T0201561

Geophysical Monograph Series

Including

IUGG Volumes
Maurice Ewing Volumes
Mineral Physics Volumes

Geophysical Monograph Series

Geophysical Monograph 160

Earth's Deep Mantle: Structure, Composition, and Evolution

Robert D. van der Hilst
Jay D. Bass
Jan Matas
Jeannot Trampert
Editors

American Geophysical Union
Washington, DC

Library of Congress Cataloging-in-Publication Data

Earth's deep mantle : structure, composition, and evolution / Robert D. van der
Hilst ... [et al.], editors.
 p. cm. -- (Geophysical monograph, ISSN 0065-8448 ; 160)
 Includes bibliographical references.
 ISBN-13: 978-0-87590-425-2
 ISBN-10: 0-87590-425-4
 1. Earth--Mantle--Research. 2. Thermochemistry--Research. 3.
Seismology--Research. 4. Heat--Convection, Natural--Research. I. Hilst, Robert
Dirk van der, 1961- II. Series.

 QE509.4.E27 2005
 551.1'16--dc22   2005028262

 ISBN-10:0-87590-425-4 (hardcover)
 ISBN-13: 978-0-87590-425-2 (hardcover)

 ISSN 0065-8448

Back Cover: Images modified after a cartoon published in F. Albarède and R. D. van der Hilst, *EOS, Transactions, American Geophysical Union*, **45**, 535–539, 1999.

CONTENTS

PREFACE

Understanding the inner workings of our planet and its relationship to processes closer to the surface remains a frontier in the geosciences. Manmade probes barely reach ~10 km depth and volcanism rarely brings up samples from deeper than ~150 km. These distances are dwarfed by Earth's dimensions, and our knowledge of the deeper realms is pieced together from a range of surface observables, meteorite and solar atmosphere analyses, experimental and theoretical mineral physics and rock mechanics, and computer simulations. A major unresolved issue concerns the nature of mantle convection, the slow (1–5 cm/year) solid-state stirring that helps cool the planet by transporting radiogenic and primordial heat from Earth's interior to its surface.

Expanding our knowledge here requires input from a range of geoscience disciplines, including seismology, geodynamics, mineral physics, and mantle petrology and chemistry. At the same time, with better data sets and faster computers, seismologists are producing more detailed models of 3-D variations in the propagation speed of different types of seismic waves; new instrumentation and access to state-of-the-art community facilities such as synchrotrons have enabled mineral physicists to measure rock and mineral properties at ever larger pressures and temperatures; new generations of mass spectrometers are allowing geochemists to quantify minute concentrations of diagnostic isotopes; and with supercomputers geodynamicists are making increasingly realistic simulations of dynamic processes at conditions not attainable in analogue experiments. But many questions persist. What causes the lateral variations in seismic wavespeed that we can image with mounting accuracy? How reliable are extrapolations of laboratory measurements on simple materials over many orders of magnitude of pressure and temperature? What are the effects of volatiles and minor elements on rock and mineral properties under extreme physical conditions? Can ab initio calculations help us understand material behavior in conditions that are still out of reach of laboratory measurement? What was the early evolution of our planet and to what extent does it still influence present-day dynamics? And how well do we know such first-order issues as the average bulk composition of Earth?

In the last decade geoscientists have begun to respond to such questions with more quantitative integrations of observations and constraints from different lines of inquiry—an approach that is changing our views on the structure, composition, and evolution of Earth's deep mantle. Studies within and across traditional disciplinary boundaries have inspired many special sessions at annual meetings of the main professional societies as well as many topical workshops. The Union Session on *Geophysical and Geochemical Models for the Structure and Composition of Earth's Mantle* at the joint EGS–EUG–AGU meeting (Nice, France, April 7–11, 2003) and a workshop on *Mantle Composition, Structure, and Phase Transitions* (Fréjus, France, April 2–6, 2003) brought together seismologists, geodynamicists, mineral physicists, and geochemists to discuss new observations and changing views on Earth's mantle. The current volume derives from these meetings, along with other invited contributions. Most contributors have combined a review component with a presentation of new developments in the respective disciplinary fields. We hope, therefore, that readers will consider this volume both as an overview of achievements made during the recent past and as a source of inspiration for new investigations.

We thank the coorganizers of the two workshops that inspired the production of this volume and Centre Nationale de la Recherche Scientific (CNRS), the Institut Nationale des Sciences de l'Univers (INSU), the French Ministry of Research and Technology, the U.S. National Science Foundation (NSF), the Consortium for Materials Properties Research in Earth Sciences (COMPRES), and the Ecole Normale Supérieure de Lyon, France, for their financial support of the workshop in Fréjus, which enabled the broad participation of students and post-docs and of scientists from Japan and the USA. We thank the authors for their contributions and also the more than 40 reviewers for their invaluable help with evaluating and improving the manuscripts.

Robert D. van der Hilst
Jay D. Bass
Jan Matas
Jeannot Trampert

Earth's Deep Mantle: Structure, Composition, and Evolution
Geophysical Monograph Series
Copyright 2005 by the American Geophysical Union
10.1029/160GM01

Earth's Deep Mantle: Structure, Composition, and Evolution—An Introduction

Robert D. van der Hilst[1], Jay D. Bass[2], Jan Matas[3], and Jeannot Trampert[4]

Here we present the general scope of the monograph and introduce the different chapters. The chapters are organized by theme instead of scientific discipline, and most combine a review of past accomplishments with discussions of results from current research. Collectively, they document the tremendous progress in understanding achieved over the past decade or so, but they also demonstrate that many controversies and challenges remain for future collaborative studies of Earth's deep interior.

GENERAL SCOPE OF MONOGRAPH

In the past decade, spectacular advances in geochemistry, theoretical and experimental mineral physics, seismic imaging, and computational geodynamics have produced new views on the inner workings of our planet. In contrast to previous single-disciplinary approaches, in the mid-1990s many investigators started to make more rigorous efforts to include constraints from other disciplines into their models or interpretations. Experimental and theoretical mineral physics research has been yielding increasingly accurate constraints on elastic parameters at pressure and temperature conditions typical for Earth's mantle. With this growing—but still vastly incomplete—data base, seismologists have begun to realize that not all inferred wavespeed variations can have a thermal origin, and evidence for lateral variations in the composition of Earth's mantle has been mounting. In a parallel effort, geodynamicists have been using the input from mineral physics research and exploiting increased computational power to simulate thermo-chemical convection, with important measures of their success being the ability to match the structure and spectra of tomographi-

cally inferred heterogeneity, reproducing essential aspects of the geochemical record of selected chemical elements, and satisfying geophysical observables such as the geoid and long wavelength gravity, dynamic topography, and heat flow.

A challenge central to all these studies has been to understand the nature, scale, and geological history of mantle convection—the slow (1–5 cm/yr) solid-state stirring that helps cool the planet by transporting radiogenic and primordial heat from Earth's interior to its surface. Evidence for a chemically heterogeneous mantle, with multiple long-lived "reservoirs", has long competed with views that convection driven by pure thermal buoyancy has effectively homogenized the mantle. The ensuing debate revolved around end-member models of either convective layering at 660 km depth (with layers above and below this depth having different—but uniform—compositions), or iso-chemical whole mantle overturn. But none of these canonical models satisfies more than what can now be considered a fairly narrow subset of available constraints. Over the past decade, however, several key discoveries and the creative interplay between various disciplines have begun to paint a picture of the lower mantle that is far more interesting—but not less enigmatic—than the one or two relatively bland shells of near-constant physical, chemical, and thermo-dynamical properties considered in the classical models. Some scientists have been trying to explain the richness of multidisciplinary constraints with hybrid models that combine the successful ingredients of the canonical models; others have been questioning the fundamental assumptions and the perceived paradoxes to produce interesting, albeit as yet dif-

[1]Massachusetts Institute of Technology, Cambridge, Massachusetts.
[2]University of Illinois, Urbana-Champaign, Urbana, Illinois.
[3]Ecole Normale Supérieure de Lyon, Lyon, France.
[4]Utrecht University, Utrecht, The Netherlands.

Earth's Deep Mantle: Structure, Composition, and Evolution
Geophysical Monograph Series 160
Copyright 2005 by the American Geophysical Union
10.1029/160GM02

ficult to validate, conceptual models of the thermo-chemical evolution of the Earth.

In the late 1990s, new types of crosscutting research began to provide important new insight. Joint interpretations of seismic images of shear- and bulk properties and then available elasticity parameters for relevant mantle compositions and *P-T* conditions produced tentative indicators for lateral variations in (major element) composition in the deep mantle (e.g., *Su and Dziewonski*, 1997; *Kennett et al.*, 1998; *van der Hilst and Kárason*, 1999). This emerging evidence inspired several groups (e.g., *Tackley*, 1998; *Kellogg et al.*, 1999; *Becker et al.*, 1999; *Davaille*, 1999) to speculate on the existence of compositionally distinct domains in the deep mantle as an explanation of some of the hitherto conflicting observations. These exciting developments followed earlier suggestions of large-scale compositional heterogeneity that were based on seismic wave speeds and the high-pressure densities of different rock lithologies, but with much less extensive elasticity and phase equilibrium data for minerals (e.g., *Bass and Anderson*, 1984; *Anderson*, 1989). In fact, the model described by *Anderson* (this volume) shares with the proposition of *Kellogg et al.* the premise of a three-layer mantle and a compositionally distinct bottom third of the mantle. The early seismological evidence for compositional heterogeneity was far from uncontroversial, however, but through detailed analysis by, for instance, *Ishii and Tromp* (1999), *Masters et al.* (2000), *Saltzer et al.* (2001, 2004), *Karato and Karki* (2001), *Anatolik et al.* (2003), and *Trampert et al.* (2004), the notion has become inescapable that much of deep mantle heterogeneity does not have a purely thermal origin.

Despite these developments, there are many uncertainties, and even first-order aspects of the nature, volume, and long-term evolution of deep mantle heterogeneity have remained enigmatic. The early thermo-chemical convection models each have their problems (see overviews by, for instance, *Tackley*, this volume), and the interpretation of the tomographic images is neither straightforward nor unique. But these are perhaps the least of all problems. Constraints as fundamental as Earth's average bulk composition remain controversial; the experimental and theoretical constraints on the temperature and pressure dependencies of elasticity mostly involve simple compositions and ignore effects of important minor elements (such as aluminum and calcium), the oxidation state of Fe (e.g., the amount of Fe^{3+} in ferromagnesian-silicates), and mineralogy; and extrapolations over large temperature and pressure ranges remain a major concern. Moreover, with few exceptions most scaling from wavespeed to temperature, composition, and mass density assume predominance of a thermal dependence, which may be incorrect. Even if the transformation from wavespeed

to major element composition is reasonable, what—if anything—does it tell us about the spatial distribution of the trace elements and noble gases used in geochemical investigations of mantle structure? For example, if the deepest mantle contains compositionally distinct domains, do they represent refractory repositories or are they depleted in heat-producing elements? Questions such as these have key implications for our understanding of the formation origin of the structures and their evolution over geological time, the early evolution of the core-mantle system, and the present-day thermal state of our planet.

OVERVIEW OF MONOGRAPH

For this monograph we have solicited contributions that relate to the general scope described above. Other important developments in the quest to understand Earth's deep mantle are not covered in detail. For instance, the lowermost mantle, including the so-called D″ region of enhanced structural and, probably, compositional heterogeneity, is not a primary target of the chapters of this volume; the progress achieved in the 1990s is reviewed elsewhere (e.g., *Gurnis et al.*, 1998; *Garnero*, 2000; *Karato et al.*, 2000). The recent discovery of a post-perovskite phase transition is covered briefly, but an overview of this rapidly evolving topic would perhaps be premature. The emphasis on the deep mantle also means that the numerous recent studies of the upper mantle and transition zone structures, including the related pressure-induced discontinuities, and the investigation of mantle "plumes" are not focused on here.

The first two chapters of this volume concern old and new constraints from geochemistry. In the past decade, many new proposals have been made and fundamental assumptions questioned. For example, the difference in mid-ocean island basalt (MORB) and ocean island basalt (OIB) trace element and noble gas distributions, which has inspired much discussion about geochemical reservoirs, can perhaps be explained by differences in sampling from a (statistically) similar source (e.g., *Anderson*, 2001; *Kellogg et al.*, 2002; *Ito and Mahoney*, 2005). This issue, however, is not addressed in detail here. *Harrison and Ballentine* (this volume) review noble gas geochemistry, the so-called helium/heat and argon "paradoxes", and the development of Earth models based on noble gas systems. They show that models based on selected systems can be misleading and that a wide range of noble gases must be considered when building robust mantle models. The changing views on the helium system are reviewed, with emphasis on the uncertainty in absolute concentrations and in the level of compatibility of parent and daughter elements. For the latest development on the latter topic we refer the reader to *Parman et al.* (2005). *Harrison and Ballentine* discuss the pitfalls of comparing systems with very different characteristic time scales and suggest that a "zero paradox" model can be constructed if (i) one relaxes the assumption that the mantle is in steady state and (ii) the noble gas concentration (e.g., ^{3}He) in the convecting mantle is higher—by a factor of 3.5—than hitherto assumed. *Albarède* (this volume) also uses characteristic time scales—here of residence times of incompatible lithophile elements—to question the basic assumption of steady state and argues that most of the noble gas and heat flux paradoxes disappear if one allows that a significant fraction (~50%) of the "missing" heat is still trapped in Earth's deep interior, perhaps in small-scale heterogeneity scattered throughout Earth's mantle.

The type of geochemical analysis presented here by *Harrison and Ballentine* and *Albarède* can provide constraints on isotope and trace element concentrations and on characteristic time scales of mixing processes, but it does not constrain the location of anomalous materials. Seismology provides the most direct probe for constraining deep mantle structure (through the mapping of 3-D variation in (an)elastic wave propagation) and on the spatial distribution of major elements (through their influence on the elastic constants and on mass density). *Trampert and van der Hilst* (this volume) review pertinent results of seismic tomography, with emphasis on lateral deviations from radially stratified (1-D) reference Earth models. Because of the non-uniqueness of joint interpretations of tomographic images along with mineral physics data *Trampert et al.* (2004) performed a statistical analysis of a wide range of possible solutions. This approach confirms and quantifies earlier indications that the propagation behavior of seismic waves is different from what can be expected from a thermal origin and shows that the emerging evidence for lateral variations in bulk composition is robust. The analysis of body wave travel times, surface wave phase velocities, and spectra of Earth's free oscillation modes suggests that lateral variations in temperature, iron, and silicon (through perovskite/magnesiowüstite partitioning) occur globally at a wide range of spatial scales and throughout the mantle, but appear stronger in the bottom 1000 km or so. In a detailed regional study based on a careful analysis of broad-band waveforms, *Helmberger and Ni* (this volume) describe rather dramatic lateral variations in shear wave speed (relative to PREM) in the deep mantle beneath southern Africa. Interestingly, the variations in *P*-wave speed are not nearly as large. Combined with the shape of the deep structure and its remarkably sharp boundaries, this observation strongly argues for a thermo-chemical origin. Detailed studies of this kind are restricted to regions of sufficiently high sampling density, and ongoing efforts to install arrays of seismograph stations in new locations may soon lead to more discoveries of thermo-chemical structures in the deep mantle.

Mantle convection experiments, both analog and numerical, play an essential role in understanding the dynamics, evolution, and structure of Earth's mantle. In particular, quantitative geodynamics provides a framework for the integration of results from different disciplines (e.g., seismology, mineral physics, geochemistry, and geodesy), and owing to vastly increased computation power, the field of quantitative geodynamical modeling of mantle convection has taken flight in the past 15 years. The mounting evidence for compositional heterogeneity implies that traditional views of mantle processes based on thermal convection may be in error, however, and strongly suggests the need for explicit thermo-chemical convection modeling. The capabilities, remaining challenges, and future of numerical modeling of thermo-chemical convection are reviewed by *Tackley et al.* (this volume). In addition to past successes of numerical modeling of deep mantle processes, *Tackley et al.* also discuss the challenge to deal with uncertainty. Fundamental uncertainties in the key physical parameters can be so large that conflicting convection behaviors seem possible. It is therefore of key importance that predictions from numerical models are tested against independent data (e.g., seismic models, geochemical records, geodynamo history). Another source of uncertainty concerns the model simplifications which reflect imperfect knowledge of many key physical processes or which are necessary to render problems tractable. This includes processes that operate on small scales but which may be key controls for the large-scale dynamical behavior. Examples include slab processes (such as hydration), melting, and plate formation. Others are too complicated or still too poorly understood, for instance, continent formation and evolution and the effect of volatiles and grain size on transport properties.

Inspired by *Kellogg et al.* (1999), *Samuel et al.* and *van Thienen et al.* investigate mantle dynamics in the presence of compositionally distinct domains in the deep mantle. *Samuel et al.* (this volume) assume that such a deep structure exists and explore the range of compositions that are consistent with the emerging evidence for compositional heterogeneity from seismic imaging. In agreement with earlier findings (e.g., *Kellogg et al.*, 1999; *Saltzer et al.*, 2004; *Trampert et al.*, 2004) they conclude that deep mantle lateral variations in iron and silicon must be invoked to explain the seismological observations. Realizing that the present-day existence of such domains begs the question as to their origin—see also *Righter* (this volume)—*van Thienen et al.* (this volume) explore possibilities for the early origin and evolution of such domains. Their analysis shows that the formation of deep, enriched, and gravitationally stable mantle domains by convective instabilities and resurfacing before the onset of modern-style plate tectonics is dynamically plausible and consistent with the geochemical record.

One must realize that the inferences on compositional heterogeneity from seismology and mineral physics described by, for example, *Trampert and van der Hilst* (this volume) and the numerical modeling of thermo-chemical convection discussed by *Tackley, Samuel et al.,* and *van Thienen* (this volume), are based on an incomplete set of highly simplified physical parameters. For example, the mineral physics parameters used to convert inferred wavespeed variations to density, temperature, and composition have been based mostly on idealized ternary compositions (FeO, MgO, and SiO_2), with the effects of calcium (CaO) and aluminum (Al_2O_3) still not well understood, let alone incorporated, and the effects on mineralogy and petrology—which can be significant, see *Speziale et al.* (this volume)—largely ignored. Thermodynamical parameters are often assumed constant, and the effects of increasing pressure are not well accounted for in most numerical simulations.

These concerns, along with those raised later in the volume, for instance by *Anderson, McCammon,* and *Speziale et al.,* highlight the critical need for more precise knowledge on the behavior of relevant mantle minerals beyond the simple compositions and mineralogies that are often used in the interpretations of seismological constraints or in the quantitative modeling of thermo-chemical convection. Although all important properties are in principle measurable, some of the complexity is out of range of current experimental capabilities. Fortunately, in the past decades, theoretical studies (that is, first-principle and ab initio calculations) have made spectacular progress owing to routine access to faster computers and more accurate approximations of the fundamental quantum-mechanics equations. *Bukowinski and Akber-Knutson* (this volume) briefly present the recent developments, the state-of-the-art, and the future challenges of theoretical mineral physics. They explain to a non-specialist in mineral physics what one can learn about the real Earth by using the extremely powerful and mature tool of quantum mechanics, and they present several examples to illustrate how one can obtain robust constraints on elasticity of mantle minerals, melting, and phase transitions that occur within the Earth's interior. For example, *Trampert and van der Hilst* (this volume) discuss how theory can help understand the seismically observed anti-correlation between bulk sound speed and shear velocity in deep mantle. Furthermore, they address various uncertainties associated with computational techniques so that the strengths and limitations of theoretical approaches can be understood. It is also clearly emphasized that any future progress goes through a constructive dialogue between the theoretical and experimental communities.

Questioning many fundamental assumptions, *Anderson* (this volume) warns against the use of variation in seismic (shear) wavespeed as a proxy for density, temperature, and composition, and suggests that simple volume scaling is best for many parameters. He presents a "zero paradox" model that, like *Kellogg et al.* (1999) and *van der Hilst et al.* (1999), considers compositional heterogeneity in the lowermost 1000 km of the mantle. However, the chemistry of *Anderson*'s deep layers is different from those suggested by *Kellogg et al.* (1999) and from the assumptions/modeling results by *Samuel et al.* and *van Thienen et al.* (this volume). Since this deep layer in *Anderson*'s model is largely isolated from shallower mantle processes, it should not play a role in explaining the geochemical record; indeed, according to *Anderson*, most of this can be explained by geodynamical processes in the top 1000 km or so of the mantle.

If some of the interpretations by *Anderson* are still speculative and provocative, there is little doubt that the interpretation of tomography and the computational modeling of thermo-chemical convection are often rather naive and—at the least—plagued by first-order uncertainties about the compositional and thermal reference state of Earth's mantle and about the elasticity data used to convert tomographically inferred wavespeed variations into estimates of temperature, density, and composition (see also *Bukowinski and Akber-Knutson*, this volume). The uncertainty in reference composition and in the elastic parameters obtained from theoretical and experimental mineral physics are the subject of several chapters of this volume. *Williams and Knittle* review the intrinsic problems of constraining Earth's bulk composition. The notion that the average bulk composition resembles that of one or more selected chondrites is not at all trivial and may lead to uncertainties that are significant, but often ignored. For a given chondritic model, the results from mass balance exercises and inversions for the average bulk composition may be very precise, but owing to uncertainty in the reference models they are perhaps not accurate to within 5% uncertainty even in major element (e.g., Fe, Mg) composition. The resulting uncertainty in density variations may leave several issues involving this important diagnostic parameter unresolved. Furthermore, a compositionally distinct (for instance, perovskite-rich) deep mantle cannot be ruled out with the data that are currently available.

But the challenges are broader and deeper than the uncertainties in compositional reference models. *Righter* (this volume) focuses on the importance of understanding the early evolution of the Earth, including the mantle-core system. An important class of chemical elements used in such studies are the so-called highly siderophile elements (HSEs), which have a very high melt/silicate partitioning coefficient. Measuring HSE concentrations is not trivial, but knowledge of their distribution can put important constraints on the conditions and timing of core formation and early mantle differentiation. Because of the high melt/silicate partitioning coefficient, equilibrium core formation would have absorbed most—if not all—HSEs into the core. A key and, as yet, unanswered question, therefore, is whether the present-day distribution of HSEs and related elements in the mantle is set by some chondritic component (with all its uncertainties; see *Williams and Knittle*, this volume) or by metal-silicate equilibrium during core formation? *Righter* argues that both are important, and he discusses the strengths and weaknesses of several hypotheses for explaining the HSE concentration in the primitive mantle. These include a late veneer (problematic), incomplete core formation (not likely), ongoing core-mantle interaction (viable but disputed), magma ocean fractionation (extent of melting uncertain); level of fractionation—which can produce perovskite-rich deep layer (*Williams and Knittle* and *Anderson*, this volume)—controversial (it may work for some elements but not for others), capability to trap metals/HSEs (controversial), and—perhaps—the occurrence of Fe-metal in the deep mantle. *Righter* dismisses the latter as insignificant because the oxygen fugacity (fO_2) may not be low enough to produce Fe-metal, but in a rigorous review of the oxidation state and oxygen fugacity in Earth's mantle, *McCammon* (this volume) shows that Fe-metal may well be present in the deep mantle. She shows that the energetics of Fe^{3+} incorporation in minerals is controlled by mineralogy and crystal chemistry at least as much as by fO_2. The oxygen fugacity generally decreases with increasing depth. In the deep mantle, however, the increase in relative proportion of Fe^{3+} (compared to all iron) is not due to changes in oxygen fugacity but because the presence of another trivalent element (Al^{3+}) in $(Mg,Fe)(Si,Al)O_3$ is balanced by iron disproportionation ($3\ Fe^{2+} \rightarrow Fe^0$–metal $+\ 2\ Fe^{3+}$). *McCammon* argues that ca. 50% of iron in the lower mantle $(Mg,Fe)(Si,Al)O_3$ occurs as Fe^{3+} (independent of fO_2).

The effects of the oxidation state (expressed by McCammon as the relative proportion of Fe^{3+}) and the need to consider $(Mg,Fe)(Si,Al)O_3$ compound the uncertainties in major element composition discussed by *Williams and Knittle* and *Righter*. The presence of trivalent elements (Fe^{3+}, Al^{3+}) and the metal phase have important implications for our understanding of major element composition, element partitioning, and mantle mineralogy (and through them, estimates of the elastic properties used in seismology and computations geodynamics), volatile solubility, and transport processes (thermal and electrical diffusivity/conductivity) in the mantle. Furthermore, *Badro et al.* (this volume) present results from high temperature/pressure experiments showing that under increasing pressure the iron in perovskite and magnesiowüstite undergoes electronic transitions—from a high-spin to a low-spin state—at lower mantle pressures (70–120 GPa). This may have important implications for the thermo-chemical state of Earth's lower mantle. The

electronic transitions in the deep mantle may enhance optic transparency and, hence, increased (radiative) thermal conduction. Whether or not this could result in more sluggish convective behavior in $(Mg,Fe)(Si,Al)O_3$-silicates at pressures in excess of 70 GPa (that is, depths larger than 1800 km) is an issue of debate and requires further research. The changing spin-states can perhaps also modify the partitioning of Fe between perovskite and magnesiowüstite, which would influence the elastic properties.

In the past decade, there has been much debate about the stability of the Mg-perovksite phase $(Mg,Fe)SiO_3$. The break down of $(Mg,Fe)SiO_3$ into compounding oxides was mentioned as a possibility by *Birch* (1952) and has been a topic of heated debate since the publication of the experimental results by *Maede et al.* (1995) and *Saxena et al.* (1996) showing that such a dissociation can indeed occur. In this volume, *Shim* reviews the experimental difficulties and concludes from X-ray diffraction studies that $MgSiO_3$ is probably stable to at least 118 GPa (2500 km depth) and perhaps over the entire pressure range of Earth's lower mantle. There are, however, several peaks in the diffraction pattern that may indicate slight changes in perovskite crystal structure. One of them occurs at 2500 K and 135–145 GPa and can be explained by the transformation to a post-perovskite phase (*Murakami et al.*, 2004). *Shim* discusses the (strong) dependence of this phase transformation on temperature and chemical composition.

The understanding and realistic description of the transition zone is undoubtedly a major challenge for the near future. The geophysical processes taking place in this region significantly influence a large number of geological records observed on the surface. The paper by *Anderson* presents a number of arguments for much of the surface record having its origin in the upper 1000 km. As noted above, for a realistic interpretation of seismic observations in the transition zone it is necessary to go beyond the relatively simple lower mantle compositions and mineralogies. The sensitivity of seismic velocities to major elements (Fe, Si, Al) can change significantly between rock types; for instance, the composition-velocity relationships are different in the harzburgite portion of the lithosphere than in the eclogite formed from subducted oceanic crust. *Ricard et al.* (this volume) attempt to bridge seismology and mineral physics and illustrate their thermodynamic approach by computing synthetic seismic images of simplified slabs of subducted lithosphere. In order to compute the seismic signature for a given rock type, they first compute the thermodynamically stable mineral assemblages that are, of course, depth dependent. They also provide a brief tutorial to the Gibbs free-energy minimization techniques. *Ricard et al.* clearly show that the contributions of composition, mineralogy and petrology, and temperature may be equally important within the transition zone. They note that the current limitations of the thermodynamic method are related to the uncertainties on the elastic and mixing properties of mantle materials under relevant mantle *P-T* conditions.

Confirming the need to consider mineralogy and lithology in the quantitative interpretation of seismologically derived wavespeed estimates, *Speziale et al.* (this volume) use recent laboratory measurements of sound velocities to examine how seismic wavespeeds are affected by the common chemical substitutions in minerals of the upper mantle and transition zone. These are the sort of systematic relationships involving velocity that enters into the questions posed in this volume by *Ricard et al.* (for slabs in the upper mantle and transition zone) and *Trampert and van der Hilst* (for the lower mantle). Significantly, *Speziale et al.* find that the velocity systematics are strongly dependent on crystal structure, and a single simple relation is not applicable to all rock types or all portions of the mantle. Since major seismic discontinuities are often accompanied by crystallographic coordination changes, these may also signal that different velocity systematics apply across such boundaries.

Finally, during the past decade we have realized that the mantle and, in particular, the transition zone, is a huge potential reservoir for water in the deep Earth. From petrologic studies of hydrous systems under extreme pressures, it was found that the high-pressure polymorphs of olivine, wadsleyite, and ringwoodite can hold up to several weight percent water. This opens the intriguing possibility for many oceans worth of water being stored at great depth (see also *Bercovici and Karato*, 2003). *Ohtani* (this volume) gives a broad overview of this subject, investigating whether subduction of oceanic lithosphere is a viable means of transporting water to the transition zone and lower mantle. An important issue is whether any mineral can hold water down to transition zone depths, or if, alternatively, there a special "choke point" beyond at which all hydrous minerals dehydrate and release the water. *Ohtani* also raises the issue of the potential seismological visibility of "water" in the deep sub-surface; answering this question will require new measurements on the high-pressure hydrous phases of the transition zone.

REFERENCES

Anderson, D. L., *Theory of the Earth*, Blackwell Scientific Publications, Oxford, 1989.

Anderson, D. L., A statistical test of the two-reservoir model for helium isotopes, *Earth Planet. Sci. Lett.*, **193**, 77–82, 2001.

Antolik M., Y. Gu, G. Ekström and A. Dziewonski. J362D28: a new joint model of compressional and shear velocity in the Earth's mantle, *Geophys. J. Int.*, **153**, 443–466, 2003.

Bass, J. D., and Anderson, D. L., Composition of the upper mantle: geophysical tests of two petrological models, *Geophys. Res. Lett.*, **11**, 237–240, 1984.

Becker, T. W., Kellogg, J. B. and R. J. O'Connell, Thermal constraints on the survival of primitive blobs in the lower mantle, *Earth Planet. Sci. Lett.*, **171**, 351–365, 1999.

Bercovi, D., and Karato S., Mantle convection and the transition-zone water filter, *Nature*, **425**, 24, 2003.

Birch, F., Elasticity and constitution of the Earth's interior, *J. Geophys. Res.*, **57**, 227–286, 1952.

Davaille, A., Simultaneous generation of hotspots and superswells by convection in a heterogenous planetary mantle, *Nature,* **402**, 756–760, 1999.

Garnero, E. J., Lower mantle heterogeneity, *Ann. Rev. Earth Planetary Sci.*, **28**, 509–37, 2000.

Gurnis, M., Wysession, M. E., Knittle, E., Buffett, B. A. (Eds.), The Core-Mantle Boundary Region, Geodynamics Series, **28**, 340 pages, AGU, 1998.

Ishii, M., and J. Tromp, Normal-mode and free-air gravity constraints on lateral variations in velocity and density of earth's mantle, *Science,* **285**, 1231–1236, 1999.

Ito, G., and Mahoney, J. J., Flow and melting of a heterogeneous mantle: 2. Implications for a chemically nonlayered mantle, *Earth Planet. Sci. Lett.*, **230**, 47–63, 2005.

Karato S.-I., Forte, A.M., Liebermann, R.C., Masters, G., Stixrude, L., (Eds.), Earth's Deep Interior: Mineral Physics and Tomography From the Atomic to the Global Scale, Geophysical Monograph Series, **117**, 289 pages, AGU, 2000.

Karato, S.-I., and B. Karki, Origin of lateral heterogeneity of seismic wave velocities and density in Earth's deep mantle, *J. Geophys. Res.*, **106**, 21771–21783, 2001.

Kellogg, L. H., B. H. Hager, and R. D. van der Hilst, Compositional stratification in the deep mantle, *Science*, **283**, 1881–1884, 1999.

Kellogg, J. B., Jacobsen, S. N., and O'Connell, R. J., Modeling the distribution of isotopic ratios in geochemical reservoirs, *Earth Planet. Sci. Lett.*, **204**, 183–202, 2002.

Kennett, B. L. N., Widiyantoro, S. and R. D. van der Hilst, Joint seismic tomography for bulk sound and shear wave speed in the Earth's mantle, *J. Geophys. Res.*, **103**, 12469–12493, 1998.

Maede, C., Mao, H.-K., and Hu, J., High-temperature phase transition and dissociation of $(Mg,Fe)SiO_3$ perovskite at lower mantle pressures, *Science*, **268**, 1743–1745, 1995.

Masters, G., G. Laske, H. Bolton, and A. Dziewonski, The relative behavior of shear velocity, bulk sound speed, and compressional velocity in the mantle: implications for chemical and thermal structure in Earth's deep interior, *AGU Monograph* **117**, 63–87, 2000.

Murakami, M., Hirose, K., Ono, S., and Ohishi, Y., Post-perovskite phase transition in $MgSiO_3$, *Science*, **304**, 855–858, 2004.

Parman, S. W., Kurz, M. D, Hart, S. R., and Grove, T. L., Helium solubility in olivine and its implications for high $^3He/^4He$ in ocean island basalts, *Nature*, in press, 2005.

Saltzer, R. L., R. D. van der Hilst, and H. Kárason, Comparing P and S wave heterogeneity in the mantle, *Geophys. Res. Lett.*, 28, 1335–1338, 2001.

Saltzer, R. L., Stutzmann, E. and R. D. van der Hilst, Poisson's ratio beneath Alaska from the surface to the Core-Mantle Boundary, *J. Geophys. Res.*, 10.1029/2003JB002712, 2004.

Saxena, S. K., *et al.*, Stability of perovskite (MgSiO$_3$) in the Earth's mantle, *Science*, **274**, 1357, 1996.

Su, W., and A. M. Dziewonski, Simultaneous inversion for 3-D variations in shear and bulk velocity in the mantle, *Phys. Earth Planet. Int.*, **100**, 135–156, 1997.

Tackley, P. J., Three-dimensional simulations of mantle convection with a thermo-chemical basal boundary layer: D″? In *The Core-Mantle Boundary Region,* M. Gurnis, M. E. Wysession, E. Knittle, B. A. Buffett (Eds), AGU, Washington DC, **28**, 231–253, 1998.

Trampert, J., Deschamps, F., Resovsky, J. and D. Yuen, Probabilistic tomography maps significant chemical heterogeneities in the lower mantle, *Science,* **306***, 853–856, 2004.*

van der Hilst, R. D. and H. Kárason, Compositional heterogeneity in the bottom 1000 kilometers of Earth's mantle: toward a hybrid convection model, *Science*, **283**, 1885–1888, 1999.

Jay D. Bass, University of Illinois, Urbana-Champaign, Urbana, Illinois 61801, USA.

Jan Matas, Ecole Normale Supérieure de Lyon, 69364 Lyon cedex 07, France.

Jeannot Trampert, Utrecht University, 3508 TA Utrecht, The Netherlands.

Robert D. van der Hilst, Massachusetts Institute of Technology, Cambridge, Massachusetts 02139, USA. (hilst@mit.edu)

Noble Gas Models of Mantle Structure and Reservoir Mass Transfer

Darrell Harrison

Institut für Isotopengeologie und Mineralische Rohstoffe, ETH Zentrum, Zürich, Switzerland

Chris J. Ballentine

School of Earth, Atmosphere and Environmental Science (SEAES), University of Manchester, Manchester, UK

Noble gas observations from different mantle samples have provided some of the key observational data used to develop and support the geochemical "layered" mantle model. This model has dominated our conceptual understanding of mantle structure and evolution for the last quarter of a century. Refinement in seismic tomography and numerical models of mantle convection have clearly shown that geochemical layering, at least at the 670 km phase change in the mantle, is no longer tenable. Recent adaptations of the mantle-layering model that more successfully reconcile whole-mantle convection with the simplest data have two common features: (i) the requirement for the noble gases in the convecting mantle to be sourced, or "fluxed", by a deep long-lived volatile-rich mantle reservoir; and (ii) the requirement for the deep mantle reservoirs to be seismically invisible. The fluxing requirement is derived from the low mid-ocean ridge basalt (MORB)-source mantle ^{3}He concentration, in turn calculated from the present day ^{3}He flux from mid-ocean ridges into the oceans ($T_{1/2} \sim 1,000$ yr) and the ocean crust generation rate ($T_{1/2} \sim 10^8$ yr). Because of these very different residence times we consider the ^{3}He concentration constraint to be weak. Furthermore, data show ^{3}He/^{22}Ne ratios derived from different mantle reservoirs to be distinct and require additional complexities to be added to any model advocating fluxing of the convecting mantle from a volatile-rich mantle reservoir. Recent work also shows that the convecting mantle ^{20}Ne/^{22}Ne isotopic composition is derived from an implanted meteoritic source and is distinct from at least one plume source system. If Ne isotope heterogeneity between convecting mantle and plume source mantle is confirmed, this result then excludes all mantle fluxing models. While isotopic heterogeneity requires further quantification, it has been shown that higher ^{3}He concentrations in the convecting mantle, by a factor of 3.5, remove the need for the noble gases in the convecting mantle to be sourced from such a deep hidden reservoir. This "zero paradox" concentration [*Ballentine et al.*, 2002] is then consistent with the different mantle source ^{3}He/^{22}Ne and ^{20}Ne/^{22}Ne heterogeneities. Higher convecting mantle noble gas concentrations also eliminate the requirement for a hidden mantle ^{40}Ar rich-reservoir and enables the heat/^{4}He imbalance to be explained by temporal variance in the different mechanisms of heat vs. He removal from the mantle system—two other key arguments for mantle layering. Confirmation of higher average convecting mantle noble gas concentrations remains the key test of such a concept.

Earth's Deep Mantle: Structure, Composition, and Evolution
Geophysical Monograph Series 160
Copyright 2005 by the American Geophysical Union
10.1029/160GM03

9

INTRODUCTION

Porcelli and Ballentine [2002], *Graham* [2002], and *Hilton and Porcelli* [2003] have recently published detailed reviews of noble gas observations and their role in developing models of the terrestrial mantle. *Ballentine et al.* [2002] and *Van Keken et al.* [2002, 2003] have further explored the role of noble gas constrained models in the light of recent seismic tomography and numerical simulations of convection in the mantle that directly contradict earlier "layered" noble gas mantle models. We provide here a summary of these works, discussing the key observational data used to constrain a variety of models, and provide an assessment of their relative robustness. We particularly highlight the observation of ^{3}He/^{22}Ne and ^{20}Ne/^{22}Ne differences resolved in samples that originate from different parts of the mantle and discuss the relevance of these features for the development of a coherent model of noble gas origin, preservation, and mixing within the mantle.

NOBLE GASES IN THE TERRESTRIAL MANTLE

The noble gases (He, Ne, Ar, Kr, and Xe) are a family of chemically inert elements with a systematic mass-dependent variation in their respective physical properties [*Ozima and Podosek*, 1983]. Their geochemical behaviour can therefore be easier to predict than the major volatile species such as carbon and nitrogen. Because they are chemically inert, they partition strongly into a mono-atomic gas phase and were therefore efficiently excluded from the "solid" planetary bodies during solar system evolution [*Pepin and Porcelli*, 2002]. These gases therefore have low abundances in the terrestrial mantle and consequently their isotopic composition is susceptible to modification by radiogenic or nucleogenic clocks, allowing the isotopic tracing of mantle reservoir evolution and mantle degassing. All the noble gas elements each have at least one isotope with a resolvable radiogenic/nucleogenic isotope component. Variations found in measured isotope ratios are a reflection of heterogeneities in the time-integrated ratio of the parent radioelement isotope to its stable daughter noble gas isotope [i.e., ^{40}K/^{40}Ar, (U + Th)/^{4}He]. We focus the discussion in this paper on He, Ne, and Ar.

NOBLE GASES AS TRACERS OF PHYSIO-CHEMICAL PROCESSES IN THE MANTLE

Despite their chemical inertness, measured noble gas elemental and isotopic ratios in samples rarely represent their respective mantle source region. This is due to modification of the mantle noble gas signature by several processes whilst on-route from the mantle to crust: in the mantle-melting

zone; during ascent from the mantle; by diffusive loss from the sample; and also by atmospheric contamination.

Although a comprehensive dataset of partition coefficients is not yet available, the most recent results [*Chamorro et al.*, 2002; *Brooker et al.*, 2003] show that during mantle melting and production of partial melts all the noble gases behave as highly incompatible elements and therefore strongly partition into the melt phase.

By far the most important process by which noble gas elemental ratios (e.g., He/Ar) may be fractionated occurs when a gas phase forms [*Moreira and Sarda*, 2000; *Burnard*, 2001; *Burnard et al.*, 2003; *Harrison et al.*, 2003]. As magma ascends though the crust, there is a decrease in pressure. A separate gas phase will form within the melt if at any point the solubility of any dissolved volatile species is exceeded. The resulting first-formed vapour phase (usually CO_2) will be preferentially enriched by the least soluble noble gases. Degassing of a silicate melt elementally fractionates the noble gases according to their relative solubilities. Solubility decreases with increasing atomic mass [*Lux*, 1987; *Carroll and Webster*, 1994; *Moreira and Sarda*, 2000]. The composition of subsequent vapour phases will continue to evolve and can be modelled by knowing the respective noble gas solubilities and estimating the initial volatile budget of the magma. This process can only fractionate relative elemental ratios; no isotope fractionation can occur.

In contrast to equilibrium solubility processes, there are some experimental and theoretical studies of noble gas diffusion that predict fractionation indices, including He isotope fractionation [*Trull and Kurz*, 1993; *Dunai and Baur*, 1995; *Trull and Kurz*, 1999; *Trieloff et al.*, 2005]. Principally this affects He because of its high diffusivity relative to the heavier gases. Theoretically it is possible that elemental fractionation may occur through diffusion (i.e., He/Ar) in both the mantle and in the crust. But perhaps more importantly the possibility that ^{3}He/^{4}He ratios may fractionate has recently been addressed [*Burnard*, 2004; *Harrison et al.*, 2004]. The basic theory is simple: Because the relative mass difference between ^{3}He and ^{4}He is large, predicted differences in their diffusivities should therefore be significant and can be defined by the following:

$$(D_{^3He}/D_{^4He}) - 1 = (M_{^4He}/M_{^3He})^{1/2} - 1 \qquad (1)$$

where D is the diffusivity and M is the relative atomic mass.

Equation (1) describes the inverse square root of mass relationship assuming vacancy diffusion mechanisms, as applicable to noble gases. ^{3}He should therefore diffuse 15% faster than ^{4}He. This is broadly consistent with experimentally determined ^{3}He and ^{4}He diffusivities in olivine and basalt glass [*Trull and Kurz*, 1993, 1999]. However, identifying analyses that show this process and being able to resolve any fractionation is difficult. But a recent model for diffusion of noble gases in the mantle (from the solid to an interstitial primary melt phase) has shown that $^3He/^4He$ ratios may be fractionated [*Burnard*, 2004], although the degree to which any ratio could be fractionated is small, up to approximately 10% (the melt would therefore have a higher $^3He/^4He$ than the residual solid). Similarly another recent study modelled He diffusion from phenocrysts stored in subsolidus magmas [*Harrison et al.*, 2004]. Fractionation of He/Ar due to preferential loss of He from the solid to a degassed melt can also fractionate ^3He from ^4He. The residual $^3He/^4He$ ratio in the phenocrysts may then be lower than the mantle source value originally trapped in the fluid inclusions.

It is however possible to model and deconvolve these various fractionation processes by utilising well-constrained radiogenic production ratios and making suitable corrections. We can predict the likely $^4He/^{40}Ar$, $^{21}Ne/^{40}Ar$, and $^{21}Ne/^4He$ production ratios in a mantle source region based on estimates of mantle K and (U + Th) compositions and nucleogenic ^{21}Ne yields. These predicted ratios are used to quantify the degree of fractionation (equilibrium or kinetic) in any given analysis, provided the production ratios are constant. A mantle reservoir isolated for 4 Ga will have an accumulated $^4He/^{40}Ar$ of $\sim$2 (present-day K/U = 12,700, Th/U = 2.6), whereas an isolation period over the last 1 Ga will result in an accumulated $^4He/^{40}Ar$ of $\approx$4 for the same K/U and Th/U [*Allègre et al.*, 1986; *Burnard et al.*, 1997]. Nucleogenic ^{21}Ne is derived from reactions of α-particles (i.e., ^4He nuclei) on ^{18}O and ^{24}Mg target nuclei. As a result, the rate of ^{21}Ne production in the mantle is determined by uranium and thorium concentrations, as is ^4He production. Because oxygen and magnesium are distributed homogeneously throughout the mantle, there should be a constant $^{21}Ne/^4He$ production ratio; recent estimates for the mantle $^{21}Ne/^4He$ production ratio range from 4.5×10^{-8} [*Yatsevich and Honda*, 1997] to 3.58×10^{-8} [*Leya and Wieler*, 1999].

The radiogenic and nucleogenic isotopes of He, Ne, and Ar are therefore powerful tools for tracing mantle processes. The $^{21}Ne/^4He$ ratio is not affected by mantle chemistry and thus traces recent gas fractionation, while the $^4He/^{40}Ar$ is in addition a function of time and of mantle chemistry. In combination these ratios can be used to identify samples containing unfractionated mantle gases and otherwise provide an estimation of the degree of, and processes controlling, fractionation.

The determination of mantle noble gas composition is also complicated by one factor that affects both isotopic and elemental compositions: For all noble gases but He, the concentrations of noble gases in the Earth's atmosphere are higher than those in mantle rocks [*Ozima and Podosek*, 1983]. A seemingly ubiquitous atmosphere-derived contaminant often clouds the mantle isotopic signatures of Ne, Ar, Kr, and Xe such that noble gas analyses appear to be mixtures of mantle and atmospheric derived gases [*Patterson et al.*, 1990, *Farley and Poreda*, 1993; *Moreira et al.*, 1998; *Harrison et al.*, 1999; *Ballentine and Barfod*, 2000; *Harrison et al.*, 2003]. Atmospheric noble gases invade the analysed gases in a number of ways: for example, magma interaction with surface equilibrated fluids to contaminate vesicle-trapped gases; post-eruptive adsorption onto the sample surface; adsorption of atmospheric gases in the mass spectrometer extraction system; and air-filled microfractures created during sample collection and preparation. Helium is largely immune to these problems due to its low abundance in the atmosphere. Fortunately there are now methodologies by which to identify atmospheric contamination and allow semiquantitative corrections [*Harrison et al.*, 2003].

ORIGIN OF NOBLE GASES IN THE MANTLE

There are two end-member models by which the Earth attained its primordial noble gas inventory during planetary formation.

1. Direct solar wind radiation implanted into accreting planetesimals (solar energetic particles; SEP). This hypothesis requires an active sun at a stage when the accreting planetesimals were small and the accretion disk relatively transparent (i.e., had lost most of its volatile components).
2. A massive proto-Earth gravitationally capturing dense solar nebula volatiles. Dissolution of solar gases into an early mantle "magma-ocean" would then allow a convecting gas-charged magma to eventually solidify at depth, trapping the noble gases in the deep Earth as a function of their relative solubilities.

Both these models make very specific predictions for the original inventory for both elemental and isotope ratios [*Moreira et al.*, 1998; *Harrison et al.*, 1999; *Trieloff et al.*, 2000; *Pepin and Porcelli*, 2002]. This area of research dealing with accretion processes, early degassing, and fractionation is very active (see review by *Porcelli and Pepin* [2004]). In the following sections we systematically describe the evolution of the noble gases in the mantle, describe different sample resources (identifying mantle reservoirs), and highlight certain aspects of the origin of these gases.

Helium

Helium isotope compositions within the Earth's mantle are a mixture of primordial He initially introduced during planetary accretion, with the addition of a time-integrated radiogenic ^{4}He component (principally from the radioactive decay of U and Th). Irrespective of the exact mechanisms of noble gas trapping in the early Earth, the primordial ^{3}He/^{4}He composition was essentially the same as the early Sun [*Wieler*, 2002]. These ancient solar gases (pre-deuterium burning stage in the Sun) have been sampled in the atmosphere of Jupiter [*Mahaffy et al.*, 2000], ^{3}He/^{4}He $\approx$ 120 R_a (R_a notation is the ^{3}He/^{4}He ratio normalised to the present-day terrestrial atmospheric ^{3}He/^{4}He value, 1.39×10^{-6}). ^{3}He is not produced in any significant quantity by nuclear processes in the mantle but can be produced in minor amounts in the crust by (n,α) reactions on lithium. The crustal He budget is dominated by radiogenic ^{4}He and has a ^{3}He/^{4}He ratio of about <0.01 R_a [*Morrison and Pine*, 1955; *Ballentine and Burnard*, 2002]. Also high ^{3}He concentrations in surface rocks have been shown to result from exposure to high-energy cosmic rays and the production of ^{3}He by in-situ spallation reactions [*Kurz*, 1986; *Niedermann*, 2002]. However, the amount of ^{3}He production in a global context is negligible [*Ballentine and Burnard*, 2002]. Furthermore, subduction of interplanetary dust particles into the mantle with a ^{3}He/^{4}He ratio similar to solar values has been proposed as a mechanism to produce mantle He isotope signatures [*Anderson*, 1993]. However, the high diffusivity of He and subsequent loss probably prevent significant subduction of He-bearing dust particles [*Stuart*, 1994].

Mantle He geochemistry is a mature subject area and there are now comprehensive He isotope datasets available from many different regions and tectonic settings. For instance, mid-ocean ridge basalts (MORB), which are sourced in the upper mantle, are characterised by a global mean ^{3}He/^{4}He = 8.75 R_a $\pm$ 2.14 (n = 658, which includes some plume-influenced ridge samples [*Graham*, 2002]), whereas some ocean island basalts (OIB), for instance, Iceland and Baffin Island, have ^{3}He/^{4}He values up to $\approx$42 R_a and $\approx$51 R_a, respectively [*Hilton et al.*, 1999; *Breddam and Kurz*, 2001; *Stuart et al.*, 2003], and were for a long time considered to be the product of "primitive" mantle plumes [*Graham*, 2002]. Helium isotopes can also be used to identify a subgroup of mantle plumes, the so-called "low-^{3}He" hotspots, as sampled at St. Helena and Tubuai, which are characterised by ^{3}He/^{4}He ratios lower than that of global mean MORB. Many of these OIB also have HIMU (high-μ or high-^{238}U/^{204}Pb) geochemical signatures that indicate ancient recycled oceanic crust within the source region. For instance the Cook–Austral Archipelago has a mean ^{3}He/^{4}He of 6.8 R_a $\pm$ 0.9 [*Hanyu and Kaneoka*, 1997]. Similarly, subcontinental lithospheric mantle (SCLM) appears to have a lower global ^{3}He/^{4}He ratio than that of MORB; global SCLM ^{3}He/^{4}He = 6.1 R_a $\pm$ 0.9 [*Gautheron and Moreira*, 2002].

Explanations for He isotope data from MORB and different OIB settings require the existence within the mantle of isolated reservoirs that have different time-integrated values for ^{3}He/(U + Th), and these reservoirs must have remained isolated for periods of time long enough for different isotopic signatures to evolve. The variability of ^{3}He/^{4}He ratios taken in conjunction with geological setting is the strongest evidence of different long-lived geochemical reservoirs in the mantle and one of the principle observations originally used to justify a geochemically layered or stratified mantle [*Kurz et al.*, 1982; *Allègre et al.*, 1986; *O'Nions*, 1987).

Neon

Neon has three isotopes: ^{20}Ne, ^{21}Ne, and ^{22}Ne. Nucleogenic ^{21}Ne is the only one produced in significant quantities in the mantle from α-particle reactions on ^{18}O and ^{24}Mg target nuclei [*Yatsevich and Honda*, 1997; *Leya and Wieler*, 1999; *Ballentine and Burnard*, 2002]. The ratio of ^{20}Ne/^{22}Ne is unaffected by in-situ nuclear processes and is therefore diagnostic of the source of Ne in the various mantle reservoirs. The atmospheric ^{20}Ne/^{22}Ne value (9.8) is not similar to any proposed accretionary source, and the precise mechanism for the formation of this value is consistent with it being the residue of early mass fractionation of a solar-like atmosphere by hydrodynamic escape or impact erosion [*Pepin and Porcelli*, 2002; *Porcelli and Ballentine*, 2002]. Solar-nebula Ne has a ^{20}Ne/^{22}Ne $\approx$ 13.8 [*Kallenbach*, 1997; *Wieler*, 2002]. This is distinct from the mixture of SEP Ne and solar wind Ne found in meteorite material irradiated by solar atoms and ions called Ne-B [*Black*, 1972]. This mixture is found in uniform proportions to give a Ne-B value of ^{20}Ne/^{22}Ne = 12.52 $\pm$ 0.18 [*Black*, 1972; *Trieloff et al.*, 2000].

Several authors have noted an apparent upper limit of ^{20}Ne/^{22}Ne = 12.5 in both OIB and MORB samples (e.g., *Farley and Poreda* [1993]; *Niederman et al.* [1997]; *Trieloff et al.* [2000]). This is seemingly distinct from a value of ^{20}Ne/^{22}Ne = 13.75 $\pm$ 0.32 reported for one aliquot in an Icelandic picrite [*Harrison et al.*, 1999] and repeated analysis of material from the Kola plume province showing a plume source ^{20}Ne/^{22}Ne $\geq$ 13.0 $\pm$ 0.2 (Figure 1; *Yokochi and Marty* [2004]). Because of the ubiquitous presence of air contamination in all basalt samples, *Ballentine et al.* [2002] argued that the highest values provide only a lower limit for the actual mantle value. In this respect, data from the Kola and Iceland systems higher than Ne-B may indicate a solar-nebula origin for gases in the Kola and possibly Iceland plume-source mantle.

More recently *Ballentine et al.* [2005] have shown that the ^{20}Ne/^{22}Ne value of the convecting mantle can be unambiguously resolved by investigating the Ne isotope systematics of

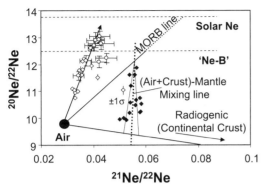

Figure 1. ^{20}Ne/^{22}Ne vs. ^{21}Ne/^{22}Ne for magmatic CO_2 well gases (solid diamonds)[*Ballentine et al.*, 2005] shown together with that from Kola plume material (open symbols)[*Yokochi and Marty*, 2004]. Well gas data define a pseudo-two-component mixing relationship between an air + crust component and the mantle end-member. Intersection of this mixing "wedge" with the air–MORB mixing line defines the convecting mantle ^{20}Ne/^{22}Ne isotopic composition to be similar to that of Ne-B, a meteorite component typical of late-accreting material exposed to extensive solar irradiation. This contrasts with the Kola data, which trend towards solar nebula values. Ne isotope heterogeneity between OIB- and MORB-mantle sources is not compatible with models that flux the convecting mantle with volatiles from the plume-mantle source.

magmatic CO_2 well gases. These magmatic fluids, instead of being contaminated with a simple atmospheric component on emplacement, have been erupted into an environment in which the groundwater containing the air contamination has accumulated crustal-radiogenic gases. This provides a unique mixing line in a three-Ne isotope plot that intersects the simple MORB–air mixing line to give a mantle end-member ^{20}Ne/^{22}Ne=12.20±0.05 (Figure 1). Resolved mantle ^{3}He/^{4}He, ^{40}Ar/^{36}Ar and the noble gas elemental pattern in these well gases are very close to the values found in mid-ocean ridge Popping rock [*Moreira et al.*, 1998] confirming that this sample resource is representative of the convecting mantle system. *Ballentine et al.* [2005] discount three mechanisms that could reduce a mantle solar-nebula value from 13.8 to 12.5: atmosphere-derived Ne recycling into the mantle, mass fractionation, and enhanced mantle nucleogenic ^{22}Ne production. Recycling and mass fractionation would have a significant impact on the other noble gases and these are not observed. Recent advances in understanding nucleogenic production of Ne [*Yatsevich and Honda*, 1997; *Leya and Wieler*, 1999; *Ballentine and Burnard*, 2002] show the huge enhancements in required mantle ^{22}Ne production rates to be highly improbable. The conclusion reached is that the Ne in the convecting mantle, with a ^{20}Ne/^{22}Ne value of ≈12.5, is dominated by solar-corpuscular–irradiated accretionary material (Ne-B).

Helium–Neon in the Mantle

The clear distinction between ^{20}Ne/^{22}Ne in the mantle (≥12.5) and that in the atmosphere (9.8) also provides a procedure for calculating the relative proportions of mantle, nucleogenic, and atmospheric Ne in any given analysis [*Honda et al.*, 1991; *Farley and Poreda*, 1993; *Graham*, 2002; *Harrison et al.*, 2003]. Because the ^{4}He production is proportional to nucleogenic ^{21}Ne production in the mantle, ^{3}He/^{4}He ratios should vary consistently and predictably with ^{21}Ne/^{22}Ne ratios, for a constant ^{3}He/^{22}Ne ratio. An introduction into the data and resolution of the relationship between He and Ne in the mantle follows.

Assuming that the Earth accreted with "solar-like" He and Ne isotope ratios, the relationship between He and Ne in the mantle can be expressed as

$$(^4He/^3He)_M + (^4He/^3He)_S = 1/(^{21}Ne/^4He)_P \times$$
$$[(^{21}Ne/^{22}Ne)_M - (^{21}Ne/^{22}Ne)_S] \times (1/^3He/^{22}Ne)_M \quad (2)$$

where P denotes production ratio, 4.5×10^{-8} [*Yatsevich and Honda*, 1997], M denotes mantle end-member ratio, corrected for atmospheric contamination [*Harrison et al.*, 2003], and S denotes initial solar composition in the early mantle: ^{4}He/^{3}He = 6000 (120 R_a) and ^{21}Ne/^{22}Ne = 0.033 [*Wieler*, 2002].

Because the ^{21}Ne/^{4}He production ratio is constant, the theoretical relationship between ^{4}He/^{3}He and ^{21}Ne/^{22}Ne is simply controlled by the ^{3}He/^{22}Ne ratio. A homogeneous ^{3}He/^{22}Ne value for the whole mantle is an implicit assumption in certain types of mantle models (as we address later in the discussion). However, several authors have previously recognised that the MORB and OIB source regions may have different ^{3}He/^{22}Ne values [*Honda and McDougall*, 1998; *Harrison et al.*, 1999; *Dixon et al.*, 2000; *Moreira et al.*, 2001; *Yokochi and Marty*, 2004]. Here we expand on this observation and model data from several localities to address this issue.

Neon isotope data from a series of hotspots and a single mid-ocean ridge sample are plotted in Figure 2. Solar starting composition is also plotted (assuming solar gas ^{20}Ne/^{22}Ne = 13.8 [*Honda et al.*, 1991; *Farley and Poreda*, 1993; *Moreira et al.*, 1998]). Using a mantle value of ^{20}Ne/^{22}Ne = 12.5 (Ne-B) does not affect any of the following arguments as long as all isotope calculations assume the same end-member value [*Harrison et al.*, 2003].

Time-integrated production of ^{21}Ne (and therefore also ^{4}He) would move data from the solar composition to the right laterally in Figure 2, as designated; this is due to variable time-integrated (U + Th)/(^{22}Ne) source ratios. The Popping rock data falls on a correlation line defined as the MORB

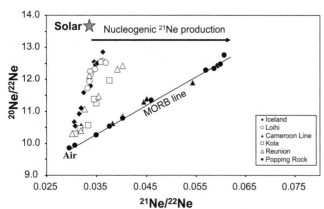

Figure 2. $^{20}Ne/^{22}Ne$ vs. $^{21}Ne/^{22}Ne$. Data from Iceland and Loihi [*Trieloff et al.*, 2000]; Cameroon [*Barfod et al.*, 1999]; Kola [*Marty et al.*, 1998]; Reunion [*Trieloff et al.*, 2002], and Popping rock [*Moreira et al.*, 1998]. All individual datasets fall on a linear array, interpreted as mixing between mantle neon and an atmospheric contaminant. As designated, a time-integrated nucleogenic ^{21}Ne (and also therefore ^{4}He) component would move an array to the right laterally. Measured $^{4}He/^{3}He$ ratios for these data are Iceland $\approx 42,000$; Loihi $\approx 29,000$; Cameroon $\approx 133,000$; Kola and Reunion $\approx 54,000$, and Popping rock $\approx 90,000$. For the plume-related data-sets, only those analyses with $^{20}Ne/^{22}Ne$ ratios ≈ 10.5 are plotted. For all subsequent calculations for all the data, only analyses with $^{20}Ne/^{22}Ne$ ratios greater than ≈ 10.5 are used. Analyses with $^{20}Ne/^{22}Ne$ ratios near to the air value (9.8) have insufficient ^{21}Ne excesses and therefore errors in estimating mantle end-member compositions are large.

line [*Sarda et al.*, 1988]. Terrestrial data from individual samples in Ne isotope plots form correlation lines that are produced by mixing of magmatic end-member compositions (high $^{20}Ne/^{22}Ne$ and high $^{21}Ne/^{22}Ne$) and air contamination (air: $^{20}Ne/^{22}Ne = 9.8$ and $^{21}Ne/^{22}Ne = 0.029$). All hotspot data fall on or between the solar and MORB compositions. The Reunion and Kola Ne data plot as intermediate between the two end-member compositions. Likewise they have interme-diate $^{3}He/^{4}He$ ratios (both $\approx 13\ R_a$). However, the Loihi and Iceland data plot on the same correlation line in Ne isotope space but have very different $^{3}He/^{4}He$ ratios ($\approx 25\ R_a$ and $\approx 17\ R_a$, respectively). In fact, the Ne-correlation line defined by these data is similar to that of solar Ne, and both hotspots would therefore be expected to have $^{3}He/^{4}He$ ratios $\geq 45\ R_a$ (this assumes an original solar $^{3}He/^{4}He$ with the addition of a time-integrated radiogenic ^{4}He component). Furthermore, the Cameroon data (a HIMU hotspot) lies on the mid-ocean ridge correlation line described by the Popping rock ($^{3}He/^{4}He \approx 8\ R_a$). The Cameroon samples have, however, a typical HIMU $^{3}He/^{4}He \approx 5.5\ R_a$, which is more radiogenic than the Popping rock. It is clear therefore that assuming mantle evolution with a constant $^{3}He/^{22}Ne$ value may be invalid. The

issue, however, is in distinguishing between two competing models that can, theoretically, both derive these discordant He and Ne data: variable primordial $^{3}He/^{22}Ne$ ratios in the mantle, or complex mixing relationships? This question has important consequences for models of mantle convection as we detail in the discussion.

The key to understanding this issue is in determining the actual primordial $^{3}He/^{22}Ne$ ratios inherent to both the MORB and hotspot source regions and identifying whether there is any variation or not. There is a broad range in the measured $^{21}Ne/^{4}He$ ratios corrected for both atmospheric contamina-tion and a solar initial value ratios, in these data (Figure 3). Moreover, these corrected nucleogenic/radiogenic $^{21}Ne/^{4}He$ ratios correlate with the primordial $(^{22}Ne/^{3}He)_M$ ratios. The

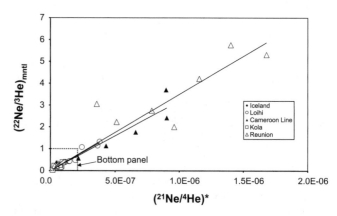

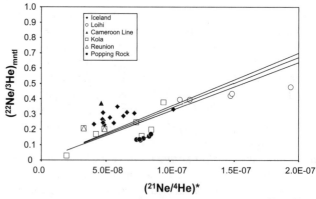

Figure 3. Correlation between radiogenic/nucleogenic $(^{21}Ne/^{4}He)^*$ and primordial $(^{22}Ne/^{3}He)_M$ ratios. $(^{21}Ne/^{4}He)^*$, the measured ratio corrected for both atmospheric contamination and a solar initial value. Top panel shows all data. Bottom panel shows an expanded view of the data in the left-hand corner of the top panel. $(^{22}Ne/^{3}He)_M$ is the mantle end-member component corrected for atmospheric contamination. Nucleogenic and primordial neon are calculated by using Figure 2. Elemental fractionation has affected most of the data. However, the correlation indicates that this frac-tionation is a recent event since it must postdate the accumulation of the radiogenic/nucleogenic isotopes. Best-fit regression lines are plotted.

high corrected $^{21}Ne/^4He$ ratios imply that most of these samples have had extensive He loss. What is more important than the mechanism that caused the Ne/He fractionation is the implication of the correlation between radiogenic and primordial Ne/He ratios. This correlation implies that the fractionation postdates the accumulation of radiogenic ^4He and nucleogenic ^{21}Ne and is therefore a recent event associated with magmatism. It is therefore possible to use the slope of the correlation lines in Fig. 3 to estimate the prefractionation $(^{22}Ne/^3He)_M$ value for these sample suites by extrapolating to the production value for corrected $^{21}Ne/^4He$. All data reviewed here, except for Iceland and Popping rock (Figure 4) fall broadly on a similar slope, including the HIMU data from the Cameroon Line. The estimated $^3He/^{22}Ne$ in the mantle source regions for these data are Loihi = 7.5 ± 1 R_a; Cameroon = 7.5 ± 1.7 R_a; Kola = 7.5 ± 2.3 R_a; Reunion = 7.5 ± 2.1 R_a. Using the same method, the $^3He/^{22}Ne$ in the Popping rock mantle source is 11.1 ± 0.5 R_a. The Iceland data have $^{21}Ne/^4He$ values near to the production value; an average of these data gives a mantle source of $^3He/^{22}Ne$ = 3.6 ± 0.5 R_a.

It is now a simple matter to test the hypothesis that the mantle contains variable $^3He/^{22}Ne$ ratios as suggested by the above results. We can calculate a predicted $^4He/^3He$ value for a given sample by using equation (2) and inputting the newly estimated $^3He/^{22}Ne$ mantle source values (Figure 3) along with the previously measured Ne data. We can then compare predicted $^4He/^3He$ results with the actual measured $^4He/^3He$ ratios for these data (Figure 4). Data from Iceland, Kola, Popping Rock, and the Cameroon Line fall on a 1:1 line of equivalence, in good agreement with the hypothesis that the mantle contains variable $^3He/^{22}Ne$ ratios. Predicted $^4He/^3He$ values for Loihi and Reunion are approximately 50% lower than the actual measured values. For simple closed system evolution, both these data sets require a mantle source $^3He/^{22}Ne$

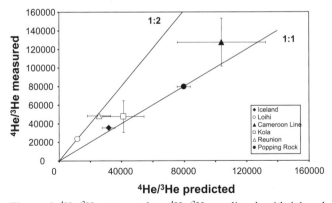

Figure 4. $^4He/^3He$ measured vs. $^4He/^3He$ predicted, with 1:1 and 1:2 lines plotted for reference. Predicted values for $^4He/^3He$ are model calculations based on the estimated mantle source $^{22}Ne/^3He$ values from Figures 3 and 4 and equation (2).

of approximately 4.5, compared to the estimated 7.5 in this model. Therefore there is excess radiogenic ^4He in these data without concomitant nucleogenic ^{21}Ne. Some unconvincing mixing calculations can be performed to explain these data, assuming two-component mixing between a plume source and an elementally fractionated upper mantle component. But where, when, and how such a mixing event could take place would be speculation and would be difficult to reconcile with the observed Ne isotopic heterogeneity. However, irrespective of the Loihi and Reunion results, we can still make a clear statement: The Earth's mantle contains heterogeneous primordial noble gas elemental ratios. The significance of this result is reiterated in the discussion.

Argon

Argon has three isotopes: ^{36}Ar, ^{38}Ar, and ^{40}Ar. Nucleogenic production of ^{38}Ar and ^{36}Ar in the mantle is insignificant. ^{40}Ar is dominantly radiogenic in origin, produced by the decay of ^{40}K [*Sarda et al.*, 1985; *Ballentine and Burnard*, 2002]. The $^{40}Ar/^{36}Ar$ ratio measured in mantle samples therefore depends on the time-integrated $^{40}K/^{36}Ar$ ratio in the source reservoir and the extent of atmospheric Ar contamination in the analysis (atmospheric $^{40}Ar/^{36}Ar$ = 295.5). Measured $^{40}Ar/^{36}Ar$ values correlate with Ne isotopes; both are a function of the same process on a local scale, mixing between magmatic and atmospheric gases. Correcting $^{40}Ar/^{36}Ar$ values for atmospheric contamination by extrapolating to the presumed mantle $^{20}Ne/^{22}Ne$ value indicates that the $^{40}Ar/^{36}Ar$ of the upper mantle is of the order 25,000–40,000 (data from the MORB Popping rock [*Moreira et al.*, 1998]. Estimates of the $^{40}Ar/^{36}Ar$ in the upper mantle source region (and by inference any sample suite) based on correlating with Ne are extremely sensitive to, and dependent on, the $^{36}Ar/^{22}Ne$ value. However, $^{36}Ar/^{22}Ne$ values in basalt samples may be variable at high $^{20}Ne/^{22}Ne$, because of the addition of a fractionated atmospheric component [*Harrison et al.*, 1999; *Trieloff et al.*, 2000; *Harrison et al.*, 2003]. It has even been suggested that the upper mantle $^{40}Ar/^{36}Ar$ value may be an order of magnitude higher than current estimates and that the value of 40,000 must be taken as a minimum (e.g., *Burnard et al.* [1999]). Recent CO_2 well gas data from New Mexico, USA, provide a strong constraint on this uncertainty by enabling the mantle $^{40}Ar/^{36}Ar$ component to be resolved with no dependency on the assumed contaminant $^{36}Ar/^{22}Ne$ value [*Ballentine et al.*, 2005]. This work limits the convecting mantle $^{40}Ar/^{36}Ar$ to be between 37,000 and 55,200.

Consistent with the He and Ne isotope modelling that shows the OIB source to be more primitive and less degassed than the MORB source, the $^{40}Ar/^{36}Ar$ of the OIB source region appears to be less radiogenic than MORB. An estimate for the $^{40}Ar/^{36}Ar$ of the mantle source supplying the Icelandic DICE samples ($^{3}He/^{4}He \approx 17R_a$) is ≈ 7300 [*Harrison et al.*, 1999; *Burnard and Harrison*, 2005] and for Loihi ($^{3}He/^{4}He$ $\sim 25 R_a$) is >8000 [*Trieloff et al.*, 2000]. But again these data must be taken as minimum values. The Ar isotopic values for mantle source regions are not well constrained due to the problems of resolving atmospheric components from analyses. It is also unclear as to whether or not there is any isotopic variation within individual mantle regions (i.e., upper convecting mantle). However, these compositions are an important constraint on mantle-degassing models.

Xenon

There are nine isotopes of xenon, a number of which are produced by radioactive decay. Xe isotope compositions in many MORB samples show resolvable excesses in ^{129}Xe and $^{131-136}Xe$ [*Staudacher and Allègre*, 1982; *Kunz et al.*, 1998; *Moreira et al.*, 1998]. Excesses in ^{129}Xe (over and above those found in the atmosphere) are derived from the short-lived radionuclide ^{129}I ($T_{1/2}$ = 15.7 Myr). Additions to the heavier isotopes are due to the spontaneous fission of ^{244}Pu ($T_{1/2}$ = 80 Myr) and also ^{238}U ($T_{1/2}$ = 4.5 Gyr). Correlations between $^{20}Ne/^{22}Ne$ and $^{129}Xe/^{130}Xe$ in the MORB sample Popping rock are consistent with an upper mantle ^{129}Xe excess >10% (over and above the $^{129}Xe/^{130}Xe$ ratio in the atmosphere [*Moreira et al.*, 1998]). Until recently, however, there was little evidence of radiogenic excesses in OIB samples despite these being an integral part of many models of mantle evolution. This is certainly due to atmospheric contributions contaminating the mantle signatures, which appears to be more prevalent in OIB samples. However, recently published Xe data from Iceland and Loihi (OIB) appear to fall on the same correlation lines as MORB in plots of multiple Xe isotopes [*Kunz et al.*, 1998; *Moreira et al.*, 1998; *Trieloff et al.*, 2000]. Xenon in these rocks may therefore be contaminated by MORB-type Xe or alternatively plume-source Xe may be similar to that in MORB, in contrast to the light noble gas isotopes (i.e., $^{3}He/^{4}He$ or $^{40}Ar/^{36}Ar$). However, the relatively imprecisely measured data preclude the resolution of any small variation in fission Xe contribution from ^{238}U and ^{244}Pu.

DISCUSSION: DEVELOPMENT OF THE NOBLE GAS MANTLE MODELS

All of the noble gas observations balancing nonradiogenic or "primitive" noble gas isotopes against radiogenic noble gas components in different mantle-derived samples can to a first approximation be exemplified by considering the He isotope system. He isotopes have the additional advantage that, unlike the heavier noble gases, the atmosphere plays little role in either contamination of the samples, or providing a possible source of recycling of He into the mantle. This is because the residence time of He in the Earth's atmosphere is on the order of 10^6 yr [*Allègre et al.*, 1986; *Torgersen*, 1989], while the atmosphere is the final sink for all other noble gases released into the atmosphere. As an example of the scale of the He depletion in the atmosphere, we can compare the present-day concentration of ^{40}Ar in the atmosphere of 0.9% to that of ^{4}He at 5 ppm. These isotopes are dominated by K and (U+Th)-derived decay, respectively, and over Earth history have been produced at a $^{4}He/^{40}Ar$ ratio of ≈ 2. If the atmosphere were closed to He loss, we would therefore expect over 3.6×10^4 times more He in the atmosphere than we see today.

We divide the models that have been proposed for the mantle system into two types: those that treat the noble gases in MORB as being the same as those that source OIB, with the addition of radiogenic noble gases, and those models that have no requirement for the OIB and MORB noble gas source to be the same (see Figure 5).

Early Flux Models

The most common form of mantle model sources the noble gases in the convecting mantle, sampled at MORB, from the mantle reservoir that provides the high $^{3}He/^{4}He$ signature of major hotspot systems such as Iceland [*Kurz et al.*, 1985; *Harrison et al.*, 1999; *Hilton et al.*, 1999] and Hawaii [*Allègre et al.*, 1983; *Kurz et al.*, 1983; *Trieloff et al.*, 2000]. There is no evidence that the various hotspots sample an isotopically homogeneous mantle source, and these probably represent a spectrum of isotopic compositions. However, it is usual to treat the OIB-source mantle as a single homogeneous reservoir for modelling purposes. In their simplest form, the observation of higher $^{3}He/^{4}He$ ratios in the OIB-source mantle compared with the convecting mantle reflects a time-averaged volatile/radioelement ratio that is higher in the OIB-source mantle [*Allègre et al.*, 1986; *Kurz et al.*, 1982; *O'Nions*, 1987]. The concentrations of ^{3}He in the MORB-source mantle combined with estimates of radioelement concentration in both the OIB and MORB-source mantles then provide the basis for calculating the noble gas concentration in the OIB-source mantle and the flux from this reservoir into the MORB-source mantle. We discuss these calculations below.

Because of eruptive degassing and other modification processes, hand samples cannot be used with any confidence to derive noble gas concentrations for either the MORB or OIB sources. The ^{3}He concentration in the MORB-source mantle used in models is almost always calculated by com-

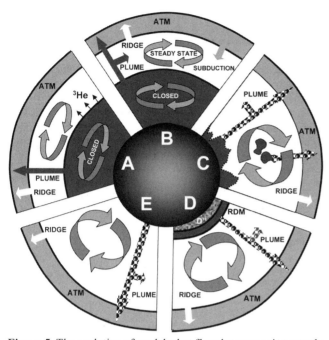

Figure 5. The evolution of models that flux the convecting mantle with noble gases from a high- $^3He/^4He$ mantle reservoir (adapted from *Porcelli and Ballentine* [2002]). See text for full description. (A) Limited interaction models (2-layer convection): A completely isolated lower mantle, with only ^3He diffusive leakage across this boundary. Plumes are entirely derived from the lower mantle. (B) Steady-state models (2-layer convection): An adaptation of (A) that has a steady-state transfer of gases from a deep, closed-system reservoir through the convecting upper mantle. Plumes are derived entirely from the lower mantle. (C) Density-contrast models (whole-mantle convection): High $^3He/^4He$ preserved in denser, deeper layers (blobs) are eroded by recycled material. Plumes are now a mix of recycled oceanic crust and high-$^3He/^4He$ mantle. (D) Residual depleted mantle model (whole-mantle convection): Recycled oceanic crust forms D″. Recycled depleted harzbugitic mantle dilutes a high-$^3He/^4He$ mantle. Varying mixtures of D″ and RDM provide HIMU to high-$^3He/^4He$ OIB plume material. (E) Noble gases from the core (whole-mantle convection): Noble gases in the convecting mantle are fluxed by a contribution from the core. Plumes are dominantly recycled material, with an admix of core-derived volatiles.

paring the present-day flux of ^3He into the oceans (^3He 1060 mol/yr [*Lupton and Craig*, 1975; *Jean-Baptiste*, 1992; *Farley et al.*, 1995, *Schlosser and Winckler*, 2002) with the rate of oceanic crust generation (20 km^3/yr [*Parsons*, 1981]). This gives an average ^3He melt concentration of 4.4×10^{-10} cm^3 (STP)/g melt. Taking an average partial melt of 10%, the ^3He concentration in this mantle source must then be 4.4×10^{-11} cm^3(STP)/g. With a $^3He/^4He$ of 8 R$_a$, typical of MORB, this gives a present-day MORB-source ^4He concentration of 3.9 $\times 10^{-6}$ cm^3 (STP)/g.

The concentration of U and Th in the mantle beneath mid-ocean ridges, estimated to be 8 and 16 ppb, respectively [*Jochum et al.*, 1983], then allows a very important calculation: Assuming closed system evolution, the U and Th concentrations predict that 1.16×10^{-5} cm^3 (STP) of ^4He per gram are produced over 4.5 Ga. This is some 3 times higher than the concentration that the flux estimate predicts. In this case, the MORB-source mantle cannot be considered as a closed system for He. Degassing of the MORB-source mantle at mid-ocean ridges provides a volatile loss mechanism, but this would also remove ^3He and result in a $^3He/^4He$ ratio far lower (more radiogenic) than observed, due to continued ^4He ingrowth. Within this framework, the only way to maintain the moderately high $^3He/^4He$ of the MORB-source mantle, but at low concentration, is to buffer the ^3He loss with input from a source with high $^3He/^4He$—the OIB-source mantle reservoir [*Kellogg and Wasserburg*, 1990; *O'Nions and Tolstikhin*, 1994; *Porcelli and Wasserburg*, 1995a, b].

Any variability in the rate of the OIB-source mantle flux to the MORB-source mantle is unconstrained, and models must make the assumption of *steady-state*: The rate of ^3He input into the MORB-source mantle is the same as its loss observed at mid-ocean ridges. In the case of steady-state ^4He, the rate of loss of ^4He at mid-ocean ridges is the same as the sum of its radiogenic production and OIB-mantle source ^4He contribution per unit time. The *residence* time of ^3He in the MORB-source mantle is then the time over which there is enough ^4He-ingrowth to bring the high $^3He/^4He$ ratio of the OIB-source mantle component (i.e., similar to ≈37 R$_a$, taken from the upper limits observed at Hawaii) to the lower $^3He/^4He$ value observed in MORB (8 R$_a$). The most complete calculations have considered a mantle layered at the 670 km discontinuity, with the OIB-source occupying the lower two-thirds (by mass) of the mantle [*Kellogg and Wasserburg*, 1990; *O'Nions and Tolstikhin*, 1994; *Porcelli and Wasserburg*, 1995a,b]. Within these models, typical residence times for ^3He in the upper mantle are calculated to be ≈1.4 Ga.

All models of this form assume that the volatile-rich mantle reservoir has been mostly isolated (or prevented from degassing) over the lifetime of the Earth. The radioelement concentration in this portion of the mantle is calculated by balancing the radioelement concentration estimated for the bulk silicate Earth (BSE) with that estimated to be in the convecting mantle and crust. In the case of the mantle layered at 670 km, the balance is made if the U + Th + K in the "lower" mantle approximates that of the BSE. It is then a simple calculation to estimate the amount of ^4He (and other radiogenic noble gases) produced in a reservoir that has been a closed system for 4.5 Ga and therefore the amount of ^3He needed in this reservoir to account for the highest

^{3}He/^{4}He ratio associated with ocean island volcanism. In this calculation, to produce a ^{3}He/^{4}He = 37 R$_a$ with BSE U = 20 ppb and Th = 76 ppb in a system closed for 4.5Ga requires a ^{3}He concentration of 3.1×10^{-9} cm^3 (STP)/g, some 70 times higher than the MORB-source reservoir. Models that require the radioelement/volatile-rich mantle volume to be smaller than the mantle below 670 km require still higher noble gas concentrations.

Early Flux Model Weaknesses

The high concentration of ^{3}He predicted for the OIB-source mantle now provides the source for testing this form of model. In steady-state, by definition, the flux of ^{3}He out of the mantle at mid-ocean ridges must equal the flux of ^{3}He into the convecting mantle. From the model estimate of ^{3}He concentration in the OIB-source mantle this provides a mass flux estimate of 3×10^{15} g/yr. This mass flux estimate is ~50 times less than the rate of ocean crust subduction. This places a strict limit on the amount of subducted material that could penetrate the OIB-source mantle. In the case of the OIB-mantle reservoir, being below the 670 km discontinuity, seismic tomography and numerical models (e.g., *van der Hilst* [1997]; *van Keken and Ballentine* [1998, 1999]) both point to a flux of material into this portion of the mantle far in excess of this value. Similarly, geophysical estimates of the plume flux ranges from 2×10^{16} to 20×10^{16} g/yr. This would mean that the ^{3}He flux at hotspots would have to be 7–70 times higher than the entire mid-ocean ridge degassing flux. This would be focused at the hotspots with the largest inferred deep mass flux (e.g., *Sleep* [1990]). It is very hard to estimate a time-averaged volatile flux at hotspots. However, such a huge focused flux at major hotspots should be observed. It is not.

It should be noted that in this framework, undegassed basalts from OIB would be expected to contain significantly more ^{3}He than undegassed MORB. A simple comparison of ^{3}He concentration in MORB and OIB shows MORB, on average, to contain higher ^{3}He concentrations. This has been called the "helium paradox" [*Anderson*, 1998] and was used to argue against a high ^{3}He concentration in the OIB-mantle source. The samples, however, are all subject to significant gas loss and in systematically different eruptive environments. The measured concentrations, therefore, reveal little about the source ^{3}He concentration.

Evolution of the Flux Models

The simple flux calculations above clearly show that high-^{3}He/^{4}He OIB cannot be entirely sourced from a deep volatile-rich and isolated reservoir. While early conceptual models

often equated the OIB to a BSE or "primitive" mantle, development of the radiogenic isotope database for OIBs showed that a large part of the OIB-source was probably related to recycled oceanic crust (e.g., *White and Hofmann* [1982]; *Zindler and Hart* [1986]; *Hauri et al.* [1994]; *Hauri* [1996]; *Eiler et al.* [1996]; *Ballentine et al.* [1997]; *Hofmann* [1997]; *Sobolev et al.* [2000]). Elongate arrays for OIBs of ^{3}He/^{4}He plotted against ^{87}Sr/^{86}Sr in particular appear to show a common origin, often called FOZO, for high ^{3}He/^{4}He samples (Figure 6). The ^{87}Sr/^{86}Sr value of FOZO shows this source to be depleted rather than primitive. An important insight into the origin of OIB is seen in a plot of ^{187}Os/^{188}Os against ^{206}Pb/^{204}Pb, where again elongate arrays trend towards a common FOZO value. Higher ^{187}Os/^{188}Os values, used to infer a recycled mafic origin (e.g., *Hauri et al.* [1994]), tend to be associated at each hotspot with lower ^{3}He/^{4}He values. Low ^{187}Os/^{188}Os, indistinguishable from the depleted convecting mantle, also appear to be characteristic of the high-^{3}He/^{4}He source (see *van Keken et al.* [2002]).

A simple calculation shows that a small amount of material from a volatile-rich high-^{3}He/^{4}He mantle source can overprint the small amount of radiogenic ^{4}He ingrowth in subducted oceanic crust (residence time = 1–2 Ga) (e.g., *Ballentine et al.* [2002]). This results in ^{3}He concentrations indistinguishable from those inferred for the Icelandic OIB-source mantle from He-Pb isotope mixing observations *Hilton et al.*, 2000]. In this form, the lack of a high-^{3}He flux at hotspots is no longer a constraint that can be used to argue against a steady-state fluxing model.

The expansion of the mantle model from the "volatile-rich" and "convecting" mantle reservoirs to a third, "recycled", reservoir has became an implicit component of more recent fluxing models that attempt to accommodate whole-mantle convection. These can be traced back to models which postulated "blobs" or "plums" of enriched material that are passive within the convecting mantle but by erosion or entrainment provide the source for OIB material [*Davies*, 1984; *Silver et al.*, 1988]. *Manga* [1996] demonstrated with numerical simulations that blobs can be preserved in a convective regime if they are 10–100 times more viscous than the surrounding mantle. Similarly, an increase in density can preserve a deep mantle layer from being homogenized by mantle convection [*Kellogg et al.*, 1999]. Applied to noble gases these take the form, for example, of density-controlled layering at 1600–1700 km depth [*Kellogg et al.*, 1999] or blobs [*Becker et al.*, 1999] within a convecting mantle that preserve a volatile-rich reservoir, while erosion of these regions by recycled material then provides a similar steady-state flux of volatile-rich material to buffer the convecting mantle system.

In a quantitative approach to this problem, *Coltice and Ricard* [1999] investigated a model in which the recycled

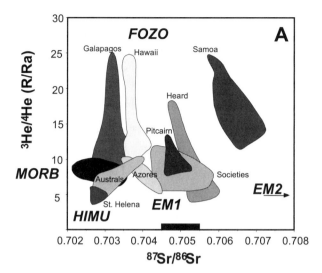

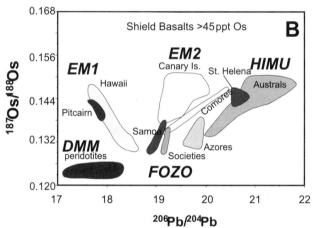

Figure 6. (A) Elongate arrays for oceanic basalts in a plot of ^{3}He/^{4}He against ^{87}Sr/^{86}Sr appear to show a common origin for the high ^{3}He/^{4}He. (B) Data arrays for ^{187}Os/^{188}Os against ^{206}Pb/^{204}Pb again converge on a high ^{3}He/^{4}He component. This high ^{3}He/^{4}He component has an elevated ^{187}Os/^{188}Os ratio indicative of recycled mafic material. Figure adapted from *van Keken et al.* [2002]; for original references see text.

component itself was divided into two components—a depleted harzburgitic residue and an altered oceanic crust. The subducted oceanic crust forms the D″ reservoir while the harzburgitic residue is mixed with a high-^{3}He/^{4}He reservoir located immediately above D″ (called the residual depleted mantle [RDM]), with starting conditions similar to that of the closed system mantle models. The importance of adding the harzburgitic material to the high-^{3}He/^{4}He reservoir is that, because it is depleted in (U + Th), it will have little impact on the ^{3}He/^{4}He yet will significantly lower the ^{3}He concentration in this reservoir. While containing a number of presently unconstrained fractionation factors for

the recycled components, this work illustrates the potential for plumes dominated by the RDM to be the source of the high-^{3}He/^{4}He reservoir. Likewise, oceanic basalts with a strong HIMU signature would be dominated by recycled oceanic crust material stored in the D″ reservoir [*Christensen and Hofmann*, 1994]. Nevertheless, the weakest aspects of these latter models [*Kellogg et al.*, 1999; *Coltice and Ricard*, 1999; *Becker et al.*, 1999] are the stringent density, size, and temperature requirements for the volatile-rich reservoirs to remain seismically invisible [*Vidale et al.*, 2001], yet still be preserved from being homogenized within the convecting mantle (e.g., *van Keken and Ballentine* [1999]).

In an attempt to reconcile whole-mantle convection without the need for seismically invisible mantle reservoirs, *Bercovici and Karato* [2003] have proposed that water-induced melting at the 440 km phase change prevents all but a small flux of noble gases and lithophile elements from being carried into the upper MORB-source mantle. Although different in physical form, this "water-filter" provides a boundary that separates a volatile and lithophile-rich lower mantle from the MORB-source upper mantle and can be considered in terms of geochemical considerations to be similar to the layered mantle fluxing models.

In a variant on the blob mantle reservoir, *Morgan and Morgan* [1999] proposed that the OIB, enriched, and high-^{3}He/^{4}He component is dispersed in much smaller scale veins, or plums, and therefore requires no density or viscosity contrast to preserve this compositional heterogeneity. These authors invoke a two-stage melting process. The first stage is a result of a thermal pulse that preferentially melts the lower melting point, enriched, veins to form the OIB. The residue is depleted in high ^{3}He/^{4}He components and has the isotopic composition of MORB. This residue is thermally buoyant and forms a layer at the top of the asthenospheric mantle. Passive decompressive melting at mid-ocean ridges then taps this layer to provide the geochemical signature of MORB. While the obvious attraction of this model is no requirement for seismically defined reservoirs, the generation of the observed uniform MORB-source ^{3}He/^{4}He from the residue of a heterogeneous OIB source has yet to be addressed. Similarly, because the model requires that all MORB melts previously supplied OIB, the ^{3}He flux from OIB volcanism would be expected to be several orders of magnitude greater than that seen at mid-ocean ridges. As discussed, this is not observed.

Extant Flux Model Weaknesses

The physical mechanism by which regions of the mantle have preserved their distinct chemical and isotopic character over the lifetime of convection within the Earth is not the focus of this paper. We discuss here the geochemical evidence that supports or refutes the concept of a volatile-rich mantle reservoir fluxing the convecting mantle system. In assessing the viability of this type of model, it is clear that He isotopes sit well with this form of conceptual model. The high concentration of ^{3}He inferred for the high ^{3}He/^{4}He source region, however, has often been presented as a problem. The closed system ^{3}He concentrations compare with the lower range observed in carbonaceous chondrites [*Mazor et al.*, 1970]. From impact degassing studies [*Ahrens et al.*, 1989], it has always been assumed that it is highly unlikely for chondritic material to preserve its volatile content during the accretionary process, and other mechanisms of volatile incorporation into the proto-Earth such as magma-ocean/massive-early-atmosphere equilibration then have to be invoked [*Mizuno et al.*, 1980; *Porcelli et al.*, 2001]. Any massive early atmosphere is predicted to have a composition similar to the solar nebula. Alternatively, solar corpuscular irradiation of pre-accretionary dust or small planetisimals may provide a high ^{3}He source and, although subject also to impact degassing, some scenarios have been discussed that may allow these to survive the accretionary process [*Trieloff et al.*, 2002; *Tolstikhin and Hofmann*, 2005]. Given the uncertainty surrounding the process(es) of volatile incorporation into the silicate portion of the Earth, the high concentration of ^{3}He in the volatile-rich mantle source cannot be regarded as a significant weakness of mantle-fluxing models.

To find a significant flaw in the fluxing models, we have to broaden our consideration to other noble gases. Unlike most other noble gas elemental ratios which are significantly and variably fractionated during eruption and magma degassing (e.g., *Sarda and Moreira* [2002]), the He/Ne value can be determined from He and Ne isotopes alone, and the He/Ne elemental ratio prior to the eruption process can be determined (see background section above and also *Honda et al.* [1991], *Honda and McDougal* [1998], and *Harrison et al.* [1999]). *Honda and McDougal* [1998] first resolved the ^{3}He/^{22}Ne difference between OIB and MORB sources, showing the MORB source to have a ^{3}He/^{22}Ne ratio up to 70% greater than that of the OIB-source mantle. Since this work, several other workers have highlighted this difference between MORB and OIB ^{3}He/^{22}Ne ratios (e.g., *Dixon et al.* [2000]). We have further demonstrated this difference in the section on He-Ne in the mantle. To maintain a difference between the flux source and the convecting mantle requires an additional complexity to be added to flux models: Either the extraction of Ne is more efficient than that of He from the convecting mantle, resulting in a shorter effective residence time for Ne in the convecting mantle, or the input of He into the convecting mantle is more efficient than that of Ne, perhaps caused by a diffusive component in the entrainment of

material from the high ^{3}He/^{4}He reservoir. While it is unlikely that the differential extraction of He and Ne is significantly different at mid-ocean ridges, again a poor understanding of the mechanism of material transfer into the convecting mantle means that this is not a fatal flaw.

To date, the only noble gas evidence that would irrefutably contradict the fluxing models is Ne isotope heterogeneity between the convecting mantle ^{20}Ne/^{22}Ne = 12.5 and evidence from plume-source volcanic material with reproducible values of ^{20}Ne/^{22}Ne > 13.0 ± 0.2 from the Kola peninsula and a single measurement from Iceland where ^{20}Ne/^{22}Ne = 13.75 ± 0.32 [*Harrison et al.*, 1999; *Trieloff et al.*, 2000; *Yokochi and Marty*, 2004; *Ballentine et al.*, 2005]. Ne isotope heterogeneity in the mantle, with a distinct difference between the value for the convecting mantle and that of the plume-source system, would rule out all models that flux the convecting mantle from a volatile-rich plume-source mantle. There is no known mechanism to reduce the Ne isotopic composition of solar nebula Ne with ^{20}Ne/^{22}Ne = 13.8 to the value of 12.5 in the convecting mantle [*Ballentine et al.*, 2005]. The assumption that all plumes have solar nebula Ne isotopic source composition is currently founded on a limited but (in the case of the Kola samples) reproducible data set. It is clear however that the fundamental change in our understanding of the relationship between the plume-source mantle and convecting mantle that this extrapolation creates makes the exploration of the character of the Ne isotopic composition of plume systems to be one of the most important short-term goals in mantle geochemistry.

Low ^{3}He Concentration Models

Anderson [1998] initiated the debate about the possibility of a mechanism that could create a volatile poor and low (U + Th) reservoir that could source high ^{3}He/^{4}He in OIB samples. *Anderson* argued that He could be incorporated within a lithosphere highly depleted in (U + Th), and could therefore preserve the ^{3}He/^{4}He at the time of trapping. This could take the form, for example, of harzburgitic residues that have been overprinted by migrating CO_2-rich fluids and accompanying noble gases. This model had the attraction of maintaining a relatively uniform mantle in a whole-mantle convective regime, while accounting for apparently lower ^{3}He concentrations in the OIB-source. However, for this shallow reservoir to source the high ^{3}He/^{4}He in oceanic intraplate volcanism, its maximum age would be on a similar timescale to that of the underlying oceanic crust. In the case of the Hawaiian hotspot this is some 80–100 Ma. In the most conservative case this would require the convecting mantle to have had ^{3}He/^{4}He = 37 R$_a$ (the highest value observed at Loihi) 100 Ma ago. Simple extrapolation of the convecting mantle ^{3}He/^{4}He from 8 R_a today, through 37 R_a at 100 Ma, soon requires the ^{3}He/^{4}He value of the convecting mantle to be far in excess of the initial solar ratio (120 R_a) at 4.5 Ga [*Porcelli and Ballentine*, 2002].

Other workers have discussed depletion of He relative to (U + Th) during the melting process. While it was generally assumed that He is more incompatible than U + Th, it has recently been suggested that He may in fact be more compatible than U + Th under certain pressure/temperature regimes, leaving a residual mantle that is more enriched in He relative to U + Th [*Graham et al.*, 1990; *Anderson*, 1998; *Helffrich and Wood*, 2001; *Brooker et al.*, 2003]. In this case the high ^{3}He/^{4}He observed in OIB could be preserved in mantle reservoirs that have been previously depleted by melting. This concept has not been extensively explored to examine the evolution of the MORB- and OIB-source reservoirs; however, the concentration of ^{3}He in a depleted source will be an important test of any such model development. For example, a 10% partial melt of a mantle source with a silicate/melt partition coefficient of 8×10^{-3} for He [*Marty and Lussiez*, 1993] will leave only 0.8% of the He in the residue. This is an upper limit [*Kurz*, 1993]. If we consider that the evidence from the Icelandic system suggests that the high ^{3}He/^{4}He source has ^{3}He concentrations higher than the MORB source (Hilton et al., 2000), then following this argument, the original source would have to have had ^{3}He concentrations 125 times higher than the current estimates of the MORB-source mantle. This does not appear to be a viable mechanism to generate high-^{3}He/^{4}He material in the mantle.

^{40}Ar Mass Balance and the Helium/Heat Imbalance

We have discussed up to this point how He (and Ne) concentration and isotope information has been used to derive a variety of conceptual models. Most of these models also address to greater or lesser extents two other observations that have been used to argue for discrete mantle domains. The first of these is the helium/heat imbalance [*O'Nions and Oxburgh*, 1986]. ^{4}He and heat are both coproducts of U and Th decay in the mantle and are produced in well-defined proportions. Even taking into account mantle heat from other sources, such as the Earth's core and secular cooling [*van Keken et al.*, 2001], the Earth today is losing proportionally more heat than He [*O'Nions and Oxburgh*, 1983]. This led *O'Nions and Oxburgh* to suggest that there is a boundary layer within the Earth that allows heat to pass conductively but prevents ^{4}He from efficiently escaping. While originally applied to the simple layered-mantle models with the boundary placed at the 670 km phase change, this applies equally well to any model that requires a boundary between the convecting mantle and a portion of the mantle with a

higher concentration of U + Th (e.g., *Kellogg et al.* [1999]; *Coltice and Ricard* [1999]; *Becker et al.* [1999]; *Bercovici and Karato* [2004]).

The ^{40}Ar mass balance is complementary to this picture: If it is assumed that the BSE has a U/K ratio similar to that of MORB (1.27×10^4; *Jochum et al.* [1983]), then only 41% of the ^{40}Ar produced since accretion is now in the atmosphere [*Turcotte and Schubert*, 1988; *Allègre et al.*, 1996]. If the MORB-source reservoir extends to the core–mantle boundary to accommodate the ^{40}Ar not in the atmosphere, the mantle should have an average ^{40}Ar concentration of 8.7×10^{-6} cm^3 (STP)/g [*Ballentine et al.*, 2002]. From the present-day ^{3}He ocean ridge flux and ocean plate generation rate, the observed MORB ^{3}He/^{4}He, and the time-integrated ^{4}He/^{40}Ar production ratio, the ^{40}Ar concentration in the MORB-source mantle is calculated to be only $\approx 2.2 \times 10^{-6}$ cm^3 (STP)/g. This difference has been used to argue for a hidden mantle reservoir with a high ^{40}Ar concentration, originally in the mantle beneath 670 km depth. If the hidden reservoir is smaller than the lower mantle, it must have proportionally higher ^{40}Ar concentrations. Because the chemical behavior of K is similar to U and Th, models that advocate a (U + Th)-enriched mantle reservoir will also place the "missing" K and therefore ^{40}Ar in this reservoir. While some workers have suggested that the total K budget of the Earth may have been overestimated [*Albarède*, 1998; *Davies*, 1999], the relative proportion of moderately volatile elements lics on a compositional trend defined by different meteorite classes [*Allègre et al.*, 1995; *Halliday and Porcelli*, 2001], and trends in Rb/Sr and K/U are compatible with a terrestrial value of K/U = 1.27×10^4 [*Halliday and Porcelli*, 2001]. Successful models must therefore account for the "missing" ^{40}Ar.

SUMMARY: TOWARDS A "ZERO PARADOX" MANTLE?

The requirement for a ^{3}He flux into the MORB-source mantle, a mantle boundary to produce the ^{4}He/heat imbalance, and a ^{40}Ar-rich reservoir are all derived from one observation—the present-day ^{3}He flux into the oceans, which in turn is used to derive the ^{3}He and related noble gas concentrations in the MORB-source mantle. This is calculated by dividing the oceanic ^{3}He flux into the average ocean crust generation rate, assuming a partial melt of 10%. It is important, however, to note that the residence time of ^{3}He in the oceans is ≈ 1000 yr and contrasts with the average age of oceanic crust (~80 Ma). Comparing systems with such different timescales is questionable. Today, the mid-ocean ridges degas significantly only along a few segments of the 65,000 km of spreading ridge [*Schlosser and Winkler*, 2002]. It would appear that the concentration estimate upon which so many of the noble gas models are based is unsound [*Ballentine et al.*, 2002].

It has been shown that if the convecting mantle has a noble gas concentration approximately 3.5 times higher than currently accepted, there is then no requirement for a ^{3}He flux into the convecting mantle source. If the ^{40}Ar concentration in the convecting mantle is higher by a similar amount, there is also no requirement for a region of the mantle to have significantly higher ^{40}Ar concentrations within a whole-mantle convective regime. Similarly, if the ^{3}He concentration is higher, the ^{4}He concentration must also be higher by a similar amount and the average degassing rate far higher over the timescale of ocean crust generation. An increase in average ^{4}He flux, again by a factor of about 3.5, then balances the amount predicted by the heat flux. This model concentration for noble gases in the convecting mantle has been called the Zero Paradox concentration [*Ballentine et al.*, 2002].

Numerical models that have investigated the He/heat imbalance also show a high frequency variance in He/heat ratios. These are dominated by variance in the He degassing flux at mid-ocean ridges compared with a more uniform heat loss throughout the oceanic crust[*van Keken et al.*, 2001]. Similarly, numerical simulations of whole-mantle degassing rates suggest that the amount of ^{40}Ar expected to be released to the atmosphere is on the order of 50%, a similar amount to that observed [*van Keken and Ballentine*, 1998, 1999]. While constraints now exist on mantle carbon concentrations from observed Nb/C correlation in MORB melt inclusions [*Saal et al.*, 2002], converting this to a ^{3}He concentration remains limited by uncertainty in convecting mantle ^{3}He/CO$_2$ ratios (0.4–6.3 × 10^9) [*Marty and Jambon*, 1987; *Graham*, 2002]. While there is no direct evidence that the average convecting mantle concentration is as high as the zero paradox concentration, there is equally little to support the concentrations derived from the ^{3}He ocean flux estimates [*Ballentine et al.*, 2002]. Nevertheless, if the zero paradox concentration is approached, the need for a ^{3}He source external to the convecting mantle is removed, and there is then no need for the noble gases that are observed in the OIB system to have a genetic connection with those in the convecting mantle—a result that sits comfortably with the observation of ^{3}He/^{22}Ne and ^{20}Ne/^{22}Ne heterogeneity in differently sourced basalt samples.

We can speculate that in such a zero paradox mantle model, convection of the whole mantle outgases only ≈50% of the mantle volatiles trapped since accretion over 4.5 Ga. All the convecting mantle noble gases can be accounted for in their entirety by a residual meteoritic (solar wind–implanted) volatile source, degassing, a small amount of recycling [*Porcelli and Wasserberg*, 1995a,b], and radiogenic noble gas production within a hetcrogeneously depleted/enriched mantle.

Homogenisation of the convecting mantle by ridge-scale melting does not remove the variance in either radioelement or volatile isotope composition completely. On the sampling scale of mid-ocean ridges this produces the observed average ^{3}He/^{4}He of 8.75 ± 2.14 R$_a$ [*Graham*, 2002]. Numerical models of whole-mantle convection show similar variance in ^{3}He/^{4}He and total outgassing of ~50% [*van Keken and Ballentine*, 1999]. Slab recycling to the D″ or core–mantle boundary results in ^{3}He-rich and high ^{20}Ne/^{22}Ne material being variably eroded from a deep reservoir. The transferred volatile-rich material, by mass balance, is very small and has little impact on the OIB radiogenic isotope tracers. Recycled material that does not acquire a significant volatile-rich component may exhibit features similar to low ^{3}He/^{4}He HIMU OIB. In this model, the primitive volatile charged plumes must efficiently transport their volatiles to the atmosphere, with negligible plume-source volatiles fluxing the convecting mantle. Numerical fluid dynamical models are required to test the efficiency of this transport and degassing mechanism. Such a model, however, would also account for the high ^{3}He/^{4}He and ^{20}Ne/^{22}Ne observed in plume material. Initial numerical modelling results suggest that a small amount of trapped, seawater-derived noble gases in the recycled material are sufficient to maintain the observed difference in ^{40}Ar/^{36}Ar and ^{129}Xe/^{130}Xe isotopic ratios between MORB and plume materials, without affecting the He or Ne isotopic composition [*Ballentine et al.*, 2004]). This provides a conceptual noble gas model of the mantle that is entirely consistent with whole-mantle convection observed by seismic tomography and numerical simulations, provides a self-consistent explanation for the convecting mantle volatile content, and does not require a major hidden reservoir within the Earth's mantle to have preserved or trapped the very high volatile concentrations required by fluxing models.

Acknowledgments. The authors thank everyone in the Isotope groups at Manchester and ETHZ as well as the AGU editorial team. We are grateful to M. Kurz for a constructive and critical review of this work.

REFERENCES

Ahrens, T. J., O'Keefe, J. D. and Lange, M. A. Formation of atmospheres during accretion of the terrestrial planets. In *Origin and evolution of planetary and satellite atmospheres*. Edited by S. K. Atreya, J. B. Pollack, and M. S. Mathews, pp. 328–385, University of Arizona press, Tucson, 1989.

Albarède, F. Time-dependent models of U-Th-He and K-Ar evolution and the layering of mantle convection. *Chem. Geol. 145*, 413–429, 1998.

Allègre, C. J., Staudacher, T., Sarda P., and Kurz, M., Constraints on evolution of Earth's mantle from rare gas systematics, *Nature*, *303*, 762–766, 1983.

Allègre, C. J., Poirier, J. P., Humler, E. and Hofmann, A. W. The Chemical Composition of the Earth. *Earth and Planet. Sci. Lett. 134 (3–4)*, 515–526, 1995.

Allègre, C. J., Staudacher, T. and Sarda, P. Rare gas systematics: formation of the atmosphere, evolution and structure of the Earth's mantle. *Earth and Planet. Sci. Lett. 87*, 127–150, 1986.

Anderson, D. L. Helium-3 from the mantle: Primordial signal or cosmic dust? *Science 261*, 170–176, 1993.

Anderson, D. L. The helium paradoxes. *Proc. Nat. Acad. Sci. USA 95 (9)*, 4822–4827, 1998.

Ballentine, C. J., and P. G. Burnard, Production, release and transport of noble gases in the continental crust, *Rev. Min. Geochem., 47*, 481–538, 2002.

Ballentine, C. J. and Barfod D. N. The origin of air-like noble gases in MORB and OIB. *Earth and Planet. Sci. Lett. 180*, 39–48, 2000.

Ballentine, C. J., Lee, D-C. and Halliday, A. N. Hafnium isotopic studies of the Cameroon line and new HIMU paradoxes. *Chem. Geol, Hofmann Volume 139*, 111–124, 1997.

Ballentine, C. J., D. Porcelli, and R. Wieler, Technical comment on 'Noble gases in mantle plumes' by Trieloff et al. (2000), *Science, 291*, 2269a, 2001.

Ballentine, C. J., Van Keken, P. E., Porcelli, D. and Hauri, E. H. Numerical models, geochemistry and the zero paradox noble-gas mantle. *Phil. Trans. R. Soc. Lond. A 360*, 2611–2631, 2002.

Ballentine, C. J., Van Keken P. E., Holland G., Hauri, E. H.and Brandenburg, J. Numerical modeling of noble gas recycling into the mantle, *http://www.agu.org/meetings/fm04/fm04-sessions/fm04_U33B.html, (abstract)* 05, 2004

Ballentine, C. J., B. Sherwood Lollar, B. Marty, and M. Cassidy, Neon isotopes constrain convection and volatile origin in the Earth's mantle, *Nature, 433*, 33–38, 2005.

Barfod, D. N., Ballentine, C. J., Halliday, A. N. and Fitton, J. G. Noble gases in the Cameroon line and the He, Ne and Ar isotopic compositions of high μ (HIMU) mantle. *J. Geophys. Res. 104*, 29509–29527,1999.

Black D. C. On the origins of trapped helium, neon and argon isotopic variations in meteorites-I. Gas-rich meteorites, lunar soil and breccia. *Geochim. Cosmochim. Acta 36*, 347–375,1972

Becker, T. W., Kellogg, J. B. and O'Connell, R. J. Thermal constraints on the survival of primitive blobs in the lower mantle. *Earth and Planet. Sci. Lett. 171*, 351–365, 1999.

Bercovici, D., and Karato, S. Whole-mantle convection and the transition-zone water filter. *Nature 425*, 39–44, 2003.

Breddam, K. and Kurz, M. D. Helium isotopic signatures of Icelandic alkaline lavas. *Eos 82 (47)*, 2001.

Brooker, R. A., Z. Du, J. D. Blundy, S. P. Kelley, N. L. Allan, B. J. Wood, E. M. Chamorro, J. A. Wartho, and J. A. Purton, The 'zero charge' partitioning behaviour of noble gases during mantle melting, *Nature, 423 (6941)*, 738–741, 2003.

Burnard, P. G. Correction for volatile fractionation in ascending magmas: noble gas abundances in primary mantle melts. *Geochim. Cosmochim. Acta 65*, 2605–2614, 2001.

Burnard, P. G. Diffusive fractionation of noble gases and helium isotopes during mantle melting. *Earth and Planet Sci. Lett. 7009*, 1–9, 2004.

Burnard, P. G., Graham D. and Turner, G. Vesicle Specific Noble Gas Analyses Of Popping Rock: Implications For Primordial Noble Gases In Earth. *Science 276*, 568–571, 1997.

Burnard P. G. and Harrison, D. Argon isotope constraints on modification of oxygen isotopes by surficial processes. *Chem. Geol. 216*, 143–156, 2005.

Burnard, P. G., Harrison, D., Turner, G. and Nesbitt, R. The degassing and contamination of noble gases in mid-Atlantic Ridge basalts. *Geochem. Geophys. Geosys.* 2002GC000326, 2003

Carroll, M. R. and Webster, J. D. Solubilities of sulfur, noble gases, nitrogen, chlorine and fluorine in magmas. In *Volatiles in Magmas, Min. Soc. America 30*, 231–271, 1994.

Chamorro, E. M., R. A. Brooker, J. A. Wartho, B. J. Wood, S. P. Kelley, and J. D. Blundy, Ar and K partitioning between clinopyroxene and silicate melt to 8 GPa, *Geochim. Cosmochim. Acta, 66*, 507–519, 2002.

Christensen, U. R. and Hofmann, A. W. Segregation of subducted oceanic-crust in the convecting mantle. *J Geophys. Res. 99 (B10)*, 19867–19884, 1994.

Coltice, C. and Ricard, Y. Geochemical observations and one layer mantle convection. *Earth and Planet. Sci. Lett. 174*, 125–137, 1999.

Davies, G. F., Geophysical and isotopic constraints on mantle convection: and interim synthesis. *J. Geophys. Res. 89*, 6016–6040, 1984.

Davies, G. F., Geophysically constrained mantle mass flows and the Ar-40 budget: a degassed lower mantle? *Earth and Planet. Sci. Lett. 166 (3–4)*, 149–162, 1999.

Dixon, E. T., Honda, M., McDougall, I., Campbell, I. H. and Sigurdsson, I. Preservation of near-solar neon isotopic ratios in Icelandic basalts. *Earth and Planet Sci. Lett. 180 (3–4)*, 309–324, 2000.

Dunai, T. J. and Baur, H. Helium, neon and argon systematics of the European subcontinental mantle: Implications for its geochemical evolution. *Geochim. Cosmochim. Acta 59 (13)*, 2767–2783, 1995.

Eiler, J. M., K. A. Farley, J. W. Valley, A. W. Hofmann, and E. M. Stolper, Oxygen Isotope Constraints On the Sources Of Hawaiian Volcanism. *Earth and Planet. Sci. Lett. 144 (3–4)*, 453–467, 1996.

Farley, K. A., Maierreimer, E., Schlosser, P. and Broecker, W. S. Constraints on mantle He-3 fluxes and deep-sea circulation from an oceanic general circulation model. *J. Geophys. Res.—Solid Earth, 100 (B3)*, 3829–3839, 1995.

Farley, K. A. and Poreda, R. J. Mantle Neon and Atmospheric Contamination. *Earth and Planet. Sci. Lett. 114 (2–3)*, 325–339, 1993.

Gautheron, C. and Moreira, M. Helium signature of the subcontinental lithospheric mantle. *Earth and Planet. Sci. Lett. 199*, 39–47, 2002.

Graham, D. W. Noble gas isotope geochemistry of mid-ocean ridge and ocean island basalts: Characterization of mantle source reservoirs. *Rev. Min. Geochem. 47*, 247–318, 2002.

Graham, D. W., Lupton, F., Albarède, F. and Condomines, M. Extreme temporal homogeneity of helium isotopes at Piton de la Fournaise, Réunion Island. *Nature 347*, 545–548, 1990.

Halliday, A. N., Porcelli, D. In search of lost planets—the paleo-cosmochemistry of the inner solar system. *Earth and Planet. Sci. Lett. 192*, 545–559, 2001.

Harrison, D., Barry, T. and Turner, G. Possible diffusive fractionation of helium isotopes in olivine and clinopyroxene phenocrysts. *European Journal of Mineralogy 16*, 213–220, 2004.

Harrison, D., Burnard, P. G., Trieloff, M. and Turner, G. Resolving atmospheric contaminants in mantle noble gas analyses. *Geochem. Geophys. Geosys. 4 (3)*, 2003.

Harrison, D., Burnard, P. G. and Turner, G. Noble gas behaviour and composition in the mantle: constraints from the Iceland Plume. *Earth and Planet. Sci. Lett. 171 (2)*, 199–207, 1999.

Hanyu, T. and Kaneoka, I. The uniform and low 3He/4He ratios of HIMU basalts as evidence for their origin as recycled materials. *Nature 390*, 273–276, 1997.

Hauri, E. H. Major Element Variability In the Hawaiian Mantle Plume. *Nature 382*, 415–419, 1996.

Hauri, E. H., Whitehead, J. A. and Hart, S. A. Fluid dynamic and geochemical aspects of entrainment in mantle plumes. *J. Geophys. Res. 99*, 24275–24300, 1994.

Helffrich, G. R., and Wood, B. W. The Earth's Mantle. *Nature 412*, 501–507, 2001.

Hilton, D. R., Gronvold, K., Macpheron, C. G. and Castillo, P. R. Extreme He-3/He-4 ratios in northwest Iceland: constraining the common component in mantle plumes. *Earth and Planet. Sci. Lett. 173 (1–2)*, 53–60, 1999.

Hilton, D. R., and Porcelli, D. Noble gases as mantle tracers. In *The mantle and core*, edited by R. W. Carlson, pp. 277–318, Elsevier, Amsterdam, 2003.

Hilton, D. R., Thirlwall, M. F., Taylor, R. N., Murton, B. J. and Nichols, A. Controls on magmatic degassing along the Reykjanes Ridge with implications for the helium paradox. *Earth and Planet. Sci. Lett. 183*, 43–50, 2000.

Hofmann, A. W. Mantle geochemistry: The message from oceanic volcanism. *Nature 385*, 219–229, 1997.

Honda, M., and McDougall, I. Primordial helium and neon in the Earth–a speculation on early degassing. *Geophys. Res. Lett. 25*, 1951–1954, 1998.

Honda, M., McDougall, I., Patterson, D. B., Doulgeris, A. and Clague, D. A. Possible solar noble-gas component in Hawaiian basalts. *Nature 349*, 149–151, 1991.

Jean-Baptiste, P. Helium-3 distribution in the deep world ocean; its relation to hydrothermal 3He fluxes and to the terrestrial heat budget. In *Isotopes of noble gases as tracers in environmental studies*, edited by H. Loosli, and E. Mazor, pp. 219–240, International Atomic Energy Agency, Vienna, Austria, 1992.

Jochum, K. P., Hofmann, A. W., Ito, E., Seufert, H. M. and White, W. M. K, U, and Th in mid-ocean ridge basalt glasses and heat production, K/U and K/Rb in the mantle. *Nature 306*, 431–436, 1983.

Kallenbach R., Ipavich F. M., Bochsler P., Hefti S., Hovestadt D., Grunwaldt H., Hilchenbach M., Axford W. I., Balsiger H., Burgi A., Coplan M. A., Galvin A. B., Geiss J., Gliem F., Gloeckler G., Hsieh K. C., Klecker B., Lee M. A., Livi S., Managadze G. G., Marsch E., Mobius E., Neugebauer M., Reiche K. U., Scholer M., Verigin M. I., Wilken B., and Wurz P. Isotopic composition of solar wind neon measured by CELIAS/MTOF on board SOHO. *J. Geophys. Res.—Space Phys. 102 (A12)*, 26895–26904, 1997.

Kellogg, L. H., Hager, B. H. and Van der Hilst, R. D. Compositional stratification in the deep mantle. *Science 283*, 1881–1884, 1999.

Kellogg, L. H., and Wasserburg, G. J. The role of plumes in mantle helium fluxes. *Earth and Planet. Sci. Lett. 99*, 276–289, 1990.

Kunz, J., Staudacher, T. and Allègre, C. J. Plutonium-fission xenon found in Earth's mantle. *Science 280*, 877–880, 1998.

Kurz, M. D. Cosmogenic helium in a terrestrial igneous rock. *Nature 320*, 435–439, 1986.

Kurz, M. D., Mantle Heterogeneity Beneath Oceanic Islands—Some Inferences from Isotopes, *Philosophical Transactions of the Royal Society of London Series A. 342*, 91–103, 1993.

Kurz, M. D., Jenkins, W. J. and Hart, S. R. Helium isotopic systematics of oceanic islands and mantle heterogeneity. *Nature 297*, 43–46, 1982.

Kurz, M. D., Jenkins, W. J., Hart, S. R. and Clague, D. Helium isotopic variation in volcanic rocks from Loihi Seamount and the island of Hawaii. *Earth and Planet. Sci. Lett. 66*, 388–406, 1983.

Kurz, M. D., Meyer, P. S. and Sigurdsson, H. Helium isotopic variation within the neovolcanic zones of Iceland. *Earth and Planet. Sci. Lett. 74*, 291–305, 1985.

Leya, I. and Wieler, R. Nucleogenic production of Ne isotopes in Earth's crust and upper mantle induced by alpha particles from the decay of U and Th. *J. Geophys. Res. Solid Earth 104(B7)*, 15439–15450, 1999.

Lupton, J. E. and Craig, H. Excess 3He in oceanic basalts, evidence for terrestrial primordial helium. *Earth and Planet. Sci. Lett. 26 (2)*, 133–139, 1975.

Lux, G. The behaviour of noble gases in silicate liquids: solution, diffusion, bubbles and surface effects, with applications to natural samples. *Geochim. Cosmochim. Acta 51*, 1549–1560, 1987.

Mahaffy, P. R., Niemann, H. B., Alpert, A., Atreya, S. K., Demick, J., Donahue, T. M., Harpold, D. N. and Owen, T. C. Noble gas abundance and isotope ratios in the atmosphere of Jupiter from the Galileo Probe Mass Spectrometer. *J. Geophys. res. Planets, 105 (E6)*, 15061–15071, 2000.

Manga, M. Mixing of heterogeneities in the mantle: Effect of viscosity differences. *Geophys. Res. Lett. 23 (4)*, 403–406, 1996.

Marty, B., I. Tolstikhin, I. L. Kamensky, V. Nivin, E. Balaganskaya, and J. L. Zimmermann, Plume-derived rare gases in 380 Ma carbonatites from the Kola region (Russia) and the argon isotopic composition in the deep mantle, *Earth Planet. Sci. Lett., 164 (1–2)*, 179–192, 1998.

Marty, B., and Jambon, A. C/3He in volatile fluxes from the solid Earth: implications for carbon geodynamics, *Earth Planet. Sci. Lett., 83*, 16–26, 1987.

Marty, B, and Lussiez, P. Constraints on Rare Gas Partition Coefficients from Analysis of Olivine-Glass from a Picritic Mid-Ocean Ridge Basalt. *Chem. Geol. 106 (1–2)*, 1–7, 1993.

Mazor, E., D. Heyamnn, and E. Anders, Noble gases in carbonaceous chondrites, *Geochim. Cosmochim. Acta, 34*, 781–824, 1970.

Mizuno, H., Nakazawa, K. and Hayashi, C. Dissolution of the primordial rare gases into the molten Earth's material. *Earth and Planet. Sci. Lett. 50*, 202–210, 1980.

Moreira, M., Breddam, K. Curtice, J. and Kurz, M. D. Solar neon in the Icelandic mantle: new evidence for an undegassed lower mantle, *Earth and Planet Sci. Lett.*, *185* (1–2), 15–23, 2001.

Moreira, M., Kunz, J. and Allègre, C. J. Rare Gas Systematics in Popping Rock: Isotopic and Elemental Compositions in the Upper Mantle. *Science 279*, 1178–1181, 1998.

Moreira, M. and Sarda, P. Noble gas constraints on degassing processes. *Earth and Planet. Sci. Lett. 176*, 375–386, 2000.

Morgan, J. P. and Morgan, W. J. Two-stage melting and the geochemical evolution of the mantle: a recipe for mantle plum-pudding. *Earth and Planet. Sci. Lett. 170 (3)*, 215–239, 1999.

Morrison, P. and Pine, J. Radiogenic origin of the helium isotopes in rock. *Ann. N. Y. Acad. Sci. 62*, 71–92, 1955.

Niedermann, S. Cosmic-Ray Produced Noble Gases in Terrestrial Rocks: Dating Tools for Surface Processes. *Rev. Min. Geochem. 47*, **731**–777, 2002.

Niedermann S., Bach W., and Erzinger J. Noble Gas Evidence For a Lower Mantle Component In MORBs From the Southern East Pacific Rise: Decoupling Of Helium and Neon Isotope Systematics. *Geochim. Cosmochim. Acta 61 (13), 2697–2715, 1997*

O'Nions, R. K. Relationships between chemical and convective layering in the earth. *J. Geol. Soc. London 144*, 259–274, 1987.

O'Nions, R. K., and Tolstikhin, I. N. Behaviour and residence times of lithophile and rare gas tracers in the upper mantle. *Earth and Planet. Sci. Lett. 124*, 131–138, 1994.

O'Nions, R. K., and Oxburgh, E. R. Heat and helium in the Earth. *Nature 306*, 429–431, 1983.

Ozima, M. and Podosek F. A. Noble gas geochemistry. *Cambridge University Press*, 1983.

Parsons, B. The rates of plate creation and consumption. *Geophys. J. R. Astron. Soc. 67*, 437–448, 1981.

Patterson D. B., Honda M., and McDougall I. Atmospheric contamination: A possible source for heavy noble gases in basalts from Loihi Seamount, Hawaii. *Geophys. Res. Lett. 17*, 705–708, 1990

Porcelli D. and Pepin R. O. The origin of noble gases and major volatiles in the Terrestrial planets. In *Treatise on Geochemsitry: The Atmosphere* (ed. R. F. Keeling) *4.* 319–342, 2004 Eselvier.,

Pepin, R. O. and Porcelli, D. Origin of Noble Gases in the Terrestrial Planets. *Rev. Min. Geochem. 47*, **191**–239, 2002.

Porcelli, D. and Ballentine, C. J. Models for the distribution of Terrestrial noble gases and evolution of the atmosphere. *Rev. Min. Geochem. 47*, 411–480, 2002.

Porcelli, D. and Wasserburg, G. J. Mass transfer of xenon through a steady-state upper mantle. *Geochim. Cosmochim. Acta 59 (10)*, 1991–2007, 1995a.

Porcelli, D. and Wasserburg, G. J. Mass transfer of helium, neon, argon, and xenon through a steady-state upper mantle. *Geochim. Cosmochim. Acta 59 (23)*, 4921–4937, 1995b.

Porcelli, D., Woolum, D. S. and Cassen, P. Deep Earth rare gases: initial inventories, capture from the solar nebula, and losses during Moon formation. *Earth and Planet Sci. Lett. 193*, 237–251, 2001.

Saal, E. S., E. H. Hauri, C. H. Langmuir, and M. R. Perfit, Vapour undersaturation in primitive mid-ocean-ridge basalt and the volatile content of Earth's upper mantle, *Nature, 419*, 451–455, 2002.

Sarda, P. and Moreira, M. Vesiculation and vesicle loss in mid-ocean ridge basalt glasses: He, Ne, Ar elemental fractionation and pressure influence. *Geochim. Cosmochim. Acta 66*, 1449–1458, 2002.

Sarda, P., Staudacher, T. and Allègre, C. J. 40Ar/36Ar ratios in MORB glasses: constraints on atmosphere and mantle evolution. *Earth and Planet Sci. Lett. 72*, 357–375, 1985.

Sarda, P., Staudacher, T. and Allègre, C. J. Neon isotopes in submarine basalts. *Earth and Planet Sci. Lett. 91*, 73–88, 1988.

Schlosser, P. and Winckler, G. Noble gases in the ocean and ocean floor. *Rev. Min. Geochem. 46*, 701–730, 2002.

Silver, P. G., Carlson, R. W. and Olson, P. Deep slabs, geochemical heterogeneity, and the large-scale structure of mantle convection—investigation of an enduring paradox. *Ann. Rev. Earth and Planet. Sci. 16*, 477–541, 1988.

Sleep, N. H. Hotspots and mantle plumes: Some phenomenology. *J. Geophys. Res. 95*, 6715–6736, 1990.

Sobolev, A. V., Hofmann, A. W. and Nikogosian, I. K. Recycled oceanic crust observed in 'ghost plagioclase' within the source of Mauna Loa lavas. *Nature 404*, 986–990, 2000.

Staudacher T and Allègre C. J. Terrestrial Xenology. *Earth and Planet Sci. Lett. 60*, 389–406, 1982.

Stuart, F. M. Speculations About the Cosmic Origin of He and Ne in the Interior of the Earth—Comment. *Earth and Planet Sci. Lett. 122 (1–2)*, 245–247, 1994.

Stuart, F. M., Lass-Evans, S., Fitton, J. G. and Ellam, R. M. High 3He/4He ratios in picritic basalts from Baffin Island and the role of a mixed reservoir in mantle plumes. *Nature 424*, 57–59, 2003.

Torgersen, T. Terrestrial helium degassing fluxes and the atmospheric helium budget: Implications with respect to the degassing processes of continental crust. *Chem. Geol. 79*, 1–14, 1989.

Tolstikhin, I. N., and A. W. Hofmann, Early crust on top of the Earth's core, *Phys. Earth Planet. Interior*, *148* (2–4), 109–130, 2005.

Trieloff, M., Kunz, J. and Allègre, C. J. Noble gas systematics of the Reunion mantle plume source and the origin of primordial noble gases in Earth's mantle. *Earth and Planet Sci. Lett. 200*, 297–313, 2002.

Trieloff, M., Kunz, J., Clague, D. A., Harrison, D. and Allègre, C. J. The nature of pristine noble gases in mantle plumes. *Science 288*, 1036–1038, 2000.

Trieloff, M., Falter, M., Buikin, A. I., Korochantseva, E. V., Jessberger, E. K. and Altherr, R. Argon isotope fractionation induced by stepwise heating. *Geochim. Cosmochim. Acta 69 (5)*, 1253–1264, 2005.

Trull, T. W. and Kurz, M. D. Experimental Measurements of He-3 and He-4 Mobility in Olivine and Clinopyroxene at Magmatic Temperatures. *Geochim. Cosmochim. Acta 57 (6)*, 1313–1324, 1993.

Trull, T. W. and Kurz, M. D. Isotopic fractionation accompanying helium diffusion in basaltic glass. *Journal of Molecular Structure 485–486*, 555–567, 1999.

Turcotte, D. L. and Schubert, G. Tectonic implications of radiogenic noble gases in planetary atmospheres. *Icarus 74*, 36–46, 1988.

Van der Hilst, R. D., Widiyantoro, S. and Engdahl, E. R. Evidence For Deep Mantle Circulation From Global Tomography. *Nature 386*, 578–584, 1997.

Van Keken, P. E. and Ballentine, C. J. Whole-mantle versus layered mantle convection and the role of a high viscosity lower mantle in terrestrial volatile evolution. *Earth and Planet. Sci. Lett. 156*, 19–32, 1998.

Van Keken, P. E. and Ballentine, C. J. Dynamical models of mantle volatile evolution and the role of phase transitions and temperature-dependent rheology. *J. Geophys. Res. Solid Earth 104 (B4)*, 7137–7151, 1999.

Van Keken, P. E., Ballentine, C. J. and Porcelli, D. A dynamical investigation of the heat and helium imbalance. *Earth and Planet. Sci. Lett. 188 (3–4)*, 421–434, 2001.

Van Keken, P. E., Hauri, E. H. and Ballentine, C. J. Mantle mixing: The generation, preservation, and destruction of chemical heterogeneity. *Annu. Rev. Earth Planet. Sci. 30*, 493–525, 2002.

Van Keken, P. E., Hauri, E. H. and Ballentine, C. J. Convective mixing in the Earth's mantle. In *The Mantle and Core*, edited by R. W. Carlson, pp. 471–492, Elsevier, New York, 2003.

Vidale J. E., Schubert G., and Earle P. S. Unsuccessful initial search for a midmantle chemical boundary with seismic arrays. *Geophys. Res. Lett. 28*, 859–862, 2001

White, W. M. and Hofmann, A. W. Sr and Nd isotope geochemistry of oceanic basalts and mantle evolution. *Nature 296*, 821–825, 1982.

Wieler, R. Noble Gases in the Solar System. *Rev. Min. Geochem. 47*, 21–63, 2002.

Yatsevich, I. and Honda, M. Production of Nucleogenic Neon In the Earth From Natural Radioactive Decay. *J. Geophys. Res. Solid Earth 102 (B5)*, 10291–10298, 1997.

Yokochi R. and Marty B. A determination of the neon isotopic composition of the deep mantle. *Earth and Planet Sci. Lett. 225*, 77–88, 2004.

Zindler, A. and Hart, S. R. Chemical Geodynamics. *Ann. Rev. Earth. Planet. Sci. 14*, 493–571, 1986.

C. J. Ballentine, School of Earth, Atmosphere and Environmental Science (SEAES), University of Manchester, Manchester M13 9PL, United Kingdom.

D. Harrison, Institut für Isotopengeologie und Mineralische Rohstoffe, ETH Zentrum, Sonneggstrasse 5, CH-8092, Zürich, Switzerland. (Darrell.Harrison@erdw.ethz.ch)

The Survival of Mantle Geochemical Heterogeneities

Francis Albarède

Ecole Normale Supérieure de Lyon, Lyon, France

The survival of mantle geochemical heterogeneities depends on the relative values of the residence times of individual elements in the mantle and of the scale-dependent mixing rate imposed by convection. Major and trace element data on fresh glasses from the PETDB data base were compiled to extract statistics on MORB compositions. It was found that the log-normal distribution provides an adequate description of the data. New means, modes, and standard deviations have been calculated after correction for crystal fractionation to 8% MgO. The residence times of the incompatible lithophile elements in the mantle were calculated as ratios of the mean mantle (Bulk Silicate Earth minus the continental crust) composition to fluxes of elements into the oceanic crust. They vary within a rather narrow range (4–9 Gy) and the larger uncertainties on U, Th, Ba, K, and Rb come from those on the continental crust composition. The mantle is therefore not at geochemical steady-state and the effect of its primordial composition on models of modern mantle geochemistry is still strong. Up to 50% of incompatible lithophile elements may never have been extracted into the oceanic crust, which expands a conclusion reached previously for ^{40}Ar. The residence time of heat in the mantle is indistinguishable from those of the incompatible lithophile elements implying that 50% of the Earth's primordial heat may still be buried at depth. This simple observation relieves the heat/helium paradox, which was used as an argument in favor of a conductive boundary layer at intermediate depth in the mantle. The balance between buoyancy flux and dissipation provides frame-independent estimates of the rates of mixing by mantle convection: Primordial geochemical anomalies with initial length scales comparable to mantle depths of plate lengths may be marginally visible at the scale of mantle melting underneath mid-ocean ridges ($\approx$50 km), but may show strongly in hot spot basalts and even more so in melt inclusions. A theory of residence time distributions of tracers in a convecting mantle with recycling of the oceanic crust is presented. It demonstrates that up to 50% primordial material may be present in the mantle, but scattered throughout as small (<10 km) domains, strongly sheared and refolded, and interbedded with younger recycled material.

1. INTRODUCTION

The demonstration by high-resolution mantle tomography of body waves that the transition zone is not an obstacle to penetration of lithospheric plates [*van der Hilst*, 1991, 1995; *Grand*, 1994; *Masters et al.*, 1996; *Li and Romanowicz*, 1996]

Earth's Deep Mantle: Structure, Composition, and Evolution
Geophysical Monograph Series 160
Copyright 2005 by the American Geophysical Union
10.1029/160GM04

has greatly strengthened the case of whole mantle convection. Further attempts by *Vidale et al.* [2001] and *Castle and van der Hilst* [2003] to identify global mid-mantle discontinuities, such as the top of the abyssal layer proposed by *Kellogg et al.* [1999], have produced no positive results. Overall seismic evidence against mantle layering was discussed by *Helffrich and Wood* [2001] and *Albarède and van der Hilst* [2002]. Mantle tomography therefore supports earlier claims based on the amplitude of the geoid and dynamic topography [e.g., *Davies and Richards*, 1992] that layered convection with a boundary in the transition zone is incompatible with the Earth's gravity field. However strong the geophysical arguments may be, they conflict with the well-entrenched geochemical perspective of a mantle made of separate reservoirs with little exchange, notably strengthened with the survival of non-radiogenic ^{3}He and solar Ne in the source of ocean island basalts [*Kurz et al.*, 1982; *Kaneoka, 1983*; *Allègre et al.*, 1983a, 1996; *Honda et al.*, 1991; *O'Nions and Tolstikhin*, 1994]. Probably the strongest argument against whole-mantle convection used to be that geochemical contrasts between the mean compositions of the upper and lower mantle would disappear in less than 1 Gy [e.g., *O'Nions and Tolstikhin*, 1996; *Albarède*, 1998]. It is only recently that whole-mantle convection models have started to explore the dispersal of geochemical anomalies using complex codes involving the various conservation equations of momentum, energy, and mass [*Gurnis and Davies*, 1986a,b; *Christensen and Hofmann*, 1994; *Schmalzl and Hansen*, 1994; *van Keken and Ballentine*, 1998, 1999; *Ferrachat and Ricard*, 1998, 2001; *Coltice and Ricard*, 1999; *Davies*, 2002; *Stegman et al.*, 2002; *Xie and Tackley*, 2004a,b].

Mixing refers to a combination of processes: (1) *stirring* by repeated streching and folding advects but preserves local geochemical (mostly isotopic) properties (2) *diffusion* alters these properties even in material at rest, and (3) *sampling* by collection of local partial melts into aggregated magmatic liquids determines some short-distance average of the geochemical properties. We will, in this work, focus on length scales (>10 m) that make the effect of diffusion negligible. The primary length scales I will consider are the characteristic dimensions of the initial geochemical heterogeneities and those of the melting regions over which melts are collected. Mixing efficiency is strongly scale-dependent, especially in 3D spherical models [*van Keken and Zhong*, 1999], and two conflicting views of mantle heterogeneity emerge from most of these models. The long-range perception, which is much akin to the view provided by the mean residence time theory (box models) so familiar to geochemists, is that the mean compositions of large domains tend to even out very rapidly. In contrast, local heterogeneities persist [e.g., *van Keken and Ballentine*, 1999; *Xie and Tackley*, 2004a,b],

sometimes in the form of long-lived layers or islands of stability [*Gurnies and Davies*, 1986a,b; *Christensen*, 1989; *Ferrachat and Ricard*, 1998]), and may survive for times in excess of the age of the Earth and therefore account for the presence of primordial ^{3}He and solar Ne in small parcels of primitive mantle strung out by the vigor of convection. This contrast is not fully unexpected from the theory of mixing [*Allègre and Turcotte*, 1986; *Olson et al.*, 1984; *Gurnis and Davies*, 1986b; *Ferrachat and Ricard*, 2001]. The number of parameters affecting convection models is very large and the uncertainties associated with these values are unclear. A clear limitation of all the tracer dispersal models so far, however, is that they deal with point-like 'particles' and not with finite-size coherent bodies that could represent real geological objects, such as oceanic plates. Our inadequate understanding of the early state of the terrestrial mantle, the uncertainties on the values of the Clapeyron slopes of phase transitions at high pressure, and of the viscosity of the lower mantle, as well as the necessity of increasing the resolution in complex 2.5 to 3D models, are only some of the major current limitations. None of the existing dynamic models have so far succeeded in reproducing even a modest fraction of the current geochemical characteristics of the modern mantle and crust to a satisfactory extent.

The correspondence between reservoir (or box) theory and standard continuum mechanics is well understood in the ocean sciences [*Keeling and Bolin*, 1967]. The residence time of an element in a particular reservoir is defined as the inverse of its probability of extraction or, equivalently, as the fraction of the element inventory extracted from the reservoir per unit time. Comparison of seawater residence times with mixing time has proven to provide an extremely powerful constraint on the chemical dynamics of the ocean [e.g., *Bolin and Rhode*, 1973]. As a rule of thumb, when the residence time of a tracer is longer than the time it takes to homogenize a fluctuation of its concentration at a given wavelength, the concentration of the tracer in the reservoir is homogeneous on the scale of that wavelength. A model of whole-mantle convection allows us to use the same approach for the terrestrial mantle. As a first approximation, I will assume that elements can only be extracted from the mantle into the oceanic crust (loosely defined to include oceanic plateaus)(Figure 1). They can further be extracted either from the oceanic crust into continental crust or recycled with the residuum into the deep mantle. Extraction of elements from the mantle wedge overlying subduction zones is restricted to what is introduced by melts and fluids lost by the subducting lithosphere. I will not assume that the mantle is homogeneous. Most of these features are those of the standard plum pudding model which has been used with a variety of nuances in the literature over the last two decades [*Davies*, 1981; *Zindler et al.*, 1984;

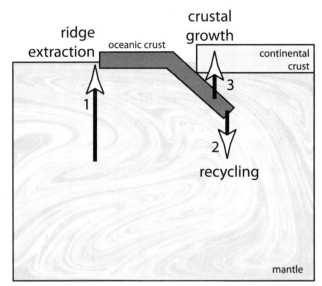

Figure 1. Model used to evaluate the residence times, their distribution, and the effect of oceanic crust recycling vs crustal growth. The residence times of each element in the mantle are calculated by dividing its total inventory in the mantle by its flux from the mantle into the oceanic crust (arrow 1).

Allègre and Lewin, 1995a; *Becker et al.*, 1999; *Kellogg et al.*, 2002; *Helffrich and Wood*, 2001; *Meibom and Anderson*, 2004]. The most important of these nuances is clearly the statement that heterogeneities must have been sheared and stretched to a very high extent through time and this insight gave rise to yet another culinary paradigm, the marble cake mantle [*Allègre and Turcotte*, 1986]. The crucial difference among models is whether separate mantle reservoirs coexist on a very large scale, typically lower vs upper mantle, over billions of years and how much transfer takes place between them [*Galer and O'Nions*, 1985; *O'Nions and Tolstikhin*, 1994]. In the present case, I simply consider the whole mantle as a single reservoir.

A number of new or improved observations makes this application of residence time theory topical. First, the modern data bases provide vastly improved estimates of the compositions of the various terrestrial reservoirs. After *Rudnick and Fountain* [1995] and *Taylor and McLennan* [1995)] who, among others, estimated a mean crustal composition, and after *Rudnick et al.* [1998], who assessed the composition of the subcontinental lithospheric mantle, the emergence of comprehensive databases, notably PETDB for Mid-Ocean Ridge Basalts (MORB) (http://petdb.ldeo.columbia.edu/petdb/) and GEOROC for Ocean Island Basalts (OIB) (http://georoc.mpch-mainz.gwdg .de/georoc/), is making it possible to estimate the residence times for a number of critical elements.

I will develop the theory of residence time distributions outlined in *Albarède* [2003] to help assess the fraction of

tracers that survives well beyond their mean residence time in the mantle. Second, the progressive isotopic mapping of the mid-ocean ridges by modern, high-throughput analytical techniques is revealing a wealth of features, such as sharp discontinuities [*Vlastelic et al.*, 2000; *Blichert-Toft et al.*, 2005] and a rich harmonic content of compositional periodograms[*Agranier et al.*, in press]. This second type of observation is first-order evidence for finite mixing times in the mantle. I will show that our improved understanding of the balance between buoyancy fluxes and viscous dissipation associated with mantle convection and plate tectonics [*Conrad and Hager*, 1999a,b] is introducing a strong constraint on mantle mixing times.

Based on a re-estimation of the fluxes of elements into the oceanic crust, the residence times of selected incompatible elements in the whole mantle will be computed. Without having to constrain the rates of mixing from geochemical data [e.g., *Allègre and Lewin*, 1995a], we can deduce them from scaling laws and then consider on which length scale the mantle composition may be heterogeneous and how primordial material can be preserved despite mantle convection.

2. A RESIDENCE-TIME TOOLBOX

Residence time is a concept introduced by *Barth* [1952] to characterize the time an atom of a particular element (or isotope) remains in a chemical system (the mantle) and describes the rate of readjustment (relaxation) of the system after a perturbation. It simply compares the outgoing flux of this element averaged over the boundary of the system to the total inventory of the element enclosed within this boundary. A convenient reference is the textbook by *Naumann and Buffham* [1983]. Let us consider a mantle of mass M containing a particular element i with a mean concentration $\overline{C_i}$. We assume that, over the time interval dt, the amount of material extracted at mid-ocean ridges and reinjected through subduction zones is Qdt, which simply amounts to assuming that the mass of the mantle does not significantly change through time. The mean concentration of the element considered in the material transferred from the mantle to the oceanic crust is $\alpha_i \overline{C_i}$, in which α_i is a coefficient measuring the efficiency of extraction for element i and can be simply related to melting equations [*Albarède*, 1998, 2001]. The flux of element i injected into the mantle is QC_i^{in}, which accounts for some fraction of element i being stored, at least temporarily, in continental crust. Mass balance requires

$$\frac{d\left(M\overline{C_i}\right)}{dt} = -Q\alpha_i \overline{C_i} + QC_i^{in} \qquad (1)$$

or

$$\frac{M}{Q\alpha_i}\frac{d\overline{C}_i}{dt}+\overline{C}_i=\tau_i\frac{d\overline{C}_i}{dt}+\overline{C}_i=\frac{C_i^{in}}{\alpha_i} \quad (2)$$

in which τ_i defined as

$$\tau_i=\frac{M}{Q\alpha_i}=\frac{M\overline{C}_i}{Q\alpha_i\overline{C}_i}=\frac{\text{mass of element } i \text{ in the mantle}}{\text{flux of element } i \text{ \underline{out} of the mantle}} \quad (3)$$

is the *relaxation time* of element i in the mantle. Conversely, the probability per unit time p_i that an atom of i is extracted from the mantle and transferred into the oceanic crust is:

$$p_i=\frac{Q\alpha_i\overline{C}_i}{M\overline{C}_i}=\frac{1}{\tau_i} \quad (4)$$

The last step is to demonstrate that, when the relaxation time τ_i remains constant, it is equal to the mean *residence time* $<t_i>$ of element i in the mantle. Let us now assume that material residence time M/Q, τ_i, and p_i are constant. Now $<t_i>$ can be calculated from the residence time of those atoms that were present in the mantle at $t=0$. Clearly, the mean concentration $\overline{C}_i^0(t)$ of these primordial atoms evolves as

$$\overline{C}_i^0(t)=\overline{C}_i^0(0)e^{-\frac{t}{\tau_i}} \quad (5)$$

We can now calculate $<t_i>$ as the mean value of the time spent in the mantle by the amount $Q\alpha_i\overline{C}_i^0(t)dt$ extracted from the mantle over the interval dt:

$$<t_i>=\frac{\int_0^\infty tQ\alpha_i\overline{C}_i^0(t)dt}{\int_0^\infty Q\alpha_i\overline{C}_i^0(t)dt}=\int_0^\infty\frac{t}{\tau_i}e^{-\frac{t}{\tau_i}}dt=\tau_i\int_0^\infty ue^{-u}du=\tau_i \quad (6)$$

Appendix A provides an alternate derivation based on Poisson processes. The theory holds that the input of element i has no effect on the probability of extracting a particular atom of i from the reservoir (as a crude analogy, adding rubidium to a system does not change the decay constant of [87]Rb). Unless the mantle is at steady-state, it is therefore incorrect to estimate the residence time of element i from its flux *into* the mantle. These linear concepts can be extended to the chemical and isotopic dynamics of interacting reservoirs and an extensive theoretical framework for the multibox theory can be found in *Albarède* [2001]. The model of 'homogenous distribution of heterogeneities' of *Helffrich and Wood* [2001] is an analogous description of a heterogeneous mantle with Poisson properties but lacks the theoretical framework provided here.

3. ABUNDANCES AND RESIDENCE TIMES OF INCOMPATIBLE ELEMENTS IN THE MANTLE

In order to evaluate the residence times of the elements in the mantle, reservoirs and fluxes must first be carefully defined. We divide the Earth into a core, a mantle, and a continental crust. The mantle, which we redefine as *bulk* mantle to minimize ambiguity, is therefore what would be left of the Earth if the core and the crust were stripped off. Alternatively, the bulk mantle is what is left of the Bulk Silicate Earth (BSE) after removal of the continental crust. The model is illustrated in Figure 1. This definition leaves oceanic crust out of the total inventory: this reservoir, however, is short-lived and makes up only a negligible part of the terrestrial budget.

I consider that an element is extracted from the mantle when it makes it into the magmatic section of the oceanic crust (basalt+dykes+gabbros). The emplacement rate of large igneous provinces over the last 200 My is 10–20% of that of mid-ocean ridges [*Coffin and Eldholm*, 1994], but the enrichment of plateau basalts with respect to normal MORB is only a factor of 2–3 [*Neal et al.*, 1997]. Neglecting plume flux may lead to residence times overestimated by 30–40%. Whether or not the oceanic crust is injected back into the mantle is irrelevant because residence times are simply inverse removal probabilities per unit time. For this reason, the oceanic lithosphere therefore remains *sensu stricto* part of the mantle and, for incompatible elements, this assumption affects only a small part of the oceanic crust inventory, a point which I will return to later. Recycling of continental crust into the mantle, as any other input, is not relevant to the residence time of elements in the mantle. The effect of recycling continental crust on mantle chemistry is not substantial, as attested to by the rare occurrence of the EM II component in oceanic basalts [*Hofmann*, 1997] and by the small abundance of atmospheric [40]Ar orphaned from continental [40]K [*Coltice et al.*, 2000].

In order to assess the composition of the oceanic crust, I downloaded more than 5000 major and trace element, and isotope analyses of MORB glasses characterized as 'fresh' from the Lamont PETDB data base. Instead of using multiple sophisticated selection criteria to evaluate a mean MORB composition as in *Salters and Stracke* [2004], I applied a brute-force procedure to the bulk of the data set. A first task was to correct the analyses for low-pressure fractional crystallization, which an inspection of element abundances vs MgO content showed was critical. This was done by calculating a least-square fit between ln $[C^i]$ and $[C^{MgO}]$ (semilog plot), where C^i represents the concentration of element i, and then extrapolating the fit to 8% MgO, which *Klein and Langmuir* [1987] argued accounts for most low-pressure fractionation (non-cumulative magmatic rocks with MgO >8% are uncommon in the oceanic crust). Such a relationship can be justified by the incremental variations (distillation) of trace elements, for which constant solid-liquid partition coefficients can be defined, whereas $[C^{MgO}]$ varies in proportion to the fraction of solid removed. For nearly all highly incompatible elements, the correlation coefficients exceeded 0.7 and, for some cases, 0.8.

It clearly shows that straightforward concentration histograms are skewed (Figure 2) and therefore that a normal probability density function is not acceptable. The histograms of concentration logarithms are shown in Figure 3. The nature of the probability density functions of concentrations and isotopic ratios has already received a great deal of attention [e.g., *Allègre and Levin*, 1995b] and we decided to use simple, but robust tests to assess which probability density functions are acceptable (Appendix B). Table 1 reports the results of these calculations together with MORB composition estimates by *Hofmann* [1988] and *Sun and McDonough* [1989]. Serious misfits are rare and mostly concern isotopic ratios, notably $^{87}Sr/^{86}Sr$ and $^{143}Nd/^{144}Nd$: although the histograms of these anomalous variables are still unimodal, they are too skewed by mixing processes to be described by normal or log-normal distributions.

The composition of the bulk mantle was estimated by removing 2.97×10^{22} kg of continental crust from 4.0×10^{24} kg of BSE. The continental crust concentrations are those from *Rudnick and Fountain* [1995] and *Taylor and McLennan* [1995], while those of the BSE are from *McDonough and Sun* [1995]. The uncertainties ensuing from different choices of continental crust composition are small except for Ta, Rb, K, U, Ba, and Pb. Likewise, except for compatible elements and those of substantial volatility, the uncertainties associated with the present choice of BSE composition can probably be neglected.

Figure 4 shows a comparison between the mean values of the log-normal deviation (geometric mean) and the mode for the available MORB trace-element data normalized to the mean mantle calculated for both *Rudnick and Fountain* [1995] and *Taylor and McLennan* [1995] compositions of the continental crust. The systematic difference between the mode and the mean is small, but robust. An analysis of the skewness (dissymmetry) and kurtosis (stockiness) of the histograms of concentration logs confirms that, although more appropriate than a normal probability density function, the log-normal probability density function is only an approximation to the data distribution. The histograms of Figure 3 unambiguously show, however, that MORB form a unique population with a single mode and that any attempt to divide them into sub-populations, such a E-MORB (enriched) and N-MORB (normal), is unsupported by the observations and

therefore is artificial and a matter of subjectivity. Although E-MORB may reveal the presence of enriched components in the mantle source [*Donnelly et al.*, 2004], they are first and foremost the incompatible element-rich tail of the MORB population. The order by which the data are displayed is somewhat arbitrary. The option chosen here was to separate elements for which the estimates in the continental crust by *Rudnick and Fountain* [1995] and *Taylor and McLennan* [1995] disagree (U, Rb, K, Th, Pb, and, to a lesser extent, Ba) from those which are agreed upon. This choice may visibly affect concentrations in the mean mantle. As expected, the modes provide MORB with a slightly more depleted character in the most incompatible elements, notably La and Ce, than the geometric mean. The apparent inconsistency of U and possibly Ba statistics calls for a re-evaluation of their distribution, but probably also their analytical procedures. The mean/mode ratio is <1 for U and Nb. The cross-over between the mean/mode ratios of Ta and Nb is surprising, but robust. The statistics for the more compatible elements, e.g., Sc, Co, Cr, Ni, are more strongly dependent on the initial assumption that the parent melt had 8% MgO.

The residence times of incompatible elements in the mantle are listed in Table 2 and, using the same order as before, depicted in Figure 5. They are all curiously similar, with most values in the range of 4–9 Gy. Such a small range of values requires that only a small fraction of the incompatible element inventory is left in the residuum. *Albarède* [1995, 2001] argued that because residence times are inversely proportional to the solid/liquid partition coefficients prevailing during MORB extraction, a positive correlation is expected between the relative standard error of the mean and the inverse of the residence time of each element. This correlation is indeed observed (Figure 6), but with an anomalous behavior of Ba, Th, and Nb. The residence times of Table 2 are very different from those proposed for some incompatible elements in the upper mantle by *Galer and O'Nions* [1985] and *O'Nions and Tosltikhin* [1995], which are all <1 Gy. A combination of different reservoir definitions and a specific depletion of the upper mantle account for the differences between the two sets of values.

4. SURVIVAL OF MANTLE GEOCHEMICAL HETEROGENEITIES

As outlined in the Introduction, modelling the survival of geochemical mantle heterogeneities requires estimates of relative values of residence and mixing times. A first method to estimate mixing times is to assess how long it takes for mantle convection to confer the distribution of an arbitrary cluster of tracers, assumed to represent initially a geodynamic object such as a plate fragment, which is treated as a random (Poisson-like) property. This method is dependent on a particular numerical model of mantle convection and involves box counting, or equivalently the calculation of a 'correlation dimension', over a range of wavelengths [*Schmalzl and Hansen* 1994; *Ferrachat and Ricard*, 1998; *Stegman et al.*, 2002]. I here use the spectral method of *Olson et al.* [1984], which provides an analytical expression of the mixing time in which both the dimension of the initial heterogeneities and the size of the sampling regions are explicitly taken into account.

Let us suppose that the mantle may be represented by a rectangular enclosure gridded with sample cubes of dimension Δl which represent the typical volume over which melting averages geochemical heterogeneities. Let us also consider that the composition of the material locally deviates from the mean value and that this chemical heterogeneity can be characterized by a typical length scale L. The mixing time at time t has been conveniently defined by *Olson et al.* [1984] as the mean time over which the variance of the compositions of all the cubes is reduced from its initial value to the current value at t. Typically, for an exponential decay of the variance, the mixing time would be the inverse of the exponential parameter. The mixing time τ_{mix} (L) depends on the length scale (wavelength) L of the heterogeneity and using *Batchelor's* [1959] original formulation, *Olson et al.* [1984] obtained the relationship:

$$\tau_{mix}(L) = \frac{\pi}{2\dot{\varepsilon}}\frac{L}{\Delta l} \qquad (7)$$

where $\dot{\varepsilon}$ is the average strain rate. For Newtonian materials, the strain rate is related to the rate of viscous dissipation Φ^{vd} through:

$$\dot{\varepsilon} = \sqrt{\frac{\Phi^{vd}}{\mu}} \qquad (8)$$

[e.g., *Schubert et al.*, 2001], where μ is the mantle viscosity. In order to evaluate τ_{mix} (L), I suggest using the frame-independent property that, because inertial effects are negligible, the rate of viscous dissipation Φ^{vd} exactly compensates the potential energy (buoyancy) flux Φ^{pe}. Today, 40 percent of this energy is used to bend the lithospheric plates [*Conrad and Hager*, 2001] and is therefore not available to mix the convective mantle. *Conrad and Hager* [1999a,b] provide an expression for the potential energy flux of subducting plates, which is

$$\Phi^{pe} = \pi^{-1/2}\rho g\alpha\Delta T v_p V_p \qquad (9)$$

where ρ is density, g is gravity, α is thermal expansion, ΔT is the temperature difference between the mantle and the sur-

face, v_p is the mean subduction velocity, and V_p is the volume of subducted plates. I assumed that plates remain identifiable until they reach a depth of 1000 km in the mantle and that they are 75 km thick. The total length of subduction zones is 36,000 km [*Forsyth and Uyeda*, 1975] and a mean rate of oceanic crust creation of 3.5 km^2 y^{-1} suggests a mean rate of subduction v_p of 10 cm y^{-1}. Using values of $\rho = 3300$ kg m^{-3} and $\alpha = 3 \times 10^{-5}$ K^{-1}, and a mantle volume of 8.9×10^{20} m^3, the power of viscous dissipation is ≈ 6 TW. The ≈ 2.8 TW buoyancy flux of the eclogitized oceanic crust [e.g., *Stacey*, 1992] should be added to that value. Studies of post-glacial rebound and dynamic topography suggest that the viscosity of the lower mantle is about 50 times higher than that of the upper mantle (see reviews in *Davies* [1999] or *Schubert et al.* [2001]). I therefore assumed a mean viscosity slightly larger (10^{22} Pa s) than the values for typical upper mantle material. The value of the mean strain rate resulting from these simple calculations is $\dot{\varepsilon} \approx 7.5 \ 10^{-16}$ s^{-1}, which is only a factor of four smaller than the value inferred by *Olson et al.* [1984] from different scaling laws. Many of these parameters are affected by substantial uncertainties and the difference with *Olson et al.*'s [1984] estimate may reflect the overall uncertainty on mixing times.

Although the relevant calculations can be found in *Olson et al.* [1984], we will assess some typical mixing times here. The value of the scaling time $\pi/2\dot{\varepsilon}$ is ≈ 66 My. If MORB sample their source over a cube with a linear dimension of 10 km, a heterogeneity with a characteristic dimension equal to the mantle depth (≈ 2900 km) will still survive after 19 Gy, while the dispersal of the D″ layer (≈ 200 km) would be largely accomplished in little more than 1.3 Gy. Ancient mantle layers are therefore expected to remain identifiable as small domains despite of convection over the entire Earth history.

5. SIMILARITIES BETWEEN THE THERMAL AND GEOCHEMICAL STRUCTURE OF THE EARTH'S MANTLE

The reasons for comparing the rate of loss and dispersal of thermal energy by mantle convection with the history of incompatible elements are profound. The secular heat inherited from terrestrial accretion and original mantle dif-

Figure 2. Histograms of concentrations and isotopic ratios of the 'fresh' MORB glass data downloaded from the Lamont-Doherty PETDB data base. The solid curve shows the theoretical normal probability density function estimated from the mean and variance of the sample. Strong differences between this curve and the histogram indicate that a normal probability density function does not adequately describe the samples.

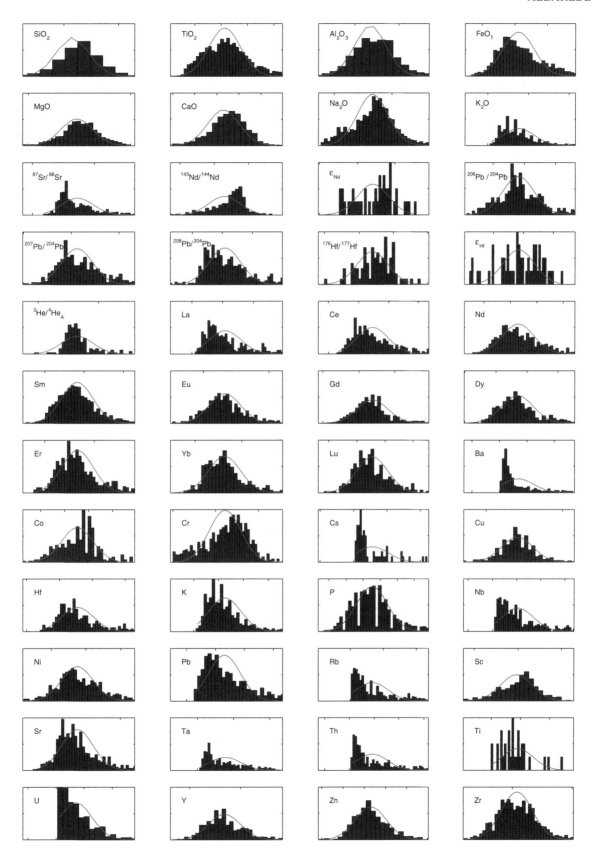

Table 1. Main Statistics of MORB Compositions Calculated From the Data Compiled in the Lamont-Doherty PETDB Data Base.

Element[1]	data #	ρ_{MgO}[2]	arith. mean[3]	geom. mean[4]	2s (rel)[5]	mode[6]	HOF88[7]	SUN89[8]
SiO_2	4838	-0.51	50.89	**50.89**	0.001	51.04	50.45	
TiO_2	4832	-0.76	1.73	**1.73**	0.017	1.84	1.615	
Al_2O_3	4832	0.53	14.47	**14.47**	0.004	14.69	15.255	
FeO_T	4143	-0.69	10.8	**10.8**	0.007	10.14	10.426	
MgO	4876		8	8		8	7.576	
CaO	4832	0.7	10.95	**10.95**	0.005	11.41	11.303	
Na_2O	4828	-0.55	2.78	**2.78**	0.007	2.91	2.526	
K_2O	4793	-0.56	0.16	**0.16**	0.047	0.17		
$^{87}Sr/^{86}Sr$	540	0.01	0.70279	0.70279	0.00006	0.70303		
$^{143}Nd/^{144}Nd$	424	0.01	0.51307	0.51307	0.00002	0.51323		
ε_{Nd}			8.5	8.4		11.5		
$^{206}Pb/^{204}Pb$	421	-0.16	18.458	18.459	0.002	18.38		
$^{207}Pb/^{204}Pb$	421	-0.1	15.518	15.518	0.0003	15.496		
$^{208}Pb/^{204}Pb$	421	-0.11	38.092	38.093	0.001	38.187		
$(^3He/^4He)_A$	180	0.2	8.1	8.1	0.047	7.7		
U	538	-0.39	0.333	**0.138**	0.112	0.233	0.0711	
Rb	661	-0.43	3.9	**2.83**	0.116	2.98	1.262	0.56
K	4798	-0.56	1465	**1328**	0.026	1375	883.7	
Th	615	-0.44	0.508	**0.383**	0.115	0.363	0.1871	0.12
Pb	406	-0.26	2.57	**0.657**	0.075	0.662	0.489	0.3
Ba	925	-0.37	42.7	**25.9**	0.093	14.4	13.87	6.3
Sr	1218	-0.13	133	**126**	0.017	115	113.2	90
La	826	-0.57	5.87	**5.35**	0.065	4.09	3.895	2.5
Ce	816	-0.77	15.9	**15.6**	0.056	12.8	12	7.5
Nd	840	-0.82	12.3	**12.3**	0.042	10.8	11.18	7.3
Sm	987	-0.8	3.9	**3.94**	0.031	3.67	3.752	2.63
Eu	805	-0.86	1.39	**1.42**	0.028	1.35	1.335	1.02
Gd	563	-0.8	5.07	**4.99**	0.029	4.73	5.08	3.68
Dy	595	-0.8	5.84	**5.7**	0.031	5.18	6.304	4.55
Y	1049	-0.69	35.7	**36.1**	0.023	32.6	35.82	28
Er	590	-0.74	3.6	**3.46**	0.032	3.14	4.143	2.97
Yb	795	-0.71	3.53	**3.44**	0.028	3.13	3.9	3.05
Lu	378	-0.78	0.552	**0.556**	0.05	0.495	0.589	0.46
P	4533	-0.68	753	**783**	0.016	860		510
Zr	1182	-0.77	116	**118**	0.031	119	104.24	74
Hf	298	-0.84	3.41	**3.54**	0.086	2.88	2.974	2.05
Ta	478	-0.53	0.428	**0.368**	0.113	0.277	0.192	0.13
Nb	778	-0.52	6	**4.89**	0.088	6.98	3.507	2.33
Co	353	0.24	43.2	**42.2**	0.01	43.8	47.07	
Cr	799	0.78	253	**191**	0.07	255		
Cu	517	0.38	73.7	**68**	0.016	66.8	74.4	
Ni	945	0.83	91.8	**74.8**	0.04	77.4	149.5	
Sc	841	0.12	40.6	**40**	0.007	41.4	41.37	
Zn	522	-0.55	92.1	**93.1**	0.025	89.7		

[1]major elements in percent, trace elements in ppm; [2]correlation coefficient between concentration or ratio and ln [MgO]; [3]arithmetic mean (normal density); [4]geometric mean (log-normal density); [5]$2 \times$ standard deviation / (geom. mean $\times \sqrt{n}$); [6]the second derivative of the 4th-degree best-fit polynomial through the cumulated distribution between 10 and 90% is set to zero; [7]N-MORB [*Hofmann*, 1988]; [8]N-MORB [*Sun and McDonough*, 1989].

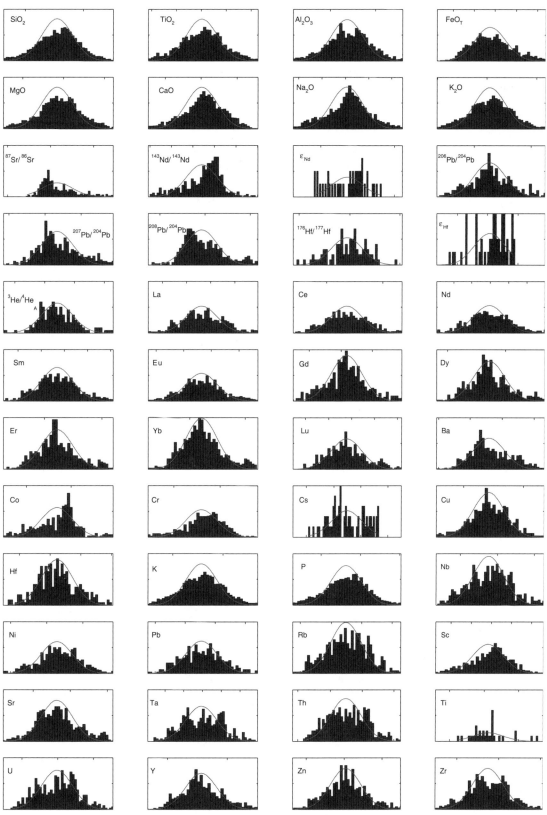

Figure 3. Same as Figure 1 but with the logarithms of concentrations and isotopic ratios.

ferentiation behaves almost exactly as primordial rare gas isotopes, e.g., ^{3}He, ^{22}Ne, and ^{36}Ar. Once extracted at the surface, it never gets recycled. The loss of secular heat is therefore closely related to the outgassing of primordial rare gas isotopes. Likewise, radiogenic heat and radiogenic isotopes of rare gases, e.g., ^{4}He and ^{40}Ar, are also closely related. *O'Nions and Oxburgh* [1983] formulated the popular paradox of heat/helium imbalance, which holds that the terrestrial loss of ^{4}He is much less than what is expected from radiogenic production and heat flow. The standard parameterization of the secular thermal evolution [e.g., *Davies*, 1980; *Turcotte*, 1980; *Stacey*, 1980] can be written:

$$\frac{d\left(mC_P T_S\right)}{dt} = -Q + m\sum_j H_0^j e^{\lambda_j(t_0-t)} \qquad (10)$$

where m denotes the mass of the Earth, C_p its specific heat, T_s the extrapolation of the adiabat to the surface, Q the total surface heat flow, t_0 the age of the Earth, H_0^j the modern heat production per mass unit of the mantle by the heat-producing isotope j (^{40}K, ^{232}Th, ^{235}U, and ^{238}U), and λ_j its decay constant. Using the residence time approach, this equation can be rewritten as:

$$\frac{dT_S}{dt} = -\frac{T_S}{\tau_Q} + \sum_j \frac{H_0^j}{C_P} e^{-\lambda_j(t_0-t)} \qquad (11)$$

where $\tau_Q = mC_p T_s/Q$ is the residence time of thermal energy in the mantle. This expression shows that the thermal evolution of the mantle is the sum of a mildly variable relaxation term and a forcing term due to radioactive heating. For C_p = 1100 J kg^{-1} K^{-1}, T_s = 1200°C, and Q = 25 TW calculated by retaining the 10% heat produced in the core and the heat produced in the crust using *Rudnick and Fountain's* [1995] estimates of U, Th, and K concentrations, we obtain a modern residence time of mantle heat $\tau_Q \approx 6.7$ (Gy).

6. DISCUSSION

The present estimates of the residence times of incompatible elements (4–9 Gy) are also a lower limit for the residence times of compatible elements, which include all the major elements. If convection overturns the whole mantle, the geochemical dynamics of the mantle-crust system is therefore not operating at steady state. It is possible that these residence times may have been shorter in the past but finding by how much is mere speculation: heat production in the past was certainly stronger and more fossil heat was present, but, since the feedback between rheology and temperature is not well understood [*Conrad and Hager*, 1999a], it is not known how much more vigorous ridge activity and plume flux should have been. The

geological record over the last 3.4 Gy lacks the dramatic shifts in magma compositions that would reflect much stronger magmatic activity. Both the major element chemistry of ancient MORB [*Abbott et al.*, 1994] and the isotopic geochemistry of inclusions from the subcontinental mantle [*Bedini et al.*, 2004] suggest that, over the last 3 Gy, mantle temperature has changed significantly less (<150°C) than predicted by standard theory.

Depending on whether the mantle is separated into isolated reservoirs, the persistence of mantle heterogeneities depends on different assumptions. If the mantle is layered with respect to mantle convection, e.g., with contrasting trace element compositions for the upper and lower mantle, or if the abyssal layer of *Kellogg et al.* [1999] does indeed exist, the two reservoirs will maintain a distinct chemistry if oceanic extraction at mid-ocean ridges and re-injection into the deep mantle at subduction zone operate at a faster pace than exchange across the reservoir boundary. For the layered mantle case, this is the equivalent to the zoned mantle convection model of *Albarède and van der Hilst* [2002]. The rate of chemical readjustment of an element i between two reservoirs labelled 1 and 2 is given by the overall relaxation time τ_R^i such that:

$$\frac{1}{\tau_R^i} = \frac{1}{\tau_1^i} + \frac{1}{\tau_2^i} \qquad (12)$$

where τ_1^i and τ_2^i are the residence times of element i in reservoirs 1 and 2, respectively [*Albarède*, 1998; 2001]. The relaxation time is therefore shorter than the individual residence times in each reservoir. The implication is that in order to keep the mean compositions of the upper and the lower mantle chemically distinct, transfer by extraction at mid-ocean ridges must be significantly faster than transfer through the transition zone by subduction. This is the essence of the canonical models of crustal growth from the upper mantle with minimum exchanges across the transition zone [*O'Nions et al.*, 1979; *Jacobsen et al.*, 1979; *Allègre et al.*, 1983b,c; *Galer and O'Nions*, 1985; *Galer et al.*, 1989; *O'Nions and Tolstikhin*, 1996]. Geophysical evidence, however, does not support this model [*van der Hilst*, 1991, 1995; *Davies and Richards*, 1992; *Grand*, 1994; *Masters et al.*, 1996; *Li and Romanowicz*, 1996].

Let us therefore assume whole-mantle convection and take the extreme view that mid-ocean ridges sample the MORB source with a probability of extraction independent of position in the mantle. Such an assumption, which holds that the mantle is evolving towards a well-mixed geochemical reservoir in a few Gy, has been successfully tested numerically [*Gurnis and Davis*, 1986a,b; *Schmalzl and Hansen*, 1994; *Ferrachat and Ricard*, 1998; *Stegman et al.*, 2002]. Although MORB appear to tap preferentially the upper mantle, the well-mixed mantle model with position-independent (Poisson) melting events is a convenient reference and the mean residence times derived from this model provide minimum values as to how long geochemical contrasts across the mantle persist. Figure 7 shows the mixing time of heterogeneities of variable wavelengths for different characteristic lengths of melting and magma mixing (50, 10, and 0.01 km). Depending on the initial size and the age of a heterogeneity, different cases may be distinguished:

• Geochemical heterogeneities with a length scale similar to mantle depths (2900 km) or to the size of large plates will be observable for times commensurate with the age of the Earth for any dimension of the melt source <50 km, and will be particularly visible in glass inclusions in olivine. Globe encircling anomalies such as the Dupal anomaly [*Dupré and Allègre*, 1983; *Hart*, 1984)] or the shear velocity anomaly under South Africa [*Ritsema et al.*, 1998; *Ni et al.*, 2003] may be extremely ancient or even attest to the presence of primordial features. The abyssal layer, which

Table 2. Composition of the Mean Mantle (in ppm) Obtained by Subtracting Continental Crust from Bulk Silicate Earth Inventories.

Element i	mean mantle conc. (RF)	(TML)	τ_i (RF) (Gy) mean	mode	τ_i (TML) (Gy) mean	mode
U	0.00988	0.0137	4.3	2.6	6.0	3.6
Rb	0.174	0.367	3.7	3.5	7.8	7.4
K	125	174	5.7	5.5	7.9	7.7
Th	0.0385	0.0541	6.1	6.4	8.5	9.0
Pb	0.0576	0.0917	5.3	5.3	8.4	8.4
Ba	3.75	4.79	8.8	15.8	11.2	20.1
Sr	17.6	18.1	8.5	9.3	8.7	9.5
La	0.519	0.534	5.9	7.7	6.0	7.9
Ce	1.38	1.44	5.3	6.5	5.6	6.8
Nd	1.11	1.14	5.5	6.2	5.6	6.4
Sm	0.380	0.383	5.8	6.3	5.9	6.3
Eu	0.146	0.147	6.2	6.6	6.3	6.6
Gd	0.521	0.524	6.3	6.7	6.3	6.7
Dy	0.653	0.652	6.9	7.6	6.9	7.6
Y	4.19	4.19	7.0	7.8	7.0	7.8
Er	0.425	0.425	7.4	8.2	7.4	8.2
Yb	0.429	0.427	7.5	8.3	7.5	8.3
Lu	0.0656	0.0658	7.1	8.0	7.2	8.0
P	84.2		6.5	5.9		
Zr	9.63	9.80	5.0	4.9	5.0	5.0
Hf	0.258	0.263	4.4	5.4	4.5	5.5
Ta	0.0293	0.0301	4.8	0.1	4.9	0.1
Nb	0.573	0.581	7.1	5.0	7.2	5.0
Co	106	106	151	146	151	146
Cr	2644	2643	839	626	839	626
Cu	30.0	29.7	27	27	26	27
Ni	1974	1974	1597	1542	1597	1541
Sc	16.2	16.1	24	24	24	23
Zn	54.9	54.8	36	37	36	37

Continental crust concentrations from *Rudnick and Fountain* [1995] (RF) and *Taylor and McLennan* [1995] (TML). Bulk silicate Earth concentrations from *Sun and McDonough* [1989]. Residence times calculated from the data in Table 1 and a production rate of oceanic rust of 66×10^{12} kg y^{-1} (3.5 km^{2}y^{-1} × 6.3 km × 3000 kg m^{-3}). These are based on a mean MgO content of the oceanic crust of 8% [e.g., *Taylor and McLennan*, 1985]. The header mean refers to the geometric mean in Table 1. The uncertainties are dominated by the simple feature of the model rather than by the spread of the data. If plume flux is taken into account, the tabulated residence times are probably overstimated by ≤30%.

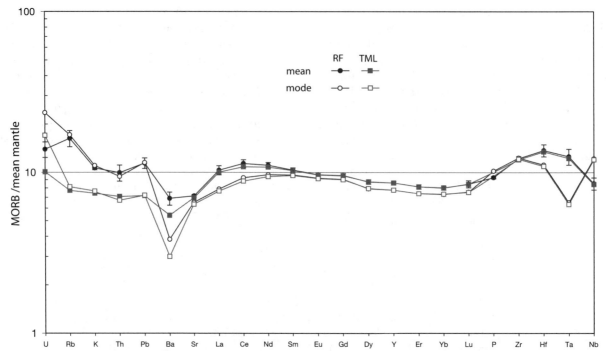

Figure 4. Geometric mean and mode of 'fresh' MORB glass concentrations of the lithophile elements normalized to the mean mantle composition. The mean mantle is obtained by subtracting continental crust, with concentrations from *Rudnick and Fountain* [1995] (RF) and *Taylor and McLennan* [1995] (TML), from the Bulk Silicate Earth, with concentrations from *Sun and McDonough* [1989]. From U to Ba, note the effect of the different estimates of the continental crust composition.

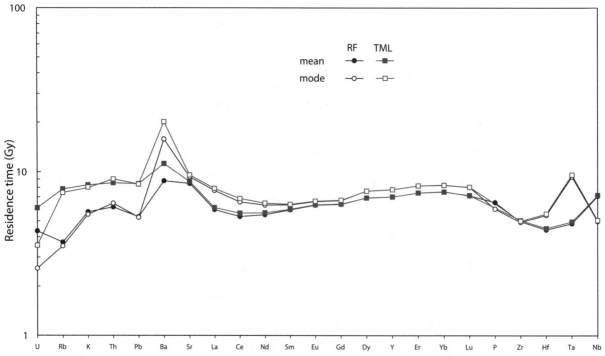

Figure 5. Residence times of the lithophile elements in the mantle (see caption of Figure 4 for the source of data).

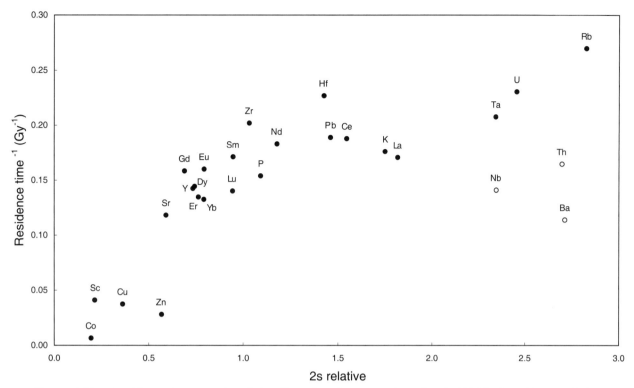

Figure 6. Plot of residence times τ_i vs the relative dispersion of concentration values in MORB. A positive correlation is expected if melt/residue fractionation controls the residence times [*Albarède*, 1995]. The misfit observed for Ba, Th, and Nb may reveal inadequate inventories or unsuspected fractionation processes. Ni and Cr, which correlate negatively with MgO, are not shown.

according to *Kellogg et al.* [1999] fills the bottom 1000 km of the deep mantle above the D″ discontinuity, should be very ancient and possibly primordial. In contrast, the abyssal domain constantly replenished by the heavier plates advocated by *Albarède and van der Hilst* [2002] [see also *Davies*, 1981] would keep its geochemical identity for billions of years without the need for a specific interface with the overlying mantle.

- In contrast, geochemical heterogeneities with a length scale similar to the thickness of the basaltic crust (≈6 km) will be observable in melt batches tapping the mantle over a characteristic length scale of ≈<60 m or will disappear within a few hundred My.

We can strengthen the conclusions derived above from the *Olson et al.*'s [1984] theory using residence time distributions. Let us now call 'tracer' a parcel of mantle significantly smaller than the length scale of any significant melting event. Once this piece of mantle is melted, the tracer and its constitutive elements are considered to have lost their primordial character: the melt is extracted while both the residue and the reinjected melt produce recycled tracers. The partial differential equations (micro- and macro-balance) ruling the

dynamics of a tracer population in a well-mixed reservoir are detailed in Appendix A. Complete solutions are provided for constant probabilities of extraction and recycling. The proportion $f_i(t,\theta)$ of tracers i with residence times in excess of θ is given by equation A11 in Appendix A:

$$f(t,\theta) = e^{-\omega_i \frac{\theta}{\tau_i}} \qquad (13)$$

where τ_i is the mean residence of element i in the mantle and ω_i the probability for a tracer i extracted into the oceanic crust to be recycled back into the mantle rather than irreversibly transferred to the continental crust (i.e., the ratio 2/(2+3) of the fluxes shown in Figure 1). This is particularly important for heat-producing elements or rare gas isotopes such as [3]He and [20]Ne (which are probably not much more incompatible than lithophile elements such as Th, U, Ba, or La) since about 50% of tracers with a mean residence time of 7 Gy will never have been extracted into the mid-ocean ridges after 4.56 Gy (Figure 8). The numbers in this figure are strikingly similar to the estimated fraction of 'missing' [40]Ar, which according to planetary inventories is still buried in the mantle [*Allègre et al.*, 1986; 1996; *Turner*, 1990].

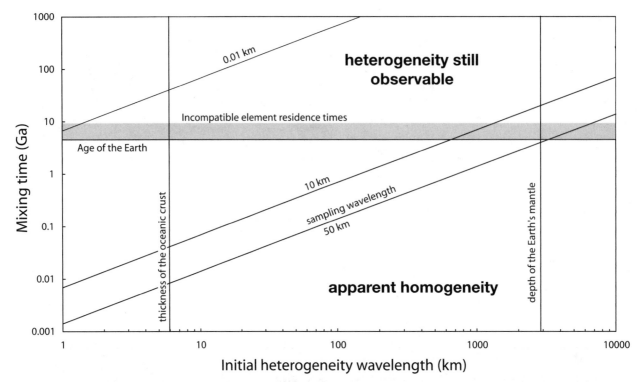

Figure 7. Relationship between the initial wavelengths of geochemical mantle heterogeneities and the time τ_{mix} required by mantle convection to erase them on the length scale of sampling used as labels on the curves. The range of residence times is derived from Figure 5. For $\tau_{mix} < \tau_i$, the heterogeneities are erased on the time scale of the mixing time.

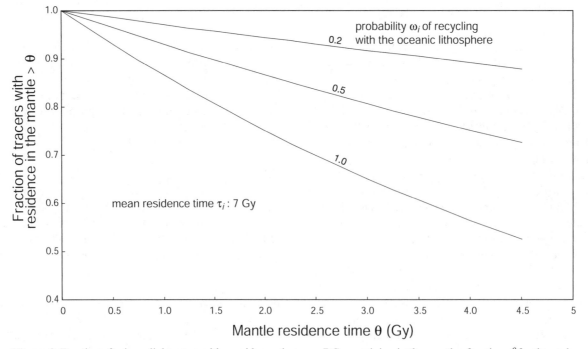

Figure 8. Fraction of primordial tracers with a residence time $\tau_i = 7$ Gy surviving in the mantle after time θ for the probabilities ω_i of recycling into subduction zones (as opposed to irreversible extraction into the continental crust) indicated as labels on the curves (equation 13 or as the ratio 2/(2+3) of the fluxes shown in Figure 1).

Gurnis and Davies [1986a] and *Xie et al.* [2004a] demonstrated numerically that primordial tracers have escaped extraction for the entire history of the planet. In order to identify the presence of primordial material in the geochemical properties of basalts, the scale of melting and mixing in their mantle source should be made as small as possible. The present results show that primordial mantle will only have survived as small domains, presumably strongly sheared and refolded and interbedded with younger recycled material (see *Helffrich and Wood*, 2001, for a discussion of how these heterogeneities relate to seismic scatterers). This is probably the main reason why ^{3}He and ^{20}Ne anomalies are restricted to hot spot basalts, for which the melt zone is not as broad as beneath mid-ocean ridges. A melt zone with a dimension <10 km is needed to make the variability inherited from a primordial mantle anomaly of ≈1000 km observable in basalts. Such melts with a primordial signature would probably be best sampled in glass inclusions in olivine. Although primordial material may indeed be present in most mantle sources, the chances of observing it as a distinctive geochemical anomaly are therefore very small. *Anderson* (2001), however, pointed out that most basalts from high ^{3}He/^{4}He hot spots actually do not show high ^{3}He/^{4}He ratios. This apparent paradox can simply be explained by the small dimension of their mantle source. A reverse situation holds for the narrow range of ^{3}He/^{4}He ratios in MORB, which probably reflects the comparatively broad melting zone present under the ridges. Because of strong concentration contrasts, rare gases may be one of the few favorable tracers. The apparent lack of anomalies associated with the extinct radioactivity of ^{146}Sm ($T_{1/2}$ = 103 My) in high-^{3}He/^{4}He hot spot basalts, such as Hawaii, was interpreted by *Boyet et al.* [2005] as a result of the dilution of the primordial mantle by younger recycled material.

Even though the combination of mixing and heat conduction eliminates essentially all thermal heterogeneities in less than 1 Gy [*Olson et al.*, 1984], regardless of their wavelength, a similar interpretation applies to the history of deep heat. Its 6.7 Gy residence time in the mantle indicates that, at the modern rate at which mantle convection is mining primordial heat, about 50% of it would still be present after 4.5 Gy. Taking into account the current uncertainties on the concentration of radioactive elements in the crust [*Rudnick and Fountain*, 1995; *Taylor and McLennan*, 1995], τ_Q is therefore indistinguishable from the residence times calculated for the incompatible elements. Such a similarity has important consequences for the apparent imbalance between the heat flow and the loss of the ^{4}He derived from the decay of U and Th [*O'Nions and Oxburgh*, 1983; *Allègre et al.*, 1996; *van Keken et al.*, 2001]. During melting, He is incompatible [*Brooker et al.*, 2003] and its residence in the

mantle must be similar to that of lithophile elements such as La or Nb. *Castro et al.* [in press] recently demonstrated that circulation in the water table at the surface of continents efficiently separates the conductive heat flow from the advective transport of ^{4}He and reproduced the observed heat/He imbalance extremely well. They further argued that seawater infiltration in the oceanic crust plays the same role. Although a conductive boundary layer is still required to separate He from heat, it need not be deep-seated or identified with the transition zone. The relative loss of ^{4}He and heat should no longer be considered as a chief argument in favor of layered-mantle convection.

As a final comment, the rather simple theory developed in the present work lends support to both the viability of whole-mantle convection and to the persistence of primordial material in the mantle. Plate subduction keeps injecting surface material into the deepest mantle. Because of intense shearing by mantle convection, heterogeneities are unlikely to survive as large 'blobs' [e.g., *Becker et al.*, 1999] but rather persist as streaks and tendrils [*Allègre and Turcotte*, 1986; *Metcalfe et al.*, 1995; *van Keken and Zhong*, 1999] of old oceanic plates in a laminated matrix of melting residuum and primordial mantle. The distribution of the refractory incompatible elements in the mantle, however, does not support a radially homogeneous mantle. *Albarède and van der Hilst* [2002] showed that the MORB source is more depleted in incompatible refractory elements than the bulk mantle, which requires that the mantle underneath the asthenosphere is correlatively enriched. In other words, the mantle is not truly 'well-mixed'. Even if the present Poisson model, as well as *Hellfrich and Wood*'s [2001], do justice to the preservation of small wavelength geochemical heterogeneities, they do not take depth of melting and ultra-deep subduction into account. The radial geochemical variability of the mantle has two essential ingredients: (i) melting is a relatively shallow process which, as argued by previous authors [*Davies and Gurnis*, 1986a; *Davies*, 1992; *Xie et al.*, 1984a], largely confines primordial component to the deep mantle, and (ii) deep subduction of oceanic plates loaded with oceanic plateaus permanently replenishes the lower mantle with more fertile components inclusive of Fe [*Albarède and van der Hilst*, 2002].

7. CONCLUSIONS

Residence time analysis of MORB geochemistry combined with an application of *Olson et al.*'s [1984] theory of mixing by whole-mantle convection provided the following results:

- The residence times of the incompatible lithophile elements in the mantle vary within a rather narrow range (4–9 Gy). The large uncertainties for U, Th, Ba, K, and

Rb derive directly from those on continental crust composition.

- The mantle is not at geochemical steady-state and the implications of its primordial composition for modern mantle geochemistry are still strong. Up to 50% of incompatible elements may never have been extracted into the oceanic crust, which generalizes a conclusion reached for ^{40}Ar [*Allègre et al.*, 1986; 1996; *Turner*, 1990].
- The balance between the buoyancy flux and dissipation provides convenient estimates of the rates of mixing by mantle convection: primordial geochemical anomalies with an initial length scale comparable to mantle depths of the order of plate lengths may marginally be visible at the scale of mantle melting beneath mid-ocean ridges. They should be more readily identifiable when the scales of melting and homogenization are shorter, such as within hot spots and melt inclusions.
- The residence time of heat in the mantle (≈ 7 Gy) is indistinguishable from the residence time of the incompatible lithophile elements and thus 50% of the primordial heat may still be buried at depth.
- If the residence time of He is similar to that of the incompatible elements, this relieves the heat/helium paradox of *O'Nions and Oxburgh* [1985], which was used as an argument in favor of a conductive boundary layer at intermediate depths in the mantle.
- A simple theory of residence time distributions shows that up to 50% primordial material may be present in the mantle; because of strong continuous shearing by mantle convection, this primordial material does not form a continuous domain but, more likely, scattered as small (<10 km) domains, strongly sheared and refolded, and interbedded with younger recycled material.
- Shallow melting and ultra-deep subduction nevertheless keep the mantle radially zoned.

APPENDIX A: RESIDENCE-TIME DISTRIBUTIONS

Let us define $C_i(\mathbf{x}, t, \theta) d\theta$ as the concentration at time t and position $\mathbf{x}$ of a passive tracer i with a residence time in the mantle comprised between θ and $\theta + d\theta$, and $p_i(\mathbf{x}, t)$ the probability at t of a tracer located at $\mathbf{x}$ to be extracted into the crust. We will separate the 'internal' coordinates embodied in vector $\mathbf{x}$ and the 'external' coordinate θ. Let the local velocity of the tracer be $v_i^{\text{ext}}(\mathbf{x}, t) = d\mathbf{x}/dt$. The conservation equation can be written as:

$$\frac{\partial C_i}{\partial t} = -p_i(\mathbf{x}, t) C_i - v_i^{\text{ext}}(\mathbf{x}, t) \nabla C_i - v^{\text{int}} \frac{\partial C_i}{\partial \theta} \quad \text{(A1)}$$

with $v^{\text{int}} = d\theta/dt = 1$. The Lagrangian 'microbalance' equation:

$$\frac{\partial C_i}{\partial t} = -p_i(\mathbf{x}, t) C_i - \frac{\partial C_i}{\partial \theta} \quad \text{(A2)}$$

describes the conservation of tracers when the observer follows the convective motion of the mantle. This equation can be simplified for constant p_i (Poisson process). Such a system is also known as well-mixed. Integrating over the volume Ω of the mantle, we get the macro-balance equation:

$$\int_\Omega \frac{\partial C_i}{\partial t} dV = \frac{d}{dt} \int_\Omega C_i dV = \frac{dn_i}{dt} = -p_i n_i - \frac{\partial n_i}{\partial \theta} + \delta(\theta) Q_i(t) \quad \text{(A3)}$$

in which $n_i(t, \theta) d\theta$ is the total number of tracers with a residence time in the mantle comprised between θ and $\theta + d\theta$ and Q_i is the input of fresh tracers, notably those recycled from the crust, either oceanic or continental. Adjusting this equation to radioactive nuclides is easily done. In the general case, this equation has no closed-form solution. So far, no assumption has been made that the p_i values and therefore the residence times are constant. Let us now investigate some illustrative cases in which these variables are constant.

Transient of primordial tracers. Let us evaluate the proportion of primordial tracers that have never been extracted at time t. Their residence time in the mantle is $\theta = t$ and the number n_i^π of primordial tracers with a residence time in the mantle comprised between θ and $\theta + d\theta$ is:

$$n_i^\pi(t, \theta) = N_i(0) \delta(t - \theta) e^{-p_i t} \quad \text{(A4)}$$

where $N_i(0)$ is the total number of tracers i in the mantle-crust system.

Transient: constant recycling rate. Let us denote ω_i the probability that a tracer i extracted at a ridge crest is recycled into the mantle. The fraction extracted into continental crust will be $1 - \omega_i$. The probability of extracting the tracers into the continental crust is $(1 - \omega_i) p_i$. For simplicity, we assume that ω_i is constant and that continental crust never gets recycled into the mantle. The total number $N_i^\Sigma(t)$ of tracers i (both primordial and recycled) present at time t in the mantle is:

$$N_i^\Sigma(t) = N_i(0) e^{-(1 - \omega_i) p_i t} \quad \text{(A5)}$$

the number $N_i^\omega(t, \theta)$ of recycled tracers is the difference between the total number and the surviving primordial tracers:

$$N_i^\omega(t, \theta) = N_i(0) \left[e^{-(1 - \omega_i) p_i t} - e^{-p_i t} \right] \quad \text{(A6)}$$

with the fraction $F_i(t)$ of recycled mantle tracers given by:

$$F_i(t) = 1 - e^{-\omega_i p_i t} \quad \text{(A7)}$$

The resetting (birth) rate $r_i(t)$ of recycled tracers i is equal to the total number of tracers present in the mantle multiplied by the probability of extraction and by the probability of recycling:

$$r_i(t) = \omega_i p_i N_i(0) e^{-(1-\omega_i)p_i t} \qquad (A8)$$

At time t, a fraction $e^{-p_i\theta}$ of the tracers recycled at $t-\theta$ will still survive. The number $n_i^\omega(t,\theta)\,d\theta$ of recycled tracers with a residence time in the mantle comprised between θ and $\theta+d\theta$ therefore is:

$$n_i^\omega(t,\theta) = \omega_i p_i N_i(0) e^{-(1-\omega_i)p_i(t-\theta)} e^{-p_i\theta}$$
$$= \omega_i p_i N_i(0) e^{-(1-\omega_i)p_i t} e^{-\omega_i p_i \theta} \qquad (A9)$$

The total number of primordial tracers $N_i^\pi(t,\theta)$ and recycled tracers $N_i^\omega(t,\theta)$ with a residence time in excess of θ is obtained by integrating equations A4 and A9 between $\theta=\theta$ and $\theta=t$:

$$N_i^\pi(t,\theta) + N_i^\omega(t,\theta) = N_i(0)e^{-p_i t}$$
$$+ N_i(0)e^{-(1-\omega_i)p_i t}\left(e^{-\omega_i p_i \theta} - e^{-\omega_i p_i t}\right)$$
$$= N_i(0)e^{-(1-\omega_i)p_i t}e^{-\omega_i p_i \theta} \qquad (A10)$$

The fraction $f(t,\theta)$ of mantle tracers which at time t has a residence time in excess of θ is obtained by dividing equation A10 by equation A5:

$$f(t,\theta) = e^{-\omega_i p_i \theta} \qquad (A11)$$

which is an exponential distribution truncated at $\theta=t$.

Steady-state. When continental crust forms irreversibly, the number of tracers i becomes vanishingly small. When all the tracers are re-injected into the mantle and their age reset upon re-injection, the steady-state solution to equation A3 is:

$$n_i^{ss}(\theta) = n_i^{ss}(0)e^{-p_i\theta} = p_i N_i(0)e^{-p_i\theta} \qquad (A12)$$

The fraction $W_i(\theta)$ (survival or washout function) of tracers which resided in the mantle for a time in excess of θ is:

$$W_i^{ss}(\theta) = \int_\theta^\infty p_i e^{-p_i\theta'}d\theta' = e^{-p_i\theta} \qquad (A13)$$

Dropping the ss superscript, the mean residence time τ_i of the tracers i in the mantle is:

$$\tau_i = \int_0^\infty \theta f_i d\theta = \frac{1}{p_i} \qquad (A14)$$

which reproduces a result arrived at in Section 2.

APPENDIX B: TESTS ON DISTRIBUTIONS

The log-normal character of the distributions was tested using quantile-quantile (Rankit) or QQ plots [e.g., *Sen and Srivstava*, 1990; *Hocking*, 2003]. The data were first converted into their logarithm, then reduced by dividing the deviations from the mean by the standard deviation and finally sorted in ascending order. The observed quantiles are obtained through the inverse normal function and compared with the quantiles predicted by the normal probability density function. Many elements provide a good 1:1 correlation between the two quantiles (Figure 9). One issue raised by some of the incompatible elements, such as Ba, is the under-representation of observed high-concentration values, which probably indicates that some analysts are reluctant to report values that may be considered to reflect sample contamination. In some cases, high blank problems clearly have depleted the low end of the concentration distributions. In order to rectify these problems, we fitted a log-normal distribution to a histogram truncated below 10% and above 90% of the values (which is conveniently done in QQ plots). This adjustment produced only fairly small shifts

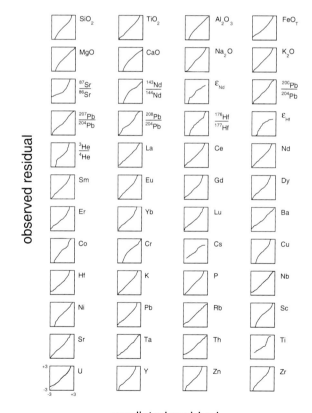

Figure 9. Quantile-quantile plots of observed vs predicted residuals assuming that concentrations and isotopic ratios follow a log-normal probability density function.

of the mean values but up to a 30% reduction of the standard deviations. The mode (most frequent value) was estimated by fitting a fourth-degree polynomial through the cumulative distribution function, again truncated below 10% and above 90% of the values. Taking the second derivative of this polynomial and equating to zero gives the mode position, which is often imprecise as a probable result of the well-known instability of numerical derivation.

Acknowledgments. Reviews by George Helffrich, Rob van der Hilst, and Bill White and careful editing of the manuscript by Janne Blichert-Toft were greatly appreciated. Discussions with Nicolas Coltice, Clint Conrad, Dan Mackenzie, Yanick Ricard, Dave Stevenson, and Paul Tackley are also gratefully acknowledged. I am particularly grateful to John Rudge who caught a significant math error in the Appendix.

REFERENCES

Abbott, D. H., L. Burgess, J. Longhi, and W. H. F. Smith (1994), An empirical thermal history of the Earth's mantle, *J. Geophys. Res., 99,* 13,835–13,850.

Agranier, A., J. Blichert-Toft, D. Graham, V. Debaille, P. Schiano, and F. Albarède (in press), Convective refolding of the mantle revealed by isotopic striations along the Mid-Atlantic Ridge, *Earth Planet. Sci. Letters.*

Albarède, F. (1995), *Introduction to Geochemical Modeling,* 543 pp., Cambridge Univ. Press, Cambridge.

Albarède, F. (1998), Time-dependent models of U-Th-He and K-Ar evolution and the layering of mantle convection, *Chem. Geol., 145,* 413–429.

Albarède, F. (2001), Radiogenic in-growth in systems with multiple reservoirs: applications to the differentiation of the mantle-crust system, *Earth Planet. Sci. Letters, 188,* 59–73.

Albarède, F., and R. van der Hilst (2002), Zone mantle convection, *Phil. Trans. R. Soc. A, 360,* 2569–2592.

Albarède, F. (2003), *Geochemistry: An Introduction,* 248 pp., Cambridge Univ. Press, Cambridge.

Allègre, C. J., T. Staudacher, P. Sarda, and M. Kurz (1983a), Constraints on evolution of Earth's mantle from rare gas systematics, *Nature, 303,* 762–766.

Allègre, C. J., S. R. Hart, and J.-F. Minster (1983b), Chemical structure and evolution of the mantle and continents determined by inversion of Nd and Sr isotopic data, I. Theoretical methods, *Earth Planet. Sci. Letters, 66,* 177–190.

Allègre, C. J., S. R. Hart, and J.-F. Minster (1983c), Chemical structure and evolution of the mantle and continents determined by inversion of Nd and Sr isotopic data, II. Numerical experiments and discussion, *Earth Planet. Sci. Letters,* 66, 191–213.

Allègre, C. J., T. Staudacher, and P. Sarda (1986), Rare gas systematics: formation of the atmosphere, evolution and structure of the Earth's mantle, *Earth Planet. Sci. Letters, 81,* 127–150.

Allègre, C. J., and D. L. Turcotte (1986), Implications of a two-component marble-cake mantle, *Nature, 323,* 123–127.

Allègre, C. J., and E. Lewin (1995a), Isotopic systems and stirring times of the earth's mantle, *Earth Planet. Sci. Letters, 136,* 629–646.

Allègre, C. J., and E. Lewin (1995b), Scaling laws and geochemical distributions, *Earth Planet. Sci. Letters, 132,* 1–13.

Allègre, C. J., A. W. Hofmann, and R. K. O'Nions (1996), The argon constraints on mantle structure, *Geophys. Res. Lett., 23,* 3555–3557.

Batchelor, G. K. (1959), Small-scale variation of convected quantities likes temperature in a turbulent fluid, 1, General discussion and the case of small conductivity, *J. Fluid Mech., 5,* 113–133.

Becker, T. W., J. B. Kellogg, and R. J. O'Connell (1999), Thermal constraints on the survival of primitive blobs in the lower mantle, *Earth Planet . Sci. Lett., 171,* 351–365.

Bedini, R.-M., J. Blichert-Toft, M. Boyet, and F. Albarède (2004), Isotopic constraints on the cooling of the continental lithosphere, *Earth Planet. Sci. Letters, 223,* 99–111.

Blichert-Toft, J., A. Agranier, M. Andres, R. Kingsley, J.-G. Schilling and F. Albarède (2005), Geochemical segmentation of the Mid-Atlantic Ridge North of Iceland and ridge-hot spot interaction in the North Atlantic, *Geochem. Geophys. Geosyst., 6,* doi:10.1029/2004GC000788.

Bolin, B., and H. Rodhe (1973), A note on the concepts of age distribution and transit time in natural reservoirs, *Tellus, 25,* 58–62.

Boyet, M., M. O. Garcia, R. Pick, and F. Albarède (2005), A search for ^{142}Nd evidence of primordial mantle heterogeneities in plume basalts, *Geophys. Res. Lett., 32,* doi:10.1029/2004GL021873.

Brooker, R. A., Z. Du, J. D. Blundy, S. P. Kelley, N. L. Allan, B. J. Wood, E. M. Chamorro, J.-A. Wartho, and J. A. Purton (2003), The 'zero charge' partitioning behaviour of noble gases during mantle melting, *Nature, 423,* 738–741.

Castle, J. C., and R. D. van der Hilst (2003), Searching for seismic scattering off mantle interfaces between 800 and 2000 km depth, *J. Geophys. Res., 108,* doi:10.1029/2001JB000286.

Castro, M. C., D. Patriarche, and P. Goblet (in press), 2-D numerical simulations of groundwater flow, heat transfer and ^{4}He transport-Implications for the He Terrestrial budget and the mantle helium-heat imbalance, *Earth Planet. Sci. Letters*

Christensen, U. (1989), Mixing by time-dependent convection, *Earth Planet. Sci. Letters, 95,* 382–394.

Christensen, U. R., and A. W. Hofmann (1994), Segregation of subducted oceanic crust in the convecting mantle, *J. Geophys. Res.,* 99, 19867–19884.

Coffin, M. F., and O. Eldhom (1994), Large igneous provinces: Crustal structure, dimensions, and external consequences, *Rev. Geophys.,* 32, 1–36.

Coltice, C., and Y. Ricard (1999a), Geochemical observations and one-layer mantle convection, *Earth Planet. Sci. Letters, 174,* 125–137.

Coltice, C., F. Albarède, and P. Gillet (2000), ^{40}K-^{40}Ar constraints on recycling continental crust into the mantle, *Science, 288,* 845–447.

Conrad, C. P., and B. H. Hager (1999a), Effect of plate bending on fault strength at subduction zones on plate dynamics, *J. Geophys. Res.,* 104, 17,551–17,571.

Conrad, C. P., and B. H. Hager (1999b), The thermal evolution of an Earth with strong subduction zones, *Geophys. Res. Lett., 26,* 3041–3044.

Conrad, C. P., and B. H. Hager (2001), Mantle convection with strong subduction zones, *Geophys. J. Int., 144,* 271–278.

Davies, G. F. (1980), Thermal histories of convective Earth models and constraints on radiogenic heat production in the Earth, *J. Geophys. Res., 85,* 2517–2530.

Davies, G. F. (1981), Earth's neodymium budget and structure and evolution of the mantle, *Nature, 290,* 208–213.

Davies, G. F. (1984), Geophysical and isotopic constraints on mantle convection: an interim synthesis, *J. Geophys. Res., 89,* 6017–6040.

Davies, G. F., and M. A. Richards (1992), Mantle convection, *J. Geol., 100,* 151–206.

Davies, G. F. (1999), *Dynamic Earth. Plates, plumes and mantle convection,* 458 pp., Cambridge Univ. Press, Cambridge.

Davies, G. F. (2002), Stirring geochemistry in mantle convection models with stiff plates and slabs, *Geochim. Cosmochim. Acta, 66,* 3125–3142.

Donnelly, K. E., S. L. Goldstein, C. H. Langmuir, and M. Spiegelman (2004), Origin of enriched ocean ridge basalts and implications for mantle dynamics, *Earth Planet. Sci. Letters, 226,* 347–366.

Dupré, B., and C. J. Allègre (1983), Pb-Sr isotope variation in Indian Ocean basalts and mixing phenomena, *Nature, 303,* 142–146.

Ferrachat, S., and Y. Ricard (1998), Regular vs. chaotic mantle mixing, *Earth Planet. Sci. Letters, 155,* 75–86.

Ferrachat, S., and Y. Ricard (2001), Mixing properties in the Earth's mantle: Effects of the viscosity stratification and oceanic crust segregation, *Geochem. Geophys. Geosyst., 2,* #2000GC000092.

Forsyth, D., and S. Uyeda (1975), On the relative importance of the driving forces of plate motion, *Geophys. J. R. Astr. Soc., 43,* 163–200.

Galer, S. J. G., and R. K. O'Nions (1985), Residence time of thorium, uranium and lead in the mantle with implications for mantle convection, *Nature, 316,* 778–782.

Galer, S. J. G., S. L. Goldstein, and R. K. O'Nions (1989), Limits on chemical and convective isolation in the Earth's interior, *Chem. Geol., 75,* 257–290.

Grand, S. P. (1994), Mantle shear structure beneath the Americas and surrounding oceans, *J. Geophys. Res., 99,* 11,591–11,621.

Gurnis, M., and G. F. Davies (1986a), The effect of depth-dependent viscosity on convective mixing in the mantle and the possible survival of primitive mantle, *Geophys. Res. Lett., 13,* 541–544.

Gurnis, M., and G. F. Davies (1986b), Mixing in numerical models of mantle convection incorporating plate kinematics, J. Geophys. Res., 91, 6375–6395.

Hart, S. R. (1984), A large-scale isotope anomaly in the Southern Hemisphere mantle, *Nature, 309,* 753–757.

Helffrich, G. R., and B. J. Wood (2001), The Earth's mantle, *Nature, 412,* 501–507.

Hocking, R. R. (2003), *Methods and Applications of Linear Models: Regression and the Analysis of Variance,* 776 pp., Wiley, New York.

Hofmann, A. W. (1988), Chemical differentiation of the Earth: the relationship between mantle continental crust, and oceanic crust, *Earth Planet. Sci. Letters, 90,* 297–314.

Hofmann, A. W. (1997), Mantle geochemistry: the message from oceanic volcanism, *Nature, 385,* 219–229.

Honda, M., I. McDougall, D. B. Patterson, A. Doulgeris, and D. Clague (1991), Possible solar noble-gas component in Hawaiian basalts, *Earth Planet. Sci. Letters, 349,* 149–151.

Jacobsen, S. B., and G. J. Wasserburg (1979), The mean age of mantle and crustal reservoirs, *J. Geophys. Res., 84,* 7411–7427.

Kaneoka, I. (1983), Noble gas constraints on the layered structure of the mantle, *Nature, 302,* 698–700.

Keeling, C. D., and B. Bolin (1967), The simultaneous use of chemical tracers in oceanic studies I. General theory of reservoir models, *Tellus, 19,* 566–581.

Kellogg, J. B., S. B. Jacobsen, and R. J. O'Connell (2002), Modeling the distribution of isotopic ratios in geochemical reservoirs, *Earth Planet. Sci. Letters, 204,* 183–202.

Kellogg, L. H., B. H. Hager, and R. D. Van der Hilst (1999), Compositional stratification in the deep mantle, *Science, 283,* 1881–1884.

Klein, E. M., and C. H. Langmuir (1987), Global correlations of ocean ridge basalt chemistry with axial depth and crustal thickness, *J. Geophys. Res., 92,* 8089–8115.

Kurz, M. D., W. J. Jenkins, and S. R. Hart (1982), Helium isotopic systematics of oceanic islands and mantle heterogeneity, *Nature, 297,* 43–47.

Masters, G., S. Johnson, G. Laske, and H. Bolton (1996), A shear-velocity model of the mantle, *Phil. Trans. R. Soc. Lond. A, 354,* 1385–1411.

Li, X.-D., and B. Romanowicz (1996), Global mantle shear velocity model developed using nonlinear asymptotic coupling theory, *J. Geophys. Res., 10*1, 22,245–22,272.

Meibom, A., and D. L. Anderson (2004), The statistical upper mantle assemblage, *Earth Planet. Sci. Letters, 217,* 123–139.

Metcalfe, G., C. R. Bina, and J. M. Ottino (1995), Kinematic considerations for mantle mixing, Geophys. Res. Lett., 22, 743–746.

McDonough, W. F., and S.-s. Sun (1995), The composition of the Earth, *Chem. Geol., 120,* 223–253.

Nauman, E. B., and B. A. Buffham (1983), *Mixing in Continuous Flow Systems,* 271 pp., Wiley, New York.

Neal, C. R., J. J. Mahoney, L. W. Kroenke, R. A. Duncan, and M. G. Petterson (1997), The Ontong Java plateau, in *Large Igneous Provinces. Continental, Oceanic and Planetary Volcanism.* Amer. Geophys. Union Monograph, ed. J. J. Mahoney, and M. F. Coffin, Amer. Geophys. Union, Washington.

Ni, S., and D. V. Helmberger (2003) Seismological constraints on the South African superplume; could be the oldest distinct structure on Earth, *Earth Planet. Sci. Lett., 206,* 119–131.

Olson, P., D. A. Yuen, and D. Balsiger (1984), Mixing of passive heterogeneities by mantle convection, *J. Geophys. Res., 89,* 525–436.

O'Nions, R. K., N. M. Evensen, and P. J. Hamilton (1979), Geochemical modeling of mantle differentiation and crustal growth, *J. Geophys. Res., 84*, 6091–6101.

O'Nions, R. K., and E. R. Oxburgh (1983), Heat and helium in the Earth, *Nature, 306*, 429–431.

O'Nions, R. K., and L. N. Tolstikhin (1994), Behaviour and residence times of lithophile and rare gas tracers in the upper mantle, *Earth Planet. Sci. Letters, 124*, 131–138.

O'Nions, R. K., and L. N. Tolstikhin (1996), Limits on the mass flux between lower and upper mantle and stability of layering, *Earth Planet. Sci. Letters, 139*, 213–222.

Ritsema, J., S. Ni, D. V. Helmberger, and H. P. Crotwell (1998), Evidence for strong shear velocity reductions and the velocity gradients in the lower mantle beneath South Africa, *Geophys. Res. Letters, 25*, 4245–4248.

Rudnick, R. L., and D. M. Fountain (1995), Nature and composition of the continental crust: a lower crustal perspective, *Rev. Geophys., 33*, 267–309.

Rudnick, R. L., W. F. McDonough, and R. J. O'Connell (1998), Thermal structure, thickness and composition of continental lithosphere, *Chem. Geol., 145*, 395–411.

Salters, V. J. M., and A. Stracke (2004), Composition of the depleted mantle, *Geochem. Geophys. Geosyst., 5*, doi:10.1029/2003GC000597.

Schmalzl, J., and U. Hansen (1994), Mixing the Earth's mantle by thermal convection: A scale dependent phenomenon, *Geophys. Res. Letters, 21*, 997–990.

Schubert, G., D. L. Turcotte, and P. Olson (2001), *Mantle Convection in the Earth and Planets*, 940 pp., Cambridge Univ. Press, Cambridge.

Sen, A., and M. Srivastava (1990), *Regression Analysis. Theory, Methods, and Applications*, 347 pp., Springer-Verlag, New York.

Stacey, F. D. (1980), The cooling Earth: A reappraisal, *Phys. Earth Planet. Int., 22*, 89–96.

Stacey, F. D. (1992), *Physics of the Earth*, 513 pp., Brookfield, Brisbane.

Stegman, D. R., M. A. Richards, and J. R. Baumgardner (2002), Effects of depth-dependent viscosity and plate motions on maintaining a relatively uniform mid-ocean ridge basalt reservoir in whole mantle flow, *J. Geophys. Res., 107*, 10.1029/2001JB000192.

Sun, S.-s., and W. F. McDonough (1989), Geochemical and isotopic systematics of oceanic basalts: implications for mantle composition and processes, *Geol. Soc. Spec. Pub., 42*, 313–345.

Taylor, S. R., and S. M. McLennan (1985), *The continental crust: its composition and evolution*, Blackwell Scientific Publications, Oxford.

Taylor, S. R., and S. M. McLennan (1995), The geochemical evolution of the continental crust, *Rev. Geophys., 33*, 241–265.

Turcotte, D. L. (1980), On the thermal evolution of the Earth, *Earth Planet. Sci. Letters, 48*, 53–58.

Turner, G. (1989), The outgassing history of the Earth's atmosphere, *J. Geol. Soc. London, 146*, 147–154.

van der Hilst, R. (1995), Complex morphology of subducted lithosphere in the mantle beneath the Tonga trench, *Nature, 374*, 154–157.

van der Hilst, R., R. Engdahl, W. Spakman, and G. Nolet (1991), Tomographic imaging of subducted lithosphere below northwest Pacific island arcs, *Nature, 353*, 733–739.

van Keken, P. E., and C. J. Ballentine (1998), Whole-mantle versus layered mantle convection and the role of a high-viscosity lower mantle in terrestrial volatile evolution, *Earth Planet. Sci. Letters, 156*, 19–32.

van Keken, P. E., and C. J. Ballentine (1999), Dynamical models of mantle volatile evolution and the role of phase transitions and temperature-dependent rheology, *J. Geophys. Res., 104*, 7137–7168.

van Keken, P., and S. Zhong (1999), Mixing in a 3D spherical model of present-day mantle convection, *Earth Planet. Sci. Letters, 171*, 533–547.

van Keken, P. E., C. Ballentine, and D. Porcelli (2001) A dynamical investigation of the heat and helium imbalance, Earth Planet. Sci Lett., 188, 421–434.

Vidale, J. E., G. Schubert, and P. S. Earle (2001), Unsuccessful initial search for a mid-mantle chemical boundary with seismic arrays, *Geophys. Res. Lett., 28*, 859–862.

Vlastelic, I., D. Aslanian, L. Dosso, H. Bougault, J. L. Olivet, and L. Geli (1999), Large-scale chemical and thermal division of the Pacific mantle, *Nature, 399*, 345–350.

Xie, S., and P. J. Tackley (2004a), Evolution of helium and argon isotopes in a convecting mantle, *Phys. Earth Planet. Int., 146*, 417–439.

Xie, S., and P. J. Tackley (2004b), Evolution of U-Pb and Sm-Nd systems in numerical models of mantle convection and plate tectonics, *J. Geophys. Res., 109*, doi:10.1029/2004JB003176.

Zindler, A., H. Staudigel, and R. Batiza (1984), Isotope and trace element geochemistry of young Pacific seamounts: implications for the scale of upper mantle heterogeneity, *Earth Planet. Sci. Lett., 70*, 175–195.

F. A. Albarède, Ecole Normale Supèrieure de Lyon, 69007 Lyon, France. (albarede@ens-lyon.fr)

Towards a Quantitative Interpretation of Global Seismic Tomography

Jeannot Trampert

Department of Earth Sciences, Utrecht University, Utrecht, The Netherlands

Robert D. van der Hilst

Department of Earth, Atmospheric, and Planetary Sciences, Massachusetts Institute of Technology, Cambridge, Massachusetts, USA

We review the success of seismic tomography in delineating spatial variations in the propagation speed of seismic waves on length scales from several hundreds to many thousands of kilometers. In most interpretations these wave speed variations are thought to reflect variations in temperature. Careful consideration of shear wave, bulk sound, and, most recently, density variations is, however, producing increasingly compelling evidence for chemical heterogeneity (that is, spatial variations in bulk major element composition) having a first-order effect on the lateral variations in mass density and elasticity of the mantle. This has profound consequences for our understanding of mantle dynamics and the thermochemical evolution of our planet. We argue that the quantitative integration of constraints from seismology, mineral physics, and geodynamics, which underlies the inference of thermochemical parameters, requires careful uncertainty analyses and should move away from emphasizing visually pleasing images and single, nonunique solutions.

1. INTRODUCTION

Knowledge of the present-day scale and nature of the solid-state convection in Earth's mantle is key to our understanding of plate tectonics—and the surface processes and hazards associated with it—and of Earth's thermochemical evolution over long periods of geological time. Indeed, whether or not mantle convection occurs in separate layers and whether or not compositionally distinct domains have survived anywhere in the convecting system have major implications for models of Earth's early geological history and subsequent development. Many integrated views on mantle dynamics and chemistry have been proposed [e.g.,

Hofmann, 1997; *Kellogg et al.*, 1999, *Helffrich and Wood*, 2001; *Albarède and van der Hilst*, 2002; *Anderson*, 2001, 2002], but firm evidence in support of any of the models is still lacking. Recently, significant progress has been made on a variety of research fronts, with seismological evidence having changed most dramatically.

Over the past two decades, global tomography, a class of inversion techniques for interpreting observations from earthquake records in terms of three-dimensional (3D) variations in Earth's elastic properties, has produced spectacular images of Earth's interior structure. Among the success stories of global tomography are the delineation of long-wavelength variations in elastic properties in Earth's mantle, which started in the early 1980s, and the detailed delineation, over the last decade or so, of trajectories of mantle convection [see reviews by, e.g., *Dziewonski and Woodhouse*, 1987; *Woodhouse and Dziewonski*, 1989;

Earth's Deep Mantle: Structure, Composition, and Evolution
Geophysical Monograph Series 160

Masters, 1989; *Romanowicz*, 1991; *Montagner*, 1994; *Masters and Shearer*, 1995; *Ritzwoller and Lavely*, 1995; *Dziewonski*, 1996; *Masters et al.*, 2000; *Kárason and van der Hilst*, 2000; *Fukao et al.*, 2001; *Romanowicz*, 2003]. It is encouraging to see that increasingly consistent information on the spatial patterns of wave speed variations is emerging from tomographic studies that use different data and/or techniques. The long-wavelength patterns are now fairly well established, and also on the issue of deep slabs there is growing consensus. Despite recent progress, however, the tomographic images of the return flow of mantle convection are still ambiguous, and there is no lack of controversy regarding the depth of origin, the morphology, the nature, and even the existence of so-called plumes.

One way of further improving spatial resolution and model accuracy is by data fusion, that is, the joint inversion of data that have different sensitivities to Earth's structure. Since these are often measured at different frequencies, it is becoming important to consider finite frequency effects, which is a topical subject of theoretical seismology [e.g., *Dahlen et al.*, 2000; *De Hoop and van der Hilst*, 2005]. The biggest challenge is to be able to perform a complete waveform inversion; we are now able to calculate exact seismograms in a full 3D Earth, but the computational resources are still lacking to apply these exciting techniques to a realistic inverse problem [*Tromp et al.*, 2005].

Although a major challenge in itself, the mere mapping of mantle structure at a wide spectrum of spatial scales is not the ultimate goal of seismic tomography. For seismology to be a key component of an inherently multidisciplinary effort aimed at understanding mantle dynamics, composition, and evolution, we must know the underlying physical or chemical causes of the variations in wave speed. Many computer simulations of mantle convection are based on instantaneous flow patterns calculated for purely thermal origins [e.g., *Schubert et al.*, 2001] and imply simple scaling relationships between seismic wave speeds and density in tomographic models. However, in recent years it has become evident that not all inferred wave speed variations are consistent with a thermal origin and that significant regional variations in major element composition exist in Earth's mantle [e.g., *Ishii and Tromp*, 1999; *van der Hilst and Kárason*, 1999]. For example, it is increasingly likely that the so-called superplumes in the deep mantle beneath Africa [see also *Helmberger and Ni*, this volume] and the Pacific reflect changes in both temperature and composition [*Trampert et al.*, 2004; see also *Samuel et al.*, this volume]. The geochemical record on Earth's differentiation over geological time, the planetary heat budget, and the secular changes in formation and subsequent evolution (stabilization) of continents also suggest that compositional heterogeneity probably exists over a wide

range of length scales [e.g., *Hofmann*, 1997; *Helffrich and Wood*, 2001; *Anderson*, 2002; *Albarède and van der Hilst*, 2002; see also the contributions to this volume by *Albarède, Anderson, Harrison and Ballentine, Righter*, and *Tackley*]. In concert with mineral physics and geodynamics, one frontier in seismological research thus concerns the detection and characterization of spatial variations in temperature and in mantle mineralogy and phase chemistry.

From existing mineral physics data, temperature and the iron and silicate content of the mantle appear to have the biggest influence on wave speeds [see also the contributions to this volume by *Badro et al.* and *Bukowinski and Akber-Knutson*], but information about *P*- and *S*-wave speeds alone is not enough to constrain these parameters. Indeed, it is important to have independent information on density variations, for instance, through analysis of gravity anomalies and quantitative integration of inferences from Earth's free oscillation frequencies. Knowledge on the nature of seismic anisotropy and attenuation would further constrain the thermochemical parameters, but their determination awaits major advances in seismic data analysis and theory.

A related challenge concerns model uncertainty. While much research effort has been put into increasing the resolution of the lateral and radial variations that can be imaged, relatively little progress has been made towards the quantitative assessment of the accuracy or uniqueness of the obtained models. Yet, this information is needed if we want to advance from a predominantly qualitative, semi-monodisciplinary to a more quantitative, multidisciplinary interpretation. Most published seismological models are solutions of an underdetermined inverse problem in the sense that the data are not sufficient to constrain independently all model parameters. In the best case, models of selected physical parameters represent an optimum fit to data, for instance, in the least-squares sense. But these fits are nonunique and often heavily influenced by a particular regularization (also referred to as damping). The same problems plague models based on mineral physics or geochemical data. The disciplinary solutions are likely to fall in different parts of the permissible model space, and one must consider uncertainty and error in order to find common ground (Plate 1).

The organization of this paper roughly follows the topics mentioned in the preceding paragraphs, and we will end with a discussion of outstanding issues and future challenges.

2. MANTLE STRUCTURE INFERRED FROM SEISMIC TOMOGRAPHY

Since the advent of seismic tomography in the late 1970s, global tomographic imaging has evolved along complementary lines, leading to the so-called high-resolution models and long-wavelength models[1]. This has several reasons. Owing to theoretical and practical considerations, seismic imaging applications have long involved relatively small subsets of the available data. On the one hand, solving the equation of motion in its most general form is not yet possible for large-scale problems, and several theoretical approximations need to be made in order to obtain tractable formulations. On the other hand, high natural noise levels in certain frequency bands and the band limitations due to instrument filters and source characteristics yield specific frequency bands for seismic analysis. Furthermore, different types of waves are sensitive to different elastic properties. And, finally, different parameterizations have been used to achieve different objectives. A seismologist thus faces a choice as to which data and, along with it, which theoretical approximations and model parameterizations, to use for a specific application.

High-resolution and long-wavelength models can have a very different appearance, but it is useful to see them as complementary rather than competing depictions of Earth's structure. Tests have shown that in regions of adequate data coverage the high-frequency body wave arrivals map long-wavelength structure correctly [e.g., Figure 2 in *van der Hilst et al.*, 1997], and with a careful spectral analysis the long-wavelength models will give a representation unbiased by small-scale structures, such as slabs in the upper mantle [*Trampert and Snieder*, 1996]. With growing data sets, the development of more powerful inversion and wave propagation theories, and increasing computational abilities, these approaches will continue to converge.

We deliberately omit discussing anisotropic and attenuation tomography. It is well established that the Earth presents distinct anisotropic regions [e.g., *Montagner*, 1998], but to date little consensus has been reached on details in existing models, except maybe in the shallowest mantle. On the interpretational side, numerous experimental data exist for anisotropic minerals, but relating observed seismic anisotropy to the chemistry of the mantle relies on many assumptions that currently cannot be unambiguously tested. Attenuation is also clearly present in the mantle [e.g., *Romanowicz and Durek*, 2000], but models strongly vary from author to author due to unresolved theoretical issues (e.g., scattering and focusing vs. intrinsic attenuation). Understanding acti-

vation processes for lower-mantle attenuation and obtaining experimental data still remain formidable challenges which make the interpretation of attenuation tomography even more speculative than that of anisotropic tomography.

2.1. Long-Wavelength Models

Long-wavelength (i.e., thousands of km) variations in wave propagation speeds have been inferred from a combination of broadband waveforms, long-period body wave arrival times, surface wave dispersion data, and splitting functions of Earth's free oscillations [e.g., *Su et al.*, 1994; *Masters et al.*, 1996, 2000; *Mégnin and Romanowicz*, 2000; *Gu et al.*, 2001; *Ritsema et al.*, 1999; *Ekström and Dziewonski*, 1998; *Li and Romanowicz*, 1996; to name but a few]. Because a combination of body- and surface waves and mode data can be used, the constraints on *S*-wave models are often better than in the case of the high resolution *P*-wave models (see below). The images are typically represented horizontally by global basis functions (surface spherical harmonics). This important class of models has continued to improve over the past two decades [e.g., *Romanowicz*, 2003], but many challenges remain. In the midmantle, substantial differences still exist between results of different research groups, in particular when different data sets are used [e.g., *Becker and Boschi*, 2002], and the magnitude of the wave speed perturbations is quite uncertain.

In the uppermost mantle, the credibility of the models has been established both by their ability to match independent waveform data and by their correlation with tectonic features. *Trampert and Woodhouse* [2000] tested all recent long-wavelength fundamental mode phase velocity models against independent waveform data and found a reassuring agreement between them. In a test against independent splitting functions, *Ritzwoller and Lavely* [1995] showed that a robust pattern of the Earth's mantle emerged at the lowest degrees, even if different data and different mapping strategies were employed. Although this quantitative comparison is now almost 10 years old, more recent long-wavelength models correlate highly with the models in the *Ritzwoller and Lavely* study.

Although smaller-scale structures can now be imaged with greater confidence, the overall view that emerged from the pioneering studies by *Masters et al.* [1982], *Woodhouse and Dziewonski* [1984], and *Dziewonski* [1984] has proved fairly robust. In the uppermost mantle, these early studies delineated the low wave speeds associated with midoceanic ridges and regions of the western Pacific characterized by back-arc volcanism (Plate 2) and demonstrated that the structure beneath continents differs from the oceanic mantle to several hundred km depth (with Precambrian cratons, shields,

---

[1] Despite popular use, references to high- vs. low-resolution models can be misleading or even incorrect. The model parameters underlying the "long-wavelength" models (i.e., the spherical harmonic coefficients) may actually be resolved better than the parameters (e.g., wave speed at nodes or in small cells) representing "high-resolution" models. In the context of this discussion, high-resolution merely indicates that—potentially—contrasts in elastic properties can be detected over smaller spatial length scales.

and platforms marked by fast seismic wave propagation to depths of 200 km and larger in some continental regions). Heterogeneity in the upper-mantle transition zone (between the seismic discontinuities near 410- and 660-km depth) and near the base of the mantle is manifest at very long wavelengths (Plate 2), with a predominance of fast wave propagation in the mantle beneath the circum-Pacific "Ring of Fire" and slow speeds beneath Africa and the central Pacific. This structure explains the long-wavelength variations in the gravity field [*Richards and Hager*, 1984; *Hager et al.*, 1985; *Cazenave et al.*, 1989] and correlates well with sites of post-Mesozoic subduction (recycling) of oceanic lithosphere [*Richards and Engebretson*, 1992; *Ricard et al.*, 1993]. The shallow and lowermost mantles are marked by relatively strong heterogeneity, but the inferred magnitude of elastic heterogeneity decreases away from these boundary regions. In the midmantle, recent models show a low-amplitude white spectrum [*Romanowicz*, 2003], but it is here that models differ most.

2.2. High-Resolution Models

In a largely separate effort, large amounts of travel time data and local basis functions (cell or grid parameterizations with a spacing of the order of 1–4°) have been used to delineate Earth's structure on a finer scale, both with *P*-data [e.g., *Fukao et al.*, 1992; *van der Hilst et al.*, 1997; *Vasco and Johnsen*, 1998; *Bijwaard et al.*, 1998; *Kennett et al.*, 1998; *Boschi and Dziewonski*, 2000; *Kárason and van der Hilst*, 2001; *Montelli et al.*, 2004] and with *S*-data [e.g., *Grand*, 1994; *Grand et al.*, 1997; *Widiyantoro et al.*, 1998]. In the *P*-wave studies, data coverage is generally not as good as in long-wavelength shear wave tomography. The uneven distribution of earthquakes and stations and the fact that surface waves provide relatively weak constraints on compressional wave speed means that high resolution is mainly restricted to seismically active regions, such as plate boundaries, or continental regions with many seismograph stations, whereas large areas beneath oceans remain without effective sampling. In modern applications, the uneven data coverage is partly balanced by the use of adaptive grids [*Abers and Roecker*, 1991; *Fukao et al.*, 1992; *Widiyantoro and van der Hilst*, 1996; *Sambridge and Gudmundson*, 1998; *Bijwaard et al.*, 1998; *Kárason and van der Hilst*, 2001; *Montelli et al.*, 2004] and, in part, remedied by the use of different data types along with 3D sensitivity kernels to account for frequency differences [e.g., *Dahlen et al.*, 2000; *Kárason and van der Hilst*, 2001; *Montelli et al.*, 2004; *van der Hilst et al.*, in preparation].

The largest single data source for this class of imaging is the *Bulletin of the International Seismological Centre*. These data are rather noisy, but the reprocessing by *Engdahl*

et al. [1998] has produced a data set of high quality and research potential. In recent years, individual efforts have been producing large data sets of travel time measurements from digital waveforms (e.g., by waveform cross-correlation [*Masters et al.*, 1996; *Ritsema et al.*, 1999]). Before too long these compilations can be expected to become the data set of choice for this class of tomographic imaging because they contain a larger proportion of later arrivals. As a next step, computational techniques are being developed that enable the use of complete seismograms [*Tromp et al.*, 2005].

Image reliability is often assessed (rather qualitatively) by model comparison and with so-called checkerboard tests, where the ability to recover a known input model is determined and used as a proxy for image reliability. Many studies have now established that the models do not critically depend on the choice of parameterization and inversion technique [e.g., *Spakman and Nolet*, 1988; *Boschi and Dziewonski*, 1999], but uneven data coverage and inconsistent data quality remain important issues.

By producing increasingly detailed images of slabs of subducted lithosphere, the most tangible trajectories of mantle flow, this class of tomographic model has made crucial contributions to the debate of layered vs. whole mantle convection, which has divided Earth scientists for almost a half century [see, e.g., recent reviews by *Hofmann*, 1997; *Helffrich and Wood*, 2001; *Albarède and van der Hilst*, 2002; *Anderson*, 2001, 2002; *Tackley*, 2002]. First on a regional [e.g., *Spakman et al.*, 1988; *van der Hilst et al.*, 1991] and later on the global scale [e.g., *van der Hilst et al.*, 1997] it was shown that slabs can penetrate across the 660-km discontinuity, implying substantial mass exchange between the upper and lower mantle. The excellent agreement between *P* and *S* images [*van der Hilst et al.*, 1997; *Grand et al.*, 1997], at least to ~1900-km depth, and the increasing similarity between slablike features in high-resolution and long-wavelength models [e.g., *Fukao et al.*, 2001; *Romanowicz*, 2003] lends further credibility to these deep structures.

But the issue is still far from being solved. Structural complexity of slab trajectories, with some slabs deflecting and apparently stagnating in the upper mantle transition zone and others penetrating to larger depths (Plate 3), demonstrates that neither strict layering at 660 km nor whole-mantle mixing with unobstructed slab penetration is a realistic flow model [*van der Hilst et al.*, 1991, 1997; *Fukao et al.*, 1992, 2001]. Combined with the emerging seismological evidence for compositional heterogeneity (see also next section)—as well as the inferences from geochemistry and Earth's heat budget—this observation inspired some of us [*van der Hilst and Kárason*, 1999; *Kellogg et al.*, 1999; *Albarède and van der Hilst*, 2002] to explore other scenarios of (thermochemical) mantle convection and evolution.

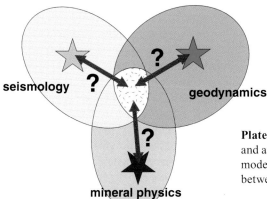

Plate 1. Optimal fits to incomplete data sets (depicted by stars) often depend on the type and amount of regularization of the inverse problem and are not unique. Unless all likely models (depicted by the ellipses) are explored and uncertainty accounted for, an overlap between models from different data sets is difficult to find, let along quantify.

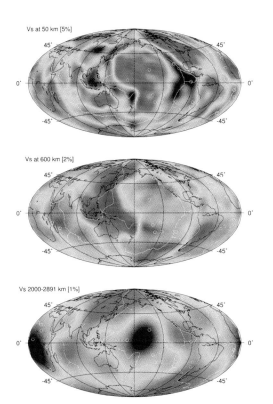

Plate 2. Models at 50- and 600-km depth obtained from the inversion of our fundamental and higher-mode surface wave phase velocity maps for variations in shear wave speed at a particular depth. The lowermost mantle layer is the most likely model from *Trampert et al.* [2004]. The perturbations are shown in percent variations with respect to PREM [*Dziewonski and Anderson*, 1981]. Negative wave speed anomalies are depicted by red colors and positive anomalies are in blue. The maximum scale is indicated beside each plot. The yellow lines indicate plate boundaries; yellow circles indicate hotspots.

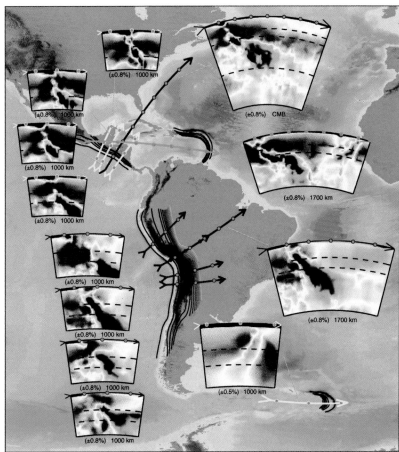

Plate 3. Cross-sections of a recent *P*-wave model [*van der Hilst et al.*, in preparation] to illustrate the likely complexity of slab structures. This model was obtained by inversion of travel time data from *P*, *PP*, *pP*, and several core phases (*PKP*, *Pdiff*), using an irregular grid parameterization, finite frequency sensitivity kernels [see also *Kárason and van der Hilst*, 2000], and corrections for crustal structure according to CRUST2.0 [http://mahi.ucsd.edu/Gabi/crust2.html]. The perturbations are shown in percent with respect to ak135 [*Kennet et al.*, 1995]. Negative anomalies are depicted by red colors and positive anomalies are in blue. The maximum scale is indicated below each cross-section together with its maximum depth. CMB, core–mantle boundary.

For technical reasons, plumes are far more difficult to image and are currently hotly debated among scientists. Recently, *Montelli et al.* [2004] described intriguing images of plumes, but further research is required to establish conclusively the existence, nature, origin depth, and volume of mantle upwellings. We speculate that over the next 5 years, improvements in theory and data coverage, especially from ocean bottom seismic experiments, will lead to significant progress on this crucial feature of mantle flow.

3. ORIGIN OF WAVE SPEED ANOMALIES: DEMISE OF THE THERMAL PARADIGM

Beyond giving insight into the present-day pattern of mantle flow, one would like to use tomographic models as input, or boundary condition, for geodynamical simulations of mantle flow. This is not trivial, however. Seismic tomography yields 3D wave speed variations, whereas convective flow is driven by lateral variations in buoyancy, i.e., mass density. It has been known since *Birch* [1961] that there are systematic relationships between *P*-wave speed and density in crustal rocks, and it thus seemed sensible to assume scaling relations between seismic wave speed and mass density for material in the deep Earth. If one makes the assumption that temperature variations within the convecting mantle are responsible for variations in seismic speed and density, results from mineral physics experiments [e.g., *Anderson et al.*, 1984] can be used to establish that deep mantle shear velocity and density are highly correlated with a ratio $\delta\ln\rho/\delta\ln V_s \approx 0.4$. This concept was used in pioneering studies in the mid-1980s. Soon after the construction of the first tomographic whole-mantle models [*Woodhouse and Dziewonski*, 1984; *Dziewonski*, 1984] it was recognized that the geoid was negatively correlated to the variations in wave speed. *Hager et al.* [1985] used lower-mantle density perturbations obtained from wave speed variations by simple scaling relationships to drive viscous flow and to generate dynamically maintained topography at the core–mantle boundary and the Earth's surface. The total gravity field due to the inferred density variations and the flow-induced topography explained the long-wavelength geoid remarkably well. The success of these and later studies led to the widespread view that mantle convection is controlled by thermal processes.

Using the growing mineral physics data base, one can test whether such a thermal interpretation of the available tomographic models is warranted. Laboratory experiments at deep mantle conditions show that heating (cooling) material reduces (increases) simultaneously the shear and bulk modulus and density of candidate minerals [e.g., *Anderson*, 1995]. The consequence of a purely thermal origin, therefore, is that bulk sound speed, shear wave speed, and density

should be perfectly correlated to one another at all depths. But this strong positive correlation appears to hold only in some parts of the Earth's mantle.

Tomographically inferred isotropic wave speed variations in the shallow mantle ($z < 200$ km) correlate well with tectonic features and expectations for the associated temperature field (Plate 2, top). Midoceanic ridges, known to be hotter than average, are seismically slower than average; the cooling of the oceanic lithosphere with age appears as progressively faster wave speeds away from the midoceanic ridge systems; and ancient continental shields are in general marked by low average heat flow and faster than average seismic propagation speed. This is consistent with analyses of mineral physics data, which show that at shallow depths in the mantle, seismic velocities are generally much less sensitive to composition than to variations in temperature [*Deschamps et al.*, 2002; *Cammarano et al.*, 2003]. However, chemical depletion of subcontinental lithosphere [e.g., the tectosphere hypothesis of *Jordan*, 1975] can lead to small detectable changes in wave speeds identifiable by considering decorrelations between *P*- and *S*-wave tomography [*Goes et al.*, 2000], combined tomography and gravity data [*Forte and Perry*, 2000; *Deschamps et al.*, 2002; *van Gerven et al.*, 2004] and seismic anisotropy [*Beghein and Trampert*, 2004].

While tomographic images are a reasonable proxy for the temperature field in the shallowest mantle, this relationship becomes increasingly more tenuous as we go deeper in the mantle. With increasing pressure the temperature sensitivity of velocities decreases and seismic wave speeds become more sensitive to composition [*Anderson*, 1989; *Trampert et al.*, 2001]. If wave speeds and mass density are not dominated by temperature effects, then it is not necessarily meaningful to assume a constant scaling between $\delta\ln\rho$ and $\delta\ln V_s$. Recent interpretations of gravity anomalies and Earth's free oscillation data [*Ishii and Tromp*, 1999, 2001, 2004] clearly show low-to-negative correlations between $\delta\ln\rho$ and $\delta\ln V_s$ in the deep mantle. Using a full-model space search technique, *Resovsky and Trampert* [2003] confirmed that the low-to-negative correlations are, indeed, highly probable, given existing seismic data throughout the mantle. While they may be consistent with geodynamic observables [e.g., *Forte and Mitrovica*, 2001], mass density anomalies derived by direct scaling of shear wave speed variations do not explain the spectral properties of the of Earth's free oscillation. Similarly, throughout the mantle, clear observations of low-to-negative correlations between $\delta\ln V_s$ and $\delta\ln V_\phi{}^2$ have been made [*Masters et al.*, 2000; *Saltzer et al.*, 2001; *Resovsky and Trampert*, 2003]. This is irrefutable observational proof that temperature alone

---

[2] In an isotropic medium the bulk sound speed V_ϕ can be derived from the P - and S-wave speed. With $V_p = \sqrt{(\kappa + 4/3\mu)/\rho}$, and $V_s = \sqrt{\mu/\rho}$, then $V_\phi = \sqrt{V_p^2 - 4/3V_s^2} = \sqrt{\kappa/\rho}$.

cannot be responsible for wave speed and density anomalies. As a consequence, the paradigm of purely thermally driven mantle flow is in need of revision, as already suggested by *Anderson* [2001], using different arguments.

4. SEISMOLOGICAL EVIDENCE FOR COMPOSITIONAL HETEROGENEITY

The anomalous frequencies of Earth's free oscillations are not the only seismological observations suggesting lateral variations in bulk composition in Earth's mantle. Other lines of evidence for compositional heterogeneity include discrepancies between the seismologically inferred and theoretically predicted ratio between relative changes in shear and compressional wave speed, the conspicuous anticorrelations between shear- and bulk sound speed, and the anomalously large range of the Poisson's ratio beneath some geographical regions. Collectively, this evidence has begun to suggest that variations in composition occur over a wide range of length scales, from local (several 100s of km) to global (e.g., the spherical harmonic degree 2 pattern).

4.1. Radial and Lateral Variation of $R = \delta \ln V_s / \delta \ln V_p$

In the seismologic and mineral physics communities, the depth dependence of the spherical average of the ratio $R = \delta \ln V_s / \delta \ln V_p$ has received considerable attention in the past 5 years or so. In combination with mineral physics data, the seismologically measured ratio is thought to indicate whether thermal and/or chemical causes are responsible for the observed lateral variations in wave speeds [*Karato and Karki*, 2001]. However, there are several caveats that one should be aware of. First, in any study based on a comparison between P and S speed, one should make sure that the sampling by P- and S-sensitive data is geographically similar [*Robertson and Woodhouse*, 1996; *Kennett et al.*, 1998; *Saltzer et al.*, 2001, 2004]. Second, there are several ways of determining R, and the result is not always the same [see, e.g., *Masters et al.*, 2000; *Saltzer et al.*, 2001, 2004]. Furthermore, the diagnostic value of R is often overstated. Scientists concur that large values of R (i.e., larger than 2.5) cannot be due to a purely thermal origin; contrary to common perception, however, the reverse is not true: While consistent with a thermal origin, small values cannot rule our compositional effects. Indeed, several seismological studies, either with travel time or normal mode data, have revealed substantial wave speed anticorrelations (suggesting nonthermal effects) in mantle regions where R values remain close to or far below values predicted from a purely thermal origin [*Beghein et al.*, 2001; *Saltzer et al.*, 2001]. Far-reaching conclusions based on the R value alone should, thus, be considered with a healthy dose of caution and skepticism.

It is generally agreed that R increases with depth, particularly close to the core [*Masters et al.*, 2000], but focusing on the radial dependence is a misleading oversimplification since lateral variations of R can be significant and important [*Masters et al.*, 2000; *Saltzer et al.*, 2001]. *Saltzer et al.* [2001] found that the strongest increase of R with depth occurs away from post-Mesozoic subduction regions, suggesting that the high values most diagnostic of compositional differences occur in the same regions where the shear wave speed is lowest, that is, in the deep mantle beneath southern Africa and the central and western Pacific. *Deschamps and Trampert* [2003] showed that the width of histograms of lateral variations of R give an unambiguous indication of the presence of chemical heterogeneities in the lowermost mantle, but that by itself the value of R can never give quantitative information on the magnitude of temperature and composition due to an invariance to arbitrary scaling.

4.2. Long-Wavelength Anticorrelations of Wave Speeds and Density

Tomographic inversion of P- and S-data reveals a strong correlation ($r \approx 0.8$) between P- and S-wave speed anomalies at all depths in the mantle [*Kennett et al.*, 1998; *Ishii and Tromp*, 1999; *Masters et al.*, 2000; *Saltzer et al.*, 2001; *Resovsky and Trampert*, 2003], although some models reveal a gradual decline in correlation beyond 2000-km depth [*Saltzer et al.*, 2004]. The high correlation is primarily due to the fact that the propagation speeds of compressional and shear waves are both dominated by changes in shear modulus or density [e.g., *Su and Dziewonski*, 1997; *Kennett et al.*, 1998; see also *Ricard et al.*, this volume]. In the deep mantle the effect of the bulk modulus (incompressibility) on P-wave speed is small and reaches a minimum near 2000-km depth [*Kennett et al.*, 1998]. While results vary [*Masters et al.*, 2000], simultaneous inversions for shear wave and bulk sound speed all reveal small or negative correlations between variations in bulk sound and shear wave speed in the deep mantle [*Su and Dziewonski*, 1997; *Kennett et al.*, 1998; *Ishii and Tromp*, 1999; *Masters et al.*, 2000]. We note that the differences between the models are most likely due to different regularizations of the inverse problem. Using a full-model space search, *Resovsky and Trampert* [2003] found that all published correlations were compatible with the available data and that negative correlations are robust (Figure 1).

Density has a different sensitivity to compositional changes than to bulk sound and shear wave speeds. *Ishii and Tromp* [1999] were the first to identify small or negative correlations between density and shear sound speed in the lower mantle. Their model was criticized as not being robust with respect to damping [*Resovsky and Ritzwoller*, 1999; *Romanowicz*,

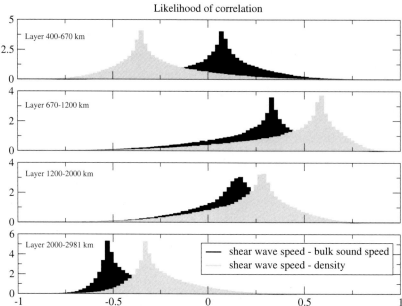

Figure 1. Likelihoods of correlations generated by randomly sampling the probability density functions for $\delta\ln V_s$, $\delta\ln V_\phi$, and $\delta\ln\rho$, after *Resovsky and Trampert* [2003]. The uppermost mantle layer is omitted because the low spherical harmonic degree expansion used does not capture the more complicated nature of Earth's structure near the surface.

2001] and initial models and parameterization [*Kuo and Romanowicz*, 2002]. While this may be true for the amplitude of the inferred density anomalies, the pattern and sign of the anomalies proved to be correct. *Resovsky and Trampert* [2003] identified all models compatible with normal mode splitting functions and fundamental mode and overtone phase velocity data. Their approach is based on forward modeling and does not require regularization; consequently, the correct amplitude can be found even in the presence of parameters with largely varying sensitivities. They confirmed that negative correlations between density and shear sound speed variations are most likely in the lowermost mantle and transition zone (Figure 1).

The amplitude of the density signal also provides compelling evidence for the presence of chemical heterogeneity. At and beyond transition zone depths, the amplitude of density anomalies are likely to be at least as large as that of shear sound speed variations [*Resovsky and Trampert*, 2003]. Existing mineral physics data show that this is not compatible with a purely thermal origin [e.g., *Karato and Karki*, 2001].

4.3. Evidence for Compositional Heterogeneity on Shorter-Length Scales

The results from the analyses of $R = \delta\ln V_s/\delta\ln V_p$, and the wave speeds and density anticorrelations mentioned in the preceding sections, all imply that variations in bulk composition deep in Earth's mantle can occur on scale lengths in excess of tens of thousands of km. Indeed, the inferred compositional heterogeneity is qualitatively consistent with the convection model proposed by *Kellogg et al.* [1999], in that the most anomalous deep mantle regions would be located away from the subducting slabs, in regions characterized by the lowest shear wave speeds. However, regional studies have begun to reveal lateral variations in composition on a much smaller length scale as well. *Saltzer et al.* [2004] inverted some 15,000 relative P and S times, obtained by broadband waveform cross-correlation, for variations in V_p, V_s, and V_ϕ beneath a great circle arc from source regions in the northwest Pacific to receivers in North America. In the deep mantle beneath Alaska, they detected regions of low- and anti-correlations between V_s and V_ϕ as small as 500 km (Plate 4). The anticorrelation and the associated large range in inferred Poisson's ratios cannot be explained by currently available data from mineral physics, and *Saltzer et al.* [2004] thus proposed that a combination of thermal and compositional changes is needed to explain the observations. Interestingly, the inferred range of variation in temperature, iron, and silicate perovskite is in excellent agreement with inferences from normal modes, which are sensitive to much longer-length scales. In combination, the normal mode and body wave studies suggest that lateral variations in bulk composition can occur over a wide range of length scales, but that a careful analysis is required to detect such heterogeneity. Knowing the scaling characteristics of compositional heterogeneity is important for understanding mantle-mixing over long periods of geological time [e.g., *Albarède*, this volume].

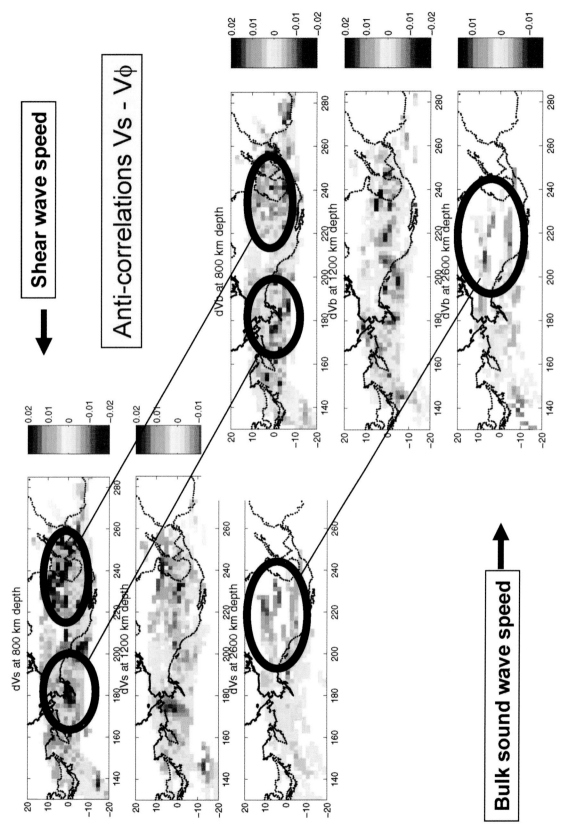

Plate 4. *S-wave and bulk sound speeds in the deep mantle beneath Alaska and the north Pacific, after Saltzer et al. [2004]. Highlighted are some clear anticorrelations for well-resolved areas of the model. Such anticorrelations are resolved at length scales of several hundred kilometers and larger. Smaller-scale variations cannot be resolved with the data and technique used.*

5. QUANTITATIVE INTERPRETATION OF TOMOGRAPHIC AND MINERAL PHYSICS DATA

For several reasons it has remained difficult to provide a quantitative interpretation of tomographic models or their related parameters described in the previous section (that is, the ratio R of the relative variation in shear and compressional wave speed, the Poisson's ratio, or the correlation—or lack thereof—between wave speeds) in terms of temperature and composition. On the one hand, we recognize outstanding issues for seismology. Constraining density variations in Earth's interior, a key ingredient for geodynamical studies, is still a formidable challenge (so far only the lowest even-numbered spherical harmonic degrees have been constrained). Furthermore, the uneven seismic source and receiver distribution produce substantial spatial variations in the reliability of the estimates of V_s, V_p, and V_ϕ and, to make matters worse, the uncertainties are different for each wave type because of differences in sampling and concomitant effects of regularization. On the other hand, we recognize outstanding issues for mineral physics research. Constraints from experimental and theoretical mineral physics exist for only some minerals and often depend critically on extrapolations over large pressure and temperature ranges and from single crystal measurements to the behavior of rock aggregates. Furthermore, experimental and ab initio results do not always agree. Moreover, the effects of, for instance, oxidation state, minor elements, and impurities on mineralogy, transport properties (such as viscosity and thermal and electric conductivity), element partitioning, and elasticity are not well known [*e.g., Badro et al.*, 2005; McCammon, this volume].

5.1. Uncertainty Analysis and Model Space Searches

Even if we do not know the origin and magnitude of all uncertainties, we must try to estimate the known uncertainties and use them in quantitative joint interpretations of seismological and mineral physics data. *Forte and Mitrovica* [2001] were the first to quantify variations in (what they called) effective temperature and composition, using tomographic models of relative bulk and shear sound variations. However, *Deschamps and Trampert* [2003] showed that realistic uncertainties in the conversion factors and in the tomographic models yielded a large uncertainty in the inferred temperature field (50%, in their study) and left composition largely unconstrained.

A major problem with the use of selected tomographic models as input data for such studies is the imprint of regularization on the solution. Tomographic problems are both overdetermined, in the sense that there are typically more data than model parameters, and underdetermined, in the sense that not all model parameters can be resolved independently.

Moreover, real data are noisy. As a consequence, the generic tomographic inversion problem does not have a unique solution. Many realizations of model parameters can produce acceptable fits to the data, and typically the seismologist forces a unique solution by imposing sensible—but subjective—constraints on the solution. One should realize that in such a case, the tomographic solution in mantle regions that are not effectively sampled by the seismic waves is entirely controlled by the combination of a priori assumptions and ad hoc regularization. Worse, even in regions that are well sampled, regularization may severely degrade the parameter estimation. If the aim is merely to produce a visually appealing image, that is, if only the geographic pattern of wave speed variations is of interest, than this may not be a serious problem. But the models produced this way (the stars in the cartoon of Plate 1) may, in fact, be sufficiently far away from the real Earth (depicted as the checkered field) to invalidate quantitative interpretations or model comparisons.

The problems associated with the use of a single model, and the bias and error due to regularization, can be avoided with techniques that explore the entire model space. Several studies have shown that the statistical properties of distributions of a large family of acceptable solutions contain different and often more meaningful information than does a single solution [*Mosegaard and Tarantola*, 1995; *Sambridge and Mosegaard*, 2002; *Beghein et al.*, 2002; *Shapiro and Ritzwoller*, 2002; *Resovsky and Trampert*, 2003; *Trampert et al.*, 2004].

5.2. Sensitivities of Wave Speeds to Temperature and Composition

Sensitivities of wave speeds to temperature and composition as a function of depth can be calculated using mineral physics data and equation of state (EOS) modeling. Many EOS models have been proposed and used for Earth studies [*Anderson*, 1989; *Poirier*, 1991; *Stacey*, 1992; *Anderson*, 1995]. The approaches that are most commonly used are either based on third-order finite strain theories or on a Mie–Grüneisen description. *Jackson* [1998] concluded that third-order Eulerian finite strain isotherms and isentropes are adequate for the range of strains encountered in the lower mantle and further showed that hot finite strain isentropes are consistent with the Mie–Grüneisen–Debye description of thermal pressure and a cold isothermal third-order compression. If we then fix a reference temperature and composition for the lower mantle using a 1D seismic reference model and choose the right mineral physics data, it is straightforward to obtain the sensitivities analytically or numerically [*Karato*, 1993; *Trampert et al.*, 2001; *Stacey and Davis*, 2004].

In reality it is not that simple. We don't really know the thermochemical reference state of the Earth [see *Williams and*

Knittle, this volume], and many different combinations of temperature and composition are compatible with a chosen seismic reference model [*Anderson*, 1989; *Stixrude et al.*, 1992; *Jackson*, 1998; *Deschamps and Trampert*, 2004]. Furthermore, both experimental and ab initio mineral physics data contain uncertainties, and there appears to be an inconsistency in the shear modulus of magnesium perovskite between experimental and ab initio data [*Deschamps and Trampert*, 2004]. In the same spirit as the model space searches mentioned above, instead of selecting one particular data set (or reference state), one can vary all parameters within their reasonable or measured bounds to obtain insight into uncertainties in the sensitivities. Such an exercise reveals that uncertainties in mineral physics data and reference state are equally important [*Trampert et al.*, 2001]. While a particular choice of mineral physics data strongly influences the inferred reference state, it has only little effect on the derivatives themselves [*Deschamps and Trampert*, 2004; *Trampert et al.*, 2004]. The importance of considering the full family of derivatives is illustrated by the fact that most published temperature derivatives fall within two standard deviations of our preferred mean derivatives [*Trampert et al.*, 2004], implying that there is no formal disagreement between any of them. Noticeable differences in tomographic interpretations arise only when considering a single realization from the wide range of plausible solutions.

5.3. Importance of Mass Density

The influence of aluminum on elastic properties is not yet well known [*Yagi et al.*, 2004], but the effect of calcium is probably small [*Shim et al.*, 2000; *Deschamps and Trampert*, 2003]. With a caveat concerning aluminum, it is thus reasonable to assume that most of the 3D structure seen by seismic data is due to spatial variations in iron, silicon (through a perovskite and magnesiowüstite ratio), and temperature. Even if we had accurate estimates of shear- and bulk sound speed variations at every point in the mantle, we would only have two constraints for these three unknowns. *Forte and Mitrovica* [2001] reduced by one the number of unknowns by inverting for an effective temperature (combination of real temperature and iron) and effective composition (combination of perovskite and iron). They showed that the effective temperature correlates highly with shear wave speed variations and argued that it is a good proxy for temperature. Realizing that knowledge of density variations would allow to distinguish between perovskite and iron effects, *Forte and Mitrovica* [2001] assumed that variations in density and shear wave speed are perfectly correlated, with a ratio between 0.1 and 0.3. This assumption is questionable, however, since several studies have now shown beyond reasonable doubt that density variations are not correlated with shear speed variations (see discussion above). Independent constraints

on density thus prove crucial to infer correct temperature and compositional variations.

5.4. Probabilistic Tomography

Probabilistic tomography can be used for the construction and, in particular, evaluation of long-wavelength models [*Beghein et al.*, 2002; *Resovsky and Trampert*, 2002, 2003]. The advantage of the approach is that all models compatible with the data are considered. The family of admissible solutions is converted into likelihoods for wave speeds and density, which can thus be regarded as a compact representation of the seismological data. The likelihoods can subsequently be narrowed by independent data, such as gravity measurements [*Resovsky and Trampert*, 2003].

Full-model space searches are very computer-intensive since the space to be explored grows exponentially with the number of unknowns. So far we have managed to search only model spaces containing up to 30 parameters. This requires a careful model parameterization. In global applications, the approach is currently useful only where data and model parameters can be expanded in spherical harmonics. Each spherical harmonic coefficient can be inverted for up to 30 depth parameters by using a separate processor. Although limiting in terms of lateral and vertical resolution, the restriction in the number of parameters does not lead to biases. Indeed, we found [*Beghein and Trampert*, 2004] that a family of models for a thick layer represents the correct statistical average of the families for finer layers. This is because all models compatible with the data are considered and the probability density functions are thus nothing more than a compact representation of the data. Caution is necessary only in the discussion of geodynamic consequences. Combined with probability density functions for sensitivities from mineral physics, the probability density functions for seismological heterogeneities can be converted into likelihoods for variations of temperature and composition in the lower mantle (Plates 5–7) [*Trampert et al.*, 2004].

For a detailed description of the concept we refer to recent papers by Trampert and co-workers. Here we summarize some of the key results that have been obtained so far. At the longest wavelengths (even degrees only): (i) The availability of an independent density constraint is essential for producing robust maps of temperature and composition; (ii) shear wave speed variations do not reflect temperature variations alone and thus should not be used to infer thermal buoyancy; (iii) throughout the mantle, significant chemical variations are required to explain seismological observations with the currently available mineral physics data. Iron variations are strongest in the lowermost mantle, but silicate variations appear equally strong throughout the lower mantle; (iv) density and

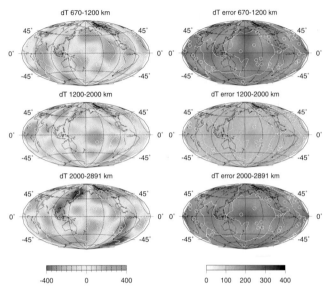

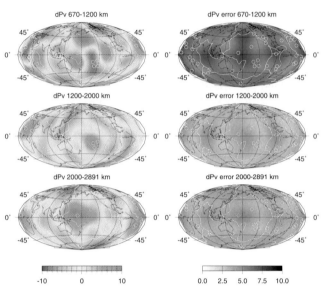

Plate 5. Relative variations, with respect to a thermochemical reference model, of temperature in the lower mantle [from *Trampert et al.*, 2004]. The exact reference model remains unknown due to nonunicity, given available constraints [e.g., *Deschamps and Trampert*, 2004]. The variations are likelihoods and are specified by Gaussian distributions. Shown are the mean (left column) and the standard deviation (right column) for different depth layers.

Plate 6. Same as for Plate 5, but for perovskite variations.

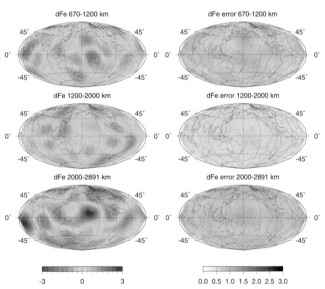

Plate 7. Same as for Plate 5, but for iron variations.

bulk sound are good proxies for iron and silicate variations, respectively; (v) compositional heterogeneity dominates buoyancy in the lowermost mantle, but even at shallower depths, its contribution to buoyancy is comparable to thermal effects; (vi) consistent with some previous suggestions [e.g., *Kellogg et al.*, 1999] the so-called superplumes in the deep mantle beneath the central Pacific and southern Africa appear compositionally distinct and have a higher intrinsic density owing to enrichment in iron and perovskite; (vii) in contrast with earlier predictions, however, these and other structures of higher than average mass density in the lowermost mantle do not stand out as anomalously hot features but appear thermally neutral or even relatively cold.

6. DISCUSSION AND CONCLUDING REMARKS

Seismic tomography has been highly successful in delineating the lateral variations of propagation speeds of elastic waves in Earth's interior. Until a few years ago the emerging tomographically inferred patterns were mostly taken as proxies of spatial variations in temperature. However, it is increasingly evident that a quantitative interpretation of the currently available seismologic and mineral physics data sets cannot be done in the framework of temperature alone and that spatial variations in major element composition must be considered. This also implies that purely thermally driven convection models cannot be representative of the Earth.

Seismology has begun to produce increasingly compelling evidence for compositional heterogeneity at all scales in the mantle, and we expect that this trend will continue. But we need more than high-quality multiparameter imaging. Indeed, a quantitative characterization of tomographically inferred heterogeneity requires a careful uncertainty analyses of both the tomographic models and the mineral physics data. Probabilistic tomography, adopted recently to the mapping of very long wavelength (>10,000 km) structures in the mantle, can achieve just that.

The preliminary results of probabilistic tomography, summarized in section 5.4, can be used to test predictions of or expectations from thermochemical models that have recently been proposed to reconcile geophysical and geochemical observations. Thermochemical convection models that put all compositional heterogeneity in small blobs [*Becker et al.*, 1999; *Helffrich and Wood*, 2001] do not explain the strong long-wavelength chemical variations unless the blobs are spatially concentrated by some (unspecified) mechanism. *Kellogg et al.* [1999] proposed a hot and intrinsically dense layer in the bottom 1000 or so km of the mantle. This layer would have a pronounced topography and be shaped rather passively through interactions with major downwellings (that is, slabs of subducted lithosphere). *Trampert et al.* [2004]

confirm that the slow shear wave speed anomalies away from downwellings are indeed due largely due to increases in mass density; in contrast to *Kellogg et al.* [1999], however, they find them to thermally neutral or even relatively cold. Preliminary results of thermochemical convection modeling using the anelastic approximation [*Tackley*, 2002] suggest that a stable stratification over long periods of time is unlikely in the lowermost mantle [*Deschamps et al.*, 2005], but the dynamics of this thermochemical system and, in particular, the implications for Earth's heat budget need further study. Furthermore, analyses of wave field distortion and scattering have so far not produced evidence for a sharp, global discontinuity between the upper-mantle transition zone and the base of the mantle [*Vidale et al.*, 2001; *Castle and van der Hilst*, 2003]; therefore, a change in bulk properties—if any—would need to be more gradual than the sharp layer boundary implied by the Kellogg et al. model.

The collective seismological evidence (summarized in section 4) and the results of probabilistic tomography [*Trampert et al.*, 2004; *Deschamps et al.*, 2005] seem most consistent with models of thermochemical mantle evolution that are characterized by a radial stratification [*Anderson*, 2001, 2002] or zonation of heterogeneity and convective mixing, as suggested by *Albarède and van der Hilst* [2002] and consistent with the evolutionary model by *van Thienen et al.* [this volume], in combination, perhaps, with slow oscillatory motion of the piles of compositionally distinct materials in the deep mantle [*Davaille*, 1999].

Probabilistic tomography is still at its early stages and lines of improvement can readily be identified. Mass density is a crucial parameter to discriminate between temperature and compositional effects, but it is poorly constrained by seismological data, especially the odd degrees. To improve the estimates on density variations, we need to incorporate more independent constraints, including geodynamical data [*Forte and Mitrovica*, 2001] and specific density–sensitive modes measured from seismic records of exceptionally strong recent earthquakes. Depth parameterization needs to be refined, but this increases the size of the model space and thus computer time. With the rapid advance of computer technology, we hope that parallel programming can effectively achieve this in the near future. It is further essential to include body wave information into the full-model space search. Classical tomography can more easily move towards data integration but needs to shy away from mere imaging, in order to improve local parameter estimation and to get error analyses on a more quantitative footing. A further target of concerted mineral physics and seismological research is the determination of the scale length of elastic and compositional heterogeneity in and between the thermochemical boundary layers of mantle convection. Refining the models including anisotropy and anelasticity will become essential to increasing the seismologic information. Careful Earth parameterization, the use of finite frequency kernels [*Dahlen et al.*, 2000], wavefield scattering and multiresolution analysis [e.g., *De Hoop and van der Hilst*, 2005], and adjoint methods [*Tromp et al.*, 2005] will no doubt prove crucial in this ambitious but exciting endeavour.

REFERENCES

Abers, G. A. and S. W. Roecker, Deep-structure of an arc-continent collision—earthquake relocation and inversion for upper mantle *P* and *S* wave velocities beneath Papua–New Guinea, *J. Geophys. Res.*, **96**, 6379–6401, 1991.

Albarède F. and R. D. van der Hilst, Zoned mantle convection, *Phil. Trans. R. Soc. Lond. A*, **360**, 2569–2592, 2002.

Anderson, D. L., *Theory of the Earth*, Blackwell Scientific Publications, Oxford, 1989.

Anderson, D. L., Top-down tectonics, *Science*, **293**, 2016-2018, 2001.

Anderson, D. L., The case for irreversible chemical stratification of the mantle, *Int. Geol. Rev.*, **44**, 97–116, 2002.

Anderson, O. L., *Equations of State of Solids for Geophysics and Ceramic Sciences*, Oxford University Press, Oxford, 1995.

Anderson, O., E. Schreiber, R. Liebermann, and N. Soga, Some elastic constant data on minerals relevant to geophysics, *J. Geophys. Res.*, **89**, 5953–5986, 1984.

Becker, T. W., Kellogg, J. B. and R. J. O'Connell, Thermal constraints on the survival of primitive blobs in the lower mantle, *Earth Planet Sci. Lett.*, **171**, 351–365, 1999.

Becker, T. W., and L. Boschi, A comparison of tomographic and geodynamic models, *Geochem. Geophys. Geosyst.*, **3**, 10.129/2001GC000168, 2002.

Beghein, C., Resovsky, J. and J. Trampert, *P* and *S* tomography using normal mode and surface wave data with a neighbourhood algorithm, *Geophys. J. Int.*, **149**, 646–658, 2002.

Beghein, C. and J. Trampert, Probability density functions for radial anisotropy: implications for the upper 1200 km of the mantle, *Earth Planet. Sci. Lett.*, **217**, 151–162, 2004.

Bijwaard, H., W. Spakman and E. R. Engdahl, Closing the gap between regional and global travel time tomography, *J. Geophys. Res.*, **103**, 30055–30078, 1998.

Birch, F., The velocity of compressional waves in rocks to 10 kilobars, Part 2., *J. Geophys. Res.*, **66**, 2199–2224, 1961.

Boschi, L. and A. M. Dziewonski, High and low-resolution images of the Earth's mantle: Implications of different approaches to tomographic modelling, *J. Geophys. Res.*, **104**, 25567–25594, 1999.

Boschi, L. and A. M. Dziewonski, Whole Earth tomography from delay times of P, Pcp, and PKP phases: lateral heterogeneities in the outer core or radial anisotropy in the mantle?, *J. Geophys. Res.*, **105**, 13675–13696, 2000.

Cammarano, F., S. Goes, Vacher P. and D. Giardini, Inferring upper-mantle temperatures from seismic velocities, *Phys. Earth Planet. Int.*, **138**, 197–222, 2003.

Castle, J. C., and R. D. van der Hilst, Searching for seismic scattering off deep mantle interfaces between 800 and 2000 km depth, *J. Geophys. Res.*, **108**, 2095, doi: 10.1029/2001JB000286, 2003.

Cazenave, A., Souriau A. and K. Dominh, Global coupling of Earth surface topography with hotspots, geoid and mantle heterogeneities, *Nature*, **340**, 54–57, 1989.

Dahlen, F. A., Hung, S. H. and G. Nolet, Frechet kernels for finite-frequency traveltimes—I. Theory, *Geophys. J. Int.*, **141**, 157–174, 2000.

Davaille, A., Simultaneous generation of hotspots and superswells by convection in a heterogenous planetary mantle, *Nature*, **402**, 756-760, 1999.

De Hoop, M. V. and R. D. van der Hilst, On sensitivity kernels for wave equation transmission tomography, *Geophys. J. Int.*, **160**, 621–633, 2005.

Deschamps, F., Trampert J. and R. Snieder, Anomalies of temperature and iron in the uppermost mantle inferred from gravity data and tomographic models, *Phys. Earth Planet. Int.*, **129**, 245–264, 2002.

Deschamps, F. and J. Trampert, Mantle tomography and its relation to temperature and composition, *Phys. Earth Planet. Int.*, **140**, 277–291, 2003.

Deschamps, F. and J. Trampert, Towards a lower mantle reference temperature and composition, *Earth Planet. Sci. Lett.*, **222**, 161–175, 2004.

Deschamps, F., Trampert, J. and P. Tackley, Thermo-chemical structure of the lower mantle: seismological evidence and consequences for geodynamics, in: Yuen, D., et al. (Editors.) *Superplume: Beyond+ Plate Tectonics*, Kluwer, in press, 2005.

Dziewonski, A. M. and D. L. Anderson, Preliminary Reference Earth Model, *Phys. Earth Planet. Int.*, **25**, 297-356, 1981.

Dziewonski, A. M., Mapping the lower mantle: determination of lateral heterogeneity in *P* velocity up to degree and order 6, *J. Geophys. Res.*, **89**, 5929–5952, 1984.

Dziewonski, A. M., Earth's mantle in three dimensions, in: Boschi, E. et al. (Editors), *Seismic modelling of Earth structure*, Instituto Natzionale di Geofisica, Rome, pp. 507–572, 1996.

Dziewonski, A. M. and J. H. Woodhouse, Global images of the Earth's interior, *Science*, **236**, 37–48, 1987.

Ekström, G. and A. M. Dziewonski, The unique anisotropy of the Pacific upper mantle, *Nature*, **394**, 168–172, 1998.

Engdahl, E. R., Van der Hilst, R. D. and R. P. Buland, Global teleseismic earthquake relocation from improved travel times and procedures for depth determination, *Bull. Seis. Soc. Am*, **88**, 722–743, 1998.

Forte, A. M. and C. A. Perry, Seismic-geodynamic evidence for a chemically depleted continental tectosphere, *Science*, **290**, 1940–1944, 2000.

Forte, A. M. and J. X. Mitrovica, Deep-mantle high-viscosity flow and thermochemical structure inferred from seismic and geodynamic data, *Nature*, **410**, 1049–1056, 2001.

Fukao, Y., Obayashi, M., Inoue, H. and M. Nenbai, Subducting slabs stagnant in the mantle transition zone, *J. Geophys. Res.*, **97**, 4809–4822, 1992.

Fukao, Y., Widiyantoro, S. and M. Obayashi, Stagnat slabs in the upper and lower mantle transition region, *Rev. Geophys.*, **39**, 291–323, 2001.

Goes, S., Govers, R. and P. Vacher, Shallow mantle temperatures under Europe from *P* and *S* wave tomography, *J. Geophys. Res.*, **105**, 11153–11169, 2000.

Grand, S. P., Mantle shear structure beneath the Americas and surrounding oceans, *J. Geophys. Res*, **99**, 11591–11621, 1994.

Grand, S. P., van der Hilst, R. D. and S. Widiyantoro, High resolution global tomography: a snapshot of convection in the Earth, *GSA Today*, **7**, 1–7, 1997.

Gu, Y. J., Dziewonski, A. M., Su, W. J. and G. Ekström, Models of the mantle shear velocity and discontinuities in the pattern of lateral heterogeneities, *J. Geophys. Res.*, **106**, 11169–11199, 2001.

Hager, B. H., Clayton, R. W., Richards, A. M., Comer R. P. and A. M. Dziewonski, Lower mantle heterogeneity, dynamic topography and the geoid, *Nature*, **313**, 541–545, 1985.

Helffrich, G. R. and B. J. Wood, The Earth's mantle, *Nature*, **412**, 501–507, 2001.

Hofmann, A. W., Mantle geochemistry: the message from oceanic volcanism, *Nature*, **385**, 219–229, 1997.

Ishii, M. and J. Tromp, Normal-mode and free-air gravity constraints on lateral variations in velocity and density of the Earth's mantle, *Science*, **285**, 1231–1236, 1999.

Ishii, M. and J. Tromp, Even-degree lateral variations in the mantle constrained by free oscillations and the free-air gravity anomaly, *Geophys. J. Int.*, **145**, 77–96, 2001.

Ishii, M. and J. Tromp, Constraining large-scale mantle heterogeneity using mantle and inner-core sensitive normal modes, *Phys. Earth Planet. Int.*, **146**, 113–124, 2004.

Jackson, I., Elasticity, composition and temperature of the Earth's lower mantle, *Geophys. J. Int.*, , 291–311, 1998.

Jordan, T., The continental tectosphere, *Rev. Geophys. Space Phys.*, **13**, 1–13, 1975.

Kárason, H. and R. D. van der Hilst, Constraints on mantle convection from seismic tomography, in: Richards, M.R., Gordon, R. and R. D. van der Hilst, (Editors), *The History and Dynamics of Global Plate Motion*, American Geophysical Union, Washington, pp. 277–288, 2000.

Kárason, H. and R. D. van der Hilst, Improving global tomography models of P-wave speed I: incorporation of differential travel times for refracted and diffracted core phases (PKP, Pdiff), *J. Geophys. Res.*, **106**, 6569–6587, 2001.

Karato, S.-I., Importance of anelasticity in the interpretation of seismic tomography, *Geophys. Res. Lett.*, **20**, 1623–1626, 1993.

Karato, S.-I. and B. Karki, Origin of lateral heterogeneity of seismic wave velocities and density in Earth's deep mantle, *J. Geophys. Res.*, **106**, 21771–21783, 2001.

Kellogg, L.H., Hager, B.H. and R. D. van der Hilst, Compositional stratification in the deep mantle, *Science*, **283**, 1881–1884, 1999.

Kennett, B. L. N., Engdahl, E. R. and R. Buland, Constraints on seismic velocities in the Earth from traveltimes, *Geophys. J. Int.*, **122**, 108–124, 1995.

Kennett, B. L. N., Widiyantoro, S. and R. D. van der Hilst, Joint seismic tomography for bulk sound and shear wave speed in the Earth's mantle, *J. Geophys. Res.*, **103**, 12469–12493, 1998.

Kuo, B. Y. and B. Romanowicz, On the resolution of density anomalies in the Earth's mantle using spectral fitting of normal mode data, *Geophys. J. Int.*, **150**, 162–179, 2002.

Li, X. D. and B. Romanowicz, Global mantle shear velocity model developed using nonlinear asymptotic coupling theory, *J. Geophys. Res.*, **101**, 22245–22273, 1996.

Masters, G., Jordan, T. H., Silver, P. G. and F. Gilbert,Aspherical earth structure from fundamental spheroidal mode data, *Nature*, **298**, 609–613, 1982.

Masters, G., Low frequency seismology and the three-dimensional structure of the Earth, *Philos. Trans. R. Soc. Lond. A*, **328**, 479–522, 1989.

Masters, G. and P. Shearer, Seismic models of the Earth: elastic and anelastic, in: Ahrens, T. J. (Editor), *Global Earth Physics: A Handbook of Physical Constants*, AGU reference shelf **1**, Amercian Geophysical Union, Washington, pp. 88–103, 1995.

Masters, G., Johnson, S., Laske, G. and H. Bolton, A shear velocity model of the mantle, *Philos. Trans. R. Soc. Lond. A*, **354**, 1385–1411, 1996.

Masters, G., Laske, G., Bolton, H. and A. M. Dziewonski, The relative behavior of shear velocity, bulk sound speed, and compressional velocity in the mantle: implications for chemical and thermal structure, in: Karato, S.-I. et al. (Editors), *Earth's Deep Interior: Mineral Physics and Tomography from the Atomic to the Global Scale*, American Geophysical Union, Washington, pp. 63–87, 2000.

Mégnin, C. and B. Romanowicz, The 3D shear velocity structure of the mantle from the inversion of body, surface and higher mode waveforms, *Geophys. J. Int.*, **143**, 709–728, 2000.

Montagner, J.-P., Can seismology tell us anything about convection in the mantle, *Rev. Geophys.*, **32**, 115–138, 1994.

Montagner, J.-P., Where can seismic anisotropy be detected in the Earth's mantle? In boundary layers ..., *Pageoph.*, **151**, 223—256, 1998.

Montelli, R., Nolet, G., Dahlen, F. A., Masters, G., Engdahl, E. R. and S.-H. Hung, Finite-frequency tomography reveals a variety of plumes in the mantle, *Science*, **303**, 338–343, 2004.

Mosegaard, K. and A. Tarantola, Monte Carlo sampling of solutions to inverse problems, *J. Geophys. Res.*, **100**, 12431–12447, 1995.

Poirier, J.-P., *Introduction to the Physics of the Earth's Interior*, Cambridge University Press, Cambridge, 1991.

Resovsky, J. S. and M. H. Ritzwoller, Regularization uncertainty in density models estimated from normal mode data, *Geophys. Res. Lett.*, **26**, 2319–2322, 1999.

Resovsky, J. S. and J. Trampert, Reliable mantle density error bars: an application of the neighbourhood algorithm to normal-mode and surface wave data, *Geophys. J. Int.*, **150**, 665–672, 2002.

Resovsky, J. S. and J. Trampert, Using probabilistic seismic tomography to test mantle velocity-density relationships, *Earth Planet. Sci. Lett.*, **215**, 121–134, 2003.

Ricard, Y., Richards, M., Lithgow-Bertelloni, C. and Y. Lestunff, A geodynamic model of mantle density heterogeneity, *J. Geophys. Res.*, **98**, 21895–21909, 1993.

Richards, M. A. and B. H. Hager, Geoid anomalies in a dynamic Earth, *J. Geophys. Res.*, **89**, 5987-6002, 1984.

Richards, M.A. and D. C. Engebretson, Large-scale mantle convection and the history of subduction, *Nature*, **355**, 437-440, 1992.

Ritsema, J., van Heijst, H. J. and J. H. Woodhouse, Complex shear wave velocity structure imaged beneath Africa and Iceland, *Science*, **286**, 4245–4248, 1999.

Ritzwoller, M. H. and E. M. Lavely, Three-dimensional seismic models of the Earth's mantle, *Rev. Geophys.*, **33**, 1–66, 1995.

Robertson, G. S. and J. H. Woodhouse, Ratio of relative S to P velocity heterogeneity in the lower mantle, *J. Geophys. Res.*, **101**, 20041–20052, 1996.

Romanowicz, B., Seismic tomography of the Earth's mantle, *Ann. Rev. Earth Planet. Sci.*, **19**, 77–99, 1991.

Romanowicz, B., Can we resolve 3D density heterogeneity in the lower mantle?, *Geophys. Res. Lett.*, **28**, 1107–1110, 2001.

Romanowicz, B., Global mantle tomography: Progress status in the past 10 years, *Ann. Rev. Earth Planet. Sci.*, **31**, 303–328, 2003.

Romanowicz, B. and J. J. Durek, Seismological constraints on attenuation in the Earth: a review, in: Karato, S.-I. et al. (Editors), *Earth's Deep Interior: Mineral Physics and Tomography from the Atomic to the Global Scale*, American Geophysical Union, Washington, pp. 161–179, 2000.

Saltzer, R. L., van der Hilst, R. D. and H. Kárason, Comparing P and S wave heterogeneity in the mantle, *Geophy. Res. Lett.*, **28**, 1335–1338, 2001.

Saltzer, R. L., Stutzmann, E. and R. D. van der Hilst, Poisson's ratio beneath Alaska from the surface to the Core–Mantle Boundary, *J. Geophys. Res.*, **109**, 10.1029/2003JB002712, 2004.

Sambridge, M. and Ó. Gudmundsson, Tomographic systems of equations with irregular cells, *J. Geophys. Res.*, **103**, 773–781, 1998.

Sambridge, M. and K. Mosegaard, Monte Carlo methods in geophysical inverse problems, *Reviews Geophys.*, **40**, 10.1029/2000RG000089, 2002.

Schubert, G., Turcotte, D. L. and P. Olsen, *Mantle Convection in the Earth and Planets*, Cambridge University Press, Cambridge, 2001.

Shapiro, N. M. and M. H. Ritzwoller, Monte-Carlo inversion for a global shear-velocity model of the crust and upper mantle, *Geophys. J. Int.*, **151**, 88–105, 2002.

Shim, S.-H., Duffy, T. S. and G. Shen, The equation of state of $CaSiO_3$ perovskite to 108 GPa at 300 K, *Phys. Earth Planet. Int.*, **120**, 327–338, 2000.

Spakman, W. and G. Nolet, Imaging algorithms, accuracy adn resolution in delay time tomography, in: Vlaar, N. J., Nolet G., Wortel M. J. R. and S. A. P. L. Cloetingh, *Mathematical Geophysics*, Reidel, Dordrecht, pp. 155–187, 1988.

Spakman, W., Wortel, M. J. R. and N. J. Vlaar, The hellenic subduction zone: A tomographic image and its geodynamic implications, *Geophys. Res. Lett.*, **15**, 60–63, 1988.

Stacey, F. D., *Physics of the Earth*, third edition, Brookfield Press, Brisbane, 1992.

Stacey, F. D. and P. M. Davis, High pressure equations of state with applications to the lower mantle and core, *Phys. Earth Planet. Int.*, **142**, 137–184, 2004.

Stixrude, L., Hemley, R. J., Fei, Y. and H. K. Mao, Thermoelasitcity of silicate perovskite and magnesiowüstite and stratification of the Earth's mantle, *Science*, **257**, 1099–1101, 1992.

Su, W. J., Woodward, R. L. and A. M. Dziewonski, Degree 12 model of shear velocity heterogeneity in the mantle, *J. Geophys. Res.*, **99**, 4945–4980, 1994.

Su, W. J. and A. M. Dziewonski, Simultaneous inversion for 3-D variations in shear and bulk velocity in the mantle, *Phys. Earth Planet. Int.*, **100**, 135–156, 1997.

Tackley, P. J., Strong heterogeneity caused by deep mantle layering, *Geochem. Geophys. Geosyst.*, **3**, 10.1029/2001GC000167, 2002.

Trampert, J. and R. Snieder, Model estimations biased by truncated expansions: Possible artifacts in seismic tomography, *Science*, **271**, 1257–1260, 1996.

Trampert J. and J. H. Woodhouse, Assessment of global phase velocity models, *Geophys. J. Int.*, **144**, 165–174, 2001.

Trampert, J., Vacher, P. and N. Vlaar, Sensitivities of seismic velocities to temperature, pressure and composition, *Phys. Earth Planet. Int.*, **124**, 255–267, 2001.

Trampert, J., Deschamps, F., Resovsky, J. and D. Yuen, Probabilistic tomography maps significant chemical heterogeneities in the lower mantle, *Science*, **306**, 853–856, 2004.

Tromp, J., Tape, C. and Q. Liu, Seismic tomography, adjoint methods, time reversals and banana-doughnut kernels, *Geophys. J. Int.*, **160**, 195–216, 2005.

van der Hilst, R. D. and H. Kárason, Compositional heterogeneity in the bottom 1000 km of Earth's mantle: towards a hybrid convection model, *Science*, **283**, 1885–1888, 1999.

van der Hilst, R. D., Engdahl, E. R., Spakman, W., and G. Nolet, Tomographic imaging of subducted lithosphere below northwest Pacific island arcs, *Nature*, **353**, 37–42, 1991.

van der Hilst, R. D., Widiyantoro, S. and E. R. Engdahl, Evidence for deep mantle circulation from global tomography, *Nature*, **386**, 578–584, 1997.

van Gerven, L., Deschamps, F. and R. D. van der Hilst, Geophysical evidence for chemical variations in the Australian Continental Mantle, *Geophys. Res. Lett.*, **31**, 10.1029/2004GL020307, 2004.

Vasco, D. W. and L. R. Johnson, Whole Earth structure estimated from seismic arrival times, *J. Geophys. Res.*, **103**, 2633–2671, 1998.

Vidale, J. E., G. Schubert, and P. S. Earle, Unsuccessful initial search for a mid-mantle chemical boundary with seismic arrays, *Geophys. Res. Lett.*, **28**, 859-862, 2001.

Widiyantoro, S. and R. D. van der Hilst, Structure and evolution of subducted lithosphere beneath the Sunda arc, Indonesia, *Science*, **271**, 1566–1570, 1996.

Widiyantoro, S., Kennett, B. L. N. and R. D. van der Hilst, Extending shear-wave tomography for the lower mantle using S and SKS arrival-time data, *Earth Planets and Space*, **50**, 999–1012, 1998.

Woodhouse, J. H. and A. M. Dziewonski, Mapping the upper mantle: Three dimensional modelling of Earth structure by inversion of seismic waveforms, *J. Geophys. Res*, **89**, 5953–5986, 1984.

Woodhouse, J. H. and A. M. Dziewonski, Seismic modelling of the Earth's large-scale three dimensional structure, *Philos. Trans. R. Soc. Lond. A*, **328**, 291–308, 1989

Yagi, T., Okabe, K., Nishiyama, N., Kubo, A. and T. Kikegawa, Complicated effects of aluminium on the compressibility of silicate perovskite, *Phys. Earth Planet. Int.*, **143–144**, 81–91, 2004.

J. Trampert, Department of Earth Sciences, Utrecht University, PO Box 80.021, 3508 TA Utrecht, The Netherlands. (jeannot@geo.uu.nl)

R. D. van der Hilst, Department of Earth, Atmospheric, and Planetary Sciences, Massachusetts Institute of Technology, Cambridge, Massachusetts 02139, USA. (hilst@mit.edu)

Seismic Modeling Constraints on the
South African Super Plume

Don V. Helmberger and Sidao Ni

Seismological Laboratory, California Institute of Technology, Pasadena, California

Tomographic studies of the structure of the lower mantle beneath South Africa reveal large-scale low velocities above the core–mantle boundary. Predicted SKS delay patterns (up to 3 s) for some of these models fit observations (Kaapvaal Array data) quite well except for magnitude level, explaining less than one-half the observed anomaly. Moreover, the sharpness in travel-time offsets and waveform complications require that nearly vertical walls separate the anomalous structure from the normal preliminary reference Earth model (PREM) mantle. We present numerous record sections along with 2D and 3D synthetics displaying multipathing of arrivals (S_d, SKS, SKKS, S, and ScS), based on a large-scale 3D structure. This kidney-shaped structure has one apex beneath the Indian Ocean (Kerguelen) and the other extending beneath the Mid-Atlantic (Cape Verde). The structure is about 1200 km wide beneath South Africa and extends upward to at least 1000 km through the lower mantle, similar to Grand's model but with an average uniform velocity decrease of about 3% relative to PREM. We have not found any evidences for ultra-low-velocity zones (ULVZ) beneath the main structure but ample evidence at some locations near the edges. We also analyzed Pd and the differentials between PcP travel times and P travel times (PcP–P) along the same great circle paths from the same events. The P-velocity is not very anomalous, perhaps –0.5%. The sharpness of the lateral boundaries (walls) and the large contrast in P and S velocities can be used in arguments for a thermochemical origin.

INTRODUCTION

The large-scale structures in the lower mantle have been well established in recent years, e.g., reviewed by *Garnero* [2000]. While the details may differ, most tomographic inversions yield a relatively fast region around the circum-Pacific belt with some complex patterns of slow velocities centered beneath the Mid-Pacific and South Africa. This picture of the mantle is consistent with the fast regions being cold and generated by past subduction [*Grand*, 2002]. The warm colors are attributed to upwelling, contain most of the world's hot-spots, and correlate with the long-wavelength geoid highs [*Hager et al.*, 1985]. Sharper seismic structures have been revealed with P-wave tomography [*van der Hilst et al.*, 1996] and with shear velocity tomography based on bodywaves [*Grand*, 1994]. The agreement between these two models for the upper 2000 km of the mantle is remarkable considering the complete independence of data and methodology [*Grand et al.*, 1997]. At greater depths, the relationship between P and S becomes more complex, indicating possible chemical influences [*van der Hilst et al.*, 2004].

The cause of the disagreement at greater depths is under investigation but the bulk sound velocity (V_c) appears to be negatively correlated with V_s within the African and Pacific anomalies [*Masters et al.*, 2000]. The most definitive data on the African structure come from the observations of

Earth's Deep Mantle: Structure, Composition, and Evolution
Geophysical Monograph Series 160
Copyright 2005 by the American Geophysical Union
10.1029/160GM06

diffracted S-waves, S_d, compared to P_d at the broadband African Array, from the same events [*Wen*, 2001]. While P_d appears PREM-like across the array, S_d develops sharp delays of up to 20 s for many samples. *Wen* [2001] was able to model many of these waveforms in 2D by introducing very low shear velocities at the base of the mantle with rapid variations at the edges of the anomaly.

The focus of this review is a detailed examination of the African superstructure as imaged by forward modeling. To develop the model, we started with Grand's tomographic model (Plate 1) and added sharp detail along line segments. Much of its sharpness is due to this style of enhancing Grand's tomography model. Note that the bottom basal layer is essentially Grand's, with a uniform 3% velocity reduction in D″. The structure above D″ is more anomalous than Grand's, extending upward to about the depth of 1700 km, where the velocity inside the structure has been uniformly decreased by 3% relative to PREM [*Dziewonski and Anderson*, 1981], but keeping Grand's model elsewhere. This simplistic model has evolved with the addition of new data, especially the recently released Kaapvaal Array data [*James et al.*, 2001]. A 2D section along a path from South America to the African Array is displayed in Plate 2B. The upper panel (Plate 2A) displays the locations of some of the events used in developing the model, including those providing the waveforms discussed in this review.

The lower panel (Plate 2C) shows the SKS delays as augmented by the African Array roughly outlining the anomalous structure [*Ni and Helmberger*, 2003a]. Note that sharp walls produce rapid jumps in SKS travel times since their ray paths (Plate 2B) can travel nearly parallel to these boundaries. These sharp features can also be effective in delaying ScS relative to S. A record section of a South American event (971128) is displayed in Figure 1 as a typical example of such (ScS–S) complexity, along with 2D synthetics for the tomographic models by *Grand* [1994] and *Ritsema et al.* [1999] and our model LVZ2, based on a low-velocity zone (LVZ) (Plate 2B). These synthetics were generated with the WKM technique introduced in *Ni et al.* [2000]. The method is basically analytical, which satisfies the wave equation assuming tomographic-type models. A comparison of synthetic seismograms for the model displayed in Figure 1, using the above semi-analytical code with a 2D numerical technique, is given in *Ni et al.* [2003d].

The beginning portion of this data set (ranges 85°) samples the westernmost edge of the model, along the dashed line in Plate 2A. There is ample evidence for a complex LVZ starting below about 40°, indicated by the green lines [*Wen et al.*, 2001; *Ni and Helmberger*, 2001]. Thus, the complexity of ScS was not modeled well by LVZ2 since we omitted this small-scale feature. However, the model predicts the separation of (ScS–S) quite well at larger distances. The LVZ2 model

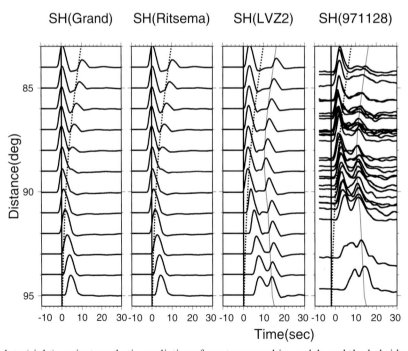

Figure 1. Comparison of data (right) against synthetic predictions from tomographic models and the hybrid model (LVZ2) displayed in Plate 2. These sections are aligned on the S-wave times predicted by PREM. The corresponding ScS times are shown as a dotted line; a heavy line indicates the observed ScS peak arrival time for comparison.

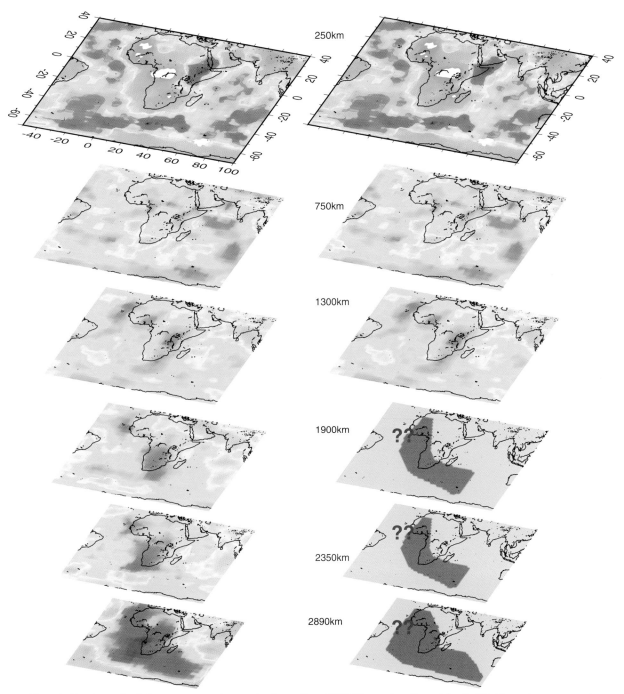

Plate 1. Panels on the left are tomography results from *Grand* [2002] as a function of depth (*Z*). The colors represent positive (blue) and negative (red) anomalies (±4%). The model on the right is Grand's model except for the solid red zone beneath Africa. We will call this hybrid model ALVS (African Low Velocity Structure), which has a uniform 3% reduction in S-velocity and PREM in P-velocity.

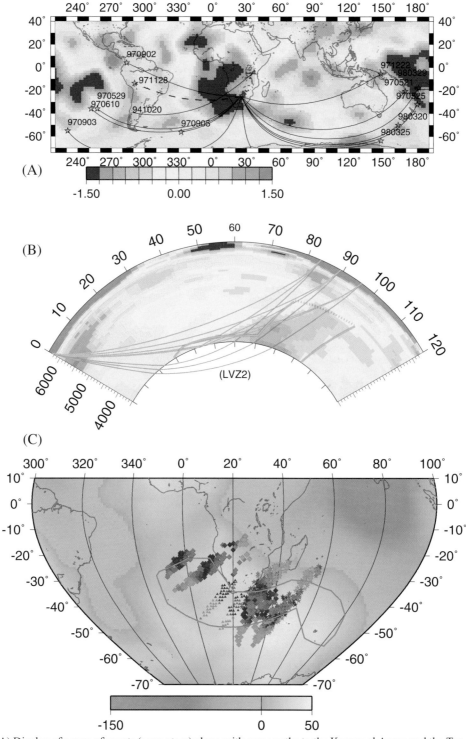

Plate 2. (A) Display of a map of events (open stars) along with some paths to the Kaapvaal Array and the Tanzania Array (solid triangles) superimposed on the tomographic results of *Ritsema et al.* [1999]. (B) Cross-section along a path from event 971128, denoted as a dashed line in the tomographic model. The hybrid model (LVZ2) is denoted by the heavy green lines. Ray paths are included for S and ScS (magenta) and SKS (blue). (C) SKS and SKKS exit points at the CMB; delays of more than 5 s are shown in magenta, those of 0 s (i.e., no delay) in blue. The colored background indicates the observed geoid in meters. The Dupal geochemical anomaly is outlined in dark blue (after *Hart* [1984]).

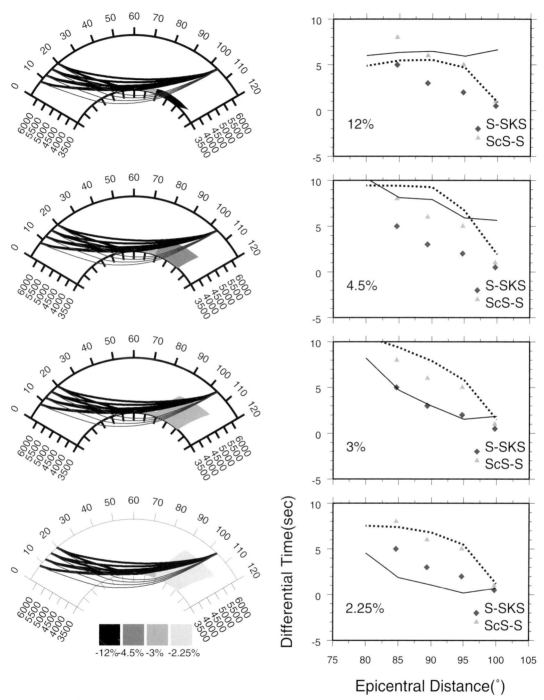

Figure 2. ALVS thickness sensitivity testing. Cross-sections with various thicknesses are presented on the left, along with ray paths where the travel times of SKS have been preserved. The basal layer on the left (D″ beneath the Atlantic) is maintained at 3%. The ALVS thickness varies from 300, 1000, 1500, and 2000 km, with velocity reductions at 12% to 2.25%. The associated differential travel time predictions (SKS-S) and (ScS-S) are displayed on the right, along with the data observed at the IRIS station LSZ (see Plate 3a for location), after *Ni and Helmberger* [2003a].

was derived from the IRIS and older WWSSN data [*Ni and Helmberger*, 2003b] before the Kaapvaal Array data were released. The model was developed largely by trial and error analysis of multi-events recorded by individual stations as displayed in Figure 2 for station LSZ. This station, located about halfway between the two arrays (Plate 3), recorded S, ScS, and SKS from many events, four events being selected as representative along a narrow South American corridor (segment A) in Plate 3 [*Ni and Helmberger*, 2003b]. Four representative models with heights constrained at 2000, 1500, 1000, and 300 km with roughly equal SKS delays of 6 s are displayed. Note that S is delayed too much for the 2000-km case and SKS–S becomes too small to fit the data, but with thinner structures SKS-S becomes too large. Thus, the combination of (ScS–S) and (SKS–S) proves quite useful in constraining possible models.

The introduction of lateral variation on a variety of scales and magnitudes allows reasonable fits to limited data but unfortunately exasperates the uniqueness question. Addressing this issue of uniqueness is one of the major efforts of this review. For example, if we reinterpret Figure 2 with the boundary moved westward, we would reach a different conclusion above the preferred height. Thus a thinner layer but with a stronger velocity drop could equally fit these data, as interpreted by *Wang and Wen* [2004]. While the above strategy produced a modification of Grand's model, *Wen et al.* [2001] explains the Kaapvaal Array data by attributing all the low-velocity structure to D″ and adopting PREM elsewhere. A rough outline of the boundary derived from this strategy is displayed in Plate 3. Much of its position is similar to that derived by *Grand* [2002; Plate 1] and is adopted into our model as a uniform (3%) slow shear-wave layer. In contrast, the D″ model of *Wen et al.* [2001] becomes highly variable where they break up array data into sectors with separate 2D models for each. Many of the synthetics produced by following their strategy fit observations very well and probably capture some of extreme variability in the real Earth. Our working approach has been to limit the number of parameters as much as possible by keeping a uniform drop in shear velocity and using numerous stations and events to set the position of the lower-mantle geography.

Our African Low Velocity Structure (ALVS) model has two branches; one extending northwestward beneath the Atlantic (Western Province) and another beneath the Indian Ocean (Eastern Province). The former has better sampling because of the richness of seismicity to the west, as displayed in Plate 2. However, there are hardly any stations in North Africa, and our model to the north becomes less defined for this reason.

The sequence of events along the great-circle path through Sandwich Island is especially well-sampled [*Ritsema et al.*, 1998a]. The addition of the Kaapvaal Array makes this corridor even more attractive for this in-depth review.

STRUCTURAL DEFINITION OF THE WESTERN PROVINCE

We begin this section with a detailed cross-section study along a great-circle path through South Africa. The events span about 80°, ranging from the East Pacific Rise (EPR) to South Sandwich Island (see Plate 3A). The stations include the two arrays, Kaapvaal and Tanzania, and the more distant IRIS stations, FURI and ATD. Events arriving from the northeast recorded along this profile are not very anomalous except for a small delay of SKS relative to S as they travel about the same distance in the ALVS [*Ritsema et al.*, 1998a].

Ray paths for the various phases are displayed in Plate 3B, including path segments for diffracted S_d (red) and SKS (blue) for two EPR events that sample the western boundary near position b. In this section, we will examine in detail the behavior of SKS delay jumps at the boundaries, followed by a discussion of S_d and P_d from an EPR event. Next, we address ScS bounce-points sampling across the western boundary at position b. Last, we examine the uniformity issue by examining S-waves turning at various depths in the ALVS model.

Jumps in SKS

The SKS ray paths are nearly vertical as they approach the array and therefore are useful in detecting horizontal velocity variations. A record section of the 970529 (EPR) events (Figure 3) displays a remarkable jump in travel time at the western boundary. Even though the records of this shallow event are complicated, they align well by applying a waveform correlation routine (column 2). The line segments included in the figure correspond to arrival times predicted by PREM from 93° to 101° (sampling outside the structure) followed by a second line segment 5 s later from 102° to 110° (inside structure). This jump is 10 times larger than array station delays (S-waves) reported by *James et al.* [2001]. The differential times between these arrivals and PREM predictions are displayed on the right, where we have subdivided the stations into two groups to indicate some small-scale feature towards the interior of the structure. The predicted times from the model are included for comparison. Other paths such as those from event 970903 (Plate 2) show a similar jump with some waveform complexity at the boundary indicative of some slow material either at the CMB or along the wall, which will be discussed later.

The SKS delays from the various events are given in map view at the CMB by projecting the rays to their piercing points (Plate 2C). Note, however, that these data do not constrain the height of the structure since travel times trade-off with velocity; rather, they simply define the edge of the boundary and magnitude of the anomaly (about 6 s) starting at position b. Note that introducing a very low velocity layer from b to c (Plate 3B) could explain these particular data with the entire anomaly within the D″ layer. To compare our results with those of *Wang and Wen* [2004], we will construct an alternative model, ULVD, specified by a thickness of 300 km with a velocity reduction of 2% at the top, grading to a 12% drop at the CMB. We will suppose that the boundary is the same as the above ALVS.

Diffracted Sd and Pd

The S_d phase from the above event (970529) is weak but this phase is quite strong for a neighboring event (970610)

located about 3° further away. These data are displayed in Figure 4, where the ray paths began to sample the top of the anomaly at about 103°, which is 3° further away than observed in Figure 3, as predicted. The delays are gradual but reach values over 8 s at the largest distances. Since the S_d has a smaller ray parameter than SKS and a longer path length in the ALVS, we would expect these phases to be delayed more than SKS. In contrast, the ULVD model remains unsampled throughout the section, with no predicted delays (see Figure 4D). However, if we placed a ULVZ in D″ starting near 105° (to the west of b), we could produce the observed delay but then it would be seen in SKS—which it is not. Note that only structures near the ray segments shown have any effect on this differential behavior since the structure beneath South America is

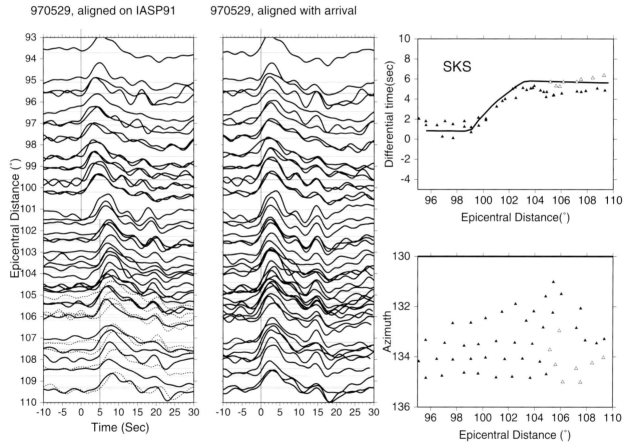

Figure 3. Displays of Kaapvaal Array data obtained from the shallow East Pacific Rise event before and after shifting, as displayed on the left and middle panels. The column on the left indicates the data plotted relative to PREM, where the upper half of the data reflect normal arrival times relative to late arrivals beyond about 100°. Heavy lines indicate the apparent offsets. Smaller-scale differences are indicated in the bottom half as a combination of solid and dashed traces, the latter indicating a small delay of a few seconds. The upper right panel displays delays in the travel time determined by cross-correlation for this event relative to PREM. Open triangles indicate the small-scale variation associated with the northernmost array stations as shown in the lower panel. Note how close these paths must be and still have different arrival times.

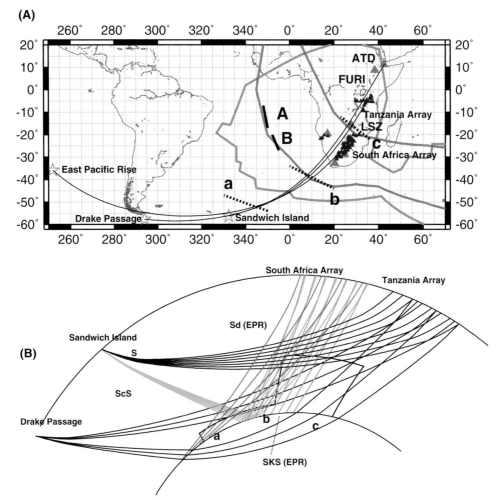

Plate 3. Upper panel: Map of events and stations (triangles) used in the construction of a 2D corridor. Lower panel: Ray paths from the various events relative to the 2D structure. Note that location (a) is at the ALVS onset where an ULVZ has been reported, and (b) and (c) indicate the upper structural boundaries. We have included the up-going S_d and SKS ray paths from two East Pacific Rise (EPR) events, after *Ni and Helmberger* [2003b]. Segments A and B in the upper map indicate the azimuths sampled by data presented in Figures 2 and 1, respectively.

common to all these S_d phases. Thus, the combination of SKS and S_d is providing some resolution on the position of the boundary and height of the structure.

We also measured P_d times with the correlation code for the above event and found only small delays, in agreement with the earlier study by *Wen et al.* [2001]. These data are rather unique in that they involve the same event and have nearly common paths (Figure 4F). The data are more extreme at the larger ranges than predicted by the ALVS model, which could mean that the midstructure is either slower or extends upward.

Differential Times Between ScS–S and PcP–P

Another excellent constraint on the western boundary is provided by ScS–S and PcP–P observations recorded by the African Array from South Sandwich Island events. Much of the data have been discussed in *Simmons and Grand* [2002]. They found that while ScS–S produces a strong anomaly, PcP–P estimates were only slightly anomalous. Their datasets involved different earthquakes so that the P and S ray paths could be slightly different. *Ni and Helmberger* [2003b,c] expanded their study once the African Array data were released. They report on an event (970602) that had high-quality P and S waves with the travel-time differentials displayed in Figure 4E. The PcP–P differentials do not change much throughout the array with perhaps a 1.5-s delay. The small shift of about 2 s relative to PREM can be attributed to the slow D'' basal layer (3%). That is, the direct P and S for this dataset have their turning depths above the anomaly (Plate 3) and have travel times similar to PREM; thus only ScS samples the structure.

Simmons and Grand [2002] interpreted their results in terms of a melt zone which would be compatible with the ULVD panel (see Figure 4B). Although this shift of the ULVD towards the west could explain the data, it now would disagree with where the SKS phases have their jump. Thus, geometry becomes extremely important in resolving these issues and in the need to investigate the whole family of S phases to provide sampling. Note that accomplishing the detailed resolution attempted here relies on the great-circle paths. To do this in 3D becomes very difficult, and using existing tomographic models as a guide is useful. We will address this issue in the next section, where we discuss the evolution in models leading to the above structure.

Structural Uniformity

One of the first PASSCAL experiments in Africa was located in Tanzania [*Nyblade et al.*, 1996]. Its purpose was to study the East African Rift System and, therefore, was not ideal for deep earth studies. However, *Ritsema et al.* [1998b] have established the station delays caused by the upper mantle, which range from 3 to 4 s for S-waves. Before these data were released, we were working on explaining the existing WWSSN (analog) data and had established some modifications of existing tomography models [*Ding*, 1997]. This model is displayed in Plate 4 and is Grand's model but with his low-velocity anomalies beneath Africa inflated to –4%. The data are plotted for two events, one beneath Sandwich Island (A) and one beneath the Drake Passage (B). The S, ScS, and SKS phases propagate through the same mantle corridor as the previously discussed events except that they have sampled more of the structure beneath South Africa. Note that the S delay generated by event A increases systematically from about 0 s at 65° to 10 s at 75°. The scatter in data is probably caused by the station delays, which are about this size. Differential times should remove these features, and the ScS–S fit is good. In contrast to ScS, SKS starts late and becomes PREM-like at the larger distances, which implies that their paths escape the slow structure by sampling to the east of the anomaly. Note that the anomalous D'' in Grand's model has been truncated along its northeastern edge to allow the SKS to become more PREM-like (Plate 1). These data can be fit better by adjusting the structure into a homogeneous block, similar to that displayed in Plate 3 [discussed in *Ritsema et al.*, 1998b].

Synthetics generated from the models in Plate 4 are a little lumpy, as are the observed waveforms, but such complexities could also be caused by the rift system. However, the model suggests a possible detachment just above D''. This feature can be tested with more distant stations (Figure 5). Again, we use a Sandwich Island event as recorded by the IRIS stations, FURI and ATD, as presented in Plate 3. We did not pay much attention to absolute time at these stations because of their well-known delays. Note that ATD is at the same WWSSN site as AAE, the most anomalous station in the network [*Dziewonski and Anderson*, 1981]. However, their differential times between S and SKS should be useful since the ray paths are quite close together near the stations. The homogeneous model (ALVS) on the left indicates that S is delayed relative to SKS, while the model on the right allows both S and SKS to miss the anomaly and be PREM-like—results that do not fit the data. In summary, the data along this corridor support our model with considerable anomalous structure protruding into the mantle. Unfortunately, we do not have such data coverage along other azimuths, although the existing (S–SKS) and ScS–S) data from IRIS and older WWSSN networks support this structure [*Ni and Helmberger*, 2003a].

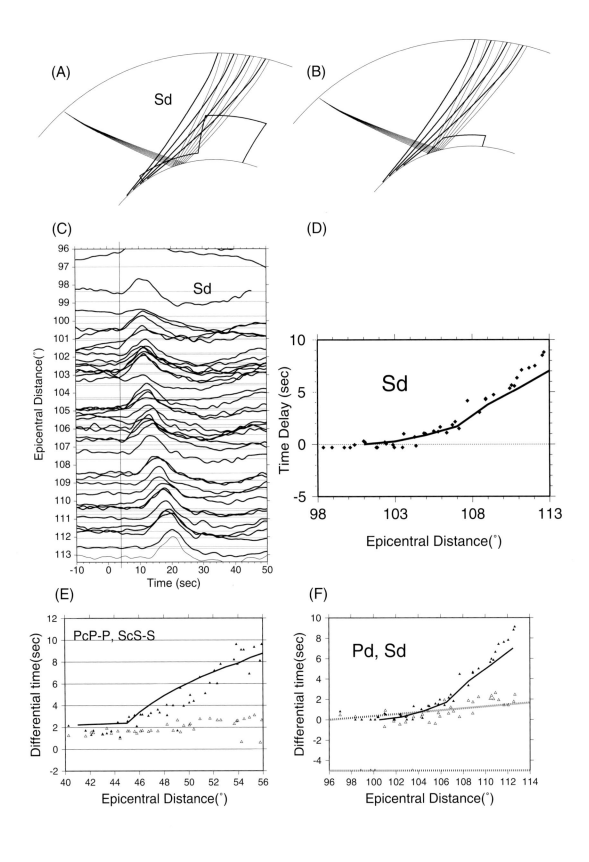

Figure 4. S_d record section produced from the Kaapvaal Array, and travel time analysis. (A) S_d ray paths (heavy lines) and ScS ray paths (thin lines) sampling a 1200-km structure with a 3% shear velocity reduction. (B) Same for a 300-km structure with a shear velocity reduction of 2% on the top and gradually increasing to 12% at the bottom (ULVD). (C) Waveform data of S_d for event 970610. S_d shows little delay before 105°, but becomes delayed up to 9 s at 113°. (D) Travel-time picks of S_d (solid symbols) determined by cross-correlation along with travel-time predictions from the two models. The 300-km model fails to explain the S_d data because no S_d ray paths sample the anomaly (the dotted line shows prediction of the 300-km model). The ALVS model (solid line) fits the S_d data quite well. Both models can explain the ScS travel time anomaly, i.e., little delay for small epicentral distances and large delay for large epicentral distances. (E) Comparison of differential PcP–P (PcP travel time minus P travel time) and ScS–S plots for event 970602. (F) Comparison of P_d and S_d for this same event, where there is a slight effect on P_d and a large effect on S_d.

STRUCTURAL DEFINITION OF THE EASTERN PROVINCE

This corridor of Kaapvaal Array data runs mostly along the structure as it bends sharply eastward beneath the Indian Ocean. The western end of the structure is easily seen in the many SKS samples in Plate 2. However, its eastern extension is more difficult to quantify. It appears that adding the SKKS phase can help, as displayed in Plate 5. This phase travels more in the shallow core and has a lower phase velocity than SKS. Thus, it crosses the core–mantle boundary (CMB) further away from the array as displayed. The actual data given in *Ni and Helmberger* [2003a] have simple waveforms, but the differential timing relative to SKS and SKKS is complicated. While the SKS phase is uniformly late, SKKS shows a pattern as displayed in Plate 5A. This feature can be explained by continuing the eastward bending nature of the structure so that the most southerly paths miss the structure as displayed in the schematic picture (Plate 5B). Several other events that yield slow SKKS values are given in Plate 2. However, this particular event (970525) is special in that the SKKS paths at distances 114°–119° lie along the same SKS paths from a closer event (980325) at 80°–85° (Plate 2A). The latter distance range encloses the S–SKS cross-over, which is easily identified, where S is delayed about the same amount as SKS [*Ni and Helmberger*, 2003a]. Since S turns about 500 km (Plate 5B) above the CMB at these ranges, we obtain direct evidence for the elevated structure seen in tomography. Completing the structural image beneath the Indian Ocean proved difficult because of the lack of stations. Thus, we assumed that the basic cross-section determined from the Western Province extends eastern to the boundar-

ies, as suggested in some tomography studies (Plate 1). We then tested this model against long-range–diffracted P and S (P_d and S_d) from events occurring in the western Pacific (Figure 6). Three events proved particularly useful, namely, the northernmost event 971222, the southern event 970904, and the middle one 990206. Record sections from these three events are displayed, plotted as conventional record sections in Figure 6 and as a function of azimuth or as a fan-array in Figure 7. The latter has proven effective in finding salt domes. Obviously, the azimuthal plots organize the data much better than the conventional record sections, proving that this structure is more vertical than horizontal. The bars indicate the relative arrivals where the data are aligned relative to PREM (shifted for range difference). The azimuths are designated with respect to the events; thus, azimuth 246° is more northerly than 202°. The interpretation is straightforward. Data in column 1 (971222) sample the northern edge, where the arrivals slow down crossing the boundary. The data in the middle column are mostly late, with a slight weakening to the south, ranging from 14 to 18 s late. The data in column 3 (970904) indicate that the southern boundary is near 214°, but the sloping travel-time curve indicates that the boundary is not parallel to the ray paths, as in Plate 5. The waveforms are also interesting in that they show multiple pulses near the boundaries, 241–246° (northern) and 212–216° (southern).

These features can be modeled in several ways. *Wen* [2001] successfully explained these data by breaking the dataset into narrow sectors (azimuthal groups) and modeling them in 2D. For the upper portion of event 971222, he modeled the second arrival as a very late ScS by invoking a very slow D″ (–12% at the CMB). This anomaly then fades away at the smaller azimuth towards the bottom, where his model of the anomaly becomes thicker but less severe.

The 3D synthetics displayed at the bottom of Figure 7 were constructed from the model presented in Plate 1. These synthetics were generated with a new code called DWKM [*Helmberger and Ni*, 2004]. The method uses WKM [*Ni et al.*, 2000] for generation of 2D synthetics and adds out-of-the-great-circle-plane arrivals by applying a diffraction operator. An example calculation is displayed in Figure 8. The column on the left displays the great-circle path contribution (2D). One then adds neighboring weighted synthetics to construct the synthetics on the right, which compare well with SEM results [*Ni et al.*, 2005]. Models containing sharp vertical walls generate distorted synthetics for about 2° to 4°, depending on frequency. Thus, the transition for the 97094 event displayed is probably more sharp than actually modeled, suggesting some horizontal change in shape. The uniqueness issue becomes more serious as we add dimensions to the model, as demonstrated above. The geometry of the ray paths

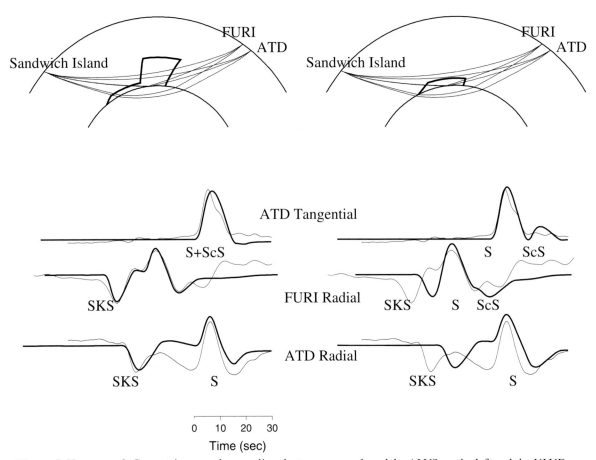

Figure 5. Upper panel: Geometric ray paths sampling the two proposed models, ALVS on the left and the ULVD on the right, at stations FURI (85°) and ATD (89°). The S-wave ray paths sample regions roughly 300 km above the CMB so that they mostly miss the ULVD structure. The observations are displayed as thin lines relative to the synthetics in heavy lines. Both synthetic S and SKS are PREM-like for the ULVD model, which does not fit the data, whereas the delay of S in ALVS is about right.

sampling the structure also becomes more dependent on the source location. This feature is demonstrated in the third column of Figure 8, where we regenerated the 3D synthetics by allowing the source to move 60 km to the north. The same effect could have been produced by rotating the ALVS structure by this amount, referred to as ALVSS in the plot. Thus, such sensitivity requires refinement in source locations to accompany structural modeling. Obviously, more data are required to nail down this portion of the structure.

Our model does not explain the edges all that well (there are some misfits between the synthetics and data); however, it does model the long-period response quite well, as discussed in *Ni et al.* [2005], where data are compared against SEM and DWKM synthetics. Predictions from tomographic models display only about one-third of the travel-time anomaly and are too smooth to produce multipathing. While the S-waves show large noticeable delays with azimuth, the

P-waves are PREM-like (also pointed out earlier by *Wen* [2001]). The remoteness of this region and the lack of Indian Ocean events make it difficult to refine the model, although other phases arriving from distant events can be used. For example, the SS phases from the above events (Figure 6) can be used to define the top of the structure, which is compatible with the earlier estimates of the western portion [*Ni et al.*, 2003c]. However, only a large sharp structure can produce such rapid changes with azimuth.

DISCUSSION

The above data analyses provide convincing evidence for a large-scale lower-mantle structure with dimensions of about 1200 km wide, 1000 km high, and 7000 km long. The walls are remarkably sharp, although the SKS delays suggest some complexities, especially along the northeastern edge of the

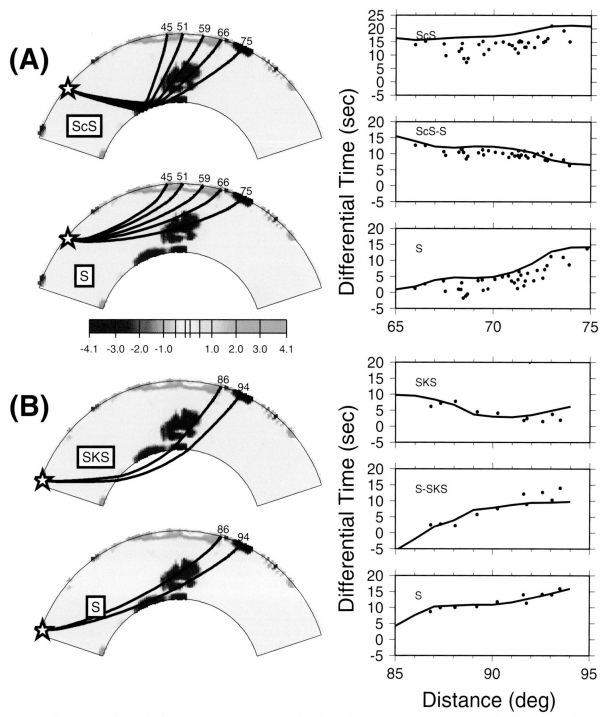

Plate 4. Cross-sections of velocity structure connecting Sandwich Island to the Tanzania Array along with ray paths appropriate for (a) ScS and S and (b) SKS and S. The corresponding travel times on the right were computed from synthetics generated for these 2D sections (solid lines) along with observed picks from the array as discussed by *Ritsema et al.* [1998b], reduced by PREM. The velocity model was derived from the tomographic images of *Grand* [1994] by applying an ad hoc enhancement scheme of all anomalies that are beneath Africa increased to 4%, after *Helmberger et al.* [2000].

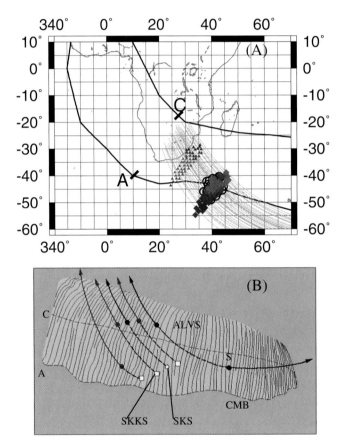

Plate 5. Upper panel: Comparison of SKS and SKKS observations from a western Pacific event (970525) showing the difference in delays apparently caused by the relative position of the CMB exit points. These points are denoted as triangles (SKS) and squares (SKKS). Magenta boxes are delayed relative to blue boxes. The open circles are SKS exit points for event 980325, which sample the same CMB region. The lower panel (schematic) indicates the unique geometry where ray paths intersect the upper surface of our proposed model (solid circles). Note that SKKS exit points for event 970525 occur at the same location of SKS for event 980325 (see Figure 2). The direct S for the latter event turns at about 500 km above the CMB and samples the ALVS for about the same amount of path length. The weight lines in the upper panel indicate the ALVS boundary and the CMB, with the Eastern Province starting nears points A and C.

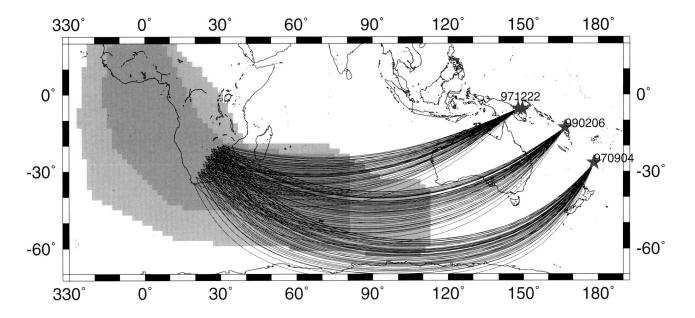

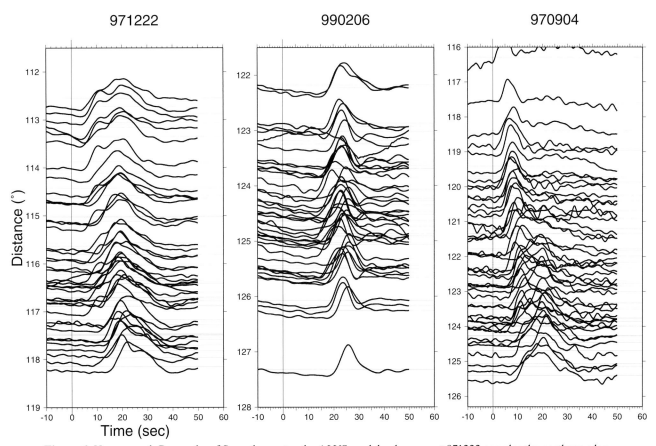

Figure 6. Upper panel: Ray paths of S_d as they enter the ALVS model, where event 971222 samples the northern edge and event 970904 samples the southern edge. Paths for event 90206 appear to enter the structure mostly along the axis. Lower panel: The three record sections assembled from the array data.

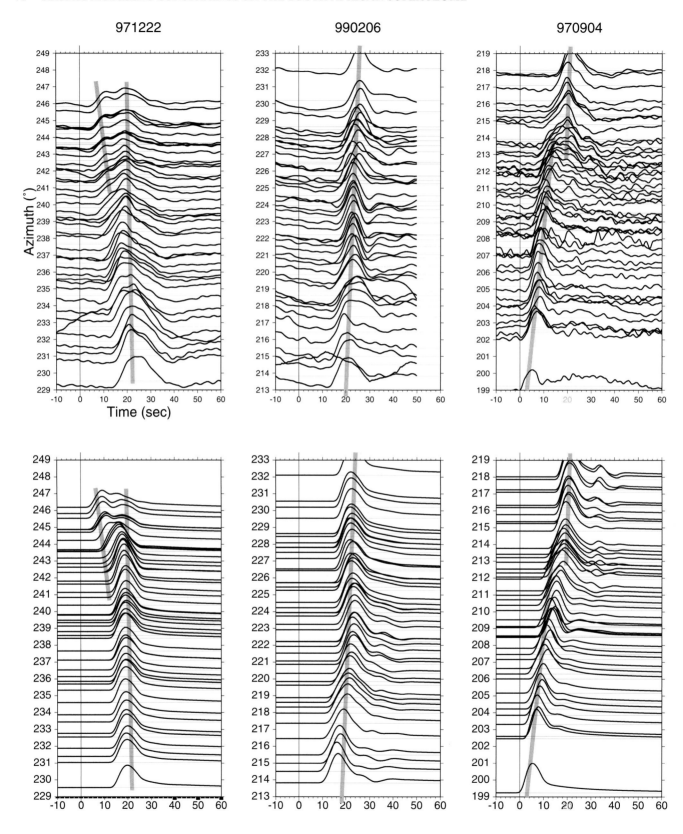

Figure 7. The upper set of observations is a replot of the same data presented in Figure 6 except as a function of azimuth. Heavy lines have been included to highlight the travel times of the multipathing along the boundaries: fast-to-slow at the northern edge (971222) and slow-to-fast at the southern edge (970904). Note that the data approach PREM outside the ALVS and reach delays of 18 s inside the ALVS. The synthetics (DWKM) based on the model are displayed in the lower panels, aligned relative to PREM predictions.

Western Province. About 10–20% variability in SKS delays suggests some small-scale extensions above the structure or perhaps some internal structure, such as enhanced percentage drop near its center or in D″. Although we favor the large structure over the thinner but more severe velocity drops proposed by *Wen* [2001] and his colleagues, we must admit that localized strong velocity drops near the edges certainly

explain some complex waveforms. However, the above 2D section discussed in detail appears well resolved and is probably representative of the Western Province. The structure beneath the Indian Ocean is less well defined, having only two good samples of thickness, namely, samples of delayed SS starting at a distance of about 125° [*Ni et al.*, 2005] and one sample of the SKS-S crossover [*Ni and Helmberger*, 2004]. The Western Province probably tilts to the east, as argued by tomographic images [*Ritsema et al.*, 1999; *Ni et al.*, 2003]. However, the sparse coverage prevents a strong conclusion, and the two models presented in Plate 3 are probably end members bracketing the true structure. Although we have not found any direct evidence for ULVZ beneath the structure, there is ample evidence for a localized ULVZ along the edges. *Wen* [2000] reports on one between Madagascar and Africa, based on precursors to PKP. Another ULVZ has

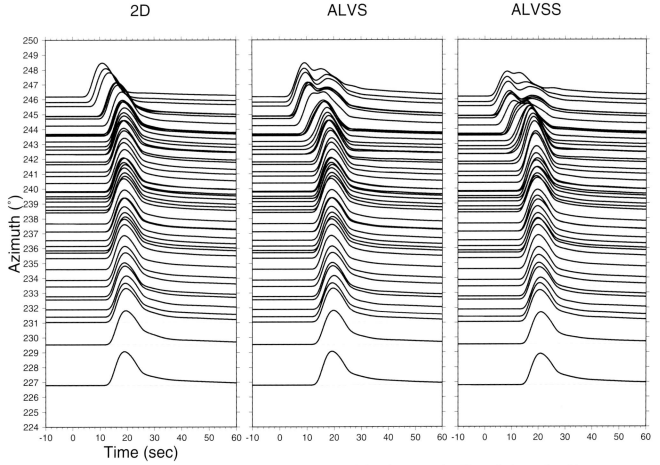

Figure 8. Synthetics generated with the DWKM method. The leftt column contains 2D synthetics produced by constructing a 2D section through the ALVS model presented in Plate 1. The middle column contains contributions from the neighboring paths, assuming an 8-s source duration. The geometry is that shown in Figure 6 for the paths along the northern edge (paths 971222). The right column displays a new set of synthetics with the structure rotated northward by 0.5° (ALVSS) or moving the source 60 km to the north.

two different types of data used to explore its location and dimensions, precursor to PKP [*Ni and Helmberger*, 2001], and SKS$_d$ [*Helmberger et al.*, 2000]. This feature is located beneath central Africa. A third ULVZ has been reported beneath Tristan by *Ni and Helmberger* [2001]. The usual interpretation of ULVZs is that they are partial melt zones [*Williams and Garnero*, 1996]. This interpretation has gained strength with the recent shockwave data [*Akins et al.*, 2003], suggesting that melt from perovskite probably sinks.

Although many interesting features of the African structure are yet to be determined, our present results have significant implications on seismological studies. For example, a common strategy followed in lower-mantle analyses is to assume that D″ is the most likely location for anomalies. It is further commonly assumed that S-wave travel times can be adequately corrected from past tomographic studies and thus ScS–S differential times become an effective means of mapping D″ (see review by *Lay et al.* [2003]). Such an approach obviously has serious problems if there are other structures extending upward into the middle mantle, as we argue in this review.

The above presentation on the seismology of this large-scale structure and its edges is relatively easy compared to explaining what this structure is and how long has it been there. Although there are many possibilities, two candidates have been addressed with considerable analyses [*Tackley*, 2000]. One is the Primitive Layer model introduced by *Kellogg et al.* [1999], who argue that this layer contains the stuff needed for "isotopic mixing" and heat production and that it has been there billions of years. Unfortunately, recent seismic studies have not indicated much evidence for a global layer [*Castle and van der Hilst*, 2003], although it does not eliminate regional structures as displayed here. A second version of the primitive layer hypothesis is that of Primitive Piles [*Tackley*, 2000]. He suggests that the two so-called megaplumes beneath the Pacific and South Africa are constructed from primitive materials that, again, are old. Convection studies on these two models suggest that this material is relatively dense, contrasts of 1–2%, along with excess temperatures, being compatible with supposed shapes [*Ni et al.*, 2002]. These density estimates are compatible with the combination of gravity anomalies and normal-mode data [*Ishii and Tromp*, 1999]. Excess temperature is more difficult to establish, but the excellent correlation between the geoid and our structure (Plate 2) would provide the ideal geometry for the Gondwanaland breakup [*Davaille*, 1999]. That is, these "piles" are expected to have hot thermal upwelling arising from their tops and perhaps edges [*Gonnermann et al.*, 2002]. Such upwelling then would explain the so-called Dupal isotope anomaly (see also Plate 2 and *Hart* [1984]).

However, the issue of a radioactive-rich layer at the base of the mantle as suggested by *Kellogg et al.* [1999] is disputed by *Anderson* [2001, 2002]. Anderson argues that such regions would overheat and overturn, delivering their radioactive elements to the upper mantle. Essentially, the accretionary process leaves the lower mantle as a drain for the denser metallic melts and refractory material. He suggests that this chemically distinct structure would have a low thermal expansivity, along with high conductivity and viscosity, and that this structure would develop irregular shapes caused by sluggish behavior (low Rayleigh number convection). This image also seems compatible with the above seismic observations. However, the top of this structure would again be hot and would supply a great deal of heat to the middle mantle.

In short, while these interpretations are all interesting, a great deal of seismology, geodynamics, and geochemical modeling is required to understand this new class of lower-mantle structure and its role in plate tectonics, if any.

Acknowledgments. We thank Lianxing Wen and Ed Garnero for their excellent reviews, Evelina Cui for her efforts in producing this manuscript, and the editors of this book, in particular, Rob van der Hilst and production coordinator Virginia Marcum. This research was supported by NSF Grant EAR-0229885. Contribution Number 9116 of the Division of Geological and Planetary Sciences, California Institute of Technology.

REFERENCES

Akins, J. A., T. J. Ahrens, and P. D. Asimow, Shock-induced melting of MgSiO$_3$ (perovskite): Implications for density of ultramafic liquids above core-mantle boundary, Abstract, *Eos Trans. AGU., 84* (46), 2771, 2003.

Anderson, D. L., Top-down tectonics, *Science, 293,* 2016, 2001.

Anderson, D. L., The case for irreversible chemical stratification of the mantle, *Int. Geol. Rev., 44,* 97–116, 2002.

Castle, J. C. and R. van der Hilst, Searching for seismic scattering off mantle interface between 800 km and 2000 km depth, *J. Geophys. Res., 108* (B2), 2095, 2003.

Davaille, A., Simultaneous generation of hotspots and superswells by convection in a heterogeneous planetary mantle, *Nature, 402,* 756–760, 1999.

Ding, Xiaoming, High resolution studies of deep earth structure, Ph.D. Thesis, California Institute of Technology, Pasadena, California, 1997.

Dziewonski, A. M. and D. L. Anderson, Preliminary Reference Earth model, *Phys. Earth Planet. Inter., 25,* 297–356, 1981.

Garnero, E. J., Lower mantle heterogeneity, *Ann. Rev. Earth Planet. Sci., 28,* 509–537, 2000.

Gonnermann, H. M., M. Manga, A. M. Jellinek, Dynamics and longevity of an initially stratified mantle, *Geophys. Res. Lett., 29,* 1029, 2002.

Grand, S. P., Mantle shear structure beneath the Americas and surrounding oceans, *J. Geophys. Res., 99,* 11,591–11,622, 1994.

Grand, S. P., R. D. van der Hilst, and S. Widiyantoro, Global seismic tomography: Global seismic tomography: A snapshot of convection in the Earth, *GSA Today, 7,* 1–7, 1997.

Grand, S. P., Mantle shear-wave tomography and the fate of subducted slabs, *Phil. Trans. R. Soc. Lond., A, 360,* 2475–2491, 2002.

Hager, B. H., R. W. Clayton, M. A. Richards, R. P. Comer, and A. M. Dziewonski, Lower mantle heterogeneity; dynamic topography and the geoid, *Nature, 313,* 541–545, 1985.

Hart, S., A large-scale isotope anomaly in the Southern Hemisphere Mantle, *Nature, 309,* 753–757, 1984.

Helmberger, D. V., S. Ni, L. Wen, and J. Ritsema, Seismic evidence for ultra low-velocity zones beneath Africa and Eastern Atlantic, *J. Geophys. Res., 105* (B10), 23865–23878, 2000.

Helmberger, D. V. and S. Ni, Approximate 3D Bodywave Synthetics for Tomographic Models, submitted to *Bull. Seismol. Soc. Am.,* 2004.

Ishii, M. and J. Tromp, Normal-mode and free-air gravity constraints on lateral variations in velocity and density of Earth's mantle, *Science, 285,* 1231–1236, 1999.

James, D. E., M. J. Fouch, J. C. VanDecar, S. van der Lee, Kaapvaal Seismic Group, Tectospheric structure beneath southern Africa, *GRL, 28,* 2485–2488, 2001.

Kellogg, L., B. H. Hager, and R. vanderHilst, Compositional Stratification in the Deep Mantle, *Science, 283,* 1881–1884, 1999.

Lay, T., E. J. Garnero, and Q. Williams, Partial Melting in a Thermo-Chemical Boundary Layer at the Base of the Mantle, submitted to *Phys. Earth Planet. Inter.,* 2003.

Masters, G., G. Laske, H. Bolton, and A. M. Dziewonski, The relative behavior of shear velocity, bulk sound speed, and compressional velocity in the mantle: Implications for chemical and thermal structure, in *Earth's Deep Interior: Mineral Physics and Tomography From the Atomic to the Global Scale,* eds. S. Karato, A. M. Forte, R. C. Liebermann, G. Masters, and L. Stixrude, *AGU,* 63–87, Washington, DC., USA., 2000.

Ni, S., X. Ding, and D. V. Helmberger, Constructing synthetics from deep earth tomographic models, *Geophys. J. Int., 140,* 71–82, 2000.

Ni, S., and D. V. Helmberger, Horizontal transition from fast (slab) to slow (plume) structures at the core-mantle boundary, *Earth Planet. Sci. Lett., 187,* 301–310, 2001.

Ni, S., E. Tan, M. Gurnis, and D. V. Helmberger, Sharp Sides to the African Super Plume, *Science, 296,* 1850–1852, 2002.

Ni, Sidao and D. V. Helmberger, Seismological Constraints on the South African Super Plume; could be the oldest distinct structure on Earth, *Earth Planet. Sci. Lett., 206,* 119–131, 2003a.

Ni, Sidao and D. V. Helmberger, Ridge-like lower mantle structure beneath South Africa, *J. Geophys. Res., 108,* 2094–2110, 2003b.

Ni, Sidao and D. V. Helmberger, Further constraints on African Super Plume Structure, *Phys. Earth Planet. Inter., 140,* 243–251, 2003c.

Ni, Sidao, Vernon F. Cormier, and D. V. Helmberger, A Comparison of Synthetic Seismograms for 2D Structures: Semianalytical versus Numerical, *Bull. Seismol. Soc. Am., 93,* No. 6, 2752–2757, 2003d.

Ni, Sidao, D. V. Helmberger, and J. Tromp, Three-dimensional structure of the African superplume from waveform modeling, *Geophys. J. Int., 161,* 283–294, 2005.

Nyblade, A. A., C. Birt, C. A. Langston, T. J. Owens, and R. Last, Seismic experiment reveals rifting of craton in Tanzania, *Eos Trans. AGU, 77,* 517–521, 1996.

Ritsema, J., S. Ni, D. V. Helmberger, and H. P. Crotwell, Evidence for strong shear velocity reductions and velocity gradients in the lower mantle beneath Africa, *Geophys. Res. Lett., 25,* 4245–4248, 1998a.

Ritsema, J., A., A. Nyblade, T. J. Owens, C. A. Langston, and J. C. VanDecar, Upper mantle seismic velocity structure beneath Tanzania, *J. Geophys. Res., 103,* 21,201–21,213, 1998b.

Ritsema, J., H. Van Heijst, and J. Woodhouse, Complex shear wave velocity structure imaged beneath Africa and Iceland, *Science, 286,* 1925–1928, 1999.

Simmons, N. A. and S. P. Grand, Partial melting in the deepest mantle, *Geophys. Res. Lett.,* 1552, 2002.

Tackley, P., Mantle Convection and Plate tectonics: Toward an integrated Physical and Chemical Theory, *Science 288,* 2002–2007, 2000.

van der Hilst, R., S. Widiyantoro, E. R. Engdahl, Evidence for deep mantle circulation from global tomography, *Nature, 386,* 578–584, 1996.

van der Hilst, R., J. Tromp, and E. Stutzman, Seismological Constraints on Mantle Structure and Composition, *Nature,* 2004.

Wang, Y. and L. Wen, Mapping the geometry and geographic distribution of a very low velocity province at the base of the Earth's mantle, *J. Geophys. Res., 109,* B10305, 2004.

Wen, L., Intense seismic scattering near the Earth's core-mantle boundary beneath the Comoros hotspot, *Geophys. Res. Lett., 27,* 3627–3630, 2000.

Wen, L., Silver P, James D and Kuehnel R., Seismic evidence for a thermo-chemical boundary at the base of the Earth's mantle, *Earth Planet. Sci. Lett., 189,* 141–153, 2001.

Wen, L., Seismic evidence for a rapidly varying compositional anomaly at the base of the Earth's mantle beneath the Indian Ocean, *Earth Planet. Sci. Lett., 194,* 83–95, 2001.

Williams, Q. and E. J. Garnero, Seismic evidence for partial melt at the base of the Earth's mantle, *Science, 273,* 1528–1530, 1996.

D. V. Helmberger and S. Ni, Seismological Laboratory, California Institute of Technology, Pasadena, California, USA. (helm@gps.caltech.edu)

Numerical and Laboratory Studies of Mantle Convection: Philosophy, Accomplishments, and Thermochemical Structure and Evolution

Paul J. Tackley[1,2], Shunxing Xie[1], Takashi Nakagawa[3], and John W. Hernlund[1]

Since the acceptance of the theory of plate tectonics in the late 1960s, numerical and laboratory studies of mantle convection and plate tectonics have emerged as a powerful tool for understanding how the solid parts of Earth and other terrestrial planets work. Here, the general philosophy of such modeling is discussed, including what can and cannot be determined, followed by a review of some of the major accomplishments and findings. Then some recent work by the authors on thermochemical convection is reviewed, with a focus on comparing the results of numerical experiments to observations from geochemistry, seismology, and geomagnetism in order to constrain uncertain physical parameters. Finally, the future of such research is discussed.

1. MODELING PHILOSOPHY AND ACCOMPLISHMENTS

Geodynamic theory plays a central role in understanding the solid Earth, as it provides a bridge or framework that links together the various other subfields, particularly seismology, mineral physics, geochemistry, geodesy, and geology. The foundation of this framework is the theory of continuum mechanics, which provides a quantitative theory linking forces within the Earth to subsequent displacements. Mineral physics plays a crucial role in geodynamic theory by providing the constitutive relations that link relevant intensive (e.g., stress, temperature, chemical potential, etc.) and extensive (e.g., strain, entropy, composition, etc.) variables. Mineral physics provides the physical properties (e.g.,

rheology, equation of state, heat capacity, thermal conductivity, etc.) necessary for a forward calculation of convection and plate tectonics, which can then predict the nature of seismic wave velocity anomalies, geochemical heterogeneity and geodetic observations (velocities, gravitational field) for comparison to observations. Alternatively, wave speed anomalies detected by seismology can be converted to buoyancy anomalies using mineral physics, then geodynamic theory can be used to compute surface gravitational anomalies, topography, and velocities.

It is useful to recognize the fundamental difference between what is obtained by observation-based fields such as geochemistry and seismology, and geodynamic theory. While observations, leading for example to seismic tomographic models, are giving us an increasingly resolved picture of the structures present inside the mantle, seismic tomography is merely a mathematical fit to the observations: describing *what* is there, analogous to Kepler's laws which give a mathematical fit to the orbits of planets round the Sun. Geodynamics, however, goes beyond this and provides the fundamental explanation of *why* these structures occur, which is analogous to Newton's laws explaining why planets move round the Sun as they do. Thus, even if seismological observations could map out three-dimensional Earth structure to a resolution of 1 km, dynamical modeling would

---

[1]Department of Earth and Space Sciences, University of California, Los Angeles, California.

[2]Institute of Geophysics and Planetary Physics, University of California, Los Angeles, California.

[3]Department of Earth and Planetary Sciences, University of Tokyo, Japan.

Earth's Deep Mantle: Structure, Composition, and Evolution
Geophysical Monograph Series 160
Copyright 2005 by the American Geophysical Union
10.1029/160GM07

still be necessary in order to understand why the observed structures occur.

In addition to explaining observations, convection investigations also offer a predictive capability, i.e., predicting what is expected to be found once observational techniques improve enough to observe them. Two prominent examples of this are as follows: (i) Based on numerical experiments of convection plus consideration of geoid and surface topography (e.g., [*Davies*, 1988]), the geodynamics community was strongly in favor of predominantly whole-mantle convection, perhaps with slabs intermittently held up above the 660-km discontinuity [*Christensen and Yuen*, 1984, 1985; *Machetel and Weber*, 1991; *Peltier and Solheim*, 1992; *Tackley*, 1995], long before seismology gave us conclusive images that supported this view [*van der Hilst et al.*, 1997]. (ii) Laboratory and numerical experiments predicted the existence and dynamics of upwelling plumes (e.g., [*Griffiths and Campbell*, 1991; *Morgan*, 1971; *Olson and Singer*, 1985; *Whitehead and Luther*, 1975]) long before seismology became capable of imaging them in the deep mantle [*Montelli et al.*, 2004]. Of course, this predictive capability is limited by uncertainties in physical parameters and model simplifications as discussed later.

Several distinctions can be made regarding different types of study. One distinction is between studies that investigate the basic fluid dynamics to understand scalings, regimes of behavior, phenomenology, etc., and studies that attempt to match specific observations, be it seismological, geochemical, geological, or geodetic. As the field progresses, there has been a progression from more emphasis on the former to more emphasis on the latter. Another distinction regards the timescale that is being modeled, which ranges from instantaneous (e.g., studies that attempt to match the geoid using flow calculated from seismic tomography), to short-term (e.g., the last 120 Ma of Earth history), to long-term—either "evolution" scenarios that represent the inherently transient evolution of Earth or another planet over 4.5 Ga, or experiments that produce a "statistically steady-state". The level to which specific observations can be matched increases as timescale is decreased, mainly because the short term or instantaneous experiments require imposing the model setup or initial conditions such that observations are matched. Each of these approaches has its place, depending on the questions being posed; ultimately, all are necessary. For example, a physical description of the mantle that is able to match some instantaneous or recent geological observation on Earth is clearly not perfect if, when run for billions of years, it produces a planet that looks nothing like Earth; the reverse also applies.

Both laboratory and numerical approaches have been used, with numerical approaches becoming more prevalent as computational technology has improved. Numerical calculations are best thought of as "experiments" in the same sense as laboratory experiments, rather than "simulations", as the latter implies an attempt to simulate the real Earth, which is still some distance from what is numerically possible. Experimental studies, whether laboratory or computational, should have a clearly defined setup, including approximations made and the extent to which results are applicable to real planets, and well-posed questions to be answered.

Many fundamental concepts, scalings, and phenomena have been established by experiments and theory. Several examples of these, a few of which (certainly not all) have become so engrained in our collective consciousness that they are taken for granted, are given below.

(i) The onset of convection in terms of Rayleigh number [*Chandrasekhar*, 1961; *Holmes*, 1931], including heating from within [*Roberts*, 1967], and the planforms of convection that appear as the Rayleigh number is increased (e.g., [*Schubert and Zebib*, 1980; *Zebib et al.*, 1983, 1980]).

(ii) The basic structure of a convection cell, with narrow boundary layers thickening approximately as a half-space solution and near-adiabatic interior (e.g., summarized in *Turcotte and Schubert* [1982]).

(iii) The scaling of heat flux and velocity with Rayleigh number for constant-viscosity convection as predicted by boundary layer theory and verified by numerical experiments (e.g., [*Jarvis and Peltier*, 1982; *Turcotte and Schubert*, 1982]).

(iv) The concept of plumes and their structure and dynamics (e.g., [*Campbell and Griffiths*, 1990; *Morgan*, 1971; *Olson and Singer*, 1985; *Parmentier et al.*, 1975]).

(v) The influence of temperature-dependent viscosity, including the transition to rigid lid convection at high viscosity contrast [*Christensen*, 1984b; *Moresi and Solomatov*, 1995; *Nataf and Richter*, 1982; *Ogawa et al.*, 1991; *Ratcliff et al.*, 1997; *Solomatov*, 1995] and its influence on thermal evolution [*Tozer*, 1972].

(vi) The basic planform of three-dimensional convection, with ridge-like boundary layer instabilities that become more plume-like as they cross the domain, and the influence of internal heating—raising the internal temperature, making the downwellings more plume-like and the interior sub-adiabatic, (e.g., [*Bercovici et al.*, 1989; *Houseman*, 1988; *Parmentier et al.*, 1994; *Travis et al.*, 1990])

(vii) The influence of non-Newtonian rheology, localizing deformation and making the system much more episodic, (e.g., [*Christensen*, 1983; *Christensen and Yuen*, 1989; *Parmentier et al.*, 1976]).

(viii) The influence of depth-dependent viscosity and thermal expansivity in making deep mantle upwellings broader and increasing the horizontal wavelength of flow (e.g., [*Balachandar et al.*, 1992; *Hansen et al.*, 1993; *Tackley*, 1996]).

(ix) The realization that plate tectonics itself is a particular manifestation of more general behavior arising from mantle processes/convection in terrestrial planets and is unique to planet Earth at the present time.

(x) The ability of plastic yielding to break a rigid lid and generate a crude approximation of plate tectonics (e.g., [*Moresi and Solomatov*, 1998; *Tackley*, 2000b; *Trompert and Hansen*, 1998]).

(xi) The influence of an endothermic phase transition at 660-km depth in promoting episodic slab penetration or layering, depending on the phase buoyancy parameter (e.g., [*Christensen and Yuen*, 1984; *Christensen and Yuen*, 1985; *Machetel and Weber*, 1991]), and possibly a secular switch from layered to whole-mantle convection [*Steinbach et al.*, 1993].

(xii) The dynamics induced by continents, including episodic movements and the inherent instability of supercontinents to break up at some characteristic time after forming [*Gurnis*, 1988; *Lowman and Jarvis*, 1993; *Zhong and Gurnis*, 1993], and the different scaling of heat flux through continents and oceans [*Lenardic*, 1998].

(xiii) The ability of viscous dissipation to cause large heating locally [*Balachandar et al.*, 1993].

(xiv) Many results related to chemical mixing and thermochemical convection that are discussed later.

Additionally, a number of studies with time evolution have attempted to match specific observations, notable examples of which are the ability of convection constrained by plate motions over the last 120 Ma to match many but not all of the features observed in global tomographic models [*Bunge et al.*, 1998] and the ability of mantle flow history to match recent uplift rates in Africa [*Conrad and Gurnis*, 2003].

1.1. Problems and Limitations

Two major problems are faced by experimental studies, both of which cause uncertainty in how well the results apply to the real Earth: (i) uncertainty in the physical properties and parameters, and (ii) the effect of model simplifications, which are often made because it is not possible to treat, either numerically or in the laboratory, the full problem complexity.

1.1.1. Parameter uncertainty. In many cases the important physical properties or parameters are uncertain enough that a range of behavior is possible. One notable example is the value of the Clapeyron slope of the spinel to perovskite + magnesiowüstite phase transition, estimates of which ranged from –2 to –6 MPa/K in the early 1990s [*Akaogi and Ito*, 1993; *Ito et al.*, 1990; *Ito and Takahashi*, 1989]—enough to cause a range in behavior from almost complete layering to basically whole-mantle convection [*Machetel and*

Weber, 1991; *Solheim and Peltier*, 1994; *Tackley et al.*, 1994]. The value is still uncertain with current estimates ranging from –0.4 to –2.0 MPa/K [*Fei et al.*, 2004; *Katsura et al.*, 2003]. Additionally, uncertainties remain in the depth of the majoritic–garnet to perovskite transition [*Ono et al.*, 2001], and in whether the ilmenite to perovskite transition [*Chudinovskikh and Boehler*, 2002] occurs in realistic mantle assemblages, both of which have significant effects on mass exchange between the upper and lower mantles.

Another notable uncertainty is the density contrast between subducted mid-ocean ridge basalt (MORB), pyrolite, and depleted residue in the deep mantle, which is large enough that MORB could be either less dense or more dense than pyrolite in the deepest mantle [*Kesson et al.*, 1998; *Ono et al.*, 2001, 2005]. This determines whether MORB settles into a layer above the core–mantle boundary (CMB), which could have a large effect on geochemical observations [*Christensen and Hofmann*, 1994; *Xie and Tackley*, 2004a,b]. Another major example of uncertainty is the viscosity profile of the mantle, estimates of which are still changing (e.g., [*Forte and Mitrovica*, 1996; *Forte and Mitrovica*, 2001; *King*, 1995]). A final example of uncertainty is in how the thermal conductivity— particularly the radiative component—quantitatively varies with temperature and pressure [*Hofmeister*, 1999], which may have important effects on heat transport and thermal evolution [*van den Berg et al.*, 2001].

Some physical properties were not even known about until recently, particularly the post-perovskite phase transition [*Murakami et al.*, 2004; *Oganov and Ono*, 2004], which has important dynamical effects [*Nakagawa and Tackley*, 2004a, 2005b], and the spin transition in perovskite [*Badro et al.*, 2004], which has been argued to have a large effect on thermal conductivity, although the quantitative effect is extremely uncertain.

1.1.2. Model simplification. Models are necessarily simpler than Earth, both from a desire to study a well-constrained system where the effect of individual complexities can be clearly identified, but also because currently available numerical methods and computers are not capable of modeling the mantle/plate system with its full complexity. A major example of this is the difficulty in obtaining realistic plate tectonics in numerical models of mantle convection, which for many years necessitated inserting plates by hand and sometimes using rules for how they evolve. It is now established that simple pseudo-plastic yielding yields a crude approximation of plate tectonics in viscous flow models [*Moresi and Solomatov*, 1998; *Richards et al.*, 2001; *Tackley*, 2000b; *Trompert and Hansen*, 1998], yet these models have several shortcomings, such as double-sided subduction and no memory of previous weak plate boundaries (i.e., weak zones revert to the strength of pristine material once they become inactive). The latter may be important on the Earth as they can later be reactivated [*Gurnis et al.*, 2000]. Most likely a full visco-elasto-plastic treatment will be necessary to get realistic behavior, as hinted at by regional models (e.g., [*Regenauer-Lieb and Yuen*, 2003; *Regenauer-Lieb et al.*, 2001]).

Some other simplifications have come into focus recently, the first of which is the role of water, which has a strong effect on the viscosity [*Hirth and Kohlstedt*, 1996]. It has been argued, for example, that the deeper melting that would take place under mid-ocean ridges in a hotter mantle would generate a thicker rheological lithosphere, causing a negative exponent in the heat flux–Rayleigh number relationship [*Korenaga*, 2003]. A physical property that has so far been ignored in all global-scale convection models is grain size, which evolves with time and has a strong effect on the viscosity for diffusion creep. It has been argued that this might even cause a hotter mantle [*Solomatov*, 2001] or hot parts of the mantle [*Korenaga*, 2005] to be more viscous than average rather than less viscous as temperature-dependence would give.

Because of the uncertainties that ill-constrained parameters and model simplification introduce, it is important that modeling studies explore the full range of possibilities in order to establish which conclusions are robust and which depend on exact choice of parameter, and to also clearly identify the limitations and give appropriate caveats about applicability. Furthermore by determining which results match observations, parameters can be constrained (a type of "inversion" by repeated forward modeling). Nevertheless, some proposed behaviors or mantle models can be ruled out even with our current knowledge, including some of the mantle models that have been proposed on the basis of geochemical observations [*Tackley*, 2000a].

Performing numerical experiments using a wide range of parameters carries an additional benefit that aids in our general understanding of Earth and its place in the pantheon of terrestrial planets. While only a particular parameter within the range studied may be relevant to Earth itself, other regions of parameter space allow us to examine how the variation of certain critical parameters gives rise to the great variety of behavior observed in the other terrestrial planets, both in our solar system and beyond, e.g., the control of lithospheric strength and convective vigor upon the occurrence of plate tectonics.

2. THERMOCHEMICAL CONVECTION

2.1. Discussion and Review

Reconciling geochemical and geophysical constraints on mantle structure and evolution remains one of the "grand challenges" in solid Earth science. There has been a recent resurgence of interest in this topic, and in thermochemical convection in general, due partly to seismological models that indicate the need for chemical variations in the deepest mantle [*Ishii and Tromp*, 1999; *Karato and Karki*, 2001; *Kennett et al.*, 1998; *Masters et al.*, 2000; *Trampert et al.*, 2004] and partly to provocative proposals regarding mantle structure [*Becker et al.*, 1999; *Kellogg et al.*, 1999].

Several approaches are being taken to understanding the dynamics of thermochemical convection and how it is related to seismic, geochemical, and other observations. One important topic is "mixing", a term that is often used loosely to mean either the stirring and stretching of heterogeneities by convection (e.g., [*Christensen*, 1989a; *Ferrachat and Ricard*, 1998; *Gurnis*, 1986; *Kellogg and Turcotte*, 1990; *Metcalfe et al.*, 1995]), or the dispersion of initially nearby heterogeneities throughout the domain (e.g., [*Gurnis and Davies*, 1986; *Olson et al.*, 1984; *Schmalzl and Hansen*, 1994; *Schmalzl et al.*, 1995, 1996]).

The present discussion, however, focuses on studies where the chemical variations are active, i.e., affect the buoyancy, and often where tracking of specific trace elements is also performed. A typical investigative approach has been to insert a layer of dense material a priori and study the subsequent dynamics as it interacts with the convection (e.g., [*Christensen*, 1984a; *Davies and Gurnis*, 1986; *Hansen and Yuen*, 1988; *Hansen and Yuen*, 1989; *Olson and Kincaid*, 1991; *Schott et al.*, 2002; *Sidorin et al.*, 1998]), although *Hansen and Yuen* [2000] started with a linear chemical stratification. In the late 1990s, the field was invigorated by a number of developments, including a series of laboratory experiments that showed various hitherto unappreciated phenomena, particularly a "doming" mode, in which domes of material from a dense lower layer rise into the upper layer then fall, a process which can be repeated several times before large-scale overturn occurs [*Davaille*, 1999a,b; *Le Bars and Davaille*, 2002]; by seismic inversions that found the bases of the large-scale "megaplumes" underneath the Pacific and Africa are dense [*Ishii and Tromp*, 1999], implying a chemical origin; and by a proposal that a dense, thick but strongly undulating layer exists in the bottom part of the mantle, helping to explain geochemical and seismological constraints [*Kellogg et al.*, 1999].

2.2. Three-Dimensional Experiments

Initial three-dimensional calculations of convection with an inserted dense layer comprising ~10% of mantle depth produced isolated "ridges" or "piles" separated by exposed CMB, with the "piles" being reminiscent of the seismi-

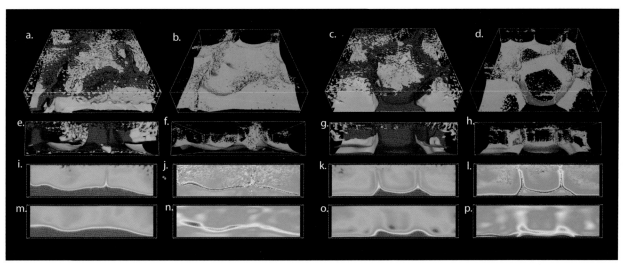

Plate 1. Three-dimensional thermochemical convection calculations [source: *Tackley*, 2002]. The left two columns show a case with a dense layer initially occupying the lower 30% of the mantle volume, while the right two columns show a case with a dense layer initially taking up 10% of the mantle volume. (a, e, c, g) Isocontours of residual temperature ±200 K; (b, d, f, h) isocontours of composition = 0.5; (i, k) vertical cross-section of temperature field, which is filtered to the nominal resolution of seismic tomography in parts (m, o), respectively; (j, l) cross-sections of $T - 0.75C$, filtered in parts (n, p), respectively.

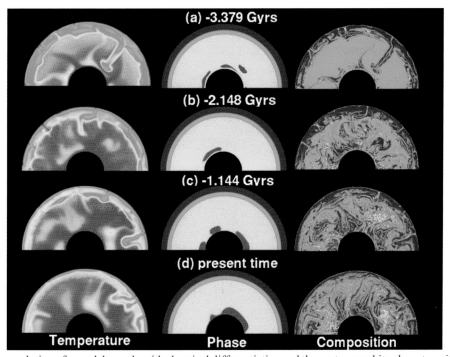

Plate 2. Time evolution of a model mantle with chemical differentiation and the post-perovskite phase transition [source: *Nakagawa and Tackley*, 2005b]. Left column: temperature. Center column: phase, where red = post-perovskite and the other colors are the usual mantle phases. Right column: composition, from red = crust to dark blue = harzburgite. Times are as indicated.

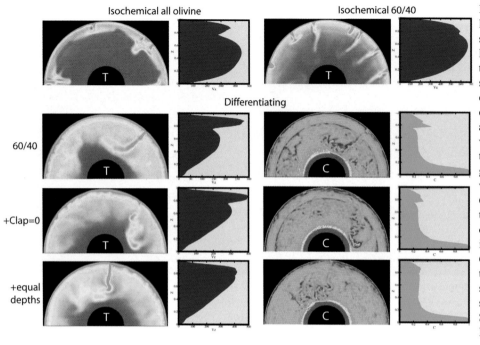

Plate 3. Effect of phase transitions on layering of flow and average composition, from two-dimensional calculations in a half cylinder. Plots show two-dimensional distributions of superadiabatic temperature (T) and composition (C) varying from red = crust to dark blue = residue; the time-averaged mean of the modulus of radial velocity (blue graphs); or the horizontally averaged composition (green graphs). Top row: isochemical cases with either (left) 100% olivine or (right) 60% olivine and 40% pyroxene–garnet; the remaining cases include chemical differentiation. Second row: 60% olivine, 40% pyroxene-garnet. Third row: 60/40 with zero Clapeyron slopes. Bottom row: 60/40 with zero Clapeyron slopes and the transition to perovskite set to an equal depth for both systems. Source: Tackley, P. J., *EOS Trans. AGU*, Fall Meeting Suppl., 2002.

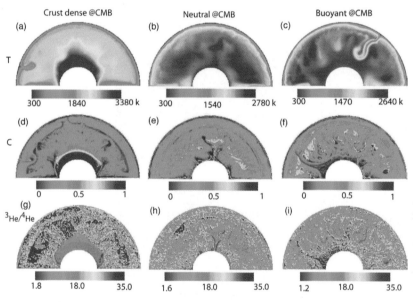

Plate 4. The effect of deep-mantle MORB density of mantle thermochemical structure including helium ratios. (a–c) temperature, (d–f) composition, and (g–i) ^{3}He/^{4}He. (a, d, g) PxGt system (hence MORB) dense at the CMB, (b, e, h) PxGt system neutrally dense at the CMB, (c, f, i) PxGt system less dense at the CMB. Source: *Xie and Tackley* [2004a].

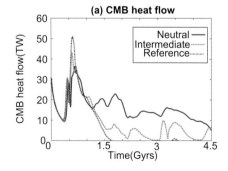
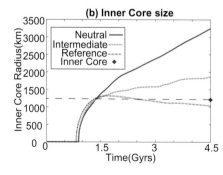

Plate 5. Time evolution of CMB heat flow and inner core size from the numerical convection calculations of *Nakagawa and Tackley* [2005a] with chemical differentiation and three different density contrasts for MORB in the deep mantle.

cally imaged megaplumes, leading to the suggestion that these structures are thermochemical in origin [*Tackley*, 1998]. This is different from the globally continuous layer comprising ~30% of mantle depth proposed by *Kellogg et al.* [1999] on the basis of geochemical mass-balance considerations. Numerical experiments with these different amounts of dense material were explored by *Tackley* [2002] and sample results are illustrated in Plate 1. When the dense material comprises 30% of mantle volume (Plate 1, left columns), a continuous undulating layer is formed, with hot upwellings rising from ridges of the dense layer. When the dense material comprises only 10% of mantle volume (Plate 1, right columns), it is swept into ridges separated by exposed patches of the CMB and again, hot upwellings rise from these ridges. The dense material becomes very hot because it is enriched in heat-producing elements (as either a primitive layer or a layer of segregated subducted oceanic crust would be) and heat must be transferred conductively across its interface (Plate 1, panels i, k, m, o). This causes very large lateral temperature variations in the deep mantle, probably causing very high amplitude seismic heterogeneity. It has been suggested that the thermal and chemical contributions to seismic velocity are of opposite sign and would approximately cancel out [*Kellogg et al.*, 1999]. The effect of this is explored in Plate 1, panels j and l. Because the chemical boundary is sharp but the thermal boundary is diffuse, the boundary of the dense material will still cause high-amplitude heterogeneity, even at the nominal resolution of seismic tomography (Plate 1, panels n, p). For intermittent ridges or piles, this high-amplitude heterogeneity is at the base of the mantle, consistent with seismological models, whereas for a global layer it is in the mid-mantle, not consistent with tomographic models, as plotted using spectral heterogeneity maps in *Tackley* [1998]. Such matters are further explored in *Deschamps et al.* [2005].

This shows how numerical modeling can be used to compute the observational signatures of proposed mantle models for comparison with observations. Recently, such models have been extended to three-dimensional spherical geometry to determine the effect of viscosity contrast on the resulting thermochemical structures [*McNamara and Zhong*, 2004] and to predict more specifically the form that chemical "piles" would take under Africa and the Pacific [*McNamara and Zhong*, 2005].

A limitation of all of these models is that a layer with a sharp interface is inserted a priori, which may or may not be reasonable as an initial condition for Earth [*Solomatov and Stevenson*, 1993]. Another method by which a layer might form is by segregation of subducted oceanic crust, if it is denser than regular mantle at deep mantle pressures.

2.3. Differentiation and Segregation

The feasibility of forming of a deep mantle layer by segregation of oceanic crust was shown by *Christensen* [1989b] and *Christensen and Hofmann* [1994]. In the latter work, the crust was generated by melting-induced differentiation of the mantle under mid-ocean ridges, something that has been seldom included in numerical studies, with some notable exceptions (e.g., [*Dupeyrat et al.*, 1995; *Ogawa*, 1997; *Ogawa*, 2000b; *Ogawa and Nakamura*, 1998; *van Thienen et al.*, 2004a,b, this volume]). However, in the last several years our group has been experimenting with such models (first reported in *Tackley and Xie* [2002]), and these will be be discussed in the remainder of this paper.

When modeling thermochemical convection, care must be taken to include reasonable relative densities of the chemically different components as a function of depth. In the models presented here, this is treated by having two reference states: one for the olivine system, and the other for the pyroxene–garnet system, the density profiles for which are illustrated in Figure 1 and discussed fully elsewhere [*Tackley and Xie*, 2003; *Xie and Tackley*, 2004a]. Of particular interest is the greater depth of the phase transition to perovskite in the pyroxene–garnet system, which causes subducted MORB to be buoyant in the range of approximately 660–720 km [*Ono et al.*, 2001], which has been proposed to have important dynamical effects [*Ringwood*, 1991]. In the experiments that follow, unmelted mantle ("pyrolite") is assumed to consist of 60% olivine and 40% pyroxene–garnet, while the extreme compositions are basaltic crust (C = 1), which

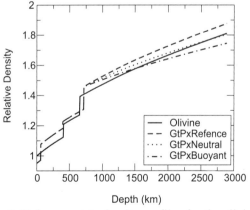

Figure 1. Reference state density profiles for the olivine and pyroxene–garnet (GtPx) components. For the latter, three different compressibilities are assumed in the lower mantle, leading to this component being either denser (GtPxReference), less dense (GtPxBuoyant), or at the same density as the olivine component. Because the slab contains a much higher fraction of the GtPx component than does pyrolite, these relationships determine the relative density of MORB, residue, and primitive material.

is pure pyroxene–garnet, and harzburgite (C = 0), which is 6:1 olivine:pyroxene–garnet. Melting occurs when the temperature exceeds a depth-dependent experimentally based solidus; then enough melt is generated to bring the temperature back to the solidus, after which the melt is immediately placed ("erupted") at the surface to form a crust. These calculations, intended to model the history of a terrestrial planet as it cools and radioactive elements decay, thus have decaying radioactive heat sources and account for heat coming out of the core, either by assuming a simple heat capacity for the core (as in *Steinbach et al.* [1993]) or by using a more sophisticated core heat balance (as in *Buffett et al.* [1996]). The models also include strongly temperature-dependent viscosity, with yielding in the lithosphere to prevent a rigid lid from forming.

2.4. Time Evolution and Post-Perovskite

Plate 2 shows a typical time-evolution for such a system with a deep mantle density contrast between harzburgite and MORB of about 2% [*Nakagawa and Tackley*, 2005b]. At early times, the upper mantle differentiates rapidly, but this is followed by whole-mantle circulation. By the end of the experiment the mantle is extremely heterogeneous, containing "blobs", stretched and folded strips of material, and a build-up of subducted crust above the CMB. This chemical structure is far more heterogeneous, and the interface of dense material at the base far less sharp, than in models in which a layer is inserted a priori. In some frames, subducted crust is trapped just below the transition zone, as also observed by *Ogawa* [2003], a topic returned to later.

These experiments also include the recently discovered post-perovskite phase transition [*Murakami et al.*, 2004; *Oganov and Ono*, 2004], which has already been shown to have important dynamical effects: destabilizing the lower boundary layer and hence resulting in more and smaller plumes and a higher mantle temperature [*Nakagawa and Tackley*, 2004a]. Regions of post-perovskite are shown in red, and many of the features predicted by *Hernlund et al.* [2005] can be seen. At early times when the core is hot, the CMB is still in the perovskite stability field, so a "double-crossing" of the phase boundary occurs, but once the core cools sufficiently, it is blanketed by post-perovskite. In hot upwelling regions the post-perovskite layer is either nonexistent (if the core is hot) or very thin. There is thus an anticorrelation between regions with a thick post-perovskite region and hot thermochemical piles. Another effect is that due to the destabilizing effect of the phase transition, a larger chemical density contrast is needed to stabilize a deep layer.

The post-perovskite phase transition provides an explanation for the seismic discontinuity at the top of D″ [*Lay and*

Helmberger, 1983] and the possible deeper discontinuity [*Thomas et al.*, 2004], and the presented models provide predictions about the relationship between these discontinuities and thermochemical piles that can be tested seismologically.

2.5. Stratification Induced by Phase Transitions

The influence of the two-system phase transitions on isochemical or differentiating convection experiments is explored in Plate 3. Almost all previously published models of phase-transition modulated mantle convection (e.g., [*Christensen and Yuen*, 1985; *Machetel and Weber*, 1991; *Peltier and Solheim*, 1992; *Tackley et al.*, 1993; *Weinstein*, 1993]) assumed a mantle made of 100% olivine, in which case (Plate 3, top left) the endothermic phase transition at 660-km depth causes a substantial inhibition of the radial flow, as indicated by the dip in the radial velocity profile. When, however, the composition is changed to pyrolite (Plate 3, top right), this inhibition becomes negligible, because the transition from garnet–majorite to perovskite is exothermic, thus largely cancelling the effect of the endothermic spinel-to-perovskite transition. A possible intermediate transition to ilmenite [*Chudinovskikh and Boehler*, 2002] would reverse this picture and increase the layering [*van den Berg et al.*, 2002], but it is doubtful whether ilmenite occurs in a pyrolite composition (S. Ono, personal communication).

When chemical variations are added, the "660" regains its dynamical importance. It was previously found that the spinel–perovskite transition on its own might act as a chemical "filter" [*Mambole and Fleitout*, 2002; *Weinstein*, 1992], but with both phase systems present, the effect is magnified (Plate 3, second row). The vertical flow again exhibits a minimum at around 660 km, and there is also a local chemical stratification around 660 km, with enriched material trapped in the transition zone and depleted material trapped at the top of the lower mantle, as also observed by *Fleitout et al.* [2000] and *Ogawa* [2000a], but curiously, not by *Christensen* [1988]. This entrapment typically does not occur when a slab first encounters the region, but rather after the crust and residue components have separated in the deep mantle and are subsequently circulating as blobs. Other features visible in the compositional field and its profile are a stable layer of segregated crust above the CMB and a thin crust at the surface.

To demonstrate that this local chemical stratification around 660-km depth is due mainly to the different depths of the perovskite phase transition rather than deflection of transitions due to their Clapeyron slopes, a case in which all Clapeyron slopes are set to zero is presented (Plate 3, third row). In this case, the stratification is still strong, though somewhat diminished, showing that the depth separation on its own can cause significant chemical stratification, which phase boundary deflection adds to. Finally, when the perovskite phase transition is set to the same depth in both systems (Plate 3, bottom row), there is no flow restriction or chemical stratification around 660-km depth.

This local chemical stratification around 660–720 km is a prediction that could be tested seismically, which would help to determine whether the currently estimated depths of these phase transitions are indeed correct.

2.6. Trace Elements and Geochemical Modeling

A long-standing challenge has been to reconcile geochemical observations with a mantle flow model, and for this purpose it is important to track trace element evolution in mantle evolution experiments so that synthetic geochemical data can be generated and compared to observations, an approach that was pioneered by *Christensen and Hofmann* [1994] and followed with other, more recent studies [*Davies*, 2002; *Ferrachat and Ricard*, 2001; *Samuel and Farnetani*, 2003]. Experiments with passive tracers carrying trace elements have also been performed [*van Keken and Ballentine*, 1998, 1999].

Here, some recently published models by *Xie and Tackley* [2004a,b] are discussed. These models build on the earlier ones of *Christensen and Hofmann* [1994] by incorporating greater realism and by including the noble gases He and Ar in addition to U-Th-Pb and Sm-Nd systems. The most important ways in which realism is increased are a 4.5-Ga run time instead of 3.6 Ga, and a secular decrease in model activity due to the decay of radiogenic heat-producing elements and cooling of the mantle and core. Plate 4 illustrates how the chemical structure after 4.5 Ga of evolution is dependent on the relative density of MORB in the deep mantle: When it is more dense (Plate 4, panels a, d, g), a layer of segregated crust is formed; when it is of equal density, a very thin transient layer of crust and residue is formed because the slab is cold (Plate 4, panels b, e, h); and when it is buoyant a layer of residue is formed together with blobs of crust in the mid mantle (Plate 4(c)(f)(i)).

The time evolution of heat flow and magmatism are plotted in Figure 2 (again from *Xie and Tackley* [2004b]). The total surface heat flow (thick solid line) decreases from about 100 TW to 20 TW over time, as expected, but conductive surface heat flow (thin solid line) is roughly constant. The discrepancy between these is made up by magmatic heat transport (dotted line), which is estimated to transport about 2 TW of heat in today's Earth, but if these models are reasonable, a lot more (up to 10s of TW) may have carried in the past. Indeed, it has even been proposed that magmatism

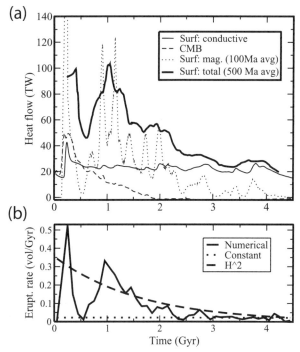

Figure 2. Time evolution of various diagnostics from a typical case in *Xie and Tackley* [2004b]. (a) Surface and CMB heat flow scaled to a spherical planet, showing CMB (dashed), surface conductive (solid), surface magmatic (dotted), and surface total (bold indicates averaging over a 500-Ma sliding window). (b) Crustal production (eruption) rate in mantle volumes per Ga, averaged over three representative cases in 100-Myr bins and compared with theoretical models (dotted and dashed lines)

transported most of the heat flux in a pre-plate-tectonic era [*Davies*, 1990, 1992; *van Thienen et al.*, 2004b], as is presently the case in Io [*Segatz et al.*, 1988]. The magmatic rate (Figure 2b) was substantially higher in the distant past, lying somewhere between the idealized models of constant at today's value (dotted line) and proportional to internal heating squared (dashed line).

The ratio ^{3}He/^{4}He, plotted as a function of position (Plate 4, panels g, h, i), shows the extremely heterogeneous nature of this model mantle. In these cases ^{3}He/^{4}He is low in former MORB because of the ingrowth of ^{4}He from U and Th (which are concentrated in MORB) and high in residue because it is assumed that He is more compatible (less incompatible) than U and Th [*Coltice and Ricard*, 1999]. However, the latter assumption is highly uncertain and its consequences are illustrated in the histograms of ^{3}He/^{4}He in Figure 3. When He is less compatible than U, both residue and crust develop a low ^{3}He/^{4}He signature, and high ^{3}He/^{4}He ratios (the peak at 35 Ra in Figure 3a) occur only in primitive, unmelted material. When, however, He is more compatible than U, the residue develops the

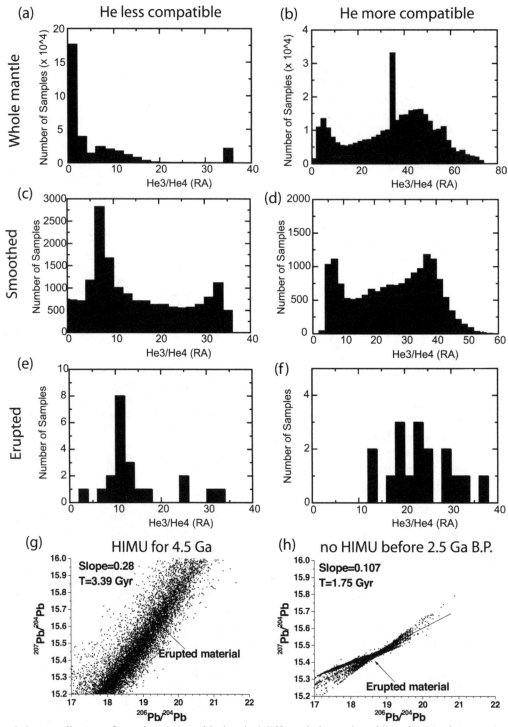

Figure 3. Isotope diagrams for various cases with chemical differentiation and tracking of trace elements. (a–f) Histograms of ³He/⁴He [*Xie and Tackley*, 2004a]. A case in which He is less compatible (more incompatible) than U is plotted in (a, c, e), while a case in which He is more compatible than U is plotted in (b, d, f). Histograms are counting either individual tracers (a, b), sampling cells (c, d)), or material erupted over the last 200 Ma (e, f)). (g, h) ²⁰⁷Pb/²⁰⁴Pb vs. ²⁰⁶Pb/²⁰⁴Pb diagrams from *Xie and Tackley* [2004b]. (g) is for a case in which high U/Pb (MU) material is subducted throughout the modeled history, whereas (h) is for a case in which HIMU material is subducted only for the last 2.5 Ga of the model time.

highest $^3He/^4He$ ratios (Figure 3b), up to 73 Ra. Figure 3(c,d) shows the effect of calculating ratios over volumetric sampling cells, rather than considering each tracer as a sample (absolute abundance matters when summing over sampling cells). The difference is most dramatic for the "less compatible" case, in which the peak is shifted to ~8 Ra, similar to that observed for MORB. Thus it is not surprising that material erupted in the last part of the experiment (Figure 3e) has a histogram similar to that observed for MORB, whereas for the "more compatible: case, the ratios are too high and too scattered (Figure 3f). *Xie and Tackley* [2004a] found that there is a trade-off with deep mantle crustal density, such that another MORB-like solution is obtained for buoyant crust and He is equally compatible. These trade-offs need to be more thoroughly investigated.

The outgassing of radiogenic ^{40}Ar in these experiments is around 50%, similar to that observed in Earth, despite over 90% of "primitive" nonradiogenic noble gases being lost [*Xie and Tackley*, 2004a]. This is possible because much of the outgassing occurs early on, when most of the final ^{40}Ar has not yet been produced.

Fractionation of U and Pb generates "HIMU" in MORB, such that Pb–Pb diagrams develop a slope that for Earth corresponds to an isochron with an age of 1.5–2.0 Ga [*Hofmann*, 1997]. In these numerical experiments, however, the slope is much larger (Figure 3g), corresponding to an isochron of 3.39 Ga. This is because the signature is dominated by early-differentiating material, both because the melting rate was much higher early on and because this material has a more extreme signature in Pb–Pb space. *Christensen and Hofmann* [1994] obtained the correct slope because their model activity was constant with time and lasted for only 3.6 Ga. *Xie and Tackley* [2004b] consider and test four hypotheses for the discrepancy between model and data: (i) HIMU is not produced early on, (ii) the melting rate is too low in these experiments, (iii) erasure of heterogeneities by stretching is not correctly treated, and (iv) the sampling length scale is inappropriate. The sampling length scale affects only the scatter not the slope; the melting rate is reasonable (Figure 2b), although a higher rate would be acceptable; and a better treatment of stretching is helpful but probably cannot account for the whole discrepancy. The most effective way to resolve the discrepancy is for HIMU to not enter the mantle prior to about 2.5 Ga before present (Figure 3h), which could be because trace elements were removed from the slab by melting prior to that time [*Martin*, 1986] or because of a change in atmosphere and ocean oxidation state [*Elliott et al.*, 1999].

2.7. Core and Geodynamo Evolution

Explaining how the heat flux out of the core can have stayed large enough over billions of years to maintain the geodynamo, without growing the inner core to a much larger size than observed, is a challenging problem. Parameterized models of core–mantle evolution in which no complexities are present predict a final inner core size that is much larger than that observed (e.g., [*Labrosse*, 2003; *Labrosse et al.*, 1997; *Nimmo et al.*, 2004]. As the heat conducted out of the core is determined by what is happening in the mantle, the time history of CMB heat flow is an important quantity to investigate with numerical models as it can provide an additional constraint on mantle models.

Nakagawa and Tackley [2004b] used a convection model similar to those already presented but with a parameterized core heat balance based on that of *Buffett et al.* [1992, 1996] to show that the presence of a layer of dense material above the CMB, either primordial or arising from segregated crust, substantially reduces the CMB heat flow and hence inner core growth. The problem is, CMB heat flow can be reduced so much that it is insufficient to facilitate dynamo action, and can even become zero or negative. Because of this, a discontinuous layer, rather than a global layer, was found to be the most promising scenario. These tradeoffs are illustrated in Plate 5 (taken from *Nakagawa and Tackley* [2005a]), which shows the time evolution of CMB heat flow and inner core size from three calculations with different deep mantle density contrasts for subducted crust. When crust is neutrally buoyant (red curves), no dense layer builds up, so CMB heat flow remains high and the inner core grows to over twice its observed radius. When crust is ~2% denser than pyrolite at the CMB (blue curves), a global layer builds up, eventually zeroing CMB heat flow and resulting in a slightly too small inner core. With an intermediate crustal density (~1% denser than pyrolite) the results are in between. *Nakagawa and Tackley* [2004b] found it difficult to match all constraints unless there is radioactive potassium in the core (e.g., *Nimmo et al.* [2004]) and mapped out the tradeoff between crustal density and core K content.

This is another example of the use of numerical experiments to constrain the values of uncertain physical parameters—in this case, the potassium concentration in the core and the density contrast of MORB in the deep mantle.

3. SUMMARY AND OUTLOOK

Mantle convection experiments play an essential role in understanding the dynamics, evolution, and structure of Earth's mantle, and many fundamental findings have been established. A major problem is parameter uncertainty, which is sometimes large enough that seemingly opposite behaviors are possible; but many mantle models can be ruled out even with existing uncertainties. Most importantly, parameters can be constrained by comparing forward models to data, as has been demonstrated in this review using comparisons with seismic models, geochemical data and geodynamo history. The presented experiments also illustrate the fundamental importance of mineral physics data, particularly on phase transitions (including the recently discovered post-perovskite transition) and the density as a function of composition and pressure.

This is an exciting time for mantle geochemistry, with new proposals being made and fundamental assumptions being questioned. For example, *Ballentine et al.* [2002] show that if current estimates for the helium flux out of the mantle are too high by a factor of three, then most of the paradoxes associated with He and Ar, including the need for a layered mantle, would vanish. The idea that MORB and ocean island basalt (OIB) could be derived from the same statistical distribution but with different sampling is explored by *Ito and Mahoney* [2005a,b] and *Meibom and Anderson* [2004]. Water may cause the transition zone to act as a trace element filter [*Bercovici and Karato*, 2003]. A new way of modeling the sampling of a chemically heterogeneous mantle with a heterogeneous size distribution has been developed [*Kellogg et al.*, 2002].

The coupled geodynamic–geochemical modeling approach reviewed here appears very promising, and there are some obvious improvements that could straightforwardly be made to the models, including three-dimensional spherical geometry, more realistic convective vigor, and higher resolution of the chemical heterogeneities (i.e., more tracers). Three-dimensional geometry would, for example, allow a distinction to be made between OIB and MORB magmatism. Additionally, the sampling algorithm by which observed compositions are calculated from that carried on tracers could be improved along the lines of *Kellogg et al.* [2002].

There are, however, several fundamental limitations that fall into the earlier-discussed category of model simplifications, even though the presented models may appear quite complex compared to previous models. Some examples are given: The slab processing (dehydration and migration of trace elements) that takes place in subduction zones may be very important for generating some geochemical signatures, but requires detailed, fine-resolution local models to treat properly (e.g., [*Ruepke et al.*, 2004; *van Keken et al.*, 2002]). The derivation of basalt composition from the source region may be more complicated than the simple partitioning assumed here [*Aharonov et al.*, 1997; *Donnelly et al.*, 2004; *Spiegelman et al.*, 2001]. Continents are geochemically important but are not included because the processes by which they form are too complicated and poorly understood to include in such global models, although some progress is being made on very early continent formation processes [*van Thienen et al.*, 2003, 2004a]. Plate tectonics itself is only crudely represented, as a full treatment of rock rheology is beyond what is presently possible in global experiments. At this point, all of these complexities are essentially absorbed into the parameterization. The arguably important effects of water and grain size [*Bercovici and Karato*, 2003; *Korenaga*, 2003; *Solomatov*, 1996] are not treated. Accurate numerical treatment of entrainment is problematic [*Tackley and King*, 2003; *van Keken et al.*, 1997] so laboratory experiments are still useful (e.g., [*Gonnermann et al.*, 2002]).

Evaluating the effect of these and other complexities on the dynamics and observational signatures will keep geodynamic researchers and their coworkers in related fields busy for long into the future.

Acknowledgments. Research was supported by NSF grant EAR0207741, the David and Lucile Packard Foundation, and IGPP Los Alamos National Laboratory. TN was also supported by 21st Century COE Program for Earth Sciences, University of Tokyo. The authors thank Arie van den Berg and two anonymous reviewers for rapid and constructive reviews.

REFERENCES

Aharonov, E., M. Spiegelman, and P. Kelemen, 3-Dimensional Flow and Reaction In Porous-Media—Implications For the Earths Mantle and Sedimentary Basins, *Journal Of Geophysical Research Solid Earth*, *102* (B7), 14821–14833, 1997.

Akaogi, M., and E. Ito, Refinement Of Enthalpy Measurement Of $MgSiO_3$ Perovskite and Negative Pressure-Temperature Slopes For Perovskite-Forming Reactions, *Geophysical Research Letters*, *20* (17), 1839–1842, 1993.

Badro, J., J. P. Rueff, G. Vanko, G. Monaco, G. Fiquet, and F. Guyot, Electronic transitions in perovskite: Possible nonconvecting layers in the lower mantle, *Science*, *305* (5682), 383–386, 2004.

Balachandar, S., D. A. Yuen, and D. Reuteler, Time-dependent 3-dimensional compressible convection with depth-dependent properties, *Geophys. Res. Lett.*, *19* (22), 2247–2250, 1992.

Balachandar, S., D. A. Yuen, and D. Reuteler, Viscous and Adiabatic Heating Effects In 3-Dimensional Compressible Convection At Infinite Prandtl Number, *Physics Of Fluids a-Fluid Dynamics*, *5* (11), 2938–2945, 1993.

Ballentine, C. J., P. E. van Keken, D. Porcelli, and E. H. Hauri, Numerical models, geochemistry and the zero-paradox noble-gas mantle, *Philosophical Transactions of the Royal Society of London Series A-Mathematical Physical & Engineering Sciences*, *360* (1800), 2611–2631, 2002.

Becker, T. W., J. B. Kellogg, and R. J. O'Connell, Thermal constraints on the survival of primitive blobs in the lower mantle, *Earth Planet. Sci. Lett.*, *171*, 351–365, 1999.

Bercovici, D., and S. Karato, Whole-mantle convection and the transition-zone water filter, *Nature*, *425* (6953), 39–44, 2003.

Bercovici, D., G. Schubert, and G. A. Glatzmaier, 3-Dimensional Spherical-Models Of Convection In the Earths Mantle, *Science*, *244* (4907), 950–955, 1989.

Buffett, B. A., H. E. Huppert, J. R. Lister, and A. W. Woods, Analytical Model For Solidification Of the Earths Core, *Nature*, *356* (6367), 329–331, 1992.

Buffett, B. A., H. E. Huppert, J. R. Lister, and A. W. Woods, On the thermal evolution of the Earth's core, *Journal of Geophysical Research*, *101* (B4), 7989–8006, 1996.

Bunge, H. P., M. A. Richards, C. Lithgowbertelloni, J. R. Baumgardner, S. P. Grand, and B. A. Romanowicz, Time Scales and Heterogeneous Structure in Geodynamic Earth Models, *Science*, *280* (5360), 91–95, 1998.

Campbell, I. H., and R. W. Griffiths, Implications Of Mantle Plume Structure For the Evolution Of Flood Basalts, *Earth and Planetary Science Letters*, *99* (1–2), 79–93, 1990.

Chandrasekhar, S., *Hydrodynamic and hydromagnetic stability*, 654 pp., Oxford University Press, London, 1961.

Christensen, U., Convection In a Variable-Viscosity Fluid—Newtonian Versus Power-Law Rheology, *Earth and Planetary Science Letters*, *64* (1), 153–162, 1983.

Christensen, U., Instability of a hot boundary layer and initiation of thermo-chemical plumes, *Annales Geophysicae*, *2* (3), 311–319, 1984a.

Christensen, U., Is subducted lithosphere trapped at the 670-km discontinuity?, *Nature*, *336* (6198), 462–463, 1988.

Christensen, U., Mixing by time-dependent convection, *Earth Planet. Sci. Lett.*, *95* (3–4), 382–394, 1989a.

Christensen, U. R., Heat transport by variable viscosity convection and implications for the Earths thermal evolution, *Phys. Earth Planet. Inter.*, *35* (4), 264–282, 1984b.

Christensen, U. R., Models of mantle convection—one or several layers, *Phil. Trans. R. Soc. London A*, *328* (1599), 417–424, 1989b.

Christensen, U. R., and A. W. Hofmann, Segregation of subducted oceanic crust In the convecting mantle, *J. Geophys. Res.*, *99* (B10), 19867–19884, 1994.

Christensen, U. R., and D. A. Yuen, The Interaction Of a Subducting Lithospheric Slab With a Chemical or Phase-Boundary, *Journal Of Geophysical Research*, *89* (B6), 4389–4402, 1984.

Christensen, U. R., and D. A. Yuen, Layered convection induced by phase transitions, *J. Geophys. Res.*, *90* (B12), 291–300, 1985.

Christensen, U. R., and D. A. Yuen, Time-Dependent Convection With Non-Newtonian Viscosity, *Journal Of Geophysical Research Solid Earth and Planets*, *94* (B1), 814–820, 1989.

Chudinovskikh, L., and R. Boehler, The $MgSiO_3$ Ilmenite-Perovskite phase boundary: Evidence for strongly negative Clapeyron slope, (submitted), 2002.

Coltice, N., and Y. Ricard, Geochemical observations and one layer mantle convection, *Earth Planet. Sci. Lett.*, *174* (1–2), 125–37, 1999.

Conrad, C. P., and M. Gurnis, Seismic tomography, surface uplift, and the breakup of Gondwanaland: Integrating mantle convection backwards in time—art. no. 1031, *Geochemistry Geophysics Geosystems*, *4*, 1031, 2003.

Davaille, A., Simultaneous generation of hotspots and superswells by convection in a heterogeneous planetary mantle, *Nature (UK)*, *402* (6763), 756–60, 1999a.

Davaille, A., Two-layer thermal convection in miscible viscous fluids, *J. Fluid Mech. (UK)*, *379*, 223–53, 1999b.

Davies, G. F., Ocean Bathymetry and Mantle Convection .1. Large-Scale Flow and Hotspots, *Journal Of Geophysical Research Solid Earth and Planets*, *93* (B9), 10467–10480, 1988.

Davies, G. F., Heat and mass transport in the early Earth, in *Origin of the Earth*, edited by H. E. Newsome, and J. H. Jones, pp. 175–194, Oxford University Press, New York, 1990.

Davies, G. F., On the Emergence Of Plate-Tectonics, *Geology*, *20* (11), 963–966, 1992.

Davies, G. F., Stirring geochemistry in mantle convection models with stiff plates and slabs, *Geochem. Cosmochem. Acta*, *66* (17), 3125–3142, 2002.

Davies, G. F., and M. Gurnis, Interaction of mantle dregs with convection—lateral heterogeneity at the core-mantle boundary, *Geophys. Res. Lett.*, *13* (13), 1517–1520, 1986.

Deschamps, F., J. Trampert, and P. J. Tackley, Thermo-chemical structure of the lower mantle: seismological evidence and consequences for geodynamics, in *Superplume: Beyond Plate Tectonics*, edited by D. A. Yuen, S. Maruyama, S. I. Karato, and B. F. Windley, pp. submitted, Springer, 2005.

Donnelly, K. E., S. L. Goldstein, C. H. Langmuir, and M. Spiegelman, Origin of enriched ocean ridge basalts and implications for mantle dynamics, *Earth & Planetary Science Letters*, *226* (3–4), 347–66, 2004.

Dupeyrat, L., C. Sotin, and E. M. Parmentier, Thermal and chemical convection in planetary mantles, *Journal of Geophysical Research, B, Solid Earth and Planets*, *100* (1), 497–520, 1995.

Elliott, T., A. Zindler, and B. Bourdon, Exploring the kappa conundrum: the role of recycling in the lead isotope evolution of the mantle, *Earth & Planetary Science Letters*, *169* (1–2), 129–145, 1999.

Fei, Y., J. Van Orman, J. Li, W. van Westeren, C. Sanloup, W. Minarik, K. Hirose, T. Komabayashi, M. Walter, and K. Funakoshi, Experimentally determined postspinel transformation boundary in Mg_2SiO_4 using MgO as an internal pressure standard and its geophysical implications, *J. Geophys. Res.*, *109* (B02305), doi:10.1029/2003JB002562, 2004.

Ferrachat, S., and Y. Ricard, Regular vs. chaotic mantle mixing, *Earth Planet. Sci. Lett.*, *155* (1–2), 75–86, 1998.

Ferrachat, S., and Y. Ricard, Mixing properties in the Earth's mantle: Effects of the viscosity stratification and of oceanic crust segregation, *Geochem. Geophys. Geosyst.*, *2*, 2000GC000092 [7490 words, 10 figures, 2 animations, 1 table], 2001.

Fleitout, L., A. Mambole, and U. Christensen, Phase changes around 670 km depth and segregation in the Earth's mantle, *EOS Trans. AGU, Fall Meeting Suppl.*, *81* (48), Abstract T12E-11, 2000.

Forte, A. M., and J. X. Mitrovica, New Inferences Of Mantle Viscosity From Joint Inversion Of Long-Wavelength Mantle Convection and Postglacial Rebound Data, *Geophysical Research Letters*, *23* (10), 1147–1150, 1996.

Forte, A. M., and J. X. Mitrovica, Deep-mantle high-viscosity flow and thermochemical structure inferred from seismic and geodynamic data, *Nature*, *410* (6832), 1049–56, 2001.

Gonnermann, H. M., M. Manga, and A. M. Jellinek, Dynamics and longevity of an initially stratified mantle—art. no. 1399, *Geophysical Research Letters*, *29* (10), 1399, 2002.

Griffiths, R. W., and I. H. Campbell, On the Dynamics Of Long-Lived Plume Conduits In the Convecting Mantle, *Earth and Planetary Science Letters*, *103* (1–4), 214–227, 1991.

Gurnis, M., Convective mixing in the Earth's mantle, Australian National University, Canberra, 1986.

Gurnis, M., Large-scale mantle convection and the aggregation and dispersal of supercontinents, *Nature*, *332* (6166), 695–699, 1988.

Gurnis, M., and G. F. Davies, The effect of depth-dependent viscosity on convective mixing in the mantle and the possible survival of primitive mantle, *Geophys. Res. Lett.*, *13* (6), 541–544, 1986.

Gurnis, M., S. Zhong, and J. Toth, On the competing roles of fault reactivation and brittle failure in generating plate tectonics from mantle convection, in *History and Dynamics of Global Plate Motions*, edited by M. A. Richards, R. Gordon, and R. van der Hilst, AGU, 2000.

Hansen, U., and D. A. Yuen, Numerical simulations of thermal-chemical instabilities at the core mantle boundary, *Nature*, *334* (6179), 237–240, 1988.

Hansen, U., and D. A. Yuen, Dynamical influences from thermal-chemical instabilities at the core-mantle boundary, *Geophys. Res. Lett.*, *16* (7), 629–632, 1989.

Hansen, U., and D. A. Yuen, Extended-Boussinesq thermal-chemical convection with moving heat sources and variable viscosity, *Earth and Planetary Science Letters*, *176* (3–4), 401–11, 2000.

Hansen, U., D. A. Yuen, S. E. Kroening, and T. B. Larsen, Dynamic consequences of depth-dependent thermal expansivity and viscosity on mantle circulations and thermal structure, *Phys. Earth Planet. Inter.*, *77* (3–4), 205–223, 1993.

Hernlund, J. W., C. Thomas, and P. J. Tackley, A doubling of the post-perovskite phase boundary and structure of the Earth's lowermost mantle, *Nature*, *434*, 882–886, 2005.

Hirth, G., and D. L. Kohlstedt, Water in the Oceanic Upper-Mantle—Implications For Rheology, Melt Extraction and the Evolution of the Lithosphere, *Earth and Planetary Science Letters*, *144* (1–2), 93–108, 1996.

Hofmann, A. W., Mantle geochemistry: the message from oceanic volcanism, *Nature*, *385* (6613), 219–29, 1997.

Hofmeister, A. M., Mantle values of thermal conductivity and the geotherm from phonon lifetimes, *Science*, *283* (5408), 1699–706, 1999.

Holmes, A., Radioactivity and Earth movements, XVII, *Trans. Geol. Soc. Glasgow*, *XVIII—Part III* (18), 559–606, 1931.

Houseman, G., The dependence of convection planform on mode of heating, *Nature*, *332* (6162), 346–349, 1988.

Ishii, M., and J. Tromp, Normal-mode and free-air gravity constraints on lateral variations in velocity and density of Earth's mantle, *Science*, *285* (5431), 1231–5, 1999.

Ito, E., M. Akaogi, L. Topor, and A. Navrotsky, Negative Pressure-Temperature Slopes For Reactions Forming $MgSiO_3$ Perovskite From Calorimetry, *Science*, *249* (4974), 1275–1278, 1990.

Ito, E., and E. Takahashi, Postspinel Transformations In the System Mg_2SiO_4-Fe_2SiO_4 and Some Geophysical Implications, *Journal Of Geophysical Research Solid Earth and Planets*, *94* (B8), 10637–10646, 1989.

Ito, G., and J. J. Mahoney, Flow and melting of a heterogeneous mantle: 1.Method and importance to the geochemistry of ocean island and mid-ocean ridge basalts, *Earth Planet. Sci. Lett.*, *230*, 29–46, 2005a.

Ito, G., and J. J. Mahoney, Flow and melting of a heterogeneous mantle: 2. Implications for a chemically nonlayered mantle, *Earth Planet. Sci. Lett.*, *230*, 47–63, 2005b.

Jarvis, G. T., and W. R. Peltier, Mantle Convection As a Boundary-Layer Phenomenon, *Geophysical Journal Of the Royal Astronomical Society*, *68* (2), 389–427, 1982.

Karato, S., and B. B. Karki, Origin of lateral variation of seismic wave velocities and density in the deep mantle, *Journal of Geophysical Research*, *106* (B10), 21771–83, 2001.

Katsura, T., H. Yamada, T. Shinmei, A. Kubo, S. Ono, M. Kanzaki, A. Yoneda, M. J. Walter, E. Ito, S. Urakawa, K. Funakoshi, and W. Utsumi, Post-spinel transition in Mg_2SiO_4 determined by high P-T in situ X-ray diffractometry, *Physics of the Earth & Planetary Interiors*, *136* (1–2), 11–24, 2003.

Kellogg, J. B., S. B. Jacobsen, and R. J. O'Connell, Modeling the distribution of isotopic ratios in geochemical reservoirs, *Earth and Planetary Science Letters*, *204* (1–2), 183–202, 2002.

Kellogg, L. H., B. H. Hager, and R. D. van der Hilst, Compositional stratification in the deep mantle, *Science*, *283* (5409), 1881–4, 1999.

Kellogg, L. H., and D. L. Turcotte, Mixing and the Distribution Of Heterogeneities In a Chaotically Convecting Mantle, *Journal of Geophysical Research Solid Earth and Planets*, *95* (B1), 421–432, 1990.

Kennett, B. L. N., S. Widiyantoro, and R. D. van der Hilst, Joint seismic tomography for bulk sound and shear wave speed in the Earth's mantle, *J. Geophys. Res. (USA)*, *103* (B6), 12469–93, 1998.

Kesson, S. E., J. D. F. Gerald, and J. M. Shelley, Mineralogy and dynamics of a pyrolite lower mantle, *Nature*, *393* (6682), 252–255, 1998.

King, S. D., The Viscosity Structure Of the Mantle, *Reviews Of Geophysics*, *33*, 11–17, 1995.

Korenaga, J., Energetics of mantle convection and the fate of fossil heat, *Geophysical Research Letters*, *30* (8), 1437, 2003.

Korenaga, J., Firm mantle plumes and the nature of the core-mantle boundary region, *Earth Planet. Sci. Lett.*, *232* (1–2), 29–37, 2005.

Labrosse, S., Thermal and magnetic evolution of the Earth's core, *Phys. Earth Planet. Int.*, *140*, 127–143, 2003.

Labrosse, S., J. P. Poirier, and J. L. Le Mouel, On cooling of the Earth's core, *Physics of the Earth and Planetary Interiors*, *99* (1–2), 1–17, 1997.

Lay, T., and D. V. Helmberger, A shear velocity discontinuity in the lower mantle, *Geophys. Res. Lett.*, *10* (1), 63–66, 1983.

Le Bars, M., and A. Davaille, Stability of thermal convection in two superimposed miscible viscous fluids, *Journal of Fluid Mechanics*, *471*, 339–63, 2002.

Lenardic, A., On the partitioning of mantle heat loss below oceans and continents over time and its relationship to the Archaean paradox, *Geophysical Journal International*, *134* (3), 706–720, 1998.

Lowman, J. P., and G. T. Jarvis, Mantle Convection Flow Reversals Due to Continental Collisions, *Geophysical Research Letters*, *20* (19), 2087–2090, 1993.

Machetel, P., and P. Weber, Intermittent Layered Convection In a Model Mantle With an Endothermic Phase-Change At 670 Km, *Nature*, *350* (6313), 55–57, 1991.

Mambole, A., and L. Fleitout, Petrological layering induced by an endothermic phase transition in the Earth's mantle–art. no. 2044, *Geophysical Research Letters*, *29* (22), 2044, 2002.

Martin, H., Effect of steeper archean geothermal gradient on geochemistry of subduction zone magmas, *Geology*, *14*, 753–756, 1986.

Masters, G., G. Laske, H. Bolton, and A. Dziewonski, The relative behavior of shear velocity, bulk sound speed, and compressional velocity in the mantle: Implications for chemical and thermal structure., in *Mineral Physics and Seismic Tomography*, edited by S.Karato et al., AGU, 2000.

McNamara, A. K., and S. Zhong, Thermochemical structures within a spherical mantle: Superplumes or piles?, *J. Geophys. Res.*, submitted, 2004.

McNamara, A. K., and S. Zhong, Thermochemical piles under Africa and the Pacific, submitted, 2005.

Meibom, A., and D. L. Anderson, The statistical upper mantle assemblage, *Earth & Planetary Science Letters*, *217* (1–2), 123–139, 2004.

Metcalfe, G., C. R. Bina, and J. M. Ottino, Kinematic Considerations For Mantle Mixing, *Geophysical Research Letters*, *22* (7), 743–746, 1995.

Montelli, R., G. Nolet, F. A. Dahlen, G. Masters, E. R. Engdahl, and S. H. Hung, Finite-frequency tomography reveals a variety of plumes in the mantle, *Science*, *303* (5656), 338–343, 2004.

Moresi, L., and V. Solomatov, Mantle convection with a brittle lithosphere—Thoughts on the global tectonic styles of the Earth and Venus, *Geophys. J. Int.*, *133* (3), 669–682, 1998.

Moresi, L. N., and V. S. Solomatov, Numerical investigation of 2D convection with extremely large viscosity variations, *Phys. Fluids*, *7* (9), 2154–2162, 1995.

Morgan, W. J., Convection plumes in the lower mantle, *Nature*, *230*, 42–43, 1971.

Murakami, M., K. Hirose, K. Kawamura, N. Sata, and Y. Ohishi, Post-perovskite phase transition in $MgSiO_3$, *Science*, *304* (5672), 855–858, 2004.

Nakagawa, T., and P. J. Tackley, Effects of a perovskite-post perovskite phase change near the core–mantle boundary on compressible mantle convection, *Geophys. Res. Lett.*, *31, L16611*, doi:10.1029/2004GL020648, 2004a.

Nakagawa, T., and P. J. Tackley, Effects of thermo-chemical mantle convection on the thermal evolution of the Earth's core, *Earth Planet. Sci. Lett.*, *220*, 107–119, 2004b.

Nakagawa, T., and P. J. Tackley, Deep mantle heat flow and thermal evolution of Earth's core based on thermo-chemical multiphase mantle convection, *Geochem., Geophys., Geosys.*, *6* (8), Q08003, doi:10.1029/2005GC000967, 2005a.

Nakagawa, T., and P. J. Tackley, The interaction between the post-perovskite phase change and a thermo-chemical boundary layer near the core-mantle boundary, *Earth Planet. Sci. Lett.*, in press, 2005b.

Nataf, H. C., and F. M. Richter, Convection Experiments In Fluids With Highly Temperature-Dependent Viscosity and the Thermal Evolution Of the Planets, *Physics Of the Earth and Planetary Interiors*, *29* (3–4), 320–329, 1982.

Nimmo, F., G. D. Price, J. Brodholt, and D. Gubbins, The influence of potassium on core and geodynamo evolution, *Geophys. J. Int.*, *156*, 363–376, 2004.

Oganov, A. R., and S. Ono, Theoretical and experimental evidence for a post-perovskite phase of $MgSiO_3$ in Earth's D″ layer, *Nature*, *430* (6998), 445–8, 2004.

Ogawa, M., A bifurcation in the coupled magmatism-mantle convection system and its implications for the evolution of the Earth's upper mantle, *Physics of the Earth and Planetary Interiors*, *102* (3–4), 259–76, 1997.

Ogawa, M., Coupled magmatism-mantle convection system with variable viscosity, *Tectonophysics*, *322*, 1–18, 2000a.

Ogawa, M., Numerical models of magmatism in convecting mantle with temperature-dependent viscosity and their implications for Venus and Earth, *Journal of Geophysical Research*, *105* (E3), 6997–7012, 2000b.

Ogawa, M., Chemical stratification in a two-dimensional convecting mantle with magmatism and moving plates, *Journal of Geophysical Research*, *108* (B12), ETG5-1-20, 2003.

Ogawa, M., and H. Nakamura, Thermochemical regime of the early mantle inferred from numerical models of the coupled magmatism-mantle convection system with the solid-solid phase transitions at depths around 660 km, *Journal of Geophysical Research*, *103* (B6), 12161–80, 1998.

Ogawa, M., G. Schubert, and A. Zebib, Numerical simulations of 3-dimensional thermal convection in a fluid with strongly temperature-dependent viscosity, *J. Fluid Mech.*, *233*, 299–328, 1991.

Olson, P., and C. Kincaid, Experiments on the interaction of thermal convection and compositional layering at the base of the mantle, *J. Geophys. Res.*, *96* (B3), 4347–4354, 1991.

Olson, P., and H. Singer, Creeping Plumes, *Journal Of Fluid Mechanics*, *158* (SEP), 511–531, 1985.

Olson, P., D. A. Yuen, and D. Balsiger, Mixing Of Passive Heterogenities By Mantle Convection, *Journal Of Geophysical Research*, *89* (B1), 425–436, 1984.

Ono, S., E. Ito, and T. Katsura, Mineralogy of subducted basaltic crust (MORB) from 25 to 37 GPa, and chemical heterogeneity of the lower mantle, *Earth Planet. Sci. Lett.*, *190* (1–2), 57–63, 2001.

Ono, S., Y. Oshishi, M. Isshiki, and T. Watanuki, In situ X-ray observations of phase assemblages in peridotite and basalt compositions at lower mantle conditions: Implications for density of subducted oceanic plate, *J. Geophys. Res.*, *110* (B02208), doi:10.1029/2004JB0003196, 2005.

Parmentier, E. M., C. Sotin, and B. J. Travis, Turbulent 3-D thermal convection in an infinite Prandtl number, volumetrically heated fluid—Implications for mantle dynamics, *Geophys. J. Int.*, *116* (2), 241–251, 1994.

Parmentier, E. M., D. L. Turcotte, and K. E. Torrance, Numerical experiments on the structure of mantle plumes, *Journal of Geophysical Research*, *80* (32), 4417–4424, 1975.

Parmentier, E. M., D. L. Turcotte, and K. E. Torrance, Studies of finite amplitude non-Newtonian thermal convection with application to convection in the Earth's mantle, *Journal of Geophysical Research*, *81* (11), 1839–1846, 1976.

Peltier, W. R., and L. P. Solheim, Mantle Phase-Transitions and Layered Chaotic Convection, *Geophysical Research Letters*, *19* (3), 321–324, 1992.

Ratcliff, J. T., P. J. Tackley, G. Schubert, and A. Zebib, Transitions in thermal convection with strongly variable viscosity, *Phys. Earth Planet. Inter.*, *102*, 201–212, 1997.

Regenauer-Lieb, K., and D. A. Yuen, Modeling shear zones in geological and planetary sciences: solid- and fluid-thermal-mechanical approaches [Review], *Earth-Science Reviews*, *63* (3–4), 295–349, 2003.

Regenauer-Lieb, K., D. A. Yuen, and J. Branlund, The initiation of subduction: Criticality by addition of water?, *Science*, *294* (5542), 578–580, 2001.

Richards, M. A., W.-S. Yang, J. R. Baumgardner, and H. P. Bunge, The role of a low viscosity zone in stabilizing plate tectonics: Implications for comparative terrestrial planetology, *Geochem. Geophys. Geosyst.*, *2*, 2000GC000115, 2001.

Ringwood, A. E., Phase-Transformations and Their Bearing On the Constitution and Dynamics Of the Mantle, *Geochimica Et Cosmochimica Acta*, *55* (8), 2083–2110, 1991.

Roberts, P. H., Convection in horizontal layers with internal heat generation. Theory., *J. Fluid Mech.*, *30*, 33–39, 1967.

Ruepke, L., J. Phipps Morgan, M. Hort, and J. A. D. Connolly, Serpentine and the subduction zone water cycle, *Earth Planet. Sci. Lett.*, *submitted*, 2004.

Samuel, H., and C. G. Farnetani, Thermochemical convection and helium concentrations in mantle plumes, *Earth and Planetary Science Letters*, *207* (1–4), 39–56, 2003.

Schmalzl, J., and U. Hansen, Mixing the Earths mantle by thermal-convection—a scale-dependent phenomenon, *Geophys. Res. Lett.*, *21* (11), 987–990, 1994.

Schmalzl, J., G. A. Houseman, and U. Hansen, Mixing Properties Of 3-Dimensional (3-D) Stationary Convection, *Physics Of Fluids*, *7* (5), 1027–1033, 1995.

Schmalzl, J., G. A. Houseman, and U. Hansen, Mixing in vigorous, time-dependent three-dimensional convection and application to Earth's mantle, *J. Geophys. Res.*, *101* (B10), 21847–58, 1996.

Schott, B., D. A. Yuen, and A. Braun, The influences of composition-sand temperature- dependent rheology in thermal-chemical convection on entrainment of the D″-layer, *Physics of the Earth & Planetary Interiors*, *129* (1–2), 43–65, 2002.

Schubert, G., and A. Zebib, Thermal convection of an internally heated infinite Prandtl number fluid in a spherical shell, *Geophys. Astrophys. Fluid Dyn.*, *15*, 65–90, 1980.

Segatz, M., T. Spohn, M. N. Ross, and G. Schubert, Tidal dissipation, surface heat flow, and figure of viscoelastic models of Io, *Icarus*, *75* (2), 187–206, 1988.

Sidorin, I., M. Gurnis, D. V. Helmberger, and D. Xiaoming, Interpreting D″ seismic structure using synthetic waveforms computed from dynamic models, *Earth and Planetary Science Letters*, *163* (1–4), 31–41, 1998.

Solheim, L. P., and W. R. Peltier, Avalanche Effects In Phase-Transition Modulated Thermal-Convection—a Model Of Earths Mantle, *Journal Of Geophysical Research-Solid Earth*, *99* (B9), 18203–18203, 1994.

Solomatov, V. S., Scaling of temperature-dependent and stress-dependent viscosity convection, *Phys. Fluids*, *7* (2), 266–274, 1995.

Solomatov, V. S., Can Hotter Mantle Have a Larger Viscosity, *Geophysical Research Letters*, *23* (9), 937–940, 1996.

Solomatov, V. S., Grain size-dependent viscosity convection and the thermal evolution of the Earth, *Earth and Planetary Science Letters*, *191*, 203–212, 2001.

Solomatov, V. S., and D. J. Stevenson, Nonfractional Crystallization Of a Terrestrial Magma Ocean, *Journal Of Geophysical Research-Planets*, *98* (E3), 5391–5406, 1993.

Spiegelman, M., P. B. Kelemen, and E. Aharonov, Causes and consequences of flow organization during melt transport: the reaction infiltration instability in compactible media, *Journal of Geophysical Research*, *106* (B2), 2061–77, 2001.

Steinbach, V., D. A. Yuen, and W. L. Zhao, Instabilities From Phase-Transitions and the Timescales Of Mantle Thermal Evolution, *Geophysical Research Letters*, *20* (12), 1119–1122, 1993.

Tackley, P. J., Mantle dynamics: Influence of the transition zone, *US National Report to the IUGG*, *1991–1994*, 275–282, 1995.

Tackley, P. J., Effects of strongly variable viscosity on three-dimensional compressible convection in planetary mantles, *J. Geophys. Res.*, *101*, 3311–3332, 1996.

Tackley, P. J., Three-dimensional simulations of mantle convection with a thermochemical CMB boundary layer: D″?, in *The Core-Mantle Boundary Region*, edited by M. Gurnis, M. E. Wysession, E. Knittle, and B. A. Buffett, pp. 231–253, American Geophysical Union, 1998.

Tackley, P. J., Mantle convection and plate tectonics: towards an integrated physical and chemical theory, *Science*, *288*, 2002, 2000a.

Tackley, P. J., Self-consistent generation of tectonic plates in time-dependent, three-dimensional mantle convection simulations Part 1: Pseudo-plastic yielding, *Geochem., Geophys., Geosys.*, *1*, 2000GC000036, 2000b.

Tackley, P. J., Strong heterogeneity caused by deep mantle layering, *Geochem. Geophys. Geosystems*, *3* (4), 10.1029/2001GC000167, 2002.

Tackley, P. J., and S. D. King, Testing the tracer ratio method for modeling active compositional fields in mantle convection simulations, *Geochem. Geophys. Geosyst.*, *4* (4), doi:10.1029/2001GC000214, 2003.

Tackley, P. J., D. J. Stevenson, G. A. Glatzmaier, and G. Schubert, Effects of an endothermic phase transition at 670 km depth in a spherical model of convection in the Earth's mantle, *Nature, 361* (6414), 699–704, 1993.

Tackley, P. J., D. J. Stevenson, G. A. Glatzmaier, and G. Schubert, Effects of multiple phase transitions in a 3-dimensional spherical model of convection in Earth's mantle, *J. Geophys. Res., 99* (B8), 15877–15901, 1994.

Tackley, P. J., and S. Xie, The thermo-chemical structure and evolution of Earth's mantle: constraints and numerical models, *Phil. Trans. R. Soc. Lond. A, 360*, 2593–2609, 2002.

Tackley, P. J., and S. Xie, Stag3D: A code for modeling thermo-chemical multiphase convection in Earth's mantle, in *Second MIT Conference on Computational Fluid and Solid Mechanics*, pp. 1–5, Elsevier, MIT, 2003.

Thomas, C., E. J. Garnero, and T. Lay, High-resolution imaging of lowermost mantle structure under the Cocos plate, *J. Geophys. Res., 109* (B08307), doi:10.1029/2004JB003013, 2004.

Tozer, D. C., The present thermal state of the terrestrial planets, *Phys. Earth Planet. Inter., 6*, 182–197, 1972.

Trampert, J., F. Deschamps, J. S. Resovsky, and D. Yuen, Probabilistic tomography maps significant chemical heterogeneities in the lower mantle, *Science, 306*, 853–856, 2004.

Travis, B., S. Weinstein, and P. Olson, 3-dimensional convection planforms with internal heat generation, *Geophys. Res. Lett., 17* (3), 243–246, 1990.

Trompert, R., and U. Hansen, Mantle convection simulations with rheologies that generate plate-like behavior, *Nature, 395* (6703), 686–689, 1998.

Turcotte, D. L., and G. Schubert, *Geodynamics: Applications of Continuum Physics to Geological Problems*, Wiley, New York, 1982.

van den Berg, A. P., M. H. Jacobs, and B. H. de Jong, Numerical models of mantle convection based on thermodynamic data for the $MgOSiO_2$ olivine-pyroxene system, *EOS Trans. AGU, Fall Meeting Suppl., 83* (47), Abstract MR72B-1041, 2002.

van den Berg, A. P., D. A. Yuen, and V. Steinbach, The effects of variable thermal conductivity on mantle heat-transfer, *Geophysical Research Letters, 28* (5), 875–8, 2001.

van der Hilst, R. D., S. Widlyantoro, and E. R. Engdahl, Evidence for deep mantle circulation from global tomography, *Nature, 386* (6625), 578–84, 1997.

van Keken, P. E., and C. J. Ballentine, Whole-mantle versus layered mantle convection and the role of a high-viscosity lower mantle in terrestrial volatile evolution, *Earth Planet. Sci. Lett., 156* (1–2), 19–32, 1998.

van Keken, P. E., and C. J. Ballentine, Dynamical models of mantle volatile evolution and the role of phase transitions and temperature-dependent rheology, *J. Geophys. Res., 104* (B4), 7137–51, 1999.

van Keken, P. E., B. Kiefer, and S. M. Peacock, High-resolution models of subduction zones: Implications for mineral hydration reactions and the transport of water into the deep mantle, *Geochem. Geophys. Geosyst., 3* (10), DOI 10.1029/2001GC000256, 2002.

van Keken, P. E., S. D. King, H. Schmeling, U. R. Christensen, D. Neumeister, and M. P. Doin, A comparison of methods for the modeling of thermochemical convection, *J. Geophys. Res., 102* (B10), 22477–95, 1997.

van Thienen, P., A. P. van den Berg, J. H. de Smet, J. van Hunen, and M. R. Drury, Interaction between small-scale mantle diapirs and a continental root, *Geochem. Geophys. Geosyst., 4* (2), doi:10.1029/2002GC000338, 2003.

van Thienen, P., A. P. van den Berg, and N. J. Vlaar, On the formation of continental silicic melts in thermo-chemical mantle convection models: implications for early Earth, *Tectonophysics, 394* (1–2), 111–124, 2004a.

van Thienen, P., A. P. van den Berg, and N. J. Vlaar, Production and recycling of oceanic crust in the early Earth, *Tectonophysics, 386* (1–2), 41–65, 2004b.

Weinstein, S. A., Induced compositional layering in a convecting fluid layer by an endothermic phase-transition, *Earth Planet. Sci. Lett., 113* (1–2), 23–39, 1992.

Weinstein, S. A., Catastrophic Overturn Of the Earths Mantle Driven By Multiple Phase-Changes and Internal Heat-Generation, *Geophysical Research Letters, 20* (2), 101–104, 1993.

Whitehead, J. A., and D. S. Luther, Dynamics of laboratory diapir and plume models, *J. Geophys. Res., 80* (5), 705–717, 1975.

Xie, S., and P. J. Tackley, Evolution of helium and argon isotopes in a convecting mantle, *Phys. Earth Planet. Inter., 146* (3–4), 417–439, 2004a.

Xie, S., and P. J. Tackley, Evolution of U-Pb and Sm-Nd systems in numerical models of mantle convection, *J. Geophys. Res., 109, B11204*, doi:10.1029/2004JB003176, 2004b.

Zebib, A., G. Schubert, J. L. Dein, and R. C. Paliwal, Character and stability of axisymmetric thermal convection in spheres and spherical shells, *Geophys. Astrophys. Fluid Dyn., 23*, 1–42, 1983.

Zebib, A., G. Schubert, and J. M. Straus, Infinite Prandtl number thermal convection in a spherical shell, *J. Fluid Mech., 97* (2), 257–277, 1980.

Zhong, S., and M. Gurnis, Dynamic feedback between a continent-like raft and thermal convection, *J. Geophys. Res., 98* (B7), 12219–12232, 1993.

J. W. Hernlund and Shunxing Xie, Department of Earth and Space Sciences, University of California, Los Angeles, California 90095-1567, USA. (hernlund@ess.ucla.edu)

Takashi Nakagawa, Department of Earth and Planetary Sciences, University of Tokyo, Japan.

P. J. Tackley, Department of Earth and Space Sciences and Institute of Geophysics and Planetary Physics, University of California, Los Angeles, California 90095-1567, USA. (ptackley@ucla.edu)

Heterogeneous Lowermost Mantle: Compositional Constraints and Seismological Observables

H. Samuel[1] and C. G. Farnetani

Laboratoire de Dynamique des Systèmes Géologiques, Institut de Physique du Globe de Paris, Paris, France

D. Andrault

Institut de Minéralogie et de Physique des Milieux Condensés, Paris, France

Several seismological observations indicate the existence of compositional heterogeneities in the lowermost mantle, in particular, the anticorrelation between bulk sound and shear wave velocity anomalies, and anomalously high values (i.e., >2.7) of the ratio $R = d\ln V_S / d\ln V_P$. Constraining the composition of such heterogeneous material is fundamental to determine its origin and its possible role on the dynamical evolution of the Earth's mantle. In this paper we propose a new approach to constrain the composition of chemically denser material in the lower mantle. Using geodynamical and seismological constraints, we show that the denser material has to be enriched in both iron and silica with respect to a pyrolitic lower mantle. The required enrichment is reduced if we consider that at high pressure Al-perovskite decreases the iron–magnesium partition coefficient between magnesiowüstite and perovskite. We then apply the estimated composition to the distribution of chemical heterogeneities calculated by our thermochemical convection model. In the deep mantle we predict broad seismic velocity anomalies and strong lateral velocity variations. Moreover, we find that areas of anticorrelation are associated with upwelling mantle flow, in agreement with tomographic studies. The calculated R ratio varies laterally and may locally have values >2.7, often associated with areas of anticorrelation. Our results compare well with seismic observations and provide a way to reconcile apparent discrepancies between global tomographic models. Finally, we suggest that only an enrichment in iron and silica in the lowermost mantle is required to explain seismological observations.

1. INTRODUCTION

Seismic tomography is a powerful tool for imaging the Earth's deep mantle structure. Despite technical differences

(i.e., data treatment procedures, parametrization, and inversion method), tomographic models display many common features, for example, the increase of the Root Mean Square (RMS) of body waves seismic velocity anomalies below a depth of 2000 km [*Grand et al.*, 1997; *Kennett et al.*, 1998; *Masters et al.*, 2000; *Mégnin and Romanowic*, 2000] and the increase of their wavelength [*Mégnin and Romanowicz*, 2000; *Su and Dziewonski*, 1991; *Li and Romanowicz*, 1996]. These observations suggest the presence of broad seismic velocity anomalies in the deep mantle, also generally

---

[1] Now at Department of Geology and Geophysics, Yale University, New Haven, Connecticut, USA.

Earth's Deep Mantle: Structure, Composition, and Evolution
Geophysical Monograph Series 160
Copyright 2005 by the American Geophysical Union
10.1029/160GM08

observed by different tomographic models [*Grand et al.*, 1997; *van der Hilst et al.*, 1997; *Masters et al.*, 2000; *Mégnin and Romanowicz*, 2000] and inferred by normal mode and free-air gravity data [*Ishii and Tromp*, 1999]. The large size of these anomalies is inconsistent with purely thermal convection and rather suggests the existence of chemical heterogeneities in the lowermost mantle. Seismic tomography provides supplementary information on the nature of deep mantle heterogeneities by using ratios and relative variations of seismic velocities for real and theoretical body waves such as (i) the ratio R of the relative variations of S-wave to P-wave velocities. Horizontally averaged R profiles show that R increases below 2000-km depth from 1.7 to values >2.7 near the Core–Mantle Boundary (CMB) [*Masters et al.*, 2000; *Saltzer et al.*, 2001; *Romanowicz*, 2001]. Such high R values cannot be explained by temperature differences alone but require the presence of compositional heterogeneities [*Masters et al.*, 2000] and possibly anelasticity [*Karato and Karki*, 2001]. (ii) The anticorrelation between bulk sound speed and shear wave velocity anomalies in the lowermost mantle is also derived from several tomographic models [*Kennett et al.*, 1998; *Masters et al.*, 2000; *Saltzer et al.*, 2001; *Antolik et al.*, 2003] and suggests the presence of chemical density heterogeneities.

While there is a general agreement on the RMS profiles and on the presence of broad seismic velocity anomalies in the lowermost mantle, high R values and the anticorrelation between V_s and V_ϕ are not commonly shown by tomographic models (see *Masters et al.* [2000] for a review). However, these differences are not necessarily contradictory. Indeed, *Saltzer et al.* [2001] found that indicators of compositional heterogeneity (i.e., $R > 2.7$ and anticorrelation between V_ϕ and V_s) can be hidden in horizontally averaged profiles because of their lateral variability. They found that areas away from slab regions present both $R > 2.5$ and anticorrelation between V_s and V_ϕ, while areas near slab regions have $R < 2.5$ and no anticorrelation. This could explain, among other things, the apparent discrepancies between different tomographic models, concerning R profiles and the presence of anticorrelation.

Tomographic models thus strongly suggest the existence of compositional heterogeneity in the deep mantle, which is also required by geochemical considerations. The large differences in trace elements and noble gases between Mid-Ocean Ridge Basalts (MORBs) and Ocean Island Basalts (OIBs) require the presence of at least two distinct reservoirs for billions of years (see *Hofmann* [1997] for a review). Laboratory experiments (e.g., *Lebars and Davaille* [2002] and references therein) and 2D–3D numerical simulations (e.g., *Tackley* [2002], *Samuel and Farnetani* [2003], and references therein) have investigated the long-term

stability and stirring of chemically denser material. A major conclusion is that even a small excess of chemical density (~1%) profoundly affects the nature of mantle convection. Under certain conditions, thermochemical convection provides a way to maintain separated reservoirs for billions of years. Assuming a relatively undegassed denser material, *Samuel and Farnetani* [2003] show that thermochemical convection can explain the observed helium ratios for MORB and OIB.

It is therefore difficult to interpret several geophysical and geochemical observations without the presence of chemically denser material in the lowermost mantle. Previous studies investigated possible mechanisms that could generate a chemical density excess, $\Delta\rho_\chi$, in the lower mantle. Considering a lower mantle assemblage of perovskite $(Fe,Mg)SiO_3$ and magnesiowüstite $(Fe,Mg)O$, *Forte and Mitrovica* [2001] and more recently *Deschamps and Trampert* [2003] suggested that $\Delta\rho_\chi$ could be due to variations of the iron and silica content as proposed by *Kellogg et al.* [1999], while *Sidorin and Gurnis* [1998] claimed the additional presence of SiO_2 stishovite to satisfy their geodynamical and seismological constraints. Furthermore, *Karato and Karki* [2001] concluded that variations in Si and Fe content alone cannot explain values of $R > 2.7$, and proposed that the presence of Ca-perovskite could satisfy this constraint. It is hard to directly compare these results, because they ensue from different reasonings: for instance, the chemical density contrast considered by *Sidorin and Gurnis* [1998] is much higher than the one required by *Forte and Mitrovica* [2001]. Anelasticity is neglected in *Sidorin and Gurnis* [1998] and *Deschamps and Trampert* [2003], contrary to *Karato and Karki* [2001] and *Forte and Mitrovica* [2001]. Finally, the constraints considered by these studies for the composition of the chemically denser material differ widely.

In this paper, we focus on the lower mantle and investigate the effect of compositional heterogeneities on seismological models and observations. Similar to *Forte and Mitrovica* [2001] we assume that the lower mantle is an assemblage of the main perovskite and magnesiowüstite phases and do not consider the effect of less abundant components such as Ca and Al. We also assume that the composition of the heterogeneous denser material in the lowermost mantle is due to variations in Fe, Mg, and Si content, relative to a pyrolitic lower mantle. First, we model thermochemical convection from 2 Gy ago (before the present; B.P.) to the present day, in order to constrain the chemical density excess $\Delta\rho_\chi$ required for the material to remain stable, and to obtain the temperature field, geometry, and distribution of the chemical heterogeneities. Second, we calculate the seismic velocity anomalies of P-waves, S-waves, and bulk sound for a wide range of compositions. Third, we use geodynamical and seismological considerations to constrain the composition of the dense material. Finally, we assign the calculated composition to the dense material in our geodynamical model. Our predicted seismic velocity anomalies, R profiles and distributions, and anticorrelation are then compared with seismological observations.

1.1. Convection Model

We use the numerical code for solid-state convection STAG3D by P. Tackley, which has been described in detail in *Tackley* [2002] and references therein. The code, in cartesian geometry, solves the equations of conservation of mass, conservation of momentum, conservation of energy, and advection of a compositional field. In our 2D convection calculations, we use 25 active tracer particles per cell to model the presence of chemically denser material. The code allows vertical compressibility; therefore, the density ρ, the thermal expansion α, and the thermal conductivity k are depth-dependent, as described in the next section.

The characteristic scales used to normalize the governing equations are the mantle depth $D = 2890$ km, the surface density ρ_0, and the superadiabatic temperature drop $\Delta T = 2500$ K. A thermal diffusion timescale is used: D^2/κ, where $\kappa = k/\rho C_p$ is the thermal diffusivity, and the specific heat $C_p = 1200$ J kg^{-1} K^{-1} is assumed constant. The internal heating is scaled over $\rho D^2/k\Delta T$. Three nondimensional numbers appear from the normalization of conservation equations: the surface dissipation number

$$Di_0 = \frac{\alpha_0 g D}{C_P}, \qquad (1)$$

the Rayleigh number, based on surface parameters

$$Ra = \frac{\rho_0 \alpha_0 \Delta T g D^3}{\kappa_0 \eta_r}, \qquad (2)$$

and the surface buoyancy number

$$B = \frac{\Delta\rho_\chi}{\rho_0 \alpha_0 \Delta T}, \qquad (3)$$

where α_0 is the surface thermal expansion, the reference viscosity $\eta_r = 1.13 \times 10^{22}$ Pa s, and the gravitational acceleration $g = 10$ m s^{-2}. $\Delta\rho_\chi$ is the chemical density excess with respect to the reference density. Therefore, temperature and compositional effect on density ρ are calculated with the linearized equation of state: $\rho = \rho_0[1 - \alpha(T - T_0) + (\Delta\rho_\chi/\rho_{0\chi})]$.

Similar to *Samuel and Farnetani* [2003], homogeneous internal heating H is linked to concentrations of ^{238}U, ^{235}U, ^{232}Th, and ^{40}K, which vary as a function of time because of

radioactive decay (using values of radioactive decay constants listed in *Turcotte and Schubert* [1982]). This yields values of H from 31 at $t = 2$ Gy B.P. to 19, corresponding to actual concentrations of U = 17 ppb (with ^{238}U/^{235}U = 135.88), ^{232}Th = 65 ppb, and ^{40}K = 25 ppb (using the heat production rates for ^{238}U, ^{235}U, ^{232}Th, and ^{40}K given in *Turcotte and Schubert* [1982]).

Our model domain is constituted of 768×128 square cells, providing the resolution of 22.6 km/cell. At the top and bottom surfaces the temperature is constant and we impose zero vertical velocity and horizontal free slip. At the sidewalls, periodic boundary conditions for temperature and velocity are imposed.

1.2. Thermodynamical Model and Parameters

1.2.1. Thermodynamical model. The code uses a thermodynamical model for depth-dependent parameters (temperature, density, and thermal expansion). The depth dependence of temperature is assumed to be adiabatic:

$$\frac{\partial T'}{\partial z'} = -DiT', \qquad (4)$$

where the primes denote nondimensional values. Density varies with depth according to the Adams–Williamson equation of state:

$$\frac{\partial \rho'}{\partial z'} = -\frac{Di}{\gamma}\rho', \qquad (5)$$

where $\gamma = \alpha K_S/\rho C_P$ is the thermodynamical Grüneisen parameter, with the assumption that $\gamma\rho$ is constant. K_S is the adiabatic bulk modulus.

Thermal expansion α varies according to the semi-empirical relation [*Anderson et al.*, 1992]:

$$\alpha = \alpha_0\, exp\left[-\frac{\delta_{T_0}}{n}\left(1 - \left(\frac{\rho_0}{\rho}\right)^n\right)\right], \qquad (6)$$

where $\delta_{T_0} = -(1/K_T)(\partial K_T/\partial T)_P$ is the isothermal Anderson–Grüneisen parameter at ambient conditions, n is a constant equal to 1.4, and K_T is the isothermal bulk modulus.

1.2.2. Thermodynamical parameters. Our objective is to investigate the effect of the presence of chemically denser material on seismic velocity anomalies in the lowermost mantle. For simplicity, we do not consider phase changes and variable viscosity. Since we are interested in lower mantle compositions, we consider only lower mantle assemblages with two phases: perovskite (Fe,Mg)SiO$_3$ and magnesiowüstite (Fe,Mg)O. The third component, CaSiO$_3$ perovskite, has elastic parameters close to (Fe,Mg)SiO$_3$ perovskite and is thus neglected, which appears to be a reasonable assumption, as shown by *Deschamps and Trampert* [2003]. Therefore, the values of the parameters chosen for our convection calculations must be consistent with respect to this assemblage. Assuming a pyrolitic composition for the reference mantle, the density of the assemblage of perovskite and magnesiowüstite at room conditions is $\rho_0 = 4160$ kg m^{-3} (using relations given in Table 1).

Surface values of ρ_0, γ_0, δ_{T_0}, and α_0, listed in Table 2, were evaluated assuming an assemblage of perovskite and magnesiowüstite. We take $\delta_{T_0} = 4.6$ for the mantle composition, which falls well within $\delta_{T_0}^{pv} = 4.1$ [*Gillet et al.*, 2000] and $\delta_{T_0}^{mw} = 6$ [*Chopelas and Boehler*, 1992] at ambient conditions. We estimated $\gamma_0 = 1.33$ by doing a Voigt average of $\gamma_0^{pv} = 1.31$ and $\gamma_0^{mw} = 1.41$ [*Jackson*, 1998] with a volumic proportion of 80% perovskite. The thermal expansion coefficient at ambient conditions for perovskite ranges between ~2 × 10^{-5} K^{-1} and 4 × 10^{-5} K^{-1}, depending on its iron content [*Karki and Stixrude*, 1999], while for magnesiowüstite, studies seem to agree for a value close to 3 × 10^{-5} K^{-1} [*Hama and Suito*, 1999], therefore we use $\alpha_0 = 2.7 \times 10^{-5}$ K^{-1}.

Using equations (4)–(6), we obtain $\rho = 5620$ kg m^{-3} at the CMB, which is in good agreement with PREM [*Dziewonski and Anderson*, 1981]. The calculated thermal expansion coefficient $\alpha = 0.9 \times 10^{-5}$ K^{-1} at the CMB is also consistent with high-pressure experiments [*Chopelas and Boehler*, 1992]. The temperature at the CMB, $T_b = 3470$ K, is about 100 K lower than a previous estimation [*Brown and Shankland*, 1981]. This choice of physical and thermodynamical parameters leads to a surface $Ra = 10^7$.

2. CALCULATION OF SEISMIC VELOCITIES

2.1. Formalism

In order to compare our thermochemical model with seismological observables, we calculate the seismic velocities for a wide range of mineralogical compositions combined with the temperature field obtained with the convection code. Making the assumption that the lower mantle is seismi-

Table 1. Iron Dependence of Physical Parameters for Perovskite and Magnesiowüsite (*Wang and Weidner* [1996] and references therein).

Parameters	Mg$_{xMg}$Fe$_{xFe}$SiO$_3$	Mg$_{xMg}$Fe$_{xFe}$O
$K_0(x_{Fe})$(GPa)	$K_0(0)$	$K_0(0) + 7.5x_{Fe}$
$\mu_0(x_{Fe})$(GPa)	-	$\mu_0(0) + 77.0x_{Fe}$
$\rho_0(x_{Fe})$ (kg/m^3)	$4108 + 1070x_{Fe}$	$3583 + 2280x_{Fe}$
$V_0(x_{Fe})$(cc/mol)	$24.46 + 1.03x_{Fe}$	$11.25 + 1.00x_{Fe}$

Table 2. Model Parameters.

Parameter	Surface value	Bottom value	Unit
T	300	3470	K
Di	0.65	0.35	-
γ	1.33	1.01	-
α	2.7×10^{-5}	0.9×10^{-5}	K^{-1}
ρ	4160	5620	kg m^{-3}
k	3	5.4	W m^{-1}

cally isotropic [*Meade et al.*, 1995], the seismic velocities of P-waves and S-waves are, respectively

$$V_P = \left(\frac{K_S + \frac{4}{3}\mu}{\rho} \right)^{1/2} \qquad V_S = \left(\frac{\mu}{\rho} \right)^{1/2}, \qquad (7)$$

where K_S is the adiabatic bulk modulus and μ the shear modulus. Another parameter commonly used in tomographic models is the theoretical bulk sound speed:

$$V_\Phi = \left(V_P^2 - \frac{4}{3}V_S^2 \right)^{1/2} = \left(\frac{K_S}{\rho} \right)^{1/2}. \qquad (8)$$

K_S, μ, and ρ depend on temperature, pressure, and composition. For our mineralogical model composed of perovskite (Fe,Mg)SiO$_3$ and magnesiowüstite (Fe,Mg)O, the composition is defined by the iron molar ratio xFe $= n$Fe/(nFe $+ n$Mg) $= 1 - x$Mg, the silica molar ratio xSi $= n$Si/(nFe $+ n$Mg), and the iron–magnesium partition coefficient:

$$KFe = \frac{(xFe/xMg)_{mw}}{(xFe/xMg)_{pv}}. \qquad (9)$$

For a given xSi, xFe composition, and KFe, we calculate $K_S^{P,T}$, $\mu^{P,T}$, and $\rho^{P,T}$ for each of the two phases considered. We can then derive $V_P(T,P)$, $V_S(T,P)$, and $V_\Phi(T,P)$ in our mantle domain, proceeding as follows:

We consider a third-order Birch–Murnaghan finite strain formalism [*Birch*, 1952], for which the pressure along an adiabat is written as follows:

$$P = \frac{3}{2} K_S^{P_0,T_S} \left[\left(\frac{\rho^{P,T_S}}{\rho^{P_0,T_S}} \right)^{(7/3)} - \left(\frac{\rho^{P,T_S}}{\rho^{P_0,T_S}} \right)^{(5/3)} \right]$$
$$\left\{ 1 + \frac{3}{4}(K_S' - 4) \left[\left(\frac{\rho^{P,T_S}}{\rho^{P_0,T_S}} \right)^{(2/3)} - 1 \right] \right\}, \qquad (10)$$

where $K_S' = (\partial K_S/\partial P)_{P=P_0}$, T_{S0} is the temperature at the top of the adiabat considered (Figure 1a), and ρ^{P_0,T_S} is approximated by:

$$\rho^{P_0,T_S} = \rho^{P_0,T_0}[1 - (T_S - T_0)\alpha^{P_0,T_0}]. \qquad (11)$$

For each phase ρ^{P_0,T_0}, the density at ambient conditions $T_0 = 300$ K and $P_0 \sim 0$ Pa is calculated using the molar mass and molar volume of each component (Fe, Si, Mg, O). This yields the relations given in *Wang and Weidner* [1996] (see Table 1). Thus, using equation (10), we calculate for each phase, ρ^{P,T_S}, the density along the adiabat.

Following *Stacey and Davis* [2004], we calculate $K_S^{P,T}$ for perovskite and magnesiowüstite at pressure P and temperature T by differentiation of equation (10) and by considering a linear dependence of K_S with temperature:

$$K_S^{P,T} = P \frac{-a(5/3)\Gamma^{5/3} + b(7/3)\Gamma^{7/3} - 3c\Gamma^3}{-a\Gamma^{5/3} + b\Gamma^{7/3} - c\Gamma^3}$$
$$+ \left(\frac{\partial K_S}{\partial T}\right)_P (T - T_S), \quad (12)$$

where $\Gamma = \rho^{P,T_S}/\rho^{P_0,T_S}$, $a = -2(K'-4)+(8/3)$, $b = -4(K'-4)+(8/3)$ and $c = b - a$.

The next step before calculating $\rho^{P,T}$ is to express $\alpha^{P,T}$, the thermal expansion coefficient at P and T. We make the reasonable assumption that T is above the Debye temperature and we use the relationship:

$$\gamma^{P,T} \rho^{P,T} = \gamma^{P_0,T_{S0}} \rho^{P_0,T_{S0}}, \quad (13)$$

with

$$\gamma = \frac{\alpha K_S}{\rho C_P}. \quad (14)$$

Combining equations (13) and (14) gives

$$\alpha^{P,T} = \frac{\gamma^{P_0,T_{S0}} \rho^{P_0,T_{S0}} C_P^P}{K_S^{P,T}}, \quad (15)$$

where we consider a pressure dependence for the specific heat C_p (see Appendix B). Then we calculate $\rho^{P,T}$, using the equivalent form of equation (11).

The pressure (or volume) dependence of shear modulus is then evaluated using the relationship [*Davies*, 1974]:

$$\mu^{P,T_0} = \left(\frac{\rho^{P,T_0}}{\rho^{P_0,T_0}}\right)^{5/3}$$
$$\left\{\mu^{P_0,T_0} + \frac{1}{2}\left[5\mu^{P_0,T_0} - 3\left(\frac{\partial\mu}{\partial P}\right)_T K^{P_0,T_0}\right]\right\}$$
$$\left[1 - \left(\frac{\rho^{P,T_0}}{\rho^{P_0,T_0}}\right)^{2/3}\right], \quad (16)$$

and its temperature dependence is expressed as:

$$\mu^{P,T} = \mu^{P,T_0} + \left(\frac{\partial\mu}{\partial T}\right)_P (T - T_0). \quad (17)$$

Finally we obtain $\rho^{P,T}$, $K_S^{P,T}$, and $\mu^{P,T}$ for pure perovskite and magnesiowüstite. We next calculate $\varphi^{P,T}$, the volumic proportion of perovskite in the assemblage of a given composition. The elastic coefficients for the assemblage are obtained by doing a Voigt–Reuss–Hill average [*Watt et al.*, 1976]; the density of the assemblage is calculated by doing a Voigt average:

$$\rho_a^{P,T} = \varphi^{P,T} \rho_{pv}^{P,T} + (1 - \varphi^{P,T}) \rho_{mw}^{P,T}. \quad (18)$$

Using equations (7) and (8), we obtain the P-wave and S-wave velocities as well as the bulk sound speed for the assemblage of perovskite and magnesiowüstite.

Note that since we do not take into account phase changes and consider only changes in the lower mantle, we take pressure values from PREM [*Dziewonski and Anderson*, 1981] instead of computing the lithostatic pressure with our densities, which would have biased the results.

The isothermal bulk and shear modulus for Mg-perovskite and periclase and their derivatives with respect to T and P, which we use in the calculations detailed above, are given in Table 3. We have chosen to use these values derived from molecular dynamic calculations [*Matsui*, 2000] because they are autocoherent and compatible with recent experimental results. We made, however, an exception for the value of $\partial\mu/\partial T = -0.019$ GPa K^{-1} for perovskite which appears to be too low compared with values proposed by other studies, which generally range between −0.027 GPa K^{-1} and −0.029 GPa K^{-1} [*Duffy and Anderson*, 1989; *Sinelnikov et al.*, 1998]. Thus, we use $\partial\mu/\partial T = -0.024$ GPa K^{-1}, the average of the extreme values. Since seismic waves propagation is adiabatic, we convert the isothermal values for bulk modulus (Table 3) to the adiabatic case (see Appendix A). For shear modulus, there is no significant difference between adiabatic and isothermal cases [*Poirier*, 1991].

In these calculations, the influence of xFe on the elastic coefficients is taken into account for $Mg_{x_{Mg}^{pv}} Fe_{x_{Fe}^{pv}} SiO_3$ perovskite and for $Mg_{x_{Mg}^{mw}} Fe_{x_{Fe}^{mw}} O$ magnesiowüstite (Table 1). Experiments have shown that the bulk modulus of perovskite [*Yeganeh-Haeri*, 1994] is not affected by the presence of iron. The influence of iron on the shear modulus of perovskite is expected to be small (see *Wang and Weidner* [1996] and references therein); however, no confirmation is available from high-pressure and temperature experiments.

2.2. The Reference Mantle

Seismic tomography displays velocity anomalies with respect to a reference model, which is often PREM, or with

Table 3. Isothermal Elastic Constants of Mg-Perovskite and Periclase and Their First-Order P and T Derivatives.

	MgSiO$_3$	MgO
$K_T^{P_0,T_0}$ (GPa)	258.1	161.0
μ^{P_0,T_0} (GPa)	176.8	131.0
$\left(\frac{\partial K_T}{\partial T}\right)_P$ (GPa K^{-1})	-0.029	-0.028
$\left(\frac{\partial K_T}{\partial P}\right)_T$	4.1	4.05
$\left(\frac{\partial \mu}{\partial T}\right)_P$ (GPa K^{-1})	-0.024*	-0.024
$\left(\frac{\partial \mu}{\partial P}\right)_T$	1.4	2.4

*Mean of extremes; all other values from *Matsui* [2000] (see text).

respect to the average velocity. Therefore, we need to choose a reference model in order to express our seismic velocities in terms of velocity anomalies. The assumed pyrolitic composition for our reference mantle is defined by xSi = 0.68 and xFe = 0.11. Although KFe is pressure dependent, its value above 40 GPa (i.e., below ~1000 km depth) seems to be constant and about 3.5 [*Guyot et al.*, 1988]. Therefore, we assume KFe = 3.5 through all our reference mantle. The reference temperature profile (shown in Figure 1a) is assumed to be adiabatic with a top adiabat at 1700 K, plus thermal boundary layers ~90 km thick at the top and bottom. The calculated reference density (Figure 1b) gives a very good fit to PREM in the lower mantle. As expected, the fit is poor in the upper mantle, since we have considered only lower mantle phases. Figure 1c shows depth profiles of thermal expansion coefficient for pure perovskite, pure magnesiowüstite and for the assemblage. At ambient conditions, the thermal expansion coefficient for perovskite and for magnesiowüstite is 2.6×10^{-5} K^{-1} and 4.1×10^{-5} K^{-1}, respectively. Such values are higher than experimental estimates for Mg-perovskite [*Fiquet et al.*, 2000; *Gillet et al.*, 2000] and magnesiowüstite (see *Hama and Suito* [1999] and references therein). However, these experiments have been performed on pure Mg-perovskite and periclase, whereas our results include the presence of iron in both phases. As pointed out by *Karki and Stixrude* [1999], measurements on Fe-bearing samples yield much higher values of thermal expansion at ambient conditions. The calculated volumic proportion of perovskite in the assemblage at ambient conditions is $\varphi^{P_0,T_0} = 82\%$, providing a thermal expansion coefficient for the assemblage $\alpha^{P_0,T_0} = 2.8 \times 10^{-5}$ K^{-1}. At high pressure and temperature (P = 135 GPa and T = 3450 K), α is 1.0×10^{-5} K^{-1} for magnesiowüstite and 0.94×10^{-5} K^{-1} for perovskite, values that compare well with experiments on Mg-perovskite [*Gillet et al.*, 2000] and MgO [*Chopelas and Boehler*, 1992].

3. RESULTS

3.1. Behavior of the Chemically Denser Material

We start the calculations at t_0 = 2 Gy B.P., which corresponds to the mean age of continental crust, and assume that the dense layer represents 25% of the whole-mantle volume. We choose this initial time to avoid the modeling of continental crust extraction.

Similar to *Tackley* [2002], we obtained our initial condition for temperature by running the calculations with the dense layer until reaching a thermal equilibrium. We affect a large value of the buoyancy number (B = 0.5) to keep the denser material in the bottom of the lower mantle with no topography. At t_0, the surface buoyancy number B is set to 0.25, which corresponds to a chemical density contrast $\Delta\rho/\rho_0$ = 1.7%.

Plate 1a shows the nondimensional potential temperature field (i.e., the temperature without the adiabatic gradient)

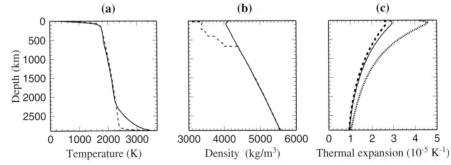

Figure 1. Depth profiles: (a) Temperature profile at present-day time for our geodynamical model (plain) and for our reference model (dashed). (b) Density profile for our reference mantle (plain) and PREM density profile (dashed). (c) Thermal expansion coefficient for our reference pyrolitic mantle, for perovskite (dashed), magnesiowüstite (dotted) and the whole assemblage (solid).

obtained after running our calculation for 2 Gy after t_0. The presence of relatively small chemical density excess $\Delta\rho/\rho$ = 1.7% greatly influences the nature of convection. The hot denser material develops a topography and is not continuous, being deflected by cold plumes (e.g., see Plates 1a and b at x/D = 0.20 and x/D = 5.5). Therefore we use the term "dense material" rather than "dense layer". The chemical density excess prevents the dense material from a complete mixing with the overlying mantle, but it is not high enough to produce a flat interface. The interface is a thermal boundary layer across which the temperature jump ranges between 300 and 800 K, as shown also in the horizontally averaged temperature profile (Figure 1a). Hot plumes are generated from domes of dense material, the height of the domes is on average 500 ± 250 km and can extend to a depth of 1500 km (e.g., see Plate 1b at x/D = 5.0). The volume of the dense material accumulated at the bottom of the lower mantle is about 15% of the whole-mantle volume; therefore, 40% of its initial volume has been brought in the overlying mantle, where stirring is efficient [*Farnetani and Samuel*, 2003; *Samuel and Farnetani*, 2003]. Filaments of heterogeneous dense material (i.e., $\chi > 0$) stretched by the convective stirring are easily visible in Plate 1b.

Finally, the value of $\Delta\rho_\chi/\rho$ = 1.7% allows a significant amount of dense material to remain stable at present-day time, with a topography. Indeed, if $\Delta\rho_\chi/\rho$ is lower than 1.7%, the dense material cannot remain stable and will be rapidly mixed with the overlying mantle, while if $\Delta\rho_\chi/\rho$ is too high, the dense material will form a flat interface, which is not observed by seismology. We remark that variable viscosity would help to stabilize the dense material [*Tackley*, 2002; *Samuel and Farnetani*, 2003]. Therefore, we can reasonably consider that the $\Delta\rho_\chi/\rho$ required to give a stable dense material with a topography ranges between 1% and 2% and perhaps less, depending on the distribution of heat-producing elements in the mantle.

In our model, the homogeneous internal heating rate is linked to the time-decreasing concentrations in heat-producing elements (due to radioactive decay; see section 1.1). Therefore, the term "present day" indicates a time for which the concentrations of heat-producing elements in our model correspond to present-day estimates of U, Th, and K mantle content, based on chondritic models [e.g., *McDonough and Sun*, 1995]. We point out that the relatively high value of the constant viscosity chosen here yields the dimensional time based on a convective time scale about 6 times smaller then the one based on the diffusive time scale. Nevertheless, even when choosing a convective time scale, we checked that the dense material could survive at least 2 Gy after our initial condition. The reason for this is that the set of parameters chosen for the convection model yields a quasi-

dynamic equilibrium of the thermochemical system, and the entrainment of dense material by thermal plumes is relatively low. When viscosity is temperature-dependent, however, one should expect a higher entrainment [*Zhong and Hager*, 2003]. This would therefore require a somewhat higher value of the chemical density contrast in order for the dense material to survive for 2 Gy.

3.2. Constraints on the Composition of the Dense Material

In the following, we consider two mineralogical end members: (i) the reference mantle, assumed to have a pyrolitic composition, and (ii) the denser material, whose mineralogical composition has to be constrained. Our approach is to use seismological models and observations as well as geodynamical considerations, to constrain the composition of the dense material. Our strategy is to calculate the seismic velocity anomalies with respect to our reference mantle for a large range of mineralogical compositions, and then to use seismology and geodynamics to constrain the composition of the dense material. For seismological constraints, we especially focus on the anticorrelation between shear wave velocity and bulk sound speed anomalies, associated with slow shear wave velocities. The second type of constraint arises from geodynamical considerations on the stability of the dense material. As previously mentioned, we can reasonably consider that the $\Delta\rho_\chi/\rho$ (at ambient conditions) required to produce a stable dense material with a dynamical topography is between 1% and 2%.

Therefore, we require the following constraints to be satisfied simultaneously:

(1) $d\ln V_\phi/d\ln V_S < 0$ associated with $d\ln V_S < 0$.
(2) $\Delta\rho_\chi/\rho$ = 1–2%.

The calculation of seismic velocity anomalies is first performed at a pressure of 100 GPa, corresponding to ~2200 km since at this depth the dense material in our numerical model starts to appear, and seismological observations indicate the existence of heterogeneity. According to our reference temperature profile (Figure 1a), the temperature at P = 100 GPa is 2200 K. Following the approach detailed in section 2.1, we calculate the seismic velocity anomalies for a large range of xFe and xSi molar ratios. Plate 2 shows the results for two cases, where we vary the temperature contrast δT between the reference mantle and the hot dense material and/or vary the iron–magnesium partition coefficient KFe between magnesiowüstite and perovskite for the dense material. In the first case (Plate 2a), δT = 400 K and KFe = 3.5 for the dense material. The blue curve corresponds to $d\ln V_S$ = 0 while the red curve corresponds to $d\ln V_\phi$ = 0. Above the blue and the red curves, $d\ln V_S$ and $d\ln V_\phi$ are negative, respectively. Therefore, areas in dark and light gray correspond to compositions that would give an anticorrelation between bulk sound speed and shear wave velocity. Moreover, the light-gray area corresponds to compositions for which anticorrelation between V_S and V_ϕ is associated with slow shear wave velocity. This corresponds to our first constraint. The composition of the dense material should therefore lie within the light-gray area. We can further constrain the composition by taking into account our geodynamical considerations (constraint 2). Isocontours of chemical density excess at ambient temperature are represented by the black lines in Plate 2: $\Delta\rho_\chi/\rho = 0\%$ (i.e., no chemical density excess), 1%, 2%, and 4%, whereas the green line is a specific isocontour $\Delta\rho_\chi/\rho = 1.7\%$ corresponding to our geodynamical model. $\Delta\rho_\chi/\rho$ were obtained by computing the densities at ambient temperature for the reference pyrolitic mantle and for a wide range of xFe and xSi compositions, as detailed in section 2.1 but at $T = 300$ K. The normalized density difference between the reference mantle and composition corresponding to various xSi and xFe values thus represents $\Delta\rho_\chi/\rho$. The compositions which satisfy both geodynamical and seismological considerations lie within the orange area since they satisfy simultaneously constraints 1 and 2. In agreement with previous work [Kellogg et al., 1999] the dense material has to be enriched in both iron and silica with respect to a pyrolitic composition (represented by the triangle in Plate 2). For $\Delta\rho_\chi/\rho = 1.7\%$, the composition that differs the least from the reference pyrolitic composition is defined by the intersection of the green line ($\Delta\rho_\chi/\rho = 1.7\%$) and the red line ($d\ln V_\phi = 0$).

We find xFe = 0.17 and xSi = 0.81 when $\delta T = 400$ K and KFe = 3.5 for the dense material (Plate 2a). For a higher temperature contrast between the reference mantle and the hot dense material, $\delta T = 800$ K (Plate 2b), a greater increase of xSi is required, while xFe remains unchanged with respect to the previous case. However, such high temperature contrast is inconsistent with the excess temperature of mantle plumes [Farnetani, 1997]. We also investigate the influence of a lower KFe for the dense material, which could be due to the presence of a small amount of aluminum in perovskite at high pressure [Andrault, 2001]. Al-perovskite is not explicitly taken into account in our calculations, since we neglect the possible effect of Al on V_0, K, and μ for perovskite. However, this simplification is probably reasonable since the amount of Al-perovskite necessary to decrease KFe is relatively low [Andrault, 2001]. We find no significant differences in silica and iron content with respect to the cases where KFe = 3.5. Table 4 summarizes the main results presented in Plates 2a and b.

Calculations have also been conducted at higher (i.e., up to 135 GPa) and lower (i.e., 25 GPa) pressures. In either case we find that the results described above remain valid.

Similarly, we investigated the effect of the reference mantle temperature and find that even a wide range of temperatures (1800–3000 K) does not affect our results significantly. Therefore, for the most plausible case, $\delta T = 400$ K, our preferred compositions for the dense material (i.e., those which differ the least from the reference pyrolitic mantle) range between 0.77 and 0.81 for xSi, and between 0.14 and 0.17 for xFe, depending on the chemical density contrast $\Delta\rho_\chi$ required (see Table 4).

3.3. Comparison With Seismological Observables

3.3.1. Check of models coherency. From further considerations, we fix the composition of the dense material to xSi = 0.81, xFe = 0.17, and KFe = 3.5, thus for $\Delta\rho_\chi/\rho = 1.7\%$. Since we use two different formalisms for the geodynamical model (equations 5 and 6) and for the postprocessing calculation of seismic velocities (see section 2.1), it is important to check that our preferred composition produces the same critical quantities obtained by the numerical convection code. From a geodynamical point of view, the critical quantity for the stability of the dense material, for a given temperature field, is the effective buoyancy number: $B_{eff} = \Delta\rho_\chi/[\rho\alpha(T - T_0)]$. In Figure 2a we compare the horizontally averaged B_{eff} obtained from STAG3D with B_{eff} from our postprocessing calculations, while in Figure 2b we compare the horizontally averaged chemical density contrast $\Delta\rho_\chi/\rho_0$ (without the effect of temperature) given by STAG3D with $\Delta\rho_\chi/\rho_0$ from our postprocessing calculations. In both figures, the small differences indicate a satisfactory coherence between the a priori thermodynamical model used in STAG3D and the postprocessing calculations.

3.3.2. Seismic velocity anomalies. We calculate seismic velocity anomalies with respect to our reference using the temperature and the compositional fields shown in Plates 1a and b. Values of xSi and xFe are linearly linked to the value of the compositional field: $\chi = 0$ corresponds to a pyrolitic composition, while $\chi = 1$ corresponds to xSi = 0.81 and xFe

Table 4. Compositions of the Dense Material That Satisfies Geodynamical and Seismological Considerations for Different Chemical Density Contrasts and Temperature Contrasts With Respect to a Reference Pyrolitic Mantle, at $P = 100$ GPa.

$\Delta\rho_\chi/\rho$	δT(K)	K Fe	xSi	xFe
1%	400	3.5	0.77	0.14
1%	800	3.5	0.80	0.14
1.7%	400	3.5	0.81	0.17
1.7%	800	3.5	0.85	0.17

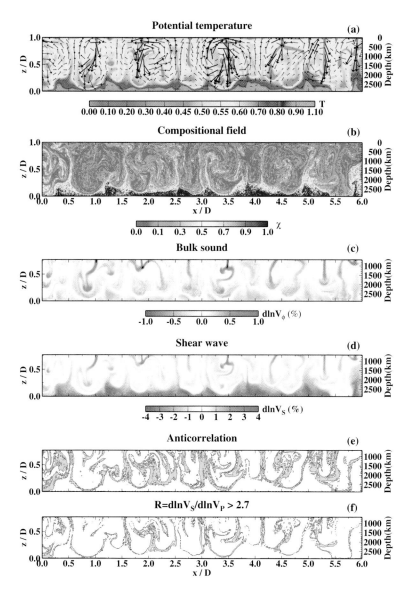

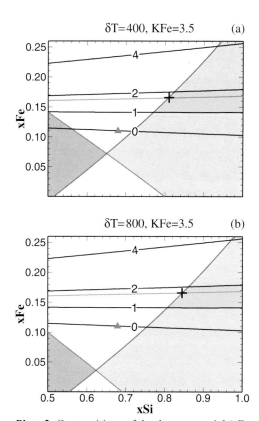

Plate 1. (a) Nondimensional potential temperature field and velocity field at present-day time; (b) corresponding compositional field, $\chi > 0$ indicates chemically dense material; (c) seismic velocity anomalies for bulk sound and (d) for shear wave. (e) Areas of anticorrelation between positive bulk sound and negative shear wave velocity anomalies (gray). (f) Areas (gray) where $R = d\ln V_S/d\ln V_P > 2.7$.

Plate 2. Compositions of the dense material (xFe and xSi). All calculations are preformed at $P = 100$ GPa and $T = 2200$ K. Blue curve: $d\ln V_S = 0$; above it, $d\ln V_S < 0$. Red curve: $d\ln V_\phi = 0$; above it, $d\ln V_\phi < 0$. Gray areas: compositions of the dense material that produces $d\ln V_\phi/d\ln V_S < 0$. Light-gray area: compositions that satisfy simultaneously $d\ln V_\phi/d\ln V_S < 0$ and $d\ln V_S < 0$ (see text). Blue triangle: assumed pyrolitic composition of reference mantle. Black lines: isocontours of chemical density contrasts at ambient temperature $\Delta\rho_\chi/\rho$ at 0%, 1%, 2%, and 4%; green line: $\Delta\rho_\chi/\rho = 1.7\%$, corresponding to our geodynamical model. Orange area: compositions that satisfy both geodynamical and seismological constraints (see text). Crosses: composition of the dense material that satisfies simultaneously $\Delta\rho_\chi/\rho = 1.7\%$, $d\ln V_\phi/d\ln V_S < 0$, and $d\ln V_S < 0$ and differs the least from a pyrolitic composition. (a) $\delta T = 400$ K and KFe $= 3.5$; (b) $\delta T = 800$ K and KFe $= 3.5$ (see text).

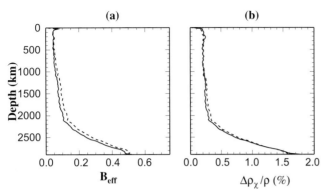

Figure 2. Comparison of horizontally averaged quantities calculated with the numerical convection code (dashed lines) and our post-processing calculations (solid lines): (a) Effective buoyancy number $B_{eff} = \Delta\rho_\chi / \rho\alpha(T - T_0)$. (b) Chemical density contrast $\Delta\rho_\chi / \rho$ at ambient temperature.

= 0.17. The obtained seismic velocity anomaly field for bulk sound (Plate 1c) and shear wave (Plate 1d) have globally a comparable shape, mainly induced by the temperature field. The amplitude of the anomalies for shear wave ranges between ±4%, while for bulk sound speed, which is less sensitive to temperature heterogeneity, the anomalies range between ±1%. Seismic velocity anomalies for P-waves (not shown) range between ±1.7% and have a shape similar to that of $d\ln V_S$ and $d\ln V_\phi$ fields. The calculated velocity anomalies compare fairly well with seismic tomography of the lower mantle [e.g., *Mégnin and Romanowicz*, 2000]. Another important feature in good agreement with local seismic studies of the lower mantle [*Bréger et al.*, 2001] is the sharp lateral variation in seismic velocity, induced by the coexistence of hotter, denser material with cold plumes. We remark that strong lateral variations in temperatures are a specific feature of thermochemical convection and cannot be generated by purely thermal convection models, for which the hot buoyant material is free to rise at shallower depths.

Although seismic velocity anomaly fields for bulk sound and shear wave have globally a comparable shape, several differences can be observed (e.g., compare Figures 1c and d at $x/D = 0.25$ and $z/D = 0-0.25$), where $d\ln V_S$ is negative while $d\ln V_\phi$ is positive. These differences produce areas of anticorrelation between V_S and V_ϕ (Plate 1e). The composition chosen for the dense material, combined with our temperature field, is responsible for this anticorrelation associated with slow shear wave velocities. We remark that these areas are not located everywhere inside the dense material because the temperature contrast δT between the two materials is too high to produce any anticorrelation (see Plate 2). Instead, anticorrelation areas are mostly located at

the interface between the hot dense material and the overlying pyrolitic mantle, where temperature (and compositional) gradients are strong. Moreover, anticorrelation areas are mainly associated with upwelling regions, while there is almost no anticorrelation in downwelling regions. This is in remarkably good agreement with recent observations by *Saltzer et al.* [2001].

The amplitudes of our calculated seismic velocity anomalies are higher than those displayed by tomographic models, mainly for two reasons: First, our model resolution of 22.6 km is much higher than the resolution of tomographic models in the lower mantle. Thus, we are able to display small structures that cannot be imaged by seismic tomography. We altered the resolution of our model from 22.6 to 180 km, using a filter (see Appendix C), and we found that the amplitudes of seismic velocity anomalies are reduced to ±3.5% for shear wave, ±1% for compressional waves, and ±0.5% for bulk sound speed. Note that even with filtering, all the results described above remain unchanged. The second reason is that tomographic models underestimate the amplitude of seismic velocity anomalies, sometimes by a factor of 3, as a result of damping [*Bréger et al.*, 1998].

3.3.3. RMS seismic velocity profiles. A common feature of tomographic models is the increase of the RMS seismic velocity anomaly below ~2000 km [*Masters et al.*, 2000; *Mégnin and Romanowicz*, 2000]. Since RMS seismic velocity gives a measure of the lateral variations of seismic velocity anomalies, the observations suggest an increase of mantle lateral heterogeneity below 2000 km. Figure 3a shows the RMS seismic velocity anomalies for shear, compression, and bulk sound. Note that we purposely exclude values corresponding to depth greater than 2800 km because our constant-temperature boundary condition at the bottom hinders lateral temperature variations below this depth. All calculated RMS profiles show an increase below ~2000 km depth, in good agreement with tomographic studies. This is due to the topography formed by the dense hotter material in the lowermost mantle. Again, we note that purely thermal convection models are unable to reproduce the observed RMS profiles.

3.3.4. The ratio R in the lower mantle. Supplementary information about mantle chemical heterogeneity can be provided by the ratio R of shear to compressional velocity anomaly:

$$R = \frac{d\ln V_S}{d\ln V_P}. \tag{19}$$

It has been argued that above a critical value of $R = 2.5-2.7$, seismic velocity anomalies cannot be explained by temperature differences alone but also point out chemical density

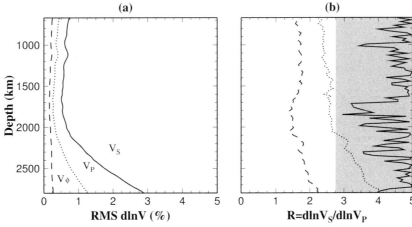

Figure 3. (a) RMS seismic velocity anomaly profiles calculated for V_S (solid line), V_P (dotted line), and V_ϕ (dashed line). (b) Horizontally averaged $R = d\ln V_S/d\ln V_P$ profile for our model: without anelastic effects (dashed line) and with anelastic effects (dotted line). The maximum R values are plotted for the case where anelastic effects are not considered (solid line). The gray area corresponds to $R > 2.7$.

differences [*Masters et al.*, 2000; *Karato and Karki*, 2001]. Global joint tomographic models [*Robertson and Woodhouse*, 1996; *Masters et al.*, 2000] have shown that, on average, R increases with depth from about 1.5–1.7 at 1000 km to values sometimes >3 near the CMB, with R becoming >2.5 at ~2000 km depth. *Saltzer et al.* [2001] have shown that the R ratio has a lateral variability, especially in the lowermost mantle, with areas where R is greater than 2.5 and others where R is less than 2, depending on the lateral position. Our calculated horizontally averaged R profile is about 1.5 at 1000 km and increases with depth to 2.2 at 2800-km depth (Figure 3b). However, R has a large lateral variability, as is clearly visible in Plate 1f, which displays areas with $R > 2.7$. These areas match fairly well areas of anticorrelation between V_ϕ and V_S (Plate 1e), emphasizing that these two features are closely related to each other. The maximum R values (Figure 3b) show that the horizontally averaged R profile hides lateral variations of $R > 2.7$, which can instead reach values up to 5. The good agreement of our results with tomographic studies, in particular, that by *Saltzer et al.* [2001], strongly supports the idea that chemical heterogeneity varies laterally as well as the anticorrelation between V_S and V_ϕ. This may be the reason why some tomographic models find relatively low R (<2.5) [*Kennett et al.*, 1998], depending on the seismic area covered.

3.3.5. Effect of anelasticity. Anelasticity as well as anharmonicity may play an important role when one tries to explain values of R greater than 2.7 [*Karato*, 1993; *Karato and Karki*, 2001]. We thus investigated the effect of anelasticity on our results, following the approach of *Karato and Karki* [2001]. Details of calculation can be found in Appen-

dix D. As shown in Figure 3b, when anelastic effects are taken into account, the averaged R profile is higher, leading to values that range between 2.7 and 4 below 2000 km, in agreement with *Karato and Karki* [2001]. However, neither anelasticity nor the presence of Ca variations seems to be necessary to explain values of $R > 2.7$, contrary to the conclusions of *Karato and Karki* [2001]. This discrepancy between our results and the study of *Karato and Karki* [2001] can be explained by the higher lateral temperature variations they considered in the lowermost mantle. Importantly, considering anelasticity does not modify our results on the composition of the dense material. This can easily be seen in Plate 2, where we determine our preferred composition for the dense material as the intersection between the two isocontours $\Delta\rho_\chi/\rho = 1.7\%$ and $d\ln V_\phi = 0$. Since anelasticity affects only S-waves but not P-waves or bulk sound speed, it will not change the position of the two isocontours $\Delta\rho_\chi/\rho = 1.7\%$ and $d\ln V_\phi = 0$ in Plate 2 and thus the composition of the dense material according to our criteria.

3.3.6. Differences between our reference model and PREM. Although it is relatively simple, our reference model gives a reasonably good fit to PREM's profiles with a mean deviation less than 1% ± 0.4% for V_S, and 1.0% ± 0.5% for V_ϕ. Actually, we did not expect the seismic velocities of our reference model to match PREM perfectly. In fact, the assumption made in PREM of a Bullen parameter

$$\eta_B = \frac{dK_S}{dP} + \frac{1}{g}\frac{d(V_\phi^2)}{dr} = 1$$

through all the lower mantle implies, among other things, that internal heating and the presence of thermal boundary layers are neglected [*Poirier*, 1991]. This assumption may not have a dramatic effect on densities, since the thermal expansion of the lower mantle is relatively small, but it can have an important effect on the elastic coefficients K and μ, and thus on V_P, V_S, and V_ϕ. Furthermore, as pointed out by *Dewaele and Guyot* [1998], PREM underestimates the anelastic effects on seismic waves. This can also be seen in Appendix D. Taking into account all these complications, we consider that our reference model is not incompatible with PREM.

3.3.7. Effect of thermodynamical parameters. The results previously described bear strong implications for understanding the structure and composition of the deep mantle. However, one should wonder to what extent our choice of model parameters affects our results. From the various laboratory studies and numerical simulations it is clear that the thermoelastic parameters and their derivatives for perovskite and magnesiowüstite are affected by significant uncertainties. To investigate the effect of these uncertainties on our results, we can consider separately the effect of each thermoelastic parameter listed in Table 3. Reasonable variations of $K_T^{P_0, T_0}$ and μ^{P_0, T_0} will not affect significantly our results since it is the *variation* of these elastic coefficients with pressure and temperature that produces seismic velocity anomalies. The effect of iron on all elastic parameters is reported to be small; therefore, the relatively small increase of the iron content that we propose for the lowermost part of the mantle does not need to be considered. Thus, we can simply investigate the effect of pressure and temperature derivatives on the slope and position of the curve $d\ln V_\phi = 0$, because it determines the composition of the dense material for a given chemical density contrast (see Plate 2). At a given depth the lateral variation of bulk sound speed is produced by temperature variations and composition at (almost) constant pressure; therefore, slight variations of $(\partial \mu / \partial P)_T$ and $(\partial K / \partial P)_T$ parameters do not affect our results. Similarly, $(\partial \mu / \partial T)_P$ does not affect the curve $d\ln V_\phi = 0$ since V_ϕ does not depend on μ. In contrast, the temperature variation of K has a considerable effect on the slope and the position of the curve $d\ln V_\phi = 0$.

For example, if $\delta T = 400$ K, KF e = 3.5, and $\Delta \rho_\chi / \rho = 1.7\%$ for dense material, an increase of $(\partial K_S / \partial T_P)$ from -0.015 GPa K^{-1} to -0.006 GPa K^{-1} for the assemblage will require a decrease of xSi from 0.81 (see Table 4 or Plate 2b) to 0.75, with almost no change in xFe for the dense material. For the same conditions, a value of $(\partial K_S / \partial T)_P = -0.026$ GPa K^{-1} for the assemblage yields xSi = 0.90. Interestingly, in that case we have observed that a low value for KFe (0.5 instead of 3.5) reduces the required Si enrichment to 0.78. The variations of $(\partial K_S / \partial T)_P$ investigated here are rather large compared to values that converge to ~ -0.015 GPa K^{-1} (see Appendix

A). Still, it appears that the xSi value of the dense material is strongly correlated to this uncertainty.

4. DISCUSSION

A robust conclusion of our study is that the dense material must be enriched in both Fe and Si with respect to a pyrolitic composition as proposed by *Kellogg et al.* [1999], in order to satisfy geodynamical and seismological constraints. Several hypotheses can be advanced for such enrichment in Si and Fe.

(1) Subducted oceanic crust is enriched in Si and Fe and has been reported to be denser than a pyrolitic mantle at lower mantle pressures [*Hirose et al.*, 1999; *Guignot and Andrault*, 2004]. Indeed, the mineralogy of a MORB-type material becomes $\sim 30\%$ mol. Al(Mg,Fe)-perovskite (with xFe ~ 0.4 and xSi ~ 0.75) and $\sim 20\%$ mol. of SiO$_2$ stishovite [*Guignot and Andrault*, 2004]. These iron and silica-rich phases could react with a pyrolitic mantle, yielding an enrichment in both Si and Fe. Following values listed in *Guignot and Andrault* [2004], xSi$_{\text{pyrolite}}/x$Si$_{\text{MORB}}$ approximates xFe$_{\text{pyrolite}}/x$Fe$_{\text{MORB}}$, where the subscript MORB refers to the Al(Mg,Fe)-perovskite and stishovite present in a MORB. To explain the highest enrichment in Si (i.e., xSi = 0.85) and Fe (i.e., xFe = 0.17; see Table 4) proposed in this study, about 30% vol. of subducted oceanic crust should be mixed with 70% vol. of pyrolitic mantle. Assuming a past subduction rate of ~ 30 km^3/yr and that the volume of the enriched material is 25% of the whole-mantle volume, the necessary amount of subducted oceanic crust would be reached after ~ 1.5 Gy of subduction.

(2) The iron enrichment could be due to the incorporation of iron from the outer core into the mantle [*Knittle and Jeanloz*, 1991], probably during the early stages of the Earth's history, when convection was more vigorous.

(3) A previous experimental study [*Guyot et al.*, 1997] proposed that Si could be incorporated in Fe$_{\text{metal}}$ at shallow depth and high temperature, during the early stages of core formation. At greater depth, the Si incorporated in Fe$_{\text{metal}}$ is no longer stable and SiO$_2$ would be formed [*Guyot et al.*, 1997]. This process could participate in the Si enrichment to the lowermost mantle.

(4) Another possibility is that the Earth's bulk composition is not derived from CI chondrites, as generally assumed, but is derived from another type of material such as Enstatite chondrites, as proposed by *Javoy* [1995]. In such a case, the bulk composition of the Earth could be depleted in Mg relative to a CI-derived bulk composition. Therefore, it would yield an Fe- and Si-enriched lowermost mantle, since the upper part of the mantle would be close to pyrolite.

5. CONCLUSIONS

Combining seismic tomography with geodynamical considerations provides a new way to constrain the nature of compositional heterogeneity in the lowermost mantle. We conducted thermochemical convection simulation, with a chemical density excess $\Delta\rho_\chi/\rho = 1.7\%$ for the dense material, starting the calculation 2 Gy B.P. After 2 Gy of convection, the dense hotter material develops a dynamical topography but remains stable in the lowermost mantle. Using a lower mantle thermodynamical model considering the assemblage of perovskite and magnesiowüstite, we constrain the composition of a dense material with respect to a reference mantle, assumed to have a pyrolitic composition. Available data and models are all compatible with the presence of a chemically denser material simultaneously enriched in Fe and Si. This enrichment is lower when the temperature contrast between the dense hotter material and the overlying mantle decreases.

We apply the composition that satisfies geodynamical and seismological constraints to the temperature and compositional fields extracted from convection simulations and compute the main seismological parameters. Our results reproduce seismological observations well, in particular:

(1) The presence of broad seismic velocity anomalies in the lowermost mantle.

(2) The increase of the RMS seismic velocity anomalies below 2000 km depth.

(3) The presence of anticorrelation between bulk sound speed and shear wave velocity anomalies, mainly located in upwelling areas.

(4) The presence of areas of $R = d\ln V_S/d\ln V_P > 2.7$ that match fairly well areas of anticorrelation between V_S and V_ϕ.

(5) The horizontally averaged R profile, which increases below 2000 km to the CMB. When anelasticity is considered, R values are higher but anelasticity is not necessary to explain $R > 2.7$. If the latter are horizontally averaged, the strong lateral variations observed for composition and temperature can hide values of $R > 2.7$.

Several mechanisms can be advanced for the origin of Si and Fe enrichment in the lowermost mantle such as the presence of subducted oceanic crust, the incorporation of Fe from the core in the early stages of Earth's mantle, chemical reactions between Si and Fe_{metal}, and/or a bulk composition of the Earth's mantle slightly different from the main canonical models.

APPENDIX A: CONVERSION FROM ISOTHERMAL TO ADIABATIC

Values for the isothermal bulk modulus K_T (Table 3) are converted to adiabatic bulk modulus K_S using:

$$K_S = K_T(1 + \alpha\gamma T). \tag{A.1}$$

The thermodynamical Grüneisein parameter γ is volume-dependent according to *Anderson* [1995]:

$$\gamma = \gamma_0 \left(\frac{\rho_0}{\rho}\right)^q, \tag{A.2}$$

where the subscript 0 indicates reference ambient P and T. Under the quasi-harmonic approximation, the derivative of equation (A.1) with respect to temperature at constant P yields

$$\left(\frac{\partial K_S}{\partial T}\right)_P = \left(\frac{\partial K_T}{\partial T}\right)_P (1 + \alpha\gamma T) + K_T\alpha\gamma\left\{1 + T\left[\alpha q + \frac{1}{\alpha}\left(\frac{\partial\alpha}{\partial T}\right)_P\right]\right\}, \tag{A.3}$$

whereas the derivative with respect to pressure at constant T yields

$$\left(\frac{\partial K_S}{\partial P}\right)_T = \left(\frac{\partial K_T}{\partial P}\right)_T (1 + \alpha\gamma T) + \gamma T\left[\frac{1}{K_T}\left(\frac{\partial K_T}{\partial T}\right)_P - \alpha q\right]. \tag{A.4}$$

For each of the two phases considered (i.e., perovskite and magnesiowüstite) we calculate the terms in equations (A.3) and (A.4) for ambient conditions using, in addition to Table 3, the following values of thermodynamical parameters: $\gamma_0^{pv} = 1.33$ [*Jackson and Rigden*, 1996], $(\partial\alpha/\partial T)_P^{pv} = 4.9 \times 10^{-8}$ K^{-2} [*Gillet et al.*, 2000], $\gamma_0^{mw} = 1.41$ [*Jackson and Rigden*, 1996], $(\partial\alpha/\partial T)^{mw} = 0.7 \times 10^{-8}$ K^{-2} [*Suzuki*, 1975]. Equation (15) provides $\alpha_0^{pv} = 2.6 \times 10^{-5}$ K^{-1} and α_0^{mw} 4.1 $\times 10^{-5}$ K^{-1}, and q is close to 1.0 for both phases. This yields $(\partial K_S)/(\partial T)^{pv}_P = -0.015$ GPa K^{-1} and $(\partial K_S)/(\partial T)^{mw}_P = -0.019$ GPa K^{-1}, in good agreement with *Jackson* [1998] and *Gillet et al.* [2000] for perovskite and with *Summino et al.* [1983] and *Isaak et al.* [1989] for magnesiowüstite. We obtain $(\partial K_S)/(\partial P)_T = 4.09$ for perovskite and $(\partial K_S)/(\partial P)_T = 4.03$ for magnesiowüstite, values that are slightly different from the isothermal case reported by *Dewaele and Guyot* [1998].

APPENDIX B: PRESSURE DEPENDENCE OF C_P

The pressure dependence of the specific heat $C_P = T(\partial S)/(\partial T)_P$ is expressed as follows:

$$C_P^P = C_P^{P_0} + \left(\frac{\partial C_P}{\partial P}\right)_T (P - P_0). \tag{B.1}$$

and can be written as

$$\left(\frac{\partial C_P}{\partial P}\right)_T = T \frac{\partial}{\partial T}\left(\left(\frac{\partial S}{\partial P}\right)_T\right)_P \quad \text{(B.2)}$$

Using the Maxwell relationship $(\partial S)/(\partial P)_T = -(\partial V)/(\partial T)_P = \alpha V$ (V being the volume), one gets:

$$\left(\frac{\partial C_P}{\partial P}\right)_T = -TV\left[\left(\frac{\partial \alpha}{\partial T}\right)_P + \alpha^2\right] \quad \text{(B.3)}$$

Therefore equation (B.3) gives the pressure dependence of C_P. For simplicity we assume that $(\partial C_P)/(\partial P)_T$ is constant and we choose the value at ambient conditions.

APPENDIX C: SEISMIC FILTERING

Our seismic filter is a moving average Gaussian operator [*Vacher et al.*, 1996]:

$$dlnV(x,z) =$$
$$dlnV(x_0, z_0)\ exp\left\{-\frac{1}{2}\left[\left(\frac{x-x_0}{L_x}\right)^2 + \left(\frac{z-z_0}{L_z}\right)^2\right]\right\},$$
$$\text{(C.1)}$$

where $dlnV(x_0, z_0)$ is the velocity anomaly at the center of the moving window. The horizontal and vertical correlation lengths, L_x and L_z, are taken as equal to 180 km.

APPENDIX D: SEISMIC ATTENUATION

At a given pressure and for a weak anelasticity, the effect of seismic attenuation on seismic velocities can be expressed as follows [*Karato*, 1993]:

$$V(w, T) = V_0(T)\left[1 - \frac{1}{2Q(w,T)}cotan\left(\frac{a\pi}{2}\right)\right].\text{(D.1)}$$

The attenuation of seismic waves is then given by the inverse of Q, the quality factor, which depends on the frequency w of seismic waves and on temperature T according to *Karato and Karki* [2001] :

$$Q(w, T) \sim w^{\alpha_a} exp\left(\frac{\alpha_a \beta T_m}{T}\right). \quad \text{(D.2)}$$

α_a and β are nondimensional constants, the latter depending on the activation enthalpy H^*, according to $H^* = \beta R_c T_m$, where R_c is the gas constant and T_m the melting temperature of the material considered. Following *Wang* [1999] the pressure dependence of T_m was estimated by integrating Lindemann's law:

$$T_m = T_{m_0} exp\left[2\gamma_S\left(1 - \frac{\rho_0}{\rho}\right) + \frac{2}{3}ln\left(\frac{\rho_0}{\rho}\right)\right], \text{(D.3)}$$

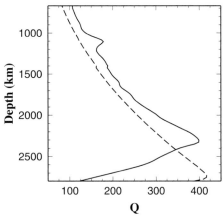

Figure 4. Horizontally averaged profile of shear quality factor, for our reference mantle (dashed line) and for our model (solid line).

where the Slater gamma $\gamma_S = \frac{1}{2}\left(\frac{dK_T}{dP} - \frac{1}{3}\right)$. We take $T_{m_0} = T_m(P_0, T_0) = 2800$ K for both perovskite and magnesiowüstite [*Wang*, 1999; *Anderson*, 1995], $w = 0.25$ Hz (i.e., a period of 4 s), and $\alpha_a = 0.2$ for the assemblage [*Karato and Karki*, 2001]. Taking $H^* = 230$ kJ/mol for magnesiowüstite and ~300 kJ/mol for perovskite (according to values listed in *Karato and Karki* [2001]), we obtain $\beta^{mw} = 9.9$ and $\beta^{pv} = 12.9$. For the assemblage of the two phases we calculate β by doing a simple Voigt average. Assuming a constant volumic proportion of perovskite $\varphi = 0.85$, we obtain $\beta = 12.5$ for the assemblage. Figure 4 shows the calculated shear quality factor average profile for our reference mantle and for our geodynamical model. Due to the decrease of Q in the lowermost mantle induced by temperature, attenuation becomes more important in the dense hotter material. Note that the shear quality factor for our reference model significantly differs from PREM, which assumes that $Q = 312$ through all the lower mantle.

Acknowledgments. We thank Laurent Guillot, François Guyot, Stéphane Labrosse, and Barbara Romanowicz for stimulating discussions and useful advice. We are grateful to Paul Tackley for providing his numerical code. Comments by Bradford Hager, Jan Matas, and an anonymous reviewer significantly improved the first version of this manuscript. Figures were made with the Generic Mapping Tools (P. Wessel and W.H.F Smith, *EOS, Trans. AGU* 76: 329, 1995).

REFERENCES

Anderson, O. L., *Equations of State of Solids for Geophysics and Ceramic Science*, vol. 81, Oxford monographs on geology and geophysics, 1995.

Anderson, O. L., H. Oda, and D. Isaak, A model for the computation of thermal expansivity at high compression and high temperatures: MgO as an example, *Geophys. Res. Lett, 19*, 1987–1990, 1992.

Andrault, D., Evaluation of (Mg, Fe) partitioning between silicate perovskite and magnesiowüstite up to 120 GPa and 2300 K, *J. Geophys. Res.*, *106*, 2079–2087, 2001.

Antolik, M., Y. J. Gu, G. Ekström, and A. M. Dziewonski, J362D28: a new joint model of compressional and shear velocity in the Earth's mantle, *Geophys. J. Int.*, *153*, 443–466, 2003.

Birch, F., Elasticity and consitution of the Earth's interior, *J. Geophys. Res.*, *57*, 227–286, 1952.

Bréger, L., B. Romanowicz, and L. Vinnick, Test of tomographic models of D" using differential travel time data, *Geophys. Res. Lett*, *25*, 5–8, 1998.

Bréger, L., B. Romanowicz, and C. Ng, The Pacific plume as seen by S, ScS, and SKS, *Geophys. Res. Lett*, *28*, 1859–1862, 2001.

Brown, J. M., and T. J. Shankland, Thermodynamic parameters in the Earth as determined from seismic profiles, *Geophys. J. R. Astron. Soc.*, *66*, 579–596, 1981.

Chopelas, A., and R. Boehler, Thermal expansivity in the lower mantle, *J. Geophys. Res.*, *19*, 1983–1986, 1992.

Davies, G. F., Effective elastic moduli under hydrostatic stress. 1. Quasi-harmonic theory, *J. Phys. Chem. Solids.*, *35*, 1513–1520, 1974.

Deschamps, F., and J. Trampert, Mantle tomography and its relation to temperature and composition, *Phys. Earth. Pl. Int.*, *140*, 277–291, 2003.

Dewaele, A., and F. Guyot, Thermal parameters of the Earth's lower mantle, *Phys. Earth. Pl. Int.*, *107*, 261–267, 1998. Duffy, T. S., and D. L. Anderson, Seismic velocities in mantle minerals and the mineralogy of the upper mantle, *J. Geophys. Res.*, *94*, 1895–1912, 1989.

Dziewonski, A. M., and D. L. Anderson, Preliminary reference Earth model, *Phys. Earth. Pl. Int.*, *25*, 297–356, 1981.

Farnetani, C. G., Excess temperature of mantle plumes : The role of chemical stratification across D", *Geophys. Res. Lett*, *24*, 1,583–1,586, 1997.

Farnetani, C. G., and H. Samuel, Lagrangian structures and stirring in the Earth's mantle, *Earth Planet. Sci. Lett.*, *206*, 335–348, 2003.

Fiquet, G., A. Dewaele, D. Andrault, M. Kunz, and T. Le Bihan, Thermoelastic properties and crystal structure of MgSiO^3 perovskite at lower mantle pressure and temperature conditions, *Geophys. Res. Lett*, *27*, 21–24, 2000.

Forte, A. M., and J. X. Mitrovica, Deep-mantle high-viscosity flow and thermochemical structure inferred from seismic and geodynamic data, *Nature*, *410*, 1049–1056, 2001.

Gillet, P., I. Daniel, F. Guyot, J. Matas, and J.-C. Chervin, A thermodynamic model for MgSiO$_3$ -perovskite derived from pressure, temperature and volume dependence of the raman mode frequencies, *Phys. Earth. Pl. Int.*, *117*, 361–384, 2000.

Grand, S. P., R. D. van der Hilst, and S. Widiyantoro, High resolution global tomography: A snapshot of convection in the Earth, *Geol. Soc. Am. Today*, *7*, 1–7, 1997.

Guignot, N., and D. Andrault, Equations of state of Na-K-Al host phases and implications for MORB density in the lower mantle, *Phys. Earth. Pl. Int.*, *143-144*, 107–128, 2004.

Guyot, F., M. Madon, J. Peyronneau, and J.-P. Poirier, X-ray microanalysis of high-pressure/high-temperature phases synthesized from natural olivine in a diamond-anvil cell, *Earth Planet. Sci. Lett.*, *90*, 52–64, 1988.

Guyot, F., J. Zhang, I. Martinez, J. Matas, Y. Ricard, and M. Javoy, P-V-T measurements of iron silicide (ε-FeSi). Implications for silicate-metal interactions in the early Earth, *Eur. J. Mineral.*, *9*, 277–285, 1997.

Hama, J., and K. Suito, Thermoelastic properties of periclase and magnesiowüstite under high pressure and temperature, *Phys. Earth. Pl. Int.*, *114*, 165–179, 1999.

Hirose, K. Y., Y. M. A. Fei, and H. K. Mao, The fate of subducted basaltic crust in the Earth's lower mantle, *Nature*, *397*, 53–56, 1999.

Hofmann, A. W., Mantle geochemistry: the message from oceanic volcanism, *Nature*, *385*, 219–229, 1997.

Isaak, D. G., O. L. Anderson, and T. Goto, Measured elastic modulus of single-crystal MgO up to 1800 K, *Phys. Chem. Miner.*, *16*, 703–704, 1989.

Ishii, M., and J. Tromp, Normal-mode and free-air gravity constraints on lateral variations in velocity and density of Earth's mantle, *Science*, *285*, 1231–1236, 1999.

Jackson, I., Elasticity, composition and temperature of the Earth's lower mantle: a reappraisal, *Geophys. J. Int.*, *134*, 291–311, 1998.

Jackson, I., and S. M. Rigden, Analysis of P-V-T data: constraints on the thermoelastic properties of high-pressure minerals, *Phys. Earth. Pl. Int.*, *96*, 85–112, 1996.

Javoy, M., The integral enstatite chondrite model of the Earth, *Geophys. Res. Lett*, *22*, 2219–2222, 1995.

Karato, S.-I., Importance of anelasticity in the interpretation of seismic tomography, *Geophys. Res. Lett*, *20*, 1623–1626, 1993.

Karato, S.-I., and B. B. Karki, Origin of lateral variation of seismic wave velocities and density in the deep mantle, *J. Geophys. Res.*, *106*, 21,771–21,783, 2001.

Karki, B. B., and L. Stixrude, Seismic velocities of major silicate and oxide phases of the lower mantle, *J. Geophys. Res.*, *104*, 13,025–13,033, 1999.

Kellogg, L. H., B. H. Hager, and R. D. van der Hilst, Compositional stratification in the deep mantle, *Science*, *283*, 1,881–1,884, 1999.

Kennett, B. L. N., S. Widiyantoro, and R. D. van der Hilst, Joint seismic tomography for bulk sound and shear wave speed in the Earth's mantle, *J. Geophys. Res.*, *103*, 12,469–12,493, 1998.

Knittle, E., and R. Jeanloz, Earth's core-mantle boundary: Results of experiments at high pressures and temperatures, *Science*, *251*, 1,438–1,443, 1991.

Lebars, M., and A. Davaille, Stability of thermal convection in two superimposed miscible viscous fluids, *J. Fluid Mech.*, *471*, 339–363, 2002.

Li, X., and B. Romanowicz, Global mantle shear velocity model developed using nonlinear asymptotic coupling theory, *J. Geophys. Res.*, *101*, 22,245–22,272, 1996.

Masters, G., G. Laske, H. Bolton, and A. M. Dziewonski, The relative behavior of shear velocity, bulk sound speed and compressional velocity in the mantle: implication for chemical and thermal structure, in *Earth's deep interior. Mineral physics and tomography from the atomic to the global scale*, vol. 117, edited by S.-I. Karato, A. Forte, R. C. Liebermann, G. Masters, and L. Stixrude, pp. 63–87, American Geophysical Union, Washington, DC, 2000.

Matsui, M., Molecular dynamics simulation of MgSiO$_3$ perovskite and the 660-km seismic discontinuity, *Phys. Earth. Pl. Int.*, *121*, 77–84, 2000.

McDonough, W. F., and S. Sun, The composition of the Earth, *Chem. Geol.*, *120*, 223–253, 1995.

Meade, C., P. G. Silver, and S. Kaneshima, Laboratory and seismological observations of lower mantle anisotropy, *Geophys. Res. Lett*, *22*, 1293–1296, 1995.

Mégnin, C., and B. Romanowicz, The three-dimensional shear velocity structure of the mantle from the inversion of body, surface and higher-mode waveforms, *Geophys. J. Int.*, *143*, 709–728, 2000.

Poirier, J.-P., *Introduction to the physics of the Earth's interior*, Cambridge University Press, 1991.

Robertson, G. S., and J. H. Woodhouse, Ratio of relative S to P heterogeneity in the lower mantle, *J. Geophys. Res.*, *101*, 20,041–20,052, 1996.

Romanowicz, B., Can we resolve 3D density heterogeneity in the lower mantle?, *Geophys. Res. Lett*, *6*, 1107–1110, 2001.

Saltzer, R. L., R. D. van der Hilst, and H. Kàrason, Comparing P and S wave heterogeneity in the mantle, *Geophys. Res. Lett*, *28*, 1335–1338, 2001.

Samuel, H., and C. G. Farnetani, Thermochemical convection and helium concentrations in mantle plumes, *Earth Planet. Sci. Lett.*, *207*, 39–56, 2003.

Sidorin, I., and M. Gurnis, Geodynamically consistent seismic velocity predictions at the base of the mantle, in *The Core–Mantle Boundary Region*, edited by M. Gurnis, M. E. Wysession, E. Knittle, and B. A. Buffett, pp. 209–230, American Geophysical Union, Washington, DC, 1998.

Sinelnikov, Y. D., G. Chen, D. R. Neuville, M. T. Vaughan, and R. C. Liebermann, Ultrasonic shear wave velocities of MgSiO^3 perovskite at 8 GPa and 800 K and lower mantle composition, *Science*, *281*, 677–679, 1998.

Stacey, F. D., and P. M. Davis, High pressure equations of state with applications to the lower mantle and core, *Phys. Earth. Pl. Int.*, *142*, 137–184, 2004.

Su, W., and A. M. Dziewonski, Predominance of long-wavelength heterogeneity in the mantle, *Nature*, *352*, 121–126, 1991.

Sunmino, Y., O. L. Anderson, and Y. Suzuki, Temperature coefficients of single cristal MgO between 80 and 1300 K, *Phys. Chem. Miner.*, *9*, 38–47, 1983.

Suzuki, I., Thermal expansion of periclase and olivine, and their anharmonic properties, *J. Phys. Earth*, *23*, 145–159, 1975. Tackley, P. J., Strong heterogeneity caused by deep mantle layering, *Geochem. Geophys. Geosyst.*, *3*(4), 10.1029/2001GC000,167, 2002.

Turcotte, D. L., and G. Schubert, *Geodynamics: Application of continuum physics to geological problems*, John Wiley, New York, 1982.

Vacher, P., A. Mocquet, and C. Sotin, Comparison between tomographic structures and models of convection in the upper mantle, *Geophys. J. Int.*, *124*, 45–56, 1996.

van der Hilst, R. D., S. Widiyantoro, and E. R. Engdahl, Evidence for deep mantle circulation from global tomography, *Nature*, *386*, 578–584, 1997.

Wang, Y., and D. J. Weidner, $(\partial \mu / \partial T)_P$ of the lower mantle, *Pure Appl. Geophys.*, *146*, 533–549, 1996.

Wang, Z. W., The melting of Al-bearing perovskite at the core mantle boundary, *Phys. Earth. Pl. Int.*, *115*, 219–228, 1999.

Watt, J., G. Davies, and R. O'Connel, The elastic properties of composite materials, *Rev. Geophys. Space Phys.*, *14*, 541–563, 1976.

Yeganeh-Haeri, A., Synthesis and re-investigation of the elastic properties of single-crystal magnesium silicate perovskite, *Phys. Earth. Pl. Int.*, *87*, 111–121, 1994.

Zhong, S., and B. H. Hager, Entrainment of a dense layer by thermal plumes, *Geophys. J. Int.*, *154*, 666–676, 2003.

D. Andrault, Institut de Minéralogie et de Physique des Milieux Condensés, Paris 6, Campus Boucicaut, Bat. 7, 140 rue de Lourmel, 75015 Paris, France. (dandrault@ipgp.jussieu.fr)

C. G. Farnetani, Laboratoire de Dynamique des Systèmes Géologiques, Institut de Physique du Globe de Paris, 4 place jussieu, B.P. 89, 75252 Paris CEDEX 05, France. (cinzia@ipgp.jussieu.fr)

H. Samuel, Department of Geology and Geophysics, Kline Geology Laboratory, Yale University, P.O. Box 208109, New Haven, Connecticut 06520-8109, USA. (henri.samuel@yale.edu)

Numerical Study of the Origin and Stability of Chemically Distinct Reservoirs Deep in Earth's Mantle

P. van Thienen[1], J. van Summeren, R. D. van der Hilst[2], A. P. van den Berg, and N. J. Vlaar

Institute of Earth Sciences, Utrecht University, Utrecht, The Netherlands

Seismic tomography is providing mounting evidence for large scale compositional heterogeneity deep in Earth's mantle; also, the diverse geochemical and isotopic signatures observed in oceanic basalts suggest that the mantle is not chemically homogeneous. Isotopic studies on Archean rocks indicate that mantle inhomogeneity may have existed for most of the Earth's history. One important component may be recycled oceanic crust, residing at the base of the mantle. We investigate, by numerical modeling, if such reservoirs may have been formed in the early Earth, before plate tectonics (and subduction) were possible, and how they have survived—and evolved—since then. During Earth's early evolution, thick basaltic crust may have sunk episodically into the mantle in short but vigorous diapiric resurfacing events. These sections of crust may have resided at the base of the mantle for very long times. Entrainment of material from the enriched reservoirs thus produced may account for enriched mantle and high-μ signatures in oceanic basalts, whereas deep subduction events may have shaped and replenished deep mantle reservoirs. Our modeling shows that (1) convective instabilities and resurfacing may have produced deep enriched mantle reservoirs before the era of plate tectonics; (2) such formation is qualitatively consistent with the geochemical record, which shows multiple distinct ocean island basalt sources; and (3) reservoirs thus produced may be stable for billions of years.

1. INTRODUCTION

Understanding mantle convection and the relationships with plate tectonic motion at Earth's surface and heat loss from Earth's deep interior remains one of the most challenging objectives in global geophysics. For many decades the discussions centered on canonical end-member models of strict layering or unhindered whole-mantle flow, but over the past decade cross-disciplinary research has gradually shifted the paradigm toward a class of hybrid models that combine certain aspects of the classical models. Reconciling constraints from different research disciplines remains difficult, however.

Geochemical analysis of mid-ocean ridge basalts (MORB) and ocean island basalts (OIBs) have been used to argue for the long term survival of compositionally distinct domains in Earth's mantle. The arguments have been reviewed extensively elsewhere [e.g., *Hofmann*, 1997; *Albarède and van der Hilst*, 2002; *Anderson*, 2002b; *Davies*, 2002; *Harrison and Ballentine*, this volume]; here we mention only a few observations. Sampling issues aside, the surprisingly uniform trace element composition of MORBs suggests their derivation

---

[1] Currently at Institut de Physique du Globe de Paris, Saint-Maur-des-Fossés, France.
[2] Also at Department of Earth, Atmospheric and Planetary Sciences, Massachusetts Institute of Technology, Cambridge, Massachusetts.

Earth's Deep Mantle: Structure, Composition, and Evolution
Geophysical Monograph Series 160
Copyright 2005 by the American Geophysical Union
10.1029/160GM09

from a well-mixed source (typically referred to as the depleted MORB mantle, or DMM). In contrast, OIB analyses have revealed a markedly more diverse pattern, resulting in a rich nomenclature that includes enriched mantle 1 and 2 (EM-1 and EM-2) and HIMU, which has high-$\mu = {}^{238}U/{}^{204}Pb$ (see Table 1) and the postulation of several distinct geochemical reservoirs [e.g., *Zindler and Hart*, 1986; *Hofmann*, 1997; *Harrison and Ballentine*, this volume]. The longevity of some of these reservoirs (apparent age up to 2.5 Gyr), and, thus, their relative stability against remixing, was inferred from lead isotope dating [*Chase*, 1978; *Tatsumoto*, 1978; *Allègre and Lewin*, 1995]. "Primitive" He ratios observed in OIBs may suggest input from an undifferentiated mantle source, but this interpretation has been questioned [*Porcelli and Halliday*, 2001; *Meibom et al.*, 2003]. Furthermore, the energy balance between heat production, heat loss, and planetary cooling, is often used to argue for "hidden" reservoirs of enhanced heat production [e.g., *Kellogg et al.*, 1999, and references therein], although there is some uncertainty about the heat flux out of the core.

The presence of distinct mantle reservoirs is consistent with cosmogenic arguments and heterogeneous accretion models of Earth's formation [e.g., *Anderson*, 2002a; *Righter*, this volume] and with mounting evidence for compositional heterogeneity in the deep mantle from seismic tomography [*Creager and Jordan*, 1986; *van der Hilst and Kárason*, 1999; *Masters et al.*, 2000; *Saltzer et al.*, 2001, 2004; *Trampert et al.*, 2004; *Trampert and van der Hilst*, this volume]. Collectively, this evidence suggests that instead of whole-mantle convection a kind of layered or zoned convection is more realistic.

Traditional layered convection models assumed that the 660 km discontinuity between upper and lower mantle formed such a reservoir boundary. However, seismic tomography shows subducting slabs crossing the 660 km discontinuity and sinking into the deeper mantle beneath several convergent margins, which suggests that the boundary between the different chemical reservoirs cannot be this discontinuity [*van der Hilst et al.*, 1997]. In an attempt to break away from canonical—and

unsuccessful—end-member models of unhindered whole-mantle convection or strict stratification at 660 km depth, *van der Hilst and Kárason* [1999] and *Kellogg et al.* [1999] postulated the existence of compositional layering in the deep lower mantle and explored its gravitational stability. Their results show that a deep layer with a relative chemical density contrast of 4% is stable, although significant topography may develop. The effective density contrast may approach zero, however, because of the higher internal temperature of the layer due to enhanced internal heating. Numerical models by *Hansen and Yuen* [1988, 2000], with the latter using a larger aspect ratio of their model domain, advecting heat sources, and strongly temperature and pressure dependent viscosity, and by *Tackley* [2002] (using a 3D model), as well as the analog experiments by *Davaille* [1999] also show the development of the deep dense layer with significant topography. Numerical experiments by *Samuel and Farnetani* [2003, and this volume], which include trace element fractionation and degassing leading to distinct helium signatures, show that a chemical density excess of only 2.4% may be sufficient to stabilize a deep layer.

Absent strong seismological evidence for a well-defined layer, several alternative configurations of these main reservoirs have been proposed [see, e.g., *Tackley*, 2002, for a review], with the part of the deep mantle that contains more primitive material being capped by enriched recycled crust [*Hofmann*, 1997; *Coltice and Ricard*, 1999, and references therein], the enriched crustal reservoir residing at the core–mantle boundary (CMB) having a layer or mounds of primitive mantle on top [*Christensen and Hofmann*, 1994; *Hansen and Yuen*, 2000; *Tackley*, 2000], a layer of enriched crustal material at the CMB with pockets of primitive mantle residing throughout the mantle [*Becker et al.*, 1999], and an unstratified but heterogeneous mantle in which depleted mantle and subducted lithosphere are continuously mixed [*Allègre and Turcotte*, 1986; *Helffrich and Wood*, 2001]. Recognizing the lack of evidence for global seismic interfaces anywhere between 660 km and the top of the so-called

Table 1. Oceanic Basalt Source Reservoirs and Their Distinguishing Isotopic Characteristics.

Source[a]	$\frac{{}^{143}Nd}{{}^{144}Nd}$	$\frac{{}^{87}Sr}{{}^{86}Sr}$	$\frac{{}^{206}Pb}{{}^{204}Pb}$	$\frac{{}^{207}Pb}{{}^{204}Pb}$	$\frac{{}^{238}U}{{}^{204}Pb}$	$\frac{{}^{187}Os}{{}^{188}Os}$	$\frac{{}^{4}He}{{}^{3}He}$	Interpretation
DMM	high	low	low	low		low		
HIMU	high (<DMM)	low	high	high	high	high		Recycled crust
EM-1	low	interm.	low	rel. low		high		Recycled (continental) crust
EM-2	low	high	low	rel. high		interm.		Recycled continental crust
FOZO/C	high	low	interm.	interm.		low	high	

[a]DMM, depleted MORB mantle; HIMU, high-μ (${}^{238}U/{}^{204}Pb$); EM, enriched mantle; FOZO, focal zone; and C, common. Information compiled from *Zindler and Hart* [1986], *Chauvel et al.* [1992], *Hart et al.* [1992], *Hauri and Hart* [1993], *Hanan and Graham* [1996], *Hofmann* [1997], and *Hauri* [2002].

D″ layer, *Albarède and van der Hilst* [2002] proposed zoned mantle convection, a modification of the model by Kellogg and coworkers, in which variable depth subduction sets up a radial gradient in mantle mixing that may leave the lowermost mantle largely exempt from convective overturn.

In addition to demonstrating conclusively its existence, its nature, and its volume, there are two fundamental questions regarding our understanding of deep mantle heterogeneity. First, when and how was it formed? Second, how did it evolve over long periods of geological time?

Several origins of compositional heterogeneity in Earth's deep mantle have been proposed. Interaction of liquid iron from the outer core with silicates in the lowermost mantle may produce a heavy phase in the lowermost mantle [*Knittle and Jeanloz*, 1991; *Guyot et al.*, 1997]. However, the effect on major element composition of present-day core–mantle interaction is likely to be confined to a relatively small region above the CMB and cannot explain the seismically observed compositional heterogeneity in the bottom 1000 km of the mantle [*van der Hilst and Kárason*, 1999; *Trampert et al.*, 2004; *Trampert and van der Hilst*, this volume]. According to *Scherstén et al.* [2004], tungsten isotopes in OIBs rule out a core contribution to their source, but their conclusion is somewhat controversial [see *Jellinek and Manga*, 2004], and evidence to the contrary also has been reported [*Humayun et al.*, 2004]. The formation of a deep reservoir by perovskite fractionation in a magma ocean, suggested by *Agee and Walker* [1988], is not compatible with geochemical evidence [*Halliday et al.*, 1995; *Blichert-Toft and Albarède*, 1997]. The sinking of subducted slabs of former oceanic lithosphere into the lower mantle may add enriched crustal material to a hypothetical reservoir somewhere in the lower mantle. *Christensen and Hofmann* [1994] showed by numerical modeling that this could produce the range of isotopic ratios found in oceanic basalts. *Albarède and van der Hilst* [2002] suggested that in a zoned mantle, an enriched reservoir, possibly of early origin, could be replenished and maintained by selective deep subduction of enriched, eclogitic oceanic plateau crust.

While plate tectonics is unlikely to have been important in a hot early Earth [*Sleep and Windley*, 1982; *Vlaar*, 1985; *Vlaar and van den Berg*, 1991; *van Thienen et al.*, 2004c], there are geochemical indications of the existence of distinct reservoirs in the early Earth. Most likely, the earliest Archean rocks found on Earth have been derived from an already depleted mantle [e.g., *Hamilton et al.*, 1983; *Patchett*, 1983; *Vervoort et al.*, 1996; *Blichert-Toft et al.*, 1999]. The age of this differentiation has been found from Lu-Hf data to be at least 4.08 Gyr before present [*Amelin et al.*, 2000]. Sm-Nd data indicate an even older mean age of differentiation: 4.460 ± 0.115 Gyr [*Caro et al.*, 2003]. Hf isotopes in carbonatites and kimberlites from Greenland and North America of up to 3 Gyr of age indicate derivation of these rocks from an enriched mantle source with an unradiogenic Hf signature [low ^{176}Hf/^{177}Hf; *Bizzarro et al.*, 2002]. *van Thienen et al.* [2004b] proposed that prior to plate tectonics the necessary fluxes could have taken place in the form of the episodic rapid sinking of complete segments of crust (>1000 km long) into the mantle.

In this paper, we describe and model two dynamic regimes, representative for the early and present-day Earth, that we think are important for producing and maintaining deep mantle heterogeneity. In this paper we do not attempt to study the transition between these regimes. Building on previous work [*van Thienen et al.*, 2004b], we investigate the possibility of an early Earth formation of a deep enriched mantle reservoir by convective instabilities (diapirism). Subsequently, we use numerical convection experiments to investigate the long-term dynamical behaviour and stability of such a reservoir and to evaluate the geochemical characteristics. In the context of our experiments we use the term "stability" to indicate that the deep layer survives on a time scale of billions of years. Processes are studied in the context of a transient mantle dynamic regime, and predictions from our models are compared with geochemical constraints.

2. NUMERICAL MODELS: GENERAL METHODOLOGY

In separate numerical experiments we investigated both the formation of enriched deep mantle reservoirs under thermal conditions representative of the early Earth and the long-term evolution and stability of such deep mantle reservoirs. Before discussing these experiments in more detail, we here describe the general numerical elements.

We use a mantle convection code based on the finite element package SEPRAN [*Segal and Praagman*, 2000; *van den Berg et al.*, 1993] to solve the energy, Stokes, and continuity equations in the extended Boussinesq approximation [e.g., *Steinbach et al.*, 1989]:

$$\rho c_p \left(\frac{\partial T}{\partial t} + u_j \partial_j T \right) - \alpha T \frac{dp}{dt} = \tau_{ij} \partial_j u_i + \partial_j (k \partial_j T)$$
$$+ \rho_0 H + \frac{\Delta S}{c_p} \frac{dF}{dt} T + \sum_k \frac{\gamma_k \delta \rho_k T}{\rho_0^2 c_p} \frac{d\Gamma_k}{dt}, \quad (1)$$

$$\partial_j \tau_{ij} - \partial_i \Delta p + \Delta \rho g_i = 0, \quad (2)$$

$$\partial_j u_j = 0. \quad (3)$$

For symbol definitions, see Tables 2 and 3.

Table 2. Symbol Definitions and Parameter Values.

Symbol	Property	Value/unit
c_p	heat capacity at constant pressure	1250 Jkg^{-1}K^{-1}
F	degree of depletion	
g	gravitational acceleration	9.8 m s^{-2}
H	radiogenic heat productivity	Wkg^{-1}
k	thermal conductivity	Wm^{-1}K^{-1}
p	pressure	Pa
Δp	nonhydrostatic pressure perturbation	Pa
ΔS	entropy change upon full differentiation	300 Jkg^{-1}K^{-1}
t	time	s
T	temperature	°C
T_0	non-dimensional surface temperature	273/ΔT
u	velocity	m s^{-1}
z	depth	m
$z_0(T)$	temperature-dependent depth of phase transition	m
γ_k	Clapeyron slope for transition k	Pa K^{-1}
Γ_k	phase function for transition k	
δz	depth range of phase transition	m
κ	thermal diffusivity	10^{-6} m^{2}s^{-1}
ρ	density	kg m^{-3}
$\delta\rho_k$	density increase of phase transition k	kg m^{-3}
τ_{ij}	deviatoric stress tensor	Pa
$\tau_{1/2}$	half-life of radiogenic heating	2.5 Gyr
Experiment I only		
h	depth scale	1200 · 10^{3}m
ΔT	temperature scale	2450°C
α	thermal expansion coefficient	3 · 10^{-5} K^{-1}
η_0	viscosity scale	10^{20} Pa s
ρ_0	reference density	3416 kg m^{-3}
$\delta\rho$	density difference:	
	peridotite upon full depletion	-226 kg m^{-3}
	basalt	-416 kg m^{-3}
	eclogite	200 kg m^{-3}
	felsic material	-800 kg m^{-3}
Experiment II only		
h	depth scale	2900 · 10^3 m
ΔT	temperature scale	4000°C
α	thermal expansion coefficient	K^{-1}
α_0	reference thermal expansion coefficient	2 · 10^{-5} K^{-1}
$\Delta\alpha$	thermal exp. contrast across mantle	5
η_0	viscosity scale	5 · 10^{20} Pa s
ρ_0	reference density	3400 kg m^{-3}

For numerical purposes, equations (1)–(3) are transformed into their non-dimensional equivalents using the following scaling:

$$T = \Delta T \cdot T', x_i = h \cdot x_i', t = \frac{h^2}{\kappa} \cdot t', u_i = \frac{\kappa}{h} \cdot u_i',$$
$$p = \frac{\eta_0 \kappa}{h^2} \cdot p', \tau = \frac{\eta_0 \kappa}{h^2} \tau' \qquad (4)$$

with the parameters as listed in Table 2; the primes indicate dimensionless parameters.

Flow is driven by density perturbations, which are related to variations of temperature, composition, and phase, and are described by the equation of state:

$$\Delta \rho = \rho_0 \left\{ -\alpha(T - T_{ref}) + \sum_k \Gamma_k \frac{\delta\rho_k}{\rho_0} + \frac{\delta\rho}{\rho_0} F \right\} \quad (5)$$

The second term inside the braces of equation (5) describes the effect of solid-state phase transitions of mantle peridotite around 670 km (in experiments I and II) and 400-km

Table 3. Parameter Definitions.

Symbol	Definition	Experiment
e_{ij}	$\partial_j u_i + \partial_i u_j$	I,II
e	$[\frac{1}{2} e_{ij} e_{ij}]^{\frac{1}{2}}$	I,II
k	$\kappa \rho c_p$	I,II
α	$\dfrac{\alpha_0 \Delta \alpha}{[(\Delta \alpha^{1/3} - 1)(1 - z) + 1]^3}$	II
Γ_k	$\frac{1}{2} [1 + \sin(\pi \frac{z - z_0(T)}{\delta z})]$	I,II
$\delta \rho$	$\frac{\partial \rho}{\partial F}$	I
$\delta \rho_k$	$\frac{\partial \rho}{\partial \Gamma_k}$	I,II
τ_{ij}	ηe_{ij}	I,II
τ	$[\frac{1}{2} \tau_{ij} \tau_{ij}]^{\frac{1}{2}}$	I,II

(experiment I) depth. This term also includes the effects of solid-state phase transitions of basaltic material. The last term in brackets describes the effect of chemical depletion due to partial melting, as discussed later.

3. EXPERIMENT I: EARLY FORMATION OF DEEP MANTLE RESERVOIRS

In this section we investigate by geodynamical modeling Earth's early differentiation and the formation of a deep reservoir enriched in incompatible trace elements by downward transport and deep storage of oceanic crust. Because of resolution requirements, we limit the computational domain to a depth of 1,200 km, thus focusing on the processes of crust production and recycling that take place at the surface and in the relatively shallow mantle.

3.1. Numerical Aspects of Experiment I

3.1.1. Model description. In experiment I (see Table 4) deformation is controlled by a composite rheology [*van den Berg et al.*, 1993], which includes diffusion creep, dislocation creep, and a stress limiter mechanism [*van den Berg et al.*, 1993; *van Hunen et al.*, 2002]. We also consider partial melting and melt segregation and their effects on diffusion and dislocation creep. As described in *van Thienen et al.* [2003], the effect of partial melting is controlled by a parameter called the degree of depletion, F, which indicates the mass fraction of melt removed from a control volume of mantle rock. The parameter values used in experiment I are given in Table 5.

Diffusion and dislocation creep are described by an Arrhenius relation [*Karato and Wu*, 1993; *van den Berg and Yuen*, 1998]:

$$\eta_i = f(F) B_i \exp \left[\frac{E_i + PV_i}{RT} \right] \tau^{1-n_i}, \qquad (6)$$

with the proportionality factor $f(F)$ representing the effect of partial melting (see below). A stress-limiter mechanism is included to approximate brittle failure of the lithosphere. It is implemented according to *van Hunen et al.* [2002]:

$$\eta_y = \tau_y \dot{\varepsilon}_y^{-1/n_y} \dot{\varepsilon}^{(1/n_y)-1}, \qquad (7)$$

where the normal yield strength τ_y is approximated in Byerlee's law with the lithostatic pressure [e.g., *Moresi and Solomatov*, 1998]:

$$\tau_y = C_0 + \mu \rho g z. \qquad (8)$$

To simulate the effect on viscosity of dehydration during partial melting of mantle material (model run I-a, Table 4), we varied $f(F)$ in equation (6) linearly between 1 and 10 for the values of F between 0.005 and 0.05 [*Karato*, 1986; *Hirth and Kohlstedt*, 1996; *Mei and Kohlstedt*, 2000a, b]. We set f to a constant value of 1 for $F < 0.005$ and to 10 for $F > 0.05$. For eclogite we use $f = 0.1$ because of the lower flow strength of omphacite [*Piepenbreier and Stöckhert*, 2001]. For the partial melting of basaltic crust (model run I-b, Table 4), we applied the solidus and liquidus of a hydrated tholeiite, using data from *Green* [1982].

In our numerical experiments, partial melting occurs when the equilibrium degree of depletion F corresponding to the local pressure and temperature conditions is higher than the actual local degree of depletion. We assume complete melt extraction but we do not account for compaction of the residual rock. Upon melt extraction an equivalent amount of basaltic material is added at the top boundary of the model (inflow boundary condition), directly above the region of partial melting, where it is assumed to be hydrated.

Upon partial melting, the local incompatible trace element concentration is adjusted according to a batch melting model with a bulk partition coefficient of 10^{-2}, comparable to that of K and somewhat greater than that of U and Th [*Henderson*, 1982; *Beattie*, 1993]. This results in a concentration of trace elements in the basaltic crust. The local rate of radiogenic heat production is proportional to the incompatible trace element concentration.

In our models, the basaltic crustal material produced by partial melting of mantle peridotite undergoes solid-state phase transitions at 30- (transformation to eclogite, with a relaxation time of 1.25 Myr) and 400-km depths. The latter transition corresponds to the decomposition of omphacite which results in an increase of the garnet content [see *Irifune and Ringwood*, 1993; *Okamoto and Maruyama*, 2004] and to the transformation of coesite into stishovite [see *Aoki and Takahashi*, 2004]. In our model these transitions are combined and assumed to occur instantaneously. Although the actual depth of the basalt to eclogite phase transition is closer to 40 km [1.2 GPa;

Table 4. Numerical Models Used in This Study and Their Characteristics. $\delta\rho_c/\rho_0$ indicates the relative intrinsic excess density of the deep layer, and H_{enr}/H_{depl} its relative enrichment in heat-producing elements relative to the rest of the mantle. The effective thermal Rayleigh numbers at the start (Ra_{start}^{eff}) and the end (Ra_{end}^{eff}) of the experiments are indicated, as are the Plates illustrating the results.

Model	Specific characteristics		Ra_{start}^{eff}	Plate
I-a	partial melting of mantle melt segregation $\rightarrow$ crustal production		$9.0 \cdot 10^8$	1,2
I-b	I-a plus partial melting of basaltic crust to form felsic material		$6.5 \cdot 10^6$	3

Model	$\delta\rho_c/\rho_0$	H_{enr}/H_{depl}	Ra_{start}^{eff}	Ra_{end}^{eff}	Plate
II-a	0.000	1	$7.2 \cdot 10^6$	$1.6 \cdot 10^6$	5a
II-b	0.005	1	$7.2 \cdot 10^6$	$2.1 \cdot 10^6$	5b
II-c	0.010	1	$7.2 \cdot 10^6$	$1.8 \cdot 10^6$	5c
II-d	0.015	1	$7.2 \cdot 10^6$	$2.3 \cdot 10^6$	5d
II-e	0.010	2	$7.2 \cdot 10^6$	$2.3 \cdot 10^6$	4

Hacker, 1996], garnet starts to be formed from about 0.7 to 0.8 GPa [*Green and Ringwood*, 1967; *Ito and Kennedy*, 1971; *Hacker*, 1996], raising the density of the assemblage above that of basalt. We note, however, that an increase of 10 km in transformation depth results in similar model dynamics and would therefore not affect the conclusions of the research reported here. We prescribe an excess density of eclogite relative to undepleted peridotite of 200 kg m^{-3}, similar to that used in earlier numerical studies [*Dupeyrat and Sotin*, 1995; *van Hunen et al.*, 2002]. A lower excess density would result in a stretching of the time scale of the resurfacing process, whereas a higher excess density would speed it up.

The density inversion between eclogite and peridotite below 670 km depth [e.g., *Ringwood and Irifune*, 1988; *Irifune and Ringwood*, 1993] is not taken into account. Numerical experiments by *Christensen* [1988] suggest that no long-term trapping of eclogite in this zone is expected. New petrological data by *Hirose et al.* [1999] show that the inferred density inversion zone is narrower than reported earlier (with the lower boundary at 720 km instead of 800 km), and the positive Clapeyron slope of the lower bounding transition would cause a further narrowing for cool subducting material. On the basis of these results, *Hirose et al.* [1999] also expect no significant trapping of basaltic material in this zone.

3.1.2. Model domain and initial and boundary conditions. Approximately 5,100 quadratic elements were used to discretize the domain, measuring 1,200 × 1,200 km, for solving the energy and momentum equations (1)–(3). The chemical evolution was evaluated by means of approximately 350,000 active particle tracers.

An initial condition was obtained by the internal heating (extreme rate of $250 \cdot 10^{-12}$Wkg^{-1}) of a pristine subsolidus mantle to high temperatures, whereby a crust was formed by partial melting of the convecting mantle. The composition and temperature field at the time when a 30-km crust had been produced was taken as the initial condition for the main numerical experiment (see Plate 1a), with the internal heating reduced to $15 \cdot 10^{-12}$Wkg^{-1}, scaled by the concentrations of local trace elements and decaying with time (see Table 2). This internal heating rate corresponds to Bulk Silicate Earth values around 3.5–4.0 Gyr before present [*van Schmus*, 1995]. Note that the layered convection in the initial situation is produced by the model and not prescribed explicitly.

The vertical boundaries have a periodic boundary condition. The top boundary has a prescribed temperature of 0ºC, a tangential stress-free condition, and a prescribed inflow velocity corresponding to the amount of basaltic crust being produced by partial melting of the underlying mantle. The lower boundary has a free-slip condition and is thermally insulated. This latter point results in an absence of plumes from the deep lower mantle. However, because deep mantle plumes can be expected to affect the positioning of sinking crust rather than its rate of accumulation, we do not expect this to significantly influence our results. It will affect the stability of the deep, dense layer once that has been formed, which will be the subject of investigation in the second series of experiments.

3.2. Large-Scale Resurfacing in Earth's Early Evolution

Results of numerical experiment I-a (Table 4) are presented in Plate 1. For each pair of panels, the ones on the left show

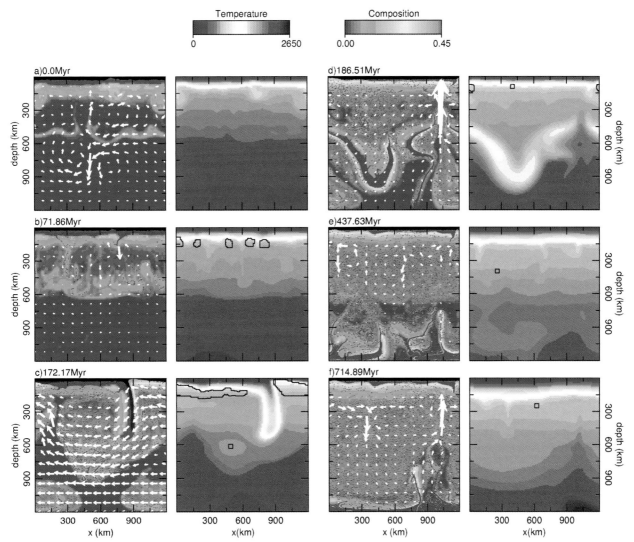

Plate 1. Secular evolution of the composition (left) and temperature (right) fields of model I-a. The color scale in the composition plots indicates the degree of depletion of mantle peridotite. Black indicates basalt, and red indicates eclogite. The arrows show the instantaneous flow field. The contours in the temperature frames indicate regions of partial melting.

Table 5. Definitions and Values of Rheological Parameters for Experiment I.

Symbol	Property	Value/unit
B_1	diffusion creep prefactor	Pa s
B_2	dislocation creep prefactor	$Pa^{n_2}s$
C_0	cohesion factor	0 Pa
E_1	diffusion creep activation energy	$270 \cdot 10^3$ Jmol^{-1}
E_2	dislocation creep activation energy	$485 \cdot 10^3$ Jmol^{-1}
e_{ij}	strain rate tensor	s^{-1}
e	second invariant of the strain rate tensor	s^{-1}
$f(F)$	composition-dependent viscosity prefactor	
n_1	diffusion creep stress exponent	1
n_2	dislocation creep stress exponent	3.25
n_y	yield exponent	10
R	gas constant	8.341 Jmol^{-1}K^{-1}
V_1	diffusion creep activation volume	$6 \cdot 10^{-6}$m^{3}mol^{-1}
V_2	dislocation creep activation volume	$7.5 \cdot 10^{-6}$m^{3}mol^{-1}
$\dot{\varepsilon}$	strainrate	s^{-1}
$\dot{\varepsilon}_y$	yield strainrate	10^{-15}s^{-1}
η	viscosity	Pa s
η_0	reference viscosity	10^{20} Pa s
η_y	yield viscosity	Pa s
μ	friction coefficient	0.03
σ_n	normal stress	Pa
τ_{ij}	deviatoric stress tensor	Pa
τ	second invariant of the deviatoric stress tensor	Pa
τ_y	yield stress	Pa

the composition field at six different times. The color scale indicates the degree of depletion of mantle peridotite, ranging from fertile (blue) to depleted (orange). Black tracers are used to monitor the basaltic material, and red tracers represent eclogite. White arrows illustrate the instantaneous flow field. The corresponding temperature field at each time, is shown in the right panel; the color scale indicates local temperature, and the contours depict regions of partial melting.

Plate 1b shows that in the first 70 Myr of model evolution small-scale delamination of the lower crust results in the mixing of small amounts of eclogitic lower crust into the upper mantle [see *van Thienen et al.*, 2004b].

Around 170 Myr (Plate 1c), a locally thickened section of crust, largely transformed into eclogite, starts to sink into the mantle; in doing so, it pulls thinner neighboring crust along with it, which initiates a resurfacing event. New crust is being produced near the sinking crustal section by partial melting of complementary upwelling mantle material (see melting contour in temperature plot). The resurfacing process continues until the sinking crust breaks off, at which time—in this model run—some 1,500 km of crust has been pulled into the mantle.

Plate 1d shows the crustal material sinking into the lower mantle, forcing lower mantle material into the upper mantle.

The subsequent frames show the eclogite body settling on the bottom boundary. Because of its high content of radiogenic trace elements, the eclogite body heats up relatively quickly (Plate 1f).

3.3. Trace Element Evolution

The secular evolution of the distribution of a (hypothetical) incompatible trace element (partition coefficient 10^{-2}) is illustrated in Plate 2. The data shown are corrected for radioactive decay, such that only the effects of fractionation are visible in this figure. The panels show snapshots of the entire computational domain. Yellow regions are depleted relative to the initial uniform composition, and gray/blue regions depict enrichment. Red and green curves indicate the location of the phase boundaries for peridotite (red) and basalt (green) at 10% (dashed), 50% (solid), and 90% (dashed) of the transition.

After ~120 Myr of model evolution, the upper mantle is somewhat depleted by the extraction of an enriched basaltic crust, with the highest degree of trace element depletion occurring in the shallow upper mantle; the lower mantle, which at this early stage in model evolution convects separately from the upper mantle, is still pristine. Continued

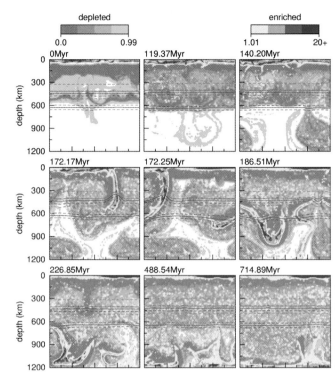

Plate 2. Incompatible trace element concentrations, corrected for radio-active decay, at nine different times in the evolution of model I-a (see also Plate 1). Pristine material is white, depleted material is yellow, and enriched material is blue/gray. Curves indicate phase boundaries at 50% (solid) and 10% and 90% (dashed) of the transition for peridotitic (red) and basaltic (green) compositions.

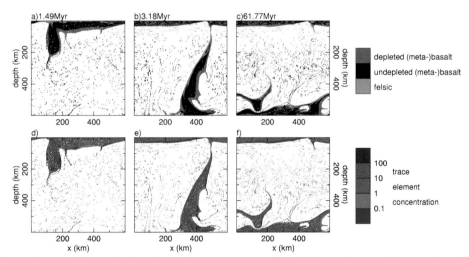

Plate 3. Nonmantle material type (a–c) and trace element concentration relative to those at pristine mantle peridotite (d–f) of model I-b, which is started from a 600 × 600 clipping from the upper right corner of the 1200 × 1200-km domain of model I-a at some time between (b) and (c) of Plate 1.

partial melting gradually depletes the upper mantle, and some exchange of material between the upper and lower mantle takes place, including small blobs of delaminated eclogite (gray).

The resurfacing event illustrated in Plate 1 is clearly visible around 170 Myr. During the 50 Myr that follow, the old, enriched crust slowly sinks to the bottom of the model mantle, where it remains stable.

The final situation, in terms of incompatible trace elements, shows (a) an enriched basaltic crust; (b) a highly depleted shallow mantle; (c) a moderately depleted deep upper mantle and lower mantle, in which small blobs of enriched eclogite (gray spots) are present; and (d) an enriched reservoir at the bottom boundary of the mantle.

3.4. The Effect of Continental Crust Production

Using the instantaneous composition and temperature field of model I-a at the onset of the resurfacing event, between (b) and (c) of Plate 1, we computed model I-b at a higher resolution in a 600 × 600-km domain [upper right quarter of model I-a; see *van Thienen et al.*, 2004a, for more details]. In this model, partial melting of the basaltic crust was included. The purpose of this numerical experiment was to investigate the effect of this additional melting process on the trace element geochemistry of the material that eventually forms the deep, dense layer. Note that reduction of the Rayleigh number by reducing the domain size (see Table 4) does not significantly affect the resurfacing event. The resulting model development is illustrated in Plate 3. Plate 3a–c shows only (meta-)basaltic and felsic material. Felsic material, indicated in green, is produced by the partial melting of hydrous (meta-)basalt. This results in depletion of the source rock (purple). Undifferentiated (meta-)basalt is indicated in black. The major part of the sinking crust consists of undepleted metabasalt, but significant amounts of felsic material and depleted metabasalt are also present. The corresponding trace element concentration plots (Plate 3d–f) show that both undepleted and felsic material are enriched in incompatible trace elements relative to undifferentiated mantle peridotite. (Meta-)basalt that has undergone partial melting shows depletion.

4. EXPERIMENT II: LONG-TERM STABILITY OF DEEP MANTLE RESERVOIRS

In numerical experiment II (see Table 4 for parameter settings of the different model runs) we consider the long-term behavior of compositionally distinct deep mantle domains. We assume that at the start of the model calculations such domains had already been formed, for instance by the mechanism dis-

cussed above. Because the formation process is not modeled, we need less resolution than in the previous numerical experiments and can include the entire depth of the mantle.

4.1 Numerical Aspects of Experiment II

4.1.1. Model description. In contrast to experiment I, partial melting and fractionation of trace elements are not included in experiment II. Since dislocation creep is predominant at relatively low pressures only [*Karato and Li*, 1992; *van den Berg and Yuen*, 1996], we ignore this deformation mechanism in our study of the long-term dynamic behavior of intrinsically dense domains in the lower mantle. Instead, we include a pressure- and temperature-dependent Newtonian viscosity:

$$\eta(T, z) = 100\eta_0 \exp\left[-\ln(20) \cdot T + \ln(100) \cdot z\right] \quad (9)$$

An endothermic phase transition at around 660-km depth is defined by using a negative Clapeyron slope of $\gamma = dp/dT = -2.5 \cdot 10^6$ PaK^{-1}. Since this is believed to be the main obstacle for material exchange between the upper and lower mantle and since we do not aim here to produce a detailed picture of the transition zone dynamics, no phase transitions other than at 660-km have been modeled.

The deep mantle reservoir has a variable excess intrinsic density (see Table 4). Table 4 also lists the various distributions of the internal heating rate between the deep reservoir and the overlying material (H_{enr} and H_{depl}, respectively). In all numerical experiments, the heat productivity integrated over the total mantle volume has the same value. This value was chosen such that the volume-averaged heat productivity has the chondritic value of $5.0 \cdot 10^{-12}$ Wkg^{-1} after a 4.5-Gyr period of radioactive decay (half-life time = 2.5 Gyr). Then it was scaled by the factor 0.629 to correct for an exaggerated heat production that would result from an overestimation of the mantle volume due to the 2D, Cartesian geometry of the model (because its mantle is much shallower, this effect is not so important in experiment I). We note that the thermal expansivity α decreases fivefold over the depth range of the model (see Tables 2 and 3).

4.1.2. Model domain and initial and boundary conditions. Our Cartesian box model scales to having a depth of 2,900 km. An aspect ratio of 2.5 is applied. The numerical mesh contains 250,000 tracers that were initially placed at random locations over a total of 11,000 finite elements.

Free-slip conditions exist on all model boundaries. A temperature of 0°C is prescribed at the surface. On the vertical boundaries, reflecting temperature boundary conditions exist. At the bottom boundary, which represents the CMB, a thermal coupling with the core is implemented by prescribing the time-dependent CMB temperature $T_{CMB}(t)$ to equal the temperature of an isothermal heat reservoir, which is cooled by the CMB heat flux, calculated from the finite-element mantle model. The heat capacity of the core heat reservoir is set to a fraction $X = 0.8$ of the mantle heat capacity.

The initial model conditions were chosen to simulate a hot early Earth, where chemical layering in the lowermost mantle is already present. A deep mantle reservoir is prescribed as a flat layer that occupies the lower 20% of the mantle volume. Initially, the temperature field represents a uniform hot mantle of 4000°C, which has been cooled for 60 Myr from the top boundary in order to prevent undesirable large temperature variations. Positive and negative temperature perturbations were added in the upper right and left corners of the domain, respectively, inducing convective circulation.

4.2. Evolution and Stability of Compositional Layering

4.2.1. Evolving flow regimes. We use model II-e to describe the most important characteristics of the models examined. The calculated composition, temperature, stream-function contours, and viscosity field are given in the snapshots at five different times in the model evolution in Plate 4. Figure 1 shows time series of the temperature, viscosity, flow velocity, heat flow, and entrainment rate (defined as the amount of primitive material that resides above the chemical interface relative to the total amount of primitive material present).

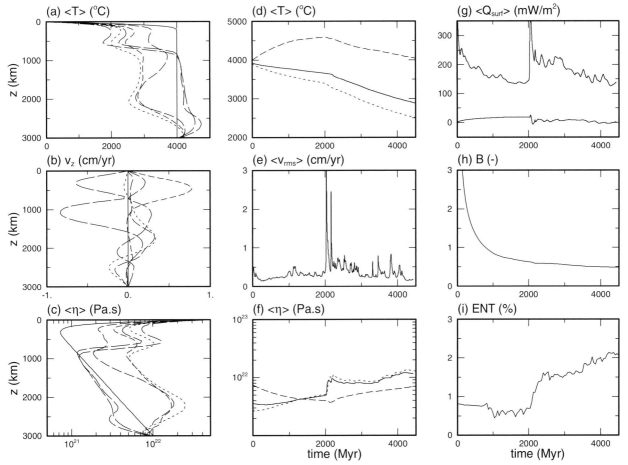

Figure 1. Time series and profiles of several global quantities for the reference model II-e. Panels (a)–(c) show vertical profiles of the horizontally averaged temperature, the vertical flow velocity at $x = 0.5$, and the horizontally averaged viscosity, respectively, at 893-Myr intervals (decreasing dash size corresponds to increasing model time). Panels (d)–(f) show time series of the volume-averaged temperature, root mean square velocity, and volume-averaged viscosity, respectively. Dashed and dotted curves in panels (d) and (f) indicate values for the enriched layer and the depleted mantle, respectively. Panels (g)–(i) show mean surface heat flux (top curve) and mantle to core heat flux (bottom curve), the buoyancy parameter B, and entrainment, respectively.

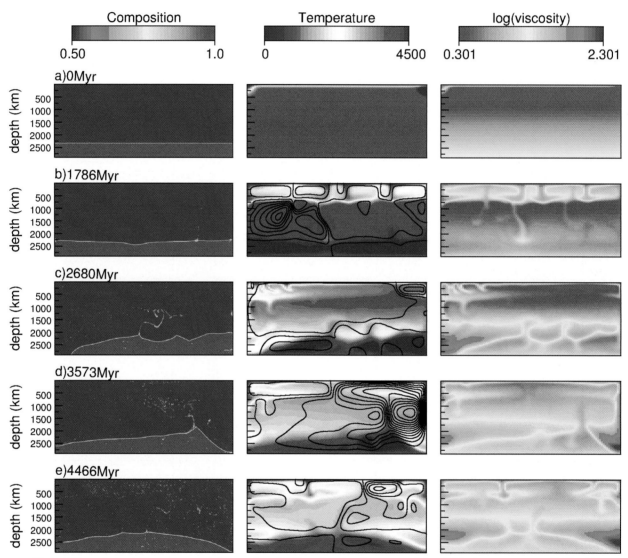

Plate 4. Snapshots at five subsequent times of model II-e, showing the composition field, the temperature field with stream lines, and the viscosity field (effective values range from 10^{21} Pa s (red) to 10^{23} Pa s (blue)).

During the first 2 Gyr of model time, convection is strictly layered at the 660-km phase transition, preventing material exchange between the upper and lower mantle (see Plate 4b). At these early times, the vigour of convection in the mid-mantle (the part of the lower mantle above the deep mantle reservoir) is too small to cause significant topography on the deep mantle reservoir.

Upon further cooling of the upper mantle (Figure 1a,d), the mantle downwellings grow larger and stronger (Plate 4b, Figure 1c,f), and some of them are now able to break through the phase transition and penetrate the lower mantle (Plate 4c). The inferred breakdown of layered mantle convection is in accord with results by *Steinbach et al.* [1993]. In our numerical experiments, the time to the change in dynamic regime depends mainly on the cooling rate of the upper mantle and varies between 1 and 2.2 Gyr: In model II-e it starts around 2 Gyr. The downwellings that penetrate the lower mantle produce pronounced flow velocities and large material fluxes across the phase transition (Plate 4b, 1b,e). They also create topography on the deep mantle reservoir, which at times is swept into isolated piles (Plate 4d,e). Plate 4e shows that after some 4.5 Gyr of model evolution, a partially layered convection regime has set in, in which some cold downwellings are temporarily obstructed by the endothermic phase transition near 670-km depth.

In this model, the cool downwelling material remains mostly on top of the deep, dense layer. The presence of the deep mantle reservoir thus limits the amount of contact between the relatively cold downwellings and the hot core, which results in thermal blanketing of the core (Figure 1a,d,g). At the relatively high temperatures in the deep mantle, the material originating from the surface heats up and eventually rises back to the surface in the form of mantle plumes (Plate 4d), entraining material from the bottom high-density reservoir. Entrainment is a possible explanation for the typical chemical signature in OIBs [*Zhong and Hager*, 2003] as was also shown in *Samuel and Farnetani* [2003], where different chemical components were modeled explicitly in a numerical evolutionary mantle model.

4.2.2. Long-term stability of the deep domains. In our computer simulations, the stability of compositional layering is controlled by a time-dependent buoyancy number, $B(t)$, which gives the ratio of the positive thermal buoyancy relative to negative compositional buoyancy:

$$B = \frac{1}{\alpha_0 \delta T(t)} \cdot \frac{\delta \rho_c}{\rho_0} \qquad (10)$$

where $\delta \rho_c$ is the compositional density difference and $\delta T(t)$ is the time-dependent temperature difference between the deep layer and the overlying mantle. We carried out systematic variations in model II-e to investigate the effects on layer stability of variations in excess compositional density and heat productivity in the deep mantle reservoir. The relevant parameters are listed in Table 4.

The time series of $B(t)$, calculated with equation (10), is shown in Figure 1h. Initially, when there is no temperature difference between the two reservoirs, the buoyancy number has a large value. When temperature differences appear, the value of $B(t)$ drops and layer stability decreases. Plate 5 shows snapshots of the composition and temperature fields after 4.5 Gyr of model time for models II-a, b, c, and d—models selected to illustrate the fate of the primitive reservoir when stability decreases.

The topography of the layer is small if the intrinsic density of the layer is large, and global layering as was proposed by *Kellogg et al.* [1999] is observed (model II-e in Plate 4). When the stability decreases, the deep reservoir is pushed aside by cold downwellings and is swept into isolated piles, as modeled by *Hansen and Yuen* [2000] and *Tackley* [2000] (model II-c in Plate 5c). Further reduction of stability leads to a regime where the primitive material is present as blobs of material floating in the lower mantle (model II-b in Plate 5b), which can be compared with the work of *Becker et al.* [1999]. Ultimately, under conditions least favoring stability, total mixing occurs (model II-a in Plate 5a).

The effects of density and heat productivity distribution on layer stability are examined separately. As a measure for stability, the mean depth of tracers initially placed in the deep mantle reservoir (with composition C_{enr}) is used. The mean depth of C_{enr} tracers is calculated at every time step as follows:

$$z_{C_{enr}} = \frac{1}{n} \sum_{i=1}^{n} z_i, \qquad (11)$$

where n is the amount of C_{enr} tracers, and z_i is depth of the ith tracer. To give a reference for interpreting this mean depth, we note that in the initial condition $z_{C_{enr}} = 0.9$ and for total mixing $z_{C_{enr}} = 0.5$.

In Figure 2a the mean depth evolution is compared for models II-a, b, c and d, which all have a uniform heat production. The density jump $\delta \rho_c / \rho_0$ varies from 0.0 to 0.015.

Figure 2a shows that the mean depth for all models starts at $z = 0.9$, as was expected. Then, after cold downwellings penetrate the lower mantle and induce layer instability, the mean depth becomes shallower. For subsequently smaller values of the density jump, the deviation from the initial condition is bigger, indicating less stable layering. In Figure 2b the mean depth is compared for models with a uniform density jump of 0.01 and a heat production ratio H_{enr}/H_{depl}

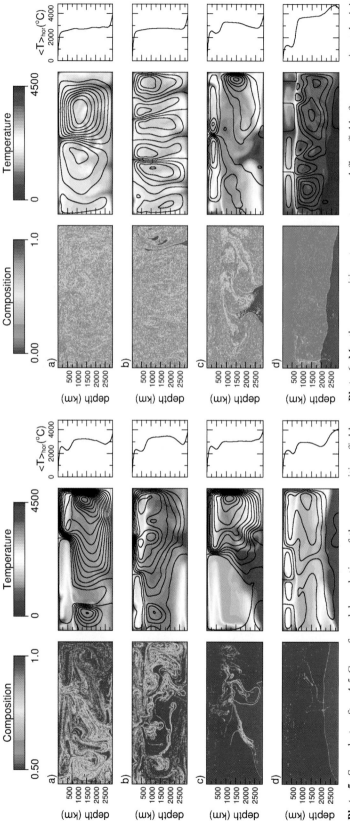

Plate 6. Mantle composition, temperature, and flow field after approximately 4.4 Gyr of evolution, for four different magnitudes of the postspinel Clapeyron slope. (a) $\gamma = 0$ MPa/K, (b) $\gamma = -1.25$ MPa/K, (c) $\gamma = -2.5$ MPa/K, (d) $\gamma = -3.75$ MPa/K.

Plate 5. Snapshots after 4.5 Gyr of model evolution of the composition field, temperature and flow field (stream lines), and horizontally averaged temperature profiles for models II-a,b,c, and d.

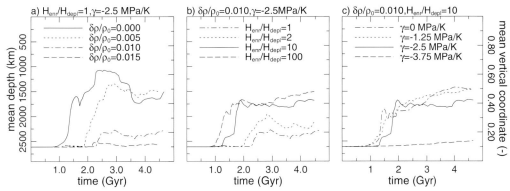

Figure 2. Mean depth/vertical coordinate of dense tracer population according to equation (11) as a function of time: (a) Uniform distribution of heat productivity; (b) dense layer enriched in heat-producing elements; (c) sensitivity to the postspinel Clapeyron slope.

that varies between 1 and 100. With increasing enrichment of the deep mantle reservoir, stability decreases as is illustrated by the shallower mean depth of the tracers.

Figure 3 shows a domain diagram indicating which models are stable or instable in the $\delta\rho_c/\rho_0 - H_{enr}/H_{depl}$ space. In agreement with previous studies (e.g., *Kellogg et al.* [1999]), the diagram shows a clear trade-off between density contrast and excess heat productivity in controlling the long-term stability of the internal layering.

4.3. Sensitivity to the Postspinel Clapeyron Slope

Most mantle convection models that include solid-state phase transitions apply a Clapeyron slope of around -2.5 MPa K^{-1} for the postspinel phase transition, consistent with several petrological [e.g., *Bina and Helffrich*, 1994; *Hirose*, 2002] and seismological [e.g., *Lebedev et al.*, 2002] studies. Recently, however, *Katsura et al.* [2003] have shown that the slope may in fact have a smaller magnitude, probably in the range of -0.4 to -2.0 MPa K^{-1}. To investigate the effect of a reduction of the magnitude of the Clapeyron slope of the postspinel phase transition on the stability of a deep dense layer, we have conducted a series of experiments for a range of Clapeyron slopes, with a fixed 1% $\delta\rho_c/\rho_0 = 0.01$ and $H_{enr}/H_{depl} = 10$. The results are presented in Plate 6 (snapshots after $\sim$ 4.4 Gyr of evolution) and Figure 2c (time series of average vertical coordinate of dense tracers).

Figure 2c shows that for low values of the Clapeyron slope, 0 and -1.25 MPa/K, respectively, the dense layer starts mixing through the mantle around 1 Gyr after the start of the models, and is more or less completely mixed after 4.5 Gyr. This picture is confirmed by Plate 6a,b, which shows that nonetheless the heavy tracers are not uniformly mixed throughout the mantle. For a stronger phase transition ($\gamma = -2.5$MPa K^{-1}), we observe the survival of part of

the deep dense layer after 4.5 Gyr in the form of a mound. Survival is near complete for the highest Clapeyron slope of -3.75 MPa K^{-1}.

Clearly, the ability of cold downwelling from the upper mantle to penetrate into the (deep) lower mantle strongly affects the mixing of the deep layer through the mantle. As illustrated by Plate 4b,c and Figure 1e, the transition of completely layered to partially layered convection is accompanied by an increase in the root mean square velocity, and cold downwellings cause significant topography in the deep layer (Plate 4d,e).

5. DISCUSSION AND CONCLUDING REMARKS

5.1. Transient Dynamical Regimes

Model series I and II both show a transient behavior. In model I-a, the initial hot phase is characterized by layered mantle convection and the formation of a thick basaltic crust. Exchange of material between the upper and lower mantle and the sinking of crustal sections into the lower mantle break this layered convection on a relatively short timescale of O(10^8 yr). The progressed cooling of the model results in the production of a thinner and more stable crust, and because of the absence of a sharp temperature contrast between upper and lower mantle due to whole-mantle convection and lack of fertile mantle material, lower mantle diapirism, inducing large-scale melting events, no longer takes place. A transition to plate tectonics would be expected here, though this model is incapable of reproducing plate tectonics. Because of the small extent of this model, characteristic time scales of cooling and thus transition between regimes are short compared to the age of Earth.

Around 1 to 2 Gyr, model II-e shows a transition from layered mantle convection to a mode of convection where

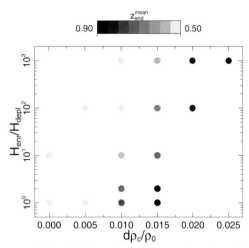

Figure 3. Stability of a deep dense layer as a function of chemical density contrast (horizontal axis) and internal heating of the layer (vertical axis). The gray scale of the dots shows the mean depth of deep layer tracers after 4.5 Gyr of model evolution, according to equation (11). Dark gray dots indicate stable configurations, light gray dots show complete mixing.

the phase transition acts as a filter to subducting slabs, which results in cold downwellings reaching the deep mantle and disturbing the compositional layering. This observation is qualitatively consistent with some interpretations of the geochemical record: On the basis of geochemical mass balances of large-ion-lithophile (incompatible) elements such as Sr, Nd, and Hf and noble gas isotope systems, *Allègre* [1997] argued that the transition to whole-mantle convection may have started some 0.5 Gyr ago.

The ultimate stability of the primitive reservoirs depends on the initially prescribed parameters, which directly and indirectly control the excess density of the deep layer. This is illustrated by Figure 3, which shows stable layering for large excess chemical densities of the lower layer and low internal heating rates and also instability for small density contrasts at high internal heating rates.

5.2. Reconciliation With Isotope Geochemistry

By numerical modeling we have explored a mechanism for the formation and deep mantle storage of enriched crustal material (see Plate 1). Plate 3 demonstrates that significant amounts of mafic (metabasalt) material, perhaps corresponding to the HIMU source, are transported to the base of the mantle. *Christensen and Hofmann* [1994] also found that, in a plate tectonic like model, significant volumes of subducted crust may settle at the base of the mantle and contribute to Pb and Nd isotope ratios which correspond to observed values in the MORB to HIMU spectrum.

Tatsumi [2000b] proposed that the EM-1 reservoir may have been formed by the accumulation of delaminated pyroxenitic restite from which continental material had been extracted, produced between 4 and 3 Ga by partial melting of arc basalts. He supported this hypothesis by calculating the corresponding isotopic evolution for Nd and Pb isotopes, assuming 45% and 60% of batch melting, which resulted in pyroxenitic and eclogitic restites, respectively. He found that a pyroxenitic restite with a 3–4% andesitic component (interpreted as retained melt in his model) produced an isotopic signature characteristic of EM-1, with low $^{143}Nd/^{144}Nd$, $^{207}Pb/^{204}Pb$, and $^{206}Pb/^{204}Pb$ ratios (see Table 1). Although subduction zones are the setting in which this model is thought to have operated [*Tatsumi*, 2000a,b], we propose that a resurfacing setting such as discussed in section 3.2 is equally valid, with a sinking crust in this context corresponding to a subducting slab in the plate tectonics sense. Significant quantities of basalt are produced in association with recycling of crust (either small-scale delamination or resurfacing), more or less analogous to arc basalts. During a resurfacing event, partial melting of this basalt may take place, equivalent to the partial melting associated with delamination in Tatsumi's model. Plate 3 shows that a significant part of the metabasalt which has sunk into the mantle has undergone (various degrees of) partial melting (pink color). No melt retention is included in our model but, as in Tatsumi's model, some retention is not unlikely.

Alternatively, our model is also consistent with the interpretation of the EM-1 and EM-2 isotopic signatures being caused by recycled continental material [*Zindler and Hart*, 1986; *Chauvel et al.*, 1992; *Hofmann*, 1997], since Plate 3 shows significant amounts of felsic material entrained in the resurfacing event and residing at the bottom of the model domain. The inhomogeneity of the deep material shown in Plate 3 is consistent with the apparent coexistence of different sources in the Earth's mantle that are interpreted for the different types of OIBs (see Table 1).

Most geochemical mantle models include, next to DMM, a reservoir of enriched recycled crust and a primitive undegassed mantle reservoir with a primitive trace element composition and a high $^3He/^4He$ ratio [see *Tackley*, 2000], but the use of He isotopes in oceanic basalts as an indicator for derivation from a primitive undegassed reservoir has been questioned [*Meibom et al.*, 2003].

In model I-a, no undifferentiated mantle peridotite survives longer than 500 million years of model evolution (see Plate 2). However, the small size of the model domain and its implications of a more rapid cooling of the system result in a reduction of the thermal Rayleigh number ($= \rho_0 \alpha g_0 \Delta T h^3 / \eta_0 \kappa$), which favors the development of single-layer mantle convection over two-layer convection [*Christensen and Yuen*, 1985]. This would reduce the probability of the survival of pristine material. However, the lower velocities associated with the reduced Rayleigh number oppose this effect.

5.3. Long-Term Stability of Deep Mantle Domains

Although they are sensitive to prior assumptions and model geometry, the results of experiment II can be used to give a qualitative insight in the dynamics of mantle evolution.

The series of model II versions show that a deep, enriched reservoir may be dynamically stable on a 4.5-Gyr time scale, which is in line with previous studies [*Kellogg et al.*, 1999; *Hansen and Yuen*, 2000; *Tackley*, 2002; *Samuel and Farnetani*, 2003; *Zhong and Hager*, 2003; *Nakagawa and Tackley*, 2004]. The relative enrichment in heat-producing elements as seen in the results of numerical experiment Ia (Plate 2) gives an estimate of the position of Earth on the vertical axis in Figure 3 (that is, between 10^1 and 10^2). An estimate of the position on the horizontal axis is more difficult, as the phase relations of basaltic material are poorly known for the deep mantle. But in order to be consistent with the presence of some chemically distinct material in the deep mantle, an excess density of at least 1.5% seems required. However, for Clapeyron slope magnitudes less than the canonical -2.5 MPa K^{-1} for the postspinel phase transition, as suggested by recent work of *Katsura et al.* [2003], cold downwellings from the upper mantle may be too strong and numerous to allow survival of a deep dense layer over billions of years. In our models, the primitive reservoir appears in several forms. For decreasing stability conditions, it first appears as a global layer on top of the CMB, transforms into isolated piles and then into blobs, before it finally mixes with the depleted mantle material.

5.4. Concluding Remarks

We have presented computer simulations of a possible early formation and the subsequent long-term evolution of compositionally distinct domains in Earth's deep interior, and we have shown that these models are in qualitative agreement with the geochemical record; that is, a deep, dense layer consisting of a geochemically inhomogeneous melange is consistent with the presence of distinct types of OIB. From such analysis alone we cannot, however, determine which of the models presented here (if any) is representative for planet Earth. Seismic tomography is producing increasingly convincing evidence for the presence of compositional heterogeneity both on large [*Masters et al.*, 2000; *Saltzer et al.*, 2001; *Trampert et al.*, 2004] and small length scales *Saltzer*

et al. [2004]; (see *Trampert and van der Hilst* [this volume] for a review), but studies of scattering of the high-frequency seismic wavefield [*Vidale et al.*, 2001; *Castle and van der Hilst*, 2003] have, so far, not produced convincing evidence for the type of global interface that one would expect for global layering. In combination, the results from geodynamical modeling, geochemistry, and seismic imaging are consistent with mantle models in which the changes in bulk composition gradually occur over a large depth range (for instance by depth-dependent subduction in the zoned convection-proposed by *Albarède and van der Hilst* [2002], or in which small-scale heterogeneity exists in the mantle above a distinct layer that has a diffuse boundary, significant short-wavelength topography, or both. These models are broadly consistent with the mantle model postulated by *Kellogg et al.* [1999]. Further narrowing the range of plausible mantle models requires (1) better seismological constraints on wavespeed and density ratios and on mantle interfaces, and their uncertainties; (2) continued quantitative integration of results from seismology, experimental and theoretical mineral physics, geochemistry, and geodynamical modeling; and (3) careful analysis of secular variations in the geological record (e.g., major element basalt chemistry) and comparison with predictions from geodynamical modeling of Earth's differentiation over long periods of geological time.

Acknowledgments. We wish to thank Jan Matas, Michael Manga, and an anonymous reviewer for constructive criticism which helped improve the paper. Peter van Thienen acknowledges the financial support provided through the European Community's Human Potential Programme under contract RTN2-2001-00414, MAGE. This work was sponsored by the Stichting Nationale Computerfaciliteiten (National Computing Facilities Foundation, NCF) for the use of supercomputer facilities, with financial support from the Nederlandse Organisatie voor Wetenschappelijk Onderzoek (Netherlands Organization for Scientific Research). Further computational facilities used were funded by ISES (Netherlands Research Centre for Integrated Solid Earth Science), which is gratefully acknowledged.

REFERENCES

Agee, C. B., and D. Walker, Mass balance and phase density constraints on early differentiation of chondritic mantle, *Earth Plan. Sci. Let., 90*, 144–156, 1988.

Albarède, F., and R. D. van der Hilst, Zoned mantle convection, *Philos. T. Roy. Soc. A, 360* (1800), 2569–2592, 2002.

Allègre, C. J., Limitation on the mass exchange between the upper and lower mantle: evolving convection regime of the Earth, *Earth Plan. Sci. Let., 150* (1–2), 1–6, 1997.

Allègre, C. J., and E. Lewin, Isotopic systems and stirring times of the earth's mantle, *Earth Plan. Sci. Let., 136*, 629–646, 1995.

Allègre, C. J., and D. L. Turcotte, Implications of a two-component marble-cake mantle, *Nature, 323*, 123–127, 1986.

Amelin, Y., D.-C. Lee, and A. N. Halliday, Early-middle Archaean crustal evolution deduced from Lu-Hf and U-Pb isotopic studies of single zircon grains, *Geochimica et Cosmochimica Acta, 64* (24), 4205–4225, 2000.

Anderson, D. L., The case for irreversible chemical stratification of the mantle, *International Geology Review, 44*, 97–116, 2002a.

Anderson, O. L., The power balance at the core-mantle boundary, *Phys. Earth Planet. Inter., 131*, 1–17, 2002b.

Aoki, I., and E. Takahashi, Density of MORB eclogite in the upper mantle, *Phys. Earth Planet. Inter., 143–144*, 129–143, 2004.

Beattie, P., The generation of uranium series disequilibria by partial melting of spinel peridotite: constraints from partitioning studies, *Earth Plan. Sci. Let., 117*, 379–391, 1993.

Becker, T. W., J. B. Kellogg, and R. J. O'Connell, Thermal constraints on the survival of primitive blobs in the lower mantle, *Earth Plan. Sci. Let., 171*, 351–365, 1999.

Bina, C. R., and G. Helffrich, Phase transition Clapeyron slopes and transition zone seismic discontinuity topography, *J. Geophys. Res., 99* (B8), 15,853–15,860, 1994.

Bizzarro, M., A. Simonetti, R. K. Stevenson, and J. David, Hf isotope evidence for a hidden mantle reservoir, *Geology, 30* (9), 771–774, 2002.

Blichert-Toft, J., and F. Albarède, The Lu-Hf isotope geochemistry of chondrites and the evolution of the mantle-crust system, *Earth Plan. Sci. Let., 148*, 243–258, 1997.

Blichert-Toft, J., F. Albarède, M. Rosing, R. Frei, and D. Bridgwater, The Nd and Hf isotopic evolution of the mantle through the Archean. Results from the Isua supracrustals, West Greenland, and from the Birimian terranes of West Africa, *Geochimica et Cosmochimica Acta, 63* (22), 3901–3914, 1999.

Caro, G., B. Bourdon, J.-L. Birck, and S. Moorbath, ^{146}Sm-^{142}Nd evidence from Isua metamorphosed sediments for early differentiation of the Earth's mantle, *Nature, 423*, 428–432, 2003.

Castle, J. C., and R. D. van der Hilst, Searching for seismic scattering off mantle interfaces between 800 and 2000 km depth, *J. Geophys. Res., 108*, 2095, doi:10.1029/2001JB000286, 2003.

Chase, C. G., Oceanic island Pb: two-stage histories and mantle evolution, *Earth Plan. Sci. Let., 52*, 277–284, 1978.

Chauvel, C., A. W. Hofmann, and P. Vidal, HIMU- EM: The French Polynesian connection, *Earth Plan. Sci. Let., 110*, 99–119, 1992.

Christensen, U. R., Is subducted lithosphere trapped at the 670-km discontinuity?, *Nature, 336*, 462–463, 1988.

Christensen, U. R., and A. W. Hofmann, Segregation of subducted oceanic crust in the convecting mantle, *J. Geophys. Res., 99* (B10), 19,867–19,884, 1994.

Christensen, U. R., and D. A. Yuen, Layered convection induced by phase transitions, *J. Geophys. Res., 99*, 10,291–10,300, 1985.

Coltice, N., and Y. Ricard, Geochemical observations and one layer mantle convection, *Earth Plan. Sci. Let., 174*, 125–137, 1999.

Creager, K. C., and T. H. Jordan, Aspherical structure of the core–mantle boundary from pkp travel times, *Geophys. Res. Let., 13* (13), 1497–1500, 1986.

Davaille, A., Simultaneous generation of hotspots and superswells by convection in a heterogeneous planetary mantle, *Nature, 402*, 756–760, 1999.

Davies, G. F., Stirring geochemistry in mantle convection models with stiff plates and slabs, *Geochimica et Cosmochimica Acta, 66* (17), 3125–3142, 2002.

Dupeyrat, L., and C. Sotin, The effect of the transformation of basalt to eclogite on the internal dynamics of Venus, *Planetary and Space Science, 43* (7), 909–921, 1995.

Green, D. H., and A. E. Ringwood, An experimental investigation of the gabbro to eclogite transformation and its petrological applications, *Geochimica et Cosmochimica Acta, 31*, 767–833, 1967.

Green, H. T., Anatexis of mafic crust and high pressure crystallization of andesite, in *Andesites*, edited by R. S. Thorpe, John Wiley and Sons, 1982.

Guyot, F., J. H. Zhang, I. Martinez, J. Matas, Y. Ricard, and M. Javoy, P-V-T measurements of iron silicide (epsilon-FeSi). Implications for silicate-metal interactions in the early Earth, *European Journal of Mineralogy, 9* (2), 277–285, 1997.

Hacker, B. R., Eclogite formation and the rheology, buoyancy, seismicity and H_2O content of oceanic crust, in *Subduction: Top to Bottom*, pp. 337–346, AGU Monogr., 1996.

Halliday, A. N., D.-C. Lee, S. Tommasini, G. R. Davies, C. R. Paslick, J. G. Fitton, and D. E. James, Incompatible trace elements in OIB and MORB and source enrichment in sub-oceanic mantle, *Earth Plan. Sci. Let., 133*, 379–395, 1995.

Hamilton, P. J., R. K. O'Nions, D. Bridgwater, and A. Nutman, Sm-Nd studies of Archaean metasediments and metavolcanics from West Greenland and their implications for the Earth's early history, *Earth Plan. Sci. Let., 62*, 263–272, 1983.

Hanan, B. B., and D. W. Graham, Lead and helium isotope evidence from oceanic basalts for a common deep source of mantle plumes, *Science, 272*, 991–995, 1996.

Hansen, U., and D. A. Yuen, Numerical simulations of thermal-chemical instabilities at the core-mantle boundary, *Nature, 334*, 237–240, 1988.

Hansen, U., and D. A. Yuen, Extended-Boussinesq thermal-chemical convection with moving heat sources and variable viscosity, *Earth Plan. Sci. Let., 176*, 401–411, 2000.

Hart, S. R., E. H. Hauri, L. A. Oschmann, and J. A. Whitehead, Mantle plumes and entrainment: isotopic evidence, *Science, 256*, 517–520, 1992.

Hauri, E. H., Osmium isotopes and mantle convection, *Phil. Trans. R. Soc. Lond., A 360*, 2371–2382, 2002.

Hauri, E. H., and S. R. Hart, Re-Os isotope systematics of HIMU and EMII oceanic island basalts from the south Pacific Ocean, *Earth Plan. Sci. Let., 114*, 353–371, 1993.

Helffrich, G. R., and B. J. Wood, The Earth's mantle, *Nature, 412*, 501–507, 2001.

Henderson, P., *Inorganic Geochemistry,* Pergamon, 1982.

Hirose, K., Phase transitions in pyrolitic mantle around 670-km depth:Implications for upwelling of plumes from the lower mantle, *J. Geophys. Res., 107* (B4), doi:10.1029/2001JB000597, 2002.

Hirose, K., Y. Fei, Y. Ma, and H.-K. Mao, The fate of subducted basaltic crust in the earth's lower mantle, *Nature, 397*, 53–56, 1999.

Hirth, G., and D. L. Kohlstedt, Water in the oceanic upper mantle: implications for rheology, melt extraction and the evolution of the lithosphere, *Earth Plan. Sci. Let., 144*, 93–108, 1996.

Hofmann, A. W., Mantle geochemistry: the message from oceanic volcanism, *Nature, 385*, 219–229, 1997.

Humayun, M., L. Qin, and M. D. Norman, Geochemical evidence for excess iron in the mantle beneath Hawaii, *Science, 306*, 91–94, 2004.

Irifune, T., and A. E. Ringwood, Phase transformations in subducted oceanic crust and buoyancy relations at depths of 600–800 km in the mantle, *Earth Plan. Sci. Let., 117*, 101–110, 1993.

Ito, K., and G. C. Kennedy, An experimental study of the basalt-garnet granulite-eclogite transition, in *The Structure and Physical Properties of the Earth's Crust*, edited by J. G. Heacock, pp. 303–314, American Geophysical Union, Washington D.C., 1971.

Jellinek, A. M., and M. Manga, Links between long-lived hot spots, mantle plumes, D'', and plate tectonics, *Rev. of Geophysics, 42* (RG3002), doi: 10.1029/2003RG000144, 2004.

Karato, S., Does partial melting reduce the creep strength of the upper mantle?, *Nature, 319*, 309–310, 1986.

Karato, S.-I., and P. Li, Diffusion creep in perovskite: implications for the rheology of the lower mantle, *Science, 255*, 1238–1240, 1992.

Karato, S.-i., and P. Wu, Rheology of the upper mantle: a synthesis, *Science, 260*, 771–778, 1993.

Katsura, T., H. Yamada, T. Shinmei, A. Kubo, S. Ono, M. Kanzaki, A. Yoneda, M. J. Walter, E. Ito, S. Urakawa, K. Funakoshi, and W. Utsumi, Post-spinel transition in Mg_2SiO_4 determined by high P-T in situ X-ray diffractometry, *Phys. Earth Planet. Inter., 136*, 11–24, 2003.

Kellogg, L., B. H. Hager, and R. D. van der Hilst, Compositional stratification in the deep mantle, *Science, 283*, 1881–1884, 1999.

Knittle, E., and R. Jeanloz, Earth's core–mantle boundary: results of experiments at high pressures and temperatures, *Science, 251*, 1438–1443, 1991.

Lebedev, S., S. Chevrot, and R. D. van der Hilst, Seismic evidence for olivine phase transitions near 410 and 660-kilometers depth, *Science, 296*, 1300–1302, 2002.

Masters, G., G. Laske, H. Bolton, and A. Dziewonski, The relative behavior of shear velocity, bulk sound speed, and compressional velocity in the mantle; implications for chemical and thermal structure, in *Earth's Deep Interior; Mineral Physics and Tomography From the Atomic to the Global Scale*, edited by S.-I. Karato, A. M. Forte, R. C. Liebermann, G. Masters, and L. Stixrude, pp. 63–86, American Geophysical Union, Washington, D.C., 2000.

Mei, S., and D. L. Kohlstedt, Influence of water on plastic deformation of olivine aggregates 1: Diffusion creep regime, *J. Geophys. Res., 105*, 21,457–21,469, 2000a.

Mei, S., and D. L. Kohlstedt, Influence of water on plastic deformation of olivine aggregates 2: Dislocation creep regime, *J. Geophys. Res., 105*, 21,471–21,481, 2000b.

Meibom, A., D. L. Anderson, N. H. Sleep, R. Frei, C. P. Chamberlain, T. H. Hren, and J. L. Wooden, Are high $^3He/^4He$ ratios in oceanic basalts an indicator of deep-mantle plume components, *Earth Plan. Sci. Let., 208*, 197–204, 2003.

Moresi, L., and V. Solomatov, Mantle convection with a brittle lithosphere: thoughts on the global tectonic styles of the Earth and Venus, *Geophys. J. Int., 133*, 669–682, 1998.

Nakagawa, T., and P. Tackley, Thermo-chemical structure in the mantle arising from a three-component convective system and implications for geochemistry, *Phys. Earth Planet. Inter.*, 2004.

Okamoto, K., and S. Maruyama, The eclogite-garnetite transformation in the MORB + H_2O system, *Phys. Earth Planet. Inter., 146*, 283–296, 2004.

Patchett, P. J., Importance of the Lu-Hf isotopic system in studies of planetary chronology and chemical evolution, *Geochimica et Cosmochimica Acta, 47*, 81–91, 1983.

Piepenbreier, D., and B. Stöckhert, Plastic flow of omphacite in eclogites at temperatures below 500 °C—implications for inter-plate coupling in subduction zones, *Int. J. Earth Sciences, 90*, 197–210, 2001.

Porcelli, D., and A. N. Halliday, The core as a possible source of mantle helium, *Earth Plan. Sci. Let., 192*, 45–56, 2001.

Ringwood, A. E., and T. Irifune, Nature of the 650-km seismic discontinuity; implications for mantle dynamics and differentiation, *Nature, 331* (6152), 131–136, 1988.

Saltzer, R., R. D. van der Hilst, and H. Kárason, Comparing P and S wave heterogeneity in the mantle, *Geophys. Res. Let., 28*, 1335–1338, 2001.

Saltzer, R. L., R. Stutzmann, and R. D. van der Hilst, Poisson's ration beneath Alaska from the surface to the Core-Mantle Boundary, *J. Geophys. Res., 109*, doi:10.1029/2003JB002712, 2004.

Samuel, H., and C. G. Farnetani, Thermochemical convection and helium concentrations in mantle plumes, *Earth Plan. Sci. Let., 107*, 39–56, 2003.

Scherstén, A., T. Elliot, C. Hawkesworth, and M. Norman, Tungsten isotope evidence that mantle plumes contain no contribution from the Earth's core, *Nature, 427*, 234–237, 2004.

Segal, A., and N. P. Praagman, The sepran package, *Tech. Rep.*, http://dutita0.twi.tudelft.nl/sepran/sepran.html, 2000.

Sleep, N. H., and B. F. Windley, Archean plate tectonics: constrains and inferences, *Journal of Geology, 90*, 363–379, 1982.

Steinbach, V., U. Hansen, and A. Ebel, Compressible convection in the Earths mantle—a comparison of different approaches, *Geophys. Res. Let., 16* (7), 633–636, 1989.

Steinbach, V., D. A. Yuen, and W. Zhao, Instabilities from phase-transitions and the timescales of mantle thermal evolution, *Geophys. Res. Let., 20* (12), 1119–1122, 1993.

Tackley, P. J., Mantle convection and plate tectonics: Toward an integrated physical and chemical theory, *Science, 288*, 2002–2007, 2000.

Tackley, P. J., Strong heterogeneity caused by deep mantle layering, *Geochem. Geophys. Geosyst., 3* (4), 1024, doi:10.1029/2001GC000167, 2002.

Tatsumi, Y., Slab melting: its role in continental crust formation and mantle evolution, *Geophys. Res. Let., 27* (23), 3941–3944, 2000a.

Tatsumi, Y., Continental crust formation by crustal delamination in subduction zones and complementary accumulation of the enriched mantle I component in the mantle, *Geochem. Geophys. Geosyst., 1*, paper number 2000GC000094, 2000b.

Tatsumoto, M., Isotopic composition of lead in oceanic basalt and its implication to mantle evolution, *Tectonophysics, 38*, 64–87, 1978.

Trampert, J., F. Deschamps, J. Resovsky, and D. A. Yuen, Probabilistic tomography maps significant chemical heterogeneities in the lower mantle, *Science, 306* (5697), 853–856, 2004.

van den Berg, A. P., and D. A. Yuen, Is the lower-mantle rheology Newtonian today?, *Geophys. Res. Let., 23* (16), 2033–2036, 1996.

van den Berg, A. P., and D. A. Yuen, Modelling planetary dynamics by using the temperature at the core–mantle boundary as a control variable: effects of rheological layering on mantle heat transport, *Phys. Earth Planet. Inter., 108*, 219–234, 1998.

van den Berg, A. P., P. E. van Keken, and D. A. Yuen, The effects of a composite non-Newtonian and Newtonian rheology on mantle convection, *Geophys. J. Int., 115*, 62–78, 1993.

van der Hilst, R. D., and H. Kárason, Compositional heterogeneity in the bottom 1000 km of the Earth's mantle: Toward a hybrid convection model, *Science, 283*, 1885–1888, 1999.

van der Hilst, R. D., S. Widiyantoro, and E. R. Engdahl, Evidence for deep mantle circulation from global tomography, *Nature, 386*, 578–584, 1997.

van Hunen, J., A. P. van den Berg, and N. J. Vlaar, On the role of subducting oceanic plateaus in the development of shallow flat subduction, *Tectonophysics, 352*, 317–333, 2002.

van Schmus, W. R., Natural radioactivity of the crust and mantle, in *Global Earth Physics. A handbook of constants*, edited by T. J. Ahrens, pp. 283–291, AGU, Washington, 1995.

van Thienen, P., A. P. van den Berg, J. H. De Smet, J. van Hunen, and M. R. Drury, Interaction between small-scale mantle diapirs and a continental root, *Geochem. Geophys. Geosyst., 4*, 10.1029/2002GC000,338, 2003.

van Thienen, P., A. P. van den Berg, and N. J. Vlaar, On the formation of continental silicic melts in thermo-chemical mantle convection models: implications for early Earth, *Tectonophysics, 394* (1–2), 111–124, 2004a.

van Thienen, P., A. P. van den Berg, and N. J. Vlaar, Production and recycling of oceanic crust in the early earth, *Tectonophysics, 386* (1-2), 41–65, 2004b.

van Thienen, P., N. J. Vlaar, and A. P. van den Berg, Plate tectonics on the terrestrial planets, *Phys. Earth Planet. Inter., 142* (1-2), 61–74, 2004c.

Vervoort, J. D., P. J. Patchett, G. E. Gehrels, and A. P. Nutman, Constraints on early Earth differentiation from hafnium and neodymium isotopes, *Nature, 379*, 624–627, 1996.

Vidale, J. E., G. Schubert, and P. S. Earle, Unsuccessful initial search for a midmantle chemical boundary with seismic arrays, *Geophys. Res. Let., 28* (5), 859–862, 2001.

Vlaar, N. J., Precambrian geodynamical constraints, in *The Deep Proterozoic Crust in the North Antlantic Provinces*, edited by A. C. Tobi and J. L. R. Touret, pp. 3–20, Reidel, 1985.

Vlaar, N. J., and A. P. van den Berg, Continental evolution and archeao-sea-levels, in *Glacial Isostasy, Sea-Level and Mantle Rheology*, edited by R. Sabadini, K. Lambeck, and E. Boschi, Kluwer, Dordrecht, Netherlands, 1991.

Zhong, S. J., and B. H. Hager, Entrainment of a dense layer by thermal plumes, *Geophys. J. Int., 154* (3), 666–676, 2003.

Zindler, A., and S. Hart, Chemical geodynamics, *Ann. Rev. Earth Planet. Sci., 14*, 493–571, 1986.

R. D. van der Hilst, Department of Earth, Atmospheric and Planetary Sciences, Massachusetts Institute of Technology, Cambridge, Massachusetts 02139, USA.

A.P. van den Berg, J. van Summeren, and N. J. Vlaar, Institute of Earth Sciences, Utrecht University, PO Box 80.021, 3508 TA Utrecht, The Netherlands.

P. van Thienen, Département de Géophysique Spatiale et Planétaire, Institut de Physique du Globe de Paris, 4 Avenue de Neptune, Saint-Maur-des-Fossés, France. (thienen@ipgp.jussieu.fr)

The Role of Theoretical Mineral Physics in Modeling the Earth's Interior

Mark S.T. Bukowinski and Sofia Akber-Knutson

Department of Earth and Planetary Science, University of California at Berkeley,
Berkeley, California

What can theoretical mineral physics tell us about the real Earth? To answer this question in a meaningful way, it is necessary to have a common understanding of the geological background, as well as the capabilities and limitations of modern computational condensed matter physics. To the uninitiated, it is difficult to assess the relative worth of two different predictions made by reputable theoretical mineral physicists. Why are there different predictions to begin with? These are fair questions that deserve an objective answer. This paper is our attempt to write a sort of tutorial, aimed at nonmineral physicists, about what mineral physics can and cannot do. While quantum mechanics is proven to provide an extremely accurate representation of microscopic physics, and while first principles computational techniques are extremely sophisticated, their practical application to complex systems such as geological minerals requires approximations that limit their accuracy. We discuss the theoretical foundations and their practical implementations, and the concomitant uncertainties. We do this through examples that we hope will illustrate how computational errors compare with the small differences among typical seismic-velocity and density models. It is our hope that the careful reader will be left with a better appreciation not only of the limitations, but also of the power of first principles methods, and the extremely bright prospects of these methods for contributions to geology in the near future.

INTRODUCTION

The Earth is a mineral assemblage—living organisms excluded, of course, but these constitute a negligible fraction of the Earth's mass. And while the liquid outer core and the water of the oceans are not minerals by standard definition, any self-respecting mineral physicist would consider them legitimate objects of study. Hence much of the Earth, and any other planet in which we develop a deep enough interest, constitutes the domain of mineral physics. Such diverse geologically relevant phenomena as equations of state, elastic velocities, phase boundaries, phase transformation mechanisms and kinetics, element partitioning, thermal and chemical diffusion, melting, rheological properties, defect mechanisms, optical and vibrational spectra, and others can in principle be related to each other, given a sufficiently accurate theoretical model of bonding. Theory is indispensable if measurements are to inform our understanding beyond the particular, and necessarily limited, experimental context, which must itself be defined via a suitable conceptual framework.

Modern mineral physics is condensed matter physics and chemistry seeking to decipher—in terms of the language of materials—the clues yielded to the probing of seismologists, geochemists, geodynamicists and every other sort of geologist. What mineral assemblage is consistent with

Earth's Deep Mantle: Structure, Composition, and Evolution
Geophysical Monograph Series 160
Copyright 2005 by the American Geophysical Union
10.1029/160GM10

geological constraints and has the observed seismic velocity? What is the nature of the Ultra Low Velocity zone above the core? The D″ zone? The anisotropy of the inner core and of the D″ zone? What accounts for the lateral heterogeneity in the lower mantle or the radial discontinuities in density and seismic velocity? What is the composition of the core and what is the structure of the iron alloy in the inner core? Is there a compositional Earth model that can make seismologists, geodynamicists, and geochemists all happy? These are some of the questions that have no answers from geological observations alone. Rather, the answers rely on comparisons of geological data with the properties of candidate Earth-forming materials.

All such properties are in principle measurable. But the Earth imposes stringent requirements on experimental equipment: pressures as high as 360 GPa and temperatures that, in some models, approach a substantial fraction of 10,000 K. Experimentalists' ingenuity, aided by modern materials, has given us static measurement techniques with the ability to almost routinely measure material densities and structures for such pressures, but readily attainable temperatures remain below those in the core. Properties requiring more than x-ray spectra are generally more restricted in pressure and temperature. Notably, elastic velocity measurements, which the interpretation of seismic velocities requires, are still limited to pressures that rarely exceed those in the middle lower mantle, or temperatures of 2500 K, but not both at once [e.g., *Jackson et al.*, 2000; *Liebermann and Li*, 1998; *Liu et al.*, 2000; *Sinogeikin and Bass*, 2000, 2002; *Sinogeikin et al.*, 2003]. And some phenomena, such as the microscopic mechanism(s) by which Fe and Al affect the properties of $MgSiO_3$ perovskite, the most abundant mineral in the lower mantle, cannot be fully understood without detailed atomistic simulations.

This is where theoretical mineral physics enters as a geological discipline. It aims to provide a conceptual and quantitative framework, within which the great variety of diverse phenomena encountered in Earth materials can be related to each other, to experimental observations, to fundamental concepts of bonding, and to geological phenomena such as those alluded to earlier. For example, measured Raman or infrared frequencies can constrain parameters in an approximate theoretical model for the energy of a crystal. The model can in turn be used to determine, among other things, the equilibrium structure of the crystal under various pressures and temperatures. If the model is good enough, it can, at least in principle, be used to compute chemical potentials, which in turn can be used to determine element partitioning among coexisting phases and the way in which partitioning is affected by pressure. One might thus find that an element that is lithophile at low pressures becomes siderophile at high

pressures. Thus, through theory, a finite set of measurements can be bootstrapped to examine a broad spectrum of physical and chemical phenomena that are relevant to geological studies.

It is not the purpose of this brief article to present a detailed review of mineral physics theory, nor of all the geological problems that it could help to resolve. Several such reviews have been written in the recent past [*Bukowinski*, 1994; *Bukowinski et al.*, 1996; *Karki et al.*, 2001a; *Stixrude*, 2000, 2001; *Stixrude et al.*, 1998]. We instead attempt to address directly the frequently asked question, "What can mineral physics tell us about the real Earth?" We have been asked similar questions often enough to suggest that the existing reviews do not provide a satisfactory answer. Although these reviews present excellent discussions of methodologies and specific examples, they perhaps fail to take into account the fact that many Earth scientists may find the formal discussions of quantum mechanical techniques somewhat less than informative or exciting. Theoretical mineral physicists, on the other hand, delight in discussing all the clever approximations that improve the agreement of computed mineral properties with experimentally determined ones. The accuracy of such results is sometimes overstated, and the deduced geological implications may be presented somewhat naïvely. To the uninitiated, it is usually far from clear which, if any, of two different predictions obtained by reputable numerical mineral modelers is correct. We therefore undertook to write a sort of tutorial, aimed at nonmineral physicists, about what theoretical mineral physics can and cannot do, especially when considered in a geophysical context, where sometimes one tries to interpret differences of, say, a few percent in seismic velocity or a fraction of a percent in density, and where these differences reflect tradeoffs among the influences of several environmental factors. We do not aim to present an exhaustive discussion; rather we use some illustrative examples that, we hope, will stimulate some readers to peruse the previous review articles and the general theoretical mineral physics literature. Before presenting the illustrative examples, however, we find it useful to review the terrestrial setting and give a quick overview of the theoretical foundations that underlie the computations. Without such foundations it is difficult to talk about uncertainties in a manner that makes sense to anyone but the experts.

THE DEEP EARTH ENVIRONMENT

Within the Earth's interior, a great variety of minerals are subjected to a large range of environmental conditions (Figure 1). One might say that all the necessary high-pressure experiments have been carried out in the Earth's inner laboratory. Although most of this laboratory is inaccessible,

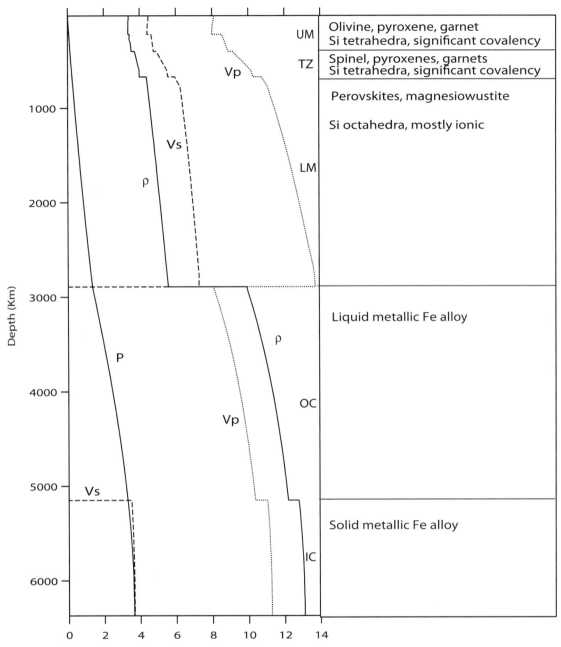

Figure 1. The spherically averaged Earth, as given by the preliminary reference Earth model [*Dziewonski and Anderson*, 1981]. Units are as follows: pressure P (Mbar); density ρ (gm/cm^3); and velocities Vs and Vp (km s^{-1}). UM, TZ, LM, OC, and IC: upper mantle, transition zone, lower mantle, outer core, and inner core, respectively. The dominant material composition and type of structure are shown next to each layer.

it does leave imprints on a variety of geological observations. For example, seismologically deduced densities and velocities constitute a form of equation of state of the mineral assemblages within the Earth. Mineral physicists strive to combine these observations with theoretical foundations to decipher the actual chemical compositions and crystallo-

graphic structures of the mineral components, the structure of the assemblage (deformation matrix) as well as environmental parameters, such as the temperature, predominant strain rate, chemical reactions, etc. Thus the conceptual framework is that of materials science, while the object of study is an incompletely constrained aggregate of minerals, some of which are yet to be observed in natural samples, and others that have not yet been discovered, either through observation or experiment, or theoretical simulation.

Beside the obvious need to understand what materials are in the Earth, and how they affect its properties and behavior, the candidate materials are interesting in their own right. Apart from the metallic core, the volumetrically significant minerals are silicates and oxides. The great majority of them are compounds of SiO_2, MgO, FeO, Al_2O_3, CaO, and Na_2O. In the upper mantle, minerals composed of these simple oxides tend to form structures that are somewhat open and, like the SiO_4 tetrahedron, are influenced by a significant amount of covalency in bonding [*Gibbs et al.*, 1998; *Pauling*, 1980]. As pressure increases with depth, the interplay between bonding and structure yields more tightly packed configurations, in which bonding tends to be more ionic. This increased ionicity reflects effects of pressure as well as changing local symmetry—a consequence of cation–anion coordination changes—on electronic levels [*Bukowinski*, 1994]. Ionic materials, whose structures are largely controlled by a balance of Coulomb interactions and short range repulsions between approximately spherical closed–shell ions, are easier to model than the more open, directionally bonded, covalent materials. This is fortunate: Since there is less mineral data at the higher pressures and temperatures of the deep mantle, the need for reliable predictive models is greater there.

Another fortunate circumstance is that the Earth is not large enough for its internal pressures to greatly modify the nature of bonding in most materials [*Bukowinski*, 1994]. From the top to the bottom of the mantle, its density increases by a factor of about 5/3. The work needed to accomplish this compression is of the order of 1 electron volt (eV) per atom. Since 1 eV is comparable to typical bond energies, one might expect mantle pressures to be sufficient to drastically alter chemical bonding and hence the nature of mantle materials. This kind of reasoning led to an early suggestion that the Earth's core may be a metallic phase of mantle oxides [*Ramsey*, 1949]. Things didn't turn out to be quite this exciting. To understand why not, we consider what is a significant pressure at the atomic scale. Atomic energies being of the order of e^2 / a_o, where a_o (0.529177 Å) is the Bohr radius and e is the electronic charge, and atomic volumes being measured in terms of a_o^3, the atomic unit of pressure is e^2 / a_o^4, which amounts to approximately 300 Mbars, or two orders of magnitude larger than the highest pressure within the Earth! Although some highly compressible compounds of no geological significance (e.g., CsI) may be metallized by Earth pressures [*Aidun et al.*, 1984], and the chemical affinity of metallic K can change from lithophile to siderophile in the mantle [*Bukowinski*, 1976a, 1979; *Ito et al.*, 1993;

Lee and Jeanloz, 2003; *Parker et al.*, 1996], most bonding changes are limited to slight shifts in the relative importance of covalency and ionicity.

Deciphering the composition of the Earth's interior is but the first step towards a physical and chemical model that is consistent with seismic, geochemical, and geodynamic observations. Such models might be sufficient to account for the spherically averaged Earth, as represented for example by the Preliminary Reference Earth Model. But to understand inner Earth processes, we need to be able to account for the deviations from spherical symmetry. Lateral density and velocity deviations are well documented, and their structure is steadily coming into focus [*Cazenave and Thoraval*, 1994; *Grand et al.*, 1997; *Kennet et al.*, 1998; *Li and Romanowicz*, 1996; *Montelli et al.*, 2004; *Su et al.*, 1994; *van der Hilst*, 1997; *Wen and Anderson*, 1997]. It is not difficult to come up with a list of processes that *might be* responsible for the observed lateral structure. But to identify the ones that *actually are* responsible requires accurate estimates of the response of the velocities and density to pressure, temperature, structural phase transformations, partial melting, and chemical composition. Hence the ability to estimate accurately the relative change in the shear velocity, in response to a change in the independent variables, may be more important than an accurate estimate of the velocity itself.

SOME THEORETICAL FOUNDATIONS

First Principles

Computations of the structure and thermoelastic properties of minerals start with the assumption that the mineral is in thermodynamic equilibrium, implying constant temperature T, pressure P, and composition $\mathbf{X}$, where $\mathbf{X}$ is a vector containing the molar fractions of all the chemical components. This is not a strictly valid assumption within the Earth, where the radial pressure gradient is of the order 0.3 kbar/km, and temperature gradients can be as high as a few tens of degrees per kilometer. However, for a 1-cm mineral grain this leads to temperature differences of the order of 10^{-3} K, and pressure differences no larger than 10^{-7} GPa across the grain, far smaller than is attainable in a typical high-pressure experiment. Although P and T gradients of this magnitude have well known effects on the large scale dynamical properties of the Earth's interior, and may affect equilibrium properties on a macroscopic scale in interesting ways [e.g., *Bina and Kumazawa*, 1993], local elastic properties at a given P and T are not affected by the small local gradients.

Given the above assumptions, the thermoelastic mineral properties may be obtained from a suitable thermodynamic potential, such as the Helmholtz free energy $F(V,T,\mathbf{X})$, or the Gibbs free energy $G(P,T,\mathbf{X})$. From thermodynamics we have

$$F(V,T,\mathbf{X}) = E(V,S,\mathbf{X}) - TS \qquad (1)$$

$$G(P,T,\mathbf{X}) = F(V,T,\mathbf{X}) + PV \qquad (2)$$

where E, S, and V are the internal energy, entropy, and volume of the material, respectively. To simplify notation, in the following discussion we suppress explicit references to the compositional variable $\mathbf{X}$.

How one proceeds from equation (1) or (2) to an actual computation depends on the kind of model adopted to describe the material. Such models range from those that treat the material as a continuum that ignores the atomistic nature of matter, to first principles theories that account explicitly for the electronic degrees of freedom. Examples of the former include finite strain theory, which deduce equations of state from the assumption that the energy can be expanded as a power series of the Eulerian strain [e.g., Birch, 1952]; the Vinet equation of state, which follows from an explicit form of the energy density of a material [Vinet.et al., 1986]; as well as various other parametric equations of state [Stacey and Davis, 2004]. Our emphasis here is on microscopic theories of materials, as deduced from first principles and the various approximations that arise from a systematic and judicious reduction of the degrees of freedom.

The internal energy E provides the most direct connection to microscopic theory. For example, to compute macroscopic properties of a mineral under conditions where its structure is unknown, we need a microscopic description in terms of the atomic positions, collectively described by the $3N$-dimensional vector $\mathbf{R}$, where N is the number of atoms. If we can compute E for such a system, then

$$G(P,T) = E(\mathbf{R}) + F_{th}(\mathbf{R},T) + PV(\mathbf{R}) \qquad (3)$$

where $E(\mathbf{R})$ is the internal energy at $T = 0$, and $F_{th}(\mathbf{R},T)$ is the thermal free energy, which includes the energy of zero-point vibrations. Note that the volume dependence of E is implicit in the atomic coordinates. By optimizing the free energy with respect to $\mathbf{R}$, we obtain the equilibrium structure, volume, and at least in principle, any other physical property at the imposed values of P and T.

The most rigorous computational approach to equation (3) is founded on first principles in which quantum theory is used to derive the total internal energy without any reference to data (except for values of fundamental physical constants such as the electronic charge and mass, Planck's constant, etc.). This quantum mechanical foundation is extremely

strong. Whenever it is possible to obtain accurate computations of properties that can be accurately measured, full agreement is obtained within the experimental precision [e.g., Goddard, 1985; Richards, 1979]. This point cannot be overemphasized: Quantum electrodynamics, which describes all microscopic electromagnetic phenomena, is the most accurate theory known to man. The errors and uncertainties encountered in computations for complex systems, such as crystals or molecules, arise not from limitations of quantum theory per se, but from approximations made to render the many-body nature of the problems computationally tractable.

The starting point for first principles theories is the exact expression for the internal energy of a system of nuclei and electrons:

$$E(\mathbf{R}) = \langle \Psi(\mathbf{x},\mathbf{R}) | \check{\mathbf{H}} | \Psi(\mathbf{x},\mathbf{R}) \rangle \qquad (4)$$

where $\mathbf{x}$ stands for the n electronic positions and their spins and, as before, $\mathbf{R}$ represents the positions of the N nuclei. Here E is expressed as the quantum mechanical expectation value of the total energy operator, or Hamiltonian, for a system of nuclei and electrons,

$$\check{\mathbf{H}} = -\frac{1}{2}\sum_{i=1}^{n}\nabla_i^2 - \frac{1}{2M_\alpha}\sum_{\alpha=1}^{N}\nabla_\alpha^2 - \sum_{i=1}^{n}\sum_{\alpha=1}^{N}\frac{Z_\alpha}{|\mathbf{r}_i - \mathbf{R}_\alpha|}$$
$$+\sum_{i<j}\frac{1}{|\mathbf{r}_i - \mathbf{r}_j|} + \sum_{\alpha<\beta}\frac{Z_\alpha Z_\beta}{|\mathbf{R}_\alpha - \mathbf{R}_\beta|} \qquad (5)$$

and $\Psi(\mathbf{r},\mathbf{R})$ is the normalized (that is, $\langle\Psi|\Psi\rangle \equiv \int d\mathbf{r}\Psi^*(\mathbf{r},\mathbf{R})\Psi(\mathbf{r},\mathbf{R}) = 1$) wavefunction of the system, obtained as a solution to Schrödinger's equation

$$\check{\mathbf{H}}\Psi = E\Psi \qquad (6)$$

which follows from (4) via variation with respect to Ψ^*, the complex conjugate of the wavefunction. Z_α and M_α are the charge and mass, respectively, of nucleus α, located at position $\mathbf{R}_\alpha$; $\mathbf{r}_i$ is the position of the ith electron. The first two terms in equation (5) are the kinetic energy operators for the electrons and nuclei, respectively. The last three terms give the contributions to the total energy of the electron–nucleus, electron–electron, and nucleus–nucleus Coulomb interactions, respectively. Equation (5) is written in atomic units, in which the charge e and mass m of the electron, and $\hbar$ (Planck's constant divided by 2π) are all set to unity. Except for relativistic effects, which are negligible for most systems of geophysical interest, equations (4)–(6) give an exact formulation for the internal energy of a material. But, as Dirac so aptly stated, "The underlying physical laws necessary for the mathematical

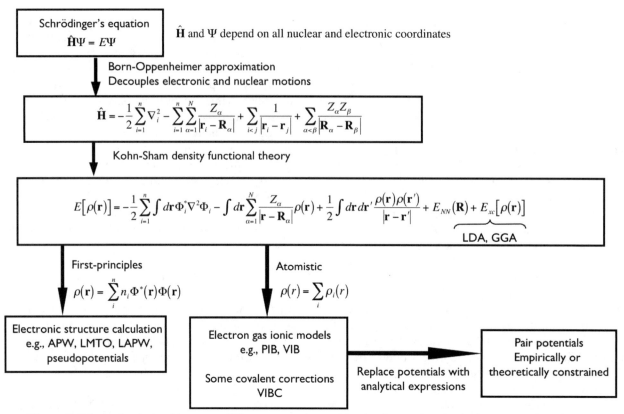

Figure 2. Schematic relationship among different approaches and levels of approximation in the computation of material properties. See text for details.

theory of a large part of physics and the whole of chemistry are thus completely known, and the difficulty is only that the exact application of these laws leads to equations much too complicated to be soluble" [*Dirac*, 1929].

The transformation of equations (4)–(6) to something that is "soluble" or, more appropriately, numerically tractable, is a difficult and subtle process of systematic reduction of the number of degrees of freedom. The starting point of all such reductions is based on the realization that, since electrons are thousands of times lighter than protons, they move that much faster than protons. The consequence of this, as proved by *Born and Oppenheimer* [1927], is that at any given time the electrons are in equilibrium with the instantaneous positions of all nuclei. This allows a factorization of the total wavefunction, $\Psi(\mathbf{x}, \mathbf{R}) = \Phi(\mathbf{x}, \mathbf{R})\Xi(\mathbf{R})$, into nuclear, $\Xi(\mathbf{R})$, and electronic, $\Phi(\mathbf{x}, \mathbf{R})$, wavefunctions, where the latter depends on the nuclear coordinates only parametrically. $\Phi(\mathbf{x}, \mathbf{R})$ is a solution of equation (6) with a Hamiltonian from which the nuclear kinetic energy term [second term in equation (5)] has been eliminated.

Although decoupling the electronic motions from those of the nuclei is a huge simplification of the equations, we are still left with something of the order of 10^{23} interact-

ing electrons in a small crystal. In crystalline materials, periodic boundary conditions allow a reduction of the total number of electrons to those present in a single unit cell, which can range from a few to many thousands. Luckily, a sort of mean field approach, in which each electron is assumed to move independently in an average potential due to the nuclei and all other electrons, provides a surprisingly good starting point for accurate solutions, in spite of the fact that the motions of the strongly coupled electrons might be expected to be highly correlated. In fact, such correlations are important, but they can be treated as part of the effective potential seen independently by each electron. The many-electron problem has thus been reduced to an effective single-electron problem.

The evolution of the single-electron approach to the solution of the many-electron Schrödinger's equation has a rich history. An overview of the physics underlying these developments is provided by [*Bukowinski*, 1994] and references found therein. Here we briefly discuss only the density-functional approach since it is arguably the most rigorous and fruitful approach to the physics of condensed matter. An approach not based on density functionals, the Hartree–Fock approximation, is the oldest successful scheme for computing electronic structures [*March*, 1975]. Methods based on the Hartree–Fock equation continue to be favored in the study of the physics and chemistry of atoms, molecules, and molecular clusters [*Gibbs*, 1982] and have also been applied to crystals [*D'Arco et al.*, 1994; *Pisani et al.*, 1988; *Sherman*, 1991, 1993a, 1993b]. However, density-functional applications dominate condensed matter computational research in general, and mineral physics in particular.

Density-functional theory rests on two facts established by *Hohenberg and Kohn* [1964]: (1) The ground state energy of a many-electron system subjected to external potentials is a unique functional of the electron density, and (2) such a functional exists and is minimized by the exact ground-state electron density. Using the Hohenberg–Kohn theorem as a foundation, *Kohn and Sham* [1965] showed that for the case where the external potential is the Coulomb attraction of nuclei, the many-electron ground state energy can be written as that of a collection of independent electrons moving in an effective potential as follows:

$$E\left[\rho\left(\mathbf{r}\right)\right] = -\frac{1}{2}\sum_{i=1}^{n}\int d\mathbf{r}\,\Phi_i^*\nabla^2\Phi_i - \int d\mathbf{r}\sum_{\alpha=1}^{N}\frac{Z_\alpha}{|\mathbf{r}-\mathbf{R}_\alpha|}\rho\left(\mathbf{r}\right)$$
$$+\frac{1}{2}\int d\mathbf{r}\,d\mathbf{r}'\frac{\rho\left(\mathbf{r}\right)\rho\left(\mathbf{r}'\right)}{|\mathbf{r}-\mathbf{r}'|} + E_{NN}\left(\mathbf{R}\right) + E_{xc}\left[\rho\left(\mathbf{r}\right)\right] \quad (7)$$

The electron density is given by $\rho\left(\mathbf{r}\right) = \sum_{i}^{n} n_i\Phi_i^*\left(\mathbf{r}\right)\Phi_i\left(\mathbf{r}\right)$, where $\Phi_i\left(\mathbf{r}\right)$ is the *i*th single electron wavefunction, n_i is its occupation number, and $\mathbf{r}$ stands for the position and spin of a single electron. Square brackets are used to indicate a functional dependence. The first term on the right hand side is the total kinetic energy of independent electrons; the second and third terms give the Coulomb interaction energy of electrons with nuclei and other electrons, respectively; and the fourth term is the corresponding interaction among nuclei. $E_{xc}\left[\rho\left(\mathbf{r}\right)\right]$, the quantum-mechanical exchange-correlation energy, is the functional whose existence is guaranteed by the Hohenberg–Kohn theorem, but whose exact form is not known. It represents the effect of two kinds of electron–electron correlations. The exchange part is a consequence of the requirement that the total electron wavefunction must be antisymmetric with respect to the exchange of any two electrons. This causes electrons of like spin to avoid each other, such that each electron is followed by a "hole" from which exactly one electron of like spin is excluded. This is the origin of the Pauli exclusion principle, which states that no two electrons can occupy the same quantum state. The correlation part stems from the additional correlation between electrons that arises due to the Coulomb repulsion between them, regardless of their spin. Both types of correlation serve to lower the Coulomb repulsion energy of the system. They also help to screen electrons from each other,

since the exchange-correlation "hole" amounts to a positively charged cloud attached to each electron, the combination "looking" effectively neutral to other electrons at distances outside the "hole."

To evaluate the total energy, it is necessary to know the single electron wavefunctions $\Phi_i\left(\mathbf{r}\right)$. This is accomplished through a variational optimization of the energy given by equation (7), which yields the electronic Schrödinger equation:

$$\left[-\frac{1}{2}\nabla^2 - \sum_{\alpha=1}^{N}\frac{Z_\alpha}{|\mathbf{r}-\mathbf{R}_\alpha|} + \int d\mathbf{r}_2\frac{\rho\left(\mathbf{r}_2\right)}{|\mathbf{r}_2-\mathbf{r}|} + V_{xc}\left(\mathbf{r}\right)\right]\Phi_i\left(\mathbf{r}\right)$$
$$= \varepsilon_i\Phi_i\left(\mathbf{r}\right) \quad (8)$$

where

$$V_{xc}\left(\mathbf{r}\right) = \frac{\delta E_{xc}\left[\rho\left(\mathbf{r}\right)\right]}{\delta\rho\left(\mathbf{r}\right)} \quad (9)$$

is the exchange-correlation potential, δ stands for functional differentiation, and ε_i is the energy associated with the single electron state $\Phi_i\left(\mathbf{r}\right)$. Strictly speaking, the ε_i values are the energies of weakly interacting "quasiparticles" made up of electrons and the exchange-correlation holes that follow them and screen them from other electrons (hence the weak interaction).

The beauty of equations (7)–(9) is that they give an exact formulation for the ground state properties of an arbitrary system of nuclei and electrons in terms of just the single electron density. Nothing needs to be known about the many-electron wavefunction! The entire complexity of the many-body aspects is contained in the density functional $E_{xc}\left[\rho\left(\mathbf{r}\right)\right]$.

The exact form of $E_{xc}\left[\rho\left(\mathbf{r}\right)\right]$ remains the "holy grail" of the condensed matter theory community. Its form is well known for a uniform electron gas in the high-density limit [*Gell-Mann and Brueckner*, 1957]. In materials of interest to us, the electron density is neither uniform nor high. It is therefore remarkable that accurate approximations can be developed by assuming that the exchange correlation interaction can be expressed in terms of the local density approximation (LDA), in which

$$E_{xc}\left[\rho\left(\mathbf{r}\right)\right] = \int d\mathbf{r}\,\rho\left(\mathbf{r}\right)\varepsilon_{xc}\left[\rho\left(\mathbf{r}\right)\right], \quad (10)$$

where $\varepsilon_{xc}\left[\rho\left(\mathbf{r}\right)\right]$ is the exchange-correlation energy per electron of a uniform electron gas with density equal to the local value of the nonuniform density $\rho\left(\mathbf{r}\right)$. To study the effects of electron spin polarization, ε_{xc} is expressed as $\varepsilon_{xc}\left[\rho\uparrow\left(\mathbf{r}\right),\rho\downarrow\left(\mathbf{r}\right)\right]$, where $\rho\uparrow\left(\mathbf{r}\right),\rho\downarrow\left(\mathbf{r}\right)$ are the spin-up and spin-down electron densities, respectively. *Lundqvist and*

March [1987] provide useful reviews of these developments. An early successful approximation is the Hedin–Lundqvist expression [*Hedin and Lundqvist*, 1971], which continues to be used in some applications. More accurate expressions can be found in some cases from Quantum Monte Carlo simulations of the electron gas [*Ceperley and Alder*, 1980]. Results of these simulations have been represented in a parametric form that obeys the known high-density limiting behavior [*Perdew and Zunger*, 1981].

The success of the various LDA approximations is demonstrated by numerous computations that yield results that are often in excellent agreement with experimental measurements. The earlier review papers provide many examples relevant to Earth studies [*Bukowinski*, 1994; *Bukowinski et al.*, 1996; *Karki et al.*, 2001a; *Stixrude*, 2000, 2001; *Stixrude et al.*, 1998]. The method is not without flaws, however. Accurate first principles computations based on LDA are believed to underestimate the volume at zero pressure due to an apparent tendency to overestimate the binding energy. Although the errors usually amount to a few percent at most, they are not easy to evaluate because many published results are for static lattices that do not include the effects of thermal expansion nor the zero-point vibration contributions. The latter increases the zero-pressure volume by a percent or so, which softens the lattice and consequently lowers the bulk modulus by several percent. When such vibrational effects are included, the LDA gives excellent equations of state but, at least in some cases, appears to now slightly underbind [*Karki and Wentzcovitch*, 2002; *Karki et al.*, 2000b; *Wentzcovitch et al.*, 2004]. A more serious shortcoming of LDA is that it does not always yield accurate relative energies for different structures of the same material. For example, LDA predicts that the stable phase of Fe at zero pressure is hexagonal close packed, rather than the observed body-centered cubic (bcc) [*Stixrude et al.*, 1997].

The principal shortcoming of the LDA is its reliance on functionals that are local and strictly valid for uniform electron gases only. Including the effects of nonuniformity has proven a rather difficult task [*Lundqvist and March*, 1987]. Particularly fruitful approaches are the generalized gradient approximations (GGAs), in which some account is taken of the local variability of the electron density, as given for example by its gradient $\nabla \rho(\mathbf{r})$. Taking advantage of the approximate locality of the exchange and correlation interactions, GGA theories use equation (10), but replace $\varepsilon_{xc}\left[\rho(\mathbf{r})\right]$ with $\varepsilon_{xc}\left[\rho(\mathbf{r}),|\nabla\rho(\mathbf{r})|\right]$, or its spin-polarized generalization. A commonly used functional was developed by Perdew and collaborators [*Perdew et al.*, 1997, 1998]. In contrast to the LDA methods, the GGA functional tends to underestimate binding somewhat, even in static lattice calculations. However, it does correctly predict that the stable phase of Fe

at ambient conditions has the bcc structure. It also appears to yield the correct phase relationships in silica [*Hamann*, 1996], but such results should be tempered by the fact that the calculations did not include zero-point vibrations, which could affect the relative energies of some of the phases.

Practical Considerations in First Principles Computations

Whether one chooses to use the LDA or GGA approximation to the exchange correlation functional, one must also choose a specific method of solution for equation (8). Names like APW, KKR, LAPW, LMTO, PAW, pseudopotential, etc., are found in the literature. These methods differ primarily in the choice of the function basis sets used to expand the electronic wavefunctions, and other approximations aimed at computational efficiency commensurate with contemporary computer capabilities. All of these methods have been used successfully to study Earth materials. The approximations traditionally associated with some of the methods can be relaxed to achieve more accuracy. For example, with modern computer speeds and memory capacities, the "muffin-tin" form of the electronic potential commonly used in the APW and KKR methods can be relaxed to the unrestricted form more commonly associated with modern methods like LAPW. If all such nonessential approximations were eliminated, then all these methods, if carefully carried out, should in principle yield the same results when applied to the same problem.

The great majority of modern first principles computations are done with the pseudopotential and LAPW methods. The latter is an "all-electron" method since it solves Schrödinger's equation for all electrons in the system. Because it also makes a minimum of simplifications, it is believed to give the most accurate properties of materials as described by current density functional theory. However, because of its heavy computational demands, it is not practical for computations on materials with structures much more complex than, say, $MgSiO_3$ perovskite. In contrast, the pseudopotential method, which replaces the nucleus and core electrons by a weak potential, can efficiently handle much more complex materials. The method is based on an exact transformation of Schrödinger's equation in which the effect of the orthogonality of valence electron states to core states is converted to an effective non-local potential [*Phillips and Kleinman*, 1959; *Pickett*, 1989]. The form of this potential is not fully specified by this orthogonality requirement, leaving an unlimited number of possible representations. This flexibility can be exploited to suit the needs of a given problem. A common choice is to require that the valence wavefunction match the all-electron wavefunction for radii outside some cutoff value and that it be nodeless inside that radius. The resulting smooth and slowly varying valence and conduction wavefunctions can be efficiently represented by a reasonable number of plane wave states. The problem is thus transformed from one in direct space to one formulated in reciprocal space, where various integrals can be handled extremely efficiently with the fast Fourier transform algorithm. Combined with a self-consistent molecular dynamics approach, the pseudopotential method yields a very efficient method of searching for equilibrium lattice structures [*Wentzcovitch and Martins*, 1991].

Atomistic Models

First principles methods provide the most rigorous means of computing physical properties of materials. Since they do not require experimental data as input, they can provide unbiased estimates of material properties under conditions not yet accessible to experimental techniques. The reliability of first principles computations should increase as pressure increases the electronic density. Hence, if computed properties agree with measurements at low pressures, they should be even more dependable under high-pressure conditions, where accurate data may not be available. Thus, when computers become sufficiently powerful, most computations will likely be carried out within the first principles framework. But until then, we will continue to need more approximate, but faster, methods that allow efficient simulations of complex materials or systems containing a large number of atoms. Besides efficiency, approximate models can offer significant insights by reducing the complex, and seemingly abstract, properties of interacting electrons to a simpler and more intuitive picture. The most successful models of this type are those based on an atomistic view of matter.

Atomistic models treat materials as being composed of atoms or ions, rather than of nuclei and electrons. The objective of such models is to reduce the total energy calculation to a sum over interatomic/ionic interaction energies. In its most general form, the energy of such models can be expressed as

$$E = \sum_i \phi_i + \sum_{i<j} \phi_2\left(r_{ij}\right) + \sum_{i<j<k} \phi_3\left(r_{ij}, r_{ik}, r_{jk}\right) + \dots \quad (11)$$

where ϕ_i is the self-energy of the ith ion (for the sake of simplicity, here and in what follows, we refer only to interacting ions, though the description applies equally well to atoms), r_{ij} is the distance between ions i and j, ϕ_2 is the interaction between these ions when no other ions are present, and ϕ_3 is the three-body interaction that measures the change in energy associated with the perturbation introduced by a third ion. ϕ_3 can also be used to represent bond-angle-bending potentials.

The self-energy ϕ_i is the energy due to the interactions among the electrons belonging to ion i, and between these electrons and the nucleus. Since interactions with other ions in the material perturb the electronic wavefunctions, these intra-ionic interactions are affected by any changes in relative ionic positions, such as those accompanying structural changes brought about by external stresses. These changes affect the total energy of the material and hence affect its properties.

To be of practical use, equation (11) should give accurate energies when the sum can be truncated at the two- or three-body terms. Any such truncation necessarily results in *effective potentials* that are likely to be useful only in a limited context. For example, pair interactions found to give accurate properties for a solid are not likely to yield accurate pair bond lengths for isolated ionic pairs.

In ab initio implementations of equation (11), all potential terms are obtained from quantum mechanics. Among these, methods based on the Gordon–Kim electron gas model of pair interactions [*Clugston*, 1978; *Cohen and Gordon*, 1975; *Gordon and Kim*, 1972; *Kim and Gordon*, 1974a, 1974b; *Waldman and Gordon*, 1979] have proven particularly useful in the study of equations of state and mineral structures. In electron gas models, the total charge density of a material is taken to be the sum of the spherical charge densities of the constituent ions. This turns out to be an excellent approximation for reasonably ionic materials, as demonstrated by the fact that the total first principles electronic density in such materials can be approximated with good accuracy by a superposition of spherical ionic charge distributions [e.g., *Mehl et al.*, 1988]. Given the total charge density of a pair of interacting ions, the energy is computed with density functional theory through equation (7), except that the kinetic energy density at a point is approximated as that of a free electron gas of the same density; hence the kinetic energy $T = \int d\mathbf{r}\, \rho(\mathbf{r}) \varepsilon_k\left[\rho(\mathbf{r})\right]$ is calulated, where $\varepsilon_k[\rho] = (3/10)\left(3\pi^2\right)^{2/3} \rho^{2/3}$ is the kinetic energy per electron. The potential energy of interaction between ions i and j is then obtained as the difference between the energy of the interacting ions minus the sum of the energies of the isolated (noninteracting) ions:

$$\phi_{ij}(r) = E\left[\rho_i + \rho_j\right] - E\left[\rho_i\right] - E\left[\rho_j\right] \quad (12)$$

where r is the distance between the ions, and ρ_i is the electron density of the ith ion. The total energy is then given by

$$E(\mathbf{R}) = \sum_{i<j} \phi_{ij}\left(r_{ij}\right) + \sum_i \phi_i \quad (13)$$

where r_{ij} is the distance between ions i and j, and as before, ϕ_i is the self-energy of the ith ion/atom.

As it turns out, to obtain accurate pair interactions, one must take into account the fact that an ion's electron density is subjected to perturbations from the other ions in the material. Such effects are particularly significant for O^{2-} ions since the doubly charged species is not stable as a free ion. Within a condensed material, the excess electrons on negatively charged O ions are stabilized by the negative Coulomb potential at its site, which is dominated by the cations in the nearest neighbor coordination shell. The consequent sensitivity of oxygen electronic densities to the crystal environment introduces many body effects. For example, the potential between an oxygen and another ion is modified when a third atom affects the oxygen's charge density and hence its interaction with the other ions.

Over the years, several models have been developed that attempt to account for the influence of the material on local charge densities. The two such models still in use approximate the effect of the crystal on an ion by a Watson sphere [*Watson*, 1958]. This is a charged sphere, centered on the ion in question and bearing a surface charge equal to the negative of the ionic charge. The potential due to the Watson sphere, which closely resembles the spherical average of the crystal potential at the ionic site, is added to the other terms on the left-hand side of equation (8) when used to calculate ionic charge densities. In the Potential Induced Breathing (PIB) model the radius of the sphere is adjusted so that the potential inside the sphere equals the crystal Coulomb potential at the site (Madelung potential) [*Boyer et al.*, 1985; *Cohen*, 1987; *Isaak et al.*, 1990]. As the crystal configuration changes, so does the Madelung potential and hence so does the electronic density of the ion. Because the oxygen ion is thus forced to slightly compress with pressure, the computed equations of state show more compressibility and are in much better agreement with data than earlier models with rigid oxygen ions. In the Variationally Induced Breathing (VIB) model, a further improvement is achieved by optimizing the total energy of the material with respect to the Watson sphere radii [*Bukowinski et al.*, 1996; *Wolf and Bukowinski*, 1988]. This effectively allows the electronic wavefunctions to react to overlap interactions as well as the Madelung potential. VIB and PIB yield similar zero-pressure volumes; in VIB, however, oxygen ions contract a little more in response to pressure since by doing so they can lower repulsive overlap interactions. The extent of this contraction is limited by negative feedback brought in by the fact that the self-energy increases with ionic contraction. This additional many-body contribution appears to account nicely for oxide and silicate energetics; VIB-calculated equations of state compare favorably with data and those obtained from first principles [*Bukowinski*, 1994; *Bukowinski et al.*, 1996; *Downs and Bukowinski*, 1997; *Inbar and Cohen*, 1995].

Electron gas type models are quite efficient. Using desktop computers, it is possible to find rather accurate fully optimized equilibrium structures for crystals with many atoms per cell and tens of structural degrees of freedom

[*Bukowinski and Downs*, 2000; *Downs and Bukowinski*, 1997]. However, there are applications where even these methods are too slow for practical use. Models based on simple approximations to interionic potentials, based on analytic expressions containing adjustable parameters, have long been recognized as powerful tools for the efficient computation of a great variety of material properties [*Catlow and Mackrodt*, 1985; *Gale*, 2001]. Computer codes have been written that can generate and analyze a great many properties of highly complex materials using a variety of potential models [e.g., *Busing*, 1981; *Gale*, 1997].

Methods based on parametric potentials differ from electron gas–type potentials in that they do not involve the use of quantum mechanics to compute interaction energies. Rather, they use mathematical expressions that fairly approximate the form of the interaction energies. For example, a popular expression for interionic potentials is

$$\phi_{ij}\left(r_{ij}\right) = \frac{I_i I_j}{r_{ij}} + b_{ij}\, e^{-\frac{r_{ij}}{\rho_{ij}}} - \frac{c_{ij}}{r_{ij}^6} \qquad (14)$$

where I_i is the charge on the ith ion, and b_{ij}, ρ_{ij}, and c_{ij} are parameters that describe a given pair of atoms. The first term is the Coulomb interaction, the second represents the short-range overlap repulsion, and the last is the van der Waals representation of the interaction between induced dipoles on the ions. In practice, the parameters in equation (14) are determined from fits to experimental data [*Bush et al.*, 1994; *Gale*, 1996; *Lewis and Catlow*, 1985; *Matsui*, 1994], or to energy surfaces computed from ab initio or approximate quantum mechanical methods [*Burnham*, 1990; *Kramer et al.*, 1991; *Tsuneyuki et al.*, 1988]. Unavoidably, such fits entail some compromises: except for the Coulomb interaction, potential forms such as (14) are approximations that work well over a limited range of interionic distances. Thus, for example, when high-pressure properties are computed with potentials constrained by low-pressure data, the results amount to extrapolations that should be regarded with some care. In contrast, electron gas potentials are recomputed at each pressure and thus account for pressure-induced changes in electron orbitals, and the consequent changes in interaction energies.

WHAT CAN THEORETICAL MINERAL PHYSICS TELL US ABOUT THE REAL EARTH?

Any property of any material can, in principle, be computed with methods built upon the foundations described above. Before about two decades ago, only simple crystals of geological interest could be studied with useful accuracy via first principles techniques [*Bukowinski*, 1976b, 1976c, 1980, 1985]. By today's standards, these early computations seem almost primitive. And yet these methods served to make some significant predictions. For example, a theoretical examination of the electronic structure of potassium found that at pressures above about 25 GPa, potassium becomes a transition metal, which presumably would render it miscible in Fe, and hence make it likely that radioactive ^{40}K is present as a heat source in the Earth's core [*Bukowinski*, 1976a, 1979]. Twenty-seven years later this was confirmed experimentally [*Lee and Jeanloz*, 2003]. More recently, a phase transformation was predicted to occur in Al_2O_3 at a pressure of around 80 GPa. Experiments to much higher pressures had failed to identify any transformations [*Jephcoat et al.*, 1988]. An LAPW calculation that assumed the experimentally observed structure for corundum predicted a phase transformation to the Rh_2O_3 type II structure [*Marton and Cohen*, 1994]. Using the pseudopotential method, *Thomson et al.* [1996] were able to fully optimize the corundum structure and showed that the x-ray spectra collected by *Jephcoat* [1988] were consistent with a mixture of the corundum and Rh_2O_3 type II structures at high pressures, thus verifying the earlier prediction by Marton and Cohen. These purely theoretical predictions were later verified experimentally with the help of high-intensity synchrotron radiation [*Funamori and Jeanloz*, 1997]. The pressure at which stishovite transforms to the $CaCl_2$-type structure was accurately determined by an LAPW calculation [*Cohen*, 1992] before it was experimentally refined [*Kingma et al.*, 1995]. There is thus no doubt that computational mineral physics can make significant predictions about simple, but geologically relevant, materials.

Modern computers, coupled with sophisticated codes, can also offer much quantitative information about complex Earth materials. Certain material properties, such as general elastic constants under deep Earth pressure and temperature conditions, continue to severely test experimental capabilities but no longer pose serious difficulties to computational mineral physics. But how reliable are these computed properties? The published literature attests to the fact that they can be quite reliable [*Bukowinski*, 1994; *Bukowinski et al.*, 1996; *Karki et al.*, 2001a; *Stixrude*, 2000, 2001; *Stixrude et al.*, 1998]. Yet, it would be naïve to believe that every sophisticated computation is accurate. Perhaps the easiest way to develop a sense of what to make of typical theoretical mineral physics claims is to examine some relevant examples. In what follows, we take a look at a few examples that illustrate both the strengths and limitations of current computational methods, as well as potential weaknesses and pitfalls.

Equations of State

Equations of state (eos) provide what are perhaps the most direct comparisons of theory with experiment at high pressures. That a model yields an accurate eos does not guarantee that it will yield other properties with equal accuracy. Nevertheless, being a derivative of the energy, the pressure as a function of the volume and temperature does offer a fairly sensitive test of the theory. It is not easy to accurately predict a material's compressibility with simple models that are not custom-designed for that material. Materials that do not undergo any internal strains upon compression require that interionic repulsions be accurately represented over a sufficiently broad range of pressure. But some materials, such as quartz, compress almost entirely through internal strains, with hardly any change in bond lengths. Any attempt to model such materials accurately requires that covalent angle-bending forces be properly accounted for [e.g., *Stixrude and Bukowinski*, 1988]. This is not an issue with first principles methods that use the full potential seen by the electrons (e.g., LAPW, pseudopotential), but electron gas–based atomistic models commonly used in mineral physics do poorly, unless some rather artificial means of approximating polarizability are added [e.g., *Jackson and Gordon*, 1988; *Lacks and Gordon*, 1993, 1995].

The published literature contains many accurate theoretical eos. The LAPW method gives the closest approximation to true LDA properties and therefore serves as a benchmark for other implementations of LDA. However, because of its high computational demands, LAPW has not been applied to highly complex silicates or oxides. Where it has been used, it typically yields volumes that differ from data by less than 4%, of which a substantial fraction must be attributed to thermal corrections, which are not included in these calculations [*Cohen*, 1991, 1992; *Mehl et al.*, 1988; *Stixrude and Cohen*, 1993]. Many other examples of LDA approximations can be found in the review papers cited earlier. Here we consider a few examples that illustrate the attainable accuracy, as well as some of the problems that can arise when modeling typical mantle minerals.

As our first example, we consider the eos of forsterite which, with 28 atoms in the unit cell and 14 free structural parameters, is complex enough to seriously test computational models. Figure 3 compares some computed eos with data. The pseudopotential static eos by *Brodholt et al.* [1996] is representative of LDA accuracy and of its tendency to overbind.

A comparison of this eos with the VIBC curves (to be discussed below) demonstrates that even if vibrational corrections were added, the LDA volumes would remain too small. However, the remaining volume error is less than 2%. Some of the risks associated with the use of pseudopotentials is illustrated by the *Wentzcovitch and Stixrude* [1997] eos. Although these authors obtained good elastic constants, the volume error is considerably larger than that of *Brodholt et*

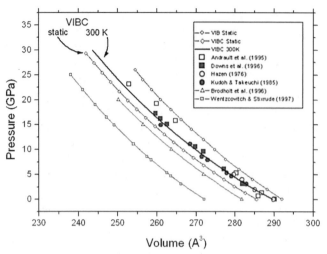

Figure 3. Experimental and computed equations of state of forsterite. The pseudopotential calculations by *Brodholt et al.* [1996] and *Wentzcovitch and Stixrude* [1997] are for static lattices. See text for other details.

al. [1996] The cause appears to be the Trouiller–Martins [*Trouiller and Martins*, 1980] pseudopotential for Mg, which assumes that the Mg 2p states can be treated as part of the Mg core. If the 2p electrons are treated as valence states, the volume of forsterite increases to something much closer to experimental and typical LDA results [*Kiefer et al.*, 1997]. Using such improved Mg pseudopotentials, *Karki et al.* [1999] obtained excellent results for the elastic properties of MgO. The lesson learned here is that, when constructing pseudopotentials, assumptions regarding which electron states can be treated as core states must always be tested.

For a comparison with more approximate methods, Figure 3 also shows eos computed with the VIB method described earlier and with VIBC, which adds a small empirically constrained covalent correction to VIB [*Akber-Knutson et al.*, 2002]. Electron gas models like VIB tend to slightly underestimate the Si–O bond length, which results in silicate lattices that are somewhat too dense and too stiff. Excellent eos can be obtained by somewhat reducing the charges of Si and O [e.g., *Bukowinski and Downs*, 2000]. To avoid ad hoc charge reductions that would need to be repeated for every different crystal and for every different pressure, we opted to introduce a weighted covalent energy term that replaces some of the Coulomb binding lost when charges are reduced. The resulting Si–O pair potential then is

$$\phi_{Si-O} = \sqrt{1 - f_{Si}f_O}\,\phi_{cv} + \phi_{coul} + \phi_{short} \qquad (15)$$

where f_i ($i = Si, O$) is the ratio of the actual ionic charge to its nominal full charge, ϕ_{cv} is the covalent energy of the bond, ϕ_{coul} and ϕ_{short} are the long-range Coulomb and short-range

overlap potentials; we followed *Urusov and Eremin* [1995] in our choice of weight for the covalent interaction. The latter is calculated as in VIB except that the charge densities are obtained for the reduced ionic charges. The total energy of the material is optimized with respect to the f_i values. In applications discussed in this review, we chose to represent ϕ_{cv} with a Morse-type expression

$$\phi_{cv} = -D\left[2e^{-\beta(r-r_0)} - e^{-2\beta(r-r_0)}\right],$$

the parameters D, β, and r_0 being chosen to reproduce the compressibility and density of stishovite [*Panero et al.*, 2001; *Weidner et al.*, 1986]. Although no data on forsterite were used for the covalent correction, the resulting eos is much better than that obtained from VIB and is considerably more accurate than the first principles results.

As our next example, we consider the eos of $MgSiO_3$ perovskite, which has received much attention from modelers using a spectrum of methods ranging from first principles to empirical pair potentials [*Bukowinski and Wolf*, 1988; *D'Arco et al.*, 1993; *Gillet et al.*, 1996, 2000; *Hama et al.*, 2000; *Karki et al.*, 2001b; *Marton et al.*, 2001; *Matsui*, 1988, 2000; *Matsui and Price*, 1992; *Matsui et al.*, 1994; *Oganov et al.*, 2001a, 2001b, 2001c; *Stixrude*, 1998; *Wentzcovitch et al.*, 1995; *Winkler and Dove*, 1992; *Wolf and Bukowinski*, 1985, 1987; *Wolf and Jeanloz*, 1985]. *Oganov et al.* [2001b] did a useful comparison of LDA and GGA pseudopotential computations, as shown in Figure 4. Although the LDA curve appears to be a good match to data, the agreement would disappear if thermal and zero-point corrections, amounting to an approximately 1% increase in volume, were added. The static GGA eos overestimates the volumes even more, in keeping with GGA's usual tendency to underbind. A very good eos is obtained by shifting the GGA 298 K eos so that the volume at zero pressure matches experiment. This brings us to the interesting observation that GGA physical properties of perovskite-structured materials match experimental data quite well when calculated at the experimental volume, rather than the equilibrium volume predicted by GGA [*Oganov et al.*, 2001b; *Vanderbilt*, 1998]. Thus, shifting the pressure as in Figure 4, the GGA physical properties of $MgSiO_3$ perovskite calculated at experimental volumes bring these properties into very good agreement with data at the observed pressure as well. One can also make the case that LDA results calculated at the experimental volume produce good agreement with data [*Karki and Wentzcovitch*, 2002; *Karki et al.*, 1999, 2000a, 2000b; *Wentzcovitch et al.*, 2004]. While this process violates the first principles underpinning of the calculations, it may be justified if it consistently yields accurate estimates of material properties.

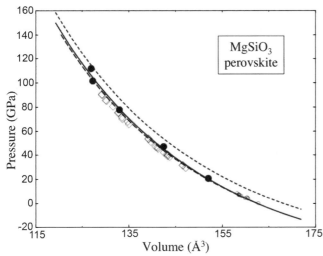

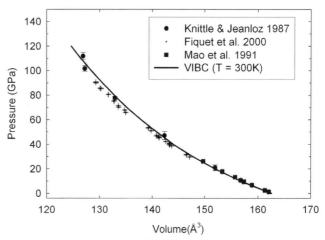

Figure 4. MgSiO$_3$ perovskite eos. Theoretical curves: long dashes, pseudopotential LDA static eos [*Karki*, 1997]; short dashes, pseudopotential GGA static eos; solid line, pseudopotential GGA eos shifted in pressure [*Oganov et al.*, 2001b]. Experimental points: solid circles are from *Knittle and Jeanloz* [1987]; small circles from *Yagi et al.* [1982]; diamonds from *Fiquet et al.* [2000]. Adapted from *Oganov et al.* [2001b].

Figure 5. Equation of state of MgSiO$_3$ perovskite. Solid line, theoretical eos obtained with the VIBC method; symbols, experimental data from studies identified in label.

Staying with the non-first-principles theme, we again find that the semiempirical VIBC (Figure 5), with an empirical correction identical to that used in the forsterite eos, yields a room temperature eos that is in better agreement with data than those derived from first principles methods, except for the pressure-shifted GGA. Computed eos parameters are also at least as accurate as the first principles results (Table 1).

While we are on the topic of eos, we should note that much progress has been made in recent years in the first principles computations of vibrational effects on equilibrium properties. Zero-point vibrations increase the volume by a percent or two and decrease the bulk modulus by a few percent [*Karki et al.*, 1999, 2000a, 2000b]. Finite temperature effects may be computed from vibrational spectra obtained through quasiharmonic lattice dynamics or from simulations such as molecular dynamics, in which the motions of the atoms in thermal equilibrium are propagated through time by solving classical equations of motion [*Frenkel and Smit*, 1996]. The quasiharmonic approximation sometimes fails to yield accurate thermal properties,

as was shown in the case of CaO [*Karki and Wentzcovitch*, 2003]. In contrast, in a study of the thermoelastic properties of MgSiO$_3$ perovskite, *Wentzcovitch et al.* [2004] found that results obtained with quasiharmonic lattice dynamics and molecular dynamics agree with each other and with the available data. Such successes give strong evidence that first principles methods either are, or soon will be, capable of generating highly reliable thermodynamic properties of earth materials.

The results we have discussed above are fairly representative. First principles methods based on LDA or GGA give material volumes with errors that rarely exceed a few percent. Electron gas potential models deliver very good accuracy with relatively minimal computing cost. This is an impressive performance for methods that either employ no data on materials at all or use a minimal amount of data to correct an otherwise ab initio method. It is not unreasonable to expect that in the near future, computed eos will be routinely as dependable as the best experimentally measured ones. Computational materials science will approach the extreme accuracy of computational quantum mechanics of simpler systems [e.g., *Goddard*, 1985; *Richards*, 1979]. It will then be possible to construct compositional models of the three-dimensional Earth that are based largely on computer-generated candidate materials. The materials' physical

Table 1. Calculated and Measured Volumes and Bulk Moduli of MgSiO$_3$ Perovskite.

Parameter	Exp	VIBC	LDAa	LDAb	LDAc	Shifted GGA	H-F
Volume, Å^3	162.45	162.5	157.5	162.47	157.87	162.4	162.12
K$_0$, GPa	264, 253	256	267.1	257.8	258.3	266.7	309

Experimental volume: *Ross and Hazen* [1989]; experimental K$_0$: single-crystal Brillouin [*Yeganeh-Haeri*, 1994]; the lower value is from a recent single and polycrystalline Brillouin measurement [*Sinogeikin et al.*, 2004]. LDAa: *Wentzcovitch et al.* [1995]; LDAb: *Karki* [1997]; LDAc and GGA: *Oganov et al.* [2001b]; H-F: *D'Arco et al.* [1993]; VIBC: Bukowinski (unpublished). The GGA eos was shifted in pressure as described in the text.

properties and their sensitivity to composition, temperature, pressure, and structure will be efficiently explored and "measured" with the help of computations that will be much faster and cheaper than experimental measurements.

But in this also lurks a significant danger. The apparent success of first principles and approximate theories of materials can lull us into a sense that all is well when it isn't. The eos of $CaSiO_3$ perovskite provides a good example. Early eos indicated that $CaSiO_3$ perovskite is at least as incompressible as $MgSiO_3$ perovskite [*Mao et al.*, 1989; *Tamai and Yagi*, 1989; *Tarrida and Richet*, 1989]. Computed eos agreed with these data, thus at once lending support to the measurements and apparently validating theory and its computational implementations [*Hemley et al.*, 1987; *Sherman*, 1993b; *Wolf and Bukowinski*, 1987; *Wolf and Jeanloz*, 1985].

Figure 6 compares the data, including more recent data obtained under quasihydrostatic conditions [*Shim et al.*, 2000; *Wang et al.*, 1996], with VIB and VIBC calculations. We now know that VIB tends to underestimate the compressibility of silicates by as much as 20%. As we saw earlier, VIBC gives a much better eos. More recent first principles calculations also give results in good agreement with the newer data [*Chizmeshya et al.*, 1996; *Karki and Crain*, 1998; *Wentzcovitch et al.*, 1995]. Of the earlier computations, all but Sherman's were similar in kind to VIB while Sherman's was based on the Hartree–Fock method, which also overestimates the bulk modulus of $MgSiO_3$ perovskite (Table 1). There is a clear lesson here: Satisfaction over good agreement between data and computation ought to be indulged in with a good measure of caution.

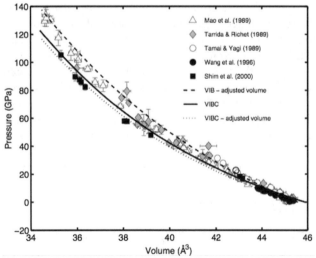

Figure 6. $CaSiO_3$ perovskite eos. Experimental data are represented by symbols; theoretical results are shown as lines. The shifts in volume in the calculated eos, illustrated by the two VIBC curves, amount to a fraction of a percent.

Elasticity and Seismic Velocities

Elastic moduli, as determined from seismic velocities, are among the most detailed and direct data on the Earth's interior. Interpretation of radial and lateral variations in velocities in terms of mineral elastic constants (as functions of pressure, temperature, and mineral composition) is essential to constraining the composition and evolution of the Earth. While the adiabatic bulk (K_S) and shear (μ) moduli, and density (ρ) have been measured experimentally, data are sparse for pressures and temperatures of the lower mantle, especially for the shear modulus [*Liebermann and Li*, 1998]. Theoretical calculations can be used not only to help extrapolate to deep Earth conditions but also to understand observed trends and identify mechanisms responsible for compositional variations (e.g., iron content and substitution mechanisms of Fe and Al into Mg, Ca, and Si sites). The calculation of first principles elastic constants was made practical by the development of an efficient structural optimization scheme that uses the Hellman–Feynman theorem to compute forces within the pseudopotential approach [*Wentzcovitch et al.*, 1993]. An excellent review of recent theoretical progress in the computation of general elastic properties of minerals is given by *Karki et al.* [2001a].

Elastic moduli may be computed by subjecting the equilibrium structure of a crystal to a small deviatoric strain ε'. The energy of the strained lattice is given by

$$F\left(V,T,\varepsilon'\right) = F_o\left(V\right) + F_{th}\left(V,T\right) + \frac{1}{2}C_{ijkl}\left(V,T\right)\varepsilon'_{ij}\varepsilon'_{kl} \quad (16)$$

where $F_0(V)$ and $F_{TH}(V,T)$ are the static and thermal contributions to the free energy, respectively, and C_{klmn} ($klmn \in \{1...3\}$) are the elements of the elastic constant tensor. The last term gives the stored elastic energy due to the small applied deviatoric strains. In most cases, the complete elasticity tensor can be obtained by computing the free energy changes brought about by a few carefully chosen small strains [e.g., *da Silva et al.*, 1997], and fitting the results to a quadratic in the strain [*Karki et al.*, 2001a]. The average bulk and shear moduli for a polycrystalline aggregate can then be estimated with the help of averaging schemes such as those of Voigt, Reuss, Voigt–Reuss–Hill, and Hashin–Shtrikman [*Watt*, 1979]. Average moduli for isotropic aggregates may also be obtained directly from eos by using approximate thermal models based on Debye theory [*Hama and Suito*, 1998; *Hama et al.*, 2000]. Modeling the response of elastic anisotropy to pressure and temperature requires the calculation of the full elastic tensor, as has been done, for example, to examine the elasticity of $MgSiO_3$ perovskite [*da Silva et al.*, 1997; *Karki et al.*, 1997; *Oganov et al.*, 2001b; *Stixrude*, 1998; *Wentzcovitch et al.*, 1998].

Table 2. Theoretical and Experimental Elastic Constants (GPa) for $MgSiO_3$ Perovskite (numbers in parentheses indicate uncertainty in last digit shown).

		$V(Å^3)$	C_{11}	C_{22}	C_{33}	C_{12}	C_{13}	C_{23}	C_{44}	C_{55}	C_{66}	K_s	μ
Theory													
	a	164.7	460	506	378	139	184	177	162	159	112	261	143
	b	165.05	531	531	425	44	143	166	237	249	136	245	192
	c		487	524	456	128	144	156	203	186	145	258	175
	d		485	560	474	130	136	144	200	176	155	259	179
	e	162.31	500	509	398	116	210	188	174	189	102	270	146
	f	162.4	493	546	470	142	146	160	212	186	149	267	179
	g	159.6	491	554	474	134	139	152	203	176	153	263	178
	h		484	542	477	146	146	162	195	172	151	267	173
	i		449	500	434	123	129	144	183	162	138	241	161
Exp													
	j	162.45	482(4)	537(3)	485(5)	144(6)	147(6)	146(7)	204(2)	186(2)	147(3)	264(5)	177(4)
	k		481(4)	528(3)	456(4)	125(3)	139(3)	146(3)	200(2)	182(2)	147(2)	253(3)	175(2)

a. *Matsui et al.* [1987], 0 K (athermal), empirical pair potentials.
b. *Cohen* [1987], 298 K (athermal), PIB.
c. *Karki et al.* [1997], 0 K (athermal), LDA from lattice strains.
d. *Wentzcovitch et al.* [1998], 0 K (athermal), LDA from lattice strains.
e. *Oganov et al.* [2001a], 300 K (thermal), pair potential, from phonon calculations.
f. *Oganov et al.* [2001b], 0 K (athermal), pressure-shifted GGA, from lattice strains.
g. *Kiefer et al.* [2002], 0 K (athermal), LDA, from lattice strains.
h. *Wentzcovitch et al.* [2004], 300K (thermal), LDA, pressure-shifted by 5 GPa.
i. *Wentzcovitch et al.* [2004], 300K (thermal), LDA, no pressure shift.
j. *Yeganeh-Haeri* [1994], single crystal Brillouin spectroscopy.
k. *Sinogeikin et al.* [2004], single and polycrystalline Brillouin spectroscopy.

Table 2 compares calculated elastic constants of $MgSiO_3$ perovskite, along with volumes and bulk and shear moduli, with experimental data (note that C_{ij}, where $ij \in \{1...6\}$, is in the Voigt notation). All LDA and GGA calculations used the pseudopotential method. Thermal (athermal) calculations do (do not) include vibrational effects. For $MgSiO_3$ perovskite at fixed pressure, the elastic constants at 0 K under static conditions are 3–5% greater than they would be at 300 K, if the same Hamiltonian were used. It is apparent that first principles methods give considerably more accurate elastic constants than methods based on parametric or ab initio pair potentials. However, it is also apparent that for no single computation do all the elastic constants agree well with experimental results. Though a few of the constants appear to be subject to considerable experimental uncertainty, some of the differences are due to limitations in the density functionals currently available.

The pressure dependence of elastic constants and velocities provides a starting point for interpreting seismically derived velocity–pressure profiles. Density functional theory is quite successful in reproducing the overall effect of pressure on elasticity: It gives accurate pressure derivatives, though the absolute values of the elastic constants do not always fully agree with measured constants [*Karki*, 1997]. A thorough assessment is not yet possible since most of the computations

are athermal. Figure 7 shows predicted elastic velocities for $MgSiO_3$ perovskite, one of the rare cases for which some thermal results are available; unfortunately, these data are only for zero-pressure conditions. Nevertheless, this example illustrates the potential predictive power of available theory. The density functional predictions are in quite satisfying agreement with the zero-pressure data. They also illustrate some of the limitations. The slightly low static GGA velocities can only decrease even further upon thermal correction. On the other hand, the pressure-shifted thermal GGA velocities are slightly high. The good agreement with data of the static LDA results will also be somewhat weakened by thermal corrections. The VIBC results illustrate a different approach. The computed thermal eos is combined with the *Hama and Suito* [1998] method to compute the elastic velocities for an isotropic aggregate. This requires the experimental zero-pressure value of the shear modulus as input. Since VIBC gives excellent eos, including the zero-pressure density, it reproduces the zero-pressure elastic velocities very accurately. The predicted high-pressure velocities take advantage of the apparent ability of VIBC to account accurately for the effects of compression, which suggests that this model may be an excellent means for extrapolation of experimental low-pressure velocities. We note that the predicted velocities, obtained through different computational models,

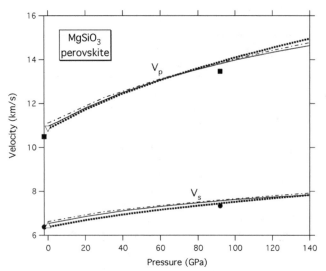

Figure 7. Elastic velocities versus pressure for MgSiO$_3$ perovskite. Solid line: 0 K, static LDA curve [*Kiefer et al.*, 2002]; dash-dot line: pressure-shifted GGA room temperature [*Oganov et al.*, 2001b]; dotted line: room temperature VIBC. Solid squares and circles: static GGA [*Kiefer et al.*, 2002]. Open triangle and inverted triangle: experimental values from *Sinogeikin et al.* [2004].

differ by at most 2% throughout the mantle's pressure range. While numerically small, such differences are comparable to those caused by significant changes in temperature and/or composition, indicating that theoretical approaches are not yet sufficiently accurate to allow unambiguous conclusions about absolute velocity values.

The density functional velocities in Figure 7 were computed independently at a number of pressures. As discussed above, the VIBC velocities were calculated via the *Hama and Suito* [1998] method, which relies on a Vinet eos fit to the calculated eos. We recalculated the velocities with a method analogous to that of Hama and Suito but based on a third-order Birch Murnaghan finite-strain eos fit [*Aidun et al.*, 1984; *Bukowinski and Aidun*, 1985]. The velocities thus computed were indistinguishable from those obtained with the Vinet eos.

The interpretation of the three-dimensional seismic structure of the Earth requires accurate thermoelastic information, such as temperature at constant-pressure derivatives of velocities and elastic constants. Because thermoelastic parameters correspond to high-order derivatives of the energy, they are not only harder to compute accurately but also difficult to validate because of the scarcity of accurate data.

As an example, let us consider the observed anticorrelation between the bulk sound speed and shear velocity in the deep lower mantle [*Masters et al.*, 2000; *Su and Dziewonski*, 1997; *Vasco and Johnson*, 1998]. This finding implies nega-

tive values of $(\partial \ln V_S / \partial \ln V_C)_P$. The anharmonic thermal contributions to lateral velocity heterogeneity are given by

$$\left(\frac{\partial \ln V_C}{\partial T}\right)_P = \frac{1}{2}\alpha(1-\delta_S), \quad \left(\frac{\partial \ln V_S}{\partial T}\right)_P = \frac{1}{2}\alpha(1-\Gamma) \quad (17)$$

where V_C and V_S are the bulk and shear elastic velocities, respectively; $\delta_S = (\partial \ln K_S / \partial \ln \rho)_P$; and $\Gamma = (\partial \ln G / \partial \ln \rho)_P$. For MgSiO$_3$ perovskite, α is not well constrained by existing data; the best that can be said is that its zero-pressure value lies somewhere on the range of (2 to 3)$\times 10^{-5}$ K^{-1} [*Chopelas*, 1996; *Funamori et al.*, 1996; *Jackson and Rigden*, 1996; *Kato et al.*, 1995; *Ross and Hazen*, 1989; *Utsumi et al.*, 1995; *Wang et al.*, 1994]. However, we don't need α to examine whether anharmonicity alone can account for the negative value of $(\partial \ln V_S / \partial \ln V_C)_P$, which is given by the ratio of the paired equations (17). Thermodynamic arguments suggest that δ_S ~ 2.7 at surface conditions [*Bukowinski and Wolf*, 1990], while measurements indicate a δ_S value of 3.3 [*Funamori et al.*, 1996] and a value of ~5.4 for Γ [*Sinelnikov et al.*, 1998]. Typical values of both parameters for lower-mantle minerals exceed 1 [*Anderson and Isaak* [1995]. Systematics indicate that Γ approaches a value of 6 in the deep mantle [*Anderson*, 1989]. It therefore appears that anharmonic effects could explain the negative correlation only if δ_S were smaller than 1 in the deep mantle. Computations show that δ_S decreases with pressure for MgO but remains above 1 for mantle pressures [*Agnon and Bukowinski*, 1990; *Isaak et al.*, 1992]. Values of δ_S have been estimated for MgSiO$_3$ perovskite from phonon spectra obtained via density functional perturbation theory [*Karki et al.*, 2001b], the δ_S dropping from 2.7 at zero pressure to 1.2 at the core–mantle boundary. Our own preliminary calculations based on VIBC give a δ_S that decreases from about 3 at the surface to 1.8 at the core–mantle boundary. Both the LDA and VIBC results are based on the quasiharmonic approximation, which may not be quite adequate for estimating anharmonic parameters. Nevertheless, these calculations suggest, but do not yet prove, that δ_S may not become small enough for thermal gradients alone to explain the negative correlation. Anelasticity may well play an important role here [*Karato*, 1993; *Karato and Karki*, 2001], as might compositional gradients or the recently discovered postperovskite phase of MgSiO$_3$ [*Iitaka et al.*, 2004; *Murakami et al.*, 2004; *Oganov and Ono*, 2004; *Tsuchiya et al.*, 2004a, 2004b]. Though first observed experimentally [*Murakami et al.*, 2004], first principles computations have played an important role in determining the exact structure and elastic properties of this new phase. Further work is needed to establish whether this transformation takes place in the mantle when Fe, Ca, and Al are included in the composition of the lower mantle.

For our last elasticity example we consider recent results regarding the effect of Al substitution into $MgSiO_3$ perovskite, which nicely illustrate the utility of parametric pair potentials in the study of thermal effects in large systems. Although the lower mantle is believed to contain a small percent of Al, the effects of this component on the density and elasticity of model mineral assemblages were often considered negligible. As the main host for Al appears to be $MgSiO_3$ perovskite [*Irifune*, 1994], a finding that Al at 5 mol % lowers the bulk modulus of $MgSiO_3$ perovskite by 11% and increases its volume slightly [*Zhang and Weidner*, 1999] got considerable attention. Subsequent experiments verified the volume increase, but there were significant discrepancies regarding the effect on the bulk modulus [*Andrault et al.*, 2001; *Daniel et al.*, 2001]. Some first principles computations also suggested a large effect on the bulk modulus [*Brodholt*, 2002]. Given that a significant change in the bulk modulus could require major adjustments in lower mantle compositional models [e.g., *Mattern et al.*, 2005], there arose a strong desire to identify the Al substitution mechanism and its effects on $MgSiO_3$ perovskite properties.

Al can dissolve into $MgSiO_3$ perovskite in a number of ways, as explored with potential models by *Richmond and Brodholt* [1998]. In the absence of other impurities, two substitution mechanisms are dominant: (1) a charge-coupled mechanism, $Si^{4+} + Mg^{2+} \rightarrow 2Al^{3+}$ and (2) an oxygen vacancy mechanism, $2Si^{4+} + O^{2-} \rightarrow 2Al^{3+}$. Experimental measurements indicate that both mechanisms are present [*Navrotsky et al.*, 2002]. First principles athermal calculations show that the charge-coupled mechanism should be heavily favored and that the effect on the bulk modulus would be small [*Yamamoto et al.*, 2003]. However, finite temperatures will cause both mechanisms to be active, their relative contributions being determined by equilibrium processes. To study these thermal effects, we used parametric pair potentials to calculate phonon spectra of large aluminous perovskite supercells [*Akber-Knutson and Bukowinski*, 2004]. The static lattice results obtained with the paired potentials are in generally good agreement with the first principles calculations of Yamamoto et al.

Configurational entropies were estimated by randomly sampling many distinct arrangements of Al and oxygen vacancies. Contrary to expectation, the configurational entropy does not give oxygen vacancies a significant advantage over charge-coupled substitution in the simulations. This is caused by partial ordering in oxygen vacancy substitution, brought about by a preference for Al, rather than Si, in the 5-coordinated polyhedra created by oxygen vacancies. The net effect on the bulk modulus and volume depends on both the Al concentration and the substitution mechanism, as illustrated in Figure 8. Charge-coupled substitution is

predicted to dominate at all pressures and temperatures in the mantle: There is less than one vacancy per 10^4 oxygen atoms at the bottom of the lower mantle. Contrary to some of the earlier studies, the bulk modulus of the resulting aluminous $MgSiO_3$ perovskite decreases by less than 2% relative to pure $MgSiO_3$, a prediction that has since been validated with Brillouin spectroscopy [*Jackson et al.*, 2004] and static compression measurements on the same sample [*Daniel et al.*, 2004].

Melting and Other Phase Transitions

The methods described in the section on theoretical foundations can in principle be used to compute any physical property of any Earth material. In this section we briefly present some examples that illustrate the ability of theoretical mineral physics to make useful predictions. Some of these predictions have since been verified by experiment, while others remain untested or unresolved due to discrepancies among experimental measurements.

The determination of melting temperatures at high pressure presents difficult experimental and theoretical challenges. Knowledge of the melting behavior of $MgSiO_3$ perovskite, with and without the addition of Fe and Al, could place an upper bound on the temperature in the lower mantle. This goal has proven highly elusive: Measured melting tem-

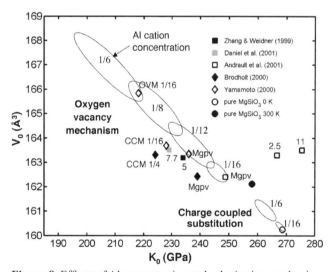

Figure 8. Effects of Al concentration and substitution mechanism on the bulk modulus and volume of aluminous $MgSiO_3$ perovskite at zero pressure. Ellipses show range of values spanned by different configurations of Al ions and O vacancies at the noted fractional Al concentrations. Decimal numbers next to experimental data points indicate Al mol %. Open and solid circles indicate the calculated thermal effect on pure $MgSiO_3$ perovskite. CCM: charge-coupled mechanism; OVM: oxygen vacancy mechanism.

peratures of $(Mg,Fe)SiO_3$ perovskites differ enough to yield extrapolated melting temperatures at the core mantle boundary that range approximately from 3000K to 7000K [*Heinz and Jeanloz*, 1987; *Knittle and Jeanloz*, 1989; *Sweeney and Heinz*, 1993; *Zerr and Boehler*, 1993]. Molecular dynamics simulations based on parametric pair potentials support the higher melting temperatures [*Chaplot et al.*, 1998; *Oganov*, 2002]. Similarly unresolved is the melting temperature of MgO. Melting temperatures obtained from classical molecular dynamics [*Vocadlo and Price*, 1996] and from VIB ab initio potentials [*Cohen and Weitz*, 1998] are in good agreement with each other but significantly exceed the temperatures obtained experimentally [*Zerr and Boehler*, 1994]. Hence melting temperatures in the mantle remain poorly characterized, and it is too early to tell whether the theoretical contributions are sufficiently reliable.

The promise of first principles estimates of melting temperatures is perhaps best illustrated with recent developments pertaining to the melting of pure Fe. Insightful discussions of the still evolving experimental phase diagram of Fe at high pressures have been published by *Poirier* [2000] and *Stixrude and Brown* [1998]. Iron melting temperatures have been examined directly and indirectly with several experimental techniques [*Anderson and Duba*, 1997; *Andrault et al.*, 2000; *Boehler*, 1993; *Boehler et al.*, 1990a, 1990b; *Brown and McQueen*, 1980; *Nguyen and Holmes*, 2004; *Saxena and Dubrovinsky*, 2000; *Shen et al.*, 1998a; *Williams et al.*, 1987; *Yoo et al.*, 1993]. The results thus far are highly inconclusive. Combining diamond cell and shock-compression techniques, *Williams et al.* [1987] found a high slope of the melting temperature that extrapolated to 7600 ± 500 K at the inner core–outer core boundary. Diamond cell measurements by *Boehler et al.* [1990] found a melting temperature at 120 GPa that was well over 1000 K below that found by *Williams et al.* A shock wave experiment by *Yoo et al.* [1993] yields an estimate that appears to be consistent with an extrapolation of the Williams et al. curve but is barely compatible with an earlier shock wave determination by *Brown and McQueen* [1980] (see Figure 9). Given the technical difficulties associated with the measurements [*Poirier*, 2000], and without additional information, we can do no better than assume that the actual melting curve lies within the roughly 2000 K range spanned by the estimated melting temperatures at the inner core boundary.

The needed additional information may well be supplied by some recent first principles computations of the melting temperature of Fe [*Alfe et al.*, 1999, 2002d; *Belonoshko et al.*, 2000; *Laio et al.*, 2000]. The work of Belonoshko et al. and Laio et al. relied on parametric potentials fitted to ab initio calculations; these potentials were then used to carry out molecular dynamics simulations of large systems containing

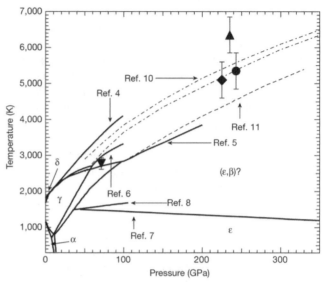

Figure 9. Phase diagram of Fe. Solid lines are from diamond anvil experiments: 4 [*Williams et al.*, 1987]; 5 [*Boehler*, 1993]; 6 [*Shen et al.*, 1998b]; 7 [*Saxena and Dubrovinsky*, 2000]; 8 [*Andrault et al.*, 2000]. Theoretical results: 10 [*Alfe et al.*, 2002d], upper (lower) dash-dotted curves are calculations with (without) free energy correction ; 11 [*Laio et al.*, 2000]. Solid symbols are from shock-compression experiments: inverted triangle, *Ahrens et al.* [2002]; diamond, *Nguyen and Holmes* [2004]; triangle, *Yoo et al.* [1993]; circle, *Brown and McQueen* [1986]. Figure taken from *Nguyen and Holmes* [2004]; reprinted with permission from *Nature*.

coexisting solid and liquid phases of Fe. Similar methods were used earlier to estimate the melting temperatures of stishovite [*Belonoshko and Dubrovinsky*, 1995]. *Alfe et al.* [1999, 2002] took a different approach: They computed first principles free energies for the solid and liquid phases of Fe and determined melting temperatures from their equality at equilibrium. They also showed that if a necessary free energy correction is made to the former authors' results, essentially exact agreement among the three calculations is found [*Alfe et al.*, 2003]. A distinguishing characteristic of the first principles calculations by Alfe et al. is their careful analysis of systematic and numerical errors. They estimate that the uncertainty of their melting temperature at the inner core surface is approximately 600 K, which is similar to the experimental uncertainties in calculated temperatures at melting observed in shock wave experiments. It is noteworthy that the most recent shock wave observation of Fe melting [*Nguyen and Holmes*, 2004] (see Figure 9) yields a temperature in excellent agreement with the computations of Alfe et al., as is also the case for the earlier shock-based estimate by *Brown and McQueen* [1980].

There is yet another testing ground for theoretical mineral physics, and it too arises at the intersection of the physics of

Fe and the geophysics of the core. In spite of the obvious lack of certainty regarding the details of iron's phase diagram, it has become a virtual tenet of geophysics that the Fe in the inner core has the hexagonal close-packed (hcp) structure, or something close to it. On the basis of their shock wave data, *Brown and McQueen* [1986] suggested that there was a solid–solid ε to γ phase transition in Fe at 243 GPa. Theoretical arguments were later used to suggest that the bcc structure of Fe would again be stable at high pressure and that what Brown and McQueen saw was the ε to bcc transformation [*Bassett and Weathers*, 1990; *Ross et al.*, 1990]. That possibility was dismissed when first principles calculations appeared to demonstrate that the bcc structure is dynamically unstable at high pressure [*Stixrude and Cohen*, 1995a] and that the hcp structure was not only stable at core pressures but could also explain the inner core's seismic anisotropy [*Steinle-Neumann et al.*, 2001; *Stixrude and Cohen*, 1995b].

The dismissal of the bcc phase may have been premature. The first principles calculations supporting the hcp hypothesis did not take into account vibrational contributions. That seemed reasonable since the ground-state enthalpy of the bcc phase exceeded that of the hcp phase by an amount significantly larger than typical core thermal energies. However, the calculation by *Steinle-Neumann et al.* [2001] showed that thermal contributions can drastically change elastic properties, and thus should have served as a warning of the dangers of neglecting the effects of temperature. Using a simple thermal model (which in retrospect might not be quite appropriate for estimating thermal effects on the elastic constant tensor), Steinle-Neumann et al.'s computations indicated that the anisotropy of Fe may be extremely sensitive to temperature—so much so that it may even change its orientation, as the calculations indicated. If so, might not temperature also affect the stability of bcc iron in the core, and might not this stability be further modified by the minor alloying elements sure to be present in the core, as suggested by a study of iron–silicon alloys [*Lin et al.*, 2002]?

Two independent theoretical simulations show that bcc Fe may well be present in at least part of the inner core [*Belonoshko et al.*, 2003; *Vocadlo et al.*, 2003b]. Using ab initio finite-temperature molecular dynamics simulations of the bcc Fe to compute its free energy under core conditions, Vocadlo et al. found that at high pressures the bcc phase becomes entropically stabilized. And although the hcp phase of pure Fe retains a slight free energy advantage over the bcc phase, addition of a few percent of S or Si stabilized the bcc phase at the expense of hcp. Belonoshko et al. used an embedded atom potential model fitted to first principles computations to perform simulations with very large unit cells (up to 64,000 atoms per cell), which the authors found to be crucial for determining the correct high temperature

structure. They also found that the bcc phase of pure Fe was stable within the inner core. Although the two simulations differ in procedural detail and specific predictions, their joint implication that the inner core, or part of it, may be composed of a bcc Fe alloy is plausible, and will remain so unless and until disproved experimentally. Meanwhile, the recent development of first principles chemical potentials is likely to play a very significant role in assessing the consequences of the structure of Fe on the detailed composition of the inner and outer core [*Alfe et al.*, 2002a, 2002b, 2002c, 2002d, 2003; *Vocadlo et al.*, 2003a].

SOME FINAL REFLECTIONS

Our sampling of computational mineral physics contributions to our understanding of Earth materials leaves little doubt that this field is close to reaching maturity. Predicted thermoelastic properties typically differ from experiment by at most a few percent. The biggest source of error in carefully conducted computations is the density functional theory itself. It is difficult to estimate the absolute errors due to the functionals, but the differences between LDA and GGA approximations are of the order of a few percent. The theory can in some case yield qualitatively incorrect results. For example, it predicts that FeO is a metal, whereas in fact it is a Mott insulator. The problem arises from inadequate treatment of the strong electron–electron correlation that tends to localize electrons on individual atomic sites when the overlap between nearest neighbor transition-metal orbitals is small. Related to this is current theories' inaccurate treatment of the energy of adding an electron to a localized orbital on an atom. The solution must await what are likely to be rather complex corrections to current mean-field approximations in density functional theory.

We mentioned earlier "carefully conducted computations." By this we mean that the calculation has been subjected to all necessary convergence tests. Since the operator acting on the wavefunction in equation (8) depends on the wavefunction (through the electron density), solutions are generally found by guessing the initial density and then solving iteratively until some convergence criteria are met, leading to so-called "self-consistent" solutions. Furthermore, as the wavefunctions are expanded in terms of some appropriate set of basis functions, it is necessary to ensure that enough such functions are included that the energy errors are below some acceptable value. In computing the total ground state energy, or the thermal energy derived from thermally excited phonons, it is necessary to approximate sums over the Brillouin zone (unit cell in wave-vector space) by some finite sampling. For example, accurate eos may require that the error from various approximations be smaller than 10^{-3} eV per atom. It is common to report on these various sources of errors and on what tests were performed to ensure that they do not influence the final results of calculations. However, as illustrated by the forsterite eos discussed earlier, errors due to the use of inadequate pseudopotentials may be more difficult, or at least more laborious, to detect. It is perhaps desirable to generate a database containing pseudopotentials for all geologically important ions/atoms that have been tested against reliable all-electron calculations (e.g., LAPW) under well-defined pressure limits.

Continued improvements of the accuracy, and in the range of problems that can be usefully simulated, require several developments. Three developments can be seen as prerequisite to a computational methodology that is complete and flexible enough to treat reliably the great diversity of material problems presented by the Earth: 1. A description of bonding that is accurate over the entire pressure range spanned by the Earth's interior; 2. efficient and reliable methods for treating atomic motions that go beyond the quasiharmonic approximation and hence remain valid at very high temperatures and near displacive phase transformations and that can treat general atomic displacements such as those responsible for diffusion; and 3. efficient and reliable sampling of phase space to allow accurate representations of thermal effects. Developments of the first kind are the most challenging, in that they require fundamental improvements in density functional theory. Quantum Monte Carlo Methods, applied to solids [*Foulkes et al.*, 2001] and to the electron gas [*Ceperley and Alder*, 1980], are most likely to inform improvements in the form of density functionals for electron–electron interactions. The second development could be effectively accomplished through first principles molecular dynamics [*Car and Parrinello*, 1985] and, at a more approximate level, through various cell methods in which the partition function is a product of single particle partition functions, each particle being considered to move in a cell defined by its static nearest neighbors. Slow processes like solid-state diffusion and related phenomena in silicates will be examined with methods of accelerated molecular dynamics [*Voter et al.*, 2002]. Another powerful method relies on thermodynamic integration to calculate the free energy change caused by turning on an accurate first principles energy correction relative to a system whose free energy is known exactly (i.e., the harmonic lattice). This approach has been successfully used to obtain the first principles melting curves for iron discussed earlier [*Alfe et al.*, 1999] and plays a significant role in first principles estimates of chemical potentials that, in turn, will be indispensable in future analyses of element partitioning among different phases in the mantle, as well as between mantle and core. The third development, concerning the accurate sampling of phase space, requires simulations of very large systems (10^5 to 10^6 atoms), as these lead to greatly

improved statistics. Progress will entail sophisticated algorithms combined with parallel processing. The algorithms are likely to include multiscale approaches in which finite element methods are combined with molecular dynamics, which in turn use potentials derived from electronic structure calculations, or ab initio electron-gas pair potentials. We believe that such methods will allow simulations of defect structures and dislocations and thus will initiate realistic studies of the physics of rheological processes at high pressures and temperatures. A promising recently developed method, based on atomic-like orbitals, requires a computational effort that scales approximately linearly with system size [*Craig et al.*, 2004] and thus promises to enable first principles simulations of very large systems. The successful applications of first principles thermodynamics to the physics and chemistry of iron will be extended to complex silicates in all their distinct phases and to the study of their mutual equilibria with respect to element partitioning. We are confident that the three challenges will be successfully met in what promises to be an exciting future for computational and theoretical mineral physics.

Acknowledgments. We thank Lars Stixrude for helpful comments and for providing numerical data for some of our plots. A detailed review by Jan Matas helped improve the clarity of some arguments. We thank Artem Oganov, Bijaya Karki, and Boris Kiefer for providing copies of plots or numerical data. Jay Bass and Jennifer Jackson kindly sent us preprints of some of their papers, and Jeff Nguyen sent us the figure on Fe melting. Comments by David Price and an anonymous reviewer helped improve the original manuscript. This work was supported by NSF Grant EAR 00-03456.

REFERENCES

Agnon, A., and M.S.T. Bukowinski, δ_s at high pressure and $d\ln V_s/d\ln V_p$ in the lower mantle, *Geophys. Res. Lett.*, *17*, 1149–1152, 1990.

Ahrens, T.J., K.G. Holland, and C.Q. Chen, Phase diagram of iron, revised core temperatures, *Geophys. Res. Let.*, *29*, 54-1–54-4, 2002.

Aidun, J., M.S.T. Bukowinski, and M. Ross, Equation of state and metallization of CsI, *Phys. Rev.*, *B29*, 2611–2621, 1984.

Akber-Knutson, S., and M.S.T. Bukowinski, A study of the substitution mechanism in Aluminum-bearing MgSiO3 perovksite at lower mantle pressures and temperatures, *Phys. Earth and Planet. Int.*, *in press*, 2004.

Akber-Knutson, S., M.S.T. Bukowinski, and J. Matas, On the Structure and Compressibility of CaSiO$_3$ perovskite, *Geophys. Res. Let.*, *29* (3), 1034, DOI:10.1029/2001GL013523, 2002.

Alfe, D., M.J. Gillan, and G.D. Price, The melting curve of iron at the pressures of the Earth's core from *ab initio* calculations, *Nature*, *401*, 462–464, 1999.

Alfe, D., M.J. Gillan, and G.D. Price, Ab initio chemical potentials of solid and liquid solutions and the chemistry of the Earth's core, *Journal of Chemical Physics, 116* (16), 7127–7136, 2002a.

Alfe, D., M.J. Gillan, and G.D. Price, Composition and temperature of the Earth's core constrained by combining ab initio calculations and seismic data, *Earth and Planet Sci Lett, 195,* 91–98, 2002b.

Alfe, D., M.J. Gillan, and G.D. Price, Thermodynamics from first principles: temperature and composition of the Earth's core, *Mineralogical Magazine, 67* (1), 113–123, 2003.

Alfe, D., M.J. Gillan, L. Vocadlo, J. Brodholt, and G.D. Price, The ab initio simulation of the Earth's core, *Philosophical Transactions of the Royal Society of London Series A-Mathematical Physical & Engineering Sciences, 360* (1795), 1227–1244, 2002c.

Alfe, D., G.D. Price, and M.J. Gillan, Iron under Earth's core conditions: Liquid-state thermodynamics and high-pressure melting curve from ab initio calculations, *Phys. Rev., 65* (16), 165118-1–11, 2002d.

Anderson, D.L., *Theory of the Earth*, Blackwell Scientific Publications, Boston, 1989.

Anderson, O.L., and A. Duba, Experimental melting curve of iron revisited, *Journal of Geophysical Research-Solid Earth, 102* (B10), 22659–22669, 1997.

Andrault, D., N. Bolfan-Casanova, and N. Guignot, Equation of state of lower mantle (Al,Fe)-MgSiO$_3$ perovskite, *Earth and Planet Sci Lett, 193,* 501–508, 2001.

Andrault, D., G. Fiquet, T. Charpin, and T. le Bihan, Structure analysis and stability of β-iron at high *P* and *T*, *Am. Mineral., 85,* 364–371, 2000.

Bassett, W.A., and M.S. Weathers, Stability of the body-centered cubic phase of iron: a thermodynamic analysis, *J. Geophys. Res, 95,* 21709–21712, 1990.

Belonoshko, A.B., R. Ahuja, and B. Johansson, Quasi-ab initio molecular dynamics study of Fe melting, *Phys. Rev. Lett., 84,* 3638–3641, 2000.

Belonoshko, A.B., R. Ahuja, and B. Johansson, Stability of the body-centred-cubic phase of iron in the Earth's inner core, *Nature, 424* (6952), 1032–1034, 2003.

Belonoshko, A.B., and L.S. Dubrovinsky, Molecular dynamics of stishovite melting, *Geochim. Cosmochim. Acta, 59,* 1883–1889, 1995.

Bina, C.R., and M. Kumazawa, Thermodynamic coupling of phase and chemical boundaries in planetary interiors, *Phys. Earth Planet. Int., 76,* 329–341, 1993.

Birch, F., Elasticity and constitution of the earth's interior, *J. Geophys. Res., 57,* 227–286, 1952.

Boehler, R., Temperatures in the Earths Core From Melting-Point Measurements of Iron At High Static Pressures, *Nature, 363* (6429), 534–536, 1993.

Boehler, R., N.v. Bargen, and A. Chopelas, Melting, thermal expansion,and phase transitions of iron at high pressures, *J. Geophys. Res., 95,* 21731–31736, 1990a.

Boehler, R., N. Vonbargen, and A. Chopelas, Melting, Thermal Expansion, and Phase Transitions of Iron At High Pressures, *Journal of Geophysical Research—Solid Earth and Planets, 95* (B13), 21731–21736, 1990b.

Born, M., and J.R. Oppenheimer, On the quantum theory of molecules, *Annalen der Physik, 84,* 457–484, 1927.

Boyer, L.L., M.J. Mehl, J.L. Feldman, J.R. Hardy, J.W. Flocken, and C.Y. Fong, Beyond the rigid-ion approximation with spherically symmetric ions, *Phys. Rev. Lett., 54,* 1940–1943, 1985.

Brodholt, J., A. Patel, and K. Refson, An ab initio study of the compressional behavior of forsterite, *Am. Mineral., 81,* 257–260, 1996.

Brodholt, J.P., Pressure-induced changes in the compression mechanism of aluminous perovskitein the Earth's mantle, *Nature, 407,* 620–622, 2002.

Brown, J.M., and R.G. McQueen, Melting of iron under core conditions., *Geophys. Res. Lett., 7,* 533–536, 1980.

Brown, J.M., and R.G. McQueen, Phase transitions, Gruneisen parameter, and elasticity for shocked iron between 77 GPa and 400 GPa, *J. Geophys. Res., 91,* 7485–7494, 1986.

Bukowinski, M.S.T., The effect of pressure on the physics and chemistry of potassium, *Geophys. Res. Lett., 3,* 491–494, 1976a.

Bukowinski, M.S.T., On the electronic structure of iron at core pressures, *Phys. Earth Planet. Int., 13,* 57–66, 1976b.

Bukowinski, M.S.T., A theoretical equation of state for the inner core, *Phys. Earth Planet. Int., 14,* 333–344, 1976c.

Bukowinski, M.S.T., Compressed potassium: a siderophile element, in *High-Pressure Science and Technology*, edited by K.D. Timmerhaus, and M.S. Barber, pp. 237–244, Plenum Press, New York, 1979.

Bukowinski, M.S.T., Effect of pressure on bonding in MgO, *J. Geophys. Res., 85,* 285–292, 1980.

Bukowinski, M.S.T., First principles equation of state of MgO and CaO, *Geophys. Res. Lett., 12,* 536–539, 1985.

Bukowinski, M.S.T., Quantum Geophysics, *Annu. Rev. Earth Planet. Sci., 22,* 167–205, 1994.

Bukowinski, M.S.T., and J. Aidun, First principles versus spherical ion models of the B1 and B2 phases of NaCl, *J. Geophys. Res., 90,* 1794–1800, 1985.

Bukowinski, M.S.T., A. Chizmeshya, G.H. Wolf, and H. Zhang, Advances in electron-gas potential models: Applications to some candidate lower mantle minerals, *Molecular Engineering, 6,* 81–112, 1996.

Bukowinski, M.S.T., and J.W. Downs, Structures, compressibilities and relative stabilities of the α, β and γ phases of Mg$_2$SiO$_4$ deduced from an electron-gas ionic Hamiltonian, *Geophys. J. Int., 143,* 295–301, 2000.

Bukowinski, M.S.T., and G.H. Wolf, Equation of state and possible critical phase transitions in MgSiO$_3$ perovskite at lower-mantle conditions, in *Structural and Magnetic Phase Transitions in Minerals*, edited by S. Ghose, J.M.D. Coey, and E. Salje, pp. 91–112, Pringer Verlag, New York, 1988.

Bukowinski, M.S.T., and G.H. Wolf, Thermodynamically consistent decompression: implications for lower mantle composition, *J. Geophys. Res., 95,* 12583–12593, 1990.

Burnham, C.W., The ionic model: perceptions and realities in mineralogy, *Am. Mineral., 75,* 443–463, 1990.

Bush, T.S., J.D. Gale, C.R.A. Catlow, and P.D. Battle, Self-consistent interatomic potentials for the simulation of binary and ternary oxides, *J. Mater. Chem., 4,* 831–837, 1994.

Busing, W.R., WMIN, a computeer program to model molecules and crystals in terms of potential energy functions, in *ORNL-5747*, Oak Ridge National Laboratory, Oak Ridge, 1981.

Car, R., and M. Parrinello, Unified approach for molecular dynamics and density-functional theory, *Phys. Rev. Lett.*, *55*, 2471–2474, 1985.

Catlow, C.R.A., and W.C. Mackrodt, Computer Simulations of Condensed Matter, North-Holland, Amsterdam, 1985.

Cazenave, A., and C. Thoraval, Mantle Dynamics Constrained by Degree-6 Surface Topography, Seismic Tomography and Geoid—Inference on the Origin of the South Pacific Superswell, *Earth and Planetary Science Letters*, *122* (1–2), 207–219, 1994.

Ceperley, D.M., and B.J. Alder, Ground state of the electron gas by a stochastic method, *Phys. Rev. Lett.*, *45*, 566–569, 1980.

Chaplot, S.L., N. Choudhury, and K.R. Rao, Molecular dynamics simulation of the phase transitions and melting in MgSiO$_3$ with the perovskite structure, *Am. Mineral.*, *83* (9–10), 937–941, 1998.

Chizmeshya, A.V.G., G.H. Wolf, and P.F. Mcmillan, First-principles calculation of the equation-of-state , stability, and polar optic modes of CaSiO$_3$ perovskite, *Geophys. Res. Lett.*, *23*, 2725–2728, 1996.

Chopelas, A., Thermal expansivity of lower mantle phases MgO and MgSiO$_3$ perovskite at high pressure derived from vibrational spectroscopy, *Phys. Earth Planet. Int.*, *98*, 3–15, 1996.

Clugston, M.J., The calculation of intermolecular forces. A critical examination of the Gordon-Kim model, *Adv. phys.*, *27* (26), 893–912, 1978.

Cohen, A.J., and R.G. Gordon, Theory of the lattice energy, equilibrium structure, elastic constants, and pressure-induced phase transformations in alkali-halide crystals, *Phys. Rev, B12*, 3228–3241, 1975.

Cohen, R.E., Elasticity and equation of state of MgSiO$_3$ perovskite, *Geophys. Res. Lett.*, *14*, 1053–1056, 1987.

Cohen, R.E., Bonding and elasticity of stishovite SiO$_2$ at high pressures: Linearized augmented Plane Wave calculations, *Am. Mineral.*, *76*, 733–742, 1991.

Cohen, R.E., First-principles predictions of elasticity and phase transitions in high pressure SiO$_2$ and geophysical implications, in *High-Pressure Research: Applications to Earth and Planetary Sciences*, edited by Y. Syono, and M.H. Manghnani, pp. 425–431, Tokyo: TERRAPUB/Washington D. C. : American Geophysical Union, 1992.

Cohen, R.E., and J.S. Weitz, The melting curve and premelting of MgO, in *Properties of Earth and Planetary Materials at High Pressure and Temperature*, edited by M.H. Manghnani, and A. Yagi, pp. 185–196, American Geophysical Union, Washington, D.C., 1998.

Craig, M.S., M.C. Warren, M.T. Dove, J.D. Gale, D. Sanchez-Portal, P. Ordejon, J.M. Soler, and E. Artacho, Simulations of minerals using density-functional theory based on atomic orbitals for linear scaling, *Phys. Chem. Minerals*, *31*, 12–21, 2004.

D'Arco, P., G. Sandrone, R. Dovesi, E. Apra, and V.R. Saunders, A quantum-mechanical study of the relative stability under pressure of MgSiO$_3$-ilmenite, MgSiO$_3$-perovskite, and MgO-periclase+SiO$_2$-stishovite assemblage, *Phys. Chem. Minerals*, *21*, 285–293, 1994.

D'Arco, P., G. Sandrone, R. Dovesi, R. Orlando, and V.R. Saunders, A quantum mechanical study of the perovskite structure type of MgSiO$_3$, *Phys. Chem. Minerals*, *20* (407–414), 1993.

da Silva, C., L. Stixrude, and R.M. Wentzcovitch, Elastic constants and anisotropy of forsterite at high pressure, *Geophysical Research Letters*, *24* (15), 1963–1966, 1997.

Daniel, I., J.D. Bass , G. Fiquet, H. Cardon, J. Zhang, and M. Hanfland, Effect of aluminium on the compressibility of silicate perovskite, *Geophys. Res. Lett.*, *31*, L15608,doi:10.1029/2004GL020213, 2004.

Daniel, I., H. Cardon, G. Fiquet, F. Guyot, and M. Mezouar, Equation of state of Al-bearing perovskite to lower mantle conditions, *Geophys. Res. Let.*, *28*, 3789–3792, 2001.

Dirac, P.A.M., Quantum mechanics of many-electron systems, *Proceedings of the Royal Society of London*, *123*, 714–733, 1929.

Downs, J.W., and M.S.T. Bukowinski, Variationally induced breathing equations of state of pyrope, grossular, and majorite garnets, *Geophys. Res. Lett.*, *24*, 1959–1962, 1997.

Dziewonski, A.M., and D.L. Anderson, Preliminary reference Earth model, *Phys. Earth Planet. Int.*, *25*, 297–356, 1981.

Fiquet, G., A. Dewaele, D. Andrault, M. Kunz, and T. Le Bihan, Thermoelastic properties and crystal structure of MgSiO3 perovskite at lower mantle pressure and temperature conditions, *Geophysical Research Letters*, *27* (1), 21–24, 2000.

Foulkes, W.M.C., L. Mitas, R.J. Needs, and G. Rajagopal, Quantum Monte Carlo simulations of solids, *Rev. Mod. Phys.*, *73*, 33–83, 2001.

Frenkel, D., and B. Smit, *Understanding Molecular Simulation—From Algorithms to Applications*, 443 pp., Academic Press Limited, San Diego, 1996.

Funamori, N., and R. Jeanloz, High-pressure transformation of Al$_2$O$_3$, *Science*, *278* (November 7), 1109–1111, 1997.

Funamori, N., T. Yagi, W. Utsumi, T. Kondo, T. Uchida, and M. Funamori, Thermoelastic Properties of MgSiO$_3$ Perovskite Determined by in Situ X Ray Observations up to 30 Gpa and 2000 K, *J. Geophys. Res.*, *101* (B4), 8257–8269, 1996.

Gale, J.D., Empirical potential derivation for ionic materials, *Phil. Mag.*, *B73*, 3–19, 1996.

Gale, J.D., GULP—A computer program for the symmetry adapted simulationod solids, *J. Chem. Soc. Faraday Trans*, *93*, 629–637, 1997.

Gale, J.D., Simulating the crystal structures and properties of ionic materials from interatomic potentials, in *Molecular Modeling Theory: Applications in the Geosciences*, edited by R.T. Cygan, and J.D. Kubicki, pp. 37–62, Geochemical Society/Mineralogical Society of America, Washigton, D.C., 2001.

Gell-Mann, M., and K.A. Brueckner, Correlation energy of an electron gas at high density, *Phys. Rev.*, *106*, 364–368, 1957.

Gibbs, G.V., Molecules as models of bonding in silicates, *Am. Mineral.*, *67*, 421–450, 1982.

Gibbs, G.V., F.C. Hill, M.B. Boisen, and R.T. Downs, Power Law Relationships between Bond Length, Bond Strength and Electron Density Distributions, *Physics & Chemistry of Minerals*, *25* (8), 585–590, 1998.

Gillet, P., I. Daniel, F. Guyot, J. Matas, and J.-C. Chervin, A thermodynamic model for $MgSiO_3$-perovskite derived from pressure, temperature and volume dependence of the Raman mode frequencies, *Phys. Earth Planet. Int.*, *117* (1–4), 361–384, 2000.

Gillet, P., F. Guyot, and Y. Wang, Microscopic anharmonicity and equation of state of $MgSiO_3$: perovskite, *Geophys. Res. Lett.*, *23*, 3043–3046, 1996.

Goddard, W.A., Theoretical chemistry comes alive: full partner with experiment, *Science*, *227*, 917–923, 1985.

Gordon, R.G., and Y.S. Kim, Theory for the forces between closed-shell atoms and molecules, *J. Chem. Phys.*, *56*, 3122–3133, 1972.

Grand, S.P., R.D. van der Hilst, and S. Widiyantoro, Global seismic tomography: Snapshot of convection in the Earth, *Geol. Soc. Am. Today*, *7* (4), 1–7, 1997.

Hama, J., and K. Suito, High-temperature equation of state of $CaSiO_3$ perovskite and its implications for the lower mantle, *Phys. Earth and Planet. Int.*, *105*, 33–46, 1998.

Hama, J., K. Suito, and O.L. Anderson, Thermoelasticity of silicate perovskites and magnesiowustite and its implications for the Earth's lower mantle, *Am. Mineral.*, *85*, 321–328, 2000.

Hamann, D.R., Generalized Gradient Theory for Silica Phase Transitions, *Physical Review Letters*, *76* (4), 660–663, 1996.

Hedin, L., and B.I. Lundqvist, Explicit local exchange-correlation potentials, *J. Phys. C*, *4*, 2064–2083, 1971.

Heinz, D.L., and R. Jeanloz, Measurement of the melting curve of $Mg_{0.9}Fe_{0.1}SiO_3$ at lower mantle conditions and its geophysical implications, *J. Geophys. Res.*, *92*, 11437–11444, 1987.

Hemley, R.J., M.D. Jackson, and G.R. G., Theoretical study of the structure, lattice dynamics, and equations of state of perovskite-type $MgSiO_3$ and $CaSiO_3$, *Phys. Chem. Minerals*, *14*, 2–12, 1987.

Hohenberg, P.C., and W. Kohn, Inhomogeneous electron gas, *Phys. Rev.*, *B 136*, 864–871, 1964.

Iitaka, T., K. Hirose, K. Kawamura, and M. Murakami, The elasticity of $MgSiO_3$ post-perovskite phase in the Earth's lowermost mantle, *Nature*, *430*, 442–445, 2004.

Inbar, I., and R.E. Cohen, High-pressure effects on thermal properties of MgO, *Geophys. Res. lett.*, *22*, 1533–1536, 1995.

Irifune, T., Absence of an aluminous phase in the upper part of the Earth's lower mantle, *Nature*, *370*, 131–133, 1994.

Isaak, D.G., O.L. Anderson, and R.E. Cohen, The relationship between shear and compressional velocities at high pressures: reconciliation of seismic tomography and mineral physics, *Geophys. Res. Lett.*, *19*, 741–744, 1992.

Isaak, D.G., R.E. Cohen, and M.J. Mehl, Calculated elastic and thermal properties of MgO at high pressures and temperatures, *J. Geophys. Res.*, *95*, 7055–7067, 1990.

Ito, E., K. Morooka, and O. Ujike, Dissolution of K in molten iron at high pressure and temperature, *Geophys. Res. Lett.*, *20* (15), 1651–1654, 1993.

Jackson, I., and S.M. Rigden, Analysis of P-V-T Data—Constraints on the Thermoelastic Properties of High-Pressure Minerals, *Physics of the Earth & Planetary Interiors*, *96* (2–3), 85–112, 1996.

Jackson, J.M., J. Zhang, and J.D. Bass Sound velocities and elasticity of aluminous $MgSiO_3$ perovskite: Implications for aluminum heterogeneity in Earth's lower mantle, *Geophys. Res. Lett.*, *31*, L10614, doi:10.1029/2004GL19918, 2004.

Jackson, M.D., and R.G. Gordon, MEG investigation of low pressure silica-Shell model for polarization, *Phys. Chem. Minerals*, *16*, 212–220, 1988.

Jackson, M.J., S.V. Sinogeikin, and J.D. Bass Sound velocities and elastic properties of γ-Mg_2SiO_4 to 873 K by Brillouin spectroscopy, *Am. Mineral.*, *85*, 296–303, 2000.

Jephcoat, A.P., R.J. Hemley, and H.K. Mao, X-ray diffraction of ruby (Al_2O_3 :Cr^{3+}) to 175 GPa, *Physica*, *B150*, 115–121, 1988.

Karato, S., Importance of anelasticity in the interpretation of seismic tomography, *Geophys. Res. Lett.*, *20*, 1623–1626, 1993.

Karato, S., and B.B. Karki, Origin of lateral variation of seismic velocities and density in the deep mantle, *J. Geophys. Res.*, *106*, 21,771–21,783, 2001.

Karki, B.B., High-pressure structure and elasticity of the major silicate and oxide minerals of the Earth's lower mantle, PhD thesis, University of Edinburgh, Edinburgh, 1997.

Karki, B.B., and J. Crain, First-principles determination of elastic properties of $CaSiO_3$ perovskite at lower mantle pressures, *Geophys. Res. Lett.*, *25*, 2741–2744, 1998.

Karki, B.B., L. Stixrude, S.J. Clark, M.C. Warren, G.J. Ackland, and J. Crain, Elastic properties of orthorhombic $MgSiO_3$ perovskite at lower mantle pressures, *Am. Mineral.*, *82*, 635–638, 1997.

Karki, B.B., L. Stixrude, and R. Wentzcovitch, High-pressure elastic properties of major materials of Earth's mantle from first principles, *Reviews of Geophysics*, *39* (4), 507–534, 2001a.

Karki, B.B., and R. Wentzcovitch, First-principles lattice dynamics and thermoelasticity of $MgSiO_3$ ilmenite at high pressure, *J. Geophys. Res.*, *107*, 2267–2273, 2002.

Karki, B.B., and R. Wentzcovitch, Vibrational and quasiharmonic thermal properties of CaO, *Phys. Rev. B*, *68*, 224304/1, 2003.

Karki, B.B., R.M. Wentzcovitch, S.d. Gironcoli, and S. Baroni, First-principles determination of elastic anisotropy and wave velocities of MgO at lower mantle conditions, *Science*, *286*, 1705–1707, 1999.

Karki, B.B., R.M. Wentzcovitch, S.d. Gironcoli, and S. Baroni, Ab initio lattice dynamics of $MgSiO_3$ perovskite at high pressure, *Phys Rev B*, *62*, 14750–14756, 2000a.

Karki, B.B., R.M. Wentzcovitch, S.d. Gironcoli, and S. Baroni, High pressure lattice dynamics and thermoelasticity of MgO, *Phys. Rev. B*, *61*, 8793–8800, 2000b.

Karki, B.B., R.M. Wentzcovitch, S.d. Gironcoli, and S. Baroni, First principles thermoelasticity of $MgSiO_3$ perovskite: consequences for the inferred properties of the lower mantle, *Geophys. Res. Lett.*, *28*, 2699–2702, 2001b.

Kato, T., E. Ohtani, H. Morishima, D. Yamazaki, A. Suzuki, M. Suto, T. Kubo, T. Kikegawa, and O. Shimomura, In Situ X Ray Observation of High-Pressure Phase Transitions of $MgSiO_3$ and Thermal Expansion of $MgSiO_3$ Perovskite at 25 Gpa by Double-Stage Multianvil System, *J. Geophys. Res.*, *100* (B10), 20475–20481, 1995.

Kennet, B.L.N., S. Widiyantoro, and R.D. van der Hilst, Joint seismic tomography for bulk sound and shear wave speed in the Earth's mantle, *J. Geophys. Res.*, *103*, 12469–12494, 1998.

Kiefer, B., L. Stixrude, and R.M. Wentzcovitch, Calculated elastic constants and anisotropy of Mg$_2$SiO$_4$ spinel at high pressure, *Geophysical Research Letters*, *24* (22), 2841–2844, 1997.

Kiefer, B., L. Stixrude, and R.M. Wentzcovitch, Elasticity of (Mg,Fe)SiO$_3$—Perovskite at high pressures—art. no. 1539, *Geophysical Research Letters*, *29* (11), 1539, 2002.

Kim, Y.S., and R.G. Gordon, Study of the electron gas approximation, *J. Chem. Phys.*, *60* (5), 1842–1850, 1974a.

Kim, Y.S., and R.G. Gordon, Theory of binding of ionic crystals: Applications to alkali-halide and alkaline-earth-dihalide crystals, *Phys. Rev.*, *B9*, 3548–3554, 1974b.

Kingma, K.J., R.E. Cohen, R.J. Hemley, and H.-k. Mao, Transformation of stishovite to a denser phase at lower-mantle pressures, *Nature*, *374*, 243–245, 1995.

Knittle, E., and R. Jeanloz, Synthesis and equation of state of (Mg,Fe)SiO$_3$ perovskite to over 100 GPa, *Science*, *235*, 669–670, 1987.

Knittle, E., and R. Jeanloz, Melting curve of (Mg,Fe)SiO$_3$ perovskite to 96 Gpa: Evidence for a structural transition in lower mantle melts, *Geohys. Res. Lett.*, *16*, 421–424, 1989.

Kohn, W., and L.J. Sham, Self-consistent equations including exchange and correlation effects, *Phys. Rev.*, *140*, 1133–1138, 1965.

Kramer, G.J., N.P. Farragher, B.W.H.v. Beest, and R.A. Santen, Interatomic force fields for silicas, aluminophosphates, and zeolites: Derivation based on ab initio calculations, *Phys. Rev, B43*, 5068–5080, 1991.

Lacks, D.J., and R.G. Gordon, Calculations of pressure-induced phase transitions in mantle minerals, *Phys. Chem. Minerals*, *22*, 145–150, 1995.

Lacks, D.L., and R.G. Gordon, *Phys. Rev., B48*, 2889, 1993.

Laio, A., S. Bernard, G.L. Chiarotti, S. Scandolo, and E. Tosatti, Physics of iron at Earth's core conditions, *Science*, *287* (5455), 1027–1030, 2000.

Lee, K.M., and R. Jeanloz, High-pressure alloying of potassium and iron: Radioactivity in the Earth's core?, *Geophys. Res. Lett.*, *30* (23), 2212, doi:10.1029/3003GL018515, 2003.

Lewis, G.V., and C.R.A. Catlow, Potential models for ionic oxides, *J. Phys. C: Sol. St. Phys.*, *18*, 1149–1161, 1985.

Li, X.D., and B. Romanowicz, Global mantle shear velocity model developed using nonlinear asymptotic coupling theory, *J. Geophys. Res.*, *101*, 22245–22272, 1996.

Liebermann, R.C., and B. Li, Elasticity at high pressure and temperature, in *Ultrahigh Pressure Mineralogy: Physics and Chemistry of the Earth's Deep Interior*, edited by R.J. Hemley, pp. 459–492, Mineralogical Society of America, Washington, D.C., 1998.

Lin, J.F., D.L. Heinz, A.J. Campbell, J.M. Devine, and G.Y. Shen, Iron-silicon alloy in Earth's core?, *Science*, *295* (5553), 313–315, 2002.

Liu, J., G. Chen, G.D. Gwanmesia, and R.C. Liebermann, Elastic wave velocities of pyrope-majorite garnets (Py$_{62}$Mj$_{38}$ and Py$_{50}$Mj$_{50}$) to 9 GPa, *Phys. Earth and Planet. Int.*, *120*, 153–163, 2000.

Lundqvist, S., and N.H. March, *Theory of the Inhomogeneous Electron Gas*, Plenum, London, 1987.

Mao, H.K., L.C. Chen, R.J. Hemley, A.P. Jephcoat, Y. Wu, and W.A. Bassett, Stability and equation of state of CaSiO$_3$—perovskite to 134 GPa,, *J. Geophys. Res.*, *94*, 17889–17894, 1989.

March, N.H., *Self-consistent Fields in Atoms-Hartree and Thomas-Fermi atoms*, 233 pp., Prgamon Press, Ltd., Oxford, 1975.

Marton, F.C., and R.E. Cohen, Prediction of a high-pressure phase transition in Al$_2$O$_3$, *Amer. Mineral.*, *79*, 789–792, 1994.

Marton, F.C., J. Ita, and R.E. Cohen, Pressure-volume-temperature equation of state of MgSiO3 perovskite from molecular dynamics and constraints on lower mantle composition, *Journal of Geophysical Research-Solid Earth*, *106* (B5), 8615–8627, 2001.

Masters, G., G. Laske, H. Bolton, and A.M. Dziewonski, The relative behavior of shear velocity, bulk sound speed, and compressional velocity in the mantle: implications for chemical and thermal structure, in *Earth's Deep Interior: Mineral Physics and Seismic Tomography From the Atomic to the Global Scale*, edited by S.K.e. al., pp. 63–87, AGU, Washington, D.C., 2000.

Matsui, M., Molecular dynamics study of MgSiO$_3$ perovskite, *Phys. Chem. Minerals*, *16*, 234–238, 1988.

Matsui, M., A transferable interatomic potential model for crystals and melts in the system CaO-MgO-Al$_2$O$_3$-SiO$_2$, *Mineralogical Magazine*, *58A*, 571–572, 1994.

Matsui, M., Molecular dynamics simulation of MgSiO$_3$ perovskite and the 660-km discontinuity, *Phys. Earth and Planet. Int.*, *121*, 77–84, 2000.

Matsui, M., A. M., and T. Matsumoto, Computational model of the structural and elastic properties of the ilmenite and perovskite phases of MgSiO$_3$, *Phys. Chem. Minerals*, *14*, 101–106, 1987.

Matsui, M., and G.D. Price, Computer simulation of the MgSiO$_3$ polymorphs, *Phys. Chem. Minerals*, *18*, 365–372, 1992.

Matsui, M., G.D. Price, and A. Patel, Comparison between lattice dynamics and molecular dynamics methods: Calculation results for MgSiO$_3$ perovskite, *Geophys. Res. Lett.*, *21*, 1659–1662, 1994.

Mattern, E., J. Matas, Y. Ricard, and J.D. Bass Lower mantle composition and temperature from mineral physics and thermodynamic modelling, *Geophys. J. Int.*, in press, 2005.

Mehl, M.J., R.E. Cohen, and H. Krakauer, Linearized Augmented Plane Wave electronic structure calculations for MgO and CaO, *J. Geophys. Res.*, *93*, 8009–8022, 1988.

Montelli, R., G. Nolet, F.A. Dahlen, G. Masters, E.R. Engdahl, and S.-H. Hung, Finite-frequency tomography reveals a variety of plumes in the mantle, *Science*, *303*, 338–343, 2004.

Murakami, M., K. Hirose, K. Kawamura, N. Sata, and Y. Ohishi, Post-perovskite phase transition in MgSiO$_3$, *Science*, *304* (5672), 855–858, 2004.

Navrotsky, A., M. Schoenitz, H. Kojitani, H. Xue, J. Zhang, D.J. Weidner, and R. Jeanloz, Aluminum in magnesium silicate perovskite: Formation, structure and energetics of magnesium-rich defect solid solutions, *J. Geophys. Res.*, *108* (B7), 1–11, 2002.

Nguyen, J.H., and N.C. Holmes, Melting of iron at the physical conditions of the Earth's core, *Nature*, *427*, 339–342, 2004.

Oganov, A.R., Computer Simulation Studies of Minerals, PhD thesis, University College London, London, 2002.

Oganov, A.R., J.P. Brodholt, and D.L. Price, The elastic constants of MgSiO₃ perovskite at pressures and temperatures of the Earh's lower mantle, *Nature*, *411*, 934–937, 2001a.

Oganov, A.R., J.P. Brodholt, and G.D. Price, Ab initio elasticity and thermal equation of state of MgSiO₃ perovskite, *Earth and Planet Sci Lett*, *184*, 555–560, 2001b.

Oganov, A.R., J.P. Brodholt, and G.D. Price, Comparative study of quasiharmonic lattice dynamics, molecular dynamics and Debye model applied to MgSiO₃ perovskite, *Phys. Earth Planet. Int.*, *122*, 277–288, 2001c.

Oganov, A.R., and S. Ono, Theoretical and experimental evidence for a post-perovskite phase of MgSiO₃ in the Earth's D″ layer, *Nature*, *430*, 445–448, 2004.

Panero, W.R., L.R. Benedetti, and R. Jeanloz, Equation of state of stishovite and interpretation of SiO₂ shock compression data, 2001.

Parker, L.J., T. Atou, and J.V. Badding, Transition element-like chemistry for potassium under pressure, *Science*, *273*, 95–97, 1996.

Pauling, L., The nature of silicon-oxygen bonds, *Am. Mineral.*, *65*, 321–323, 1980.

Perdew, J.P., K. Burke, and M. Ernzerhof, Generalized Gradient Approximation Made Simple (Vol 77, Pg 3865, 1996), *Physical Review Letters*, *78* (7), 1396, 1997.

Perdew, J.P., K. Burke, and Y. Wang, Generalized Gradient Approximation for the Exchange-Correlation Hole of a Many-Electron System (Vol 54, Pg 16 533, 1996), *Physical Review B-Condensed Matter*, *57* (23), 14999, 1998.

Perdew, J.P., and A. Zunger, Self-interaction correction to density—functional approximations for many-electron systems, *Phys. Rev.*, *B23*, 5048–5079, 1981.

Phillips, J.C., and L. Kleinman, New method for calculating wave functions in crystals and molecules, *Phys. Rev.*, *116*, 287–294, 1959.

Pickett, W.E., Pseudopotential methods in condensed matter applications, *Comp. Phys. Comm.*, *9*, 115–197, 1989.

Pisani, C., R. Dovesi, and C. Roetti, *Hartree-Fock ab initio treatments of crystalline systems*, Springer-Verlag, Berlin, 1988.

Poirier, J.P., *Introduction to the Physics of the Earth's Interior*, 312 pp., Cambridge University Press, Cambridge, 2000.

Ramsey, W.H., On the nature of the earth's core, *Mon. Not. Roy. Astr. Soc., Geophys. Suppl.*, *5*, 409, 1949.

Richards, G., Third age of quantum chemistry, *Nature*, *78*, 507, 1979.

Richmond, N.C., and J.P. Brodholt, Calculated role of aluminum in the incorporation of ferric iron into magnesium silicate perovskite, *Am. Mineral.*, *83* (9–10), 947–951, 1998.

Ross, M., D.A. Young, and R. Grover, Theory of the iron phase diagram at earth core conditions, *J. Geophys. Res.*, *95*, 21713–21716, 1990.

Ross, N.L., and R.M. Hazen, Single crystal X-ray diffraction study of MgSiO₃ perovskite from 77 to 400 K, *Phys. Chem. Minerals*, *16*, 415–420, 1989.

Saxena, S.K., and L.S. Dubrovinsky, Iron phases at high pressures and temperatures: Phase transition and melting, *Am. Mineral.*, *85*, 372–375, 2000.

Shen, G., H.-k. Mao, R.J. Hemley, T.S. Duffy, and M.L. Rivers, Melting and crystal structure of iron at high pressures and temperatures, *Geophys. Res. Lett.*, *25*, 373–376, 1998a.

Shen, G.Y., H.K. Mao, R.J. Hemley, T.S. Duffy, and M.L. Rivers, Melting and crystal structure of iron at high pressures and temperatures, *Geophys. Res. Lett.*, *25* (3), 373–376, 1998b.

Sherman, D.M., Hartree-Fock band structure equation of state, and pressure-induced hydrogen bonding in brucite, Mg(OH)₂, *Am. Mineral.*, *76*, 1769–1772, 1991.

Sherman, D.M., Equation of state and high-pressure phase transitions of stishovite (SiO₂): Ab initio (periodic Hartree-Fock) results, *J. Geophys. Res.*, *98*, 11,865–11,874, 1993a.

Sherman, D.M., Equation of state, elastic properties, and stability of CaSiO₃ perovskite—first-principles (periodic Hartree-Fock) results, *J. Geophys. Res.*, *98*, 19795–19805, 1993b.

Shim, S.-H., T.S. Duffy, and G. Shen, The equation of state of CaSiO₃ perovskite to 108 GPa and 300 K, *Phys. Earth Planet. Int.*, *120*, 327–338, 2000.

Sinelnikov, Y.D., G. Chen, D.R. Neuville, M.T. Vaughan, and R.C. Liebermann, Ultrasonic shear velocities of MgSiO₃ perovskite at 8 GPa and 800 K and lower mantle composition, *Science*, *281*, 677–679, 1998.

Sinogeikin, S.V., and J.D. Bass Single-crystal elasiticity of pyrope and MgO to 20 GPa by Brillouin scattering in the diamond cell, *Phys. Earth and Planet. Int.*, *120*, 43–62, 2000.

Sinogeikin, S.V., and J.D. Bass Elasticity of majorite and a majorite–pyrope solid solution to high pressure: implications for the transition zone, *Geophys. Res. Lett.*, *29*, **4**-1-**4**-4, 2002.

Sinogeikin, S.V., J.D. Bass , and T. Katsura, Single-crystal elasticity of ringwoodite to high pressures and high temperatures: Implications for the 520 km seismic discontinuity, *Phys. Earth and Planet. Int.*, *136*, 41–66, 2003.

Sinogeikin, S.V., J. Zhang, and J.D. Bass Elasticity of Single Crystal and Polycrystalline MgSiO₃ Perovskite by Brillouin Spectroscopy, *Geophys. Res. Let.*, 2004.

Stacey, F.D., and P.M. Davis, High pressure equations of state with applications to the lower mantle and core [Review], *Physics of the Earth & Planetary Interiors*, *142* (3–4), 137–184, 2004.

Steinle-Neumann, G., L. Stixrude, R.E. Cohen, and O. Gulseren, Elasticity of iron at the temperature of the Earth's inner core, *Nature*, *413* (6851), 57–60, 2001.

Stixrude, L., Elastic constants and anisotropy of MgSiO₃ perovskite, peroclase, and SiO₂ at high pressure, in *The Core-Mantle Region*, edited by M. Gurnis, M. Wysession, E. Knittle, and B. Buffett, pp. 83–96, American Geophysical Union, Washington, D.C., 1998.

Stixrude, L., Density functional theory in mineral physics, in *Physics Meets Mineralogy: Condensed-Matter Physics in Geosciences*, edited by H. Aoki, Y. Syono, and R.J. Hemley, pp. 21–43, Cambridge University Press, Cambridge, 2000.

Stixrude, L., First principles theory of mantle and core phases, in *Molecular Modeling Theory: Applications in the Geosciences*, edited by R.T. Cygan, and J.D. Kubicki, pp. 319–343, Geochemical Society/Mineralogical Society of America, Washigton, D.C., 2001.

Stixrude, L., and J.M. Brown, The Earth's core, in *Ultrhigh-Pressure Mineralogy: Physics and Chemistry of the Earth's Deep Interior*, edited by R.J. Hemley, pp. 261–282, Mineralogical Society of America, Washington, D.C., 1998.

Stixrude, L., and M.S.T. Bukowinski, Simple covalent potential models of tetrahedral SiO2: Applications to a-quartz and coesite at pressure, *Phys. Chem. Minerals, 16*, 199–206, 1988.

Stixrude, L., and R.E. Cohen, Stability of orthorhombic $MgSiO_3$ perovskite in the Earth's lower mantle, *Nature, 364*, 613–616, 1993.

Stixrude, L., and R.E. Cohen, Constraints on the crystalline structure of the inner core: Mechanical instability of bcc iron at high pressure, *Geophys. Res. Lett., 22*, 125–128, 1995a.

Stixrude, L., and R.E. Cohen, High-pressure elasticity of iron and anisotropy of Earth's inner core, *Science, 1972–1975*, 1995b.

Stixrude, L., R.E. Cohen, and R.J. Hemley, Theory of minerals at high pressure, in *Ultrahigh-Pressure Mineralogy: Physics and Chemistry of the Earth's deep Interior*, edited by R.J. Hemley, pp. 639–671, Mineralogical Society of America, Washington D. C., 1998.

Stixrude, L., E. Wasserman, and R.E. Cohen, Composition and temperature of the Earth's inner core, *J. Geophys. Res., 102*, 24729–24739, 1997.

Su, W.-J., and A.M. Dziewonski, Simultaneous inversion for 3-D variations in shear and bulk velocity in the mantle, *Phys. Earth Planet. Int., 100*, 135–156, 1997.

Su, W.-j., R.L. Woodward, and A.M. Dziewonski, Degree 12 model of shear velocity heterogeneity in the mantle, *J. Geophys. Res., 99* (B4), 6945–6980, 1994.

Sweeney, J.S., and D.L. Heinz, Melting of iron-magnesium silicate perovskite, *Geophys. Res. Lett., 20*, 855–858, 1993.

Tamai, H., and T. Yagi, High-pressure and high-temperature phase relations in $CaSiO_3$ and $CaMgSi_2O_6$ and elasticity of perovskite-type $CaSiO_3$, *Phys. Earth Planet. Int., 54*, 370–377, 1989.

Tarrida, H., and P. Richet, Equation of state of $CaSiO_3$ perovskite to 96 GPa, *Geophys. Res. Lett., 16*, 1351–1354, 1989.

Thomson, K.T., R.M. Wentzcovitch, and M.S.T. Bukowinski, Polymorphs of alumina predicted by first principles: Putting pressure on the ruby pressure scale, *Science, 274*, 1880–1882, 1996.

Trouiller, N., and J.L. Martins, Efficient pseudopotentials for plane-wave calculations, *Phys. Rev., B 43*, 1993–2006, 1980.

Tsuchiya, T., J. Tsuchiya, K. Umemoto, and R. Wentzcovitch, Elasticity of post-perovskite $MgSiO_3$, *Geophys. Res. Let., 31*, doi 10.1029/2004GL020278, 2004a.

Tsuchiya, T., J. Tsuchiya, K. Umemoto, and R. Wentzcovitch, Phase transition in $MgSiO_3$-perovskite in the Earth's lower mantle, *Earth & Planetary Science Letters, 224*, 241–248, 2004b.

Tsuneyuki, S., M. Tsukada, H. Aoki, and Y. Matsui, First-principles interatomic potential of silica applied to molecular dynamics, *Phys. Rev. Lett., 61*, 869–872, 1988.

Urusov, V.S., and N.N. Eremin, Energy minimum criteria in modeling structures and properties of minerals, *Phys. Chem Minerals, 22*, 151–158, 1995.

Utsumi, W., N. Funamori, T. Yagi, E. Ito, T. Kikegawa, and O. Shimomura, Thermal expansivity of $MgSiO_3$ perovskite under high pressures up to 20 GPa, *Geophys. Res. Lett., 22*, 1005–1008, 1995.

van der Hilst, R.D., Evidence for deep mantle circulation from global tomography, *Nature, 386*, 578–589, 1997.

Vanderbilt, D., First-principles theory of structural phase transitions in cubic perovskites, *J. Korean Phys. Soc., 32*, S103–S106, 1998.

Vasco, D.W., and L.R. Johnson, Whole Earty structure estimated from seismic arrival times, *J. Geophys. Res., 103*, 2633–2672, 1998.

Vinet, P., J. Ferrante, J.R. Smith, and J.H. Rose, A universal equation of state for solids, *J. Phys. C, 19*, L467–L473, 1986.

Vocadlo, L., D. Alfe, M.J. Gillan, and G.D. Price, The properties of iron under core conditions from first principles calculations, *Phys. Earth and Planet. Int., 140* (1–3), 101–125, 2003a.

Vocadlo, L., D. Alfe, M.J. Gillan, I.G. Wood, J.P. Brodholt, and G.D. Price, Possible thermal and chemical stabilization of body-centred-cubic iron in the Earth's core, *Nature, 424* (6948), 536–539, 2003b.

Vocadlo, L., and G.D. Price, The melting of MgO—computer calculations via molecular dynamics, *Phys. Chem. Minerals, 23*, 42–49, 1996.

Voter, A.F., F. Montalenti, and T.C. Gemann, Extending the time scale in atomistic simulations of materials, *Annu. Rev. Mater. Res., 32*, 321–346, 2002.

Waldman, M., and R.G. Gordon., Scaled electron gas approximation for intermolecular forces, *J. Chem. Phys., 71*, 1325–39, 1979.

Wang, Y., D. Weidner, and F. Guyot, Thermal equation of state of $CaSiO_3$ perovskite, *J. Geophys. Res., 101*, 661–672, 1996.

Wang, Y., D.J. Weidner, R.C. Liebermann, and Y. Zhao, *P-V-T* equation of state of $(Mg,Fe)SiO_3$ perovskite: constraints on composition of the lower mantle, *Phys. Earth Planet. Int., 83* (1), 13–40, 1994.

Watson, R.E., Analytic Hartree-Fock solutions for O^{-2}, *Phys. Rev., 111*, 1108–1110, 1958.

Watt, J.P., Hashin-Shtrikman bounds on the effective elastic moduli of polycrystals with orthorhombic symmetry, *J. Appl. Phys., 50* (10), 6290–6295, 1979.

Weidner, D., J.D. Bass , A.E. Ringwood, and W. Sinclair, The single-crystal elastic moduli of stishovite, *J. Geophys. Res., 87* (B6), 4740–4746, 1986.

Wen, L., and D.L. Anderson, Slabs, hotspots, cratons and mantle convection revealed from residual seismic tomography in the upper mantle, *Phys. Earth Planet. Int., 99*, 131–143, 1997.

Wentzcovitch, R., B.B. Karki, S. Karato, and C.R.S. da Silva, High pressure elastic anisotropy of $MgSiO_3$ perovskite and geophysical implications, *Earth Planet. Sci. Lett., 164*, 371–378, 1998.

Wentzcovitch, R., and J.L. Martins, First principles molecular dynamics of Li: test of a new algorithm, *Sol. Stat. Comm., 78*, 831–834, 1991.

Wentzcovitch, R., N.L. Ross, and G.D. Price, Ab initio study of $MgSiO_3$ and $CaSiO_3$ perovskites at lower-mantle pressures, *Phys. Earth and Planet. Int., 90*, 101–112, 1995.

Wentzcovitch, R.M., B.B. Karki, M. Coccocioni, and S.d. Giron-coli, Thermoelastic properties of MgSiO$_3$-perovskite: Insights on the nature of the Earth's lower mantle, *Phys. Rev. Lett.*, *92*, 018501-1–018501-4, 2004.

Wentzcovitch, R.M., J.L. Martins, and G.D. Price, Ab initio molecular dynamics with variable cell shape: applications to MgSiO$_3$, *Phys. Rev. Lett.*, *70*, 3947–3950, 1993.

Wentzcovitch, R.M., and L. Stixrude, Crystal chemistry of forsterite: A first-principles study, *Am. Mineral.*, *82*, 663–671, 1997.

Williams, Q., R. Jeanloz, J. Bass, R. Svendsen, and T.J. Ahrens, The melting curve of iron to 2.5 Mbar: first experimental constraint on the temperature at the earth's center, *Science*, *236*, 181–182, 1987.

Winkler, B., and M.T. Dove, Thermodynamic properties of MgSiO$_3$ perovskite derived from large scale molecular dynamics simulations, in *Phys. Chem. Minerals*, pp. 407–415, 1992.

Wolf, G.H., and M.S.T. Bukowinski, Ab initio structural and thermoelastic properties of orthorhombic MgSiO$_3$ perovskite, *Geophys. Res. Lett.*, *12*, 809–812, 1985.

Wolf, G.H., and M.S.T. Bukowinski, Theoretical study of the structural properties and equations of state of MgSiO$_3$ and CaSiO$_3$ perovskites: Implications for lower mantle composition, in *High-Pressure Research in Mineral Physics*, edited by M.H. Manghnani, and Y. Syono, pp. 313–331, Terra Scientific Company Tokyo/American Geophysical Union, D.C., 1987.

Wolf, G.H., and M.S.T. Bukowinski, Variational stabilization of the ionic charge densities in the electron-gas theory of crystals: Applications to MgO and CaO, *Phys. Chem. Minerals*, *15*, 209–220, 1988.

Wolf, G.H., and R. Jeanloz, Lattice dynamics and structural distortions of CaSiO$_3$ and MgSiO$_3$ perovskites, *Geophys. Res. Lett.*, *12*, 413–416, 1985.

Yagi, T., H.K. Mao, and P.M. Bell, Hydrostatic compression of perovskite-type MgSiO$_3$, in *Advances in Physical Geochemistry*, edited by S.K. Saxena, pp. 317–325, Springer Verlag, Berlin, 1982.

Yamamoto, T., D. Yuen, and T. Ebisuzaki, Substitution mechanism of Al ions in MgSiO3 perovskite under high pressure conditions from first-principles calculations, *Earth and Planet Sci Lett*, *206*, 617–625, 2003.

Yeganeh-Haeri, A., Synthesis and re-investigation of the elastic properties of single-crystal magnesium silicate perovskite, *Phys. Earth Planet. Int.*, *87*, 111–122, 1994.

Yoo, C.S., N.C. Holmes, M. Ross, D.J. Webb, and C. Pike, Shock Temperatures and Melting of Iron At Earth Core Conditions, *Phys. Rev. Lett.*, *70* (25), 3931–3934, 1993.

Zerr, A., and R. Boehler, Melting of (Mg,Fe)SiO$_3$-perovskite to 625 kilobars: Indications of a high melting temperature in the lower mantle, *Science*, *262*, 553–555, 1993.

Zerr, A., and R. Boehler, Constraints on the melting temperatures of the lower mantle from high-pressure experiments on MgO and magnesiowustite, *Nature*, *371*, 506–508, 1994.

Zhang, J., and D. Weidner, Thermal equation of state of aluminum-enriched silicate perovskite, *Science*, *284*, 782–784, 1999.

S. Akber-Knutson and M.S.T. Bukowinski, Department of Earth and Planetary Science, University of California at Berkeley, Berkeley, California 94720, USA. (markb@socrates.berkeley.edu)

Self-Gravity, Self-Consistency, and Self-Organization in Geodynamics and Geochemistry

Don L. Anderson

Seismological Laboratory, Caltech, Pasadena, California

"...it is a privilege to see so much confusion."
–Marianne Moore, *The Steeple-Jack*

The results of seismology and geochemistry for mantle structure are widely believed to be discordant, the former favoring whole-mantle convection and the latter favoring layered convection with a boundary near 650 km. However, a different view arises from recognizing effects usually ignored in the construction of these models, including physical plausibility and dimensionality. Self-compression and expansion affect material properties that are important in all aspects of mantle geochemistry and dynamics, including the interpretation of tomographic images. Pressure compresses a solid and changes physical properties that depend on volume and does so in a highly nonlinear way. Intrinsic, anelastic, compositional, and crystal structure effects control seismic velocities; temperature is not the only parameter, even though tomographic images are often treated as temperature maps. Shear velocity is not a good proxy for density, temperature, and composition or for other elastic constants. Scaling concepts are important in mantle dynamics, equations of state, and wherever it is necessary to extend laboratory experiments to the parameter range of the Earth's mantle. Simple volume-scaling relations that permit extrapolation of laboratory experiments, in a thermodynamically self-consistent way, to deep mantle conditions include the quasiharmonic approximation but not the Boussinesq formalisms. Whereas slabs, plates, and the upper thermal boundary layer of the mantle have characteristic thicknesses of hundreds of kilometers and lifetimes on the order of 100 million years, volume-scaling predicts values an order of magnitude higher for deep-mantle thermal boundary layers. This implies that deep-mantle features are sluggish and ancient. Irreversible chemical stratification is consistent with these results; plausible temperature variations in the deep mantle cause density variations that are smaller than the probable density contrasts across chemical interfaces created by accretional differentiation and magmatic processes. Deep-mantle features may be convectively isolated from upper-mantle processes. Plate tectonics and surface geochemical cycles appear to be entirely restricted to the upper ~1,000 km. The 650-km discontinuity is mainly an isochemical phase change but major-element chemical boundaries may occur at other depths. Recycling laminates the upper mantle and also makes it statistically heterogeneous, in agreement with high-frequency scattering studies. In contrast to standard geochemical models and recent modifications, the deeper layers need not be accessible to surface volcanoes. There is no conflict between geophysical and geochemical data, but a physical basis for standard geochemical and geodynamic mantle models, including the two-layer and whole-mantle versions, and qualitative tomographic interpretations has been lacking.

Earth's Deep Mantle: Structure, Composition, and Evolution
Geophysical Monograph Series 160
Copyright 2005 by the American Geophysical Union
10.1029/160GM11

1. INTRODUCTION

The "classical" models of mantle structure and evolution are the whole-mantle convection, or pyrolite, model [*Ringwood*, 1975] with deep slab penetration, and the two-reservoir model [e.g., *DePaolo and Wasserburg*, 1976; *Allegre*, 1982] with an undegassed primitive lower mantle. The recognition of fundamental problems with these models gave rise to an alternative chemically stratified mantle model with essentially all of the incompatible and heat-producing elements residing in the crust and inhomogeneous upper mantle [*Anderson*, 1983, 1989a]. In this model, most of the mantle, by mass balance, is infertile and large-ion-lithophile (LIL)-depleted and the deep mantle is inaccessible because of its high intrinsic density. This model (the Theory of the Earth [TOE] model), essentially the inverse of the above models, is justified and described elsewhere [*Anderson*, 1989a, 2002a] and will not be part of the present conversation until the other models—the classical and the so-called standard models—are dealt with. The recent widespread recognition—or re-recognition [e.g., *Zindler and Hart*, 1986]—of problems with the two-reservoir model has led a proliferation of modifications and a return to variants of the whole-mantle convection scheme of *Ringwood* [1975]. These new hybrid models are designed to explain one or two features but often conflict with other observations. It is not always clear that these complex conceptual models are any more physically realistic than the classical models [e.g., *van Keken et al.*, 2002]. There is now a feeling of crisis and frustration amongst specialists in mantle geochemistry. Francis Albarède, in his Plenary Lecture of the European Union of Geosciences meeting in 2001, put it well (www.theconference.com/JConfAbs/6/Albarede.html):

The paradigm of layered mantle convection was established nearly 20 years ago, mostly based on geochemical mass balance and heat budget arguments. It is now stumbling over the difficulty imposed by convection models to maintain a sharp interface in the mantle at mid-depth and by overwhelming tomographic evidence that at least some of the subducting lithospheric plates are currently reaching the core-mantle boundary. Discontinuities in the deep mantle...remain elusive...The present situation...remains frustrating because the reasons why the layered convection model was defended in the first place are still there and do not find a proper answer with the model of homogeneous mantle convection ...the imbalance between heat flow and heat production requires that the deep mantle is rich in U, Th, and K... the imbalance of some refractory lithophile elements between the composition of the Earth estimated with a homogeneous

mantle and the composition of chondrites leaves a number of 'paradoxes' unresolved...convective mixing should take place with a characteristic time of less than 1 Gy and should essentially wipe out mantle isotopic heterogeneities. In addition, frustrating evidence that the lower mantle hides a geochemical 'black box', with a non-primitive composition and hardly accessible to observation, is mounting.

Albarède and Boyet [2003] hint at the power of a false color image in upsetting a paradigm that 20 years of paradoxes and more quantitative developments in geophysics had not:

For more than a decade, conflicting evidence between seismic tomography and isotope geochemistry of rare gases has thwarted the construction of a unifying convection model and blurred our vision of lower mantle chemistry and mineralogy. All body wave models vividly depict lithospheric plates penetrating the 660 km discontinuity...

2. PRELUDE—ASSERTIONS & SEMANTICS

What is the conflicting evidence? Are the paradoxes real, or are they the result of nonphysical assumptions or data selection? The purpose of this paper is to review the evolution of ideas in mantle geochemistry and dynamics, to test these ideas for physical plausibility, and to address the concerns raised above and, repeatedly, elsewhere [e.g., *Albarède and van der Hilst*, 1999, 2002a,b; hereafter referred to as *AvdH*]. The models mentioned above are motivated almost entirely by one-dimensional (1D) box-model constructs, body-wave travel-times, vivid 2D color images, and purely thermal interpretations of these images and of melting anomalies in the mantle. Many unstated assumptions behind the assertions will be brought up front, reassessed, and dropped. The idea that the mantle is subdivided into a small number of large, isotopically distinct, homogeneous, and accessible reservoirs, separated by major seismic discontinuities, has dominated recent thinking in mantle geochemistry and geodynamics, but it has never had a sound theoretical underpinning, nor was it a unique interpretation of the data [e.g., *Zindler and Hart*, 1986]. Furthermore, it conflicted—and still conflicts—with a wide variety of evidence in many disciplines, including isotope geochemistry itself.

The paradigm that has been called the standard model of mantle geochemistry [e.g., *Hofmann*, 1997] is a two-box model with a semi-isolated leaky lower layer and a well-stirred and homogenized upper layer—above 650-km depth. The original standard model attributed nonridge basalts—so-called melting anomalies—to mixtures between depleted and primitive mantle, with no enriched components.

The two-layer model has generated a number of paradoxes and problems, including the current "crisis", and is responsible for a proliferation of increasingly complex, contrived, and contradictory models for mantle structure, evolution, and convection. The idea of a homogeneous, depleted (i.e., stripped of LIL or crustal elements), well-stirred, upper mantle ('the convecting mantle') and an undepleted ("primitive", with chondritic abundances of all elements, including the noble gases) gas-rich lower mantle (PM), starting at the major mantle phase boundary, is a direct consequence of assumptions, not of data or calculation. The model itself has elements of the Maxwell Demon paradox.

The isotope-based box-models replaced the petrology-based homogeneous pyrolite (deep slab penetration) model of *Ringwood* [1966, 1975], although both involve an unprocessed primitive mantle. These models assume major element homogeneity, or do not address bulk chemistry at all, or assume relationships between major elements (tomography) and isotopes.

2.1. Reservoirs vs. Components

Magmas from mid-ocean ridges (MORB) and ocean islands (OIB) are chemically distinct and are traditionally attributed to separate, isolated reservoirs, e.g., the two-layer model. It is assumed that the MORB source is the "depleted upper mantle" (DUM), the residue after removal of the continental crust (CC) from PM; CC, DUM, and PM are the only three reservoirs. This was first modeled by *Jacobsen and Wasserburg* [1979], using calculations commonly termed "box models". Other magmas were treated as simple mixtures between melts from DUM and PM (lower mantle). Some OIB were considered as pure PM from the deep mantle [*O'Nions et al.*, 1980]; the upper mantle, by definition, had no enriched material. Terrestrial heat flow and estimates of the abundances of U and Th cannot be explained by a mantle composed entirely of MORB and residual peridotites; the remaining heat sources are usually hidden at great depths in isolated reservoirs, either in the unprocessed PM or in a thin radioactive layer.

Refined mass-balance calculations, which include major elements, are consistent with a mantle that is almost entirely depleted in LIL (80–90% infertile peridotite and pyroxenite, and about 7% basalt such as MORB; see Chapter 8 in *Anderson*, 1989a). The estimated K content of the Earth is 151 ppm and the ^{40}Ar in the atmosphere represents 77% of that produced over 4.5 Ga; this observation alone rules out the standard model. A very small fraction of enriched material (EM), such as CC, enriched-MORB (EMORB), OIB, and kimberlite is all that is needed to give chondritic abundances of the refractory trace elements, including U and Th. For example, kimberlitic magma need amount to only 0.085% of the mantle to account for the missing heat-producing elements. This amount of EM—which is complementary to depleted material (DM) and CC—could be stored beneath continents, below oceanic plates, in the perisphere, or elsewhere in the upper mantle or transition zone (TZ). A layer of kimberlite only 2 km thick suffices, or a layer 20 km thick with 10% of the LIL concentrations of kimberlite. Such amounts of enriched material in the shallow mantle cannot be ruled out; if they exist, there is no need for a vast unprocessed primitive reservoir or deep U-rich layers. Enriched magmas such as EMORB and OIB erupt in the ocean basin, and continental flood basalts are associated with continental break-up or rifts. It is only by assumption that these are excluded from upper mantle inventories and attributed to a leaky deeper reservoir. Additionally, a large fraction of the surface heat flow is due to secular cooling, not radioactive decay.

Reservoir—or layered—models are not the only way to satisfy geochemical constraints [*Zindler and Hart*, 1986]. *Meibom and Anderson* [2003] propose an alternative hypothesis: Rather than being divided into isolated reservoirs, the mantle is fertilized, filtered, and sampled by various plate-tectonic processes; large-scale averaging at ridges explains why extreme compositions are apparently missing there. Removal of small-degree melts and other EM at subduction zones and elsewhere creates DM and complementary shallow enriched or metasomatized regions.

2.2. Conflicts and Paradoxes

There are a large number of geochemical paradoxes (sometimes called enigmas, dilemmas, conundrums, or surprises) associated with what is nominally a model based on geochemistry—a sure sign that the model is wrong. The notion developed among some workers that there is a conflict between geophysics and geochemistry [e.g., *Silver et al.*, 1988; *van der Hilst et al.*, 1997] rather than simply strong evidence against the standard 1D box model of geochemistry [e.g., *Zindler and Hart*, 1986]. The view of conflict and paradox is widespread [e.g., *Becker et al.*, 1999; *Coltice and Ricard*, 1999; *Helffrich and Wood*, 2001; *Ballentine et al.*, 2002] although it basically goes back to a single thermal interpretation of a 2D seismic image on the one hand, and a nonunique box model on the other. All of these authors make modifications to the standard models in order to satisfy a given observation. Few authors itemize or evaluate the "conflicting" geophysical or geochemical evidence (see Appendix 1) but refer only to models that are mainly a result of assumptions, unphysical scaling relations and ad hoc interpretations. The assumptions underlying the standard models have remained intact.

Allegre [1997] stated that it is increasingly accepted that the Earth's mantle has a two-layer structure. *Kamber and Collerson* [1999] also claimed that "growing geophysical and geochemical evidence" supported the original "standard" two-layer model. At the same time, *van der Hilst et al.* [1997 and elsewhere] stated their view, which they call a consensus, that slabs of subducted lithosphere sink deep into the "lower mantle", contradicting the two-layer model, and that present-day mantle convection is "therefore" predominated by whole-mantle flow. This was also a contested view in 1979 [e.g., *Anderson*, 1979; *Elsasser et al.*, 1979]. *Ringwood* [1975] thought that eclogite was denser than pyrolite, at all depths, and argued for deep slab penetration and a chemically homogeneous mantle, whereas *Anderson* [1979, 1989b] argued that eclogite would be trapped in the transition zone (see Figure 1). What is meant by deep penetration is often confusing because the classical lower-mantle boundary is at 1,000-km depth [*Birch*, 1952], not at 650 km. These models will be referred to separately as the two-layer, or standard, model and the whole-mantle model, and together as the standard—or classical—models. *Becker et al.* [1999] argue that *"neither end-member scenario is tenable"*. Both of these models are "whole-mantle" in the sense that material from all depths in the mantle is assumed to be accessible to surface volcanoes. Also in 1997 *Hofmann* [1997] recognized the problems and summarized attempts to modify and fix the standard models, mainly by changing the geometry of the reservoirs. However, certain key assumptions were retained; that is, primordial undegassed or less-degassed regions survive in the deep mantle, "high" ^{3}He/^{4}He ratios imply high ^{3}He contents and an undegassed reservoir, enriched components do not exist in "the convecting mantle" (e.g., the MORB reservoir), all regions of the mantle are accessible, temperature is the main variable in controlling physical properties and melting anomalies, and solid-state convective stirring—rather than recycling, magma blending, and sampling theory—is responsible for magma heterogeneity or homogeneity and the scales of heterogeneity. The standard models thus have basically been replaced by a standard set of inviolate assumptions. Various paradoxes can be traced directly to these assumptions—geometric details are less important—and to things that are ignored, such as physics (e.g., effects of pressure, thermodynamic consistency) and statistics (e.g., sampling theory and the central limit theorem).

Some very recent modifications to the standard models [e.g., *van Thienen et al.*, this volume] retain the ideas that oceanic ridges tap the shallowest mantle; ocean island volcanoes tap deep primitive reservoirs; there must be deep, radioactive-rich layers; and oceanic crust—including thick-crust plateaus—must sink deeply into the lower mantle.

Since, in all these models, critical processes occur deep in the mantle, the effects of pressure are of prime importance but are almost always ignored. The standard models are built on geochemistry and seismic body-wave travel times and do not address physical plausibility or the effects of pressure or scale. Other geophysical techniques and physical considerations have been brought to bear on the problem, with quite different results.

2.3. Seismological Layers

The classical 1D seismological models of the mantle included a TZ between 400- and 1,000-km depth that was attributed to phase changes in mantle minerals [*Birch*, 1952]. The Repetti discontinuity near 1,000-km depth was the single abrupt feature between Moho and the core–mantle boundary. By 1970 seismologists had identified about six seismic discontinuities in the mantle: a major one near 630-km depth and others at 280, 520, 940, 410, and 1,250 km [e.g., *Whitcomb and Anderson*, 1970; for updates on mantle discontinuities see *Bina*, 1991; *Revenaugh and Jordan*, 1991; *Deuss and Woodhouse*, 2002]. Although the attention of geochemists and geodynamicists has been focused on the better known 410- and 650-km features, about 10 discontinuities in the mantle have been identified by a variety of high-resolution or correlation techniques and are, roughly, in the depth intervals of 60–90, 130–170, 220, 280–320, 400–415, 500–560, 630–670, 800–940, 1,250–1,320, and 2,500–2,700 km. Negative velocity jumps have been reported from depths of 380, 410, 450, 610, and 720 km [*Fee and Dueker*, 2004; *Song et al.*, 2004]. Some of these discontinuities are probably chemical in nature and some are apparently highly variable in depth (the reports of reflections or scatterers between 800- and 1,320-km depth may all be due to a few highly irregular interfaces). There are numerous reflectors and scatterers in the shallow mantle above about 220-km depth [e.g., *Deuss and Woodhouse*, 2002; *Levander and Niu*, 2003; *Rost and Garnero*, 2004]. Thus, the mantle is not as simple as it appears from global tomography or as implied by the standard models.

Interpretation of the 410- and ~630-km discontinuities as phase changes had been firmly established by 1970. Thermochemical calculations showed that the depths and the abrupt increases in seismic velocity could be explained by phase changes in olivine [*Akimoto*, 1969; *Anderson*, 1967a, 1970]. The phase change interpretation received strong support from static high-pressure experiments on analog compounds in Australia and shock wave experiments on silicates in the United States (see *Ahrens et al.*, 1969; *Ringwood*, 1975). The predicted postspinel breakdown to rock salt and perovskite structures was achieved in several labs in

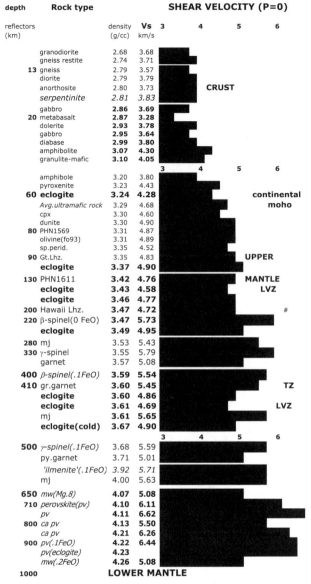

Figure 1. Density and shear velocity of crustal and mantle minerals and rocks, at standard *T* and *P*, tabulated and arranged according to increasing density. This approximates the situation in an ideally chemically stratified mantle. *P* and *T* effects may change the ordering and the velocity and density jumps. Eclogite can settle to various levels, depending on composition; the deeper eclogite bodies have lower velocity than other rocks of similar density. Eclogite has a much lower melting point than peridotites and will eventually heat up and rise, even if it is not in a TBL. The ilmenite form of garnet and enstatite is stable at low temperature but will convert to more buoyant phases as it warms. Velocity decreases do not necessarily imply hot mantle. *Song et al.* [2004] and *Fee and Dueker* [2004] found LVZs at various depths above 720 km. Also shown are the approximate depths of prominent or robust reflections from the mantle; they are not necessarily due to the adjacent lithologies.

1976–1978, confirming the thermochemical calculations and inferences from shock waves and analog compounds. The predicted negative Clapyron slope was confirmed by calorimetric measurements. The phase change hypothesis for both the 410-km and the 650-km discontinuities was generally accepted by 1970, and there were numerous candidates for the other discontinuities. The alternate view—that major chemical changes were responsible for the major seismic discontinuities—came later; it was not the initial explanation. The numerous minor reflectors, however, may be due to chemical interfaces (Figure 1). Subducting slabs may be trapped at various depths, including phase changes, and may be responsible for complicating reflections from the phase change boundaries.

2.4. Isotopic Reservoirs

Between 1979 and 1987, isotope geochemists developed a two-reservoir model, with continuous continental growth and, initially, no recycling. They assumed that the still growing CC was being extracted from some fraction of PM, leaving behind a depleted (upper) and an undifferentiated (deeper) reservoir. They did not challenge the phase change hypothesis; they seemed unaware of it.

Some early mass-balance calculations suggested that about 30% of the mantle could be depleted, assuming 100% efficiency, to form the CC, and that this corresponded to the volume of the upper mantle. Estimates of errors [*Allegre et al.*, 1983; *Anderson*, 1983; *Zindler and Hart*, 1986] cast doubt on this. Some other early calculations suggested perhaps 50% of the mantle was depleted [*DePaolo*, 1980; *O'Nions et al.*, 1980]. Refined calculations permitted much more extensive processing of the mantle, up to 90% [e.g., *Zindler and Hart*, 1986]. It is now clear that at least half, and perhaps 80%, of the original LIL budget of the mantle resides in the CC, lithosphere, upper mantle, and (in the case of argon) atmosphere, thereby reversing the original conclusions that motivated the standard model. Some mass-balance calculations—including the ^{40}Ar in the atmosphere—required >70% of the mantle to be depleted [*Anderson*, 1983; *Zindler and Hart*, 1986], which, following the box-model logic, should be equated with the lower mantle. Mass-balance calculations are consistent with a LIL-depleted and infertile lower mantle that is complementary to the crust and to the olivine-rich shallow mantle [*Anderson*, 1983].

Numerous other geochemical components were identified, and the box model concept soon required five or six isolated reservoirs or layers instead of two, plus recycling. These components have similarities to surficial reservoirs (sediments, crust, delaminated lower crust, lithosphere, seawater, air) but they were also placed in the deepest mantle, by assumption. The isotopic evidence for various components in the mantle did not imply a layered mantle or deep sources, but "the depleted reservoir" was invariably associated with the upper mantle and the others were placed in the lower mantle. DM—and the other "reservoirs"—could have been placed anywhere in the mantle, or distributed; isotopes and mass-balance cannot constrain this [*Zindler and Hart*, 1986]. The bottom line: Geochemical data do not constrain the depths or configurations of reservoirs, and the early isotope papers did not claim that they were; no reference was made to seismological discontinuities. Geochemical data are not responsible for the perceived crisis.

2.5. Geodynamic Constraints

Early geodynamic modelers assumed that mantle convection was mainly restricted to the low-viscosity upper mantle or asthenosphere. *Elsasser et al.* [1979], however, argued that if the plates are the upper boundary layer of mantle convection, then the depth of the system should be comparable to the sizes of plates, or thousands of kilometers. There have been very few simulations of layered convection and none with realistic geometries or physical properties. Nevertheless, the few that exist are instructive. Two-layer simulations of convection and shallow return-flow models are able to satisfy geophysical observations [*McKenzie and Richter*, 1981; *Cizkova and Matyska*, 2004] including heat flow and apparent slab penetration into the deeper layer. A layered model with a major element chemical interface near 900 km at the base of Bullen's TZ explains both the geoid and the dynamic topography [*Wen and Anderson*, 1997], which a boundary at 650 km does not. Chemically heterogeneous or chemically stratified models in general cannot be ruled out. What models can be ruled out, however, are the original mass-balance and primitive undegassed lower mantle models; models that ignore recycling [e.g., *DePaolo and Wasserburg*, 1976; *Jacobsen and Wasserburg*, 1979]; radioactivity-rich lower mantle models; whole-mantle convection schemes based on qualitative tomographic interpretations [e.g., *van der Hilst et al.*, 1997]; geoid models that do not satisfy long-wavelength topography; layered models that consider the 650-km phase boundary to be primarily a chemical or isotope boundary or the only plausible barrier to convection; thermal models that ignore secular cooling; and layered models with shear-coupling across layers. None of this implies a conflict between geochemistry and geophysics. Mantle convection and stirring schemes based on heating from below, 2D geometry (Cartesian or axi-symmetric boxes), and the Boussinesq approximation (ignoring pressure effects on physical properties) should not be viewed as definitive [*Bunge et al.*, 2001; *Tackley*, 1998]; nonetheless,

they have been quoted as contributing to the crisis. There is still a large universe of models, including more physically plausible ones, that cannot be ruled out.

Different scenerios, consistent with high-temperature accretion, a heterogeneous mantle, and gravitational stratification, have been proposed using petrological, major element, and classical physics considerations [e.g., *Birch*, 1952; *Agee*, 1990; *Agee and Walker*, 1988; *Anderson*, 1987b, 1989a,b; *Gasparik*, 1997; *Mattern et al.*, 2005] and based on quantitative analyses of tomography [e.g., *Scrivner and Anderson*, 1992; *Wen and Anderson*, 1995; *Gu et al.*, 2001; *Anderson*, 2002a; *Ishii and Tromp*, 2004; *Trampert et al.*, 2004], sampling theory [*Meibom and Anderson*, 2003], and dynamic topography and plate reconstructions [*Wen and Anderson*, 1997]. Geophysical and thermodynamic data taken as a whole are consistent with a chemically and convectively inhomogeneous mantle that is quite different from the standard models currently being modified and debated.

The evidence for a major phase-change boundary at 650-km depth is not new but apparently it helped precipitate the crisis. As recently as 2002, *van Keken et al.* [2002] state that "the strong seismic reflector at 670-km depth has generally been assumed to be a flow barrier...a major chemical discontinuity" and according to *Helffrich and Wood* [2001], "The seismic discontinuities at 410 and 660 km depth initially appeared to be likely boundaries for the compositional layering". This is not correct.

The early isotope mass balance and two-layered convection papers made no reference to seismic discontinuities. The initial discovery of these discontinuities by seismologists 10 years earlier was immediately followed by thermodynamic calculations that showed that their depths and other characteristics—such as sharpness—were generally consistent with isochemical phase changes, the 410-km one being primarily due to the olivine–spinel phase change and the 650-km one due to a spinel–postspinel phase change with a negative Clapyron slope [*Anderson*, 1967b]. (The 650-km mantle discontinuity is widely referred to as the "660" or "670" km discontinuity, but the observed range in depths is at least 30 km and the global average depth is 650 km.) Previously, the 400–1,000 km depth interval was attributed to a spread-out phase change [e.g., *Bullen*, 1947; *Birch*, 1952; *Ringwood*, 1966]. The phase change interpretation of the upper mantle seismic discontinuities [*Akimoto*, 1969; *Anderson*, 1970] has been firmly established for some time and neither was involved in early geodynamic and geochemical models. Neither mass balance nor seismology requires a chemical change at 650 km, and convection simulations do not imply whole-mantle convection or rapid mixing. Geodynamic data do not conflict with geochemical and geophysical data, although some convection models do.

2.6. Mantle Filters

Tomographic studies suggesting that some slabs cross 650-km depth do not imply that all do, or that key trace elements and isotopes avoid subduction zone processes and are recycled into the deep mantle. The TZ may act as a petrological filter [e.g., *Zindler and Hart*, 1986]. Recycled material can also be trapped at other depths, both deeper and shallower; thick, cold slabs can sink further and take longer to warm up; younger slabs or those with thick crust tend to underplate continents. Slabs dehydrate and sediments get stripped off at shallow depths. Small-degree melts and LIL are also removed from the mantle in high thermal-gradient regions of the lithosphere, and on the wings of ocean ridge melting domains.

Subduction zone processing is the most severe filter; sediment melting, eclogitization, and slab erosion remove LIL from the slab. Dehydration and partial melting filter out LIL. The dry and depleted residual phases—peridotites and eclogites—equilibrate at various depths, and the removed material metasomatizes the shallow mantle (the mantle wedge, the perisphere, and the plate). Young oceanic plates, subducted seamount chains, ridges, and plateaus thermally equilibrate and melt at depths different from those for older, thicker plates.

The 650-km discontinuity, with its negative Clapyron slope, is a temporary barrier to cold sinking material of the same composition; if the conditions are right, however, such material can eventually break through. A different material, with higher-pressure phase changes, e.g., eclogite, can be stranded by phase changes in peridotite. Eclogite can density-equilibrate at depths above 600 km (Figure 1). Chemical discontinuities, even those with very small density jumps, can be a barrier—or filter—to through-going convection. The high-viscosity mesosphere may also play a role in stratifying the mantle [*Cizkova and Matyska*, 2004].

Fertile depleted material—e.g., non-MORB source—is not necessarily the shallowest mantle; it appears in purest form at mature and fast-spreading ridges, not at the onset of continental breakup, or the initiation of spreading, or even on the flanks of spreading centers. Metasomatic fluids and magmas infiltrate the shallow mantle and are more likely to be sampled at the onset—or termination—of spreading than at mature ridges. Delaminated continental crust is a particularly potent source of mantle heterogeneity and melting anomalies; it starts out warmer and equilibrates faster than subducted oceanic crust. Large, fertile, low-melting point blobs can be responsible for "melting anomalies". These recycling, filtering, and sampling processes can explain many geochemical observations while avoiding the pitfalls associated with isolated mantle reservoirs and deep penetration of all slabs and all components.

2.7. Why a Crisis Now?

Fundamental problems with the standard model—and the assumptions underlying it—had been identified early on [e.g., *Armstrong*, 1981; *Allegre*, 1982; *Anderson*, 1982a, 1982b; *Zindler et al.*, 1984] and it has been essentially abandoned by isotope geochemists for some time [*Hofmann*, 1997]. *Armstrong* [1991] reviewed the early history of the standard model, which he called a "myth". Completely different chemically layered models were developed that overcame the problems with the two-layer and Ringwood–Elsasser models [e.g., *Anderson*, 1979, 1989a; *Agee*, 1990]. A large number of models are presumably consistent with both the geophysical and the geochemical data, even if the standard models are not [*Zindler and Hart*, 1986]. Seismic data have very little to say about isotopes, and isotopes cannot constrain mantle stratigraphy.

Why then do we suddenly have a conflict or a crisis? What happened? In addition to the rediscovery of the 650-km phase boundary and the nonuniqueness of mass-balance calculations, isotope geochemists–who previously supported the standard model [e.g., *Ballentine et al.*, 2002]—most commonly refer to a paper by *van der Hilst et al.* [1997] and a widely reproduced dramatic color cross-section from this milestone paper. Visual, or intuitive, interpretations of these images, qualitative chromotomography (QCT), and the association of color with temperature has had enormous cross-disciplinary impact, more so than quantitative and statistical analysis of tomographic inversions. The visual impression of a tomographic image, the assumed implications regarding isotopic as well as major element recycling, and the plausibility or implausibility of layering and survival of heterogeneities in a convecting mantle are now at the heart of the perceived crisis. These are issues related to mineral properties, physics, and scaling to large systems but not to isotope geochemistry or even seismology.

2.8. Overwhelming Tomographic Evidence?

The view expressed by van der Hilst, Albarède, and others, although widely held in the isotope geochemistry community, is not a consensus view of seismologists, primarily because body-wave tomography is a powerful but imperfect tool. Travel-time tomography, used alone, is particularly limited. The results depend crucially on the ray geometry, which is constrained by the geometry of earthquakes and seismic stations [*Vasco et al.*, 1994] and, to a lesser extent, by the details of the mathematical techniques employed [*Spakman and Nolet*, 1988; *Spakman et al.*, 1989; *Shapiro and Ritzwoller*, 2004]. Seismic ray coverage is sparse and spotty. Moreover, the visual appearance of displayed results depends on the color scheme, reference model, cropping, and cross-sections chosen. The resulting images contain artifacts that appear convincing [*Spakman et al.*, 1989]. Even further difficulties result from the limitations of present algorithms, which cannot correct completely for finite frequency, source, and anisotropic effects and can certainly not constrain regions where there are little data. Global tomographic models have little detailed resolution. High-frequency refection, scattering, and coda studies paint a much more complex picture of the mantle than is available from long-period waves [e.g., *Thybo and Perchuc*, 1997; *Shearer and Earle*, unpubl.].

Color 2D cross-sections are particularly ambiguous. Although certainly vivid—and impressive to nonspecialists—they are not "overwhelming, compelling or convincing" evidence; they can be overinterpreted. They cannot do justice to the information content of a typical tomographic study. There are also issues of physical, geodynamic, and petrological interpretations: For example, are "blue" regions of the "lower mantle"—even if real— unambiguous indicators of cold, dense material that started at the Earth's surface? This again is a mineral physics and scaling issue. Can global tomography–or midocean ridge basalts–reveal the true heterogeneity of the mantle or is it averaged out? These are sampling issues.

2.9. Ways out of the Crisis, if There Is One

Any inverse problem, including seismic tomography, must deal with limitations dictated by the distribution and quality of the data, sampling theory, and trade-offs between diverse structures within the Earth; it is likely that the Earth possesses a substantial component in the null-space of any mix of data [e.g., *Shapiro and Ritzwoller*, 2004]. This is also true for isotope data and box models. Methods are available to control the overinterpretation of sparse data, but these guarantee neither physical acceptability of the resulting model nor a model that resembles the real Earth. These limitations are fundamental. Producing realistic, physically plausible Earth models and interpretations requires physical constraints to be applied [*Shapiro and Ritzwoller*, 2004] and an appreciation of what sparse coverage, smoothing, and overparameterization can do [*Vasco et al.*, 1994; *Boschi and Dziewonski*, 1999].

Tomographic images are often interpreted in terms of an assumed velocity–density–temperature correlation; e.g., high shear velocity (blue) is attributed to cold, dense slabs, and low shear velocity (red) is interpreted as hot, rising, low-density blobs. Many factors control seismic velocity, and some do not involve temperature or density [see, for example, Chapters 5–7 and 13–16 in *Anderson*, 1989a]. Cold, dense regions of the mantle can have low shear velocities [e.g.,

Trampert et al., 2004; *Presnall and Gudfinnsson*, 2005]. Changes in composition or crystal structure can lower the shear velocity and increase the bulk modulus and/or density, as can verified by checking any extensive tabulation of elastic properties and densities of minerals (see Figure 1). *Ishii and Tromp* [2004], for example, found negative correlations between velocity and density in the upper mantle. If one drops the purely thermal and visual interpretations of tomographic cross-sections, a resolution of some of the paradoxes is apparent. The tomographic interpretations that precipitated the perceived crisis are not unique.

For reviews of the situation regarding seismic modeling, including uncertainties and limitations, see *Ritsema et al.* [1999], *Vasco et al.* [1994], *Boschi and Dziewonski* [1999], *Dziewonski* [2004, 2005], *Julian* [2004], *Shapiro and Ritzwoller* [2004], *Lay* [2005], and *Ritsema* [2005]. Bottom line: The mantle is not similar to the one- and two-layer 1D structures that underlie the standard models, and temperature is not the sole parameter controlling seismic velocities.

3. PHILOSOPHY AND GROUND RULES

The present contribution is the flip-side of a series of recent review, comment, and synthesis papers that defend—or present modifications to—the standard models, without changing the basic assumptions [e.g., *van der Hilst et al.*, 1997; *AvdH*; *Kellogg et al.*, 1999; *Helffrich and Wood*, 2001; *DePaolo and Manga*, 2003; *van der Hilst*, 2004]. These represent one particular train of thought—and a limited parameter and assumption range—regarding mantle structure and evolution, with an emphasis on mass-balance, heat budget, visual tomography, and convection and mixing simulations. For a review of the pros and cons of this class of models, see *van Keken et al.* [2002]. Recognition of the nonuniqueness of the tomographic and geochemical models is the first step in resolution of the crisis. Other elements that have been missing in recent discussions of mantle structure and evolution are physical plausibility, initial conditions, petrology and mineral physics, and to some extent, statistics—particularly the central limit theorem. Physics of materials at high pressure, thermodynamics, the initial state of the mantle, scaling, and sampling theory are discussed later. These topics can be grouped into themes of self-gravitation (pressure), self-consistency (thermodynamics), and self-organization (thermochemical convection in a gravity field). Pressure can make gravitational stratification irreversible and the deeper layers inaccessible. This possibility alone can resolve many of the geochemical paradoxes; that is, there can be hidden and inaccessible regions. Sampling theory vs. convective mixing has been discussed elsewhere [*Meibom and Anderson*, 2003].

A major advance has recently been made by *Trampert et al.* [2004], using probabilistic tomography and the integration of mineral physics and seismology. According to these authors, and Steve Grand [personal communication, 2005], temperature variations are much weaker than—or even the opposite of—those inferred from shear velocity by using conventional scaling relations. In the deep mantle, the correlation between shear velocity and temperature is weak, and large features of low shear velocity may be due to variations in silicon and iron content or to the spin-state of iron but not to high temperatures [see also *Ishii and Tromp*, 2004]. These features are not buoyant—or strongly buoyant—as was assumed in the past. The size of these features and the lack of a strong thermal signature, which is not predicted by the standard models, can be understood by considering the effect of pressure on thermal properties.

4. THEMES

4.1. Pressure and Chemical Stratification

Pressure decreases interatomic distances, which has strong nonlinear effects on such properties as thermal expansion, conductivity, melting point, and viscosity; on the interpretation of seismic images; and on the possibility of chemical stratification of the mantle. The range of plausible lithologies in both the upper mantle and lower mantle is such that there may be only very small differences in the seismic velocities [e.g., *Zhao and Anderson*, 1994; *Lee*, 2003; see also Figure 1], particularly at high pressure. Even eclogite and peridotite can have similar seismic velocities over some depth intervals. Various candidate lower mantle assemblages have almost identical elastic properties. Thus, compositional changes and interfaces are hard to detect—one reason why some investigators have assumed they do not exist. Some arguments for whole-mantle convection are based on the absence of obvious seismological evidence for layering and thermal boundary layers (TBLs), or on the presence of broad high-velocity bands below 650-km depths (see Appendix 1). Note that chemical discontinuities can have similar shear velocity on both sides and that the shear velocity can drop across a density jump discontinuity.

It has recently become possible to resolve density from seismic data and to separate the effects of temperature and composition [*Trampert et al.*, 2004; *Ishii and Tromp*, 2004]. The basic assumption in many tomographic interpretations—that low seismic velocity is always a proxy for high temperature and low density—is not valid. There is no correlation in properties between the upper mantle, midmantle, and lower mantle and no evidence for either deep slab penetration or continuous plume-like low-velocity upwellings

[*Becker and Boschi*, 2002; *Ishii and Tromp*, 2004, *Trampert et al.*, 2004]—consistent with chemical stratification [*Wen and Anderson*, 1995, 1997]. Whether the stratification is primordial or stable requires information outside the realm of seismology. Chemical stratification of the mantle—both reversible and irreversible—depends on the initial conditions and the variation of physical properties with temperature and pressure. These, plus evidence for stratification, are treated below.

4.2. The Initial State

Although a homogeneous mantle with constant properties is the simplest imaginable assumption, no one has simply explained how the mantle may have arrived at such a state, except by slow, cold, homogeneous accretion. This is an unstated assumption in the standard models. The accretion of Earth was more likely to have been a violent high-temperature process that involved repeated melting and vaporization, and the probable end result was a hot, gravitationally differentiated body.

The assumed starting composition for the Earth is most plausibly based on cosmic abundances, e.g., the refractory parts of carbonaceous, ordinary, or enstatite chondrites [*Ringwood*, 1966; *Anderson*, 1989a; *Agee*, 1990; *Javoy*, 1995]. These compositions predict that the lower mantle has more silicon than the olivine-rich buoyant shallow mantle and that only a small fraction [~7%] of the mantle can be basaltic [e.g., *Anderson*, 1989a, Chapter 8]. The pyrolite model initially assumed 20–25% basalt. Isotope-based models do not constrain the major elements.

The volatile components still in the Earth were most likely added as a late veneer after the planet had cooled to the point where it could retain volatiles. The other choices for starting compositions—considering the standard models—are (1) undegassed volatile-rich ("primordial") components with abundances of both refractory and volatile elements, including noble gases, similar to unfractionated undegassed carbonaceous chondrites [*Kellogg and Wasserburg*, 1990], and (2) whole-mantle compositions dictated by upper-mantle peridotites (the pyrolite and the whole-mantle convection model of *Ringwood* [1975]). The first option evolves to the standard model upon degassing and crustal extraction from the upper mantle; the lower mantle remains fertile and gas-rich. The second option is equivalent to a particularly fertile, whole-mantle convection model, with slabs sinking readily into the deep mantle.

Melting and gravitational differentiation during accretion sets the stage for mantle evolution, including the distribution of radioactive elements. This step is often overlooked in geochemical and geodynamic models, which often start with a homogeneous mantle, or even a cold, gas-rich mantle. The initial temperatures may have been forgotten, but the stratification of major and radioactive elements can be permanent. An excellent summary of initial conditions from a petrological point of view is given by *Ringwood* [1966, 1979].

4.3. Petrological Models

Mantle models based on accretional differentiation and petrology [e.g., *Birch*, 1952; *Anderson*, 1983, 1987a; *Agee*, 1990; *Armstrong*, 1991; *Gasparik*, 1997] are more complex than one- and two-layer models [e.g., *Coltice and Ricard*, 1999; *Helffrick and Wood*, 2001], hybrid models [e.g., *AvdH*], and convection models that ignore petrology, the effects of pressure, and early and irreversible gravitational differentiation. High-resolution seismological techniques involving reflected and converted phase and scattering are starting to reveal the real complexity of the mantle.

In addition to the neglected issues of pressure, petrology, and initial conditions, some other assumptions also distinguish the standard models from alternative models. Some of these are discussed below.

4.4. Sampling or Stirring?

Assumption: The "convecting" upper mantle is efficiently stirred and homogenized.
Corollary: Non-MORB and enriched magmas come from the lower mantle.

The survival of layers, blobs, and reservoirs and the possible entrainment and homogenization by vigorous convective stirring are issues in all current models of mantle structure and evolution. At high Rayleigh numbers, inhomogeneities in the mantle may be stretched, thinned, folded, and stirred, the usual explanation for homogeneous MORB. Inefficient stirring is used to explain "anomalous" basalts in the single layer and mantle-blob schemes of mantle convection [*Becker et al.*, 1999; *Coltice and Ricard*, 1999]. The usual assumption in these calculations is whole-mantle convection, uniform density and viscosity, very high Rayleigh number (10^7–10^8), no pressure effects, steady-state, unidirectional stirring, Newtonian rheology, long stirring times, and no plates; i.e., chaotic mixing is encouraged. Often the calculations are done in 2D Cartesian coordinates with uniform surfaces, no continents, and no internal heating [e.g., *van Keken et al.*, 2002]; the fluid and the surface are often not free to self-organize. The experimenter has a great deal of control on the outcome. An important issue, therefore, is, "What is the Rayleigh number of the mantle and is it really so easily homogenized by convection?"

Mantle convection is most often treated with the simplest possible assumption, the Boussinesq equations. These assume that volume (V) is a function only of temperature (T) and that all other properties are independent of T, V, and pressure (P), even those that explicitly depend on V. Results of such simulations, and of laboratory experiments, are of pedagogic interest but are not relevant to the behavior of the mantle. But scaling relations can be used to predict general behavior such as layering vs. mixing. For example, if the intrinsic density jump across a chemical interface is greater than the density change that can be created by thermal expansion at that depth, then the stratification will probably be stable and irreversible. If the effects of pressure on material properties and the Rayleigh number—and original density stratification—are ignored, it is more likely that the mantle can be well-stirred and homogenized. In a convectively layered mantle with volume dependent properties, the effective Rayleigh number is low and chemical differentiation (gravitational stratification) rather than homogenization must be considered a possible outcome. In the standard two-box model, stratification is due to removal of the crust from the upper layer and is unrelated to accretion or density stratification.

The assumption that the entire upper mantle is efficiently stirred and homogenized, or that whole-mantle convection must destroy chemical heterogeneities, underlies some of the geochemical paradoxes and provides the rationale for recent hybrid models. Left to itself and allowed the necessary degrees of freedom—the essence of self-organization—the mantle may behave in a way inconsistent with imposed boundary conditions, stirring history, and material parameters. Even if the mantle is convecting, the isotopic diversity of magmas may not be an issue of solid-state stirring, layering, or entrainment [*Zindler et al.*, 1984; *Gerlack*, 1990; *Meibom and Anderson*, 2003]. Some stirring calculations give mixing times much greater than sampling times [*Olsen et al.*, 1984] and much greater than the characteristic time of 1 Ga assumed by *Albarède* [2001]. Another standard assumption is that homogeneous basalts require a homogeneous source. The alternative to homogenization by convection is magma blending during the sampling and eruption stage [*Anderson*, 1989a, p. 231] or the extraordinarily powerful central limit sampling theorem [*Anderson*, 2000, 2001; *Meibom and Anderson*, 2003], which explains why chemical diversity is rare along spreading ridges.

4.5. Distribution of Radioactive Elements

Assumption: The lower mantle is primordial, undegassed, or less degassed compared with the upper mantle; it is enriched or undepleted in U, Th, and K.

Mass-balance and thermal constraints are consistent with the view that the radioactive elements are strongly concentrated into the crust and upper mantle [*Birch*, 1952; *Clark and Turekian*, 1979; *Anderson*, 1989a, 2002a]. Note: The upper mantle—a seismological concept—is not equivalent to "the MORB-source", "the convecting mantle", "the depleted upper mantle" (DUM) of the standard models; the "upper mantle" of recent geochemical models—above 670 km—is not the same as the classical upper mantle of seismology: above 1,000-km depth [*Bullen*, 1947; *Birch*, 1952].

There is a rapid decrease in concentrations of radioactive elements from the upper crust (U = 2.8 ppm) through the midcrust (1.6 ppm) and the lower crust (0.2 ppm) to lithospheric xenoliths (0.04–0.12 ppm) [*Rudnick and Fountain*, 1995]. *Clark and Turekian* [1979] suggested that a general exponential decrease with depth in the mantle with a characteristic scale of 1,000 km would satisfy the thermal constraints and that essentially no radioactive elements may be present in the deep mantle. This is a plausible but not unique interpretation, but it does contradict the common view that heat flow and heat production *requires* that the deep mantle is rich in U, Th, and K. The "missing" U, Th, and K may be in the upper mantle: in EMORB components, in kimberlite, and in other enriched components, some recycled [*Anderson*, 1989a; Chapter 8]. Others have proposed deep radioactive-rich layers [*Kellogg et al.*, 1999], assuming that the MORB source (depleted MORB endmember plus U- and Th-free peridotite) fills up 'the upper mantle'. Some enriched components in the mantle—kimberlites, for example—are so enriched that if they occupy only a small fraction of a percent of the upper mantle, they can account for the missing elements and, at the same time, account for the fact that MORB is not exactly complementary to the continental crust.

The conjecture that there are no enriched components in the upper mantle—that the whole upper mantle is uniform and depleted—is the source of some of the geochemical paradoxes. Paradoxes in mantle dynamics and thermal evolution can be traced to the assumption that the lower mantle has high concentrations of U, Th, and K [e.g., *Bunge et al.*, 2001].

A plausible alternative layered model—with none of the above paradoxes—has almost all of the radioactivities in the crust and upper mantle [e.g., Chapter 8 in *Anderson*, 1989a]. Secular cooling also occurs mainly in the upper mantle; the existence of deep and large TBLs depends on the balance of heating and cooling of the upper layers. The standard model with strong heating from the core, little heating in the upper layer, and strong heating and little secular cooling in the deep layer, will develop a strong TBL at the base of "the upper-mantle" and possibly one at the core–mantle boundary; this is a necessary but not sufficient condition for plume formation.

The amount of ^{4}He escaping through the surface is only about 5% of that thought to be generated by radioactive decay [*O'Nions and Oxburgh*, 1983]. More than 20% of the Earth's ^{40}Ar is not in the atmosphere [*Anderson*, 1989a, p. 152]. Where are the missing noble gases stored? Some rocks dredged along the mid-Atlantic ridge are extraordinarily rich in gas, including ^{3}He, ^{4}He, and ^{40}Ar [*Sarda and Graham*, 1990, and references therein]. Furthermore, such rocks show little indication of gas loss or contamination. Most MORB show fractionated noble gas patterns, consistent with degassing that occurred before sampling. If the gas-rich rocks are typical of undegassed MORB and if they represent ~7% of the mantle, then they can account for the missing Ar and He and for the helium–heat flow paradox. Even normal degassed MORB have higher ^{3}He contents than OIB [*Anderson*, 1998].

4.6. Distribution of Major Elements

Assumption: The mantle is chemically homogeneous in the major elements, allowing whole-mantle convection.

Seismological and mineralogical properties can be compared in a variety of ways [e.g., *Sammis et al.*, 1970; *Anderson and Bass*, 1984; *Anderson*, 1989a; *Duffy and Anderson*, 1989; *Zhao and Anderson*, 1994; *Stacey and Isaak*, 2000]. Conclusions using this 1D equation-of-state (EOS) approach are not robust; a variety of compositional models and temperatures for the lower mantle are consistent with the data [e.g., *da Silva et al.*, 2000; *Bunge et al.*, 2001; *Stacey and Isaak*, 2001; *Kiefer et al.*, 2002; *Lee et al.*, 2004], but this in itself explains some of the paradoxes such as the apparent absence—or invisibility—of compositional discontinuities. Equations-of-state modeling (1D) and inspection of a tomographic cross-section (2D) are tools that are much too blunt to "prove" that the lower mantle has the same (or different) chemistry as the upper mantle, even for the major elements. Mass-balance and EOS calculations with major elements are permissive of bulk chemistry changes at 400-, 650-, or 1,000-km depth but do not require these.

Recent studies imply—or are consistent with—an iron-rich high-temperature lower mantle [*Stacey and Isaak*, 2000; *Lee et al.*, 2004; *Mattern et al.*, 2005] or a lower layer that is chemically different from the upper mantle. The possibilities that iron is concentrated into a postperovskite phase and a low-spin iron-rich phase complicate matters still further and have not yet been incorporated into EOS modeling. High thermal conductivity—lattice and/or radiative—in the deep mantle may allow core heat to be conducted through the lower TBL without plume formation. This depends on the local Rayleigh number at the base of the mantle, an approach that is developed below.

4.7. Self-Consistency and Self-Organization

Assumption: Convection models cannot maintain sharp interfaces in the mantle...convective mixing should take place with a characteristic time of less than 1 Gy and should essentially wipe out mantle heterogeneities.

Thermal convection is a self-organized thermodynamic system far from equilibrium in a gravity field and is sensitive to initial conditions and slight changes in conditions or parameters [*Tackley*, 2000; *Anderson*, 2002a,b]. The mantle is a 3D object, and convection and plate tectonics are 3D processes. Tomographic cross-sections and convection in Cartesian boxes or with enforced axes of symmetry probably cannot reveal mantle processes. It is dangerous to generalize from a few oversimplified simulations [see, for example, *Phillips and Bunge*, 2005]. The above assumptions about the mantle depend on material properties, the mode of heating, and even the presence or not of a continent on the surface. The plausibility of chemical stratification depends on how—or if—pressure is treated. The role of mineral physics in mantle geodynamics is to provide a way to test hypotheses and to assure physical consistency in tomographic and convection studies. The parameters that control natural convecting systems at high-temperature and pressure are interrelated. Thermodynamic variables are often indiscriminately assumed to be constant, or to vary independently, ignoring thermodynamic constraints that preclude such assumptions [*Schubert et al.*, 2001].

5. SCALING

The important parameters in mantle convection are density and expansivity. Because seismology measures seismic velocities, going from a tomographic model to a convective model requires scaling relations. Such relations, and the bridges they provide from mineral physics and petrology to seismology and geodynamics, are developed in the following sections. No scaling relations exist between isotopic properties and seismology, or between the colors on a tomographic cross-section and the direction of mantle flow, such as were assumed in developing the standard models—two-layer or whole-mantle.

5.1. Scaling Relations in Seismology

The relative behaviors of density, shear velocity, and bulk sound speed in the mantle are now being determined, and the results have implications for chemical and thermal structure [e.g., *Trampert et al.*, 2001, 2004]. Prior to these developments it was common in seismology to attempt to infer

density from the shear velocity and sometimes to assume or infer a relationship between seismic velocity and temperature. This approach was responsible for the initial revival of the whole-mantle convection hypothesis (see Appendix 1). It is still necessary, however, to use self-consistent thermodynamic relations.

5.2. Scaling Relations in Geodynamics

A measure of the vigor of convection and the distance from static equilibrium is the Rayleigh number, Ra. This is also the scaling parameter for the effect of size on the system. Estimates of Rayleigh numbers for the mantle often do not take into account the effect of pressure on physical properties or of layering. In a spherical shell, convection occurs spontaneously when Ra exceeds about 2,000 [*Chandrasekhar*, 1961]. Chaotic convection and efficient mixing—implicit in the "convecting mantle" scenario of the standard model—are thought to require Ra of $>10^7$, which is very far from static or conductive equilibrium. Can realistic mantle models yield such values?

Mantle convection and mixing calculations with Ra of 10^7 or higher [e.g., *Bunge et al.*, 2001] are, in effect, assuming that the zero-pressure values of thermal and transport properties are maintained throughout the mantle and that whole-mantle convection operates. If one instead uses values estimated for the base of the mantle [*Tackley*, 1998], one derives a value of only about 4,000 for whole-mantle convection, barely twice the critical value. The Rayleigh number depends on the thickness of the convection layer cubed. Therefore, a chemically (or convectively) layered mantle will have a lower effective Rayleigh number than will a chemically homogeneous mantle with whole-mantle convection, and stirring will be inefficient. If the lower 1,000 km of the mantle is convectively isolated, Ra for that region drops to 500. It is thus possible that the deep mantle convects sluggishly, episodically, or not at all. In systems involving a pressure gradient, a single Rayleigh number is not an adequate description of convective style.

5.3. Volume as a Scaling Parameter

The main effects of pressure, temperature, and phase changes on physical properties are volume changes. The thermal, elastic, and rheological properties of solids depend on interatomic distances, or lattice volumetric strain, and are relatively indifferent as to what causes the strain (*T*, *P*, or crystal structure) [*Birch*, 1952, 1961; *Anderson*, 1967a, 1987a, 1989; *Anderson and Anderson*, 1970]. Intrinsic temperature effects are those that occur at constant volume. Quasiharmonic approximation is widely used in mineral physics but not in seismology or geodynamics, where less physically sound relationships are traditionally used. This is ironic since some of the earliest and most fundamental contributions to this field were made by geophysicists [e.g., *Thomsen*, 1972; *Davies*, 1974].

A parameter that depends on P, T, phase (ϕ), and composition © (within limits, e.g., constant mean atomic weight) can be expanded as

$$M (P, T, \phi, ©) = M (V) + \varepsilon$$

where ε represents higher-order intrinsic effects at constant molar V. This is the basis of Birch's Law [*Birch*, 1961], the seismic equation of state [*Anderson*, 1967a, 1987a], laws of corresponding states, and quasiharmonic approximations. Lattice dynamic parameters and thermodynamic and anharmonic parameters are interrelated by way of volume. In some cases, the intrinsic effect of temperature is important, but the quasiharmonic approximation is a step away from the Boussinesq and related approximations that ignore the effects of volume change on most physical properties or that combine pressure and temperature effects in thermodynamically inconsistent ways.

5.4. Beyond Boussinesq

The effect of volume changes on thermodynamic properties are determined by dimensionless parameters. Scaling parameters for volume-dependent properties [*Anderson*, 1987a, 1989a] can be written as power laws or as logarithmic volume derivatives about the reference state:

Lattice thermal conductivity $\quad d \ln \kappa_L / d \ln V \sim 4$
Bulk modulus $\quad d \ln K_T / d \ln V \sim 4$
Thermal expansivity $\quad d \ln \alpha / d \ln V \sim -3$
Viscosity $\quad d \ln \nu / d \ln V \sim 40\text{–}48$

These are semiempirical, numerical values estimated from laboratory and geodynamic measurements. Viscosity is an activated process and the above scaling is partly the volume effect on the preexponential and partly the effect on the activation volume [see *Anderson*, 1989a]. Estimates of viscosity can also be based on the homologous temperature assumption and the effect of compression on the melting point.

In the upper mantle, T and phase changes mainly control V variations, whereas in the deep mantle it is P, crystal structure, and © (including low-spin iron). T is particularly important in the upper mantle where the coefficient of thermal expansion is large and increases with temperature and therefore with depth.

5.5. Things That Do Not Scale With Volume

Shear velocity, rigidity, viscosity, radiative conductivity, and seismic attenuation have intrinsic temperature, compositional, or structural dependencies in addition to volume-dependent terms. They are not simple functions of density (see Figure 1). Shear velocity has an anelastic term that depends on frequency and microstructure. Rigidity has a strong intrinsic temperature term and is affected by iron substitution; the bulk modulus and molar volume are relatively insensitive to iron [Table 6.2 in *Anderson*, 1989a]. The pressure and temperature derivatives for bulk modulus fall into a narrow range for most minerals, whereas those for rigidity do not [Table 6.7 in *Anderson*, 1989a].

Viscosity is one of the most important—but most uncertain and most variable—parameters in mantle dynamics. However, the buoyancy parameter $\alpha \delta T$ is more important in discussions of chemical stratification. It takes a temperature perturbation of about 1,000 K to overcome a 1% density jump in the deep mantle. Viscosity decreases strongly over the depth intervals 1,000–1,400 km and 2,000–2,500 km [e.g., *Forte and Mitrovica*, 2001]. The former interval also has anomalous thermal and FeO gradients [*Mattern et al.*, 2005] and seems to be a fundamental boundary, even barrier, in the mantle [e.g., *Wen and Anderson*, 1997].

6. APPLICATIONS

The specific volume at the base of the mantle is 64% of that at the top [*Dziewonski and Anderson*, 1981]. Compression, composition, and phase changes, and to some extent temperature, are all involved. Although volume scalings such as the Debye theory and the quasiharmonic approximation are strongly grounded in classical physics, they have not been implemented in mantle convection codes. The scaling theory reviewed above will be applied to a few situations relevant to the deep mantle: the Rayleigh number, the thickness and growth time of a deep thermal boundary layer, the "detectability" of chemical interfaces, and the possibility of irreversible chemical layering.

6.1. Rayleigh Numbers

The TBL thickness of a fluid cooled from above or heated from below grows as follows:

$$h \sim (\kappa \, t)^{1/2}$$

where κ is thermal diffusivity, and t is time. The TBL becomes unstable and detaches when the local or sublayer Rayleigh number

$$Ra_c = \alpha\, g\, (\delta T)\, h^3/\kappa\nu$$

exceeds about 1,000 [*Howard*, 1966; *Elder*, 1976], where *g* is acceleration due to gravity; ν is kinematic viscosity, and δT is the temperature increase across the TBL.

The combination $\alpha/\kappa\nu$ decreases with *V* decreases, thereby lowering Ra_c at high *P* or low *T*. Compressible flow, temperature-dependent viscosity, and internal heating make δT across a deep TBL less than δT at the surface [*Tackley*, 1998; *Lenardic and Kaula*, 1994], much less if there are intervening boundary layers. Because $\alpha\delta T$ decreases with depth, irreversible density stratification is favored, even for small intrinsic density contracts. The viscosity at the base of the mantle is probably higher than in the asthenosphere, and the drop in viscosity across the TBL is less.

The local Rayleigh number controls TBL thickness. At the surface the issue is complicated because of water [*Hirth and Kohlstedt*, 1996], intrinsic buoyancy, and the fact that α, κ and ν are strongly *T*-dependent. For parameters appropriate for the top of the mantle, treated as a constant viscosity fluid, the surface TBL becomes unstable at a thickness of about 100 km [*Elder*, 1976]. The time-scale is about 10^8 years, approximately the lifetime of surface oceanic plates.

6.2. Tomography

The implication of the volume scaling plus the role of anelasticity is that temperature effects are much less important in the deep mantle than at the surface. At high-pressure, temperature has little effect on density and other properties that depend on density. Chemical- and phase-changes, anisotropy, and fluids should dominate seismic velocity variations; this changes the usual visual and thermal interpretations of seismic images.

The changes that control upper mantle properties are melting, anelastic, mineralogical, chemical, and thermal [e.g., *Goes et al.*, 2000; *Lee*, 2003; *Perry et al.*, 2003]. Upper mantle chemical and temperature gradients create a low-velocity, low-viscosity, high-attenuation, and low-conductivity zone, usually equated with the asthenosphere, at depths of ~100 to 200 km. At greater depths, the temperature gradient and the effect of temperature on *V* and elastic properties become small but pressure continues to increase. At very high pressures, thermal expansivity is low, and composition and mineralogy control lattice volume. Composition, silicate and metallic melts, and phase changes (including the low-spin transition in FeO and the postperovskite phase change) become the important controls on *V*, conductivity, buoyancy, and seismic parameters. Activated and quantum effects have additional (intrinsic) temperature dependencies. Purely thermal upwellings are expected to have low bulk modulus, low

compressional velocity, and low density. Such is not the case for the large lower-mantle features [*Ishii and Tromp*, 2004; *Trampert et al.*, 2004]; they have the appropriate dimensions to be thermal in nature but resemble more a chemically dense layer at high pressure, i.e., large-scale marginally stable domes with large relief. These domes may have neutral density or may be slowly rising or sinking. In any case, they will affect the geoid, the dynamic surface topography, and the relief on other chemical boundaries, even if they are isolated. D″ may be a very dense—probably iron-rich—layer, and the overlying "layer", D′, would be a less dense region trapped between D″ and the rest of the lower mantle. Stratification may have been established during accretional melting of the Earth [*Anderson*, 1989a; *Agee*, 1990; *Armstrong*, 1991] by downward drainage of dense melts, residual refractory phases, and iron partitioning into phases that may include postperovskite [*Murakami et al.*, 2004], low-spin iron-rich oxides and sulfides [*Gaffney and Anderson*, 1973; *Li et al.*, 2004], and intermetallic compounds. The large, low-shear-velocity features are more appropriately called "domes", a geologically descriptive term, than "megaplumes", which implies a thermal, active upwelling with low-density and low-bulk modulus. The existence of large lateral changes in chemistry and lithology makes suspect all attempts to infer composition and temperature from 1D mantle models such as PREM.

6.3. Thermal Boundary Layers

A homogeneous fluid with constant properties, heated from below, will develop symmetric upper and lower thermal boundary layers. Downwellings and upwellings from the boundaries have the same dimensions and time constants. However, because the upper, middle, and lower thirds of the mantle have quite different viscosities, spectral and spatial characteristics, and correlations with plate reconstructions, the concept has developed of a tripartite mantle [*Anderson*, 2002a] in which the upper 1,000 km is the active and accessible layer. The boundaries are much more subtle than phase change discontinuities. Whether these distinctively different tomographic regions differ in intrinsic chemistry and whether they can exchange material are matters of current debate. The domes in the lower third of the mantle—the abyss—have dimensions much larger than upper mantle slabs, consistent with the scaling. The small ultra-low-velocity zones [*Garnero and Helmberger*, 1995] are here interpreted as dense regions containing core or mantle fluids or reaction products; they are probably very iron-rich.

If the whole lower TBL goes unstable, we have a diapiric plume. Scaling relations show that these will be huge, slow to develop, and long-lived. When a thin low-viscosity, low-density layer at the CMB feeds a plume head, we have cavity plumes with large bulbous heads and narrow tails. Temperature-dependence of viscosity is required. Mantle flow driven by cooling of the top TBL and internal heating makes it difficult for cavity plumes to form—this is one of the plume paradoxes [*Nataf*, 1991]. Pressure broadens considerably the dimensions of diapir plumes and cavity plume heads. It is this pressure-broadening that makes intuitive concepts about plumes [e.g., *DePaolo and Manga*, 2003], based on unscaled laboratory simulations, implausible for the mantle.

A solution to this plume paradox was investigated by *Lenardic and Kaula* [1994]. The solution itself is paradoxical. The condition for a cavity plume is created by destabilizing the upper TBL, which sinks, thins, and cools the lower TBL. Plume formation is triggered from above rather than from an instability in the boundary layer itself as in normal plume theory. In effect, plumes are "splashed out" by impacting cold slabs. However, in general, the effects of temperature are reversed by pressure, and only broad domes or diapirs can form at depth.

6.4. The Lower Thermal Boundary Layer

The presence of a lower TBL and the need to power the dynamo do not require that plumes exist or that they have the properties assigned to plumes in the current literature. The key questions are (1) Are the lattice and radiative conductivities of the lower mantle high enough to conduct heat out of the core without formation of TBL instabilities? (2) Are the dimensions and timescales of these upwellings, if they exist, of the order of hundreds of kilometers and tens of millions of years? (3) Do they rise to the surface? (4) Can they create the sorts of melting anomalies seen at the surface? and (5) Is the lower TBL stabilized by high intrinsic density? The neglect of pressure and scaling effects is responsible for the widely held misconception that narrow rapidly upwelling cylinders are *required* by boundary layer and dynamo theory [e.g., *DePaolo and Manga*, 2003]. This misconception is maintained by unscaled laboratory injection experiments and Boussinesq computer simulations [e.g., *Cordery et al.*, 1997]. The core can get rid of its heat by mechanisms other than ~200-km-wide plumes rapidly rising to the surface. A TBL can be stable if its Rayleigh number is subcritical or if the layer is dense enough.

The critical dimension of lower-mantle thermal instabilities, ignoring radiative transfer, is predicted from the above considerations to be about 10 times larger than at the surface, or about 1,000 km. This is consistent with seismic tomography [*Hager et al.*, 1985; *Hager and Clayton*, 1989; *Tanimoto*, 1990; *Gu et al.*, 2001] and with compressible flow calculations from depth-dependent properties [*Tackley*, 1998]. If there is an appreciable radiative component to the conductivity, or a chemical component to the density (i.e., chemical stratification), then the scale-lengths can be much greater. Large-scale thermochemical features have been inferred from seismology [*Ishii and Tromp*, 2004; *Trampert et al.*, 2004] and simulated with non-Boussinesq convection calculations [*Tackley*, 1998].

The timescale of deep thermal instabilities scaled from the upper mantle value is $\sim 3 \times 10^9$ years. Radiative transfer and other effects may increase thermal diffusivity, further increasing the time required to form an instability. The surface TBL cools rapidly and becomes unstable quickly. The same theory, scaled for the density increase across the mantle, predicts large and long-lived features above the core. This—plus chemical stratification—is the most dramatic effect of pressure and volume scaling. Further implications are that the long-wavelength geoid and the rotation axis are very stable over time [e.g., *Anderson*, 1989a, Chapter 12] and that long-lived hot and cold regions of the lower mantle are more likely to influence the temperature of the mesosphere and upper mantle than vice versa. There is no evidence that the large low-shear-velocity features cause the TZ to thin (one indication of heating)

If the abyss represents one-third of the mantle, it will have an reduction in Ra due to this effect alone by a factor of 27 relative to a reference state of whole-mantle convection. A similar reduction is accomplished by viscosity increase alone. Together, these decrease the deep mantle Ra by about 10^3, compared with whole-mantle, constant-property values. There may also be other chemical boundaries in the mantle [*Anderson*, 1979, 2002a] that would further reduce Ra and the δT across TBLs. The predicted large-scale longevity and sluggishness of deep mantle features are not entirely due to high viscosity. Low α at high P means that intrinsically dense layers may be permanently trapped; moderate jumps (~1%, depending on δT) in density between layers in the mantle can stabilize chemical layering [*Tackley*, 1998; *Anderson*, 2002a]. Unreasonably high mantle temperatures do not occur in these trapped layers if most of the radioactivity is in the crust and upper mantle [*Anderson*, 1989b; *Anderson*, 2002a]. Heat can also be conducted more efficiently at high P.

6.5. Relation to Plume Heads

The scales of lower mantle thermal diapiric instabilities are much larger than those quoted for plume heads. The volumes of plume heads are often assumed to be related to the sizes of large igneous provinces—a circular argument—or to the thickness of D″, assuming it is purely thermal. The Boussinesq approximation also favors small plume heads. If the thickness of D″ is controlled by compositional layering or phase changes [Lay et al., 1998] then one must look elsewhere for the scale of lower mantle thermal instabilities. Da Silva et al. [2000] found that the thermal gradient in much of the lower mantle is superadiabatic. If this is interpreted as evidence for a TBL, the layer is more than 1000 km thick, consistent with the scaling relations, and must have taken a long time to form. Heat is supplied to the CMB at less than 0.10 the rate at which heat is removed from the surface boundary layer [e.g., Christensen and Tilgner, 2004], so deep TBL instabilities are slow to form. Nevertheless, if there is little radioactivity in the deep mantle [e.g., Anderson, 1983], the bottom heating may dominate.

7. STRATIFIED MANTLE?

The conflict, confrontation, and crisis that permeate the isotopic literature may be a chimera. The whole-mantle convection/deep-slab penetration interpretation of tomographic cross-sections is neither unique nor robust [Boschi and Dziewonski, 1999; Cizkova et al., 1999; Davaille, 1999; Hamilton, 2002; Cizkova and Matyska, 2004; Dziewonski, 2004, 2005] and it is inconsistent with geophysical evidence more broadly defined. Petrology-based models tend to be gravitationally stratified, with buoyant olivine-rich regions at the top, dense perovskite- and iron-rich features at depth, and low-melting-point eclogitic materials—among other things—at intermediate depths [e.g., Anderson, 1983; Agee, 1990] (see Figure 1). In these models, the midmantle, upper mantle, and crust are chemically complementary to the deep mantle and to each other. Abrupt seismic discontinuities are not necessarily isotope or reservoir boundaries. What distinguishes this class of models from the others is gravitational stratification by density—which is reversible in the upper mantle—and upward concentration of volatile and LIL, including U, Th, and K. Another distinction is that the deeper layers are not necessarily accessible to surface volcanoes. Plate tectonics and geochemical cycles may be entirely restricted to the upper ~1,000 km, where thermal expansion is high and melting points, viscosity, and thermal conductivity are low.

This class of layered models does not have the paradoxes associated with the standard models or with QCT interpretations, and is consistent with the effects of volume changes discussed in this chapter. The presence of inaccessible (residual) regions is consistent with various mass-imbalance calculations that are paradoxical in the standard models, or with the standard assumptions. Even the crust and the continental lithosphere have managed to stratify themselves by their intrinsic density [Rudnick and Fountain, 1995; Lee et al., 2003].

The seismic velocities of plausible materials in the mantle differ little from one another, even if the density contrasts are adequate to permanently stabilize the layering against convective overturn [see Anderson, 2002a, and eclogite vs. peridotite in Table 1 of Chapter 3 in Anderson, 1989a]. Since chemical discontinuities can be almost invisible to 1D seismology, compared to phase-changes, and since even small chemical density contrasts can stratify the mantle, the possibility must be kept in mind that there may be multiple chemical layers in the mantle, some of which may be subtle (e.g., see Figure 1). The major seismic discontinuities in the mantle are due to mineralogical and phase-changes, not chemical changes, but this does not rule out a chemically heterogeneous mantle.

7.1. Chemical Discontinuities?

The claimed lack of evidence for global seismic interfaces between 650-km depth and D" is sometimes taken as evidence against chemical stratification. Actually, a systematic search for mantle reflections [Deuss and Woodhouse, 2002] found a continuum of robust reflectors between 750- and 1,200-km depth. Their reflection histogram can be explained by three phase-changes (at average depths of 400, 520, and 650 km with variations of 40 km) and three chemical boundaries (at average depths of 220, 850, and 1,100 km with variations of 100 km). But the evidence for chemical stratification can also be subtle. Chemical discontinuities are sometimes discounted because of the perception that a fortuitous cancellation of chemical and thermal effects is required to explain their apparent absence. But small effects are predicted, as are large undulations in topography, consistent with the observations [e.g., Deuss and Woodhouse, 2002].

Phase-change discontinuities are, in general, easier to detect with seismology than chemical discontinuities. Different methods must be used for the latter. The evidence for large-scale stratification of the mantle includes tomographic patterns and spectra [e.g., Tanimoto, 1990; Scrivner and Anderson, 1992; Ray and Anderson, 1994; Wen and Anderson, 1995; Gu et al., 2001; Anderson, 2002a; Becker and Boschi, 2002; Trampert et al., 2004], dynamic topography and evidence for slab flattening, and apparent pile-ups [Fukao et al., 1992, 2001]. Support for a complication, discontinuity, or barrier near 900-km depth is widespread [Whitcomb and Anderson, 1970; Tanimoto, 1990; Revenaugh and Jordan, 1991; Kawakatsu and Niu, 1994; Ritzwoller and Lavely, 1995; Wen and Anderson, 1995; Forte and Mitrovica, 2001; Shen et al., 2003; Mattern et al., 2004]. The tomographic structure of the mantle above the Repetti discontinuity—the upper mantle proper—correlates with present and past plate tectonics [Wen and Anderson, 1995; Becker *and Boschi*, 2002] and behaves as expected for fluid with a moderate Rayleigh number: cooled from above and driven by the plates and lithospheric architecture.

A variety of evidence therefore suggests there might be an important geodynamic boundary, possibly a barrier to convection at midmantle depths. This does not rule out thermal-coupling and apparent continuity across such boundaries [*Cizkova and Matyska*, 2004].

The lowermost 700–1,000-km layer is rich in seismological 3D detail and differs from the overlying mantle in all respects [e.g., *Garnero and Helmberger*, 1995; *Lay et al.*, 1998a; *da Silva et al.*, 2000; *Gu et al.*, 2001; *Anderson*, 2002a; *Trampert et al.*, 2004]. A dense, trapped layer is most consistent with the observations and the physics. This part of the mantle is probably stabilized against convective overturn by the effects of pressure and composition and possibly by high thermal conductivity and by the low-spin transition in FeO [*Badro et al.*, 2003]. The idea that the upper and lower thirds of the mantle might be chemically distinct and isolated is contested [e.g., *Helffrich and Wood*, 2001; *DePaolo and Manga*, 2003; *van der Hilst*, 2004] and is inconsistent with whole-mantle convection, deep slab penetration, plume, and homogeneous mantle ideas based on 1D Earth models, visual impressions of tomographic cross-sections through QCT, and Boussinesq or laboratory simulations. These approaches are becoming increasingly hard to defend.

8. SUMMARY

The petrological and mineral physics case for an inhomogeneous mantle and some sort of convective or chemical stratification is strong [e.g., *Duffy and Anderson*, 1989; *Agee*, 1990; *Javoy*, 1995; *Gasparik*, 1997; *Wen and Anderson*, 1997; *Anderson*, 2002a; *Meibom and Anderson*, 2003; *Lee et al.*, 2004]. The seismological evidence is equally strong (Appendix 1), even though detail is washed out in the kind of frequencies and path-lengths used in global tomography. EOS modeling is ambiguous and nonunique and does not demand a homogeneous mantle. The scattering of high-frequency waves suggests a shallow heterogeneous mantle.

The kind of global chemical stratification that seems to be most consistent with all geochemical and geophysical evidence is essentially the inverse of the standard model and recent modifications of it [e.g., *van der Hilst*, 2004]. It involves a refractory, barren, inaccessible deep mantle (but not primordial, undegassed, or highly radioactive), with irregular chemical boundaries near 1,000- and 2,000-km depths, and a passively convecting, but not mixing, upper mantle. Homogenization is achieved by sampling and melting, not by vigorous solid-state convection. The upper mantle is heterogeneous—both radially and laterally—in

isotope geochemistry and fertility because of recycling and delamination. Most of the mantle is depleted in LIL and barren; only parts of the upper mantle are depleted and fertile. Magmas are homogenized at ridges where large volumes of mantle are processed continuously [*Meibom and Anderson*, 2003]. Large fertile blobs in the upper mantle, however, can create melting anomalies, even at ridges. The "missing" He and Ar may be in the upper mantle.

This kind of chemically layered model removes the objections that have been raised against some layered models and erroneously extended to all such models [e.g., *Hager et al.*, 1985; *Davies*, 1988; *Wen and Anderson*, 1997; *Coltice and Ricard*, 1999; *Helffrich and Wood*, 2001; *Schubert et al.*, 2001]. It appears to be a zero-paradox model.

Acknowledgements. I thank Adrian Lenardic, Russell Hemley, Jun Korenaga, and Anne Hofmeister for incisive reviews of an early draft and Raymond Jeanloz for discussions of related points. Frank Stacey, Geoffry Davies, Shijie Zhong, Robert Liebermann, and Yaoling Niu gave important advice for improvement and in correcting mistakes. Scott King's comments on the penultimate and ultimate versions were invaluable. Gillian Foulger, Seth Stein, Anders Meibom, Kanani Lee, and Jeroen Ritsema reviewed, and provided valuable input for, the revised version. Conversations with David Stevenson, Rob van der Hilst, Jeroen Ritsema, Jeroen Tromp, Bradford Hager, Richard O'Connell, Thomas Duffy, and Thorne Lay helped me identify problems and consolidate ideas, but mistakes and oversights, of course, are mine.

APPENDIX 1. THE GEOPHYSICAL DATA

The evidence against the standard model is not evidence against chemical stratification in general. However, because of the perceived crisis, some investigators have argued for a return to one-layer mantle models, ignoring a large body of other geophysical evidence. Geophysical data that have been cited in support of whole mantle convection include the following:

1. The long-wavelength tomographic structure of the lower mantle, and whole-mantle convection models with plausible velocity–density scalings and viscosity models, successfully explain the geoid [*Hager et al.*, 1985; *Hager and Clayton*, 1989].
2. The bathymetry of the seafloor is explained by conductive cooling of the plate and whole-mantle convection [*Davies*, 1988].
3. Selected tomographic cross-sections show a few high-velocity features in the mantle below 650-km depth. Intuitive scaling relations suggest that these may be cold, dense slabs [*van der Hilst et al.*, 1997].
4. Scattering of high-frequency seismic energy is thought to be consistent with slab fragments in the lower man-

tle and with whole-mantle convection [*Helffrich and Wood*, 2000].

5. The imbalance between heat flow and heat production is thought to require a deep mantle, rich in U, Th, and K.

6. The seismic properties of the lower mantle are consistent with pyrolite, the prototype upper-mantle rock.

Several of these interpretations rely on scaling relations and assumed mineral properties at high pressure. Problems with these interpretations are the following:

1. The geoid represents the combined effects of density variations in the interior of the mantle and the accompanying distortion of the boundaries (termed "dynamic topography") including the surface, the core-mantle boundary and any internal interfaces [e.g., *Hager et al.*, 1985]. Both geoid and dynamic topography must be explained by the same model. Layered convection with a chemical boundary near 900 km deep, can explain both datasets [*Wen and Anderson*, 1997].

2. The inability of some layered models to explain bathymetry is due to the assumption that most surface heat flow is from radioactivity in the lower mantle. A chemically stratified mantle with most of the radioactive elements in the crust and upper mantle [*Anderson*, 1989a] does not suffer from this problem or a from a problem with lower mantle overheating. On the other hand, the large dynamic topography associated with whole-mantle convection affects the square-root age bathymetry relation. In the model of Wen and Anderson, dynamic topography is generated by density variations in the upper mantle and is of low amplitude. Ocean-floor bathymetry is dominated by cooling of the plate. Anomalous bathymetry is primarily due to shallow variations in density, not necessarily high temperature.

3. If the mantle is layered, tomography and the geoid can rule out shear-coupling between layers, but thermal or topographic coupling, or accidental correlations, cannot be ruled out. In layered convection simulations, thermal-coupling induces structures that visually resemble downwellings that penetrate the interface, but are not [e.g., *Cizkova and Matyska*, 2004]. Quantitative analysis of tomographic models and the history of plate subduction confirm the importance of a barrier near the depth of 900 km [*Wen and Anderson*, 1995].

4. Recent seismic scattering studies have been able to better isolate the source of the scattering. Most of the scattered energy comes from the upper mantle [*Shearer and Earle*, unpubl.; *Baig and Dahlen*, 2004] and can be attributed to slab fragments [e.g., *Meibom and Anderson*, 2003] or small-scale layering in the upper mantle. There are robust reflectors throughout the upper 1,200-km of the mantle [*Deuss and Woodhouse*, 2002].

5. The imbalance between heat productivity and heat flow is a result of secular cooling and time lags associated with heat transport to the surface. Layered mantle models have larger time lags. Mass-balance calculations are consistent with most of the radioactivity, and with other LIL elements, being in the crust and upper mantle [*Anderson*, 1989a, Chapter 8; *Rudnick and Fountain*, 1995].

6. The seismic properties of the lower mantle are consistent with rocks quite different from pyrolite [*Lee et al.*, 2004; *Mattern et al.*, 2005], including rocks that are similar in major element chemistry to meteorites, and to meteorite compositions after the crust and upper mantle are removed.

REFERENCES

Agee, C. B., A new look at differentiation of the Earth from melting experiments on the Allende meteorite, Nature 346, 834–837, 1990.

Akimoto, S., High pressure transformations, Intern. Symp. Phase Transition Earth's Interior, Canberra, 1969.

Albarède, F., Plenary Lecture, European Union of Geosciences, 2001. (www.theconference.com/JConfAbs/6/Albarède.html)

Albarède, F., and M. Boyet, A watered-down primordial lower Mantle, AGU, 84(46), Fall Meeting Suppl., Abstract, T21A-06, 2003.

Albarède, F., and R. D. van der Hilst, New mantle convection model may reconcile conflicting evidence, Eos 8(45), pp. 535, 537–539, 1999.

Albarède, F., and R. D. van der Hilst, Zoned mantle convection, Geochimica Cosmochimica Acta 66 (15A), A12–A12 Suppl., 2002a.

Albarède, F., and R. D. van der Hilst, Zoned mantle convection, Philos. Trans. Roy. Soc. A 360(1800), 2569–2592, 2002b.

Allegre, C. J., Chemical geodynamics, Tectonophysics 81, 109–32, 1982.

Allegre, C.J., Limitation on the mass exchange between the upper and lower mantle: The evolving convection regime of the Earth, Earth Planet. Sci. Lett. 150, 1–6, 1997.

Allegre, C., S. R. Hart, and J. F. Minster, Chemical structure and evolution of the mantle and continents determined by inversion of Nd and Sr isotopic data. I. Theoretical methods, Earth Planet. Sci. Lett. 66, 177–90, 1983.

Anderson, D. L., A seismic equation of state, Geophys. Jour. R. Astron. Soc. 13, 9–30, 1967a.

Anderson, D. L., Phase changes in the upper mantle, Science 157, 1165–1173, 1967b.

Anderson, D. L., Petrology of the mantle, Mineralog. Soc. America Spec. Paper, 3, 85–93, 1970.

Anderson, D. L., Chemical stratification of the mantle, Jour. Geophys. Res., v. 84, no. B11, p. 6297–6298, 1979.

Anderson, D. L., Isotopic evolution of the mantle; the role of magma mixing, Earth Planetary Sci. Lett., 57, 1–12, 1982a.

Anderson, D. L., The chemical composition and evolution of the mantle, in Advances in Earth and Planet. Sci., v. 12, High-pressure research in geophysics, edited by S. Akimoto and M. H. Manghnani, 301–318, 1982b.

Anderson, D. L., Chemical composition of the mantle, Jour. Geophys. Res., v. 88 supplement, p. B41–B52, 1983.

Anderson, D. L., A seismic equation of state II. Shear properties and thermodynamics of the lower mantle, Phys. Earth Planet. Interiors 45, 307–323, 1987a.

Anderson, D. L., Thermally induced phase changes, lateral heterogeneity of the mantle, continental roots and deep slab anomalies, Jour. Geophys. Res. 92, 13,968–13,980, 1987b.

Anderson, D. L., Theory of the Earth, Blackwell Scientific Publications, Boston, 366 pp., 1989a. (http://resolver._Caltech.edu/CaltechBOOK:1989.001)

Anderson, D. L., Where on Earth is the crust?, Physics Today 38–46, 1989b.

Anderson, D. L., The helium paradoxes, Proc. National Acad. Sci. USA 95, 4822–4827, 1998.

Anderson, D. L., The statistics of helium isotopes along the global spreading ridge system, Geophys. Res. Lett. 27 (16), 2401–2404, 2000.

Anderson, D. L., A statistical test of the two reservoir model for helium isotopes, Earth Planetary Science Letters 193, 77–82, 2001.

Anderson, D. L., The case for the irreversible chemical stratification of the mantle, International Geology Review 44, 97–116, 2002a.

Anderson, D. L. Plate tectonics as a far-from-equilibrium self-organized system, in Plate Boundary Zones, AGU Geodynamics Series, 30, 411–425, 2002b.

Anderson, D. L., and Anderson, O. L., The bulk modulus-volume relationship for oxides, Jour. Geophys. Res. 75(17), 3494–3500, 1970.

Anderson, D. L., and Bass, J. D., Mineralogy and composition of the upper mantle: Geophys. Res. Lett. 11, 637–640, 1984.

Anderson, D. L., Sammis, C. G., and Jordan, T. H., Composition and evolution of the mantle and core, Science 171, no. 3976, p. 1103–1112, 1971.

Anderson, O. L., Equations of state of solids for geophysics and ceramic science, Oxford University Press, New York, 405 pp., 1995.

Armstrong, R. L., Radiogenic isotopes: The case for crustal recycling on a near-steady-state no-continental-growth Earth, Royal Society of London Philosophical Transactions, v. 301, 443–472, 1981.

Armstrong, R. L., The persistent myth of crustal growth, Australian Journal of Earth Sciences 38, 613–630, 1991.

Baig, A. M. and F. A. Dahlen, Traveltime biases in random media and the S-wave discrepancy, Geophysical Journal International 158, 922–938, 2004, doi:10.1111/j.1365-246X.2004.02341.x

Badro, J., G. Fiquet, F. Guyot, J.-P. Rueff, V. V. Struzhkin, G. Vankó, and G. Monaco, Iron partitioning in Earth's mantle: Toward a deep lower-mantle discontinuity, Science, 300, 789, 2003.

Ballentine, C. J., D. C. Lee, and A. N. Halliday, Hafnium isotopic studies of the Cameroon line and new HIMU paradoxes, Chem. Geol. 139, 111–124, 1997.

Ballentine, C. J., Van Keken, P. E., Porcelli, D., Hauri, E. H., Numerical models, geochemistry and the zero-paradox noble-gas mantle. Philos Transact Ser A Math Phys Eng Sci. 360(1800):2611–31, 2002.

Becker, T. W., and L. Boschi, A comparison of tomographic and geodynamic mantle models, Geochem. Geophys. Geosyst., 3, 2001GC000168, 2002.

Becker, T. W., Kellogg, J. B., and O'Connell, R. J., Thermal constraints on the survival of primitive blobs in the lower mantle, Earth and Planetary Science Letters 171, 351, 1999.

Bina, C. R., Mantle discontinuities, Rev. Geophys. Suppl., 783–793, 1991.

Birch, F., Elasticity and constitution of the Earth's Interiors, J. Geophys. Res. 57, 227–286, 1952.

Birch, F., Composition of the earth's mantle, Geophys. J. R. Astron. Soc. 4, 295–311, 1961.

Boschi, L., and A. Dziewonski, High- and low-resolution images of the Earth's mantle, J. Geophys. Res. 104, 25,567–25,594, 1999.

Bullen, K., An Introduction to the Theory of Seismology, Cambridge University Press, Cambridge, 276 pp, 1947.

Bunge, H.-P., Ricard, Y., and J. Matas, Non-adiabaticity in mantle convection, Geophys. Res. Lett., 28, 879–882, 2001.

Chandrasekhar, S., Hydrodynamic and Hydromagnetic Stability, Dover Publishing, New York, 654, 1961.

Christensen, U. R., and A. Tilgner, Power requirement of the geodynamo from ohmic losses in numerical and laboratory dynamos, Nature 429, 169–171, 2004, doi:10.1038/nature02508.

Cizkova, H., and C. Matyska, Layered convection with an interface at a depth of 1000 km: Stability and generation of slab-like downwellings, Phys. Earth Planet. Int. 141, 269–279, 2004.

Clark, S. P., and K. K. Turekian, Thermal constraints on the distribution of long-lived radioactive elements in the Earth, Phil. Trans. R. Soc. Lond. 291, 269–275, 1979.

Coltice, C., and Y. Ricard, Geochemical observations and one-layer mantle convection, Earth Planet. Sci. Lett. 174, 125–137, 1999.

Cordery, M. J., Davies, G. F., and Campbell, I. H., Genesis of flood basalts from eclogite-bearing mantle plumes, J. Geophys. Res. 102, 20,179–20,197, 1997.

da Silva, C, R. S., R. M. Wentzcovitch, A. Patel , G. D. Price , and S. I. Karato , The composition and geotherm of the lower mantle: Constraints from the elasticity of silicate perovskite, Physics of the Earth and Planetary Interiors 118, 103–109, 2000.

Davaille, A., Two-layer thermal convection in miscible viscous fluids, J. Fluid Mech. 379, 223–253, 1999.

Davies, G. F., Effective elastic-moduli under hydrostatic stress.1. Quasi-harmonic theory, Journal of Physics and Chemistry of Solids 35(11), 1513–1520, 1974.

Davies, G. F., Ocean bathymetry and mantle convection 1. Large-scale flow and hotspots, J. Geophys. Res. 93, 10,467–10,480, 1988.

Davies, G. F., Mantle plumes, mantle stirring and hotspot chemistry, Earth Planet. Sci. Lett. 99, 94–109, 1990.

DePaolo, D. J., Crustal growth and mantle evolution: Inferences from models of element transport and Nd and Sr isotopes, Geochim. Cosmochim. Acta 44, 1185–96, 1980.

DePaolo, D., and G. J. Wasserburg, Inferences about magma sources and mantle structure from variations of $^{143}Nd/^{144}Nd$, Geophys. Res. Lett. 3, 743–746, 1976.

DePaolo, D. J., and M. Manga, Deep origin of hot spots—the mantle plume model, Science 300(5621), 920–921, 2003.

Deuss, A., and J. H. Woodhouse, A systematic search for mantle discontinuities using SS-precursors, Geophys. Res. Lett. 29, 10.1029/2002GL014768, 2002

Dubuffet, F., Yuen, D. A., and Rainey, E. S. G., Controlling thermal chaos in the mantle by positive feedback from radiative thermal conductivity, Nonlinear Processes in Geophysics 9, 311–323, 2002.

Duffy, T. S., and D. L. Anderson, Seismic velocities in mantle minerals and the mineralogy of the upper mantle, Jour. Geophys. Res. 94, 1895–1912, 1989.

Dziewonski, A. M., The robust aspects of global seismic tomography, in Plates, Plumes & Paradigms, Foulger, G. R., Natland, J. H., Presnall, D. C, and Anderson, D. L., eds., Boulder, CO, Geological Society of America, 2005.

Dziewonski, A. M., Global seismic tomography: What we really can say and what we make up, 2004. (www.mantleplumes.org/Penrose/PenPDFAbstracts/DziewonskiAdam_abs.pdf)

Dziewonski, A. M., and D. L. Anderson, Preliminary Reference Earth Model, Phys. Earth Planet. Interiors 25, 297–356, 1981.

Elder, J., The Bowels of the Earth, Oxford Univ. Press, Oxford, 1976.

Elsasser, W. M., P. Olson, and B. D. Marsh, The depth of mantle convection, Jour. Geophys. Research 84, 147–154, 1979.

Fee, D., and K. Dueker, Mantle transition zone topography and structure beneath the Yellowstone hotspot, Geophys. Res. Lett. 31, L18603, doi:10.1029/2004.

Forte, A. M., and J. X. Mitrovica, Deep-mantle high-viscosity flow and thermochemical structure inferred from seismic and geodynamic data, Nature 402, 1881–1884, 2001.

Fukao, Y., Widiyantoro, S., and Obayashi, M., Stagnant slabs in the upper and lower mantle transition region, Rev. Geophy. 28, 291–323, 2001.

Gaffney, E. S., and D. L. Anderson, The effect of low-spin Fe on the composition of the lower mantle, J. Geophys. Res. 78, 7005–7014, 1973.

Garnero, E., and D. Helmberger, A very slow basal layer underlying large-scale low-velocity anomalies in the lower mantle beneath the Pacific, Phys. Earth Planet. Interiors 91, 161–176, 1995.

Gasparik, T., A model for the layered upper mantle, Phys. Earth Planet. Interiors 100, 197–212, 1997.

Gerlack, D. C., Eruption rates and isotopic systematics of ocean islands: Further evidence for small-scale heterogeneity in the upper mantle, Tectonophysics 172, 273–289, 1990.

Goes, S., R. Govers, and P. Vacher, Shallow mantle temperatures under Europe from P and S wave tomography, J.Geophys. Res. 105, 11,153–11,169, 2000.

Gu, Y., Dziewonski, A. M., Weijia, S., and Ekstrom, G., Models of the mantle shear velocity and discontinuities in the pattern of lateral heterogeneities, J. Geophys. Res. 106, 11,169–11,199, 2001.

Hager, B. W., R. W. Clayton, M. A. Richards, R.P . Comer, and A. Dziewonski, Lower mantle heterogeneity, dynamic topography and the geoid, Nature 313 (6003), 541–545, 1985.

Hager, B. H., and R. W. Clayton, Constraints on the structure of mantle convection using seismic observations, flow models, and the geoid, in Mantle Convection, R. W. Peltier. Gordon and Breach. New York, pp. 657–763, 1989.

Hamilton, W. B., The closed upper-mantle circulation of plate tectonics, in Plate Boundary Zone, S. Stein, ed., AGU Monograph, 2002.

Helffrich, G. R., and Wood, B. J., The Earth's mantle, Nature 412, 501–507, 2001.

Hirth, G., and D. L.Kohlstedt, Water in the oceanic upper mantle: Implications for rheology, melt extraction and evolution of the lithosphere, Earth Planet. Sci. Lett. 144, 93–108, 1996.

Hofmann, A. W., Mantle geochemistry: The message from oceanic volcanism, Nature 385, 219–229, 1997.

Hofmeister, A. M., Mantle values of thermal conductivity and a geotherm from phonon lifetimes, Science 283, 1699–1706, 1999.

Howard, L. N., Convection at high Rayleigh number, in Proceedings of the 11th Congress of Applied Mechanics, Munich (Germany), H. Gortler, ed., pp. 1109–1115, Springer-Verlag, Berlin, 1966.

Ishii, M., and J. Tromp, Constraining large-scale mantle heterogeneity using mantle and inner-core sensitive modes, Physics Earth Planet. Interiors 146, 113–124, 2004.

Ita, J. J., and S. D. King, The influence of thermodynamic formulation on simulations of subduction zone geometry and history, Geophys. Res. Lett. 125, 1463–1466, 1998.

Jacobsen , S. B., and G. J. Wasserburg, The mean age of mantle and crustal reservoirs, J. Geophys. Res. 84, 7411–7427, 1979.

Javoy, M., The integral enstatite chondrite model of the Earth. Geophys. Res. Lett. 22, 2219–2222, 1995.

Jordan, T. H., Lithospheric slab penetration into the lower mantle beneath the sea of Okhotsk, J. Geophys. Res. 43, 473–496, 1977.

Julian, B. R., Seismology: The hunt for plumes, 2004. (http://www.mantleplumes.org/Seismology.html)

Kawakatsu, H., and F. Niu, Seismic evidence for a 920-km discontinuity in the mantle, Nature 371, 301–305, 1994.

Kellogg, L. H., and G. J. Wasserburg, The role of plumes in mantle helium fluxes, Earth Planet. Sci. Lett. 99, 276–289, 1990.

Kellogg, L. H., B. H. Hager, and R. D. van der Hilst, Compositional stratification in the deep mantle, Science 410, 1049–1056, 1999.

Kiefer, B., L. Stixrude, and R. M. Wentzcovitch, Elasticity of $(Mg,Fe)SiO_3$-perovskite at high pressures, Geophys. Res. Lett., 29, art. no. 014683, 2002.

Lay, T., The deep mantle thermo-chemical boundary layer: the putative mantle plume source, 2005. (http://www.mantleplumes.org/TopPages/TheP3Book.html)

Lay, T., Q. Williams, and E.Garnero, The core–mantle boundary layer and deep Earth dynamics: Nature 392, 461–468, 1998.

Lee, C.-T., Compositional variation of density and seismic velocities in natural peridotites at STP conditions: Implications for seismic imaging of geochemical heterogeneities in the upper mantle, J. Geophys. Res. 108, 2003, doi:10.1029/2003JB002413.

Lee, K. K. M., B. O'Neill, W. R. Panero, S.-H. Shim, L. R. Benedetti, and R. Jeanloz, Equations of state of the high-pressure phases of a natural peridotite and implications for the Earth's lower mantle, Earth Planet. Sci. Lett. 223(3–4), 381–393, 2004.

Lenardic, A., and W. M. Kaula, Tectonic plates, D″ thermal structure, and the nature of mantle plumes, J. Geophys. Res., 99, 15,697–15,708, 1994.

Levander, A., and F. Niu, Receiver-function imaging of the crustal and mantle structure beneath South Africa: A comparison of stacking and migration, EOS Trans. 84, 46, 2003.

Li, J., V. Struzhkin, H.-K. Mao, J. Shu, R. J. Hemley, Y. Fei, B. Mysen, P. Dera, V. Prakapenka, and G.Shen, Electronic spin state of iron in lower mantle perovskite, Proc. Natl. Acad. Sci. USA 101, 14,027–14,030, 2004.

Loper, D. E., and F. D. Stacey, The dynamical and thermal structure of deep mantle plumes, Phys. Earth Planet.. Interiors 33, 304–317, 1983.

Lubimova, H., Thermal history of the Earth with consideration of the variable thermal conductivity of the mantle, Geophys. J Roy. Astron. Soc. 1, 115–134, 1958.

Mattern, E., J. Matas, Y. Ricard, and J. Bass, Lower mantle composition and temperature from mineral physics and thermodynamic modeling, Geophys. J. Int. 160, 973–990, 2005.

McKenzie, D. P., and F. Richter, Parameterized thermal convection in a layered region and the thermal history of the Earth, J. Geophys. Res. 86, 11,667–11,680, 1981.

Meibom, A., and Anderson, D. L., The statistical upper mantle assemblage, Earth Planet Sci. Lett. 217, 123–139, 2003. (www.mantleplumes.org/Preprints.html)

Mitrovica, J. X., and Forte, A. M., Radial profile of mantle viscosity: Results from the joint inversion of convection and postglacial rebound observable, J. Geophys. Res. 102, 2751–2769, 1997.

Murakami, M., Hirose, K., Kawamura, K., Sata, N., and Ohishi, Y., Post-perovskite phase transition in MgSiO₃, Science 304, 855–858, 2004.

Nataf, H.-C., Mantle convection, plates, and hotspots, Tectonophysics 187, 361–377, 1991.

Olsen, P. L., D. A. Yuen, and D. S. Balsiger, Mixing of passive heterogeneities by mantle convection, J. Geophys. Res. 89, 425–36, 1984.

O'Nions, R. K., Evensen, N. M., and Hamilton, P. J., Differentiation and evolution of the mantle, Philos. Trans. Roy. Soc. Lond. A 297, 479–93, 1980.

O'Nions, R. K., and R. Oxburgh, Heat and helium in the Earth. Nature 306, 429–431, 1983.

Perry, H. K. C., A. M. Forte, and D. W. S. Eaton, Upper mantle thermochemical structure below North America from seismic geodynamic flow model, Geophys. J. Int. 154, 279–299, 2003.

Phillips, B. R,. and H-P. Bunge, Heterogeneity and time dependence in 3D spherical mantle convection models with continental drift, Earth Planet. Sci. Lett. 233(1–2), 121–135, 2005.

Presnall, D. C., and G. H. Gudfinnsson, Carbonatitic melts in the oceanic low-velocity zone and deep upper mantle, in Plates, Plumes & Paradigms, G. R. Foulger, J. H. Natland, D. C. Presnall, and D. L. Anderson, eds., Geological Society of America, 2005 (in press).

Ringwood, A. E., Composition and phases of the mantle, in Advances in Earth Sciences, P. M. Hurley, ed., MIT Press, Cambridge, Mass., 502 pp., 1966.

Ringwood, A. E., Compositions and Petrology of the Earth's Mantle, McGraw-Hill, New York, 618 pp., 1975.

Ringwood, A. E., Origin of the Earth and Moon, Springer-Verlag, New York, 1979.

Ritsema, J., Global seismic maps, Web supplement, 2005. (http://www.mantleplumes.org/TopPages/TheP3Book.html)

Ritsema, J., van Heijst, H. J., and Woodhouse, J. H., Complex shear wave velocity structure imaged beneath Africa and Iceland, Science 286, 1925–1928, 1999.

Rost, S., and E. J. Garnero, Array seismology advances research into Earth's Interior, Eos Trans. 85, 301, 305–306, 2004.

Rudnick, R. L., and D. M. Fountain, Nature and composition of the continental crust: A lower crustal perspective, Rev. Geophysics 33, 267–309, 1995.

Sammis, C. G., D. L. Anderson, and T. H. Jordan, Application of isotropic finite strain theory to ultra-sonic and seismological data, Jour, Geophys. Res. 75(23), 478–480, 1970.

Sarda, P., and D. W. Graham, Mid-ocean ridge popping rocks: Implications for degassing at ridge crests, Earth Planet. Sci. Lett. 97, 268–289, 1990.

Schubert, G., Turcotte, D. L., and Olsen, P., Mantle Convection in the Earth and Planets, Cambridge University Press, Cambridge, 2001.

Scrivner, C., and Anderson, D. L., The effect of post Pangea subduction on global mantle tomography and convection, Geophys. Res. Lett. 19, 1053–1056, 1992.

Shapiro, N. M., and M. H. Ritzwoller, Thermodynamic constraints on seismic inversions, Geophys. J. Int. 157, 1175–1188, doi:10.1111/j.1365-246X.2004.02254.x, 2004.

Shearer, P. M., and Earle, P. S., Modeling the global short-period wavefield with a Monte Carlo seismic phonon method, submitted, 2004.

Silver, P. G., R. W. Carlson, and P. Olson, Deep slabs, geochemical heterogeneity and the large-scale structure of mantle convection: Investigation of an enduring paradox, Ann. Rev. Earth Planet. Sci. 16, 477–541, 1988.

Song, T. R. A., D. V. Helmberger, and S. Grand, Low velocity zone atop the 410 seismic discontinuity in the northwestern US, Nature 427, 530–533, 2004.

Spakman, W., and G. Nolet, Imaging algorithms, accuracy and resolution in delay time tomography, in Mathematical Geophysics, Vlaar et al. (eds.), Reidel, pp. 155–188, 1988.

Spakman, W., S. Stein, R. D. Van der Hilst, and R. Wortel, Resolution experiments for NW Pacific subduction zone tomography, Geophys. Res. Lett. 16, 1097–1100, 1989.

Stacey, F. D., Properties of a harmonic lattice, Phys. Earth Planet. Interiors 78, 19–22, 1993.

Stacey, F. D., and Loper, D. E., The thermal boundary-layer interpretation of D″ and its role as a plume source, Phys. Earth Planet. Interiors 33, 45–55, 1983.

Stacey, F. D., and D. G. Isaak, Extrapolation of lower mantle properties to zero pressure: Constraints on composition and temperature, Amer. Mineralogist 85, 345–353, 2000.

Stein, C., J. Schnalzl, and U.Hansen, The effect of rheological parameters on plate behavior in a self-consistent model of mantle convection, Phys. Earth Planet. Interiors 142, 225–255, 2004.

Tackley, P., Mantle convection and plate tectonics: Toward an integrated physical and chemical theory, Science 288, 2002–2007, 2000.

Tackley, P., Three-dimensional simulations of mantle convection with a thermo-chemical basal boundary layer: D″, in Gurnis, M., Wysession, M. E., Knittle, K., and Buffett, B., eds., The Core–Mantle Boundary Region, American Geophysical Union, Washington, D.C., pp. 231–253, 1998.

Tanimoto, T., Predominance of large-scale heterogeneity and the shift of velocity anomalies between the upper and lower mantle, J. Phys. Earth 38, 493, 1990.

Thybo, H., and E. Perchuc, The seismic 8-degree discontinuity and partial melting in the continental mantle, Science 275, 1626–1629, 1997.

Trampert, J., P. Vacher, and N. Vlaar, Sensitivities of seismic velocities to temperature, pressure and composition in the lower mantle, Phys. Earth Planet. Interiors 124, 255–267, 2001.

Trampert, J., F. Deschamps, J. Resovsky and D. Yuen, Probabalistic tomography maps chemical heterogeneities throughout the lower mantle, Science.306, 853–856, 2004.

van der Hilst, R. D., Widiyantoro, S., and Engdahl, E. R., Evidence of deep mantle circulation from global tomography, Nature 386, 578–584, 1997.

van der Hilst, R. D., Changing views on Earth's deep mantle, Science 306, 817–818, 2004.

van Keken, P., E. H. Hauri, and C. J. Ballentine, Mantle mixing: The generation, preservation, and destruction of chemical heterogeneity, Annual Rev. Earth Planet. Sci. 30, 493–525, 2002.

Vasco, D. W., L. R. Johnson, R. J. Pulliam, and P. S. Earle, Robust inversion of IASP91 travel time residuals for mantle P and S velocity structure, earthquake mislocations, and station corrections, J. Geophys. Res. 99, 13,727–13,755, 1994.

Wen, L., and D. L. Anderson, The fate of slabs inferred from seismic tomography and 130 million years of subduction, Earth Planet Sci. Lett. 133, 185–198, 1995.

Wen, L., and D. L. Anderson, Layered mantle convection: A model for geoid and topography, Earth Planet. Sci. Lett. 146, 367–377, 1997.

Whitcomb, J. H., and D. L. Anderson, Reflection of P'P' seismic waves from discontinuities in the mantle, Jour. Geophys. Res. 75(29), 5713–5728, 1970.

Zhao, Y., and D. L. Anderson, Mineral physics constraints on the chemical composition of the Earth's lower mantle, Phys. Earth Planet. Interiors 85, 273–292, 1994.

Zindler, A., and S. R. Hart, Chemical geodynamics, Ann. Rev. Earth Planet Sci. 14, 493–571, 1986.

Zindler, A., H. Staudigel, and R. Batiza, Isotope and trace element geochemistry of young Pacific seamounts: Implications for the scale of mantle heterogeneity. Earth Planet. Sci. Lett. 70, 175–195, 1984.

Don L. Anderson, Seismological Laboratory, California Institute of Technology, MC 252-21, Pasadena, CA 91125 (dla@gps.caltech.edu)

The Uncertain Major Element Bulk Composition
of Earth's Mantle

Q. Williams and E. Knittle

Department of Earth Sciences, University of California–Santa Cruz, Santa Cruz, California, USA

We examine the bulk mantle composition generated by assuming that the mantle undergoes a transition (at some depth) to a geophysically feasible but usually geochemically disregarded composition: a nearly pure perovskite deep mantle. This composition yields a bulk chemistry of the mantle that approaches chondritic values, and can resolve the well-known discrepancy between chondritic Mg/Si ratios and those inferred for the Earth's upper mantle. The uncertainties in the major element concentrations generated by such a shift in deep mantle composition are large: The silica content of the mantle is raised from that inferred from upper-mantle-based determinations of bulk silicate Earth composition by as much as ~5 wt%, while the MgO content is depressed by a roughly comparable amount. If the deep mantle is compositionally distinct from the overlying mantle, then the depth at which the mantle shifts composition becomes critically important for our understanding of the bulk composition of the planet. Experimental studies of the chemistry of high-pressure melts in chondritic charges indicate that a complementary relationship might exist between primitive upper mantle compositions and a perovskite-enriched or near-chondritic deep mantle. Assessing whether such a complementary relationship exists hinges on the precise aluminum and, to a lesser extent, calcium content of the deep mantle. Inversions for bulk Earth composition that assume a primitive upper mantle composition describes the chemistry of the entire mantle are often highly precise yet, because of the assumption of a chemically homogeneous mantle, may be inaccurate at the 5 wt% level for MgO and SiO_2. Subsidiary effects of oxidation state and volatile content on inversions for bulk silicate Earth composition are also noted.

INTRODUCTION

The bulk composition of the Earth has been extensively examined from a range of geochemical perspectives [e.g., *Ringwood*, 1979; *Wanke*, 1981; *Anderson*, 1982, 1983, 1989; *Allègre et al.*, 1995; *McDonough and Sun*, 1995; *Jones*, 1996; *Javoy*, 1999; *Drake and Righter*, 2002; *Burbine and O'Brien*, 2004]. Obviously, were the compositions of the two dominant regions of the planet, the Earth's core and

mantle, each well known from direct sampling, then the problem of Earth's composition would be solved. In the case of the Earth's core, comprising ~32% of the planet's mass, the primary compositional uncertainty lies in the identity of its ~10 wt% lighter alloying component: Sulfur, oxygen, silicon, and prospectively even hydrogen and carbon have all been proposed. However, the extremely limited number of samples derived from depths below about 200 km within the planet's interior necessitates indirect approaches to estimating the major element chemistry of the Earth as a whole.

The major element chemistry of Earth's mantle is thus constrained from a suite of diverse inputs. These include (1) the cosmochemical abundance of elements and the cor-

Earth's Deep Mantle: Structure, Composition, and Evolution
Geophysical Monograph Series 160
Copyright 2005 by the American Geophysical Union
10.1029/160GM12

responding average compositions of primitive meteorites, as the likely building blocks of the proto-Earth [e.g., *Palme and O'Neill,* 2003]; (2) xenolith compositions, as nominally representative samples of the upper mantle and, in perhaps some instances, the transition zone [e.g., *Jagoutz et al.,* 1979]; (3) exposed/obducted samples of Earth's subcrustal mantle, which often represent samples of mantle material from which partial melt has been extracted [e.g., *Allègre and Turcotte,* 1986]; (4) the compositions of mantle-derived magmas, as these indirectly reflect the chemistry (and occasionally the compositional heterogeneity) of their mantle source region [e.g., *Ringwood,* 1979; *Salters and Stracke,* 2004]; and finally (5) normal mode and travel time-derived seismic constraints on the elastic properties and density of the mantle coupled with the known (or inferred) elastic moduli and densities of likely mantle materials at the temperatures and pressures of Earth's interior [e.g., *Jeanloz and Knittle,* 1989; *Jackson and Rigden,* 1998; *Mattern et al.,* 2005]. Additionally, indirect compositional probes such as electromagnetic sounding techniques are occasionally used to assess variations in the abundance of elements such as iron.

Each of these avenues of inquiry into bulk mantle composition has significant weaknesses and limitations; yet taken together, they can provide an integrated and moderately consistent picture of the major element chemistry of Earth's mantle. The weaknesses of each approach are easily identifiable. For example, cosmochemical element abundances and the chemistry of chondritic meteorites are frequently viewed as the initial building blocks of the planet and thus representative of the bulk major element composition of the Earth. The subsequent compositional modification from this initial major element chemistry occurs through the differentiation processes that may happen during planetary accretion. These include preferential accretion of refractory/easily accreted components and volatilization and loss of material in large impact events during the highly energetic latter stages of the accretionary process. In addition, within the first tens of millions of years of Earth history, elements that alloy readily with iron-rich material may also be depleted within the silicate Earth through core formation. In this sense, the abundance of different light alloying elements within Earth's core ultimately becomes pivotal for inferring the likely bulk chemistry of Earth's mantle, if it assumed to be derived from representative chondritic parent material (with the recognition that chondrites themselves span a broad compositional range).

Yet, it remains controversial whether the Earth's initial composition is accurately represented by the chemistry of any known meteorite [e.g., *Burbine and O'Brien,* 2004; *Drake and Righter,* 2002]. Chondritic meteorites span a broad range of chemistries, from those with Mg/Si

ratios approaching pyrolitic (the CV-chondrites) to the Ordinary and Enstatite chondrites, which have substantially lower Mg/Si ratios [e.g., *Palme and O'Neill,* 2003]. From a dynamical standpoint, simulations of terrestrial planet accretion appear to incorporate material from wide zones within the proto-planetary accretionary disk, implying that the terrestrial planets sampled a broad range of meteorite chemistries during accretion [e.g., *Chambers and Wetherill,* 1998; *Morbidelli et al.,* 2000]. Indeed, such mixing may be required to explain the oxygen isotope compositions of Earth, Mars, the moon and Vesta [*Lodders,* 2000]. Inversions of a combination of the estimated elemental and oxygen isotope composition of the Earth compared with the composition of different meteorite groups (with oxygen isotope values providing a principal discriminant) yield a bulk Earth composition dominated by the reduced, enstatite-rich (and not very common) EL chondrite family [*Burbine and O'Brien,* 2004]. The related EH enstatite chondrite family had been previously proposed as a primary source material for the Earth [*Javoy,* 1995]. Alternatively, an unsampled "Earth chondrite" has been invoked as the principal building block of the planet [*Drake and Righter,* 2002]. The key question with such meteorite-based analyses of bulk Earth composition is: How closely does an available family (or suite) of meteorites represent Earth's composition? A different way of asking the same question is: How much compositional zonation or variation (induced by phenomena such as temperature variations and solar wind) existed within the portion of the proto-planetary accretion disk sampled by the growing Earth? Some have advocated that there was little or no chemical zonation in the region of accretion of the terrestrial planets [*Palme,* 2000], but the level of understanding of the bulk chemistry of the other inner planets may be insufficient to firmly reach this conclusion. Indeed, others have proposed precisely the opposite—that compositionally distinct reservoirs were present within the inner solar system during accretion [*Drake and Righter,* 2002]. The lack of consensus on the degree of compositional zonation (or lack thereof) in the earliest solar system provides a fundamental backdrop of uncertainty to forward modeling exercises that use solar or meteoritic compositions to constrain the composition of the Earth.

By the same token, analyzing the chemistry of the xenolith suite [e.g., *Jagoutz et al.,* 1979], or combining depleted mantle and magma compositions (as pioneered by Ringwood through his pyrolite model [e.g., *Ringwood,* 1979]) yields at best a reliable characterization of the composition of only the shallow mantle (likely no deeper than 670-km depth); whether it reflects the composition at deeper depths hinges on the amount of convective exchange between the shallower and deeper portions of the mantle. Indeed, the pyrolite model may constrain the composition of only those zones of the upper mantle that are either upwelling or juxtaposed with upwellings capable of either generating melt or entraining material. This shortcoming has allowed considerable latitude for proposing that the composition of the mantle outside of regions sampled by upwellings may differ from those regions sampled by xenoliths and magmas: an interesting avenue of inquiry, but not one that can definitively locate the unsampled regions [e.g., *Anderson, 1989*]. *Jones* [1996] proposed a somewhat different idea relative to the xenolith suite: that the actual bulk composition of the mantle may be represented by a few rare but extremely fertile xenoliths that lie close in major element composition to CV-chondrites.

Traditionally, geophysical approaches to constraining mantle composition have been the least important in what have largely been geochemically based estimates of bulk mantle composition. The rationale for this is twofold. First, the knowledge of the elastic properties of the main mineralogic constituents of the lower mantle has been viewed as insufficiently precise to provide an accurate constraint on the major element chemistry of the lower mantle, which is the dominant portion of the silicate Earth. Second, the results of geophysical approaches to lower mantle composition have tended to yield disparate results. Some models have favored a compositional stratification, often proposed as perovskite-enrichment in the lower mantle [e.g., *Jeanloz and Knittle, 1989; Anderson, 1989; Stixrude et al., 1992; Zhao and Anderson, 1994; Karki and Stixrude, 1998*], and others permit fairly broad compositional ranges in the deep mantle (with significant trade-offs with temperature) [*Deschamps and Trampert, 2004; Mattern et al., 2005*], while other bulk composition models are fully consistent with a homogeneous pyrolitic chemistry throughout the mantle [*Jackson and Rigden, 1998; Bina, 2003*], but perhaps with a superadiabatic temperature gradient [*Cammarano et al., 2005*]. Similarly, normal mode constraints permit excess densities of ~0.4% in the deep mantle [*Masters and Gubbins, 2003*]. Importantly, the concept that the 670-km discontinuity *must* be linked to any proposed compositional layering within the mantle has largely been eliminated [e.g., *Kellogg et al., 1999*]. Such a linkage *may* exist, but compositional layering might instead exist at mid- or deep lower mantle depths: Either instance is difficult to definitively seismically detect. Moreover, the apparent behavior of slabs within the mantle does not *mandate* (nor does it absolutely preclude) penetration of slabs to core–mantle boundary (or even deep mantle) depths or commensurate whole-mantle mixing through subduction [e.g., *Fukao et al., 2001*]. Indeed, the lack of obvious slab continuity into the deep mantle has partially motivated suggestions of deep mantle chemical layering [*Kellogg et al., 1999; Anderson, 2002*].

From a broad perspective, the coarse and often ambiguous compositional resolution of mineral physics interpretations of seismic properties (particularly relative to, for example, the precise geochemical determination of the chemistry of pyrolite), have resulted in Occam's Razor frequently being applied to mantle composition [*Javoy, 1999; Allègre et al., 2001*]. The default position has been to simply treat the mantle as chemically homogeneous: That such a state was generated and maintained through Earth's accretion, solidification, and subsequent evolution is simply assumed. On its face, the assumption of mantle chemical homogeneity appears not necessarily compatible with isotopic and rare gas results that indicate at most a limited mixing between chemical reservoirs assumed to be the upper and lower mantle of the planet [*Allègre, 1997*]. Thus, if the major element composition of the mantle is assumed to be homogeneous, then the existence of distinct radiogenic isotopic reservoirs must not be associated with variations in their major element chemistry [*Drake and Righter, 2002*]—a possible, but perhaps contrived, situation. In short, the processes through which separate isotopic reservoirs are produced (which might include magma ocean differentiation, crustal generation and recycling, continental/cratonic root genesis, formation of depleted residues, and preservation of chemically primitive zones) must have induced insignificant major element variations in the planet as a function of depth. If limited mixing between the upper and lower mantle occurs, the differing exposure of the shallow and deep mantle to processes that generate isotopic heterogeneities must, in turn, also produce little or no major element heterogeneity.

The solution of the issue of mantle layering through a comparison of mineral physics and seismic data continues to hinge on a variety of as-yet poorly resolved parameters, including the bulk modulus of aluminous perovskite and its pressure derivative [e.g., *Zhang and Weidner, 1999; Andrault et al., 2001; Jackson et al., 2004; Walter et al., 2004*], and the precise values of pressure–temperature cross-derivatives of elastic moduli and thermal expansion. Indeed, as noted by *Deschamps and Trampert* [2004], the issue of the effect of aluminum on the bulk modulus of silicate perovskite is of critical importance in assessing the degree of chemical stratification of Earth's mantle. Whether ~5 wt% Al_2O_3 dissolved into perovskite (near the value likely present within Earth's lower mantle) weakly increases the bulk modulus [*Andrault et al., 2001*] of the perovskite phase, decreases it [*Zhang and Weidner, 1999; Walter et al., 2004*], or has little effect [*Jackson et al., 2004*] remains unclear. Substantial shifts in the elasticity of the major phase of the lower mantle, induced by defect substitution, would produce a shift in both the preferred and permissible compositions within the lower mantle, with the principal impact being on the amount of magnesiowüstite (and thus the deep mantle MgO/SiO_2 ratio) inferred to be present at depth.

The key observation is that the issue of mantle stratification is not yet resolved and, as discussed by *Javoy* [1999], this ambiguity in lower mantle composition contributes a primary uncertainty to not only the composition of the mantle, but also to that of the Earth. Indeed, because the size of the lower mantle dominates the major element chemistry of the silicate portion of the planet, rather modest shifts in the ratio of $(Mg,Fe)(Si,Al)O_3$-perovskite to $(Mg,Fe)O$-magnesiowüstite in the lower mantle can, for example, produce significant changes in the planetary Mg/Si ratio. It has long been recognized that the Mg/Si ratios of the shallow and deep mantle could be altered by processes related to early, extensive melting of the planet: Both olivine flotation and magnesium perovskite fractionation could each deplete the shallow mantle in silicon [e.g., Agee and Walker, 1988]. The precise geochemical role of deeper magmatic segregation, through either melt descent or partial melting of the mantle at depths greater than ~750 km, has remained substantially more obscure.

Here, we review the constraints and uncertainties on the major element chemistry of Earth's mantle. We emphasize major element constraints in preference to trace element constraints simply because the partition coefficients of almost all trace elements are poorly known at deep Earth pressure and temperature conditions. Indeed, possible deep mantle polymorphism of silicate perovskite, the unknown importance of magmatic differentiation in the deep mantle [*Williams and Garnero*, 1996; *Lay et al.,* 2004; *Rost et al.,* 2005], and the ill-constrained minor mineral chemistry of the deep mantle each render the utility of trace elements in inferring the bulk chemistry of the planet potentially prone to misadventure. For example, the respective roles in trace element partitioning of near-solidus phases such as $CaSiO_3$-perovskite; possible couplings between the different substitutional mechanism of aluminum and possibly ferric iron in magnesium silicate perovskite with partitioning behavior [*Walter et al.,* 2004]; the efficiency and dynamics of mineral-melt segregation processes within deep portions of an early, partially molten mantle [*Solomatov and Stevenson,* 1993a]; the efficiency of subsequent convective mixing; and the possible sequestration of incompatible elements in the lowermost mantle each have ill-constrained effects on the ultimate trace element signature of deep magmatic processes. The degree to which such poorly understood processes have (or have not) participated in the production of an upper mantle with a generally chondritic trace element signature remains entirely conjectural. Indeed, the likely importance and persistence of magmatic processes in the deep mantle, coupled with possible continued core interaction, make the usage

of trace element constraints on bulk mantle composition potentially similar to attempting to infer the trace element characteristics of the planet without considering the presence of the continental crust. The difficulties of using trace element constraints on magmatic differentiation between the deep and shallow mantle are illustrated by the recognition that differentiation of a magma ocean almost certainly will not proceed by simple fractional crystallization [*Solomatov and Stevenson,* 1993b]; this has not, however, prevented such simple fractional crystallization models from being applied to the problem of mantle differentiation and chemical evolution [e.g., *Corgne and Wood,* 2002]. Moreover, it is well worth noting that the available information on mineral/melt partitioning of trace elements is mostly confined to pressures less than 30 GPa and temperatures of less than 2800 K [e.g., *Kato et al.,* 1988; *McFarlane et al.,* 1994; *Knittle,* 1998; *Hirose et al.,* 2004]—far short of the 135 GPa and ~4000 K conditions present at the base of the mantle, and perhaps not even on the mineral phases of principal relevance to the deeper portions of the mantle [*Shim et al.,* 2002; *Murakami et al.,* 2004]. The situation with major elements is not much better, but at least there are some constraints that allow the effects of different processes during planetary differentiation to be evaluated [*Ito et al.,* 2004; *Asahara et al.,* 2004]. Our approach thus follows on earlier mass balance approaches to mantle composition that utilized large-scale melting at depth (often via a magma ocean) coupled with fractionation of major mineral phases to generate upper mantle peridotite [e.g., *Ohtani,* 1985, 1988; *Takahashi,* 1986; *Herzberg et al.,* 1988; *Agee and Walker,* 1988]. Our intent is to reevaluate the possible role played by deep magmatic effects in mantle composition, in light of both the recent appreciation of the complex major element chemistry of deep mantle minerals and the improved dataset on high-pressure melt chemistry.

COSMOCHEMICALLY DERIVED COMPOSITIONS

The idea underpinning cosmochemical constraints on major element Earth composition is straightforward: MgO, SiO_2, Al_2O_3, CaO, and FeO collectively comprise ~99% of the weight percentage of the oxidized Earth, with the balance being primarily sodium, manganese, chromium, nickel and titanium. So, one can simply write [e.g., *Palme and O'Neill,* 2003]:

$$wt\% \, MgO + wt\% \, SiO_2 + wt\% \, Al_2O_3 + wt\% \, CaO + wt\% \, FeO = 98.5 \qquad (1)$$

If one can impose a priori constraints on four abundance ratios of these oxides (say, the Si/Mg, Ca/Mg, Al/Mg, and Fe/Mg ratios) from solar or meteoritic chemistries, the problem of mantle composition is not simply straightforward; it is trivial. However, several uncertainties emerge in applying equation (1). The first of these is the widely recognized issue of core chemistry: The iron abundance in the core is reasonably well-constrained, but whether the core abundantly sequesters any of the other cations present in equation (1) is unclear. The most widely proposed cation in equation (1) that might reside in the core is silicon (although its presence in the outer core is controversial, as the elastic properties of iron–silicon alloys may not accurately reproduce the pressure dependence of the bulk sound velocity in the outer core [*Williams and Knittle,* 1997; *Hirao et al.,* 2004]. The relationship between iron alloy elasticity and the chemical composition of the outer core is, however, controversial [*Sanloup et al.,* 2004]. Moreover, the oxygen fugacities required to incorporate significant quantities of silicon in the core are extremely reducing: For example, conditions 2 log units more reducing than the iron-wüstite buffer generate only ~0.1 wt% Si in iron in equilibrium with silicates at 2200°C [*Malavergne et al.,* 2004; *Gessmann et al.,* 2001]. It has also been proposed that aluminum might be incorporated into the core via core-mantle chemical reactions, but this suggestion is largely unquantified [*Dubrovinsky et al.,* 2001]. The second (and usually dismissed: *Palme and O'Neill,* 2003) uncertainty in equation (1) lies in the assumption that iron is completely divalent within Earth's mantle. This assumption is not fully correct even within Earth's upper mantle, as the formation of Fe^{3+} at a range of oxygen fugacities (such as in the presence of metallic iron) has been known for decades (e.g., *Bowen and Schairer,* 1932). Within lower mantle phases, significant trivalent iron exists under reducing conditions, as well [e.g., *McCammon et al.,* 2004]. Indeed, characteristic amounts of trivalent iron within aluminous perovskite are near 30% of total iron, and the trivalent iron within magnesiowüstite is near 10%. Accordingly, equation (1) should be expressed as

$$wt\% \ MgO + wt\% \ SiO_2 + wt\% \ Al_2O_3 + wt.\% \ CaO + \{1 + [Fe^{3+}/(Fe^{2+}+ Fe^{3+})](0.11)\}wt.\% \ FeO = 98.5 \quad (2)$$

Here, 0.11 represents the weight correction between trivalent oxidized iron and divalent oxidized iron. The key point about equation (2) is that the oxygen content of the deep mantle is less certain than it might appear from a typical oxide mass balance: This uncertainty in oxygen is independent of whether one expresses equation (2) in terms of atomic percentages [e.g., *Jagoutz et al.,* 1979] or oxide weight percentages [e.g., *Palme and O'Neill,* 2003]. The magnitude of the uncertainty in oxygen content is small, but not negligible—on the order of a few tenths of a weight percent—for iron oxide contents near 7 wt%. Indeed, when this is coupled

with the large uncertainty in the oxygen content of the planet's core, the uncertainty in the oxygen content of the planet expands to several weight percent of the planet's mass.

The third often-disregarded uncertainty involves volatiles: Water is possibly present in the lower mantle at the ~0.2 wt% level or higher [*Litasov et al.,* 2003; *Murakami et al.,* 2002; *Williams and Hemley,* 2001], while the abundance of carbon is largely unconstrained. Clearly, each of these cosmochemically highly abundant, yet highly volatile, elements becomes substantially less volatile at high pressures: If they can be retained in abundance in the deep Earth through the processes of accretion, magma ocean circulation, and convection-associated degassing, then the right-hand side of the oxide mass balance equations (1) and (2) could be decreased by ~1%. As an aside, both our iron valence-state–induced modification in equation (2) and the inclusion of volatiles will not impact the ratios of major elements to one another, with the important exception of those involving iron. Both modifications do, however, shift the inferred absolute quantities of these oxide components, and thus the estimated bulk chemical composition of the planet.

Ignoring these uncertainties, the difficulty with applying the purely cosmochemical approach to Earth composition is first-order: The Mg/Si ratios derived from estimates of Earth's upper mantle composition do not agree with those of most likely primitive constituents of the Earth. In particular, the Mg/Si ratio of the upper mantle substantially exceeds that of the CI chondrites [e.g., *Hart and Zindler,* 1986]. Confronted with this mismatch, the response has been generally to alter the chondritic ratio either through selective volatilization during accretion [*McDonough and Sun,* 1995] or through partitioning of silicon into the core [*Allègre et al.,* 1995]. Alternatively, the Earth has been proposed to have formed from a different family of meteorites than is represented in our modern meteoritic record—a so-called Earth chondrite or Earth achondrite [*Drake and Righter,* 2002]. A generalized description of a frequently used algorithm for determining bulk-silicate Earth composition is shown in Figure 1 [*McDonough and Sun,* 1995; *Allègre et al.,* 1995; *Drake and Righter,* 2002]. Variants of this approach do exist, but are often focused on the minor elements [*Helffrich and Wood,* 2001; *Meibom and Anderson,* 2004]. For example, to bring the radiogenic element content (K, U, Th) into accord with a bulk silicate Earth composition, ubiquitous salting of the mantle with the dregs of ancient subduction, including a component of recycled continental material, has been proposed [*Helffrich and Wood,* 2001]. The idea that extensive long-term storage of subducted material occurs at depth has long been a result of isotopic inversions of mantle evolution [e.g., *Chase and Patchett,* 1988]. However, the relationship between such isotopic and/or trace element–based

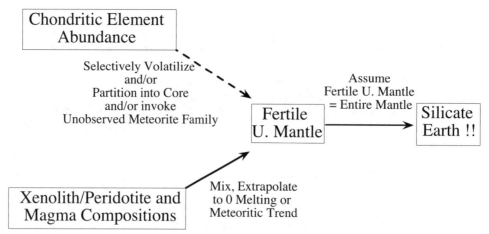

Figure 1. Flow chart of a commonly used geochemical algorithm for estimating bulk mantle composition. The dotted line indicates that the process(es) linking the two compositions is itself unconstrained. To explain deviations between features such as the chondritic Mg/Si ratio and that of the upper (U) mantle, *Allègre et al.* [1995] invoke silicon partitioning into Earth's core; *McDonough and Sun* [1995] prefer selective volatilization; and *Righter and Drake* [2002] view Earth composition as being indicative of planetary accretion from a meteorite family that has not been observed.

approaches to mantle composition and variations in, and depth dependence of, major element composition is often unclear. Moreover, the idea of distributing a calculated quantity of subducted material throughout the mantle is similar in underlying concept to the algorithm of Figure 1—namely, that the sole process responsible for mantle geochemical heterogeneity is the segregation of Earth's crust, particularly basalt petrogenesis. For comparison, *Anderson* [1983, 1989] coupled a combination of chondritic ratios of refractory elements with the compositions of five petrologic/geochemical components to solve for the relative abundances of different components: His regression yielded a distinctly compositionally zoned planet with a perovskite-enriched lower mantle.

In the following sections of this paper, we examine the implications for the major element composition of the mantle if the two families of assumptions shown in the flowchart of Figure 1 are discarded. In particular, if the assumption that the composition of fertile upper mantle is equal to that of the entire mantle is incorrect, we can consider solutions for bulk mantle chemistry that are compatible with geophysical inversions but do not require a chemically homogeneous mantle. The question we test then becomes: How closely might the Earth approximate an iron-depleted (due to core segregation) chondritic composition, thus negating the need for the suite of possible accretion/differentiation–induced alterations to a chondritic composition? Moreover, the issue of what geochemical processes might give rise to a silicon-enriched deep mantle and a Si-depleted shallower mantle is worth reconsidering as more experimental evidence is garnered on the chemical effects of melting at high pressures [e.g., *Ito et al.*, 2004]. Whether such differentiation

could have occurred during accretion/planetary growth, or through subsequent high-pressure magmatic alteration during early Earth history, is unknown. In this respect, there are provocative geochemical indications, derived from inverting for the likely residua of high-Mg melt formation in the Archean, that the upper mantle may have become progressively depleted in iron and silicon over geologic time [*Francis*, 2003]; metasomatism might, however, play a role in generating these ancient samples [*Carlson et al.*, 2005]. While the idea of a near-chondritic or perovskite-only lower mantle is certainly not new [e.g., *Anderson and Bass*, 1986; *Ohtani et al.*, 1986; *Agee and Walker*, 1988; *Anderson*, 1989; *Jeanloz and Knittle*, 1989; *Stixrude et al.*, 1992], the concept has often (with exceptions: e.g., *Javoy*, 1999; *Burbine and O'Brien*, 2004) been summarily dismissed in inversions for bulk Earth composition [*Allègre et al.*, 1995; *McDonough and Sun*, 1995; *Drake and Righter*, 2002].

SHALLOW MANTLE COMPOSITION

The assumption of mantle homogeneity that is often used in constructing bulk Earth compositional models may be stated another way: The only process that has altered the major element chemistry of geochemically significant portions of the mantle is basalt petrogenesis. Figures 2 and 3 illustrate this effect through well-known major element igneous fractionation trends, defined in this instance between the end-members of highly depleted harzburgitic mantle and basalt. These serve to illustrate the strength of Ringwood's concept of pyrolite: The primordial mantle composition of PRIMA [*Allègre et al.*, 1995], a pyrolitic fer-

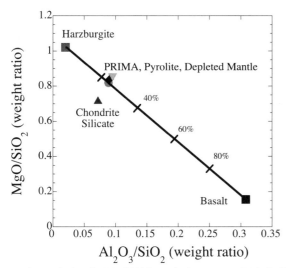

Figure 2. Variation in MgO/SiO_2 ratio in terms of Al_2O_3/SiO_2 ratio. The bold line is the terrestrial magmatic fractionation trend. Harzburgite end-member composition is from *Danchin* [1979], while the basalt end-member is from *Pearce* [1976]. The chondritic silicate composition is from *Hart and Zindler* [1986] (see Table 1). Compositions of PRIMA (circle), pyrolite (diamond), and depleted mantle (inverted triangle) are from *Allègre et al.* [1995], *McDonough and Sun* [1995], and *Salters and Stracke* [2004], respectively.

tile mantle composition [*McDonough and Sun*, 1995], and even an average "depleted mantle" composition (defined as a mantle capable of generating MORB, without enriched components [*Salters and Stracke*, 2004]) are each in close accord with one another. In effect, the extrapolation of peridotite compositions to zero melting/no basalt extraction (or, in the case of some elements/studies, the same intent is accomplished by determining the intersection of the peridotitic trend with the meteoritic trend) is quite robust and yields a highly reproducible composition. Treating this composition as the bulk mantle composition thus produces a highly precise, but possibly inaccurate, measure of the composition of the silicate Earth. The precision with which primitive mantle/pyrolitic compositions have been reported varies widely from 1% or better in the case of the major elements [*Palme and O'Neill*, 2003; *Allègre et al.*, 1995], to a "subjective" but, as we show below, likely realistic estimate of 10% uncertainty in the different major element abundances by *McDonough and Sun* [1995]. For comparison with the primordial/upper mantle-derived compositions, the chondrite mantle composition of *Hart and Zindler* [1986] is poorer in magnesium (Figure 2) but lies close to the terrestrial fractionation trend for calcium and aluminum contents and near the primitive upper mantle compositions (Figure 3).

To assess the effect of a non-PRIMA/pyrolite lower mantle on bulk Earth composition, we create a hypothetical deep mantle of pure perovskite composition. In doing so, we draw an important distinction between describing this material as a deep mantle rather than as a lower mantle composition: The latter has a well-defined depth extent, while the former does not. The composition we choose has a chemistry of $(Mg_{0.88},Fe_{0.12})SiO_3$, after the best-fit model of *Stixrude et al.* [1992], with the addition of chondritic/upper mantle abundances of CaO and Al_2O_3 of approximately 3 and 4 wt%, respectively (Table 1). The implicit rationale for choosing these weight ratios of calcium and aluminum oxides is that the chondritic and the upper mantle compositional estimates are generally similar in their contents of these elements. This does not imply that these ratios are correct; rather, they simply represent a justifiable (albeit ad hoc) estimate. We have not attempted to optimize the calcium and aluminum contents to make our hypothetical deep mantle composition fit into a complementary relationship with upper mantle and chondritic compositions. However, as is shown below, the precise amounts of calcium and aluminum within the deep mantle composition become critically important for assessing whether such complementary relationships exist: In this sense, our ad hoc estimates function as a starting point for evaluating how sensitive any possible complementary relationships are to the deep mantle Ca and Al content. From a mineralogic standpoint, the Al_2O_3 content of our hypothetical deep mantle composition could fully dissolve into single-phase magnesium silicate perovskite [e.g., *Andrault et al.*, 2001; *Ito et al.*, 2004; *Walter et al.*, 2004]. The amount of

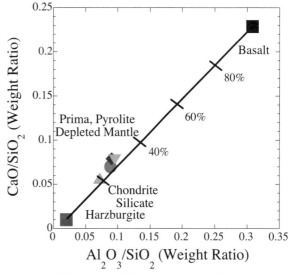

Figure 3. Variation in CaO/SiO_2 ratio in terms of Al_2O_3/SiO_2 ratio. The bold line is the terrestrial magmatic fractionation trend. Data are from the same sources as in Figure 2.

Table 1. Major Element Chemistries of Different Possible Model Mantle Compositions.

Wt. %	PRIMA	Chondrite silicate	Pyrolite	Depleted mantle	Pv-only lower mantle	Pv+PRIMA 670 km bdry	Pv+PRIMA 1000 km bdry
Al_2O_3	4.09	3.56	4.09	4.28	4.00	4.02	4.04
CaO	3.23	2.82	3.55	3.50	3.00	3.06	3.09
MgO	37.77	35.68	37.80	38.22	31.18	32.93	33.80
SiO_2	46.10	49.52	45.00	44.90	52.83	51.05	50.17
FeO	7.49	7.14	8.05	8.07	7.58	7.56	7.54

PRIMA composition is from *Allegre et al.* [1995]; Chondrite silicate is from *Hart and Zindler* [1986]; Pyrolite is from *McDonough and Sun* [1995]; Depleted mantle is from *Salters and Stracke* [2004]; Pv + PRIMA columns report a mixture of a perovskite-only lower mantle with a PRIMA composition, with the chemical boundary (bdry) lying at 670 and 1000 km, respectively.

calcium is likely modestly in excess of that observed in magnesium silicate perovskite [*Trønnes*, 2000; *Ito et al.*, 2004], and a small percentage of $CaSiO_3$-perovskite might thus also be present in our "perovskite-only" composition. A weight fraction of 1.4 wt% in the hypothetical pure perovskite deep mantle is consigned to other components, particularly sodium, chromium, manganese, nickel, and titanium oxides. Such minor element levels are in general accord with both the chondritic and upper mantle compositional models. The mixing lines between this lower mantle composition and the suite of upper mantle compositions are shown in Figures 4. The interesting aspect of Figure 4 is that the estimated bulk composition of the mantle depends crucially on where the

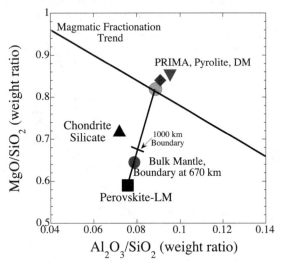

Figure 4. Compositional relationship between a possible perovskite-only deep mantle (Table 1), chondrite silicate, and upper mantle compositions for the MgO/SiO_2 ratio relative to the Al_2O_3/SiO_2 ratio. The bold line shows the magmatic fractionation trend. Average bulk silicate Earth compositions are shown for possible chemical boundaries between shallow mantle compositions (represented by PRIMA) and a deep, perovskite-only mantle at depths of 670 km (lower circle) and 1000 km depth (bar). Notably, the closest approach to the Mg–Si ratio of a chondritic silicate occurs for a compositional boundary near 1500 km in depth.

boundary lies between a perovskite-enriched deep mantle and the PRIMA/pyrolite shallow mantle. If a chemical boundary were present in the 1000–1500-km depth range [*Kellogg et al.*, 1999; *Anderson*, 2002], the bulk Mg/Si ratio of the mantle would rather closely match that of a chondritic chemistry mantle. Thus, a rather deeply chemically layered mantle could produce a major element chemistry that closely simulates that of CI chondrite silicate. Notably, the match between chondritic chemistry and the deep mantle–shallow mantle trend of Figure 4 would have been perfect had we assumed a slightly lower Al_2O_3 content (by ~1 wt%) for our perovskite-enriched lower mantle, rather than an intermediate value between the rather close alumina contents of the upper mantle and chondritic compositions. Figure 5 emphasizes the comparative closeness of the calcium and aluminum contents of the different mantle chemistries and their resulting sensitivity to small variations in the assumed initial composition of the deep mantle. In this expanded scale view, the inferred bulk mantle compositions differ little from either those of the upper mantle assemblages or those of the chondrite or deep mantle perovskite. Yet, as with Figure 4, a 1 wt% lower Al_2O_3 content of the deep mantle reservoir (corresponding to an Al_2O_3/SiO_2 ratio near 0.06) would have produced a trend significantly closer to chondritic.

The key question is whether there exists a means through which the upper mantle compositions and the deep mantle, geophysically devised chemistry of Figures 4 and 5 could arise as complements to one another? One possible mechanism is that the elevated Mg/Si ratio of the upper mantle may have been generated by olivine flotation within a primordial magma ocean [e.g., *Ohtani*, 1985, 1988; *Agee and Walker*, 1988; *Anderson*, 2002]. Moreover, characterizations of the chemistry of melts (and coexisting solids) produced by melting of chondritic materials at deep transition zone and shallow upper mantle depths have recently been conducted [*Asahara et al.*, 2004; *Ito et al.*, 2004]. Figure 6 shows a sequence of compositional vectors that may bear on the genesis of any shallow-/deep-mantle dichotomy. Clearly, olivine addition, while highly effective at raising the Mg/Si ratio, has

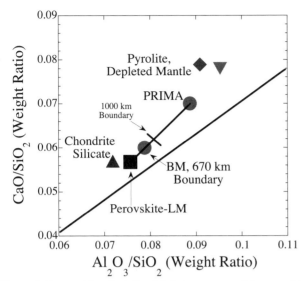

Figure 5. Compositional relationship between a possible perovskite-only deep mantle (Table 1), chondrite silicate, and upper mantle compositions for the CaO/SiO$_2$ ratio relative to the Al$_2$O$_3$/SiO$_2$ ratio. The bold line is the magmatic fractionation trend. Note the markedly expanded scale relative to Figure 4. Different possible bulk silicate Earth compositions are shown for chemical boundaries occurring at 670 and 1000 km, as in Figure 4.

a dilutional effect on the aluminum content of the shallow mantle; indeed, its effect is almost precisely identical to that of segregating aluminous perovskite into the deep mantle. Magnesiowüstite addition or subtraction is not shown on Figure 6, as its effect is simply a vertical translation: We do not preclude magnesiowüstite from playing a role in the early geochemical evolution of the mantle, as there are indications that magnesiowüstite either is the liquidus phase of chondritic compositions at shallow lower mantle depths [*Agee et al.,* 1995] or may simply lie close to the liquidus [*Trønnes,* 2000]. However, the dependence of the liquidus phase on composition and pressure, and the manner in which the liquidus phase may (or may not) preferentially differentiate in a magma ocean [*Solomatov and Stevenson,* 1993], are each complex issues. The key point illustrated by Figure 6 is that not only is the upper mantle enriched in magnesium relative to chondritic or perovskititic compositions, but also its ratio of alumina/silica is higher. Thus, if a complementary relationship exists, the upper mantle requires enrichment in magnesium coupled with depletion in silica and/or enrichment in alumina (Table 1).

The chemistries of magmas observed in the charges of *Asahara et al.* [2004] and *Ito et al.* [2004] are of particular interest. The melt evolved by *Asahara et al.* at 25 GPa is essentially indistinguishable from the terrestrial fractionation trend: Its chemical effect would be nearly identical to that

produced by basalt petrogenesis. In contrast, the results of *Ito et al.* [2004] show a considerable effect of pressure on melt chemistry, and the compositional vector is rotated towards upper mantle compositions. From a historical viewpoint, pressure-induced shifts in mantle-melting relations have long been viewed as important for the genesis of the upper mantle [e.g., *Herzberg and O'Hara,* 1985; *Walker,* 1986]. As *Ito et al.* [2004] note, their results provide indications that higher pressures could produce a complementary relationship between a chondritic source and the fertile upper mantle [*Ito et al.,* 2004]. Indeed, further pressure-induced rotation of the magma compositional vector in Figure 6 could produce a complementary relationship with a slightly alumina-depleted or magnesiowüstite-enriched perovskite-dominated lower mantle. An alternative view is that a combination of olivine

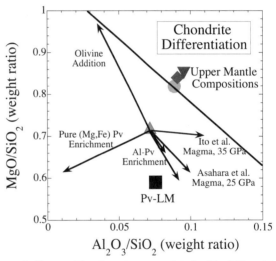

Figure 6. Compositional vectors associated with different types of fractionation processes, relative to a chondritic initial composition for the MgO/SiO$_2$ ratio relative to that of Al$_2$O$_3$/SiO$_2$. The bold line is the magmatic fractionation trend. Clearly, a modest amount of dissolution of Al$_2$O$_3$ into perovskite markedly alters the compositional trend associated with perovskite enrichment: The Al-pv vectors, derived from the experiments of *Ito et al.* [2004] (the shorter vector) and those of *Asahara et al.* [2004] (longer vector), represent the chemistry of perovskite in equilibrium with magma. The magmatic vectors are also derived from the studies of *Ito et al.* [2004] and *Asahara et al.* [2004] chondritic compositions at high pressures. The olivine addition and perovskite enrichment arrows simply represent directions; their endpoints lie off of the plot. Notably, each vector can be reversed to represent depletion of a given component. If the deep mantle region did, in fact, contain modest amounts of magnesiowüstite (less than required to generate a pyrolitic composition), its composition would be translated markedly towards that of the chondritic silicate and would more closely approach a simple fractionation trend with the upper mantle samples. This reflects that any magnesiowüstite fractionation would translate compositions vertically on this plot.

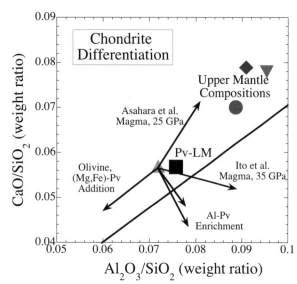

Figure 7. Compositional vectors associated with different types of fractionation processes, relative to a chondritic initial composition for the CaO/SiO$_2$ ratio relative to that of Al$_2$O$_3$/SiO$_2$. The bold line is the magmatic fractionation trend. For these ratios, the effect of olivine or (Mg,Fe)perovskite segregation is purely dilutional: The principal effect driving enrichment in these elements is likely to be associated with melting. In this instance, evolving upper mantle compositions from a chondritic precursor could be associated with extensive polybaric melting.

flotation, melting derived from shallow lower mantle depths, and associated perovskite-enrichment in the deep mantle within an Earth with close-to-chondritic major element abundances could generate the observed chemistry of the upper mantle. The key issue that emerges in such a threefold scenario is the aluminum content of the residual perovskite: Variations in the Al-content of perovskite produce a dramatic rotation in the compositional vectors associated with Al-perovskite depletion (and enrichment: Figures 6 and 7).

Figure 7 shows the corresponding set of vectors for the calcium/silica and aluminum/silica ratios. In the case of these less abundant elements, olivine addition has a purely dilutional effect, and the importance of olivine flotation for Ca and Al abundance in the upper mantle is minor relative to the role of added melts. In this instance, polybaric melting in the shallow lower mantle could be sufficient to generate the enhanced Ca and Al ratios of the upper mantle although, as with Figure 6, depletion in Al-poor perovskite may play a major role. Notably, low-pressure melting phenomena, as illustrated by the low-pressure fractionation trends on Figures 6 and 7, appear completely unable to generate any complementary relationship between chondritic compositions, our fictive deep mantle composition, and the shallow mantle compositions. The critical point of Figures 6 and 7

is thus that high pressure melting relations may bear little resemblance to those at low pressures, and inferring the bulk chemistry of the planet (or whether or not complementary relationships exist between major element reservoirs) solely from low-pressure phase relations may be fraught with misadventure.

CONCLUSIONS

The major element composition of the mantle is surprisingly poorly constrained: Precise estimates of the bulk composition exist, but their accuracy hinges on the assumption that the mantle is isochemical. The net possible variability in the mantle's MgO and SiO$_2$ contents are near 5 wt%—far in excess of the uncertainties associated with some inversions for primitive mantle composition [e.g., *Allègre et al.*, 1995; *Palme and O'Neill*, 2003]. A complementary relationship between fertile shallow mantle compositions and a perovskite-rich deep mantle may exist [e.g., *Ohtani*, 1985; *Agee and Walker*, 1988; *Anderson*, 1989]: the net bulk composition of the mantle could lie close to chondritic values. From a petrologic standpoint, the compositional effect of melting on mantle stratification hinges on the rate and degree to which magma compositional vectors rotate between pressures of 25 and 35 GPa (and at higher pressures). Moreover, the precise calcium and aluminum content of residual solids is critical for determining what (if any) role is played by perovskite fractionation in the major element chemistry of the upper mantle. Ultimately, determining whether such a complementary relationship exists hinges on both substantial improvements in the understanding of the chemistry of deep mantle melting processes, coupled with dynamical insights into how magmatic differentiation processes occur in both an early, extensively melted mantle and within the deeper reaches of the mantle throughout Earth history. Finally, the complicated oxidation state of iron and the volatile content in the deep mantle each provide uncertainties in bulk mantle composition that could approach 1 wt%.

Acknowledgments. We thank D.L. Anderson, K. Righter, D. Walker, and three anonymous reviewers for thorough comments that greatly improved the manuscript. Work supported by the U.S. National Science Foundation.

REFERENCES

Agee, C. B., and D. Walker, Mass balance and phase density constraints on early differentiation of chondritic mantle, Earth Planet. Sci. Lett., 90, 144–156, 1988.

Agee, C. B., J. Li, M. C. Shannon, and S. Circone, Pressure-temperature phase diagram for the Allende meteorite, *J. Geophys. Res., 100*, 17225–17740, 1995.

Allègre, C. J., Limitation on the mass exchange between the upper and the lower mantle: The evolving convection regime of the Earth, *Earth Planet. Sci. Lett., 150,* 1–6, 1997.

Allègre, C., G. Manhes, and E. Lewin, Chemical composition of the Earth and the volatility control on planetary genetics, *Earth Planet. Sci. Lett., 185,* 49–69, 2001.

Allègre, C. J., J. P. Poirier, E. Humler, and A. W. Hofmann, The chemical composition of the Earth, *Earth Planet. Sci. Lett., 134,* 515–526, 1995.

Allègre, C. J., and D. L. Turcotte, Implications of a two-component marble-cake mantle, *Nature, 323,* 123–127, 1986.

Anderson, D. L., The chemical composition and evolution of the mantle, in *High Pressure Research in Geophysics,* eds. S. Akimoto and M.H. Manghnani, pp. 301–318, Center for Academic Publishing, Tokyo, 1982.

Anderson, D. L., Chemical composition of the mantle, *J. Geophys. Res. (suppl.), 88,* B41–B52, 1983.

Anderson, D. L., The case for irreversible chemical stratification of the mantle, *Int. Geol. Rev., 44,* 97–116, 2002.

Anderson, D. L., *Theory of the Earth,* Blackwell Scientific Publications, Boston, 1989.

Anderson, D. L. and J. D. Bass, Transition region of the Earth's upper mantle, *Nature, 320,* 321–328, 1986.

Andrault, D., N. Bolfan-Casanova and N. Guignot, Equation of state of lower mantle (Al,Fe)-$MgSiO_3$ perovskite, *Earth Planet. Sci. Lett., 193,* 501–508, 2001.

Asahara, Y., T. Kubo, and T. Kondo, Phase relations of a carbonaceous chondrite at lower mantle conditions, *Phys. Earth Planet. Inter., 143–144,* 421–432, 2004.

Bina, C. R., Seismological constraints upon mantle composition, in *Treatise on Geochemistry, Vol. 2,* R. Carlson, Ed., Elsevier, Amsterdam, 2003.

Bowen, N. L., and J. F. Schairer, The system, $FeO-SiO_2$, *Am. J. Sci., 141,* 177–213, , 1932.

Burbine, T. H. and K. M. O'Brien, Determining the possible building blocks of the Earth and Mars, *Meteorit. Planet. Sci., 39,* 667–681, 2004.

Cammarano, F., S. Goes, A. Deuss, and D. Giardini, Is a pyrolitic adiabatic mantle compatible with seismic data?, *Earth Planet. Sci. Lett., 232,* 227–243, 2005.

Carlson, R. W., D. G. Pearson, and D. E. Jones, Physical, chemical, and chronological characteristics of continental mantle, *Rev. Geophys., 43,* RG1001, doi:10.1029/2004RG000156, 2005.

Chambers, J. E., and G. W. Wetherill, Making the terrestrial planets: N-body integrations of planetary embryos in three dimensions, *Icarus, 136,* 304–327, 1998.

Chase, C. G., and P. J. Patchett, Stored mafic/ultramafic crust and early Archean mantle depletion, *Earth Planet. Sci. Lett., 91,* 66–72, 1988.

Corgne, A., and B. J. Wood, $CaSiO_3$ and $CaTiO_3$ perovskite-melt partitioning of trace elements: Implications for gross mantle differentiation, *Geophys. Res. Lett., 29,* 1933–1937, 2002.

Danchin, R. V., Mineral and bulk chemistry of garnet lherzolite and garnet harzburgite xenoliths from the Premier Mine, South Africa, in *Proc. of the 2nd International Kimberlite Conference,* F. R. Boyd, H. O. A. Meyer, Eds., pp. 104–126, AGU, Washington, D.C., 1979.

Deschamps, F., and J. Trampert, Towards a lower mantle reference temperature and composition, *Earth Planet. Sci. Lett., 222,* 161–175, 2004.

Drake, M. J. and K. Righter, Determining the composition of the Earth, *Nature, 416,* 39–44, 2002.

Dubrovinsky, L., H. Annerstin, N. Dubrovinskaia, F. Westman, H. Harryson, O. Fabrichnaya and S. Carlson, Chemical interaction of Fe and Al_2O_3 as a source of heterogeneity at the Earth's core-mantle boundary, *Nature, 412,* 527–529, 2001.

Francis, D., Cratonic mantle roots, remnants of a more chondritic Archean mantle?, *Lithos, 71,* 135–152, 2003.

Fukao, Y., S. Widiyantoro and M. Obayashi, Stagnant slabs in the upper and lower mantle transition region, *Rev. Geophys., 39,* 291–323, 2001.

Gessmann, C. K., B. J. Wood, D. C. Rubie and M. R. Kilburn, Solubility of silicon in liquid metal at high pressure: Implications for the composition of the Earth's core, *Earth Planet. Sci. Lett., 184,* 367–376, 2001.

Hart, S. R. and A. Zindler, In search of a bulk-Earth composition, *Chem. Geol., 57,* 247–267, 1986.

Helffrich, G. R. and B. J. Wood, The Earth's mantle, *Nature, 412,* 501–507, 2001.

Herzberg, C. T., M. Feigenson, C. Skuba and E. Ohtani, Majorite fractionation in early mantle differentiation, *Nature, 322,* 823–826, 1988.

Herzberg, C. T., and M. J. O'Hara, Origin of mantle peridotite and komatiite by partial melting, *Geophys. Res. Lett., 12,* 541–544, 1985.

Hirao, N., E. Ohtani, T. Kondo and T. Kikegawa, Equation of state of iron-silicon alloys to megabar pressures, *Phys. Chem. Minerals, 31,* 329–336, 2004.

Hirose, K., N. Shimizu, W. van Westrenen, and Y. Fei, Trace element partitioning in Earth's lower mantle and implications for geochemical consequences of partial melting at the core-mantle boundary, *Phys. Earth Planet. Inter., 146,* 249–260, 2004.

Ito, E. A. Kubo, T. Katsura and M. J. Walter, Melting experiments of mantle materials under lower mantle conditions with implications for magma ocean differentiation, *Phys. Earth Planet. Inter., 143–144,* 397–406, 2004.

Jackson, I. and S. M. Rigden, Composition and temperature of the Earth's mantle: Seismological models interpreted through experimental studies of Earth materials, in *The Earth's Mantle: Structure, Composition and Evolution—The Ringwood Volume,* I. Jackson, Ed., Cambridge U. Press, Cambridge, pp. 405–460, 1998.

Jackson, J. M., J. Zhang and J. D. Bass, Sound velocities and elasticity of aluminous $MgSiO_3$ perovskite: Implications for aluminum heterogeneity in Earth's lower mantle, *Geophys. Res. Lett., 31,* L10614, doi:10.1029/2004GL019918, 2004.

Jagoutz, E., H. Palme, H. Baddenhausen, K. Blum, M. Cendales, G. Dreibus, B. Spettel,, V. Lorenz, and H. Wanke, The abundances of major, minor and trace elements in the Earth's mantle as derived from primitive ultramafic nodules, *Proc. 10th Lunar Planet. Sci. Conf.,* 2031–2050, 1979.

Javoy, M., The integral enstatite chondrite model of the Earth, *Geophys. Res. Lett., 22,* 2219–2222, 1995.

Javoy, M., Chemical Earth models, *C.R. Acad. Sci. Paris, 329,* 537–555, 1999.

Jeanloz, R. and E. Knittle, Reduction of mantle and core properties to a standard state by adiabatic decompression, *Adv. Phys. Geochem., 6,* 275–309, 1986.

Jeanloz, R. and E. Knittle, Density and composition of the lower mantle, *Phil. Trans. R. Soc. Lond. A, 328,* 377–389, 1989.

Jones, J. H., Chondrite models for the composition of Earth's mantle and core, Phil, Trans. R. Soc. Lond. A354, 1481–1494,1996.

Karki, B. B. and L. Stixrude, Seismic velocities of major silicate and oxide phases of the lower mantle, *J. Geophys. Res., 104,* 13025–13033, 1999.

Kato, T., A. E. Ringwood, and T. Irifune, Experimental determination of element partitioning between silicate perovskites, garnets and liquids: Constraints on early differentiation of the mantle, *Earth Planet. Sci. Lett., 89,* 123–145, 1988.

Kellogg, L., B. H. Hager and R. D. van der Hilst, Compositional stratification in the deep mantle, *Science, 283,* 1881–1884, 1999.

Knittle, E., The solid/liquid partitioning of major and radiogenic elements at lower mantle pressures: Implications for the core-mantle boundary region, in *The Core Mantle Boundary Region,* Eds. M. Gurnis, M.E. Wysession, E. Knittle and B. Buffett, AGU, Washington, D.C., pp. 119–130, 1998.

Lay, T., E. Garnero and Q. Williams, Partial melting in a thermochemical boundary layer at the base of the mantle, *Phys. Earth Planet. Inter., 146,* 441–467, 2004.

Litasov, K., E. Ohtani, F. Langenhorst, H. Yurimoto, T. Kubo, and T. Kondo, Water solubility in Mg-perovskite and water storage capacity in the lower mantle, *Earth Planet. Sci. Lett., 211,* 189–203, 2003.

Lodders, K., An oxygen isotope mixing model for the accretion and composition of rocky planets, *Space Sci. Rev., 92,* 341–354, 2000.

Malavergne, V., et al., Si in the core? New high-pressure and high-temperature experimental data, *Geochim. Cosmochim. Acta, 68,* 4201–4211, 2004.

Masters, G. and D. Gubbins, On the resolution of density within the Earth, *Phys. Earth Planet. Inter., 140,* 159–167, 2004.

Mattern, E., J. Matas, Y. Ricard, and J. Bass, Lower mantle composition and temperature from mineral physics and thermodynamic modeling, *Geophys. J. Int., 160,* 973–990, 2005.

McCammon, C. A., S. Lauterbach, F. Seifert, F. Langenhorst and P. A. van Aken, Iron oxidation state in lower mantle mineral assemblages. I. Empirical relations derived from high-pressure experiments, *Earth Planet. Sci. Lett., 222,* 435–449, 2004.

McDonough, W. F. and S.s. Sun, The composition of the Earth, *Chem. Geol., 120,* 223–253, 1995.

McFarlane, E. A., M. J. Drake and D. C. Rubie, Elemental partitioning between Mg-perovskite, magnesiowustite, and silicate melt at conditions of the Earth's mantle, *Geochim. Cosmochim. Acta, 58,* 5161–5172, 1994.

Meibom, A. and D. L. Anderson, The statistical upper mantle assemblage, *Earth Planet. Sci. Lett., 217,* 123–139, 2004.

Morbidelli, A., J. Chambers, J. I. Lunine, J. M. Petit, F. Robert, G. B. Valsecchi, and K. Cyr, Source regions and timescales for the delivery of water to Earth, *Meteorit. Planet. Sci., 35,* 1309–1320, 2000.

Murakami, M., K. Hirose, K. Kawamura, N. Sata and Y. Ohishi, Post-perovskite phase transition in MgSiO₃, Science, 304, 855–858, 2004.

Murakami, M., K. Hirose, H. Yurimoto, S. Nakashima and N. Takafuji, Water in Earth's lower mantle, *Science, 295,* 1885–1887, 2002.

Ohtani, E., The primordial terrestrial magma ocean and its implications for stratification of the mantle, Phys. Earth Planet. Inter., 38, 70–80, 1985.

Ohtani, E., Chemical stratification of the mantle formed by melting in the early stage of the terrestrial evolution, *Tectonophys. 154,* 201–210, 1988.

Ohtani, E., T. Kato, and H. Sawamoto, Melting of a model chondritic mantle to 20 GPa, *Nature, 322,* 352–353, 1986.

Palme, H., Are there chemical gradients in the inner solar system?, *Space Sci. Rev., 92,* 237–262, 2000.

Palme, H., and H. S. C. O'Neill, Cosmochemical estimates of mantle composition, in *Treatise on Geochemistry, Vol. 2,* Ed., K. Turekian, pp. 1–38, 2003.

Pearce, J. A., Statistical analysis of major element patterns in basalt, *J. Petrol., 17,* 15–43, 1976.

Ringwood, A. E., *Origin of the Earth and Moon,* Springer-Verlag, Berlin,1979.

Rost, S., E. J. Garnero, Q. Williams and M. Manga, Seismic evidence of plume genesis at the core-mantle boundary, *Nature, 435,* 666–670, 2005.

Salters, V. J. M., and A. Stracke, Composition of the depleted mantle, *Geochem. Geophys. Geosys.,* 5, Q05004, doi:10.1029/ 2003GC000597, 2004.

Sanloup, C., G. Fiquet, E. Gregoryanz, G. Morard and M. Mezouar, Effect of Si on liquid Fe compressibility: Implications for sound velocity in core materials, *Geophys. Res. Lett., 31,* L07604, doi:10.1029/2004GL019526, 2004.

Shim, S. H., R. Jeanloz and T. Duffy, Tetragonal structure of CaSiO₃ perovskite above 20 GPa, *Geophys. Res. Lett., 29,* 2166–2169, 2002.

Solomatov, V. S., and D. J. Stevenson, Suspension in convecting layers and style of differentiation of a terrestrial magma ocean, *J. Geophys. Res., 98,* 5375–5390, 1993a.

Solomatov, V. S., and D. J. Stevenson, Nonfractional crystallization of a terrestrial magma ocean, *J. Geophys. Res., 98,* 5391–5406, 1993b.

Stixrude, L., R. J. Hemley, Y. Fei and H. K. Mao, Thermoelasticity of silicate perovskite and magnesiowustite and stratification of the Earth's mantle, *Science, 257,* 1099–1101, 1992.

Takahashi, E., Melting of a dry peridotite KLB-1 up to 14 GPa: Implications on the origin of peridotitic upper mantle, *J. Geophys. Res., 91,* 9367–9382, 1986.

Trønnes, R. G., Melting relations and major element partitioning in an oxidized bulk Earth model composition at 15–26 GPa, *Lithos, 53,* 233–245, 2000.

Walker, D., Melting equilibria in multicomponent systems and liquidus/solidus convergence in mantle peridotite, *Contrib. Mineral. Petrol., 92,* 303–307, 1986.

Walter, M. J., A. Kubo, T. Yoshino, J. Brodholt, K. T. Koga and Y. Ohishi, Phase relations and equation–of-state of aluminous Mg-silicate perovskite and implications for Earth's lower mantle, *Earth Planet. Sci. Lett., 222,* 501–516, 2004.

Wanke, H., Constitution of terrestrial planets, *Phil. Trans. R. Soc. Lond., 303,* 287–302, 1981.

Williams, Q., and E. Garnero, Seismic evidence for partial melt at the base of Earth's mantle, *Science, 273,* 1528–1530, 1996.

Williams, Q. and R. J. Hemley, Hydrogen in the deep Earth, *Ann. Rev. Earth Planet. Sci., 29,* 365–418, 2001.

Williams, Q. and E. Knittle, Constraints on core chemistry from the pressure dependence of the bulk modulus, *Phys. Earth Planet. Inter.,* 100, 49–59, 1997.

Zhang, J. and D. J. Weidner, Thermal equation of state of aluminum-enriched silicate perovskite, *Science, 284,* 782–784, 1999.

Zhao, Y. and D. L. Anderson, Mineral physics constraints on the chemical composition of the Earth's lower mantle, *Phys. Earth Planet. Inter., 85,* 273–292, 1994.

E. Knittle and Q. Williams, Department of Earth Sciences, University of California–Santa Cruz, Santa Cruz, California 95064, USA. (quentw@emerald.ucsc.edu)

Highly Siderophile Elements: Constraints on Earth Accretion and Early Differentiation

Kevin Righter

NASA Johnson Space Center, Houston, Texas

Highly siderophile elements (HSE: Re, Au, and the PGEs) prefer FeNi metal and sulfide phases over silicate melts and minerals (olivine, pyroxene, feldspar, etc.). In addition, three HSE—Re, Pt, and Os—are involved in radioactive decay schemes: $^{187}Re \rightarrow {}^{187}Os$ (beta decay) and $^{192}Pt \rightarrow {}^{188}Os$ (alpha decay). As a result, they have provided constraints on the conditions during establishment of the primitive upper mantle, and the conditions and timing of later mantle differentiation and evolution. Hypotheses proposed to explain HSE elemental and isotopic compositions in the primitive upper mantle include mantle–core equilibrium, outer core metal addition, inefficient core formation, and late accretion (the late veneer). All of these scenarios have problems or unresolved issues. Here a hybrid model is proposed to explain the HSE concentrations in the primitive mantle, whereby Au, Pd, and Pt concentrations are set by high-pressure and temperature metal–silicate equilibrium, and Re, Ru, Rh, Ir, and Os concentrations are set by late accretion of chondritic material that is added via oxidized vapor following a giant impact (post-core formation). Processes affecting the later HSE evolution of the mantle include (1) layering caused by fractionation and/or flotation of mantle phases such as olivine, chromite, and garnet, (2) addition of metal from the outer core, and (3) recycling of oceanic crust. Uncertainties about differences in composition between the upper and lower mantle make evaluation of processes in the first category uncertain, but both the second and third processes can explain some aspects of mantle Os isotope geochemistry. This is a review of the field over the past decade and reports not only progress in the field but also highlights areas where much work remains.

1. INTRODUCTION

Determining the conditions and materials out of which the Earth formed is one of the most interdisciplinary studies in science, drawing on astronomy, astrophysics, geology, geochemistry, planetary science, and organic geochemistry. To obtain a solid understanding of this chapter in Earth's history, as many tools as possible must be applied to the

problem. Siderophile elements are those which are "iron-loving" and reside mainly in Earth's core. Because finite quantities of these elements also were left behind in Earth's mantle, they hold information about the conditions in the early Earth.

The siderophile elements encompass over 30 elements and are defined as those elements that have D (bulk distribution coefficient: wt% element in phase A/wt% element in phase B) in metal/silicate > 1, and are useful for deciphering the details of core formation. This group of elements is commonly divided into several subclasses, including the slightly siderophile elements (1 < D < 10), moderately siderophile

Earth's Deep Mantle: Structure, Composition, and Evolution
Geophysical Monograph Series 160

elements (MSE; $10 < D < 10,000$), and highly siderophile elements (HSE; $D > 10,000$). Because these three groups encompass a wide range of partition coefficient values, they can be very useful in trying to determine the conditions under which metal may have equilibrated with the mantle (or a magma ocean). HSE (Re, Au, Pt, Pd, Rh, Ru, Ir, Os; Figure 1A) are of great interest to economic geologists and their behavior in magmatic processes has been studied in detail. However, they are also useful and sensitive indicators of both metal–silicate separation and accretion of chondritic materials to terrestrial planets.

The ^{187}Re–^{187}Os and ^{190}Pt–^{186}Os isotope systems provide important constraints on the origin of HSE in the mantles of the Earth and other planetary bodies. The three elements that compose these two radiogenic isotope systems are HSE, and consequently, the determination of Os isotopic compositions can be used to monitor the long-term elemental ratios of Re/Os and Pt/Os in materials derived from planetary mantles. ^{187}Re decays to ^{187}Os via a $^-\beta$ transition, with a decay constant of $1.666 \times 10^{-11}a^{-1}$. In contrast, ^{190}Pt decays to ^{186}Os via an α transition, with a decay constant of $1.543 \times 10^{-12}a^{-1}$. The Re–Os system is more diagnostic for studies of long term relative abundances of HSE in mantles because ^{187}Re is the main isotope of Re (62.60%) and has a relatively large decay constant. The Pt–Os system is less precise in monitoring long-term Pt/Os because ^{190}Pt is a minor isotope of Pt (0.0129%), and its decay constant is relatively small. With advances in high-precision techniques for measuring ^{186}Os [*Walker et al.*, 1997a], the system has significant utility in placing constraints on the long-term Pt/Os of the Earth's upper mantle.

This review will discuss the chondritic relative abundances of HSE in the primitive upper mantle, and proposed models to explain these patterns. In addition, evidence for nonchondritic HSE, including Re/Os and Pt/Os ratios will be summarized and discussed.

2. HSE IN THE PRIMITIVE UPPER MANTLE AND EARTH ACCRETION MODELS

2.1. Primitive Upper Mantle Measurements

The most common samples of Earth's mantle are peridotite xenoliths (xeno = foreign; lith = rock; chunks of rock found in volcanic deposits formed during eruptions) and peridotite-bearing ultramafic massifs. Careful mineralogical, petrological and geochemical studies have identified certain xenolith and massif peridotites as primitive, meaning that they represent material that is likely to be most similar to the original mantle composition at ~4.4 Ga. Such compositional similarities are illustrated by the elements Mg, Al and Si; the most fertile (primitive) peridotites lie close to the intersec-

tion of the chondrite and meteorite trends on Mg/Si vs. Al/Si diagrams [e.g., *Jagoutz et al.,* 1979]. Thermobarometry studies of these xenoliths demonstrate that they represent samples from the upper mantle [*Finnerty and Boyd*, 1984]; the lower mantle may have a different composition than these samples, but this is currently poorly understood (see section 3). It is also important to note that many of these xenoliths are not isotopically chondritic, but rather exhibit evidence for ancient melting events [e.g., *Zindler and Hart*, 1987]. As a result, they cannot be considered mantle samples that have survived 4.4 Ga. Nonetheless, these same peridotite samples also have chondritic ratios of many refractory lithophile elements, and globally similar abundances of moderately (e.g., Ni/Co) and highly siderophile elements (Figure 1B).

Morgan et al. [1981], *Morgan* [1986] and *Chou et al.* [1983] determined that these samples of the primitive upper mantle have near-chondritic relative HSE abundances (e.g., Figure 2). Although more recent studies of primitive peridotite suites (both massif and xenolith sample) have identified some deviations in the near-chondritic ratios of HSE [e.g., *Pattou et al.,* 1996], there is clearly a convergence of many HSE ratios (e.g., Rh/Pd, Os/Ir, Au/Pt, Rh/Pt) to chondritic values at alumina contents appropriate for the primitive upper mantle (Figure 2B, 2C). The variation of Pd/Ir, Rh/Pd, and Os/Ir as a function of alumina content is a reflection of the relative compatibilities of each of these elements during melting. For example, Pd is incompatible relative to Ir, so that in low alumina content peridotite (which has presumably undergone extensive melting), Pd is low and Ir is high. However, Rh is compatible relative to Pd, so that in low alumina content peridotite (which has presumably undergone extensive melting), Rh is high and Pd is low. In the case of Os/Ir, both elements are equal in compatibility during melting and thus they form a horizontal trend with alumina. Several suites are characterized by Pd/Ir and Ru/Ir ratios demonstrably higher than chondritic (Figure 2A); whether this is a primary feature of the primitive mantle, or due to a later secondary process is still debated and uncertain [e.g., *Pattou et al.,* 1996; *Lorand et al.,* 1999]. This is an important issue to resolve, because it bears directly on whether the mantle values were set by a chondritic component or later chemical equilibration between metal and silicate.

2.2. Experimental and Analytical Considerations

Experimental work on these elements has been hampered by two difficult problems. First, HSE-rich metallic nuggets, sometimes sub-microscopic, are stable in some experimental samples, and thus interfere with analysis that would lead to an understanding of their equilibrium behavior. Second, standard and more common analytical techniques such as

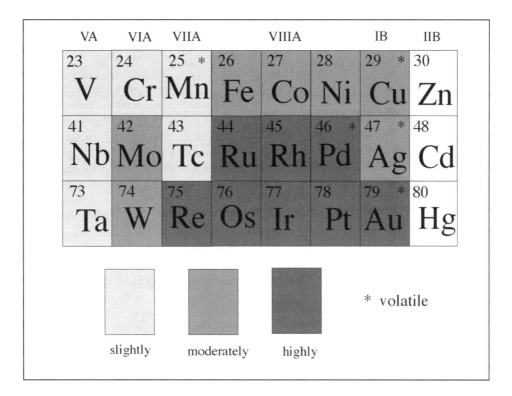

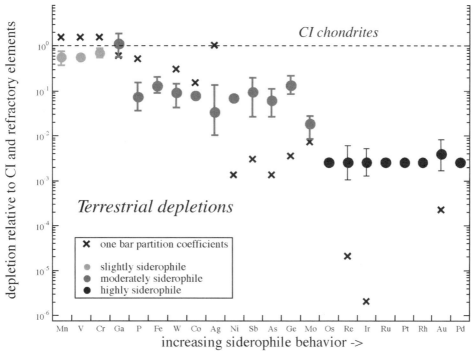

Figure 1. (Upper) Portion of the periodic table showing the highly siderophile elements Ru, Rh, Pd, Re, Os, Ir, Pt, Au. (Lower) Highly, moderately, and slightly siderophile elements and their depletions in the primitive upper mantle, along with calculated depletions using partition coefficients determined at 1 bar pressure [after *Righter and Drake*, 1997]. The latter illustrates the mismatch between concentrations expected for core–mantle equilibrium (using low-pressure D values) compared with observed concentrations.

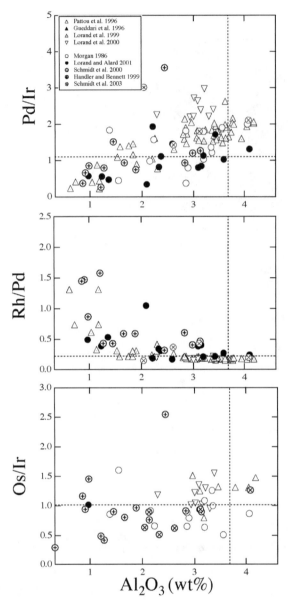

Figure 2. Pd/Ir, Rh/Pd, and Os/Ir vs. Al₂O₃ for a wide variety of mantle peridotites: massif (triangles) and xenolith (circles) peridotites. References are in the legend; horizontal dashed CI chondritic lines are from the compilation of *McDonough and Sun* [1995]; vertical dashed lines represent the alumina content of the primitive upper mantle (3.7 wt%). The variation of Pd/Ir, Rh/Pd, and Os/Ir as a function of alumina content is a reflection of the relative compatibilities of each of these elements during melting. For example, Pd is incompatible relative to Ir, so that in low-alumina-content peridotite (which has presumably undergone extensive melting), Pd is low and Ir is high. However, Rh is compatible relative to Pd, so that in low-alumina-content peridotite (which has presumably undergone extensive melting), Rh is high and Pd is low. In the case of Os/Ir, both elements are equal in compatibility during melting and thus form a horizontal trend with alumina.

electron microprobe analysis cannot be used for silicate or oxide analysis due to the very low solubilities of these elements in such materials.

2.2.1. Nugget effect. The nugget effect is a well-known, though not well-understood, problem in HSE research. It is the contamination of a trace element analysis by tiny (submicron) undissolved particles containing the element of interest as a major constituent. Risk of nugget contamination increases greatly with decreasing solubilities of the element. When the concentrations of interest are as low as 10s to 100s of parts per billion the risk of contamination by undissolved metallic particles of the highly siderophile element dispersed throughout the silicate sample must be given serious consideration. This kind of contamination is difficult to detect a priori because the number and size of nuggets are small; even careful selection of apparently clear fragments of glass under the optical microscope may not be a sufficient safeguard when the concentration is under 10 ppb. At higher bulk concentrations, say, 1 ppm, the presence of nuggets is easier to detect by optical examination, e.g., a micron-sized particle per mm³ would produce 1 ppm. Similarly, *Hervig et al.* [2004] looked for nuggets in experimental samples using transmission electron microscopy, and showed that these must be <0.1 μm in glasses where secondary ion mass spectrometry (SIMS) measurements indicated <10 ppm. One reason for conducting experiments under conditions of higher solubility is to detect the problem in advance of analysis. To control the nugget effect one must be able to reliably detect its existence. But detection and prevention are separate issues. Part of the problem with the nugget effect is that it does not appear to be entirely reproducible and this is undoubtedly a source of some of the erratic results reported in the literature.

Proposed causes of the nugget effect are many and varied. Some studies have suggested that they form during quench [*Cottrell and Walker*, 2002]. *Borisov and Palme* [1997] suggested that nuggets could form from flakes of metal breaking off the HSE metal into the melt. They used alloys of HSE (Ir–Pt) with increased ductility to reduce potential flaking. The reduced thermodynamic activity of Ir may also help ameliorate the problem if it is associated with transient melt phenomena, such as quenching or mechanical redistribution. Others have suggested that nuggets are fostered by or nucleated on trace impurities (Bi, Sb, etc.) in glasses [*Borisov et al.,* 1994]. Alternatively, it may be that nugget formation is actually a significant aspect of HSE geochemistry and that the variability observed at extreme dilution is a consequence of solvated clusters of HSE atoms [e.g., *Tredoux et al.,* 1995; *Amosse et al.,* 1990]. Whatever the cause of nuggets, this phenomenon must be addressed and limits experimental and analytical approaches to HSE studies.

2.2.2. Laser ablation inductively coupled plasma mass spectrometry (ICP-MS) microprobe. Measurement of low concentrations of HSE (<100 ppm) in metals, oxides and silicates is possible by laser ablation ICP-MS, and has been used successfully previously on a variety of natural and experimental samples [e.g., *Campbell and Humayun*, 1999; *Campbell et al.*, 2001b; *Righter et al.*, 2002, 2003a, 2004]. Analyses using laser ablation techniques on experimental spinel-melt and olivine-melt pairs indicate that the nugget effect (discussed above) can be avoided, and high sensitivity can be achieved, by using 25-, 50-, and 100-μm laser-spot sizes on the spinel, olivine, and glass, respectively [*Righter et al.*, 2004]. For analyses on metal diffusion couples, 15 μm spot sizes can be used since the concentrations of HSE in the metals are much higher [*Righter et al.*, 2002, 2003b].

2.2.3. SIMS. The advantage of SIMS is the fine-scale resolution (10–20 μm) of the primary ion beam. Successful measurement of Re, Ru, Au, Pd, Pt, and Rh in silicates and oxides was made by *Capobianco et al.* [1994], *Righter and Hauri* [1998], *Hill et al.* [2000], *Righter et al.* [2004], and *Hervig et al.* [2004], and by *Hsu et al.* [2000] on meteoritic metals. However, there can be many molecular interferences and several of the HSE do not ionize very efficiently, making SIMS generally less applicable than laser ablation-ICP-MS.

2.3. Scenarios

Four different scenarios have been proposed to explain the near-chondritic depletions of HSE in the terrestrial primitive upper mantle (PUM) (Figure 3), and each will be discussed below.

2.3.1. Mantle–core equilibrium. Attributing mantle HSE patterns to metal–silicate equilibria has been problematic because experimentally determined metal/silicate partition coefficients for the HSE are all large (>10^5) and different at 1 bar (Figure 4). Such values would produce an HSE depleted mantle and have values that are not chondritic relative. Thus the HSE concentrations could not have been set by metal–silicate equilibrium in the early Earth, and many [e.g., *Newsom*, 1990] argued against equilibrium scenarios for the HSE. However, these arguments were based on experiments carried out at 1 bar, in a temperature range of 1,200–1,600 °C, on basaltic melts, and usually in Fe-free systems [e.g., *Walter et al.*, 2000]—all different conditions from the possible high pressure and temperature, FeO-bearing peridotite melts that may have been present during core formation in the early Earth.

Experimental studies have demonstrated that D values for HSE are dependent upon oxygen fugacity, temperature, pressure, melt composition, and metallic liquid composition.

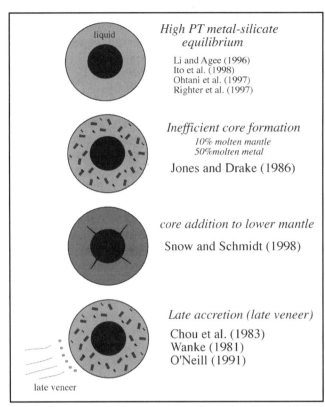

Figure 3. Four models for explanation of near-chondritic HSE contents in Earth's PUM (from top to bottom): high-PT metal/silicate equilibrium, inefficient core formation, core metal addition, and late chondritic veneer.

Despite this recognition, there remain many open questions about HSE metal–silicate partitioning, and many of these parameters have not been quantified for every element. For instance, there remains uncertainty about the valences of some of the HSE at high temperature and pressure; some experimental evidence suggests that many of these elements are stable at lower than expected valences such as 1+ or 2+ [*Borisov and Palme*, 1997; *Borisov and Palme*, 1996]. There have been many experiments completed on S-bearing systems, but the effects of other alloying light elements, such as O and C, are virtually unknown, yet relevant to planetary core formation. There has been only one systematic study of the element Rh, and this at fixed temperature, pressure, silicate melt, and metal composition [*Ertel et al.*, 1999].

2.3.1a. Temperature and oxygen fugacity. The last decade has seen efforts to define the role of variables such as oxygen fugacity (fO_2), temperature, pressure, and silicate and metallic melt compositions. Progress on the effect of temperature and oxygen fugacity has been steady. For example, the solubilities of all the HSE have been studied across a wide fO_2 range (Figure 4), from air down to near the iron–wüstite (IW) buffer (see review of *Walter et al.*, [2000]). Some work-

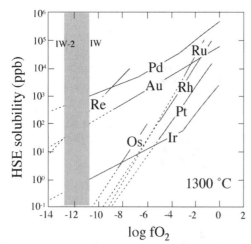

Figure 4. Summary of HSE solubility in diopside-anorthite system (Di-An) eutectic melt at 1,300 °C, 1 bar, as a function of oxygen fugacity. Shaded area (IW and "IW-2"): the oxygen fugacities at the IW buffer and at 2 log fO_2 units lower. Figure redrawn from *Walter et al.* [2000].

ers have studied the effect of temperature on HSE solubility [*Borisov and Palme* 1997; *Fortenfant et al.*, 2003].

2.3.1b. Sulfur. The role of sulfur in Fe–FeS melts on HSE partitioning has been evaluated by the work of *Fleet et al.* [1991a, 1999], and *Fleet and Stone* [1991]. Sulfide melt–silicate melt partitioning has been studied experimentally by several different groups [*Fleet et al.* 1991b, 1996; *Stone et al.*, 1990; *Crocket et al.*, 1992, 1997; *Peach et al.*, 1990, 1994]. After numerous studies involving multiple elements, it has become clear that sulfide liquid/silicate liquid D values are very large, 10^3–10^5 for most of the HSE, but still smaller than in sulfur-free systems. Despite this first order observation, there are several lingering questions regarding these experiments. First is, Why should the D values for such a geochemically diverse set of elements all be so similar over a range of conditions? This may be due to the contrasting behavior of the HSE in S-bearing systems, as compared to S-free. That is, the metal/sulfide D values are very large (100–1,000) for those elements (Ir, Os, Ru) that have large metal/silicate Ds, whereas metal/sulfide D values are small (<100) for those elements (Au, Pd, Rh) that have smaller metal/silicate Ds [*Fleet et al.*, 1991a; *Fleet and Stone*, 1991; *Chabot et al.*, 2003]. As a result, the effect of sulfur will be to equalize metal/silicate D values. Second, why is there scatter in the values of D, sometimes as much as two order of magnitude [e.g., *Fleet et al.*, 1996]? This may be due to the presence of micronuggets of HSE-rich alloys in the silicate melts, interfering with analysis of the silicate phase (see discussion above). Higher sulfur in metallic liquids decreases D(M/S) and helps to reduce D(M/S), perhaps even to values that are consistent with an equilibrium model.

2.3.1c. Pressure and melt composition. There has been little work on the effect of pressure or melt composition on HSE metal/silicate partitioning. Therefore, evaluating mantle concentrations of HSE in light of all the conditions relevant to a high-pressure and temperature (high-PT) magma ocean scenario, for instance, has not been possible yet. *Righter and Drake* [1997] proposed that mantle Re concentrations were consistent with a high-PT magma ocean scenario, but this was based on limited experimental database. Their calculations also relied upon a large melt composition and pressure effect, both of which were poorly characterized. Since this proposal, several studies have attempted to isolate a pressure effect on the partitioning of HSE. For instance, *Holzheid et al.* [2000] studied Pd and Pt partitioning between metal and K-rich silicate melt to 160 kb, and found little or no change in the solubilities. Their experiments were at oxygen fugacities believed to be high enough to avoid the nugget effect in the glass analyses, yet there was still considerable analytical scatter in the data. Their results suggested that pressure had no large effect on reducing the value of D(M/S), and concluded that Pd and Pt were not set by high-PT metal/silicate equilibrium. In a study of Pt solubility in Fe-free basaltic melts, *Ertel et al.* [2001] argue that pressure causes a huge reduction (10^3) in the value of D between 1 bar and 10 kb, but then has only a very small effect at higher pressures. However, it was not clear in these studies whether the high pressure runs were free of the nugget effect that plagues experimental glasses at low oxygen fugacities. *Danielson et al.* [2003] studied Au solubility and chose chondritic starting materials. Their focus on Au also had the advantage of a large nugget-free range of fO_2 (nuggets don't appear until <IW) making any pressure effects easier to define, and use of an ion microprobe to analyze small regions (15–20 μm). Their studies have shown that D(Au) decreases to values that are consistent with metal–silicate equilibrium.

A summary of D(M/S) data (including high pressure) for Pd (Figure 5) shows that partition coefficients for this HSE are lowered to values of ~400-600 at high-PT conditions of a peridotite magma ocean. Such values are consistent with an equilibrium scenario, as suggested by the calculations of *Righter and Drake* [1997] for Re. Data for Au and Pt also indicate that a high-PT peridotite magma ocean scenario may reduce partition coefficients sufficiently to be consistent with equilibrium. These elements of course need additional study, as do Rh, Ru, Ir and Os because their solubility at high pressures and their dependence upon silicate melt composition is unknown, and hence consistency with a high-PT equilibrium scenario cannot yet be made. Clearly, additional work will be necessary to fully evaluate whether HSE concentrations can be explained by a high-PT equilibrium scenario.

2.3.2. Outer core additions to the lower mantle. To explain the HSE concentrations in the primitive mantle, *Snow and Schmidt* [1998] argued that outer core metallic liquid was added to the lower mantle, ultimately making it to the shallow mantle and surface. If Pt/Os and Re/Os ratios in the outer core are the same as H5 (ordinary chondrite) metal, then ^{186}Os and ^{187}Os isotopic anomalies measured in basaltic and komatiitic rocks from Siberia, Gorgona, Colombia and Hawaii could be a result of core additions [*Brandon et al.*, 1998, 1999, 2003; *Meibom and Frei*, 2002; *Walker et al.*, 1995, 1997a]. Experimental evidence for this possibility was cited by *Shannon and Agee* [1998] who found that metallic liquids are mobile through a perovskite matrix at lower mantle conditions, and could perhaps be drawn out of the core by capillary action. In addition, *Humayun et al.* [2004] show that Fe enrichments and Fe/Mn measured in Hawaiian lavas are consistent with a small amount of oxidation of Fe-rich metal added to their source from the core.

However, this scenario is not without difficulties and challengers. *Righter and Hauri* [1998] point out that addition of 1% outer core material to the mantle would boost the concentrations of HSE to values far above what are observed or calculated for ocean island basalt (OIB) sources. *Righter and Hauri* [1998] also point out that addition of metal to the mantle implies an oxygen fugacity close to the IW buffer,

which is lower than any measured terrestrial mantle samples [e.g., *Wood et al.*, 1990]. As a way around this problem, *Puchtel and Humayun* [2000] argue that the radiogenic Os isotopic signature measured by many groups could be imparted on the mantle by diffusive, isotopic exchange, rather than bulk additions to the mantle. Finally, W isotopic data from Hawaiian and South African rocks indicate that a core signature is absent, despite the fact that siderophile W might be expected to yield low ^{182}W relative to bulk Earth values [*Schersten et al.*, 2004]. The lack of a core tungsten signature led *Schersten et al.* [2004] to conclude that Mn-rich sediments must have produced the radiogenic Os isotopic signature. However, the Fe/Mn measurements of *Humayun et al.* [2004] demonstrate that Mn-rich sediments might explain Os, but create major element problems in that Fe/Mn ratios are inconsistent. Additional work on W isotopes will undoubtedly shed light on this issue, but the core-mantle interaction model has survived much detailed criticism and remains a strong and viable hypothesis.

Constraints from experimental petrology are not yet mature enough to lend weight to this hypothesis. The composition of such outer core melts will of course be dependent upon the nature of HSE partitioning between solid iron-rich metal and liquid iron-rich metal. Part of this argument rests on the degree to which Re, Pt, and Os partition between solid and liquid metal (SM and LM). Most of the Re, Os, and Pt partitioning modeling is based on studies of meteoritic materials that equilibrated at low pressures. Experimental work has the potential to isolate the effects of temperature, pressure, and composition. There has been a lot of progress in the area of SM/LM HSE partitioning, and it is summarized here. Solid metal/liquid metal partition coefficients for Ru and Re were measured by *Lazar et al.* [2004]; *Chabot et al.* [2003] and *Fleet et al.* [1999] report D(SM/LM) for Re, Os, Au, Pd, Pt, Ru, Os, and Ir (Figure 6A). A summary (Figure 6A) shows that Pt, Re, and Os vary strongly with S content of metallic liquid. Partition coefficients for all eight HSE increase with increasing sulfur content, but Pd and Au have much lower values than the others (Figure 6A).

It is important to note that partition coefficients for Re, Pt and Os in the Fe-Ni-S are not in agreement with values used in the modelling of core-mantle interaction [e.g., *Brandon et al.*, 1998, 2000]. For instance, the values used in the modelling are D(Pt) = 2.9, D(Re) = 18-26, and D(Os) = 28-44. There is no wt% S value in Figure 6B or 6D which satisfies all three of these values. This could mean that the core-mantle interaction hypothesis is flawed, or it could be that we don't have sufficient experimental data to make a full assessment. There are two main variables that must still be explored for Re, Pt, and Os—effects of other light elements such as C and O, and pressure.

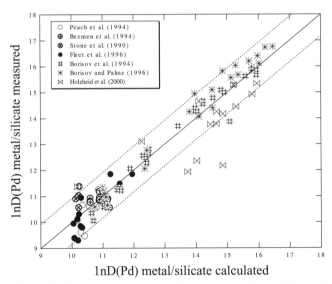

Figure 5. Comparison of calculated and measured lnD(Pd) metal/silicate. Studies included are shown in the legend. Dashed lines are 2 σ error on the regression. The equation for calculated values of lnD follows the form: a lnfO_2 + b/T + cP/T + d(NBO/t) + e(1–X_s) + f. Using this equation it can be shown that D(Pd) metal/silicate is approximately 500 under the conditions of a deep magma ocean: IW-1, 2000 °C, 250 kb, X_s = 0.12 and, peridotite (NBO/t = 2.7). Scatter in the data from *Holzheid et al.* [2000] is present in the original work.

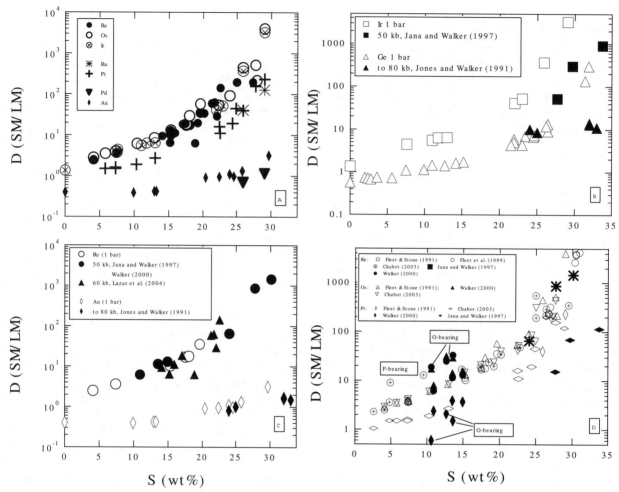

Figure 6. (A) Summary of recent work on all the HSE, determined at 1 bar pressure (data from *Fleet et al.* [1999]; *Chabot et al.* [2003]). (B) D(Ir) and D(Ge) solid metal/liquid metal (SM/LM) at 1 bar and 50–80 kb, showing a slight pressure effect. (C) D(Re) and D(Au) SM/LM at 1 bar and 50–80 kb, showing no pressure effect. (D) Summary of data for Re, Pt, and Os at 1 bar, higher pressures, and with O- and P-bearing metallic liquids. References cited in legend.

The effects of other light elements [*Hillgren et al.,* 2000; *Li and Fei,* 2003] such as C, P, O, H, Si, on SM/LM partitioning, is not known well, but should be a focus of future efforts since we know that the light element in Earth's core is only partly S [*Dreibus and Palme,* 1996]. For instance, dissolved C affects some elements differently than does dissolved S [*Chabot et al.,* 2004], and the effect of O and P is known to increase the difference between D(Pt), and D(Re) and D(Os). For instance, several D(Re) and D(Os) experiments fall above the low pressure trends, and this is attributed to higher O and P contents of the metallic liquids (Figure 6D). The effect of O and P on Pt partitioning is the opposite from Re and Os, as D(Pt) decreases with addition of O and P to the metallic liquid. It is unclear whether Pt is affected by pressure, or the lower D values of *Jana and Walker* [1997] are instead attributable to higher O or P con-

tents (Figure 6D). Clearly, light elements have a strong effect on the magnitude of D.

Finally, the effect of pressure has not been completely addressed yet. Although pressure does not affect the partitioning of some elements (Re and Au; Figure 6C), the partitioning of others may be affected by pressure (e.g., Ir and Ge in Figure 6B). Also, so far experimental work has involved the low pressure form of solid FeNi, whereas the inner core may have a different structure and therefore different partitioning behavior [e.g., *Fei et al.,* 1997; *Wood,* 1993; *Li et al.,* 2002; *Li and Fei,* 2003].

Clearly, full evaluation of the hypothesis that core metal has been added or entrained in the mantle awaits integration of a wide variety of important information including the effect of light elements and elevated pressure on D(SM/LM), and the possibility that inner core FeNi may have a different structure than low-pressure FeNi, on which all research has been done so far.

2.3.3. Inefficient core formation. Proposed by *Jones and Drake* [1986] as a way of explaining the HSE content of the PUM, this idea holds that the early Earth was not substantially molten and thus Fe–FeS melt was trapped in the upper mantle. Because Fe–FeS melt/silicate melt partition coefficients for HSE are so large, nearly the entire mantle budget of HSE would be tied up in the trapped melt. This budget would also be chondritic relative, since all HSE are delivered from chondritic materials and being trapped in the mantle. Geochemically, this idea is simple and would easily explain the HSE concentrations in the PUM. The most challenging questions this idea poses, however, are whether the early Earth was cool enough to avoid melting, and under what specific conditions can an Fe–FeS melt be trapped in a planetary mantle.

The early thermal state of the Earth has been considered from a number of perspectives. *Urey* [1952] and *Ringwood* [1960] both argued that the Earth's interior would likely be entirely molten due to a runaway thermal process caused by energy derived from gravitational separation of core and mantle. However, *Birch* [1965] and *Flaser and Birch* [1973] and *Ringwood* [1974] argued that this would not produce planet-wide melting, but still extensive melting of the outer part of the Earth (1600 to 2000 °C). The work of *Shaw* [1978, 1979] concluded similarly, that because the runaway core formation process took so much heat to the core, that some of the mantle was unaffected. Nonetheless extensive melting was produced. Recognition of the importance of impacts in the accretion process led *Kaula* [1979] to conclude that Earth would be extensively melted. Similarly, *Coradini et al.* [1983] showed that the energy derived during accretion of planets from planetesimals is large enough to produce an outer shell of melt. A number of relatively minor heat sources are discussed by *Hostetler and Drake* [1980] and include solar luminosity, radioactive decay of ^{26}Al and other short-lived isotopes, adiabatic compression, tides, and induction. In summary, despite geochemical arguments by *Jones and Palme* [2000] against a hot early Earth, physical arguments suggest there is plenty of heat in a large terrestrial planet like the Earth and avoiding melting is difficult [*Sasaki et al.,* 1986]. Because metal is delivered to a growing Earth on the same time frame as the energy of accretion is available, it would be difficult to trap metal in Earth's mantle since even as little as 15–30% silicate melt would allow segregation of metal into the core [*Taylor*, 1992].

Whether metallic liquids can move through a solid mantle matrix has been the topic of much investigation. *Rushmer et al.* [2000] summarized the difficulties in moving metal through a solid silicate mantle that is static. Metallic liquids will only interconnect at low pressures if the liquid contains a large fraction of non-metallic anions such as S, C, or O (Figure 7). Olivine, garnet, and pyroxene-bearing matrices all yield the same result; however, high pressures or high-pressure phases may allow a low enough dihedral angle [*Shannon and Agee* 1998]. Additionally, metallic liquids can interconnect if there is a significant % of silicate melt present, that helps to wet the phase boundaries [e.g., *Takahashi*, 1983; *Taylor*, 1992].

A dynamic environment may also allow interconnectivity of metallic liquid in a solid matrix [*Rushmer et al.* 2000]. *Bruhn et al.* [2000] show that molten Au becomes interconnected in a dynamic olivine-bearing matrix. In addition, *Yoshino et al.* [2003] show that Fe–S liquid can move through a peridotite matrix by permeable flow. Experimentation in metal–silicate systems in dynamic environments is in its infancy, and might provide new, more quantitative, insights into core formation processes.

It seems clear that metal will be trapped most easily in a setting in which there is little to no silicate melt present, in the low–non-metal anion content, and in an environment that is static rather than dynamic. Most of these conditions are at odds with the hot, dynamic environment that is likely for the early Earth, and thus argue against the inefficient core formation hypothesis.

2.3.4. Late accretion ("late veneer"). Accretion of material to Earth after core formation has been invoked to explain the acquisition of water [*Delsemme* 1997] and the terrestrial mantle HSE budget [*Chou et al.* 1983]. Post-core formation addition of chondritic material to an HSE depleted mantle, followed by oxidation, would lead to the 0.7% HSE depletions observed in the mantle. Although these "late veneer" scenarios are attractive explanations in their simplicity; they are problematic upon closer inspection.

Osmium isotopic measurements of chondritic materials have shown that any late veneer would have to be supplied by a relatively reduced material, such as enstatite or ordinary chondrite [*Meisel et al.* 2001]. Addition of such reduced chondritic materials to a mantle buffered at an oxygen fugacity below the IW buffer results in the metal staying reduced. Because metal is the host phase for the HSE, the fate of the HSE depends on the stability of the metal. If the mantle is hot and molten, the metal will sink to the core, taking the HSE with it, because metal droplets move rapidly through molten silicate [*Rubie et al.*, 2003; *Stevenson*, 1990]. If the mantle is cool and solid, metal will become trapped in the mantle, because low volume fraction liquid metal does not connect easily in a solid silicate matrix [*Yoshino et al.*, 2004; *Rushmer et al.*, 2000]. This, too, is problematic because

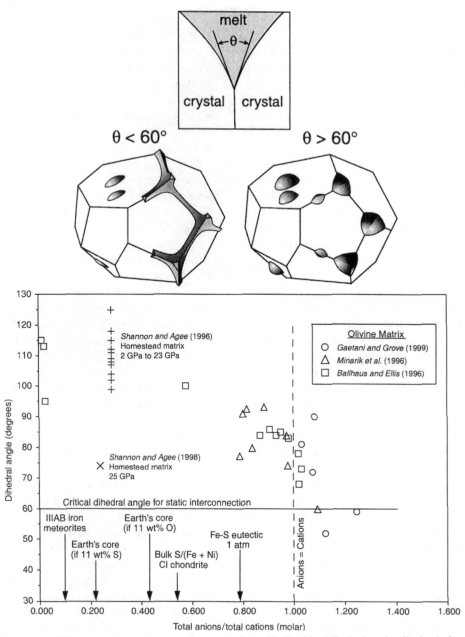

Figure 7. Summary of dihedral angles and nonmetal content of metallic liquids. Dihedral angles display is from *Rushmer et al.* [2000]. © 2000 The Arizona Board of Regents; reprinted by permission of the University of Arizona Press.

there is no metal in the current or Archean mantle [*Arculus* 1985, *Eggler and Lorand* 1992], and HSE will only become dispersed if the metal becomes oxidized or transferred to sulfide. HSE are stable as alloys even up to high oxygen fugacities [*Borisov and Palme*, 2000; Figure 8]. Even if the metal can be oxidized, it must be mixed into the mantle by 3.9 Ga, as spinel peridotites from Greenland have chondritic Os isotopic values [*Bennett et al.* 2001]. However, mixing times in the upper mantle can be as long as 0.5 Ga, pushing

the addition of such chondritic material back to the earliest part of Earth's history (~4.4 Ga).

It is clear that that name "late veneer" is a misnomer in that it must have occurred directly after core formation (not really late in the sense of the late heavy bombardment at 3.9 Ga), and it had to be mixed into at least the upper mantle (not really a veneer). Any solution to the HSE budget of Earth's upper mantle requires appropriate oxidation and mixing mechanisms.

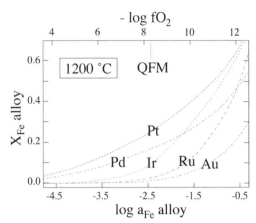

Figure 8. Summary of HSE alloy stability at QFM and 1200 °C (redrawn from *Borisov and Palme*, [2000]). $X_{HSE} > 0.5$ at QFM for many Fe–HSE alloys.

2.3.5. A hybrid model? The idea of late oxidized accretion was used explain the MSE (Ni, Co, Mo, and W), because their D(M/S) values were so dissimilar, and they exhibit such diverse behavior in sulfide and silicate systems [e.g., *Wanke*, 1981; *O'Neill*, 1991]. But similar and near-chondritic relative depletions of HSE have been shown to be a result of high pressure and temperature metal–silicate equilibrium rather than due to addition of chondritic materials [e.g., *Li and Agee*, 1996; *Walter et al.*, 2000]. This may also be true for the HSE; very different D(M/S) values, coupled with their very dissimilar sulfide and silicate partitioning would lead one to believe that a late accretion event is the only way to bring their values to near-chondritic in the mantle. However, the grouping of these eight elements together under one heading (HSE), is perhaps unjustified. After all, elements such as REE or HFSE are grouped together due to relatively well understood geochemical characteristics or behavior. This is not necessarily true of the HSE, since their valences are all different and their metal/silicate D values are variable. This, together with the persistent occurrence of nonchondritic HSE ratios in mantle peridotite [e.g., Pd/Ir, Rh/Ir; e.g., *Pattou et al.*, 1996], have motivated an alternative model to explain the HSE abundances in Earth's PUM.

If a deep magma ocean set the abundances of the MSE in Earth's PUM, then such conditions would also affect the HSE. Although most HSE have very large metal–silicate partition coefficients ($>10^7$), Re, Pt, Pd and Au are much lower. Given our current state of knowledge of these four elements (effect of pressure, temperature, fO_2, and composition), a high-PT scenario is consistent with their depletion in the PUM. Partition coefficients for Ir, Os, Ru and Rh, although poorly known at high-PT conditions, are likely to be too high to be consistent with a high-PT scenario. Instead, the chondritic relative depletions of these four elements could

be attributed to late accretion onto Earth with a relatively oxidized early atmosphere. It has been shown that many of the HSEs can be hosted by magnesioferrite spinels [*Righter and Downs*, 2000]. Late accretion of material onto an Earth with a primitive H_2O-CO_2 atmosphere [*Abe et al.*, 2000] may have resulted in oxidation of PGEs into the post-impact vapor. Later condensation of the PGEs into spinels [e.g., *Ebel and Grossman*, 1999] and mixture into the metal-free mantle would result in a chondritic relative Ir, Os, Ru, Re and Rh depletions. This would provide a way to keep these five HSEs oxidized in the mantle, while Pd, Au, and Pt (all incompatible in magnesioferrite spinel) were set by earlier metal–silicate equilibrium. This is called a hybrid model because it calls on two processes (rather than one single event) to explain the ultimate HSE pattern in the PUM.

2.3.6. Summary of HSE constraints on Earth accretion. The five models discussed above have their strengths and weaknesses. There is additional work in various areas that could help to discriminate between these models. For instance, additional work on metal/silicate partitioning at high pressures and temperatures, and on peridotitic silicate melts would be beneficial. Also, the issue of whether there is a critical silicate melt fraction for Fe–FeS–FeX liquid mobility in a mantle could be resolved with additional experiments and modeling. Finally, the effects of multiple light elements on HSE partitioning between solid and liquid metal will be useful for a thorough understanding of inner–outer core HSE distribution.

3. EARLY EARTH LAYERING OR FRACTIONATION: A CONTROL ON MANTLE OS ISOTOPIC VALUES?

3.1. Os Isotopic Values in Mantle

Osmium isotopic and bulk HSE data have been produced on a great variety of samples from oceanic basalt (MORB, OIB), to peridotite (xenolith, massif, abyssal peridotite), to inclusions in diamonds, and to iridosmine nuggets from placer deposits [e.g., *Shirey and Walker*, 1997; *Righter et al.*, 2000]. When coupled with previous isotopic constraints from Pb, Nd, Sr, and Hf [*Zindler and Hart*, 1986; *Hofmann*, 1997], this database forms the basis for models of early terrestrial layering and mantle heterogeneities. Radiogenic Os measured in samples places constraints on Re/Os and Pt/Os on various parts of the mantle (Figure 9). In most studies of mantle-derived rocks, a chondritic Re/Os source has been surmised [e.g., *Shirey and Walker*, 1997], but there are some important exceptions. For example, $^{187}Os/^{188}Os$ measurements of mantle-derived rocks have shown that parts of the mantle are superchondritic with respect to Re/Os;

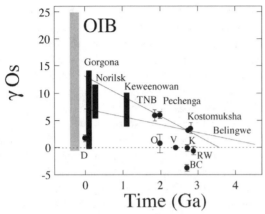

Figure 9. Gamma Os vs. time before present (in giga-annum) illustrating superchondritic Os isotopic values (γ Os > 0) in mantle materials. Figure redrawn from *Walker and Nisbet,* [2001]; OIB field from *Hauri and Hart* [1993] and references therein. TNB = Thompson Nickel Belt, Canada; D = Deccan, India; O = Onega, Russia; V = Vetreny Belt, Russia; K = Kambalda, Australia; RW = Ruth Well, Australia; BC = Boston Creek, Canada.

Kostomuksha, Belingwe, Pechenga, Keweenawan, Norilsk, and Gorgona are among the localities where such nonchondritic Re/Os ratios have been observed [*Walker and Nisbet,* 2001]. Because some of these provinces are Archean in age, it points to the possibility that there were nonchondritic Re/Os portions of the mantle early in Earth history. Some scenarios involve partitioning and distribution of HSE in the deep interior of the Earth, but since high pressure partitioning data are scarce for the HSE, only several scenarios will be discussed.

3.2. Fractionation, Layering, and Flotation

Crystallization of a magma ocean may lead to layering, due to crystal sorting during convection and cooling [*Tonks and Melosh,* 1990; *Solomatov and Stevenson,* 1994; *Solomatov,* 2000]. This may have lead to segregation of perovskite into the lower mantle, causing compositional differences between the upper and lower mantle, for example. Although geophysicists have argued that whole-mantle convection likely has homogenized any compositional differences between the upper and lower mantle [e.g., *Bunge et al.* 1998, *Young and Lay,* 1987], there is some geophysical and geochemical evidence that there is a primitive mantle reservoir within the lower mantle [*Kellogg et al.* 1999; *van der Hilst and Karason,* 1999]. Some more recent work questions the global extent of such deep mantle heterogeneities [*Castle and van der Hilst,* 2003a, 2003b; *Karason and van der Hilst,* 2001]. In addition, because silicate melts are more compressible than crystals, high pressure will tend to promote flotation.

This could have lead to olivine flotation at modest depths in the terrestrial mantle. Such an idea has been proposed as an explanation for Earth's superchondritic Mg/Si upper mantle [*Agee and Walker,* 1988].

3.2.1. Olivine and chromite. The suggestion by *Agee and Walker* [1988] that the terrestrial upper mantle has a superchondritic Mg/Si due to olivine flotation solves one problem, but unfortunately creates others. This idea can be tested using olivine/liquid partition coefficients for Ni, Co, Au, Ir, Pd, Re and Os. *Drake* [1989] showed that olivine flotation will create superchondritic Ir/Au and Ni/Co in the upper mantle, but such ratios are not observed. Similarly using new olivine melt partition coefficients [*Brenan et al.,* 2003; *Righter et al.,* 2004], it is clear that Ir/Pd and Os/Re will also become superchondritic with olivine flotation (Figure 10, 11). The situation becomes worse if chromite is added to the system, as chromite is a likely liquidus phase at these depths as well (Figure 10).

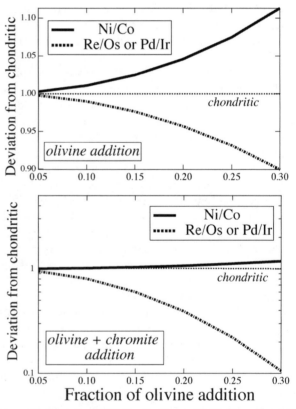

Figure 10. Deviation of Ni/Co, Re/Os, and Pd/Ir from chondritic values by olivine or olivine + chromite fractionation. Concentrations and ratios are calculated by using D(Re, Os, Pd, Ir) oliv/liq, and chromite/liq from *Brenan et al.* [2003] and *Righter et al.* [2004].

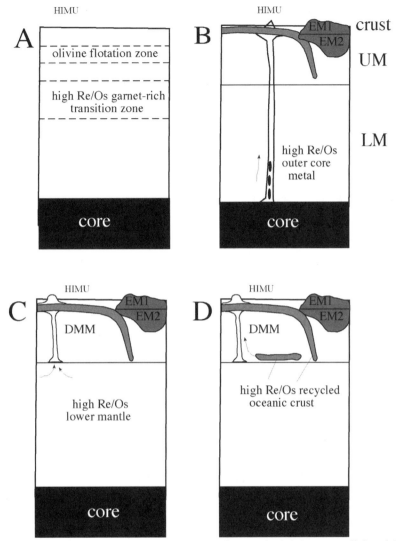

Figure 11. Four models for explanation of radiogenic Os in Earth's mantle: (A) Garnet or olivine-rich layers in the early Earth, (B) outer core metal additions to the upper mantle, (C) lower mantle differing in composition from the upper mantle, and (D) recycled oceanic crust. Enriched mantle 1 (EM1), enriched mantle 2 (EM2), high-mu mantle (HIMU), and depleted modern mantle (DMM) portions are from *Hofmann* [1997].

3.2.2. Garnet. A garnet-rich transition zone has been suggested by some [*Anderson*, 1983]. If Re is compatible in garnet, but Os is not, there may be regions of the transition zone that have Re/Os higher than the PUM (Figure 11A). Garnet is a host phase for Re [*Righter and Hauri, 1998; Hill et al.*, 2000], but there has not yet been any work on Os partitioning in garnet-bearing systems.

3.3. Core Additions

The same issues that pertain to the possibility of core additions to the mantle producing chondritic relative HSE patterns, also apply to the issue of nonchondritic HSE (see

section 2.3.2; Figure 11B). A solid understanding of the role of light elements and pressure will be very important in evaluating models for the generation of mantle sources with nonchondritic Re/Os, Pt/Os or otherwise (Pd/Ir).

3.4. A Lower Mantle Different From the Upper Mantle?

It has also been suggested by some that superchondritic Os isotopic signatures measured in plume-related and other basalts could have originated in a high Re/Os lower mantle [Figure 11C; *Walker et al.,* 1996,1997b]. Recent evidence for this possibility comes from a study by *Fortenfant et al.* [2003], in which they show that Re is much more soluble in magnesiowüstite than is Os, giving this phase the potential to elevate the Re/Os ratio of parts of the lower mantle (Figure 11C).

3.5. Recycled Oceanic Crust

Recycled oceanic crust is an essential component of mantle geochemical models (e.g., *Hofmann* [1997]; *Hauri and Hart* [1997]; Figure 11D). Even those data that argue for core metal addition to the mantle [e.g., *Brandon et al.,* 2000; *Bennett*, 1996] also require a crustal recycling component to explain all of the Os isotopic data. Although early Re-Os data for eclogites yielded low Os concentrations [*Pearson et al.,* 1995], suggesting eclogites are a weak lever on the Os isotopic composition of the mantle into which they are recycled, more recent data from *Menzies et al.* [2003] and *Barth et al.* [2002] have yielded higher Os values. In addition, Pt is mildly compatible in clinopyroxene [*Righter et al.,* 2004] and may be concentrated into clinopyroxene-bearing rocks such as eclogites and potentially producing a high Pt/Os reservoir that could yield radiogenic Os over time.

4. SUMMARY AND FUTURE

Highly siderophile elements have the potential to tell us much about terrestrial mantle structure, core-mantle exchange and mantle dynamics. With the addition of more experimental partitioning data at high pressures and temperatures, as well as the identification of more nonchondritic Pt/Os or Re/Os reservoirs, it may be possible to discriminate between the numerous models that currently exist. In particular, it is a challenge to discriminate between metal and silicate reservoirs for HSE. The most effective approach will be to synthesize all available data—isotopic, bulk rock, and experimental—in order to construct the most realistic models.

Acknowledgments. This paper was possible through support of NSF grant EAR0074036 and a NASA RTOP. Discussions with A. Brandon, L. Danielson and R. Rudnick and comments of two anonymous reviewers were beneficial to presentation of the ideas in this paper.

REFERENCES

Abe Y, Ohtani E, Okuchi T, Righter K, Drake MJ., Water in the early Earth. In (R.M. Canup and K. Righter, eds.) *Origin of the Earth and Moon*, Univ. of Arizona Press, Tucson, 413–434, 2000.

Agee, C. B. and Walker, D., Mass balance and phase density constraints on early differentiation of chondritic mantle, *Earth Planet. Sci. Lett.* 90, 144–156, 1988.

Agee, C. B. and D. Walker, Olivine flotation in mantle melt, *Earth Planet. Sci. Lett. 114,* 315–324, 1993.

Arculus, R.J., Oxidation status of the mantle: past and present, *Ann. Rev. Earth Planet. Sci.* 13, 75–95, 1985.

Amosse, J., Alibert, M., Fischer, W. and Piboule, M., Experimental study of the solubility of platinum and iridium in basic silicate melts—implications for the differentiation of the platinum group elements during magmatic processes, *Chem. Geol.* 81, 45–53, 1990.

Ballhaus, C. and Ellis, D. J., Mobility of core melts during Earth's accretion, *Earth Planet. Sci. Lett.* 143, 137–145, 1996.

Barth, M. G., Rudnick, R. L., Carlson, R. W., Horn, I. and McDonough, W. F., Re-Os and U-Pb geochronological constraints on the eclogite-tonalite connection in the Archean Man Shield, West Africa, *Precambrian Res.* 118, 267–283, 2002.

Bennett, V. C., Esat, T. M., and Norman, M. D., Two mantle-plume components in Hawaiian picrites inferred from correlated Os-Pb isotopes, *Nature* 381, 221–225, 1996.

Bennett, V. C., Nutman, A. P., Esat, T. M., Constraints on mantle evolution from ^{187}Os/^{188}Os isotopic compositions of Archean ultramafic rocks from southern West Greenland (3.8 Ga) and Western Australia (3.46 Ga), *Geochim. Cosmochim. Acta* 66, 2615–2630, 2002.

Bezmen, N. I., Asif, M., Brügmann, G. E., Romanenko, I.M. and Naldrett, A.J., Distribution of Pd, Rh, Ru, Ir, Os, and Au between sulfide and silicate metals, *Geochim. Cosmochim. Acta* 58, 1251–1260, 1994.

Birch, F., The energetics of core formation, *Jour. Geophys. Res.* 70, 6217–6221, 1965.

Borisov, A. and H. Palme, Experimental determination of the solubility of Au in silicate melts, *Mineral. Petrol.* 56, 297–312, 1996.

Borisov, A. and H. Palme, Experimental determination of the solubility of platinum in silicate melts, *Geochim. Cosmochim. Acta 61*, 4349–4357, 1997.

Borisov, A. and H. Palme, Solubilities of noble metals in Fe-containing silicate melts as derived from experiments in Fe-free systems, *Amer. Mineral.* 85, 1665–1677, 2000.

Borisov, A., Palme, H., and Spettel, B., Solubility of palladium in silicate melts: implications for core formation in the Earth, *Geochim. Cosmochim. Acta* 58, 705–716, 1994.

Brandon, A., Walker, R. J., Morgan, J. W., Norman, M. D. and Prichard, H.M., Coupled ^{186}Os and ^{187}Os evidence for core-mantle interaction, *Science 280*, 1570–1573, 1998.

Brandon, A. D., Norman, M. D., Walker, R. J. and Morgan, J. W., ^{186}Os-^{187}Os systematics of Hawaiian picrites, *Earth Planet. Sci. Lett.* 174, 25–42, 1999.

Brandon, A. D., Snow, J. E., Walker, R. J., Morgan, J. W., Mock, T. D., ^{190}Pt-^{186}Os and ^{187}Re-^{187}Os systematics of abyssal peridotites, *Earth Planet. Sci. Lett.* 177, 319–335, 2000.

Brandon, A., Walker, R. J., Puchtel, I. S., Becker, H., Humayun, M. and Revillon, S., ^{186}Os-^{187}Os systematics of Gorgona Island komatiites: implications for early growth of the inner core, *Earth Planet. Sci. Lett.* 206, 411–426, 2003.

Brenan, J. M., McDonough, and W. F. Dalpe, C., Experimental constraints on the partitioning of rhenium and some platinum-group elements between olivine and silicate melt, *Earth Planet. Sci. Lett.,* 212, 135–150, 2003.

Bruhn D., Groebner N., Kohlstedt D. L., An interconnected network of core-forming melts produced by shear deformation, *Nature* 403, 883–86, 2000.

Bunge H. P., Richards M. A., Lithgow-Bertelloni C., Baumgardner J. R., Grand S. P., Romanowicz B. A., Time scales and heterogeneous structure in geodynamic Earth models, *Science* 280, 91–95, 1998.

Campbell, A. J., and Humayun, M., Trace element microanalysis in iron meteorites by laser ablation ICP-MS, *Anal. Chem.* 71, 939–946, 1999.

Campbell A. J., Humayun M., Meibom A. and Krot A. N., Origin of Zoned Fe,Ni Metal Grains in QUE 94411, *Geochim. Cosmochim. Acta* 65, 163–180, 2001.

Capobianco, C. J., R. L. Hervig and M. J. Drake, Experiments on crystal/liquid partitioning of Ru, Rh and Pd for magnetite and hematite solid solutions crystallized from silicate melt, *Chemical Geology* 113, 23–43, 1994.

Castle, J. C. and van der Hilst, R. D., Using ScP precursors to search for mantle structures beneath 1800 km depth, *Geophys. Res. Lett.* 30, 1422, 2003.

Castle, J. C. and van der Hilst, R. D., Searching for seismic scattering off mantle interfaces between 800 km and 2000 km depth, *Jour. Geophys. Res.* 108, 2095, 2003.

Chabot, N. L., Campbell, A. J., Jones, J. H., Humayun, M. and Agee, C.B., An experimental test of Henry's Law in solid metal-liquid metal systems with implications for iron meteorites, *Met. Planet. Sci.* 38, 181–196, 2003.

Chabot, N. L., Campbell, A. J. and Humayun, M., Solid metal–liquid metal partitioning of Pt, Re, and Os: the effect of carbon, *Lunar Planet. Sci.* XXXV, 1008, 2004.

Chou, C. L., Fractionation of siderophile elements in the Earth's upper mantle, *Proc. Lunar Planet. Sci. Conf. 9th,* 219–230, 1978.

Chou, C.-L., D. M. Shaw, Crocket, J. H., Siderophile trace elements in the Earth's oceanic crust and upper mantle, *Jour. Geophys. Res. 88 (Supplement),* A507–A518, 1983.

Coradini, A., Federico, C. and Lanciano, P., Earth and Mars: early thermal profiles, *Phys. Earth Planet. Int.* 31, 145–160, 1983.

Cottrell, E. A. and Walker, D., A new look at Pt solubility in silicate liquid, *Lunar Planet. Sci.* XXXIII, #1274, 2002.

Crocket, J. H., Platinum-Group Elements in mafic and Ultramafic Rocks: A Survey, *Canadian Mineralogist 17,* 391–402, 1979.

Crocket, J. H., M. E. Fleet, and Stone, W. E., Experimental partitioning of osmium, iridium and gold between basalt melt and sulphide liquid at 1300°C, *Austral. Jour. Earth Sci.* 39, 427–432, 1992.

Crocket, J. H., M. E. Fleet, and Stone, W. E., Implications of composition for experimental partitioning of platinum-group elements and gold between sulfide liquid and basalt melt: The significance of nickel content, *Geochim. Cosmochim. Acta 61,* 4139–4149, 1997.

Danielson, L. R., Sharp, T. G., and Hervig, R. L., Implications for core formation from high pressure and temperature partitioning studies of Au, *Earth Planet. Sci. Lett.,* in review.

Delsemme A., The origin of the atmosphere and the oceans. In *Comets and the Origin and Evolution of Life,* eds. P. J. Thomas, C.F. Chyba, C.P. McKay, New York: Springer-Verlag. 296 pp, 1997.

Drake, M. J., Geochemical constraints on the early thermal history of the Earth, *Z. Naturforsch.* 44a, 883–890, 1989.

Dreibus, G. and Palme, H., Cosmochemical constraints on the sulfur content in the Earth's core, *Geochim. Cosmochim. Acta* 60, 1125–1130, 1996.

Ebel, D. S. and Grossman, L., Condensation in a Model Chicxulub Fireball, *Lunar Planet. Sci. Conf.* XXX, #1906, 1999.

Eggler, D. H. and Lorand, J. P., Sulfides, diamonds and mantle fO_2, *Proc. 5th International Kimberlite Conf.* 160–169, 1995.

Ertel, W., O'Neill, H. St. C., Sylvester, P. J. and Dingwell, D. B., Solubilities of Pt and Rh in a haplobasaltic silicate melt at 1300 °C, *Geochim. Cosmochim Acta* 63, 2439–2449, 1999.

Ertel, W., Walter, M. J., Sylvester, P. J., and Drake, M.J., Experimentally determined solubilities of Pt up to 90 kb and 1850 °C, *Lunar Planet. Sci.* XXXII #1011, 2001.

Fei, Y., Bertka, C. M. and Finger, L. W., High pressure iron-sulfur compound, Fe_3S_2, and melting relations in the Fe-FeS system, *Science* 275, 1621–1624, 1997.

Finnerty, A. A. and Boyd, F. R., Evaluation of thermobarometers for garnet peridotites, *Geochim. Cosmochim Acta* 48, 15–27, 1984.

Flasar, F. M. and Birch, F., Energetics of core formation: A correction, *Jour. Geophys. Res.* 78, 6101–6104, 1973.

Fleet, M. E., R. G. Tronnes, and Stone, W. E., Partitioning of Platinum Group Elements in the Fe-O-S System to 11 GPa and Their Fractionation in the Mantle and Meteorites, *Jour. Geophys. Res. 96,* 21,949–21,958, 1991a.

Fleet, M. E., W. E. Stone, and Crocket, J. H., Partitioning of palladium, iridium, and platinum between sulfide liquid and basalt melt: Effects of melt composition, concentration, and oxygen fugacity, *Geochim. Cosmochim. Acta* 55, 2545–2554, 1991b.

Fleet, M. E. and W. E. Stone, Partitioning of platinum-group elements in the Fe-Ni-S system and their fractionation in nature, *Geochim. Cosmochim. Acta* 55, 245–253, 1991.

Fleet, M. E., J. H. Crocket, and Stone, W. E., Partitioning of platinum-group elements (Os, Ir, Ru, Pt, Pd) and gold between sulfide liquid and basalt melt, *Geochim. Cosmochim. Acta 60,* 2397–2412, 1996.

Fleet, M. E., Liu, M. and Crocket, J. H. Partitioning of trace amounts of highly siderophile elements in the Fe-Ni-S system and their fractionation in nature, *Geochim. Cosmochim Acta* 63, 2611–2622, 1999.

Fortenant S., Gunther, D., Dingwell D. B., and Rubie D. C., Temperature dependence of Pt and Rh solubilities in a haplobasaltic melt, *Geochim. Cosmochim Acta* 67, 123–131, 2003.

Fortenfant S., Rubie D. C., Reid J., Dalpe C., Capmas F., and Gessmann C.K., Partitioning of Re and Os between liquid metal and magnesiowustite at high pressure, *Phys. Earth Planet. Int.* 139, 77–91, 2003.

Gaetani, G. A. and Grove, T. L., Wetting of mantle olivine by sulfide melt: implications for Re/Os ratios in mantle peridotite and late-stage core formation, *Earth Planet. Sci. Lett.* 169, 147–163, 1999.

Gueddari, K., Piboule, M., and Amosse, J., Differentiation of platinum group elements (PGE) and of gold during partial melting of peridotites in the lherzolitic massifs of the Betico-Rifean range (Ronda and Beni Bousera), *Chem. Geol. 134*, 181–197, 1996.

Handler, M. and Bennett, V. C., Behavior of platinum group elements in the subcontinental mantle of eastern Australia during variable metasomatism and melt depletion, *Geochim. Cosmochim. Acta 63*, 3597–3618, 1999.

Hauri, E. H. and S. R. Hart, Re-Os isotope systematics of HIMU and EMII oceanic island basalts from the south Pacific Ocean, *Earth Planet. Sci. Lett. 114*, 353–371, 1993.

Hauri, E. H. and S. R. Hart, Rhenium abundances and systematics in oceanic basalts, *Chem. Geol. 139*, 185–205, 1997.

Hill, E., Wood, B. J., and Blundy, J. D., The effect of Ca-Tschermaks component on trace element partitioning between clinopyroxene and silicate melt, *Lithos 53*, 203–215, 2000.

Hillgren, V. J., Geßmann, C. K. and Li, J., An experimental perspective on the light element in Earth's core, In (R. M. Canup and K. Righter, eds.) *Origin of the Earth and Moon*, Univ. of Arizona Press, Tucson, 245–264, 2000.

Hofmann, A. W., Mantle geochemistry: the message from oceanic volcanism, *Nature 385*, 219–229, 1997.

Holzheid, A. Sylvester, P. J., O'Neill, H. St. C., Rubie, D. C., and Palme, H., Evidence for a late chondritic veneer in the Earth's mantle from high pressure partitioning of palladium and platinum, *Nature 406*, 396–399, 2000.

Hostetler, C. J. and Drake, M. J., On the early global melting of the terrestrial planets, *Proc. Lunar. Planet. Sci. Conf. 11th*, 1915–1929, 1980.

Hsu, W., Huss, G. R., and Wasserburg, G. J., Ion probe measurements of Os, Ir, Pt, Au in individual phases of iron meteorites, *Geochim. Cosmochim. Acta 64*, 1133–1147, 2000.

Humayun, M., Qin, L, and Norman, M. D., Geochemical evidence for excess iron in the mantle beneath Hawaii, *Science 306*, 91–94, 2004.

Ito, E. Katsura, T. and Suzuki, T., Metal/silicate partitioning of Mn, Co, and Ni at high pressures and high temperatures and implications for core formation in a deep magma ocean, in: M. H. Manghnani (eds.), *Properties of Earth and Planetary Materials at High Pressure and Temperature, Geophysical Monograph 101*, AGU, Washington, DC, p. 215–225, 1998.

Jagoutz E., Palme H., Baddenhausen H., Blum K., and Cendales M., The abundances of major, minor and trace elements in the Earth's mantle as derived from primitive ultramafic nodules, *Proc. Lunar Planet. Sci. Conf. 10th*, 2031–50, 1979.

Jana, D. and Walker, D., The influence of sulfur on partitioning of siderophile elements, *Geochim. Cosmochim. Acta 61*, 5255–5277, 1997.

Jones, J. H. and Drake, M. J., Geochemical constraints on core formation in the Earth, *Nature 322*, 221–228, 1986.

Jones, J. H. and Palme, H., Geochemical constraints on the origin of the Earth and Moon, In (R. M. Canup and K. Righter, eds.) *Origin of the Earth and Moon*, Univ. of Arizona Press, Tucson, 197–216, 2000.

Jones, J. H. and Walker, D., Partitioning of siderophile elements in the Fe-Ni-S system: 1 bar to 80 kb, *Earth Planet. Sci. Lett. 105*, 127–133, 1991.

Karason, H. and van der Hilst, R.D., Tomographic imaging of the lowermost mantle with differential times of refracted and diffracted core phases (PKP, Pdiff), *Jour. Geophys. Res.* 106, 6569, 2001.

Kaula, W. M., Thermal evolution of Earth and Moon growing by planetesimal impacts, *Jour. Geophys. Res.* 84, 999–1008, 1979.

Kellogg L. H., Hager B. H., Van der Hilst R. D., Compositional stratification in the deep mantle, *Science* 283, 1881–84, 1999.

Lazar, C., Walker, D. and Walker, R. J., Experimental partitioning of Tc, Mo, Ru, and Re between solid and liquid during crystallization in Fe-Ni-S, *Geochim. Cosmochim. Acta* 68, 643–651, 2004.

Li, J., and Agee, C. B., Geochemistry of mantle–core formation at high pressure, *Nature 381*, 686–689, 1996.

Li, J. and Fei, Y., Experimental Constraints on Core Composition, in *Mantle and Core*, vol. 2, Treatise on Geochemistry, Elsevier, New York, 521–546, 2003.

Li, J., Mao, H. K., Fei, Y., Gregoryanz, E., Eremets, M. and Zha, C.S., Compression of Fe$_3$C to 30 GPa at room temperature, *Phys. Chem. Mineral.* 29, 166–169, 2002.

Lorand, J.-P. and Alard, O., Platinum group element abundances in the upper mantle: new constraints from in situ and whole rock analyses of Massif Central xenoliths (France), *Geochim. Cosmochim. Acta* 65, 2789–2806, 2001.

Lorand, J.-P., Pattou, L. and Gros, M., Fractionation of platinum group elements in the upper mantle: a detailed study in Pyrenean orogenic lherzolites, *Jour. Petrol.* 40, 957–981, 1999.

Lorand, J.-P., Schmidt, G., Palme, H. and Kratz, K.-L., Highly siderophile elements geochemistry of the Earth's mantle: new data for the Lanzo (Italy) and Ronda (Spain) orogenic peridotite bodies, *Lithos 53*, 149–164, 2000.

Mathez, E. A. and Peach, C. L., Geochemistry of platinum group elements in mafic and ultramafic rocks, In *Ore Deposition Associated with Magmas* (J.A. Whitney and A.J. Naldrett, eds.), Reviews in Economic Geology 4, 33–41, 1990.

McDonough, W. F. and Sun, S.-s., The composition of the Earth. *Chemical Geology* 120, 223–253, 1995.

Meibom, A. and Frei, R., Evidence for an ancient osmium isotopic reservoir in Earth, *Science* 296, 516–519, 2002.

Meisel T., Walker R. J., Irving A. J., Lorand J-P., Osmium isotopic compositions of mantle xenoliths: a global perspective, *Geochim. Cosmochim. Acta* 65, 1311–23, 2001.

Menzies, A. H., Carlson, R.W., Shirey, S. B., and Gurney, J. J., Re-Os systematics of diamond-bearing eclogites from the Newlands kimberlite, *Lithos* 71, 323–336, 2003.

Minarik, W. G., Ryerson, F. J. and Watson, E. B., Textural entrapment of core-forming melts, *Science* 272, 530–532, 1996.

Morgan, J. W., Ultramafic xenoliths: clues to Earth's late accretionary history, *Jour. Geophys. Res.* 91, 12375–12387, 1986.

Morgan, J. W., Wandless, G. A., Petrie, R. K. and Irving, A. J., Composition of the earth's upper mantle. I—Siderophile trace elements in ultramafic nodules, *Tectonophysics 75*, 47–67, 1981.

O'Neill, H. St. C., The origin of the Moon and the early history of the Earth—A chemical model. Part 2: The Earth, *Geochim. Cosmochim. Acta 55*, 1159–72, 1991.

Ohtani, E. Yurimoto, H., and Seto, S., Element partitioning between metallic liquid, silicate liquid, and lower-mantle minerals: implications for core formation of the Earth, *Phys. Earth Planet. Int.* 100, 97–114, 1997.

Pattou, L., Lorand, J., and Gros, M., Nonchronditic platinum group element ratios in the Earth's mantle, *Nature 379,* 712–715, 1996.

Peach, C. L., E. A. Mathez, and Keays R. R., Sulfide melt-silicate melt distribution coefficients for noble metals and other chalcophile elements as deduced from MORB: Implications for partial melting, *Geochim. Cosmochim. Acta 54*, 3379–3389, 1990.

Peach, C. L., Mathez, E. A., Keays, R. R., and Reeves, S. J., Experimentally determined sulfide-melt silicate melt partition coefficients for iridium and palladium, *Chem. Geol. 117,* 361–377, 1994.

Pearson, D. G., Snyder, G. A., Shirey, S. B., Taylor, L. A., Carlson, R. W., and Sobolov, N. V., Archean Re-Os age for Siberian eclogites and constraints on Archean tectonics, *Nature 374*, 711–713, 1995.

Puchtel, I. S. and Humayun, M., Platinum group elements in Kostomuksha komatiites and basalts: implications for oceanic crust recycling and core-mantle interaction, *Geochim. Cosmochim. Acta 64*, 4227–4242, 2000.

Righter K., and Downs R. T., Crystal structures of Re- and PGE-bearing magnesioferrite spinels: implications for accretion, impacts and the deep mantle, *Geophys. Res. Lett.* 28, 619–22, 2001.

Righter, K., Drake, M. J., Metal–silicate equilibrium in a homogeneously accreting Earth: new results for Re, *Earth Planet. Sci. Lett.*146, 541–553, 1997.

Righter, K. and E. H. Hauri, Compatibility of Rhenium in Garnet During Mantle Melting and Magma Genesis, *Science 280*, 1737–1741, 1998.

Righter, K. Drake, M. J. and Yaxley, G., Prediction of siderophile element metal—silicate partition coefficients to 20 GPa and 2800 °C: the effect of pressure, temperature, fO_2 and silicate and metallic melt composition, *Phys. Earth Planet. Int. 100*, 115–134, 1997.

Righter, K., Walker, R. J. and Warren, P. H., Significance of Highly Siderophile Elements and Osmium Isotopes in the Lunar and Terrestrial Mantles, In Origin of the Earth and Moon (eds. R. M. Canup and K. Righter), University of Arizona Press, Tucson, 291–322, 2000.

Righter, K., Campbell, A. J., Humayun, M., Diffusion in metal: application to zoned metal grains in chondrites, *Goldschmidt Conference Abstracts*, A640, 2002.

Righter, K., Campbell, A. J., Humayun, M., Diffusion of siderophile elements in Fe metal: application to zoned metal grains in chondrites, *Lunar Planet. Sci. Conf.* XXXIII, abstract 1373, [CD-ROM], 2003.

Righter, K., Campbell, A. J., Humayun, M. and Hervig, R. L., Partitioning of Ru, Rh, Pd, Re, Ir and Au between Cr-bearing spinel, olivine, pyroxene and silicate melts, *Geochim. Cosmochim. Acta 68*, 867–880, 2004.

Ringwood, A. E., Some aspects of the thermal evolution of the Earth, *Geochim. Cosmochim. Acta 20*, 241–259, 1960.

Ringwood, A. E., Composition and Petrology of the Earth's mantle, McGraw-Hill, New York, 618 pp., 1974.

Rubie, D. C., Melosh, H. J., Reid, J. E., Liebske, C., Righter, K., Mechanisms of metal–silicate equilibration in the terrestrial magma ocean, *Earth Planet. Sci. Lett.* 205, 239–255, 2003.

Rushmer T., Minarik W. G., Taylor G. J., Physical processes of core formation, In (R.M. Canup and K. Righter, eds.) *Origin of the Earth and Moon*, Univ. of Arizona Press, Tucson, 227–43, 2000.

Schersten, A., Elliot, T., Hawkesworth, C. and Norman, M., Tungsten isotope evidence that mantle plumes contain no contribution from Earth's core, *Nature 427*, 234–238, 2004.

Schmidt, G., Palme, H., Kratz, K.-L., Kurat, G., Are highly siderophile elements (PGE, Re, Au) fractionated in the upper mantle of the Earth? New results on peridotites from Zabargad, *Chem Geol.* 163, 167–188, 2000.

Schmidt, G., Witt-Eickschen, G., Palme, H., Seck, H., Spettel, B., and Kratz, K.-L., Highly siderophile elements (PGE, Re and Au) in mantle xenoliths from the West Eifel volcanic field (Germany), *Chem. Geol.* 196, 77–105, 2003.

Shannon M. C. and Agee C. B., Percolation of core melts at lower mantle conditions, *Science* 280, 1059–61, 1998.

Shannon, M. C. and Agee, C. B., High pressure constraints on percolative core formation, *Geophys. Res. Lett.* 23, 2717–2720, 1998.

Shaw, G. H., Effects of Core formation, *Phys. Earth Planet. Int.* 16, 361–369, 1978.

Shaw, G. H., Core formation in terrestrial planets, *Phys. Earth Planet. Int.* 20, 42–47, 1978.

Shirey, S. B. and R. J. Walker, The Re-Os isotope system in cosmochemistry and high-temperature geochemistry, *Ann. Rev. Earth Planet. Sci. 26*, 423–500, 1997.

Snow, J. E. and G. Schmidt, Constraints on Earth accretion deduced from noble metals in the oceanic mantle, *Nature 391*, 166–169, 1998.

Solomatov, V. S., Fluid dynamics of a terrestrial magma ocean, In (R.M. Canup and K. Righter, eds.) *Origin of the Earth and Moon*, Univ. of Arizona Press, Tucson, 323–338, 2000.

Solomatov, V. S. and Stevenson, D. J., Nonfractional crystallization of a terrestrial magma ocean, *Jour. Geophys. Res.* 98, 5391–5406, 1993.

Stone, W. E., J. H. Crocket, and Fleet, M. E., Partitioning of palladium, iridium, platinum, and gold between sulfide liquid and basalt melt at 1200 °C, *Geochim. Cosmochim. Acta 54*, 2341–2344, 1990.

Takahashi, E., Melting of a Yamato L3 chondrite (Y-74191) up to 30 kb, *Proc. NIPR 8*, 168–180, 1983.

Taylor, G. J., Core formation in asteroids, *Jour. Geophys. Res.* 97, 14717–14726, 1992.

Tonks, W. B. and Melosh, H. J., The physics of crystal settling and suspension in a turbulent magma ocean, In: The Origin of the Earth, H. Newsom and J. H. Jones, eds., pp. 151–174, Oxford Press, London, 1990.

Tredoux, M., Lindsay, N. M., Davies, G. and McDonald, I., The fractionation of platinum group elements in magmatic systems, with the suggestion of a novel causel mechanism, *S. Afr. J. Geol.* 98, 157–167, 1995.

Urey, H.C., *The Planets.* Yale Univ. Press. New Haven, CT, 245 pp., 1952.

Van der Hilst, R. D. and Karason, H., Compositional heterogeneity in the bottom 1000 kilometers of Earth's mantle: toward a hybrid convection model, *Science 283*, 1885–1888, 1999.

Walker, D., Core participation in mantle geochemistry: Geochemical Society Ingerson Lecture, GSA Denver, October 1999, *Geochim. Cosmochim. Acta 64*, 2897–2911, 2000.

Walker, R.J. and Nisbet, E., [187]Os isotopic constraints on Archean mantle dynamics, *Geochim. Cosmochim. Acta 66*, 3317–3325, 2002.

Walker, R. J., Morgan, J. W. and Horan, M. F., Osmium-187 enrichment in some plumes: evidence for core-mantle interaction?, *Science 269*, 819–822, 1995.

Walker, R. J., E. J., Hanski, J. Vuollo and J. Liipo, The Os isotopic composition of Proterozoic upper mantle: evidence for chondritic upper mantle from the Outokumpu ophiolite, Finland, *Earth Planet. Sci. Lett. 141*, 161–173, 1996.

Walker, R. J., J. W. Morgan, Beary, E. S., Smoliar, M. I., Czamanske, G.K. and Horan, M.F., Applications of the [190]Pt-[186]Os isotope system to geochemistry and cosmochemistry, *Geochim. Cosmochim. Acta 61*, 4799–4807, 1997a.

Walker, R. J., J. W. Morgan, and Hanskii, E., Re-Os systematics of Early Proterozoic ferropicrites, Pechenga Complex, northwestern Russia: Evidence for ancient [187]Os-enriched plumes, *Geochim. Cosmochim. Acta 61*, 3145–3160, 1997b.

Walker, R. J., M. Storey, and Kerr, A.C., Implications of [187]Os isotopic heterogeneities in a mantle plume: evidence from Gorgona Island and Curacao, *Geochim. Cosmochim. Acta 63*, 713–728, 1999a.

Walter, M. J., Newsom, H. E., Ertel, W. and Holzheid, A., Experimental and physical constraints on core formation: behavior of moderately and highly siderophile elements, In (R. M. Canup and K. Righter, eds.) *Origin of the Earth and Moon*, Univ. of Arizona Press, Tucson, 265–290, 2000.

Wänke, H., Constitution of the terrestrial planets, *Phil. Trans. Roy. Soc. Lon Ser. A 303*, 287–302, 1981.

Wood, B. J., Carbon in the core, *Earth Planet. Sci. Lett.* 117, 593–607, 1993.

Wood, B. J., Bryndzia, L. T. and Johnson, K. E., Mantle oxidation state and its relationship to tectonic environment and fluid speciation, *Science* 248, 337–345, 1990.

Yoshino, T., Walter, M. J., and Katsura, T., Core formation in planetesimals triggered by permeable flow, *Nature* 422, 154–157, 2003.

Yoshino, T., Walter, M. J. and Katsura, T., Connectivity of molten Fe alloy in peridotite based on in situ electrical conductivity measurements: implications for core formation in terrestrial planets, *Earth Planet. Sci. Lett.* 222, 625–643, 2004.

Young C.J., Lay T., The core-mantle boundary, *Ann. Rev. Earth Planet. Sci.* 15, 25, 1987.

Zindler, A. and Hart, S. R., Chemical geodynamics, *Ann. Rev. Earth Planet. Sci.* 14, 493–571, 1987.

Kevin Righter, Mail Code KT, NASA Johnson Space Center, 2101 NASA Parkway, Houston, Texas. (kevin.righter-1@nasa.gov)

Mantle Oxidation State and Oxygen Fugacity: Constraints on Mantle Chemistry, Structure, and Dynamics

Catherine A. McCammon

Bayerisches Geoinstitut, Universität Bayreuth, Bayreuth, Germany

This review of mantle oxidation state, based on Mössbauer measurements of mantle xenoliths and synthetic high-pressure phases, suggests relatively low $Fe^{3+}/\Sigma Fe$ in upper mantle and transition zone phases but high $Fe^{3+}/\Sigma Fe$ in lower mantle $(Mg,Fe)(Si,Al)O_3$ perovskite, even under reducing conditions, with significant implications for physical and chemical properties. Whole-rock Fe_2O_3 concentrations for pyrolite mantle are calculated to be low in the upper mantle and transition zone (ca. 0.3 wt% Fe_2O_3), but high in the lower mantle (ca. 4 wt% Fe_2O_3). High Fe_2O_3 concentrations are probably balanced according to the iron disproportionation reaction $3\,Fe^{2+} = Fe^0 + 2\,Fe^{3+}$. Oxygen fugacity is relatively high at the top of the upper mantle (near the fayalite–magnetite–quartz buffer) due to the concentration of Fe^{3+} in modally minor phases, but probably low in the transition zone and lower mantle (close to metal equilibrium). Minerals in subducting slabs may be more oxidised, however, with higher Fe_2O_3 concentrations. Coupled redox reactions between iron oxidation and the reduction of oxidised species such as carbonate may contribute to the genesis of diamonds in the lower mantle. Disproportionation of iron in the lower mantle to produce ca. 1 wt% Fe-rich metal may have occurred in the early Earth, with important consequences for mantle geochemistry. If the metal phase does not segregate appreciably, material can move across the transition zone–lower mantle boundary without a net enrichment or depletion in oxygen.

1. INTRODUCTION

The oxidation state of the Earth's mantle is related to the oxygen fugacity (fO_2) through mineral equilibria, and both oxidation state and fO_2 play an important role in many chemical and physical processes, including speciation of volatiles, magma genesis, metasomatism, and evolution of the Earth's atmosphere [e.g., *Kasting et al.*, 1993]. The oxidation state of the mantle as measured by $Fe^{3+}/\Sigma Fe$ is well constrained for primitive upper mantle where direct samples are available [e.g., *O'Neill et al.*, 1993a; *Canil et al.*, 1994] but must rely on estimates from high-pressure experiments

in deeper regions. *O'Neill et al.* [1993a] calculated Fe^{3+} concentrations for the upper mantle, transition zone, and lower mantle based on data existing at the time, which provided a foundation for mantle models over the following decade. However, the report that $(Mg,Fe)(Si,Al)O_3$ perovskite likely contains significantly more Fe^{3+} than previously estimated [*McCammon*, 1997], coupled with subsequent high-pressure experiments on the lower-mantle assemblage [*Lauterbach et al.*, 2000; *Frost and Langenhorst*, 2002; *McCammon et al.*, 2004b; *Frost et al.*, 2004], motivates a reexamination of mantle oxidation state in light of the new data.

This chapter reviews our knowledge of mantle oxidation state, focusing particularly on data collected since the analysis of *O'Neill et al.* [1993a], and considers pyrolite mantle, midocean ridge basalt (MORB), and hydrous phases. The link between $Fe^{3+}/\Sigma Fe$ and fO_2 is elucidated, with implica-

Earth's Deep Mantle: Structure, Composition, and Evolution
Geophysical Monograph Series 160

tions for the chemistry, structure, and dynamics of the mantle. The overall aim is to provide an overview of the current picture of fO_2 and oxidation state within the Earth at a level that is accessible to geochemists and geophysicists.

2. RELATION BETWEEN OXIDATION STATE AND fO_2

The oxidation state of the mantle has been interpreted in different ways but is perhaps most accurately defined in terms of the oxidation state of its constituent elements. The bulk composition of the Earth is broadly chondritic in composition, although no existing meteorites share all of the Earth's inferred compositional characteristics [Drake and Righter, 2002]. Various approaches have been used to estimate primitive mantle composition, which overall show similar general features (Table 1).

Iron is by far the most abundant element with a variable oxidation state. The oxidation state of the mantle, therefore, is often quantified in terms of the relative abundance of the three most common oxidation states of iron: Fe^0, Fe^{2+}, and Fe^{3+}. Commonly the $Fe^{3+}/\Sigma Fe$ ratio (expressed as an atomic fraction) is used to indicate the degree of oxidation of either individual minerals or the bulk assemblage.

It is worthwhile at this point to explore the differences between Fe^{2+} and Fe^{3+}. It involves more than simply the removal of a d electron, since both chemical and physical properties of the mineral phases can be affected. For example, when Fe^{2+} is substituted for Mg^{2+}, the charge balance remains the same. But if Fe^{3+} substitutes for Mg^{2+}, there is a charge excess that can only be balanced by effects such as coupled substitutions (e.g., 2 Fe^{3+} for Mg^{2+}–Si^{4+}), the creation of defects, or the addition or loss of volatiles such as hydrogen. Properties highly sensitive to such effects include sub- and supersolidus phase relations, trace element partitioning, electrical conductivity, diffusivity, elasticity, rheology, and the nature of volatile species.

Other elements may be used to quantify oxidation state, depending on the composition of the system. Carbon, for example, exists in both oxidised (carbonate) and reduced (graphite/diamond, carbon monoxide, or methane) forms, as do hydrogen and sulphur. However this chapter focuses on iron as the primary measure of oxidation state because of its high abundance.

fO_2 is a more elusive parameter. Oxygen is the most abundant element in the mantle, and although its abundance can be measured directly, it is more commonly expressed as a chemical potential. On its own, oxygen is a highly reactive gas but readily combines with other elements to produce the oxides and silicates that make up the bulk of the Earth. fO_2 is a measure of oxygen reactivity in mineral assemblages, even

Table 1. Some Chemical Models for the Mantle.

Type	Primitive mantle	Primitive mantle	Primitive mantle Modified chondritic	Bulk Earth Modified chondritic
Model	Xenolith	Pyrolite	chondritic	chondritic
O	445000	440000	447900	297000
Mg	222300	224500	227767	154000
Si	214800	209300	215567	163000
Fe	**58900**	**65300**	**58182**	**319000**
Ca	25300	25730	23099	17100
Al	22200	23600	21646	15900
Cr	3011	2935	2600	4700
Na	2889	2545	2671	1800
Ni	2108	1890	1964	18220
Ti	1350	1280	1349	810
Mn	1021	1080	1154	720
K	231	240	280	160
Co	105	105	–	880
V	82.1	82	–	95
S	80	350	–	6345
P	64.5	95	–	1100
Zn	48.5	56	–	40
C	46.2	250	–	730
Cu	28.5	30	–	60
H	–	–	–	260
Reference	[1]	[2]	[3]	[4]

All compositions are given in ppm.
–: not estimated.
References: [1] Wänke et al. [1984]; [2] Ringwood [1975]; [3] Allègre et al. [1995]; [4] McDonough [1999].

though oxygen is not present as a free gas, and can be roughly described as the oxidising potential of the phase assemblage.

The iron oxidation state and fO_2 are closely linked. In simple systems such as Fe-O, an increase in fO_2 favours the reaction

$$2FeO + \tfrac{1}{2} O_2 = Fe_2O_3, \qquad (1)$$

which stabilises iron in higher oxidation states, whereas low fO_2 stabilises iron in lower oxidation states. This concept can be applied to melt and glass systems, where various empirical relations have been derived to link fO_2 with Fe_2O_3 concentration [e.g., Kress and Carmichael, 1991]. In complex crystalline systems, however, mineral compositions also play an important role in the link between oxidation state and fO_2. This means that changes in fO_2 are not necessarily reflected by changes in iron oxidation state, and vice versa. In the mantle, where crystalline phases dominate, it is therefore necessary to consider both iron oxidation state and fO_2.

3. MEASURING OXIDATION STATE

There are numerous methods for determining major, minor, and trace element concentrations in minerals, but few of these methods are able to distinguish between different oxidation states, e.g., Fe^{2+} and Fe^{3+}. Wet chemistry has traditionally been used but generally requires large amounts of sample. Currently, spectroscopic methods offer the best possibilities to determine oxidation state.

The most common spectroscopic method in use is Mössbauer spectroscopy, which has the added advantage that structural and dynamic information can also be obtained [see instructional reviews by *Hawthorne*, 1988; *McCammon*, 2000, 2004; *Amthauer et al.*, 2004]. Significant advances in spatial resolution allow the analysis of samples as small as 100 μm [*McCammon et al.*, 1991; *McCammon*, 1994, 2004], which has motivated many recent studies of mantle phases. Higher spatial resolution can be obtained from other methods in current use, including electron energy loss spectroscopy (EELS), X-ray absorption spectroscopy, X-ray photoelectron spectroscopy, and X-ray emission spectroscopy. Advantages and disadvantages of all of these methods are reviewed by *McCammon* [1999].

4. MANTLE OXIDATION STATE

There are two approaches to determining Fe^{3+} concentrations in mantle minerals. The first is to examine natural samples from the mantle, which provides the most direct determination. The depth of sampling is limited to roughly 200 km, however, although there are a few reports of rocks from greater depths. Mineral inclusions in diamonds also provide a source of mantle samples, but the extent to which they represent bulk mantle chemistry remains an open question. A second method is to synthesise samples in the laboratory under appropriate conditions of pressure, temperature, and fO_2. Since the latter parameter is not well known in all parts of the Earth, it should be varied between minimum and maximum plausible values, e.g., between Fe metal equilibrium and several log-bar units above the fayalite–magnetite–quartz (FMQ) buffer. The mineralogical constitution of the Earth's mantle provides a road map for experiments (Figure 1).

In the remainder of this section I describe the extent of Fe^{3+} substitution in iron-bearing mantle minerals from measurements of both natural and synthetic samples. The intention is not to list all available data, but rather to provide a general picture from which conclusions regarding the concentration of Fe^{3+} in different parts of the mantle can be drawn. The data are summarised in Table 2.

4.1. Upper Mantle Phases

Upper-mantle peridotite is accessible through abyssal peridotites, peridotite massifs intruded into the continental

crust, and xenoliths entrained in alkali basalts and kimberlites. The dominant iron-bearing peridotite minerals are olivine, orthopyroxene, clinopyroxene, and garnet (Figure 1), whereas spinel is stable at shallower depths. Complementing the xenolith suite are mineral inclusions in diamonds, sometimes entrained in the same eruptive process. In addition, high-pressure laboratory studies of synthetic samples have investigated aspects such as the solubility of Fe^{3+} under different conditions. Figure 2 illustrates Mössbauer spectra from minerals contained in upper-mantle xenoliths.

Olivine incorporates negligible Fe^{3+} according to Mössbauer studies of samples from the upper mantle [e.g., *Canil et al.*, 1994; *Canil and O'Neill*, 1996], which is supported by crystal chemical arguments [*O'Neill et al.*, 1993a] and the results of thermogravimetric experiments [*Nakamura*

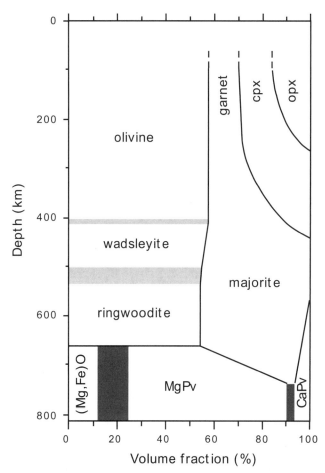

Figure 1. Relative mineral proportions below 80-km depth from *Bina* [1998] based on a pyrolite mantle composition [*Ringwood*, 1975]. Gray regions indicate divariant phase transformations in olivine polymorphs; dark regions denote uncertainties in the modal abundance of ferropericlase and $CaSiO_3$ perovskite [after *Bina*, 1998]. opx: orthopyroxene; cpx: clinopyroxene; MgPv: $(Mg,Fe)(Si,Al)O_3$ perovskite; CaPv: $CaSiO_3$ perovskite.

Table 2. $Fe^{3+}/\Sigma Fe$ for Typical Compositions of Mantle Minerals.

	$Fe^{3+}/\Sigma Fe$ at minimum fO_2	Influence of increasing fO_2	Reference
Upper mantle			
olivine	< 1%	Fe_3O_4 exsolution	[1,2]
orthopyroxene	< 1%	incr $Fe^{3+}/\Sigma Fe$	[3]
clinopyroxene	< 1%	incr $Fe^{3+}/\Sigma Fe$	[4]
spinel	< 1%	incr $Fe^{3+}/\Sigma Fe$	[5]
garnet	< 1%	incr $Fe^{3+}/\Sigma Fe$	[6]
Transition zone			
wadsleyite	2–3%	incr $Fe^{3+}/\Sigma Fe$	[7,8]
ringwoodite	2–4%	incr $Fe^{3+}/\Sigma Fe$	[7,8]
majorite	5–7%	incr $Fe^{3+}/\Sigma Fe$	[7,8]
Lower mantle			
$(Mg,Fe)(Si,Al)O_3$ perovskite	10–60%	none	[9,10]
ferropericlase	< 1%	incr $Fe^{3+}/\Sigma Fe$	[11,12]
$CaSiO_3$ perovskite	100%?	none?	[13]

References: [1] *Nitsan* [1974]; [2] *Nakamura and Schmalzried* [1983]; [3] *Annersten et al.* [1978]; [4] *Luth and Canil* [1993]; [5] *Wood and Virgo* [1989]; [6] *Gudmundsson and Wood* [1995]; [7] *O'Neill et al.* [1993b]; [8] *McCammon et al.* [2004a]; [9] *Lauterbach et al.* [2000]; [10] *McCammon et al.* [2004b]; [11] *McCammon et al.* [1998]; [12] *Frost et al.* [2001]; [13] *Bläß et al.* [2004].

and Schmalzried, 1983]. Studies of modally metasomatised xenoliths have shown higher Fe^{3+} concentrations [*McGuire et al.,* 1991], but transmission electron microscopy showed that this was due to the presence of laihunite lamellae, and not Fe^{3+} in the olivine crystal structure itself [*Banfield et al.,* 1992]. The negligible solubility of Fe^{3+} in olivine is a crucial point in understanding the mantle redox state, as it demonstrates that a phase with extremely low $Fe^{3+}/\Sigma Fe$ can be in equilibrium with nominally oxidised minerals and/or fluids.

Orthopyroxene incorporates amounts of Fe^{3+} ranging from ca. 4-10% $Fe^{3+}/\Sigma Fe$ according to Mössbauer studies of mantle samples [e.g., *Dyar et al.,* 1989; *Luth and Canil,* 1993; *Canil et al.,* 1994; *Canil and O'Neill,* 1996]. In the spinel peridotite facies there is a slight positive correlation between orthopyroxene Fe^{3+} concentration and equilibration temperature that, coupled with the observation that whole-rock Fe_2O_3 is independent of temperature, suggests the positive correlation may be due to the decrease in the modal abundance of spinel with increasing temperature [*Canil and O'Neill,* 1996].

Clinopyroxene incorporates more Fe^{3+} than orthopyroxene, with amounts ranging from ca. 10% to 40% $Fe^{3+}/\Sigma Fe$ according to Mössbauer studies of mantle samples [*Dyar et al.,* 1989; *Luth and Canil,* 1993; *Canil et al.,* 1994; *Canil and O'Neill,* 1996]. Fe^{3+} concentrations in ortho- and clinopyroxene are strongly correlated in both the spinel and garnet peridotite facies [*Canil and O'Neill,* 1996], and Fe^{3+}

concentration in clinopyroxene varies with fO_2 [*Luth and Canil,* 1993].

High-pressure experiments have shown that orthopyroxene and clinopyroxene undergo progressive mutual solid solution with increasing depth in the upper mantle to form a single pyroxene phase (space group $C2/c$), and although the pressure interval over which this occurs is dependent on bulk mantle composition, it seems relatively well established

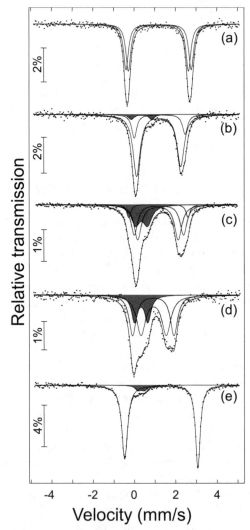

Figure 2. Room-temperature Mössbauer spectra of upper-mantle minerals from mantle xenoliths: (a) olivine and (b) orthopyroxene (Pali-Aike garnet lherzolite, Argentina); (c) clinopyroxene (Roberts Victor eclogite, South Africa); (d) spinel (Kilbourne Hole spinel lherzolite, USA); (e) garnet (Jericho garnet peridotite, Canada). $Fe^{3+}/\Sigma Fe$ is determined based on the relative areas of Fe^{3+} doublets (dark shaded areas). The abscissa (v) gives the energy shift according to the Doppler effect ($E = E_0\, v/c$, where E_0 is the initial energy of the gamma ray and c is the speed of light), and is calibrated relative to α-Fe.

that orthopyroxene disappears at pressures no greater than 10 GPa [e.g., *Zhang and Herzberg*, 1994]. (Mg,Fe)SiO$_3$ clinopyroxene (*C2/c*) contains no Fe^{3+} under highly reducing conditions [*O'Neill et al.*, 1993b], but greater Fe^{3+} concentrations are expected at higher *f*O$_2$. Fe^{3+} may be incorporated into the structure as an M^{2+}Fe^{3+}$_2$SiO$_6$ component (M=Ca^{2+}, Mg^{2+}, Fe^{2+}), similar to the inferred behaviour of orthopyroxene and clinopyroxene at low pressure [*Canil and O'Neill*, 1996]. *Huckenholz et al.* [1969] reported up to 28 mol% of this M^{2+}Fe^{3+}$_2$SiO$_6$ component in hedenbergite (CaFe^{2+}Si$_2$O$_6$) at 1175°C and 1 bar under oxidising conditions, while *Woodland and O'Neill* [1995] found no more than 4 mol% at higher pressure (up to 8.5 GPa) at conditions that were probably more reducing. Although there are no Mössbauer studies of high-pressure clinopyroxene phases with compositions relevant to the deep upper mantle, based on the above results and the nature of the high-pressure clinopyroxene structure [e.g., *Hugh-Jones et al.*, 1994; *Tribaudino et al.*, 2003], it seems reasonable to assume that the ability of mantle pyroxene to incorporate Fe^{3+} will not diminish with depth during the progressive dissolution of orthopyroxene into clinopyroxene.

Spinel, because of its prominence in oxybarometers [e.g., *Mattioli and Wood*, 1986], is probably the most extensively studied mantle mineral with regard to Fe^{3+} concentration. Numerous Mössbauer studies have been carried out, and the high iron concentration coupled with the low abundance of high-valence elements such as Si allow the determination of Fe^{3+} based on stoichiometric calculations, provided that previously calibrated standards for Fe^{3+}/ΣFe are used [e.g., *Wood and Virgo*, 1989]. Spinel is one of the major repositories for Fe^{3+} in the top of the upper mantle, with amounts ranging from ca. 15% to 35% Fe^{3+}/ΣFe [e.g., *Dyar et al.*, 1989; *Canil et al.*, 1994; *Canil and O'Neill*, 1996]. Spinel is stable within the peridotite assemblage at the top of the upper mantle, but at higher pressures it is replaced by garnet. The transition pressure and temperature depend on bulk composition [e.g., *O'Neill*, 1981; *Webb and Wood*, 1986], and for a pyrolite composition, the transition occurs at ca. 4 GPa along a typical continental geotherm.

Garnet has also been the subject of numerous studies to determine Fe^{3+} due to its importance in mineral oxybarometers [e.g., *Luth et al.*, 1990; *Gudmundsson and Wood*, 1995]; unlike spinel, however, Fe^{3+} in garnet cannot be quantified through calibrated electron microprobe measurements [e.g., *Canil and O'Neill*, 1996], and all data have thus come from Mössbauer studies. Garnet incorporates less Fe^{3+} than spinel, with amounts ranging from ca. 3% to 13% Fe^{3+}/ΣFe [e.g., *Luth et al.*, 1990; *Canil et al.*, 1994; *Canil and O'Neill*, 1996; *Woodland and Koch*, 2003]. Garnet peridotite shows a positive correlation between garnet Fe^{3+} concentration

and temperature [*Luth et al.*, 1990; *Canil and O'Neill*, 1996; *Woodland and Koch*, 2003]; coupled with the lack of correlation between whole-rock Fe$_2$O$_3$ and temperature, this correlation has been interpreted to indicate a temperature-dependent partitioning of Fe^{3+} between garnet and clinopyroxene [*Canil and O'Neill*, 1996].

Garnet is stable within the peridotite assemblage above pressures of ca. 4 GPa with compositions predominantly in the ternary system pyrope–almandine–grossular. A progressive solid solution of clinopyroxene into garnet becomes observable at pressures above 8–9 GPa, continuing with depth until the disappearance of clinopyroxene within the transition zone [e.g., *Ito and Takahashi*, 1987; *Irifune and Ringwood*, 1987]. High-pressure experiments have shown that the solubility of the Fe^{3+}-component Fe^{2+}$_3$Fe^{3+}$_2$Si$_3$O$_{12}$ in almandine (Fe^{2+}$_3$Al$_2$Si$_3$O$_{12}$) increases with increasing pressure, forming a complete solid solution above 10 GPa [*Woodland and O'Neill*, 1993]. One can reasonably assume, therefore, that the ability of upper-mantle garnet to incorporate Fe^{3+} will not diminish with increasing pressure.

4.2. Transition Zone Phases

The transition zone is not generally accessible to sampling except as rare inclusions in diamonds [e.g., *Stachel et al.*, 2000a; *Stachel*, 2001] and as ultradeep xenoliths [*Haggerty and Sautter*, 1990; *Sautter et al.*, 1991; *Collerson et al.*, 2000], although this last report is still controversial [*Neal et al.*, 2001]. The most comprehensive information on Fe^{3+} in transition zone minerals comes from laboratory studies, where composition, pressure, temperature, *f*O$_2$, and water fugacity can all be varied. Figure 3 illustrates Mössbauer spectra of transition zone phases synthesised in high-pressure experiments.

O'Neill et al. [1993b] developed a method to synthesise phases at their minimum *f*O$_2$ stability limit, i.e., in equilibrium with Fe metal and stishovite according to the reaction

$$Fe_nSiO_{2+n} = nFe + SiO_2 + (n/2) O_2. \qquad (2)$$

This defines the minimum Fe^{3+} concentration for the phase and is important for assessing the probable Fe^{3+} concentration in the mantle.

Wadsleyite incorporates significantly more Fe^{3+} than olivine. At its minimum *f*O$_2$ stability limit, *O'Neill et al.* [1993b] found 3% Fe^{3+}/ΣFe in (Mg,Fe)$_2$SiO$_4$ wadsleyite. A subsequent study using a Re capsule (i.e., under likely more oxidising conditions) showed a higher Fe^{3+}/ΣFe value [*Fei et al.*, 1992]. These results are consistent with high-pressure experiments reporting a wadsleyite-structured phase in the system Fe$_2$SiO$_4$–Fe$_3$O$_4$, where the presence of Fe^{3+} is crucial

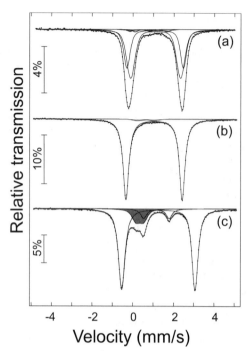

Figure 3. Room-temperature Mössbauer spectra of transition zone phases synthesised at high pressure: (a) $(Mg,Fe)_2SiO_4$ wadsleyite and (b) $(Mg,Fe)_2SiO_4$ ringwoodite [*McCammon et al.*, 2004a]; (c) $(Mg,Fe)SiO_3$ majorite [*McCammon and Ross*, 2003]. $Fe^{3+}/\Sigma Fe$ is determined based on the relative areas of Fe^{3+} doublets (shaded dark gray) and $Fe^{2.5+}$ doublets (shaded light gray). Energy calibration is the same as for Figure 2.

in stabilising the first Fe-silicate reported to crystallise in the wadsleyite structure [*Woodland and Angel*, 1998]. Atomistic computer simulations have shown that the linked tetrahedra in the wadsleyite structure are able to distort more easily to accommodate Fe^{3+} than are the isolated SiO_4 tetrahedra in olivine [*Richmond and Brodholt*, 2000].

Ringwoodite also incorporates measurable Fe^{3+}. The iron end-member, Fe_2SiO_4, forms a complete solid solution with Fe_3O_4 above 8 GPa [*Woodland and Angel*, 2000], and high-pressure experiments on $(Mg,Fe)_2SiO_4$ ringwoodite at its minimum fO_2 stability limit found 4% $Fe^{3+}/\Sigma Fe$ [*O'Neill et al.*, 1993b], although smaller values have also been reported [*Suito et al.*, 1984].

Majorite is probably the primary repository for Fe^{3+} in the transition zone. At its minimum fO_2 stability limit, *O'Neill et al.* [1993b] found 7% $Fe^{3+}/\Sigma Fe$ for $(Mg,Fe)SiO_3$ majorite, and a subsequent high-pressure study showed that $Fe^{3+}/\Sigma Fe$ could reach 22% under more oxidising conditions as a function of both total iron content and fO_2 [*McCammon and Ross*, 2003]. Fe^{3+} concentrations in high-pressure samples containing Al are lower due to the competition between Fe^{3+} and Al for the same octahedral site in the structure [*McCammon and Ross*, 2003].

4.3. Lower Mantle Phases

The lower mantle is accessible only as rare inclusions in diamonds [e.g., *Harte et al.*, 1999; *Stachel et al.*, 2000b; *McCammon*, 2001]; hence, the bulk of available information on Fe^{3+} in lower-mantle minerals comes from laboratory studies. Figure 4 illustrates Mössbauer spectra of lower mantle phases synthesised in high-pressure experiments.

$(Mg,Fe)(Si,Al)O_3$ perovskite is the most abundant phase in the lower mantle and hence dominates its physical and chemical properties. Initial studies of Fe^{3+} concentration in Al-free $(Mg,Fe)SiO_3$ perovskite under silica-saturated conditions showed 9–16% $Fe^{3+}/\Sigma Fe$ [*McCammon et al.*, 1992; *Fei et al.*, 1994; *McCammon*, 1998], depending on fO_2. Subsequent studies of compositions incorporating Al found significantly higher concentrations—up to 82% $Fe^{3+}/\Sigma Fe$ [*McCammon*, 1997; *Lauterbach et al.*, 2000; *Frost and Langenhorst*, 2002]. These high concentrations, which have been observed by both Mössbauer spectroscopy and EELS, arise from the favourable energetics of coupled Fe^{3+}–Al substitution into $MgSiO_3$ perovskite [e.g., *Richmond and Brodholt*, 1998; *Navrotsky*, 1999]. There is a strong positive correlation between $Fe^{3+}/\Sigma Fe$ and Al concentration in $(Mg,Fe)(Si,Al)O_3$ perovskite [*Lauterbach et al.*, 2000], with evidence for a change in substitution mechanism at higher Al concentrations [*Frost and Langenhorst*, 2002]. The relation between $Fe^{3+}/\Sigma Fe$ and Al concentration is independent of fO_2, and high concentrations of Fe^{3+} are found even under highly reducing conditions [*Lauterbach et al.*, 2000;

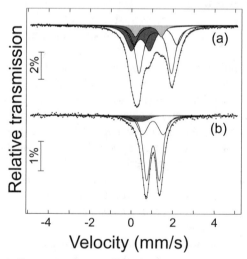

Figure 4. Room-temperature Mössbauer spectra of lower mantle phases synthesised at high pressure: (a) $(Mg,Fe)(Si,Al)O_3$ perovskite [*Lauterbach et al.*, 2000]; (b) $(Mg,Fe)O$ ferropericlase [*McCammon et al.*, 1998]. $Fe^{3+}/\Sigma Fe$ is determined based on the relative areas of Fe^{3+} doublets (dark gray) and $Fe^{2.5+}$ doublets (light gray). Energy calibration is the same as for Figure 2.

McCammon et al., 2004b; *Frost et al.*, 2004]. Charge balance at low fO_2 occurs through disproportionation (3 Fe^{2+} = Fe^0 + 2 Fe^{3+}), where iron metal is formed in discrete blebs [*Lauterbach et al.*, 2000; *Frost et al.*, 2004].

The effect of Al on the oxidation state of iron in $(Mg,Fe)(Si,Al)O_3$ perovskite can be explained simply in terms of cation size and charge balance. Mg^{2+} and Si^{4+} occupy two different sites in the perovskite structure in the end-member composition $MgSiO_3$. Fe^{2+} substitutes for Mg^{2+} on the same site, maintaining charge balance in the structure. Oxidation of Fe^{2+} to Fe^{3+} is hindered, however, because the excess positive charge limits the amount of Fe^{3+} that can be incorporated in $(Mg,Fe)SiO_3$ perovskite. In contrast, addition of the slightly smaller cation Al^{3+} enables the isovalent substitution of Mg^{2+} and Si^{4+} by the pair (Fe^{3+}, Al^{3+}) without changing the charge balance in the structure. Also, the energy associated with substitution of the trivalent cations on the octahedral site, with the creation of oxygen vacancies to balance charge, appears to be lower when both Fe and Al are present. Evidence for the presence of oxygen vacancies in $(Mg,Fe)(Si,Al)O_3$ has been found in high-pressure experiments [*Lauterbach et al.*, 2000; *Xu and McCammon*, 2002].

At the top of the lower mantle, $(Mg,Fe)_2SiO_4$ ringwoodite transforms to $(Mg,Fe)SiO_3$ perovskite plus $(Mg,Fe)O$. In a pyrolite composition, the perovskite phase contains ca. 0.08 Al per formula unit (p.f.u.), which increases to ca. 0.1 Al p.f.u. at the point where the majorite–perovskite transformation is complete [e.g., *Irifune*, 1994; *Wood*, 2000; *Hirose*, 2002; *Nishiyama and Yagi*, 2003]. These concentrations of Al in $(Mg,Fe)(Si,Al)O_3$ perovskite imply 50–60% $Fe^{3+}/\Sigma Fe$ according to current experimental data, independent of fO_2 [as shown above, $Fe^{3+}/\Sigma Fe$ in $(Mg,Fe)(Si,Al)O_3$ perovskite is determined primarily by its Al concentration, and is independent of fO_2].

The high inferred concentrations of Fe^{3+} in $(Mg,Fe)(Si,Al)O_3$ perovskite are supported by observations of inclusions in diamonds believed to have originated in the lower mantle. Although no minerals with the perovskite structure have been recovered, $(Mg,Fe)(Si,Al)O_3$ phases show compositional trends that are consistent with existence at lower-mantle conditions in the perovskite structure, with subsequent transformation within the diamond during exhumation [*Harte et al.*, 1999; *Stachel et al.*, 2000b]. $Fe^{3+}/\Sigma Fe$ appears to be preserved during exhumation, and shows a trend with Al concentration that is consistent with results from high-pressure experiments [*McCammon et al.*, 2004c].

Ferropericlase can incorporate Fe^{3+} at high pressure and temperature, depending on the iron concentration and the prevailing fO_2, with amounts ranging from 0 to 10% $Fe^{3+}/$

ΣFe for compositions relevant to the lower mantle [e.g., *McCammon et al.*, 1998; *Frost et al.*, 2001]. This is significantly less than at atmospheric pressure [e.g., *Speidel*, 1967] due to a high-pressure phase transformation in the system Fe_3O_4–$MgFe_2O_4$ [e.g., *Andrault and Bolfan-Casanova*, 2001]. A recent Mössbauer study of ferropericlase inclusions in diamonds showed a nearly linear variation of trivalent cation abundance with monovalent cation abundance, suggesting a substitution of the form $Na_{0.5}M^{3+}_{0.5}O$ (M = Fe^{3+}, Cr^{3+}, Al^{3+}) [*McCammon et al.*, 2004c].

$CaSiO_3$ perovskite incorporates only small amounts of iron, e.g., 0.01 atoms p.f.u. in a peridotite bulk composition [*Irifune*, 1994] and 0.06 atoms p.f.u. in a MORB bulk composition [*Hirose et al.*, 1999]. Nevertheless, there are indications that a significant proportion of iron may occur as Fe^{3+}. *Wang and Yagi* [1998] reported the high-pressure synthesis of $(Ca,Fe^{3+})(Si,Fe^{3+})O_3$ perovskite from andradite, and all synthesis runs in the system $CaSiO_3$–$CaFe^{3+}O_{2.5}$ that were examined using Mössbauer spectroscopy contained only Fe^{3+} [*Bläß et al.*, 2004]. In addition, ab initio calculations showed that Fe^{3+} substitution accompanied by oxygen vacancies was favoured in $CaSiO_3$ perovskite below 30 GPa, and coupled Fe^{3+}–Al substitution above 30 GPa [*Chaplin et al.*, 1999].

4.4. Effect of Water in Nominally Anhydrous Minerals

Studies have shown that water can be incorporated into a number of mantle minerals that are considered to be nominally anhydrous [e.g., *Bell and Rossman*, 1992]. This is relevant to Fe^{3+} concentrations because (1) some suggested mechanisms of dehydrogenation during ascent of mantle material involve oxidation of Fe^{2+} to Fe^{3+} [e.g., *Ingrin and Skogby*, 2000] and (2) variation in water concentration within the mantle may cause variations in Fe^{3+} concentration, depending on the link between structurally incorporated OH and Fe^{3+}.

Significant water concentrations have been reported in upper-mantle xenolith minerals: clinopyroxene (100–1,300 wt ppm), orthopyroxene (50–650 wt ppm), garnet (0–200 wt ppm), and olivine (0–150 wt ppm) [reviewed by *Ingrin and Skogby*, 2000]. Concerns have been raised, however, regarding the extent to which these measured water contents represent actual mantle values. During the ascent of mantle xenoliths, dehydrogenation reactions may take place, where the most plausible mechanism involves the oxidation of Fe^{2+} to Fe^{3+}. This implies that Fe^{3+} concentrations could potentially be higher in recovered samples than in the mantle region from which they were derived. However, *Bell and Rossman* [1992] argued that most water concentrations measured in minerals probably reflect those of their source region, given observed correlations between water content and petrological environment. Also, a recent study examined correlations between water concentration and major-element compositional data from sub-arc mantle wedge xenoliths, and concluded that the measured water concentrations probably represent unperturbed mantle values [*Peslier et al.*, 2002], which would imply that Fe^{3+} is also unchanged. On the other hand, *Ingrin and Skogby* [2000] argued that the kinetics of the dehydrogenation reaction is sufficiently fast for some minerals to have lost part of their initial water concentration during ascent from the mantle. This would apply primarily to pyroxene, since Fe^{3+} concentrations in olivine are below the detection limit, and garnet does not show the inverse correlation that would be expected between water content and Fe^{3+} concentration if significant dehydrogenation were to have occurred. While the question regarding water loss (and consequent oxidation of Fe^{3+}) during ascent still remains open, for the purpose of this review the range of Fe^{3+} concentrations measured for upper mantle samples are assumed to represent upper mantle values.

Crystal structure refinements of nominally anhydrous high-pressure phases that can incorporate water have shown that the principal hydration mechanism is by octahedral site vacancy [e.g., *Smyth and Kawamoto*, 1997; *Smyth et al.*, 2003, 2005]. An octahedral site vacancy ($\square$) can also be formed in the absence of hydrogen through oxidation of Fe^{2+} to Fe^{3+}:

$$2\ Fe^{2+} + Mg^{2+} + \tfrac{1}{2}\ O_2 = 2\ Fe^{3+} + \square + MgO, \qquad (3)$$

where higher fO_2 would favour a higher concentration of vacancies. The coupling of these two mechanisms was suggested by *Kurosawa et al.* [1997], who found a positive correlation between trivalent and monovalent (including hydrogen) cations in olivine. They suggested a structural model where octahedral vacancies are balanced by protonation of oxygen on one side, and oxidation of Fe^{2+} to Fe^{3+} on the other. Hydration under oxidising conditions could therefore stabilise large concentrations of Fe^{3+}. Such large Fe^{3+} concentrations have indeed been found in Mössbauer studies of olivine, wadsleyite, and ringwoodite synthesised at high pressure under hydrous conditions at high fO_2 [*McCammon et al.*, 2004a]. While such water-saturated, oxidising conditions are probably not relevant for the bulk mantle, they may be applicable to regions near subduction zones. These data are summarised in Figure 5.

The solubility of water in lower-mantle minerals remains a controversial issue [*Bolfan-Casanova et al.*, 2000; cf. *Litasov et al.*, 2003]. Nevertheless, EELS measurements of hydrous $(Mg,Fe)(Si,Al)O_3$ perovskites show $Fe^{3+}/\Sigma Fe$ values of ca. 60%, which are at least as large as values observed for anhydrous samples [*Litasov et al.*, 2003].

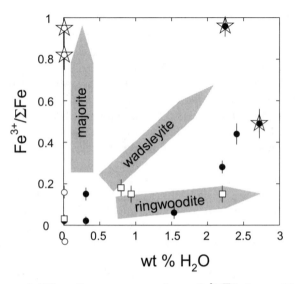

Figure 5. Effect of water concentration on $Fe^{3+}/\Sigma Fe$ in transition zone phases synthesised in the multianvil press. Symbols represent wadsleyite (solid circles), ringwoodite (open squares), and majorite (open circles). Stars indicate the runs carried out under extremely oxidising conditions, which suggests a relation between $Fe^{3+}/\Sigma Fe$ and fO_2. The trend of $Fe^{3+}/\Sigma Fe$ with H_2O concentration is indicated for each mineral [after *McCammon et al.*, 2004a].

5. IMPLICATIONS FOR PHYSICAL AND CHEMICAL PROPERTIES

The result that ca. 50% of iron in lower-mantle $(Mg,Fe)(Si,Al)O_3$ perovskite occurs as Fe^{3+}, independent of fO_2, is astonishing news for geophysicists and geochemists who have traditionally assumed that all iron occurs as Fe^{2+}. Depending on the mechanism by which Fe^{3+} substitutes into the crystal structure, the enhanced concentration of Fe^{3+} in $(Mg,Fe)(Si,Al)O_3$ perovskite can have a profound influence on physical and chemical properties.

The substitution of Fe^{3+} into the perovskite structure can occur in a number of different ways, the most likely being:

$$Fe_2O_3 + Mg^{2+} + Si^{4+} \rightarrow Fe^{3+}{}_A + Fe^{3+}{}_B + MgO + SiO_2 \quad (4)$$

$$Fe_2O_3 + 2Si^{4+} + O^{2-} \rightarrow 2\ Fe^{3+}{}_B + \square + 2SiO_2, \qquad (5)$$

where Fe^{3+} can substitute on the A (8–12-coordinated) and/or the B (octahedral) site, and $\square$ indicates an oxygen vacancy. Atomistic computer simulations have shown that the energies associated with these substitution mechanisms are similar, and that they are decreased when coupled substitutions with Al are considered [*Richmond and Brodholt*, 1998]. Chemical analysis of high-pressure run products combined with Mössbauer and/or EELS data has shown that both substitution mechanisms occur in (Mg,Fe)(Si,Al)O_3 perovskite [*Lauterbach et al.*, 2000]; whereas equation 4 appears to be favoured at Al concentrations above 0.1 cation p.f.u. [*Frost and Langenhorst*, 2002].

The physical and chemical properties inferred for the lower mantle traditionally have been based on measurements of MgSiO_3 perovskite. There was generally good agreement between different laboratories, and the effect of Fe^{2+} was relatively well understood. When the idea spread towards the end of the 1990s that Al and Fe^{3+} could have potentially large effects despite their low abundance, laboratory measurements of lower-mantle perovskite were extended to include Al. Many of the results were surprising and have stimulated an ongoing effort to redefine the nature of lower-mantle perovskite.

5.1. Elastic Properties

The combination of laboratory measurements of elastic properties with seismic data from geophysical measurements has been the foundation for current mantle models. The thermoelastic properties of MgSiO_3 and (Mg,Fe)SiO_3 perovskite have been measured by numerous laboratories, and values for the zero-pressure isothermal bulk modulus (K_{0T}) of MgSiO_3 and (Mg,Fe)SiO_3 generally fall within the range 250–260 GPa [e.g., *Funamori et al.*, 1996; *Fiquet et al.*, 1998, 2000; *Sinogeikin et al.*, 2004]. However, *Zhang and Weidner* [1999] found that the addition of 5 mol% Al$_2$O$_3$ to MgSiO_3 perovskite reduced K_{0T} to 234 GPa, i.e., by 5–10%. This result was confirmed by subsequent elastic property measurements of Mg(Si,Al)O_3 perovskite [*Kubo et al.*, 2000; *Daniel et al.*, 2001; *Yagi et al.*, 2004].

The most plausible explanation for the decrease in bulk modulus is the presence of oxygen vacancies in the structure caused by the substitution of trivalent cations. *Richmond and Brodholt* [1998] found the oxygen vacancy mechanism to be favoured when only Al substitutes into MgSiO_3 perovskite, and transmission electron microscopy revealed the presence of stishovite in the *Zhang and Weidner* [1999] sample [*Langenhorst and McCammon*, unpublished data], which is consistent with equation 5. Moreover, single-crystal measurements have revealed a 25% softening of the bulk modulus of CaTiO_3 perovskite when Fe^{3+} substitutes according to the same mechanism [*Ross et al.*, 2002], and ab initio calculations have shown that perovskites with oxygen vacancies have lower bulk moduli [*Brodholt*, 2000].

In contrast, several high-pressure experiments have shown a different result, namely, that the substitution of trivalent cations into (Mg,Fe)(Si,Al)O_3 perovskite either has no effect on the bulk modulus or causes an increase in the bulk mod-

ulus [*Andrault et al.*, 2001; *Ono et al.*, 2004; *Yagi et al.*, 2004; *Jackson et al.*, 2004] (Figure 6). Possible explanations suggested for the difference in elastic behaviour include variation in oxygen vacancy concentrations and/or ordering, partial amorphisation after sample synthesis, or as yet unidentified factors [*Andrault et al.*, 2001; *Ono et al.*, 2004; *Yagi et al.*, 2004].

Clearly the first priority for elastic property measurements is to define the elastic parameters of (Mg,Fe)(Si,Al)O_3 perovskite that are relevant for the bulk mantle. A second priority, however, is to define the properties that are relevant at the top of the lower mantle, which remains an enigmatic region. Comparison of such properties with seismic data may ultimately provide the best constraints on the chemistry of the region.

5.2. Transport Properties

Electrical conductivity models provide constraints complementary to those from seismology on the temperature and chemistry of the lower mantle. Laboratory efforts have focused on both (Mg,Fe)SiO_3 perovskite and ferropericlase in order to determine their electrical conductivities under the relevant conditions. *Xu et al.* [1998] found a threefold increase in electrical conductivity of (Mg,Fe)(Si,Al)O_3 compared to Al-free (Mg,Fe)SiO_3 perovskite, which can be attributed to a greater density of charge carriers. *Dobson and Brodholt* [2000] derived revised temperature profiles for the lower mantle based on the new conductivity data that

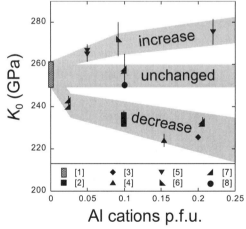

Figure 6. Effect of Al$_2$O$_3$ concentration (expressed as Al cations per formula unit) on the zero-pressure isothermal bulk modulus (K_{0T}) of (Mg,Fe)(Si,Al)O_3 perovskite. References: [1] *Fiquet et al.* [1998]; *Fiquet et al.* [2000]; *Sinogeikin et al.* [2004]; [2] *Zhang and Weidner* [1999]; [3] *Kubo et al.* [2000]; [4] *Daniel et al.* [2001]; [5] *Andrault et al.* [2001]; [6] *Ono et al.* [2004]; [7] *Yagi et al.* [2004]; [8] *Jackson et al.* [2004].

suggest a relatively cool lower mantle if Al is concentrated in the perovskite phase.

The substitution mechanism of Fe^{3+} into $(Mg,Fe)(Si,Al)O_3$ perovskite has a strong influence on its transport properties. The dominant conduction mechanism in $(Mg,Fe)SiO_3$ perovskite is inferred to be electron hopping [e.g., *Peyronneau and Poirier*, 1989], but the creation of oxygen vacancies through coupled (Al,Fe^{3+}) substitution according to equation 5 raises the possibility of ionic conduction. In situ experiments on $(Mg,Fe)(Si,Al)O_3$ perovskite show evidence for ionic conduction at high temperature [*Xu and McCammon*, 2002], and Na-doped $MgSiO_3$ perovskite has been found to exhibit extrinsic ionic conduction at temperatures up to 1800 K [*Dobson*, 2003]. Such observations are analogous to behaviour in the system $CaTiO_3$–$CaFeO_{2.5}$, where ionic conductivity increases with both temperature and oxygen vacancy concentration [*Marion et al.*, 1999]. The mobility of oxygen vacancies can be altered, however, by the nature of vacancy ordering in the structure. Ionic conductivity is high when vacancies are isolated and randomly distributed, but decreases with the tendency of vacancies to order into chains. At the point where vacancies become fully ordered in extended sheets, electrical conductivity decreases by more than one order of magnitude [*Becerro et al.*, 2000].

The data of *Dobson* [2003] suggest that extrinsic oxygen ionic conduction might be the dominant electrical conduction mechanism at the top of the lower mantle. Electrical conductivity would therefore be sensitive to composition and temperature through the concentration of oxygen vacancies and the nature of their ordering. As indicated by *Dobson* [2003], geomagnetic data with periods corresponding to skin-depths of 1000 km would be sensitive to the composition of the shallow lower mantle, and could provide new information on this enigmatic region.

5.3. Other Properties

Measurements of physical and chemical properties of $(Mg,Fe)(Si,Al)O_3$ perovskite that differ from those of $MgSiO_3$ perovskite have only just scratched the surface of relevant possibilities. Since the presence of even small amounts of trivalent cations affects the electrostatic charge balance and equilibrium defect concentration, influence on other properties could also be expected. For example, *Litasov et al.* [2003] found that the solubility of water in $(Mg,Fe)(Si,Al)O_3$ perovskite was a function of the bulk composition, which controls the nature of Fe^{3+} substitution. Major element partitioning (particularly Fe/Mg) is known to vary with Al concentration in the $(Mg,Fe)(Si,Al)O_3$ perovskite-ferropericlase assemblage [e.g., *Wood and Rubie*, 1996], and *Liebske et al.* [2005] report that crystal-melt partition coefficients for trace

elements in $(Mg,Fe)(Si,Al)O_3$ perovskite systems vary with Al concentration.

6. MANTLE FE_2O_3 CONCENTRATION

The concentration of Fe_2O_3 in mantle rocks is a useful indicator of the oxygen buffering capacity of iron [e.g., *Canil et al.*, 1994]. Fe_2O_3 concentration can be determined from (1) analysis of whole rock samples using a bulk method such as wet chemistry; or (2) Mössbauer measurements of Fe^{3+} in individual minerals multiplied by modal abundance and total iron content. The latter method has been applied to a wide range of different xenoliths [*O'Neill et al.*, 1993a; *Canil et al.*, 1994; *Canil and O'Neill*, 1996; *McCammon and Kopylova*, 2004], and is generally considered to give more accurate results than whole rock wet chemical analysis due to effects such as oxidation of samples during acid digestion, and the presence of non-primary oxidised material along grain boundaries [*O'Neill et al.*, 1993a; *Canil et al.*, 1994]. Method (2) can also be applied to deeper regions of the mantle based on laboratory measurements of $Fe^{3+}/\Sigma Fe$ at the relevant conditions.

6.1. Bulk Pyrolite Mantle

Fe_2O_3 concentration varies with rock chemistry, where the samples most depleted by partial melting generally show the lowest values, reflecting the preference of Fe^{3+} for the melt phase compared to the residue [e.g., *Canil et al.*, 1994]. Observed trends were used to determine that (1) depleted peridotite from ocean basins or from trenches associated with subduction zones should contain ≤ 0.1 wt% Fe_2O_3; (2) average continental lithospheric mantle should contain ≤ 0.2 wt% Fe_2O_3; and (3) primitive upper mantle should contain ~ 0.3 wt% Fe_2O_3 [*Canil et al.*, 1994]. These estimates are consistent with subsequent Fe_2O_3 determinations in mantle rocks [*Canil and O'Neill*, 1996; *McCammon and Kopylova*, 2004], and examples for spinel and garnet lherzolite are listed in Table 3. Differences in the magnitude of individual contributions to total Fe_2O_3 are a consequence of trade-offs between $Fe^{3+}/\Sigma Fe$, modal abundance, and total iron content.

These estimates can be extended to deeper parts of the mantle based on results of high-pressure experiments combined with mineralogical models for the mantle. At the top of the transition zone, the bulk mantle consists primarily of wadsleyite and majorite (Figure 1), and based on Fe^{3+} studies at high-pressure [*O'Neill et al.*, 1993b; *McCammon and Ross*, 2003] and phase chemistry [*Irifune and Isshiki*, 1998], the Fe_2O_3 concentration is calculated to be similar to undepleted upper mantle values (Table 3). These values are consistent with previous estimates [*O'Neill et al.*, 1993a],

Table 3. Whole-Rock Fe_2O_3 Abundance Calculated From $Fe^{3+}/\Sigma Fe$ Measurements or Estimates for Component Minerals.

Sample type	sp lherz	gt per	bulk TZ	bulk LM	bulk LM	bulk LM	MORB	MORB	MORB	MORB	MORB
# or P	Fr-1	23-5	16 GPa	23.6 GPa	24 GPa	30 GPa	10 GPa	16 GPa	25 GPa	30 GPa	37 GPa
Phases present	ol,opx, cpx,sp	ol,opx, cpx,gt	wads mj	rg,MgPv, CaPv,mj	MgPv,fp CaPv,mj	MgPv, fp,CaPv	cpx gt	mj st	mj,CaPv AlPh,st	MgPv,fp CaPv,st	MgPv,fp CaPv,st
Modal mineral abundance (vol%)											
ol,wads,rg	52(1)	60(2)	59(3)	62(3)							
opx,MgPv	28(1)	15(1)		15(2)	76(4)	77(4)				31(4)	33(5)
cpx, mj	17(1)	5(1)	41(3)	17(2)	1(1)		55(5)	90(5)	70(5)		
sp,gt	3(1)	19(1)					45(5)				
CaPv				6(1)	7(2)	8(2)			16(2)	23(2)	24(3)
fp,AlPh					16(2)	15(2)			4(1)	23(3)	24(4)
st								10(2)	10(2)	23(2)	19(3)
$Fe^{3+}/\Sigma Fe$ for individual mineral phases (at%)											
ol,wads,rg	0(1)	0(1)	3(1)	2(1)							
opx,MgPv	5(2)	7(3)		49(11)	58(11)	60(11)				75(20)	75(20)
cpx, mj	19(3)	25(10)	5(3)	5(3)	5(3)		10(5)	7(3)	7(3)		
sp,gt	26(3)	11(2)					5(3)				
CaPv				50(50)	50(50)	50(50)			50(50)	50(50)	50(50)
fp,AlPh					1(1)	1(1)			50(50)	50(50)	50(50)
st								50(50)	50(50)	50(50)	50(50)
FeO abundance for individual mineral phases (wt%)(all iron assumed to be Fe^{2+})											
ol,wads,rg	10(1)	10(1)	9(1)	11(1)							
opx,MgPv	7(1)	6(1)		6(1)	8(1)	8(1)				21(1)	21(1)
cpx, mj	3(1)	3(1)	6(1)	5(1)	3(1)		2(1)	8(1)	10(1)		
sp,gt	11(1)	8(1)					9(1)				
CaPv				0.5(5)	0.5(5)	0.5(5)			3(1)	0.4(5)	0.7(5)
fp,AlPh					17(1)	14(1)			3(1)	8(1)	7(0)
st								0.1(5)	0.2(5)	0.3(5)	0.2(5)
Contribution of each mineral phase to whole-rock Fe_2O_3 abundance (wt%)											
ol,wads,rg	0.00(3)	0.00(3)	0.17(6)	0.15(8)							
opx,MgPv	0.10(4)	0.07(3)		0.47(13)	3.7(8)	3.9(8)				5.4(16)	5.7(18)
cpx, mj	0.11(3)	0.05(2)	0.14(9)	0.05(3)	0.00(0)		0.15(8)	0.59(26)	0.52(23)		
sp,gt	0.09(2)	0.19(4)					0.24(14)				
CaPv				0.02(2)	0.02(3)	0.02(3)			0.22(23)	0.05(8)	0.09(12)
fp,AlPh					0.03(3)	0.02(2)			0.07(7)	1.0(10)	0.89(91)
st								0.01(3)	0.01(3)	0.04(7)	0.02(6)
Whole-rock wt% Fe_2O_3	0.30(6)	0.31(6)	0.31(10)	0.7(2)	3.8(8)	3.9(8)	0.38(16)	0.6(3)	0.8(3)	6.5(19)	6.7(20)
Reference	[1]	[2]	[3]	[4]	[5]	[5]	[6]	[6]	[7]	[8]	[8]

ol: olivine; wads: wadsleyite; rg: ringwoodite; opx: orthopyroxene; MgPv: $MgSiO_3$ perovskite; cpx: clinopyroxene; mj: majorite; sp: spinel; gt: garnet; CaPv: $CaSiO_3$ perovskite; fp: ferropericlase; AlPh: aluminous calcium ferrite type phase; st: stishovite; lherz: lherzolite; per: peridotite; TZ: transition zone; LM: lower mantle; MORB: midocean ridge basalt.

References: [1] *Canil et al.* [1994]; [2] *McCammon and Kopylova* [2004]; [3] *Irifune and Isshiki* [1998]; [4] *Hirose* [2002]; [5] *Nishiyama and Yagi* [2003]; [6] *Irifune et al.* [1986]; [7] *Irifune and Ringwood* [1993]; [8] *Ono et al.* [2001].

and indicate that the fertile upper mantle and transition zone are approximately isochemical with respect to Fe_2O_3 concentration.

The top of the lower mantle is marked by the transition of ringwoodite to $(Mg,Fe)SiO_3$ perovskite plus ferroperi-clase. In a pyrolite bulk composition, $(Mg,Fe)(Si,Al)O_3$ perovskite co-exists with ringwoodite and majorite when it first appears [e.g., *Irifune*, 1994; *Hirose*, 2002], and incorporates sufficient Al to require $Fe^{3+}/\Sigma Fe$ of at least 50% (see Section 4.3). The rapid increase in the proportion of $(Mg,Fe)(Si,Al)O_3$ perovskite with depth as the remainder of ringwoodite breaks down causes a sharp increase in the Fe_2O_3 concentration to a level that is more than ten times the value determined for primitive upper mantle (Table 3). This high concentration is inescapable based on the dependence of $Fe^{3+}/\Sigma Fe$ on Al concentration in $(Mg,Fe)(Si,Al)O_3$ perovskite, independent of the prevailing fO_2. The contribution of ferropericlase to the bulk Fe_2O_3 concentration in the lower mantle is negligible, since even under oxidising conditions the maximum value of $Fe^{3+}/\Sigma Fe$ in ferropericlase is not likely to exceed 10% (Table 3).

An important question is whether the increase in Fe_2O_3 concentration associated with the transition to $(Mg,Fe)(Si,Al)O_3$ perovskite is balanced by the reduction of other species. A lower mantle enriched in oxygen is not consistent with the idea of whole mantle convection, since the low Fe_2O_3 concentration in the upper mantle could not be maintained through geologic time. A more plausible scenario, therefore, is charge balance either through the reduction of volatile species, or the disproportionation of iron.

Volatile species incorporating carbon, hydrogen and sulphur have all been considered possible candidates for buffering mantle fO_2 [e.g., *Canil et al.*, 1994]. The possibility of coupled redox reactions to balance iron oxidation involving these elements can be evaluated based on their mantle abundance. For example, the reactions

$$2\,FeO + C + H_2O = CH_4 + Fe_2O_3 \qquad (6)$$

or

$$4\,FeO + S_2 + 2\,H_2O = 2\,H_2S + 2\,Fe_2O_3 \qquad (7)$$

could be considered as possible candidates. However, the volatile concentrations required for a net 3.5 wt% increase of Fe_2O_3 concentration in pyrolite are roughly 4,000 wt ppm carbon, 2,500 wt ppm water, or 7,000 wt ppm sulphur, all of which greatly exceed estimations for bulk pyrolite mantle [e.g., *Canil et al.*, 1994; *Wood et al.*, 1996].

A more likely possibility to balance charge during iron oxidation is the disproportionation reaction

$$3\,FeO = Fe + Fe_2O_3. \qquad (8)$$

The amount of Fe metal required to balance the creation of 3.5 wt% Fe_2O_3 is ca. 1 wt% [*Frost et al.*, 2004]. There is strong experimental evidence for disproportionation of Fe^{2+}. Synthesis of $(Mg,Fe)(Si,Al)O_3$ perovskite at low oxygen fugacity produces iron metal in the multianvil press [e.g., *Lauterbach et al.*, 2000; *Frost et al.*, 2004] and in the diamond anvil cell [e.g., *Miyajima et al.*, 1999].

6.2. Midocean Range Basalt (MORB)

So far attention has focused on the bulk mantle, but in order to evaluate the potential role of Fe_2O_3 concentration in mantle dynamics, it is important to consider other regions of the mantle such as subduction zones. As a starting point, the results of high-pressure experiments on MORB will be considered. Although there are no direct Mössbauer data from such assemblages, the results discussed in Section 4 can be used to estimate the ability of individual minerals to incorporate Fe^{3+}.

At upper-mantle pressures and temperatures, MORB consists primarily of garnet and clinopyroxene [e.g., *Irifune et al.*, 1986]. Based on Mössbauer studies of these phases, the estimated bulk Fe_2O_3 concentration of a garnet-clinopyroxene assemblage at 10 GPa at reducing conditions does not differ significantly from that of the bulk upper mantle (Table 3). Fe_2O_3 concentrations would be higher if conditions were more oxidising within the subducting slab, and the presence of water could increase Fe_2O_3 concentration as described in Section 4.4.

In the transition zone MORB consists almost exclusively of majorite [e.g., *Irifune and Ringwood*, 1987]. Based on Mössbauer studies that show an enhanced stability of Fe^{3+} in majorite, even under reducing conditions, Fe_2O_3 concentrations are estimated to increase with depth as the proportion of majorite increases (Table 3). The appearance of $CaSiO_3$ perovskite with increasing depth will likely increase Fe_2O_3 concentration further, since even though reported iron concentrations are small, the high proportion of Fe^{3+} expected even under reducing conditions will contribute to the overall Fe_2O_3 concentration.

In the lower mantle, MORB is reported by most studies to transform to a mixture of $(Mg,Fe)(Si,Al)O_3$ perovskite, $CaSiO_3$ perovskite, stishovite, and an Al-rich calcium ferrite-type phase [e.g., *Irifune and Ringwood*, 1993; *Ono et al.*, 2001]. The high bulk Al concentration and hence high Al concentration in $(Mg,Fe)(Si,Al)O_3$ perovskite implies high $Fe^{3+}/\Sigma Fe$, and Mössbauer analysis of high-pressure experiments in the system $MgO-FeO-CaO-Al_2O_3$ suggests that the Al-rich calcium ferrite-type phase probably consists predominantly of Fe^{3+} [*McCammon, Frost, and Liebske*, unpublished data]. As a consequence, Fe_2O_3 concentrations are calculated to be high and insensitive to fO_2. The higher Fe_2O_3 concentrations in MORB relative to those in the bulk mantle may be balanced by an increase in Fe metal according to the iron disproportionation reaction (equation 8), or may be coupled with reduction of oxidised phases in the subducting slab. This point is explored further in Section 8.2.

In summary, Fe_2O_3 concentrations in a bulk pyrolite mantle are estimated to be relatively constant with depth throughout the upper mantle and transition zone, while Fe_2O_3 concentrations in MORB may be higher compared to the bulk mantle.

At the top of the lower mantle there is a dramatic increase in Fe_2O_3 concentration in the bulk mantle due to the transformation to $(Mg,Fe)(Si,Al)O_3$ perovskite, and an even greater increase in Fe_2O_3 concentration in MORB due to the higher bulk concentration of Al (Figure 7a).

7. MANTLE fO_2

Mantle fO_2 can be determined directly from equilibrium assemblages using mineral oxybarometers. In spinel peridotite, for example, the following reaction describes the exchange of Fe^{2+} and Fe^{3+} between coexisting minerals:

$$6\ Fe_2SiO_4 + O_2 = 2\ Fe^{2+}Fe^{3+}{}_2O_4 + 6\ FeSiO_3$$
$$\text{ol} \qquad\qquad \text{sp} \qquad\qquad \text{opx} \qquad (9)$$

[e.g., *Wood et al.*, 1990]. At equilibrium

$$\log fO_2(P,T) = -\frac{\Delta G^0(P,T)}{2.303RT} - 6\log a^{ol}_{Fe_2SiO_4}$$
$$+ 3\log a^{opx}_{FeSiO_3} + 2\log a^{sp}_{Fe_3O_4}, \qquad (10)$$

where ΔG^0 is the free energy of reaction 9, and the a terms refer to the activities of the indicated components in the given minerals. Hence fO_2 is related to both the activity of magnetite in spinel (which is related to spinel $Fe^{3+}/\Sigma Fe$) and the activity of Fe^{2+} components in olivine and orthopyroxene. The determination of such activities requires activity–composition relations for the relevant solid solutions [e.g., *O'Neill and Wall*, 1987; *Wood*, 1990; *Ballhaus et al.*, 1991]. Uncertainties in the determination of fO_2 are therefore related not only to the accuracy of $Fe^{3+}/\Sigma Fe$ measurements but also to the reliability of the activity–composition models within the system under study.

The fO_2 is a strong function of temperature and pressure, which means that comparisons of absolute fO_2 are meaningful only if they refer to the same conditions. To address this problem, oxygen fugacities are commonly normalised to a common buffer, such as FMQ. This has the added advantage of minimising errors in $\Delta\log fO_2$ (FMQ) due to pressure and temperature uncertainties.

7.1. Upper Mantle

Efforts over the past 20 years have produced much data concerning upper-mantle fO_2, and although agreement is not universal, a relatively consistent picture is emerging [see reviews by *Arculus*, 1985; *Wood et al.*, 1990; *Ballhaus*, 1993; *O'Neill et al.*, 1993a]. The fO_2 measured for lithospheric upper mantle is heterogeneous on a scale of at least

four log units, where trends in fO_2 have been noted to occur with metasomatism [e.g., *McCammon et al.*, 2001], partial melting [e.g., *Bryndzia and Wood*, 1990; *Kadik*, 1997], and tectonic environment [e.g., *Wood et al.*, 1990; *Ballhaus*, 1993; *Parkinson and Arculus*, 1999]. Temporal variations have also been studied [e.g., *Ballhaus*, 1993; *Canil*, 1997; *Lee et al.*, 2003]. Generally the lowest values of fO_2 have been recorded for suboceanic abyssal peridotites and xenoliths from zones of continental extension (FMQ to FMQ-1), while more oxidised values have been recorded for xenoliths from regions of recent or active subduction (FMQ+2).

The relatively high fO_2 of lithospheric mantle contrasts with the relatively low concentrations of Fe_2O_3 (Table 3). This paradoxical relation between oxidation state and fO_2 in the upper mantle arises because the dominant iron-bearing phase, olivine, essentially excludes Fe^{3+} from its structure, and the next most abundant phase, orthopyroxene, incorporates only small amounts. Fe^{3+} is therefore concentrated in the minor phases clinopyroxene and spinel, which leads to

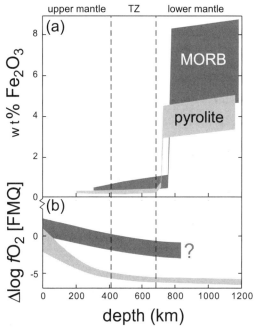

Figure 7. (a) Calculated variation of Fe_2O_3 concentration with depth for bulk pyrolite (light gray) and midocean ridge basalt (MORB) (dark gray) compositions. Shaded regions are constructed from the data and uncertainties listed in Table 3. Note that the increase in Fe_2O_3 concentration is likely balanced by iron disproportionation ($3\ Fe^{3+} \rightarrow Fe^0 + 2\ Fe^{3+}$) in the bulk mantle or reduction of oxidised species in MORB; the lower mantle is probably not enriched in oxygen relative to the upper mantle. (b) Highly schematic estimated variation of fO_2 (relative to FMQ) with depth for bulk pyrolite (light gray) and MORB (dark gray) compositions. Shaded regions are constructed from the literature data as discussed in Section 7.

relatively high oxygen fugacities according to Equation 10. An equilibrium reaction involving Fe^{3+} in clinopyroxene has also been calibrated for upper-mantle assemblages, and gives essentially the same results as equation 9 [*Luth and Canil*, 1993]. The volume change of the solids in Equation 9 is approximately half the value for the FMQ buffer, which means that at fixed composition, $\Delta \log fO_2$ (FMQ) will decrease slightly with increasing pressure [e.g., *Ballhaus*, 1995; *Wood et al.*, 1996]. Such decreases are not always observed, however, as there is no reliable barometer for spinel peridotite and pressure uncertainty in the spinel facies is therefore relatively high.

The transition to the garnet facies in the upper mantle changes the determination of fO_2 to equilibria largely dominated by garnet, e.g.:

$$4 \, Fe_2SiO_4 + 2 \, FeSiO_3 + O_2 = 2 \, Fe^{2+}_3Fe^{3+}_2Si_3O_{12} \quad (11)$$
$$\text{ol} \qquad\quad \text{opx} \qquad\qquad\qquad \text{gt}$$

[*Gudmundsson and Wood*, 1995]. The volume change of the solids in Equation 11 is negative, which means that pressure enhances the incorporation of Fe^{3+} into garnet. At fixed composition, $\Delta \log fO_2$ (FMQ) will therefore decrease strongly with increasing pressure, which has been observed in redox profiles of the Kaapvaal [*Woodland and Koch*, 2003] and Slave [*McCammon and Kopylova*, 2004] cratons. Both datasets show that $\Delta \log fO_2$ (FMQ) values decrease to less than FMQ-4 at ca. 6 GPa.

The relatively strong reduction in relative fO_2 with depth raises an interesting question regarding redox conditions in the deep upper mantle. A simple extrapolation of global values to the 410 km discontinuity leads to redox conditions below the iron–wüstite (IW) buffer [e.g., *Woodland and Koch*, 2003]. Such redox conditions would probably lie below the Ni precipitation curve, where Ni is predicted to precipitate as metal instead of residing in silicate phases [*O'Neill and Wall*, 1987]. The IW buffer has therefore been generally assumed to represent a lower limit for mantle fO_2 based on Ni concentrations measured for peridotitic mantle olivine in xenoliths and mineral inclusions in diamond [*O'Neill and Wall*, 1987]. On the other hand, relative oxygen fugacities more reducing than IW in the deep upper mantle have been suggested to account for platinum-group element fractionation between partial melt and mantle residue [*Ballhaus*, 1995].

The mineralogy of the upper mantle changes with depth in the deep part of the upper mantle (Figure 1). If garnet remains the dominant Fe^{3+}-bearing phase, fO_2 might be controlled by a modified form of equation 11:

$$4 \, Fe_2SiO_4 + 2 \, FeSiO_3 + O_2 = 2 \, Fe^{2+}_3Fe^{3+}_2Si_3O_{12}, \quad (12)$$
$$\text{ol} \qquad\quad \text{cpx} \qquad\qquad\qquad \text{gt}$$

where $FeSiO_3$ is the clinoferrosilite component in clinopyroxene. Note that this component is denser than the ferrosilite component in orthopyroxene [*Hugh-Jones et al.*, 1994], which reduces the zero-pressure volume change of the solids in equation 12 compared to FMQ, thereby reducing the effect of pressure in decreasing relative fO_2 with depth. However, an accurate assessment of fO_2 at the high pressures of the deep upper mantle (8–13.5 GPa) using equation 12 should include a volume term that accounts for the effects of both thermal expansion and compressibility, which were neglected in the original formulation of *Gudmundsson and Wood* [1995] on the basis that the resulting errors were negligible at low pressures. Another source of uncertainty is the pressure effect on the reciprocal reactions describing cation exchange in the garnet structure. *Woodland et al.* [1999] calculated that a more accurate assessment of this pressure effect for garnet peridotite assemblages up to 6.5 GPa leads to differences in $\Delta \log fO_2$ of +0.3 to -0.8 log-bar units compared to the original formulation of *Gudmundsson and Wood* [1995]. Uncertainties would be even greater at the higher pressures of the deep upper mantle.

In summary, redox conditions in the deep upper mantle are likely reduced relative to FMQ as already suggested by previous studies [e.g., *Wood et al.* 1990; *Ballhaus* 1995; *Wood et al.* 1996; *Woodland and Koch* 2003], but a definitive determination from oxybarometry of whether redox conditions lie above or below the Ni precipitation curve is still lacking.

7.2. Transition Zone

There are no natural samples from the transition zone with which to determine fO_2 using mineral oxybarometers; hence estimates must rely on the known Fe^{3+} behaviour in individual minerals. As already noted in the literature, the increased affinity of all transition zone minerals for Fe^{3+} compared to those in the upper mantle means that Fe^{3+} will not be concentrated in the modally minor phases as in the upper mantle, but will be distributed throughout the major phases, thereby lowering its chemical potential. Thus, fO_2 is also lowered, probably close to conditions of metal saturation [*O'Neill et al.*, 1993a,b]. Since such reduced conditions are also predicted to exist in the deep upper mantle (see Section 7.1), there is no compelling reason for a redox gradient across the 410-km discontinuity. A similar conclusion was reached for whole rock Fe_2O_3 concentration (see Section 6.1).

7.3. Lower Mantle

The fO_2 of the lower mantle is still an open question. As shown in Section 4.3, high $Fe^{3+}/\Sigma Fe$ in $(Mg,Fe)(Si,Al)O_3$ perovskite does not imply high fO_2. Disproportionation of iron to balance charge in $(Mg,Fe)(Si,Al)O_3$ perovskite would produce an Fe-rich metal phase that could buffer the lower mantle at reducing conditions.

Ferropericlase is the only major phase in the lower mantle where $Fe^{3+}/\Sigma Fe$ varies with fO_2 (see Section 4.2). It is possible in principle, therefore, to derive fO_2 indirectly by considering its effect on another property which is known for the lower mantle. Such an approach was used by *Wood and Nell* [1991], who determined the effect of fO_2 on the electrical conductivity of ferropericlase, and then compared the results with geophysically measured values to infer the likely fO_2 of the lower mantle. This approach is valid if the conductivity of the lower mantle is dominated by ferropericlase, but subsequent studies suggest that $(Mg,Fe)(Si,Al)O_3$ perovskite may be the dominant conductor [e.g., *Katsura et al.*, 1998; *Xu et al.*, 1998; *Xu and McCammon*, 2002]. Nevertheless, the possibility of constraining fO_2 based on the properties of ferropericlase still holds promise and could eventually provide better constraints on bulk lower-mantle fO_2.

Redox conditions in the lower mantle may not be homogeneous. Analyses of Kola carbonatites that show a plume source contribution suggest variations in refractory/volatile ratios in different regions of the mantle that are probably associated with differences in fO_2 [*Marty et al.*, 1998].

7.4. Subduction Zones

Magmas and peridotite xenoliths derived from areas of recent or active subduction are generally more oxidised than those associated with mid-ocean ridges [e.g., *Wood et al.*, 1990; *Carmichael*, 1991; *Ballhaus*, 1993]. While the actual mechanism remains controversial [e.g., *Brandon and Draper*, 1996; *Frost and Ballhaus*, 1998; *Brandon and Draper*, 1998], analyses of spinel peridotite from a range of localities suggest that the source of oxygen which oxidises the mantle wedge is from the subducted slab [*Parkinson and Arculus*, 1999; *Peslier et al.*, 2002]. The oxidation of Fe^{2+} to Fe^{3+} through hydrothermal alteration of oceanic crust probably also contributes to the downward flux of oxygen at subduction zones [e.g., *Rea and Ruff*, 1996; *Lécuyer and Ricard*, 1999].

More direct evidence of redox conditions associated with subduction in the deeper mantle may come from $Fe^{3+}/\Sigma Fe$ measurements of inclusions in diamond. The tantalising possibility that diamonds might contain tiny inclusions of deep material was recognized two decades ago by *Scott-Smith et al.* [1984], but it took more than a decade for more convincing proof to be found [*Harte et al.*, 1999]. Over the past 5 years many new inclusion suites have been described—some inferred to have originated in the transition zone [*Stachel et al.*, 2000a; *Stachel*, 2001; *Harte et al.*, 2002] and some in the lower mantle [*Harte et al.*, 1999; *Davies et al.*, 1999;

Stachel et al., 2000b; *Kaminsky et al.*, 2001]. *McCammon* [2001] reviewed evidence for their deep origin and discussed implications for mantle geochemistry and dynamics.

The current consensus is that deep diamonds are likely associated with subduction. The eclogitic nature of inclusions from the transition zone is consistent with a subduction-related origin [*Stachel*, 2001]. Most former $MgSiO_3$ perovskite inclusions contain relatively low concentrations of Al (<1.7 wt% Al_2O_3), suggesting origin of diamonds containing these inclusions near the top of the lower mantle, since majorite garnet would normally incorporate the major portion of mantle Al at these depths [e.g., *Irifune and Ringwood*, 1987]. Trace element analyses show highly enriched compositions for majorite garnet inclusions [*Stachel*, 2001; *Harte et al.*, 2002] and former $CaSiO_3$ perovskite inclusions [*Harte et al.*, 1999; *Stachel et al.*, 2000b; *Kaminsky et al.*, 2001], while former $MgSiO_3$ perovskite inclusions show generally less than chondritic compositions. This is consistent with high-pressure partitioning experiments at lower-mantle conditions that show a preference of rare earth elements for $CaSiO_3$ perovskite [e.g., *Kato et al.*, 1988, 1996]. Former $CaSiO_3$ perovskite inclusions show Eu anomalies, which are interpreted to reflect a source rock containing feldspar and therefore crustal origin [*Harte et al.*, 1999; *Stachel et al.*, 2000b]. The combination of the above factors has led to suggestions of a subduction-related origin for transition zone and lower-mantle diamonds [*Harte et al.*, 1999; *Stachel et al.*, 2000b; *Kaminsky et al.*, 2001; *Hutchison et al.*, 2001; *Stachel*, 2001].

Mössbauer measurements of majorite garnet inclusions in diamonds from the transition zone [*Harte et al.*, 2002] suggest that they were formed at fO_2 conditions more oxidising than the bulk mantle, which is inferred to be close to metal saturation (see Section 7.2). *Harte et al.* [2002] found $Fe^{3+}/\Sigma Fe$ values up to 32% in majorite garnet inclusions, which are significantly higher than the values predicted from high-pressure experiments for such compositions at metal saturation (<10%) [*McCammon and Ross*, 2003]. Mössbauer measurements of ferropericlase inclusions in diamonds from the lower mantle suggest that they were formed at a range of redox conditions, some more oxidising than metal equilibrium [*McCammon et al.*, 2004c]. These results suggest a redox gradient between subducted material and the bulk mantle (Figure 7b), similar to the gradient in Fe_2O_3 concentration predicted between subducted slabs and the surrounding mantle, and may play an important role in the genesis of deep diamonds (see Section 8.2).

8. DISCUSSION

The overall picture of fO_2 within the bulk mantle is a general decrease with depth, with reduced conditions throughout the transition zone and the lower mantle. In contrast, the bulk concentration of Fe^{3+} is low throughout the upper mantle and (likely) the transition zone but increases dramatically near the top of the lower mantle as $(Mg,Fe)(Si,Al)O_3$ perovskite becomes stable. The increased Fe^{3+} abundance is likely balanced by a corresponding amount of Fe metal through disproportionation of Fe^{2+}. These contrasts in oxidation state and fO_2 have important implications for mantle chemistry, structure and dynamics, and a few are discussed below.

8.1. Implications for Mantle Properties and Dynamics

As shown in Section 4.3, the Fe^{3+} concentration in $(Mg,Fe)(Si,Al)O_3$ perovskite depends on the bulk Al concentration, not on fO_2, and $(Mg,Fe)(Si,Al)O_3$ perovskite with lower-mantle Al concentration is *always* found to contain significant quantities of Fe^{3+} (ca. 60%) in high-pressure experiments, even under conditions of low oxygen fugacity where metallic iron is stable [*Lauterbach et al.*, 2000; *Frost et al.*, 2004]. Iron therefore occurs in three coexisting valence states: Fe^{3+}, Fe^{2+}, and Fe^0. Such behaviour is already well known at ambient pressure in phases such as $Fe_{1-x}O$, where the energetics of Fe^{3+} substitution into the structure favour nonstoichiometry in $Fe_{1-x}O$ (i.e., the presence of Fe^{3+}), even in equilibrium with metallic iron. In $(Mg,Fe)(Si,Al)O_3$ perovskite, the free energy is lower for trivalent cation substitution (either as a coupled isovalent substitution or involving creation of an oxygen vacancy) than for divalent substitution, even under reducing conditions. The constraints of charge balance therefore require a corresponding reduction reaction to balance the oxidation of Fe^{2+} to Fe^{3+} during the transformation to $(Mg,Fe)(Si,Al)O_3$ perovskite. Laboratory experiments conducted under reducing conditions show the formation of metallic iron resulting from Fe^{2+} disproportionation [e.g., *Miyajima et al.*, 1999; *Lauterbach et al.*, 2000; *Frost et al.*, 2004].

As discussed in Section 6.1, disproportionation is the only plausible mechanism to balance charge during the transformation to $(Mg,Fe)(Si,Al)O_3$ perovskite if the lower mantle is not significantly enriched in oxygen. Direct evidence for disproportionation in the lower mantle could be obtained, in principle, from geophysical methods or geochemical analysis of lower-mantle material. Unfortunately seismic waves would likely not be sensitive to the small amount of metal needed to balance charge (ca. 1 wt.% Fe), and electrical conductivity measurements would not be able to detect small amounts of a noninterconnected phase. The only samples available from the lower mantle, inclusions in diamond, were likely formed in a region where redox conditions were different from the bulk mantle, and charge may have been balanced by

mechanisms other than iron disproportionation, for example reduction of oxidised species (see Section 8.2).

Iron disproportionation in the bulk mantle has a significant influence on the geochemical evolution of the Earth [*Frost et al.*, 2004]. The formation of silicate perovskite in the early Earth, either through solid-state transformation during accretion or by crystallisation from a magma ocean, would be accompanied by disproportionation of Fe^{2+} because the precursor material would have been poor in Fe^{3+}. Fe-rich metal would have been present from the moment that silicate perovskite formed, and could not have segregated substantially without enriching the lower mantle significantly in oxygen. Physical evidence that the metal phase remains dispersed comes from high-pressure experiments that show finely dispersed small particles of metal throughout $(Mg,Fe)(Si,Al)O_3$ perovskite following iron disproprtionation [*Lauterbach et al.*, 2000; *Frost et al.*, 2004]. Since cation diffusion in silicate perovskite is extremely slow [*Yamazaki et al.*, 2000; *Holzapfel et al.*, 2001] and ferropericlase is likely weaker than silicate perovskite [*Yamazaki and Karato*, 2001], iron-rich metal particles probably require extremely long timescales to form larger grains. During mantle convection, iron-rich metal in upwelling material could recombine with Fe^{3+} to form Fe^{2+}, thus preserving the bulk oxygen content.

If a solid lower mantle were present before core formation was complete, the Fe-rich metal could have incorporated a significant concentration of siderophile elements that would have otherwise been taken up by the core [*Frost et al.*, 2004]. The subsequent redistribution of these elements throughout the mantle as a consequence of convection could then explain the apparent overabundance of siderophile elements in the mantle [e.g., *Ringwood*, 1966] and remove the need for a "late veneer" addition of siderophile elements during accretion [*O'Neill*, 1991, and references therein]. The segregation of ca. 10% of the disproportionated metal phase to the core would cause a net increase in oxygen content of the lower mantle, and following convection could account for the increase in Fe_2O_3 concentration of the upper mantle to its present day level [*Frost et al.*, 2004].

8.2. Implications for Deep Diamond Formation

One of the enigmas of deep diamonds is the mechanism of their formation. Since diamond growth is generally favoured in regions of redox gradients [*Deines*, 1980], the redox contrast between subducted material and the bulk mantle may provide a clue to diamond formation in the lower mantle. As discussed in Section 6.1, the mechanism to balance the oxidation of iron to Fe^{3+} during the transformation to $(Mg,Fe)(Si,Al)O_3$ perovskite is likely iron disproportionation, because the abundance of oxidised species in the bulk mantle that could be reduced is not sufficient to balance charge. In subducted material, however, the abundance of oxidised species is likely greater.

Magnesite ($MgCO_3$) is known to be stable at conditions throughout the transition zone and lower mantle [*Biellmann et al.*, 1993; *Fiquet et al.*, 2002; *Isshiki et al.*, 2004], so charge balance of iron oxidation could occur according to the following reaction involving subducted carbonates:

$$4\,FeO + MgCO_3 = MgO + C + 2\,Fe_2O_3. \quad (13)$$

For example, an increase from ca. 0.4 to 0.8 wt% Fe_2O_3 in the transition zone (Figure 7a) would involve only 0.1 wt% magnesite ($MgCO_3$), which requires only ca. 150 ppm of carbon. A further increase of 6 wt% Fe_2O_3 associated with the garnet–perovskite transformation could be accomplished by the reduction of ca. 1.5 wt% magnesite, involving ca. 2,000 ppm of carbon. Such quantities are easily accommodated by current estimates of sediment subduction [e.g., *Rea and Ruff*, 1996]. The origin of lower-mantle diamonds might therefore be linked to equation 13, which could account for the high proportion of ferropericlase in lower-mantle diamond inclusions—more than half of worldwide occurrences [*McCammon*, 2001]—and also observed trends in carbon isotopic compositions of lower-mantle diamonds [*Stachel et al.*, 2002; *McCammon et al.*, 2004c].

9. SUMMARY

This paper has reviewed current knowledge of iron oxidation state and fO_2 in different regions of the mantle. The important results can be summarised as follows:

(1) The energetics of Fe^{3+} incorporation in minerals is controlled by crystal chemistry at least as much as by fO_2. Some minerals, e.g., olivine, incorporate negligible Fe^{3+}, while others, such as $(Mg,Fe)(Si,Al)O_3$ perovskite, incorporate substantial amounts of Fe^{3+}, even under reducing conditions (balanced, for example, by disproportionation), which has important implications for physical and chemical properties. The link between iron oxidation state (as measured by $Fe^{3+}/\Sigma Fe$) and fO_2 is therefore not a simple one in mineral systems.

(2) The fO_2 is relatively oxidising in shallow lithospheric mantle (near FMQ) due to the concentration of Fe^{3+} in modally minor phases such as spinel but decreases significantly with depth due to the increased affinity of garnet for Fe^{3+} to near IW in the deep lithospheric mantle. The ability of all transition zone phases to accommodate Fe^{3+}, even under reducing conditions, suggests that fO_2 is near metal equilibrium in the transition zone and also near metal equilibrium in the lower mantle due to the suggested presence of iron-

rich metal resulting from iron disproportionation. The fO_2 is not homogeneous throughout the mantle, however, since subducted slabs appear to contain more oxidised material.

(3) The whole-rock Fe_2O_3 concentration of the mantle is estimated from mantle xenolith and high-pressure experimental data to be ca. 0.3 wt% Fe_2O_3 in the upper mantle and transition zone in agreement with previous estimates, but at least 10 times greater in the lower mantle due to the preference of $(Mg,Fe)(Si,Al)O_3$ perovskite for Fe^{3+}, even under reducing conditions. Analogous estimates for MORB show an even greater increase in Fe_2O_3 concentration in the lower mantle.

(4) In the bulk mantle, the increase in Fe_2O_3 concentration associated with the transition to $(Mg,Fe)(Si,Al)O_3$ perovskite is likely balanced by disproportionation of iron ($3\,Fe^{3+} \rightarrow Fe^0 + 2\,Fe^{3+}$), since oxygen enrichment of the lower mantle is inconsistent with geochemical constraints, and the mantle does not contain sufficient volatiles to oxidise the amount of iron required. The presence of a metal phase in the lower mantle has important implications for mantle geochemistry, and if the metal phase does not segregate appreciably, material can move across the lower mantle boundary without a net enrichment or depletion in oxygen.

(5) In subducting slabs there may be sufficient material (carbonates, for example) to oxidise the amount of Fe^{3+} required to satisfy requirements for $(Mg,Fe)(Si,Al)O_3$ perovskite. Such coupled redox reactions could be related to the origin of deep diamonds.

Acknowledgments. I am pleased to acknowledge the contributions of my collaborators B. Harte, J. Harris, M. Hutchison, T. Stachel M. Kopylova, H. O'Neill, N. Ross, F. Langenhorst, S. Lauterbach, F. Seifert, and P. van Aken in many of the projects described in this paper. The manuscript was significantly improved by the efforts of the editor and two anonymous reviewers.

REFERENCES

Allègre, C. J., J. P. Poirier, E. Humler, and A. W. Hofmann, The chemical composition of the Earth, *Earth Planet. Sci. Lett.*, *134*, 515–526, 1995.

Amthauer, G., M. Grodzicki, W. Lottermoser, and G. J. Redhammer, Mössbauer spectroscopy: Basic Principles, in *Spectroscopic Methods in Mineralogy*, edited by A. Beran, and E. Libowitsky, pp. 345–367, Eötvös University Press, Budapest, 2004.

Anderson, D. L., Chemical composition of the mantle, *J. Geophys. Res.*, *88*, B41–B52, 1983.

Andrault, D., and N. Bolfan-Casanova, High-pressure phase transformations in the $MgFe_2O_4$ and Fe_2O_3-$MgSiO_3$ systems, *Phys. Chem. Minerals*, *28*, 211–217, 2001.

Andrault, D., N. Bolfan-Casanova, and N. Guignot, Equation of state of lower mantle (Al,Fe)-$MgSiO_3$ perovskite, *Earth Planet. Sci. Lett.*, *193*, 501–508, 2001.

Annersten, H., M. Olesch, and F. A. Seifert, Ferric iron in orthopyroxene: a Mössbauer spectroscopic study, *Lithos*, *11*, 301–310, 1978.

Arculus, R. J., Oxidation status of the mantle—Past and present, *Ann. Rev. Earth Planet. Sci.*, *13*, 75–95, 1985.

Ballhaus, C., Redox states of lithospheric and asthenospheric upper mantle, *Contrib. Mineral. Petrol.*, *114*, 331–348, 1993.

Ballhaus, C., Is the upper mantle metal-saturated?, *Earth Planet. Sci. Lett.*, *132*, 75–86, 1995.

Ballhaus, C., R. F. Berry, and D. H. Green, High pressure experimental calibration of the olivine-orthopyroxene-spinel oxygen barometer: implications for the oxidation state of the upper mantle, *Contrib. Mineral. Petrol.*, *107*, 27–40, 1991.

Banfield, J. F., M. D. Dyar, and A. V. McGuire, The defect microstructure of oxidised mantle olivine from Dish Hill, California, *Amer. Mineral.*, *77*, 977–986, 1992.

Becerro, A. I., F. Langenhorst, R. J. Angel, S. Marion, C. A. McCammon, and F. Seifert, The transition from short-range to long-range ordering of oxygen vacancies in $CaFe_xTi_{1-x}O_{3-x/2}$ perovskites, *Physical Chemistry Chemical Physics*, *2*, 3933–3941, 2000.

Bell, D. R., and G. R. Rossman, Water in Earth's mantle: The role of nominally anhydrous minerals, *Science*, *255*, 1391–1397, 1992.

Biellmann, C., P. Gillet, F. Guyot, J. Peyronneau, and B. Reynard, Experimental evidence for carbonate stability in the Earth's lower mantle, *Earth Planet. Sci. Lett.*, *118*, 31–41, 1993.

Bina, C. R., Lower mantle mineralogy and the geophysical perspective, in *Ultrahigh-Pressure Mineralogy: Physics and Chemistry of the Earth's Deep Interior*, edited by R.J. Hemley, pp. 205–240, Mineralogical Society of America, Washington DC, 1998.

Bläß, U. W., F. Langenhorst, T. Boffa-Ballaran, F. Seifert, D .J. Frost, and C. McCammon, A new oxygen deficient perovskite phase $Ca(Fe_{0.4}Si_{0.6})O_{2.8}$ and phase relations along the join $CaSiO_3$–$CaFeO_{2.5}$ at transition zone conditions, *Phys. Chem. Minerals*, *31*, 52–65, 2004.

Bolfan-Casanova, N., H. Keppler, and D. C. Rubie, Water partitioning between nominally anhydrous minerals in the MgO-SiO_2-H_2O system up to 24 GPa: Implications for the distribution of water in the Earth's mantle, *Phys. Earth Planet. Int.*, *182*, 209–221, 2000.

Brandon, A. D., and D. S. Draper, Constraints on the origin of the oxidation state of mantle overlying subduction zones: An example from Simcoe, Washington, USA, *Geochim. Cosmochim. Acta*, *60*, 1739–1749, 1996.

Brandon, A. D., and D. S. Draper, Reply to the comment by B.R. Frost and C. Ballhaus on "Constraints on the origin of the oxidation state of mantle overlying subduction zones: an example from Simcoe, Washington, USA", *Geochim. Cosmochim. Acta*, *62*, 333–335, 1998.

Brodholt, J. P., Pressure-induced changes in the compression mechanism of aluminous perovskite in the Earth's mantle, *Nature*, *407*, 620–622, 2000.

Bryndzia, L. T., and B. J. Wood, Oxygen thermobarometry of abyssal spinel peridotites: the redox state and C-O-H volatile composition of the Earth's sub-oceanic upper mantle, *Amer. J. Sci.*, *290*, 1093–1116, 1990.

Canil, D., Vanadium partitioning and the oxidation state of Archean komatiite magmas, *Nature*, *389*, 842–845, 1997.

Canil, D., and H. S. C. O'Neill, Distribution of ferric iron in some upper-mantle assemblages, *J. Petrol.*, *37*, 609–635, 1996.

Canil, D., H. S. C. O'Neill, D. G. Pearson, R. L. Rudnick, W. F. McDonough, and D. A. Carswell, Ferric iron in peridotites and mantle oxidation states, *Earth Planet. Sci. Lett.*, *123*, 205–220, 1994.

Carmichael, I. S. E., The oxidation state of basic magmas: a reflection of their source regions? *Contrib. Mineral. Petrol.*, *106*, 129–142, 1991.

Chaplin, T. D., J. P. Brodholt, N. C. Richmond, and N. L. Ross, Computer simulation of the incorporation of aluminium and ferric iron into calcium silicate perovskite, *Eos Trans. AGU*, *80*, Fall Meet. Suppl., Abstract V22A-04, 1999.

Collerson, K. D., S. Hapugoda, B. S. Kamber, and Q. Williams, Rocks from the mantle transition zone: Majorite-bearing xenoliths from Malaita, Southwest Pacific, *Science*, *288*, 1215–1223, 2000.

Daniel, I., H. Cardon, G. Fiquet, F. Guyot, and M. Mezouar, Equation of state of Al-bearing perovksite to lower mantle pressure conditions, *Geophys. Res. Lett.*, *28*, 3789–3792, 2001.

Davies, R., W. L. Griffin, N. J. Pearson, A. S. Andrew, B. J. Doyle, and S.Y. O'Reilly, Diamonds from the deep: Pipe DO-27, Slave Craton, Canada, in *The J.B. Dawson Volume, Proc. VII Inter. Kimberlite Conf.*, edited by J.J. Gurney, J.L. Gurney, M.D. Pascoe, and S.H. Richardson, pp. 148–155, Red Roof Design, Cape Town, South Africa, 1999.

Deines, P., The carbon isotopic composition of diamonds: Relationship to diamond shape, color, occurrence and vapor composition, *Geochim. Cosmochim. Acta*, *44*, 943–961, 1980.

Dobson, D., Oxygen ionic conduction in $MgSiO_3$ perovskite, *Phys. Earth Planet. Int.*, *139*, 55–64, 2003.

Dobson, D. P., and J. P. Brodholt, The electrical conductivity and thermal profile of the Earth's mid-mantle, *Geophys. Res. Lett.*, *27*, 2325–2328, 2000.

Drake, M. J., and K. Righter, Determining the composition of the Earth, *Nature*, *416*, 39–44, 2002.

Dyar, M. D., A. V. McGuire, and R. D. Ziegler, Redox equilibria and crystal chemistry of coexisting minerals from spinel lherzolite mantle xenoliths, *Amer. Mineral.*, *74*, 969–980, 1989.

Fei, Y., H. K. Mao, J. Shu, G. Parthasarathy, W. A. Bassett, and J. Ko, Simultaneous high-P, high-T X ray diffraction study of β-$(Mg,Fe)_2SiO_4$ to 26 GPa and 900 K, *J. Geophys. Res.*, *97*, 4489–4495, 1992.

Fei, Y., D. Virgo, B. O. Mysen, Y. Wang, and H. K. Mao, Temperature dependent electron delocalization in $(Mg,Fe)SiO_3$ perovskite, *Amer. Mineral.*, *79*, 826–837, 1994.

Fiquet, G., D. Andrault, A. Dewaele, T. Charpin, M. Kunz, and D. Haüsermann, *P-V-T* equation of state of $MgSiO_3$ perovskite, *Phys. Earth Planet. Int.*, *105*, 21–31, 1998.

Fiquet, G., A. Dewaele, D. Andrault, M. Kunz, and T. Le Bihan, Thermoelastic properties and crystal structure of $MgSiO_3$ perovskite at lower mantle pressure and temperature conditions, *Geophys. Res. Lett.*, *27*, 21–24, 2000.

Fiquet, G., F. Guyot, M. Kunz, J. Matas, D. Andrault, and M. Hanfland, Structural refinements of magnesite at very high pressure, *Amer. Mineral.*, *87*, 1261–1265, 2002.

Frost, B. R., and C. Ballhaus, Comment on "Constraints on the origin of the oxidation state of mantle overlying subduction zones: an example from Simcoe, Washington, USA" by A.D. Brandon, D.S. Draper, *Geochim. Cosmochim. Acta*, *62*, 329–331, 1998.

Frost, D. J., and F. Langenhorst, The effect of Al_2O_3 on Fe-Mg partitioning between magnesiowüstite and magnesium silicate perovskite, *Earth Planet. Sci. Lett.*, *199*, 227–241, 2002.

Frost, D. J., F. Langenhorst, and P. A. van Aken, Fe-Mg partitioning between ringwoodite and magnesiowüstite and the effect of pressure, temperature and oxygen fuagcity, *Phys. Chem. Minerals*, *28*, 455–470, 2001.

Frost, D. J., C. Liebske, F. Langenhorst, C. A. McCammon, R. Trønnes, and D. C. Rubie, Experimental evidence for the existence of iorn-rich metal in the Earth's lower mantle, *Nature*, *428*, 409–411, 2004.

Funamori, N., T. Yagi, W. Utsumi, T. Kondo, T. Uchida, and F. M, Thermoelastic properties of $MgSiO_3$ perovskite determined by *in situ* X ray observations up to 30 GPa and 2000 K, *J. Geophys. Res.*, *101*, 8257–8269, 1996.

Gudmundsson, G., and B. J. Wood, Experimental tests of garnet peridotite oxygen barometry, *Contrib. Mineral. Petrol.*, *119*, 56–67, 1995.

Haggerty, S. E., and V. Sautter, Ultradeep (greater than 300 kilometers), ultramafic upper mantle xenoliths, *Science*, *248*, 993–996, 1990.

Harte, B., J. W. Harris, M. T. Hutchison, G. R. Watt, and M. C. Wilding, Lower mantle mineral associations in diamonds from São Luiz, Brazil, in *Mantle Petrology: Field Observations and High-Pressure Experimentation: A Tribute to Francis R. (Joe) Boyd*, edited by Y. Fei, C.M. Bertka, and B.O. Mysen, pp. 125–153, Geochemical Society, USA, 1999.

Harte, B., J. W. Harris, M. C. Wilding, V. Sautter, and C. McCammon, Eclogite-garnetite inclusions in diamonds from the São Luiz area, Brazil, *18th General Meeting Inter. Mineral. Assoc. Abstract Volume*, 74, 2002.

Hawthorne, F. C., Mössbauer spectroscopy, in *Spectroscopic Methods in Mineralogy and Geology*, edited by F.C. Hawthorne, pp. 255–340, Mineralogical Society of America, Washington, D.C., 1988.

Hirose, K., Phase transitions in pyrolitic mantle around 670-km depth: Implications for upwelling of plumes from the lower mantle, *J. Geophys. Res.*, *107*, 10.1029/2001JB000597, 2002.

Hirose, K., Y. W. Fei, Y. Z. Ma, and H. K. Mao, The fate of subducted basaltic crust in the Earth's lower mantle, *Nature*, *397* (6714), 53–56, 1999.

Holzapfel, C., D. C. Rubie, D. J. Frost, and F. Langenhorst, Fe-Mg interdiffusion in $(Mg,Fe)SiO_3$ perovskite, *Eos Trans. AGU*, *82(47), Fall Meet. Suppl.*, Abstract V51A-0967, 2001.

Huckenholz, H. G., J. F. Schairer, and H. S. Yoder, Synthesis and stability of ferri-diopside, *Mineral. Soc. Am. Spec. Pap.*, *2*, 163–177, 1969.

Hugh-Jones, D. A., A. B. Woodland, and R. J. Angel, The structure of high-pressure C2/c ferrosilite and crystal chemistry of high-pressure C2/c pyroxenes, *Am. Mineral.*, *79*, 1032–1041, 1994.

Hutchison, M. T., M. B. Hursthouse, and M. E. Light, Mineral inclusions in diamonds: associations and chemical distinctions around the 670-km discontinuity, *Contrib. Mineral. Petrol.*, *142*, 119–126, 2001; *142*, 260, 2001.

Ingrin, J., and H. Skogby, Hydrogen in nominally anhydrous upper-mantle minerals: Concentration levels and implications, *Eur. J. Mineral.*, *12*, 543–570, 2000.

Irifune, T., Absence of an aluminous phase in the upper part of the Earth's lower mantle, *Nature*, *370*, 131–133, 1994.

Irifune, T., and M. Isshiki, Iron partitioning in a pyrolite mantle and the nature of the 410-km seismic discontinuity, *Nature*, *392*, 702–705, 1998.

Irifune, T., and A. E. Ringwood, Phase transformations in primitive MORB and pyrolite compositions to 25 GPa and some geophysical implications, in *High-Pressure Research in Mineral Physics*, edited by M.H. Manghnani, and Y. Syono, pp. 231–242, Terra Scientific Publishing Company, Tokyo, 1987.

Irifune, T., and A. E. Ringwood, Phase transformations in subducted oceanic crust and buoyancy relationships at depths of 600–800 km in the mantle, *Earth Planet. Sci. Lett.*, *117*, 101–110, 1993.

Irifune, T., T. Sekine, A. E. Ringwood, and W.O. Hibberson, The eclogite-garnetite transformation at high pressure and some geophysical implications, *Earth Planet. Sci. Lett.*, *77*, 245–256, 1986.

Isshiki, M., T. Irifune, K. Hirose, S. Ono, Y. Ohishi, T. Watanuki, E. Nishibori, M. Takata, and M. Sakata, Stability of magnesite and its high-pressure form in the lowermost mantle, *Nature*, *427*, 60–63, 2004.

Ito, E., and E. Takahashi, Ultrahigh-pressure phase transformations and the constitution of the deep mantle, in *High-Pressure Research in Mineral Physics*, edited by M. H. Manghnani, and Y. Syono, pp. 221–229, Terra Scientific Publishing Company, Tokyo, 1987.

Jackson, J. M., J. Zhang, and J. D. Bass, Sound velocities and elasticity of aluminous $MgSiO_3$ perovskite: Implications for aluminum heterogeneity in Earth's lower mantle, *Geophys. Res. Lett.*, *31*, doi:10.1029/2004GL019918, 2004, 2004.

Kadik, A., Evolution of Earth's redox state during upwelling of carbon-bearing mantle, *Phys. Earth Planet. Int.*, *100*, 157–166, 1997.

Kaminsky, F. V., O. D. Zakharchenko, R. Davies, W. L. Griffin, G.K. Khachatryan-Blinova, and A.A. Shiryaev, Superdeep diamonds from the Juina area, Mato Grosso State, Brazil, *Contrib. Mineral. Petrol.*, *140*, 734–753, 2001.

Kasting, J. F., D. H. Eggler, and S. P. Raeburn, Mantle redox evolution and the oxidation state of the Archean atmosphere, *J. Geol.*, *101*, 245–257, 1993.

Kato, T., A. E. Ringwood, and T. Irifune, Experimental determination of element partitioning between silicate perovskites, garnets and liquids: constraints on early differentiation of the mantle, *Earth Planet. Sci. Lett.*, *89*, 123–145, 1988.

Kato, T., E. Ohtani, Y. Ito, and K. Onuma, Element partitioning between silicate perovskites and calcic ultrabasic melt, *Phys. Earth Planet. Int.*, *96*, 201–207, 1996.

Katsura, T., K. Sata, and E. Ito, Electrical conductivity of silicate perovskite at lower-mantle conditions, *Nature*, *395*, 493–495, 1998.

Kress, V. C., and I. S. E. Carmichael, The compressibility of silicate liquids containing Fe_2O_3 and the effect of composition, temperature, oxygen fugacity and pressure on their redox states, *Contrib. Mineral. Petrol.*, *108* (1–2), 82–92, 1991.

Kubo, A., T. Yagi, S. Ono, and M. Akaogi, Compressibility of $Mg_{0.9}Al_{0.2}Si_{0.9}O_3$ perovskite, *Proc. Japan Acad.*, *76B*, 103–107, 2000.

Kurosawa, M., H. Yurimoto, and S. Sueno, Patterns in the hydrogen and trace element compositions of mantle olivines, *Phys. Chem. Minerals*, *24*, 385–395, 1997.

Lauterbach, S., C.A. McCammon, P. van Aken, F. Langenhorst, and F. Seifert, Mössbauer and ELNES spectroscopy of $(Mg,Fe)(Si,Al)O_3$ perovskite: A highly oxidised component of the lower mantle, *Contrib. Mineral. Petrol.*, *138*, 17–26, 2000.

Lécuyer, C., and Y. Ricard, Long-term fluxes and budget of ferric iron: implication for the redox states of the Earth's mantle and atmosphere, *Earth Planet. Sci. Lett.*, *165*, 197–211, 1999.

Lee, C. A., A. D. Brandon, and M. Norman, Vanadium in peridotites as a proxy for paleo-fO_2 during partial melting: Prospects, limitations, and implications, *Geochim. Cosmochim. Acta*, *67*, 3045–3064, 2003.

Liebske, C., A. Corgne, D. J. Frost, D. C. Rubie, and B. J. Wood, Compositional effects on element partitioning between Mg-silicate perovskite and silicate melts, *Contrib. Mineral. Petrol.*, in press.

Litasov, K., E. Ohtani, F. Langenhorst, H. Yurimoto, T. Kubo, and T. Kondo, Water solubility in Mg-perovskites and water storage capacity in the lower mantle, *Earth Planet. Sci. Lett.*, *211*, 189–203, 2003.

Luth, R. W., and D. Canil, Ferric iron in mantle-derived pyroxenes and a new oxybarometer for the mantle, *Contrib. Mineral. Petrol.*, *113*, 236–248, 1993.

Luth, R. W., D. Virgo, F. R. Boyd, and B. J. Wood, Ferric iron in mantle-derived garnets. Implications for thermobarometry and for the oxidation state of the mantle, *Contrib. Mineral. Petrol.*, *104*, 56–72, 1990.

Marion, S., A. I. Becerro, and T. Norby, Ionic and electronic conductivity in $CaTi_{1-x}Fe_xO_{3-\Delta}$ (x=0.1–0.3), *Ionics*, *5&6*, 385–392, 1999.

Marty, B., I. Tolstikhin, I. L. Kamensky, V. Nivin, E. Balaganskaya, and J.-L. Zimmerman, Plume-derived rare gases in 380 Ma carbonatites from the Kola region (Russia) and the argon isotopic composition in the deep mantle, *Earth Planet. Sci. Lett.*, *164*, 179–192, 1998.

Mattioli, G. S., and B. J. Wood, Upper mantle oxygen fugacity recorded by spinel lherzolites, *Nature*, *322*, 626–628, 1986.

McCammon, C., Deep diamond mysteries, *Science*, *293*, 813–814, 2001.

McCammon, C. A., A Mössbauer milliprobe: Practical considerations, *Hyper. Inter.*, *92*, 1235–1239, 1994.

McCammon, C. A., Perovskite as a possible sink for ferric iron in the lower mantle, *Nature*, *387*, 694–696, 1997.

McCammon, C. A., The crystal chemistry of ferric iron in $Mg_{0.95}Fe_{0.05}SiO_3$ perovskite as determined by Mössbauer spectroscopy in the temperature range 80–293 K, *Phys. Chem. Minerals*, *25*, 292–300, 1998.

McCammon, C. A., Methods for determination of $Fe^{3+}/\Sigma Fe$ in microscopic samples, in *The P.H. Nixon Volume, Proc. VII Inter. Kimberlite Conf.*, edited by J.J. Gurney, J.L. Gurney, M.D. Pascoe, and S.H. Richardson, pp. 540–544, Red Roof Design, Cape Town, South Africa, 1999.

McCammon, C. A., Insights into phase transformations from Mössbauer spectroscopy, in *Transformation Processes in Minerals*, edited by S. Redfern, and M. Carpenter, pp. 241–264, Mineralogical Society of America, Washington, DC, 2000.

McCammon, C. A., Mössbauer spectroscopy: Applications, in *Spectroscopic Methods in Mineralogy*, edited by A. Beran, and E. Libowitsky, pp. 369–398, Eötvös University Press, Budapest, 2004.

McCammon, C. A., and M. G. Kopylova, A redox profile of the Slave mantle and oxygen fugacity control in the cratonic mantle, *Contrib. Mineral. Petrol.*, *148*, 55–68, 2004.

McCammon, C. A., and N. L. Ross, Crystal chemistry of ferric iron in $(Mg,Fe)(Si,Al)O_3$ majorite with implications for the transition zone, *Phys. Chem. Minerals*, *30*, 206–216, 2003.

McCammon, C. A., V. Chaskar, and G. G. Richards, A technique for spatially resolved Mössbauer spectroscopy applied to quenched metallurgical slags, *Meas. Sci. Technol.*, *2*, 657–662, 1991.

McCammon, C. A., D. C. Rubie, C. R. Ross II, F. Seifert, and H. S. C. O'Neill, Mössbauer spectra of $^{57}Fe_{0.05}Mg_{0.95}SiO_3$ perovskite at 80 and 298 K, *Amer. Mineral.*, *77*, 894–897, 1992.

McCammon, C. A., M. Hutchison, and J. Harris, Ferric iron content of mineral inclusions in diamonds from São Luiz: A view into the lower mantle, *Science*, *278*, 434–436, 1997.

McCammon, C. A., J. P. Peyronneau, and J.-P. Poirier, Low ferric iron content of (Mg,Fe)O at high pressures and high temperatures, *Geophys. Res. Lett.*, *25*, 1589–1592, 1998.

McCammon, C. A., W. L. Griffin, S. H. Shee, and H. S. C. O'Neill, Oxidation during metasomatism in ultramafic xenoliths from the Wesselton kimberlite, South Africa: Implications for the survival of diamond, *Contrib. Mineral. Petrol.*, *141*, 287–296, 2001.

McCammon, C. A., D. J. Frost, J. R. Smyth, H. M. Laustsen, T. Kawamoto, N. L. Ross, and P. A. van Aken, Oxidation state of iron in hydrous mantle phases: Implications for subduction and mantle oxygen fugacity, *Phys. Earth Planet. Int.*, *143–144*, 157–169, 2004a.

McCammon, C. A., S. Lauterbach, F. Seifert, F. Langenhorst, and P.A. van Aken, Iron oxidation state in lower mantle mineral assemblages I. Empirical relations derived from high-pressure experiments, *Earth Planet. Sci. Lett.*, *222*, 435–449, 2004b.

McCammon, C. A., T. Stachel, and J. W. Harris, Iron oxidation state in lower mantle mineral assemblages II. Inclusions in diamonds from Kankan, Guinea, *Earth Planet. Sci. Lett.*, *222*, 423–434, 2004c.

McDonough, W. F., Earth's core, in *Encyclopedia of geochemistry*, edited by C. P. Marshall and R. W. Fairbridge, pp. 151–156, Kluwer Academic Publishers, Dordrecht, 1999.

McGuire, A. V., M. D. Dyar, and J. E. Nielson, Metasomatic oxidation of upper mantle peridotite, *Contrib. Mineral. Petrol.*, *109*, 252–264, 1991.

Miyajima, N., K. Fujino, N. Funamori, T. Kondo, and T. Yagi, Garnet-perovskite transformation under conditions of the Earth's lower mantle: an analytical transmission electron microscopy study, *Phys. Earth. Planet. Int.*, *116*, 117–131, 1999.

Morgan, J., and W. E. Anders, Chemical composition of Earth, Venus and Mercury, *Proc. Nat. Acad. Sci.*, *77*, 6973–6977, 1980.

Nakamura, A., and H. Schmalzried, On the stoichiometry and point defects of olivine, *Phys. Chem. Minerals*, *10*, 27–37, 1983.

Navrotsky, A., A lesson from ceramics, *Science*, *284*, 1788–1789, 1999.

Neal, C. R., S. E. Haggerty, and V. Sautter, "Majorite" and "Silicate Perovskite" mineral compositions in xenoliths from Malaita, *Science*, *292*, 1015, 2001.

Nishiyama, N., and T. Yagi, Phase relation and mineral chemistry in pyrolite to 2200°C under the lower mantle pressures and implications for dynamics of mantle plumes, *J. Geophys. Res.*, *108*, 10.1029/2002JB002216, 2003.

Nitsan, U., Stability field of olivine with respect to oxidation and reduction, *J. Geophys. Res.*, *79*, 706:711, 1974.

O'Neill, H. St. C., The transition between spinel lherzolite and garnet lherzolite, and its use as a geobarometer, *Contrib. Mineral. Petrol.*, *77*, 185–194, 1981.

O'Neill, H. St. C., The origin of the moon and early history of the Earth—a chemical model. Part 2: The Earth, *Geochim. Cosmochim. Acta*, *55*, 1159–1172, 1991.

O'Neill, H. St. C., and V. J. Wall, The olivine-orthopyroxene-spinel oxygen geobarometer, the nickel precipitation curve, and the oxygen fugacity of the Earth's upper mantle, *J. Petrol.*, *28*, 1169–1191, 1987.

O'Neill, H. St. C., D. C. Rubie, D. Canil, C. A. Geiger, C. R. Ross II, F. Seifert, and A. B. Woodland, Ferric iron in the upper mantle and in transition zone assemblages: Implications for relative oxygen fugacities in the mantle, in *Evolution of the Earth and Planets*, edited by T. Takahashi, R. Jeanloz, and D. C. Rubie, pp. 73–88, American Geophysical Union, Washington D.C., 1993a.

O'Neill, H. St. C., C. A. McCammon, D. C. Canil, D. C. Rubie, C. R. Ross II, and F. Seifert, Mössbauer spectroscopy of transition zone phases and determination of minimum Fe^{3+} content, *Amer. Mineral.*, *78*, 456–460, 1993b.

Ono, S., E. Ito, and T. Katsura, Mineralogy of subducted basaltic crust (MORB) from 25 to 37 GPa, and chemical heterogeneity of the lower mantle, *Earth Planet. Sci. Lett.*, *190* (1–2), 57–63, 2001.

Ono, S., T. Kikegawa, and T. Iizuka, The equation of state of orthorhombic perovskite in a peridotitic mantle composition to 80 GPa: implications for chemical composition of the lower mantle, *Phys. Earth Planet. Int.*, *145*, 9–17, 2004.

Parkinson, I. ., and R. J. Arculus, The redox state of subduction zones: Insights from arc-peridotites, *Chem. Geol.*, *160*, 409–423, 1999.

Peslier, A. H., J. F. Luhr, and J. Post, Low water contents in pyroxenes from spinel-peridotites of the oxidized, sub-arc mantle wedge, *Earth Planet. Sci. Lett.*, *201*, 69–86, 2002.

Peyronneau, J., and J. P. Poirier, Electrical conductivity of the Earth's lower mantle, *Nature*, *342*, 537–539, 1989.

Rea, D. K., and L. J. Ruff, Composition and mass flux of sediment entering the world's subduction zones:Implications for global sediment budgets, great earthquakes, and volcanism, *Earth Planet. Sci. Lett.*, *140*, 1–12, 1996.

Richmond, N. C., and J. P. Brodholt, Calculated role of aluminum in the incorporation of ferric iron into magnesium silicate perovskite, *Amer. Mineral.*, *83* (9–10), 947–951, 1998.

Richmond, N. C., and J. P. Brodholt, Incorporation of Fe^{3+} into forsterite and wadsleyite, *Amer. Mineral.*, *85*, 1155–1158, 2000.

Ringwood, A. E., Chemical evolution of the terrestrial planets, *Geochim. Cosmochim. Acta*, *30*, 41–104, 1966.

Ringwood, A. E., *Composition and Petrology of the Earth's Mantle*, McGraw-Hill, New York, 1975.

Ross, N. L., R. J. Angel, and F. Seifert, Compressibility of brownmillerite ($Ca_2Fe_2O_5$): Effect of vacancies on the elastic properties of perovskites, *Phys. Earth. Planet. Int.*, *129*, 145–151, 2002.

Sautter, V., S. E. Haggerty, and S. Field, Ultradeep (> 300 kilometers) ultramafic xenoliths—Petrological evidence from the transition zone, *Science*, *252*, 827–830, 1991.

Scott-Smith, B. H., R. V. Danchin, J. W. Harris, and K. J. Stracke, Kimberlites near Orrroroo, South Australia, in *Kimberlites and Related Rocks*, edited by J. Kornprobst, pp. 121–142, Elsevier, 1984.

Sinogeikin, S. V., J. Zhang, and J. D. Bass, Elasticity of single crystal and polycrystalline $MgSiO_3$ perovskite by Brillouin spectroscopy, *Geophys. Res. Lett.*, *31*, doi:10.1029/2004GL019559, 2004, 2004.

Smyth, J. R., and T. Kawamoto, Wadsleyite II: A new high pressure hydrous phase in the peridotite-H_2O system, *Earth Planet. Sci. Lett.*, *146*, E9–E16, 1997.

Smyth, J. R., C. M. Holl, D. J. Frost, S. D. Jacobsen, F. Langenhorst, and C. A. McCammon, Structural systematics of hydrous ringwoodite and water in Earth's interior, *Amer. Mineral.*, *88*, 1402–1407, 2003.

Smyth, J. R., C. M. Holl, F. Langenhorst, H. M. Laustsen, G. R. Rossman, A. Kleppe, C. A. McCammon, and T. Kawamoto, Crystal chemistry of wadsleyite II and water in the Earth's interior, *Phys. Chem. Minerals*, 31, 691–705, 2005.

Speidel, D. H., Phase equilibria in the system $MgO-FeO-Fe_2O_3$: The 1300°C isothermal section and extrapolations to other temperatures, *J. Amer. Ceram. Soc.*, *50*, 243–248, 1967.

Stachel, T., Diamonds from the asthenosphere and the transition zone, *Eur. J. Mineral.*, *13*, 883–892, 2001.

Stachel, T., G. P. Brey, and J. W. Harris, Kankan diamonds (Guinea) I: from the lithosphere down to the transition zone, *Contrib. Mineral. Petrol.*, *140*, 1–15, 2000a.

Stachel, T., J. W. Harris, G. P. Brey, and W. Joswig, Kankan diamonds (Guinea) II: lower mantle inclusion parageneses, *Contrib. Mineral. Petrol.*, *140*, 16–27, 2000b.

Stachel, T., J. W. Harris, S. Aulbach, and P. Deines, Kankan diamonds (Guinea) III: $\Delta^{13}C$ and nitrogen characteristics of deep diamonds, *Contrib. Mineral. Petrol.*, *142*, 465–475, 2002.

Suito, K., Y. Tsutsui, S. Nasu, A. Onodera, and F. E. Fujita, Mössbauer effect study of the γ-form of Fe_2SiO_4, in *High Pressure in Science and Technology (Proc. 9th AIRAPT Conf.)*, edited by C. Homan, R.K. MacCrone, and E. Whalley, pp. 295–298, Elsevier Science Publ. Co., New York, 1984.

Tribaudino, M., F. Nestola, C. Meneghini, and G. D. Bromiley, The high-temperature $P2_1/c$-$C2/c$ phase transition in Fe-free Ca-rich $P2_1/c$ clinopyroxenes, *Phys. Chem. Minerals*, *30*, 527–535, 2003.

Wang, Z., and T. Yagi, Incorporation of ferric iron in $CaSiO_3$ perovskite at high pressure, *Mineral. Mag.*, *62*, 719–723, 1998.

Wänke, H., G. Dreibus, and E. Jagoutz, Mantle chemistry and accretion history of the Earth, in *Archean Geochemistry*, edited by A. Kröner, G.N. Hanson, and A.M. Goodwin, pp. 1–24, Springer-Verlag, Berlin, 1984.

Webb, S. A., and B. J. Wood, Spinel-pyroxene-garnet relationships and their dependence on Cr/Al ratio, *Contrib. Mineral. Petrol.*, *92*, 471–480, 1986.

Wood, B. J., An experimental test of the spinel peridotite oxygen barometer, *J. Geophys. Res.*, *95*, 15845–15851, 1990.

Wood, B. J., Phase transformations and partitioning relations in peridotite under lower mantle conditions, *Earth Planet. Sci. Lett.*, *174*, 341–354, 2000.

Wood, B. J., and J. Nell, High-temperature electrical conductivity of the lower-mantle phase (Mg,Fe)O, *Nature*, *351*, 309–311, 1991.

Wood, B. J., and D. C. Rubie, The effect of alumina on phase transformations at the 660-kilometer discontinuity from Fe-Mg partitioning experiments, *Science*, *273*, 1522–1524, 1996.

Wood, B. J., and D. Virgo, Upper mantle oxidation state: Ferric iron contents of lherzolite spinels by ^{57}Fe Mössbauer spectroscopy and resultant oxygen fugacities, *Geochim. Cosmochim. Acta*, *53*, 1277–1291, 1989.

Wood, B. J., L. T. Bryndzia, and K. E. Johnson, Mantle oxidation state and its relationship to tectonic environment and fluid speciation, *Science*, *248*, 337–345, 1990.

Wood, B. J., A. Pawley, and D. Frost, Water and carbon in the Earth's mantle, *Phil. Trans. R. Soc. Lond. A*, *354*, 1495–1511, 1996.

Woodland, A. B., and H. S. C. O'Neill, Synthesis and stability of $Fe_3^{2+}Fe_2^{3+}Si_3O_{12}$ garnet and phase relations with $Fe_3Al_2Si_3O_{12}$-$Fe_3^{2+}Fe_2^{3+}Si_3O_{12}$ solutions, *Amer. Mineral.*, *78*, 1002–1015, 1993.

Woodland, A. B., and H. S. C. O'Neill, Phase relations between $Ca_3Fe_2^{3+}Si_3O_{12}$-$Fe_3^{2+}Fe_2^{3+}Si_3O_{12}$ garnet and $CaFeSi_2O_6$-$Fe_2Si_2O_6$ pyroxene solid solutions, *Contrib. Mineral. Petrol.*, *121*, 87–98, 1995.

Woodland, A. B., and R. J. Angel, Crystal structure of a new spinelloid with the wadsleyite structure in the system Fe_2SiO_4-Fe_3O_4 and implications for the Earth's mantle, *Am. Mineral.*, *83*, 404–408, 1998.

Woodland, A. B., and R. J. Angel, Phase relations in the system fayalite-magnetite at high pressures and temperatures, *Contrib. Mineral. Petrol.*, *139*, 734–747, 2000.

Woodland, A. B., and M. Koch, Variation in oxygen fugacity with depth in the upper mantle beneath the Kaapvaal craton, Southern Africa, *Earth Planet. Sci. Lett.*, *214*, 295–310, 2003.

Woodland, A. B., R. J. Angel, M. Koch, M. Kunz, and R. Miletich, Equations of state for $Fe_3^{2+}Fe_2^{3+}Si_3O_{12}$ "skiagite" garnet and Fe_2SiO_4-Fe_3O_4 spinel solid solutions, *J. Geophys. Res.*, *104*, 20049–20058, 1999.

Xu, Y., and C. A. McCammon, Evidence for ionic conductivity in lower mantle (Mg,Fe)(Si,Al)O_3 perovskite, *J. Geophys. Res.*, *107*, doi:10.1029/2001JB000677, 2002.

Xu, Y., C. A. McCammon, and B. T. Poe, The effect of alumina on the electrical conductivity of silicate perovskite, *Science*, *282*, 922–924, 1998.

Yagi, T., K. Okabe, A. Kubo, and T. Kikegawa, Complicated effects of aluminium on the compressibility of silicate perovskite, *Phys. Earth Planet. Int.*, *143–144*, 81–91, 2004.

Yamazaki, D., and S. Karato, Some mineral physics constraints on the rheology and geothermal structure of Earth's lower mantle, *Amer. Mineral.*, *86*, 385–391, 2001.

Yamazaki, D., T. Kato, H. Yurimoto, E. Ohtani, and M. Toriumi, Silicon self-diffusion in $MgSiO_3$ perovskite at 25 GPa, *Phys. Earth Planet. Int.*, *119*, 299–309, 2000.

Zhang, J., and C. Herzberg, Melting experiments on anhydrous peridotite KLB-1 from 5.0 to 22.5 GPa, *J. Geophys. Res.*, *99*, 17729-17742, 1994.

Zhang, J., and D. J. Weidner, Thermal equation of state of aluminum-enriched silicate perovskite, *Science*, *284*, 782–784, 1999.

C. A. McCammon, Bayerisches Geoinstitut, Universität Bayreuth, D-95440 Bayreuth, Germany. (catherine.mccammon@uni-bayreuth.de)

Thermochemical State of the Lower Mantle: New Insights From Mineral Physics

James Badro, Guillaume Fiquet, and François Guyot

Laboratoire de Minéralogie Cristallographie, Institute de Physique du Globe de Paris, Paris, France

We report recent findings in the field of high-pressure mineral physics with important implications for Earth's lower mantle. We show that the two main constituents of the lower mantle, namely, $(Mg,Fe)SiO_3$—magnesium silicate perovskite—and $(Mg,Fe)O$—ferropericlase—undergo electronic transitions at lower mantle pressures (70 and 120 GPa), in which iron transforms from the high-spin state to the low-spin state. The transformations modify the thermochemical state of Earth's lower mantle. Minerals bearing high-spin iron have characteristic absorption lines in the near-infrared, hindering radiative conductivity at lower-mantle temperatures. These absorption lines shift to the visible (green to violet) range in the low-spin state, and their intrinsic intensities decrease; the minerals thus become increasingly transparent in the near-infrared and their radiative and total thermal conductivities rise. Thus, the heat conductivity of the lowermost mantle could be higher than previously thought. Moreover, the spin-driven partitioning of iron between the two mineral phases can explain large-scale chemical heterogeneities in the mantle that are driven by regional temperature variations. It is noteworthy that the transition pressures correspond to the bottom third of the lower mantle (70 GPa, 1700-km depth), and to the last 300 km above the core–mantle boundary (120 GPa, 2600-km depth); these regions have very special geophysical signatures, as chemical heterogeneities have been reported by seismology in the first case and the bottom 300 km of Earth's mantle constitutes the D″ layer. Our observations provide a mineral physics basis for these features in Earth's lower mantle.

INTRODUCTION

Seismic wave analysis is the main tool for studying Earth's structure [*Jeffreys and Bullen*, 1940], whether on the shallow (crustal seismology, oil exploration) or the deep (Earth's inner structure) scale [*Birch*, 1952]. Radial seismological models, such as PREM [*Dziewonsky and Anderson*, 1981], obtained from inversion of seismic travel times and normal mode spectra, give the density, bulk and shear moduli, compressional and shear sound velocities in the Earth as a function of depth. In terms of seismic wave travel times, structurally different entities are determined on the basis of discontinuities in density, bulk, and shear properties [*Birch*, 1952; *Jeffreys and Bullen*, 1940]. Seismological profiles show without any doubt that the most severe discontinuity in material properties occurs at a depth of 2890 km; this is the core–mantle boundary (CMB) that separates the dense iron-rich core below from the lighter silicate-rich mantle above. In the mantle, the sharpest discontinuity occurs at a depth of 660 km, followed by a broader one at 410 km; the 660-km discontinuity separates the lower mantle below from the upper mantle above, whereas the 410-km discon-

Earth's Deep Mantle: Structure, Composition, and Evolution
Geophysical Monograph Series 160
Copyright 2005 by the American Geophysical Union
10.1029/160GM15

tinuity further separates the uppermost mantle above from the transition zone below. In one of its major achievements, high-pressure mineral physics has shown that these seismic discontinuities are a direct consequence of phase transformations in candidate mantle minerals (e.g. *Ringwood* [1975]; *Jackson and Rigden* [1998]; and a review article by *Fiquet* [2001] and references therein). Although the question of a difference in chemical composition between the upper and lower mantle is still a matter of debate, the mineralogy of these two reservoirs is in any case fundamentally different; the upper mantle is mainly constituted of fourfold-coordinated silicate framework minerals (olivine, pyroxene, garnet, and their high-pressure modifications), whereas the lower mantle consists mainly of sixfold coordinated silicates— magnesium silicate perovskite (Mg-*pv*) and calcium silicate perovskite (Ca-*pv*)—and dense oxides such as ferropericlase (*fp*) [*Fiquet*, 2001; *Liu*, 1974].

Although apparent changes in the lower mantle cannot be observed using radial (one-dimensional) seismological models, the recent developments [*Fukao et al.*, 2001; *Grand*, 2002; *Romanowicz*, 2003; *Trampert and van der Hilst*, this volume] of seismic tomography as well as high-precision modelling of seismic anisotropy [*Garnero and Lay*, 2003] have revealed more subtle phenomena in the deepest parts of the lower mantle. For instance, lateral velocity anomalies at depths around 1700 km have been ascribed to chemical heterogeneities [*Karato and Karki*, 2001; *van der Hilst and Kárason*, 1999], as thermal heterogeneities alone cannot reasonably explain the magnitude of these anomalies. Such effects cannot be addressed by mineral physics by probing structural transformations, because the absence of sharp discontinuity withdraws all chances for the phenomenon to be related to a structural phase transformation. Of course, past reports of chemical reactions in the lower mantle, such as the breakdown of Mg-*pv* into silica and *fp* [*Saxena et al.*, 1996] could have brought a logical mineral-chemical basis to these observations. However, it appears today that this breakdown was not observed in any of the numerous studies performed since [*Fiquet*, 2001; *Fiquet et al.*, 2000; *Kesson et al.*, 1998; *Serghiou et al.*, 1998; *Shim and Jeanloz*, 2002], and it is widely accepted today that Mg-*pv* is chemically stable (i.e. it does not break down into its simple constituents) under the pressure and temperature (P–T) conditions of Earth's mantle. Given that geophysical observations tend to show a chemical origin for lower-mantle heterogeneities, a deeper insight in the chemistry of minerals in the lower mantle is necessary. This has led us to investigate directly the properties that control the chemistry in these compounds, namely, their electronic structure, by carrying out x-ray emission spectroscopy (XES) measurements on the lower mantle's most abundant minerals [*Badro et al.*, 2003, 2004] at high pressure.

A mysterious and uncharacterised geophysical entity, the D″ layer, is a 300-km thick layer at the bottom of the mantle that sits right above the CMB [*Lay*, 1983]. The highest resolution seismic studies even locally suggest another very thin layer (on the order of tens of km) characterized by very low seismic wave velocities between the D″ layer and the CMB: the ultra-low velocity zone (ULVZ). If the top of the D″ layer represents a discontinuity, it could have a structural signature in terms of mineral transformations. But as very little is known about the chemical balance (and therefore about the chemistry) at those depths, with the chemical input of cold downwelling slabs and the chemical output through upwelling plumes, there is a possibility that D″ and the ULVZ constitute a chemical layer different from the surrounding mantle and core; in this case again, and this time because the chemistry of the reservoir (and therefore the mineralogical composition) is not known, little can be obtained from structural studies stricto sensu to help address this issue.

Until very recently, high pressure and high temperature (HP–HT) experiments performed in the laser-heated diamond anvil cell in the P-T conditions of Earth's lower mantle have shown the stability of lower-mantle phases (Mg-*pv*, *fp*, and Ca-*pv*) up to 110 GPa (2450-m depth) [e.g. *Fiquet*, 2001], with possible small distortions (and minor phase changes) of the Mg-*pv* phase [*Shim et al.*, 2001]. However, a recent x-ray diffraction study showed that Mg-*pv* undergoes a transformation to a denser high-pressure modification named post-perovskite phase (Mg-*pp*), at the P–T conditions of the D″ layer [*Murakami et al.*, 2004a]. This mineral physics observation directly addresses the question of the origin of the D″ layer in terms of a structural mineralogical transformation. In order to constrain and provide input for geophysical and geochemical models, and also to cross-check the validity of the new material's properties with geophysical observations, various physical and chemical properties (beyond crystal structure) of this new phase need to be known (Clapeyron slope, electronic and transport properties, etc. We have performed a XES study that shows that the properties of Mg-*pv* are severely modified in the same pressure range as the Mg-*pv* to Mg-*pp* transformation [*Badro et al.*, 2004], partly addressing issues related to the properties of that phase.

In this article, we present a study of the electronic properties of lower-mantle minerals subject to extreme conditions, using x-ray spectroscopy to gain insight in important physical and chemical properties that cannot otherwise be obtained. The rationale lies in the fact that the strong changes in the properties of the lower mantle's main minerals, Mg-*pv*, *fp*, and Ca-*pv*, should likely be determined by their iron (and to a lesser degree aluminium) content. Indeed, the alkali-earth end-member minerals MgO (periclase), $MgSiO_3$-*pv* and $CaSiO_3$-*pv* do not seem to exhibit any physical or chemical peculiarity with increasing pressure and temperature. They remain mostly transparent broad-band-gap ionically bonded insulators. In the presence of iron oxide, periclase and Mg-*pv* readily form solid solutions by Mg/Fe exchange, giving rise to electronically complex compounds (Mg,Fe)O-*fp* and (Mg,Fe)SiO$_3$-*pv*, whereas Ca-*pv* remains essentially iron-free. Because Mg is a more abundant major element than Fe in Earth's mantle, all the present (oxidized) iron is accommodated in the solid solution *fp* and Mg-*pv* phases, and no pure (oxidized) iron end-member minerals should exist.

The specificity of iron comes from the fact that it is a transition metal; unlike the other major elements (Mg, Ca, Si, Al) of the bulk silicate Earth [BSE; *McDonough and Sun*, 1995], it has partially filled *3d* orbitals that give rise to a series of possible energy configurations depending on its atomic environment. Notably, iron adopts different valences, for instance, ferrous (Fe^{2+}) and ferric (Fe^{3+}) iron, and different electronic configurations, for instance, high-spin and low-spin states. The energy differences involved in *3d* orbitals are small compared to those in *s* and *p* orbitals, implying that the sensitivity of iron-bearing compounds to energy changes induced by pressure and temperature is more dramatic. Properties such as electrical, thermal, and radiative conductivity can vary significantly with pressure or temperature in such compounds, whereas they remain almost unperturbed in the iron-free end-members. The physical and chemical complexity of lower-mantle mineralogical assemblages is undoubtedly linked to the presence of iron.

ELECTRONIC STRUCTURE OF IRON— THERMODYNAMICS

Iron is a *3d* transition metal, and its ionic electronic structure is (Ar)$3d^6$ for ferrous iron (Fe^{2+}) and (Ar)$3d^5$ for ferric iron (Fe^{3+}). The *3d* orbitals are constituted of a subset of five orbitals with different symmetry. Three of these (d_{xy}, d_{xz}, d_{yz}) make up a set of orbitals named t$_{2g}$ and the two others (d_{z^2} and $d_{x^2-y^2}$) make up a set of orbitals named e_g. Iron can be in the high-spin (HS) or low-spin (LS) states, as defined by the occupation of that outermost *3d* orbital. Their energetic fine structure is shown in Figure 1 in the case of an octahedral (sixfold coordination) field, within the framework of crystal field theory [*Burns*, 1993]. The latter is a simple yet sound theoretical framework, for estimating the electronic contributions to internal energy and entropy of a HS or LS state. All thermodynamic calculations presented in this paper are based on this simple theory, and should only be considered as first-order approximations. More precise theoretical modelling is required to obtain increased precision or insight into more complex effects. Depending on the values of the pairing energy (E_p) and the t$_{2g}$–e_g splitting (known as 10 Dq or Δ)—often referred to as the crystal field stabilisation parameter—the lowest energy configuration can either be the HS state (high E_p/Δ ratio or weak field) where the spin quantum number is maximum or the LS state (low E_p/Δ ratio or strong field). The crystal field stabilisation parameter (CFSP) Δ increases with decreasing iron–oxygen bond-length, and thus with increasing density, whereas the pairing energy remains essentially unchanged; in this respect, an iron-bearing mineral consisting of HS iron at low pressure (at the surface of the globe) can undergo a HS to LS (or spin pairing) transition at higher pressure (at depth in the Earth). Obviously, such a transformation has an effect on various properties linked to the energetics (energy and entropy) of the compound. A striking example of the effects of spin transitions is the colour change of haemoglobin, the substance in red blood cells that bonds oxygen to its octahedral ferrous iron site, making it the oxygen carrier in the body: iron in this molecule undergoes a HS to LS transition upon release of the oxygen atom (and the opposite upon capture) and transforms in color from red (oxygen-rich arterial blood) to purple (oxygen-depleted venous blood). Other band-structure effects can induce spin-pairing transitions (for instance, an increase of e_g and t$_{2g}$ bandwidth), but these only apply in

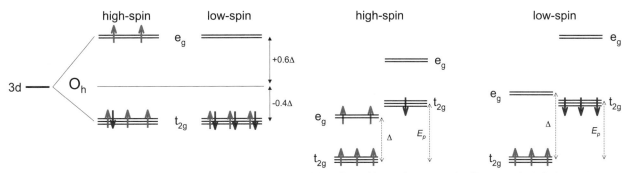

Figure 1. Electronic fine-structure of ferrous iron in the high-spin and low-spin states. The figure on the left shows the population of both states in terms of orbital symmetry (t$_{2g}$ and e_g), whereas the figure on the right shows the energetic diagram as a function of the crystal field splitting energy Δ and coulomb repulsion energy E_p.

the case of band-like systems, which is irrelevant to the case of diluted iron in lower-mantle minerals.

The internal energy in the HS and LS states of Fe^{2+} is obtained by summing the energies of all six electrons, depending on their energy level and pairing, which yields (see Figure 1)

$$U_{HS} = 4 \times (-0.4\Delta) + 2 \times (0.6\Delta) + E_p = -0.4\Delta + E_p \quad (1a)$$

$$U_{LS} = 6 \times (-0.4\Delta) + 3E_p = -2.4\Delta + 3E_p \quad (1b)$$

and similarly for Fe^{3+}, one obtains

$$U_{HS} = 3 \times (-0.4\Delta) + 2 \times (0.6\Delta) = 0 \quad (1c)$$

$$U_{LS} = 5 \times (-0.4\Delta) + 2E_p = -2.0\Delta + 2E_p \quad (1d)$$

The (configurational) entropy is given by $S = k_B \ln(n \cdot (2s + 1))$ where s is the total spin, n the orbital degeneracy, and k_B the Boltzmann constant [Sherman, 1991]. For ferrous iron (Fe^{2+}), the total spin is $s_{HS} = 2$ and $s_{LS} = 0$; the degeneracy term is $n_{LS} = 1$ in the LS state and $n_{HS} = 3$ in the HS state but should be considered $n_{HS} = 1$ at low temperatures because the degeneracy could be lifted. Therefore, the entropy of LS state is $S_{LS} = 0$ (there is a single microstate), and the entropy of the HS state is $S_{HS} = k_B \ln 5$ at high temperature and $S_{HS} = k_B \ln 5$ at low temperature. For ferric iron (Fe^{3+}), the total spin is $s_{HS} = 2.5$ and $s_{LS} = 0.5$; the degeneracy term is $n_{HS} = 1$ in the HS state, and $n_{LS} = 3$ in the LS state at high temperature but should be considered $n_{LS} = 1$ at low temperatures because, once again, the degeneracy could be lifted. Therefore, the entropy of HS state is equal to $S_{HS} = k_B \ln 6$, and the entropy of the LS state is equal to $S_{LS} = k_B \ln 6$ at high temperature and to $S_{LS} = k_B \ln 2$ at low temperature. It is interesting to note that $\Delta S = 0$ at high temperature for the spin-pairing transition in ferric iron.

A spin transition occurs when the overlap in electronic orbitals is such that Δ reaches a critical value thermodynamically favouring the LS state. Δ increases with compression following pressure according to Burns [1993]:

$$\Delta(V) = \Delta^0 \cdot \left(\frac{V_0}{V}\right)^{5/3} \quad (2)$$

and reaches a critical value Δ_c, defined by thermodynamic equilibrium in the canonical *(T,V)* ensemble, minimising the Helmholtz free energy $F = U - TS$. This is what we will call hereafter a *density-driven transition*, although the term may not be perfectly correct because it doesn't reflect the entropy dependence but rather has the meaning that the transition is due to increasing electronic overlap, which is a sole function of density. Therefore, at the spin transition

$F_{HS} = F_{LS}$, and incorporating the terms detailed above, one obtains at the transition

$$\Delta_c = E_p + \frac{1}{2} k_B T \ln 15 \quad (3a)$$

at mantle temperatures and

$$\Delta_c = E_p + \frac{1}{2} k_B T \ln 5 \quad (3b)$$

at room temperature (for ferrous iron).

In a system at a fixed pressure and temperature [canonical *(P,T)* ensemble], a spin transition is thus followed by a local volume collapse around the concerned ionic species. This is of course accompanied by an increase in Δ according to equation (2), and this reduced volume for iron in the LS state with respect to the HS state affects partitioning of iron between various phases coexisting in the HS and LS state. It is also partially responsible for the important changes expected in the transport properties as indicated above. Whether this volume collapse, following the spin transition, should be accounted for in the equilibrium condition ($\Delta G = 0$) is still a matter of debate and depends on whether the transition is cooperative in character or not. Therefore, two cases will be discussed below: (i) taking this volume change into account at equilibrium (a first-order transition with phase coexistence) and (ii) considering the volume collapse occurs only afterwards (a cooperative first-order transition with no phase coexistence). Of course, one must keep in mind that we are dealing with a multicomponent system and that our one-component thermodynamic treatment is not rigorously exact.

X-RAY EMISSION SPECTROSCOPY

XES is a spectroscopic technique that allows probing of the structure of the low-energy outermost electronic levels or orbitals of a chemical element [Jenkins, 1999]. It is a chemically sensitive local technique in the sense that it probes only the local environment around a given chemical species. The process is schematically described in Figure 2. When a primary x-ray excitation source strikes a sample, the x-ray can either be absorbed by the atom or scattered through the material. The process in which an x-ray is absorbed by the atom by transferring all of its energy to an electron is called the "photoelectric effect" [Ashcroft and Mermin, 1976; Jenkins, 1999]. During this process, if the primary x-ray had sufficient energy, electrons are ejected from the inner shells (e.g. 1s electrons), creating core-holes (vacancies). These holes present an unstable condition for the atom. As the atom returns to its stable condition, electrons from the outer shells are transferred to the inner shells and, in

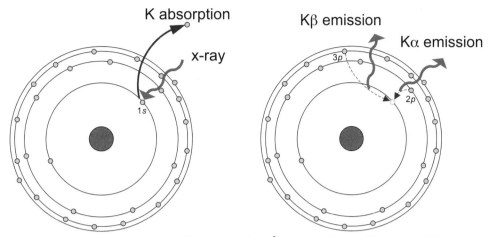

Figure 2. Schematic representation of the shell structure of a Fe^{2+} ion as well as an illustration of the x-ray absorption (left) followed by the Kα and Kβ emission processes (right).

the process, give off a characteristic x-ray whose energy is the difference between the two binding energies of the corresponding shells. Because each element has a unique set of energy levels, each element produces x-rays at a unique set of energies, allowing one to nondestructively measure the elemental composition of a sample. The process of emissions of characteristic x-rays is called x-ray fluorescence or x-ray spectroscopy. Their analysis is called x-ray fluorescence spectroscopy, or x-ray emission spectroscopy (XES).

The characteristic x-rays are labelled K, L, M, or N to denote the shells from which they originated. Another designation, α, β, or γ, is made to mark the x-rays that originated from the transitions of electrons from higher shells. Hence, a Kα x-ray is produced from a transition of an electron from the L to the K shell (or $2p$ to $1s$), and a Kβ x-ray is produced from a transition of an electron from the M to a K shell (or $3p$ to $1s$), etc. Given that within the shells there are multiple orbits of higher and lower binding-energy electrons, a further designation is made as $α_1$, $α_2$ or $β_1$, $β_2$, etc. to denote transitions of electrons from these orbits into the same lower shell. Now, the hole left in the upper levels interacts with the outermost electronic shell ($3d$ in the case of iron), an exchange interaction that gives rise to small energy shifts (on the order of a few electron volts) of the emission line [*Jenkins*, 1999]. These shifts, and the lineshapes accompanying them, depend on the specific electronic configuration and structure of that outermost shell; their analysis in turn allows determining these, which as we said earlier, is responsible for the electronic complexity of transition metals.

The Kβ emission line originates from the $3p$ to $1s$ decay. Kβ spectra can be interpreted using atomic multiplet calculations [*Hermsmeier et al.*, 1988] and configuration interactions. The spectral shape of Kβ emission line in $3d$ transition metal compounds is dominated by final state interaction

between the $3p$ core-hole and the electrons of the partially filled $3d$ shell (see Figure 3). Qualitatively, the main effect is due to the exchange interaction between the core-hole and the local moment, which results in the splitting of the Kβ spectrum into HS and LS final states. This simple picture also predicts that the energy separation between the two peaks is given by the product of the exchange integral J and $(2S + 1)$, where S is the total spin of the $3d$ shell; and that the intensity ratio between the two is given by $S/(S + 1)$ [*Tsutsumi et al.*, 1976]. Both the energy splitting and intensity ratio are modified when configuration interaction is taken into account

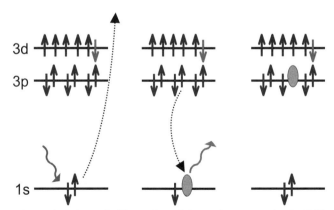

Figure 3. Schematic illustration of the various states in the Fe^{2+} Kβ x-ray emission process: the initial state (left panel) is a K-edge x-ray absorption process ejects an electron (arrow), which leaves an empty core-hole (circle) in the $1s$ shell. This is followed by a Kβ x-ray emission process in the intermediate state (middle panel), which is associated with the collapse of a $3p$ electron onto the $1s$ shell to fill the core-hole. The final state (right panel) corresponds to a core-hole in the $3p$ shell, which strongly interacts with the $3d$ magnetic moment. That interaction in the final state, in turn. allows the determination of the structure of that $3d$ shell.

[*Hermsmeier et al.*, 1988]. These detailed calculations show that the $\underline{3p}^13d^\uparrow$ final state is characterized by a single peak, which constitutes the most of the intensity of the main emission line. On the other hand, the $\underline{3p}^\uparrow 3d^\uparrow$ final state is further split into two components, one at significantly lower energy than predicted by simple theory and one at slightly lower energy than the main emission line ($\underline{3p}^13d^\uparrow$), which appears as a shoulder to the main emission line. However, the simplified picture does point out the qualitative changes one would expect as the $3d$ electrons go from HS state to LS state, namely, smaller energy splitting between the main peak and the satellite as well as reduction of the intensity ratio of the satellite peal to the main peak. Figure 4 reports two Kβ emission spectra from iron in wüstite (FeO) and pyrite (FeS$_2$) at room conditions [*Badro et al.*, 1999], which are archetypes of HS- and LS-bearing iron, respectively. The difference is quite striking in terms of lineshape. Indeed, the emission spectrum of HS Fe is characterized by a main peak K$\beta_{1,3}$ with an energy of 7058 eV and a satellite peak Kβ' located at lower energy appearing as a result of the $3p$ core-hole–$3d$ exchange interaction in the final state as described above. Given that the LS state of Fe^{2+} (d^6 configuration) is characterized by a total magnetic moment equal to zero, it should lead to the disappearance of the low-energy satellite, a very clear spectral signature that can be seen in Figure 4.

As we will see later, not only do the intensities on the satellite and widths of the main line change between the HS and LS configurations, but so does the position of the main line [*Peng et al.*, 1994]. This has been given less attention in recent studies because of the difficulty of making absolute energy measurements. Indeed, in most studies, the position of the main line is taken as a reference and is set to the default value of 7058 eV. This has the advantage of not needing to calibrate the geometry of the x-ray spectrometer for every measurement, but its drawback is that one cannot use that additional information (main line position) to detect, constrain, and quantify the spin transition.

EXPERIMENTAL DETAILS

In order to probe the electronic properties of lower-mantle minerals in the pressure and temperature conditions of the lower mantle, all our samples were loaded in specially designed diamond anvil cells. These had radial openings with respect to the compression axis, through which the XES spectra were measured. The incoming x-rays were monochromatised with a high-resolution (approximately 100-meV bandwidth) fixed exit monochromator at 15 keV for *fp* and 14 keV for Mg-*pv*. The beam was focused into the sample through one of the diamond anvils, and the emitted x-rays were measured at 90° through the radial openings in the diamond anvil cell, thanks to x-ray–transparent high-strength beryllium gaskets. Pressure was measured using the ruby fluorescence technique, and one of the samples (Mg-*pv*) was heated with the single-mode TEM$_{00}$ fundamental harmonic of a Nd:YAG infrared laser. Both the very small size of the samples (50 μm diameter, 10 μm thickness) and their low-iron content are responsible for the very weak signals. In turn, this rules out the use of a pressure medium, in order to maximise signal. Even in these conditions, every spectrum required accumulation times of 6 h for *fp* to 18 h for Mg-*pv*.

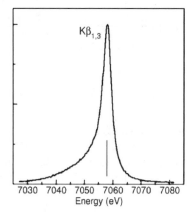
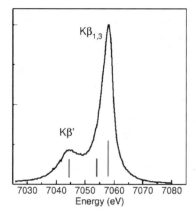

Figure 4. Example of two Kβ x-ray emission spectra of a low-spin (left) and a high-spin (right) compound, namely, FeS$_2$ (pyrite) and FeO (wüstite), respectively. The characteristic lineshape of the low-spin compound consists of a single main line (position marked by a long vertical line), whereas that of a high-spin compound consists of two smaller lines (positions marked by short lines) in addition to that same main line. One of them is at significantly lower energy and constitutes the Kβ' satellite, whereas the second is a weak feature at slightly lower energy than the main line and appears as a shoulder in the peak.

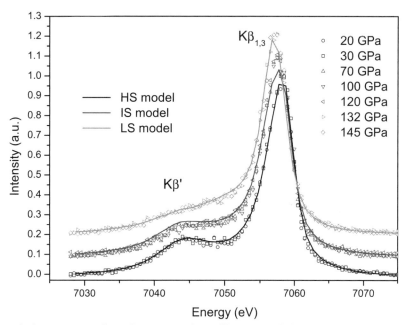

Figure 5. X-ray emission spectra collected on magnesium silicate perovksite ($Mg_{0.9}Fe_{0.1}$)SiO_3 between 20 and 145 GPa. The presence of a satellite structure ($K\beta'$ line) on the low-energy side of the iron main emission line ($K\beta_{1,3}$ line) is characteristic of a high spin $3d$ magnetic moment. The spin state of iron transforms twice at 70 and 120 GPa, as indicated by the changes in $K\beta'$ line intensity. Moreover, the position of the $K\beta_{1,3}$ line shifts at each transition, and by a total of -0.75 eV between 20 and 145 GPa, which is in agreement with a HS–LS transition in iron. The spectra have been vertically shifted (each group separately) for clarity. The first (bottom) group is characteristic of the HS state, the second (middle) is characteristic of the mixed state (mixture of HS and LS iron), and the third (top) is characteristic the LS state. Note that all spectra within one group are close to identical in terms of both $K\beta$ intensity and $K\beta_{1,3}$ position. The solid lines are models constructed from reference molecular compounds (see text and Figure 7), and are not fitted to the data; in this respect, the agreement with the data is excellent and brings a totally independent confirmation of the three-state scenario with a fully HS state, a mixed state, and a fully LS state.

The spectrometer for the x-ray emission measurement is an 80-cm Rowland circle instrument laid out in the horizontal plane. Emission was measured using the (531) reflection of a spherically bent (focusing) silicon single-crystal analyzer. The incoming x-ray beam was focused using a pair of Kirkpatrick–Baez mirrors to increase photon density on the microscopic sample.

LOWER MANTLE MINERALS

(Mg,Fe)SiO₃ Perovskite

We probed the spin state and measured the spin magnetic moment of iron in ($Mg_{0.9} Fe_{0.1}$)SiO_3 perovskite from 20 to 145 GPa using high-resolution $K\beta$ XES. To release stresses and to avoid any presence of disordered or amorphous phases, the sample was heated with a Nd:YAG laser operating in TEM_{00} mode (as indicated above) at each pressure point between 30 and 120 GPa. At higher pressure, laser radiation did not

couple with the sample and heating of the sample was no longer possible.

The spectra (Figure 5), which have been vertically shifted by groups for clarity, reveal two transitions associated with a decrease of $K\beta'$ peak intensity (Figure 6a) and shifting of the $K\beta_{1,3}$ peak to lower energy (Figure 6b), occurring around 70 and 120 GPa, respectively. The intensity decrease of the $K\beta'$ peak (Figures 5 and 6a) is directly proportional to the relative abundances of the HS and LS species (Figure 6a, right scale). Moreover, the energy shift of the $K\beta_{1,3}$ line (by -0.75 eV) measured in Mg-pv between 20 and 145 GPa (Figure 6b) is consistent with predicted values for a HS to LS transition (and also linearly scales to the relative abundance of the species as shown in Figure 6b); indeed, the transfer of spectral weight to the satellite region in the HS state, compared to the LS state in which the satellite is largely reduced, is balanced by a displacement of the main peak in the opposite direction, to keep fixed the centre of mass of the emission line [*Peng et al.*, 1994]. The pressure dependence of the iron magnetic

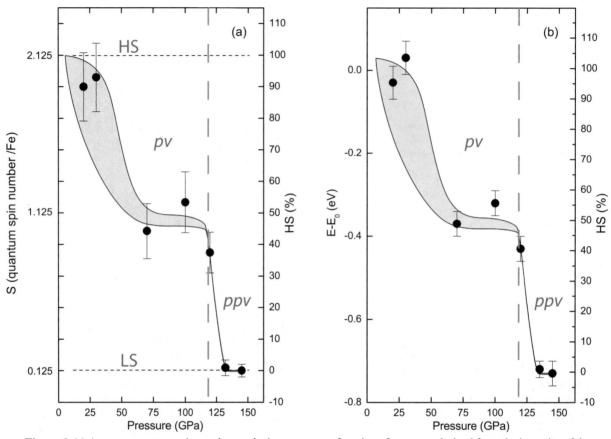

Figure 6. (a) Average quantum spin number on the iron atom as a function of pressure, derived from the intensity of the Kβ line. It indicates that two transitions occur in Mg-*pv* at 70 and 120 GPa. The surface curves are guides to the eye. In the lower pressure range, the two curves represent tentative upper and lower bounds of the iron magnetic moment. The corresponding abundance x of iron atoms in the HS state is represented on the right vertical axis, and the abundance of the LS atoms is $(1-x)$. (b) K$\beta_{1,3}$ line position as a function of pressure. This figure shows again that the two transitions occur at 70 and 120 GPa. Note that the surface curves (guides to the eye) are identical in both figures and show that the system evolves in three independent states: the HS state below 70 GPa, the mixed state between 70 and 120 GPa, and the LS state above 120 GPa. The vertical dashed line separates the perovskite from the post-perovskite phases.

state derived from the data analysis (Figures 6a and 6b) shows the two successive transitions at around 70 and 120 GPa. Unfortunately, we cannot state whether the first transition is sharp or gradual due to the limited pressure sampling in the range of the first transition (70 GPa), although recent studies have shown that it is a gradual transition that ends at 70 GPa [*Jackson et al.*, 2005]. However, the precision at the onset of the second transition (120 GPa) clearly indicates a sharp transition. Interestingly, our measurements show a nearly constant Kβ' peak intensity (Figure 6a) and K$\beta_{1,3}$ peak position (Figure 6b) as well as a similar lineshape (Figure 5) for all spectra in a given spin-state (HS, mixed state, LS). Therefore, iron in a HS state at low pressure and around 70 GPa undergoes a first transition that is completed to a "mixed" state where the relative amount of the HS and

LS iron species is roughly 55% and 45% (Figures 6a and 6b, right scale). At 120 GPa, a second sharp transition occurs, transforming the remaining HS iron to the LS state; above this pressure, all iron in Mg-*pv* is in the LS state.

Iron in Mg-*pv* can be in the ferrous (Fe^{2+}) or ferric (Fe^{3+}) valence-state and can occupy one of two crystallographic sites [*McCammon*, 1997]. Although the exact partitioning behaviour of ferrous and ferric iron between these two sites is poorly characterized, current interpretation [*McCammon*, 1997] indicates that the large "dodecahedral" A site (actually a bicapped trigonal prism, but referred to as dodecahedral hereafter for consistency with the generally accepted nomenclature) is the host of ferrous iron, whereas the smaller octahedral B site is the host of ferric iron. In the conditions of synthesis and with 2 mol% iron [Fe/(Fe + Mg) = 0.1)] in our sample, Mg-*pv* should contain about 80% Fe^{2+} and 20% Fe^{3+} [*Fei et al.*, 1994], although it has been shown recently [*Jackson et al.*, 2005] that a similar synthesis yields 60% ferrous iron and 40% ferric iron.

The changes up to 70 GPa could be related to a recently reported transition [*Jackson et al.*, 2005] concerning ferric iron and observed by nuclear forward scattering (a synchrotron time-resolved Mössbauer spectroscopy technique), which shows that Fe^{3+} undergoes a gradual spin-pairing transition between room pressure and 70 GPa. Given that we observe the transition of 45% of the Fe ions at that pressure, this could indicate that the amount of Fe^{3+} in our Mg-*pv* sample is actually higher than we expected and closer to that measured by *Jackson et al.* [2005] than that by *Fei et al.* [1994]. Unfortunately, we have no analysis of the redox state of our sample to address this issue, and relating these populations, whether on a site-specific basis or on a valence-specific basis, is beyond the scope of this study; it should, however, be considered in the future. A careful characterisation of the samples has to be undertaken in order to ascribe the transitions to either valences or sites. Studying the electronic properties of Al-bearing Mg-*pv*

[*Li et al.*, 2004] is a way to address this issue, because the Fe^{3+}/Fe ratio increases with Al content [*McCammon*, 1997]. Although this study [*Li et al.*, 2004] does not cover the entire lower-mantle pressure range, it indicates that the transitions in Al-bearing Mg-*pv* could be shifted to higher pressures. On the other hand, the presence of large stresses could have increased the observed pressure of transition, or even smeared the pressure range in which it occurred (the samples in that study were not annealed at high pressure). The work by *Li et al.* [2004] nevertheless shows that it is important to gage the effect of aluminium on the transition pressure, as one should bear in mind that the transition will lose all geophysical relevance if its pressure shifts for any given reason above 136 GPa (CMB).

With this in hand, support for our interpretation of the two transitions can be obtained from a direct comparison (i.e. without fitting, Figures 5 and 7) with reference spectra of molecular spin crossover compounds [*Vankó et al.*, 2002] containing ferrous or ferric iron in the HS and LS states. In these systems, which are used as references in XES [*Vankó et al.*, 2002], iron is in a localized state and adopts a band-free atomic behaviour, as in Mg-*pv*, which

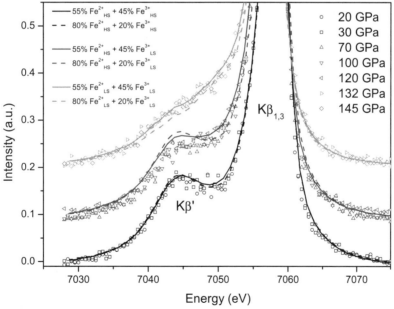

Figure 7. Fit of the a priori models to the x-ray emission spectra collected on magnesium silicate perovksite $(Mg_{0.9}Fe_{0.1})SiO_3$ between 20 and 145 GPa. These models, constructed from reference molecular spin crossover compounds, are not fitted to the data, but only overlain. The spectra (and models) have been vertically shifted (each group separately) for clarity; the first (bottom) group is characteristic of the HS state, the second (middle) is characteristic of the mixed state (mixture of HS and LS iron), and the third (top) is characteristic of the LS state. The solid lines are models based on 55% Fe^{2+} and 45% Fe^{3+} (close to the composition found in *Jackson et al.* [2005]) and the dashed lines are models based on 80% Fe^{2+} and 20% Fe^{3+} (close to the composition found in *Fei et al.* [1994]). The agreement with the data suggests that the prior (55% Fe^{2+} and 45% Fe^{3+}) is more self-consistent than the latter. The larger discrepancy in the intermediate "mixed" model suggests that more complex phenomena might occur in that state.

contains only 10 mol% of iron. Given that x-ray emission spectroscopy is a locally sensitive technique (depending on local environment of atoms, and not on the general structure), such model spectra can be used to interpret those obtained in Mg-*pv*. Composite spectra can thus be constructed from these reference spectra by linear combination of the spectra of the four components (ferrous HS and LS, ferric HS and LS) weighted by the fractions of the individual states. Using two models with different Fe^{3+}/Fe^{2+} ratios, one with 80% Fe^{2+} and 20% Fe^{3+} (according to *Fei et al.* [1994], dashed lines), and one with 55% Fe^{2+} and 45% Fe^{3+} (close to *Jackson et al.* [2005], solid lines), we created model spectra for a fully HS (low pressure), a fully LS (high pressure), and a mixed state (Fe^{2+} HS and Fe^{3+} LS) Mg-*pv*. These models (Figure 7) agree (intensity of the $K\beta'$ line and position of the $K\beta_{1,3}$ line) with the measured spectra and confirm that the LP spectra have the signature of a full HS state and that the highest pressure spectra have the signature of a full LS state. Comparison of the spectra in the "mixed" state do not show as good an agreement as the two extreme states, indicating that more complex phenomena may be taking place, but strengthens the idea (this can already be seen in the LS state) that the model with 55% Fe^{2+} and 45% Fe^{3+} composition shows a better agreement with the experiment (Figure 7) than does the model with the 80% Fe^{2+} and 20% Fe^{3+}. This could confirm the transition in Fe^{3+} at 70 GPa (as shown in *Jackso, et al.* [2005]) and that in Fe^{2+} at 120 GPa.

(Mg,Fe)O Ferropericlase

The case of *fp* is simpler, because it contains one species of iron (i.e. ferrous iron) and one site in the structure which is the standard octahedral site in the cubic *fcc* structure. We monitored the spin state and measured the spin magnetic moment of iron in $(Mg_{0.83} Fe_{0.17})O$ *fp* as a function of pressure, from 0 to 80 GPa, by high-resolution $K\beta$ XES. As described above, two features are identified in the emission spectrum as a main peak ($K\beta_{1,3}$) and a satellite peak ($K\beta'$).

The spectra of the micrometer-sized samples were measured at different pressures (Figure 8) and the spectrum of a larger sample at ambient pressure was measured outside the cell. The spectra show that there is a decrease of the satellite peak intensity at 49 GPa and that it vanishes at 75 GPa, indicating a HS to LS transition in iron in *fp* between those pressures. Upon decompression, the spectrum at 64 GPa shows a partial reversal from HS to LS (a mixture of phases) and at 41 GPa, the spectrum shows the presence of only the HS state (indicating a complete transformation). The sample was probably subject to large radial pressure gradients in the absence of a pressure-transmitting medium and without

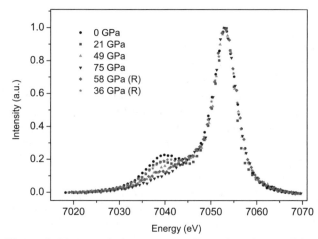

Figure 8. X-ray emission spectra collected on ferropericlase $(Mg_{0.83}Fe_{0.17})O$ at different pressures. The presence of a satellite structure ($K\beta'$ line) on the low-energy side of the iron main emission line ($K\beta_{1,3}$ line) is characteristic of a high-spin *3d* magnetic moment. This structure collapses at high pressure upon compression, and then reforms upon pressure decrease (spectra annotated with letter "R" in the legend). The system is in the HS state at 36 GPa, in a HS–LS mixture at 49 and 58 GPa, and in the LS state at 75 GPa. The pure LS component appears between 58 and 75 GPa. Note here that, unlike the experimental data collected on Mg-*pv*, the absolute energy was not calibrated, and the main line in each spectrum is fixed at an arbitrary value. This does not allow us to use in this case the energy shift (but only the satellite peak intensity) at the spin transition to constrain and quantify the transition, as was shown in the case of Mg-*pv*.

laser-annealing. Whether the transition is gradual (occurring over a large pressure range) or whether this observation is an artefact due to pressure gradients is still an open question. Measurements will need to be carried out either in a hydrostatic environment or along with laser annealing of the sample (e.g. in the experiment in Mg-*pv*) in order to fully address this important issue.

Similar measurements on FeO [*Badro et al.*, 1999] have shown there is no transition to a low-spin state to at least 143 GPa. Very recent experimental work [*Lin et al.*, 2005] shows that the transition pressure decreases with decreasing iron content. This has also been suggested by theoretical calculations [*Brodholt*, 2003]. The transition could thus be accompanied by phase separation between an iron-rich high-spin and a magnesium-rich low-spin *fp*, as shown by some experimental reports [*Dubrovinsky et al.*, 2000, 2001], although opposing claims have also been reported [*Lin et al.*, 2003]. Recent high-temperature equation of state measurements [*Fei et al.*] show a volume collapse at around 60 GPa, which is possibly a signature of the spin transition. Whether phase separation occurs at the onset of this transition at high temperature remains to be elucidated.

CRYSTAL-FIELD SPLITTING AND TEMPERATURE EFFECT

Although the effect of temperature still needs to be probed directly in order to transfer these laboratory observations to the real-Earth system, one can estimate that effect by using simple thermodynamic arguments. Recent experimental work [*Lennartsson et al.*, 2004] has shown the way with the in situ study of *fp* and needs to be extended to higher pressures and temperatures and other systems (e.g. Mg-*pv*). As explained above, electronic transitions are most of the time density-driven transition, i.e. they occur at a given critical density. Therefore, the effect of temperature, apart from entropy contributions that have been estimated earlier, would be to increase the transition pressure so as to counterbalance the effect of thermal expansion with that of isothermal compressibility. At a given pressure, the thermal expansion is given by:

$$\alpha = \frac{1}{V}\left(\frac{\partial V}{\partial T}\right)_P = \left(\frac{\partial \ln V}{\partial T}\right)_P \qquad (4)$$

and at a given temperature, the isothermal compressibility is given by:

$$\chi_T = -\frac{1}{V}\left(\frac{\partial V}{\partial P}\right)_T = -\left(\frac{\partial \ln V}{dP}\right)_T = \frac{1}{K_T}. \qquad (5)$$

Counterbalancing the effect of thermal expansion with that of pressure means that

$$d\ln V = \left(\frac{\partial \ln V}{\partial T}\right)_P dT + \left(\frac{\partial \ln V}{\partial P}\right)_T dP \equiv 0 \qquad (6)$$

hence obtaining

$$\frac{dP}{K_T} - \alpha \cdot dT = 0,$$

the differential equation which can be expressed in an integral form as follows:

$$\int_{P_0}^{P} \frac{1}{K_T} \cdot dP = \int_{T_0}^{T} \alpha \cdot dT.$$

(Mg,Fe)O Ferropericlase

Given that this compound contains only one species of iron in one site, estimating the temperature-dependence is simpler in the case of *fp* than for Mg-*pv*. Δ^0 in this compound has been measured by spectroscopy and is equal to 1.34 eV [*Burns*, 1993] at room conditions. At room temperature, the molar volume V_c at the transition (70 GPa and 300 K) is 9.15 cm^3/mol [*Knittle*, 1995]. Therefore

$$\left(\frac{V_0}{V_c}\right) = 1.30$$

(for *fp* as well as for the octahedral site in *fp*) and

$$\Delta(V) = \Delta^0 \cdot \left(\frac{V_0}{V}\right)^{5/3} = 2.07 \text{ eV}$$

at the transition. One can therefore estimate the value of E_p by using equation (3b), because the transition occurs at low temperature (300 K), yielding $E_p = 2.05$ eV, which is in excellent agreement with the accepted value in dense oxides and silicates of 2.09 eV in *fp* [*Burns*, 1993].

At 2500 K (average lower-mantle geotherm at 70 GPa [*Brown and Shankland*, 1981]), the transition should occur for a CFSP

$$\Delta_c = E_p + \frac{1}{2}k_B T \ln 15 = 2.34 \text{ eV}.$$

Using equation (2), we obtain

$$\left(\frac{V_0(T_0)}{V_c(T)}\right) = 1.40.$$

The pressure corresponding to this compression can be calculated by using the thermal equation of state

$$P(T) = \frac{3}{2}K_0(T) \cdot \left[\left(\frac{V(T)}{V_0(T)}\right)^{-\frac{7}{3}} - \left(\frac{V(T)}{V_0(T)}\right)^{-\frac{5}{3}}\right] =$$
$$\frac{3}{2}K_0(T) \cdot \left[\left(\frac{V(T)}{V_0(T_0)}\frac{V_0(T_0)}{V_0(T)}\right)^{-\frac{7}{3}} - \left(\frac{V(T)}{V_0(T_0)}\frac{V_0(T_0)}{V_0(T)}\right)^{-\frac{5}{3}}\right] \qquad (7)$$

where

$$V_0(T) = V_0(T_0)\exp\left(\int_{T_0}^{T}\alpha(T)dT\right),$$
$$\alpha(T) = a_0 + a_1 T + a_2 T^{-2}, \qquad \text{and}$$
$$K_0(T) = K_0 + \frac{\partial K_0}{\partial T}\Delta T$$

[*Fei*, 1995; *Knittle*, 1995]. At 2500 K ($\Delta T = 2200$ K), this is equal to 104 GPa and yields a Clapeyron slope

$$\frac{dP}{dT} = \frac{\Delta S}{\Delta V} = 16 \text{ MPa/K}.$$

Another way to estimate the temperature effect is to calculate directly the Clapeyron slope from the entropy and volume variation at the transition. The entropy change for one iron atom, that is, one octahedron, is $\Delta S = -k_B \ln 15$, and the volume change can be calculated by using the tabulated contraction of the Fe^{2+} ionic radius in an octahedral site between a HS and LS transition, which is 0.17 Å. The octahedral volume is 13.44 Å^3 at ambient pressure and reaches 9.60

Å³ at the transition pressure. That volume corresponds to a Fe–O bond length of 1.93 Å. When iron transforms to the LS state, the Fe–O bond length shortens by 0.17 Å to reach 1.76 Å, corresponding to an octahedral volume of 7.30 Å³, and the volume contraction is therefore $\Delta V = V_{LS} - V_{HS}$–2.3 Å³ per octahedron, or 1.39 cm³/mol. This gives a Clapeyron slope

$$\frac{dP}{dT} = \frac{\Delta S}{\Delta V} = 16.2 \text{ MPa/K},$$

which is very close to the value obtained above.

(Mg,Fe)SiO₃ Perovskite

As there are two transitions in Mg-*pv*, there have to be two sites/species with two different values of Δ. Information on the electronic structure of iron in the perovskite structure is very scarce [*Burns*, 1993; *Keppler et al.*, 1994]. Given that crystal field theory is a first-order approximate theory, we will consider from here on an "average" model that has the following characteristics: Two quasi-octahedral sites in Mg-*pv* contain ferric and ferrous iron, respectively, and undergo spin pairing at 70 and 120 GPa, respectively. The use of a quasi-octahedral model for the "dodecahedral" site is justified by its bi-capped trigonal prism geometry and can be understood in terms of Mg–O bond-lengths in that site [*Wentzcovitch et al.*, 1995]: At 100 GPa, there are six bonds that have an average length of 1.90 Å, two with a length of 2.213 Å, and four others with lengths longer than 2.5 Å. This means that the close oxygen environment is quasi-octahedral, with the bonds at 1.90 Å responsible for most of the electronic overlap on the atom in the site.

For Fe^{2+}, at the spin transitions (70 and 120 GPa, 300 K),

$$\Delta_c = E_p + \frac{1}{2} k_B T \ln 5 = 2.07 \text{ eV}$$

and

$$\left(\frac{V_0}{V_{c,B}}\right) = 1.20, \text{ and } \left(\frac{V_0}{V_{c,A}}\right) = 1.31$$

yielding

$$\Delta_{pv,B}^0 = \Delta_c \cdot \left(\frac{V_0}{V_{c,B}}\right)^{-5/3} = 1.53 \text{ eV}$$

and

$$\Delta_{pv,A}^0 = \Delta_c \cdot \left(\frac{V_0}{V_{c,A}}\right)^{-5/3} = 1.32 \text{ eV},$$

the latter being surprisingly close to the value of 1.34 eV in *fp*. The A and B subscripts to the CFSP denote that these could

be related to sites A and B of the structure. Nevertheless, there is no direct proof of this yet, and this nomenclature can also represent two virtual sites "A" and "B".

The effect of temperature is obtained from the Clapeyron slope

$$\frac{dP}{dT} = \frac{\Delta S}{\Delta V}$$

of the transition, where entropy and volume changes need to be estimated as shown above. For ferrous iron, ΔS between the LS state and the HS state is per atom [*Sherman*, 1991] at high temperature (restored orbital degeneracy), whereas for ferric iron, $\Delta S = 0$. Thus the Clapeyron slope for a spin-pairing transition involving ferric iron (possibly that at 70 GPa) at high temperature is null, and the transition pressure should not vary with temperature. The volume variation of an octahedral site at the transition can be approximated as follows. Using tabulated values, the ionic radius difference between octahedrally coordinated HS and LS Fe^{2+} is 0.17 Å. In the HS phase, the volume of the A and B sites at 100 GPa [*Wentzcovitch et al.*, 1995] is 15.73 cm³/mol (26.1 Å³ per site) and 3.73 cm³/mol (6.19 Å³ per site), respectively, corresponding to average Fe–O bond lengths of 1.97 Å (average for the six closest oxygen atoms) in the A site and 1.67 Å in the B site. The 0.17 Å contraction yields ΔV of 3.74 and 1.02 cm³/mol for the A and B sites, respectively. With 75% Fe in the dodecahedral site and 25% in the octahedral site, the average ΔV is 3.06 cm³/mol. Finally one obtains a Clapeyron slope

$$\frac{dP}{dT} = 7.4 \text{ MPa/K},$$

which is obtained by taking into account only Fe^{2+}; the effect of Fe^{3+} will be to decrease that slope because it contributes to ΔV but not to ΔS. But it can be estimated only if the proportion of Fe^{3+} is known.

A linear extrapolation to 2500 K (ΔT=2200 K) gives a pressure correction of 15 GPa. Whether this value is to be added to the experimental transition pressure of 120 GPa determined remains to be elucidated; because the LS spin state was quenched from a high-temperature annealing, the actual pressure could reflect the transition at high temperature. Ideally, such measurements should be preformed in situ at high pressures and temperatures [*Lennartsson et al.*, 2004].

DISCUSSION

In this study, we showed that iron in both of these phases undergoes spin-pairing transitions at lower-mantle depths. Of the many aspects these transitions can bear, two seem of utmost importance: the heat conductivity of the minerals and mineral assemblages, and the partitioning or iron between the two phases.

Thermal Conductivity in the Lower Mantle

Heat can be transported in the mantle by conduction, radiation, or convection. Convection in the lower mantle is initiated only if the other two processes fail to transfer heat produced from the secular cooling of the core and by radioactive decay, through the lower mantle, e.g. if the ratio of heat transport through convection to heat transport through conduction and radiation, is high enough. Changes in the conduction or radiation properties of lower-mantle mineral assemblages will therefore strongly affect lower-mantle dynamics [*Clark*, 1957; *Dubuffet et al.*, 2000; *Goto et al.*, 1980; *Kellogg et al.*, 1999; *Shankland et al.*, 1979; *Tackley*, 2002; *van den Berg et al.*, 2001]. We have shown that iron in both major phases of the lower mantle undergoes spin-pairing transitions between 70 and 120 GPa [*Badro et al.*, 2003, 2004]. One of the main and intrinsic characteristics of LS iron-bearing minerals resides in the blue-shift [*Burns*, 1993; *Sherman*, 1991] of iron absorption bands (the absorption bands initially in the infrared [IR] and red region shift to the green–blue region).

A HS–LS transition should increase the intrinsic radiative thermal conductivity of Mg-*pv* and *fp* [*Burns*, 1993; *Sherman*, 1991]. Indeed, radiative conductivity in HS iron-bearing lower-mantle minerals is hindered by light absorption in the near IR spectral region, due to intraband transitions [*Clark*, 1957; *Goto et al.*, 1980; *Keppler et al.*, 1994; *Shankland et al.*, 1979; *Williams et al.*, 1990]. An HS–LS transition of iron in these minerals is accompanied by a blue-shift of these absorption bands, from the near IR to the visible (blue-green) spectrum, as has been shown for octahedral Fe^{2+}-bearing minerals [*Burns*, 1993] and is likely in the case of dodecahedral coordination. In turn, light with wavelengths in the red to near IR spectral range should radiate through the assemblages with a longer mean absorption length. This spectral range corresponds to the maximum of blackbody emission for temperatures between 2000 K and 3000 K (970 nm at 3000 K), meaning that most of the radiative energy emitted from the core and lowermost mantle should be affected by the HS–LS transition.

Figure 9 shows the normalized blackbody thermal radiation as a function of wavelength at three different temperatures (2000, 2500, and 3000 K) that can be considered as bounds for the lower-mantle geotherm

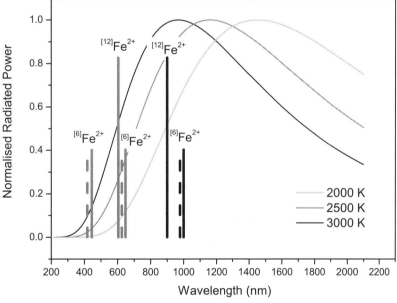

Figure 9. Normalized blackbody thermal radiance as a function of wavelength at three different temperatures (2000, 2500, and 3000 K) that can be considered as bounds for the lower-mantle geotherm. The absorption bands of $(Mg_{0.9}Fe_{0.1})SiO_3$ perovskite are reported in the HS state (solid black vertical line) and in the LS state (solid gray vertical line), whereas those of *fp* are similarly represented by dashed black and gray lines. In the HS state, the most intense absorption bands are in the near IR region, where blackbody radiation is maximal at these temperatures; this makes these compounds bad radiative thermal conductors. In the LS state, these same bands undergo a blue-shift to the visible region, where blackbody radiation is weaker. The total radiative power throughput in both minerals should significantly increase after the transition.

[*Brown and Shankland*, 1981]. The absorption bands of $(Mg_{0.9}Fe_{0.1})SiO_3$ perovskite are reported in the HS state (black sticks [vertical bars]) according to pressure-corrected measurements in *Keppler et al.* [1994] performed at room conditions, and in the LS state (gray bars according to the model proposed by *Burns* [1993]. The same applies for *fp*, whose absorption lines are represented by dashed bars. In the HS state, the most intense absorption bands are in the near IR region [*Sherman*, 1991], where blackbody radiation is maximal at these temperatures; this makes these compounds bad radiative thermal conductors. In the LS state, these same bands undergo a blue-shift to the visible region, where blackbody radiation is weaker. The total power throughput in both minerals should significantly increase after the transition, because IR radiation travels with a larger mean free path. This contributes to an enhanced radiative thermal conductivity in this state [*Sherman*, 1991] and is characteristic of the mineral assemblages of Earth's lowermost mantle. Of course, effects of spin changes on thermal conductivities will be significant only if they are not overprinted by other effects such as light scattering by grain boundaries. The question of the actual grain size in the lowermost mantle and of its effect of physical properties deserves further studies.

Our Mg-*pv* sample was heated by using an IR laser at each pressure point to anneal the sample and to release stresses generated at such high pressures. Heating using the 1064 nm radiation of a Nd:YAG laser is achieved in metal-bearing silicates through the absorption by crystal field absorption bands [*Boehler*, 2000]; that energy is then converted to heat, and the technique has been used to reach temperatures in excess of 5000 K in the diamond anvil cell [*Boehler*, 2000]. However, above 120 GPa, when all the iron in Mg-*pv* transformed to the LS state, it was no longer possible to heat the sample with the radiation of the Nd:YAG laser (1064 nm) if operating at the same power. This provides an additional indication that above 120 GPa, absorption by Mg-*pv* in the IR is hindered, and the mean free path for photons increases.

With both Mg-*pv* and *fp* becoming increasingly transparent to red and IR radiation at depths between the bottom third of the mantle and the core–mantle boundary, the transitions could have a strong dynamical signature as inferred from geodynamical modelling [*Dubuffet et al.*, 2000; *Kellogg et al.*, 1999; *van den Berg et al.*, 2001]; an increase in thermal conductivity will result in a decrease of the Rayleigh number and hence may hinder convection and favour layering in the lowermost mantle. Direct measurements of absorption spectra in the visible and IR, which need to be carried out at high pressures and high temperatures, are required to quantify the effect of these transitions and hence to precisely address this issue.

Iron Partitioning in the Lower Mantle

We have seen that both *fp* and Mg-*pv* undergo a spin-pairing transition in the 70 GPa range. Although our study does not offer the resolution required to accurately determine which compound undergoes the transition first, there is necessarily a pressure region where the two samples are in different states, i.e. one of them is in the LS state and the other in the HS state. Even if the transition pressures were identical to start with, the Clapeyron slopes are different enough so that lateral temperature variations at depth in the mantle would create such conditions.

Considering this eventuality is important, because a spin-pairing transition in only one of the minerals should alter the chemical behaviour of the assemblage as a whole, in the sense that the partition coefficient (of that species undergoing the transition, e.g. iron) between the minerals of the assemblage should be affected [*Burns*, 1993; *Malavergne et al.*, 1997]. This is due to the fact that the LS iron atom occupies a smaller volume than the HS species, and that at high pressure, the $P\Delta V$ energy term is all the stronger and favours iron in the LS site: One therefore expects an exchange of Fe with Mg between the phases in order to minimize the system's Gibbs free energy.

Crystal field theory [*Burns*, 1993] is a sound theoretical framework for estimating the electronic contributions to enthalpy and entropy of a HS or LS state, which in turn provides the Gibbs free energy change for the substitution reaction of Mg and ferrous Fe between Mg-*pv* and *fp* given by:

$$(Mg^{2+})_{fp} + (Fe^{2+})_{pv} \leftrightarrows (Mg^{2+})_{pv} + (Fe^{2+})_{fp} \qquad (8)$$

At thermodynamic equilibrium, this Gibbs free energy is proportional to the logarithmic partition coefficient of iron between the two phases defined by

$$\ln(K) = \ln\left(\frac{x_{Fe}}{x_{Mg}}\right)_{fp} - \ln\left(\frac{x_{Fe}}{x_{Mg}}\right)_{pv} = -\frac{\Delta G}{RT} \qquad (9)$$

where $R = Nk_B$ is the universal gas constant, and x the molar fractions. As proposed above, we will consider two thermodynamic models for the transition.

1. Cooperative phase transition. The first case we consider is that of a cooperative phase transition, as developed in *Badro et al.* [2003]. In this case, the Gibbs free energy of the system varies dramatically during the spin transition, and the effective ΔG between the two phases is finite in a similar fashion as Landau-type transitions. This effective free energy difference can be estimated as $\Delta G = \Delta F_{electronic} + P\Delta V$. The Helmholtz free energy term can be readily calculated from the equations given above, and we estimate the $P\Delta V$ term hereafter.

We will consider only the case of ferrous (Fe^{2+}) iron, because it is the only species we followed in both compounds. As the ferric iron content of our *fp* sample in the conditions of synthesis is almost null, the transition occurring in *fp* can be attributed to that of ferrous iron. The partitioning of ferric iron according to the relationship above introduces an additional term like the creation of a hole to accommodate the charge balance, as well as that of a defect, which are both unknown and too complex to account for here. For all these reasons, we shall limit the scope of this chapter to ferrous iron in the A site (which undergoes the spin-pairing transition at 120 GPa), keeping in mind that the framework used here can only be considered qualitatively, not quantitatively.

At 70 GPa,

$$\left(\frac{V}{V_0}\right)_{fp} = 0.77$$

and

$$\left(\frac{V}{V_0}\right)_{pv} = 0.83$$

which gives CFSP values $\Delta_{fp} = 2.07$ eV and $\Delta_{pv,A} = 1.80$ eV according to equation (2). This implies, using equations (1a) and (1b), that

$$U_{fp}^{HS} = 1.222 \text{ eV} \quad U_{pv,A}^{HS} = 1.330 \text{ eV}$$
$$U_{fp}^{LS} = 1.182 \text{ eV} \quad U_{pv,A}^{LS} = 1.830 \text{ eV} \quad (10)$$

and allows us to estimate the partition coefficient with the Gibbs free energy change upon iron substitution given by

$$\Delta G = G_{Fe,fp} - G_{Fe,pv} = U_{fp} - U_{pv} - T(S_{fp} - S_{pv}) + P(V_{fp} - V_{pv}) \quad (11)$$

where the volume difference can be calculated as shown above to be $\Delta V = -1.39$ cm³/mol. We see that the estimated

partition coefficient of ferrous iron between *fp* and Mg-*pv*'s A site at 70 GPa and 2500 K (approximately 1700 km depth) is 1.7 in the case where iron is in the HS in both phases phase (ΔS and ΔV are both identically 0, and partitioning is driven only by internal energy minimisation), and is in agreement with measurements [*Kesson, et al.*, 2002; *Mao et al.*, 1997; *Murakami et al.*, 2004b] performed on recovered samples. This partition coefficient increases to 13.3 when Fe^{2+} undergoes the transition in *fp*, enriching that phase in iron with respect to (and to the detriment of) Mg-*pv*. The numerical values are condensed in Table 1.

At 120 GPa (approximately 2600 km depth),

$$\left(\frac{V}{V_0}\right)_{fp} = 0.70$$

and

$$\left(\frac{V}{V_0}\right)_{pv} = 0.765$$

gives CFSP values $\Delta_{fp} = 2.43$ eV and $\Delta_{pv,A} = 2.07$ eV according to equation (2). It is at that pressure that Mg-*pv* becomes fully LS. This implies, using equations (1a) and (1b), that

$$U_{fp}^{HS} = 1.078 \text{ eV} \quad U_{pv,A}^{HS} = 1.222 \text{ eV}$$
$$U_{fp}^{LS} = 0.318 \text{ eV} \quad U_{pv,A}^{LS} = 1.182 \text{ eV} \quad (12)$$

The partition coefficients are also reported in Table 1. We see that the partition coefficient increases to 28 when both compounds are in the LS state and remains 15 times higher than in the top part (fully HS) of the lower mantle.

2. Noncooperative phase transition. Here we consider that the HS–LS phase transition is noncooperative in nature and that both phases coexist in a finite pressure range. Thermodynamic equilibrium is classically defined by minimization of the Gibbs free energy. In this case, the transition is defined by $G_{HS} = G_{LS}$, and there is no discontinuity of the Gibbs free energy at the transition, meaning that the parti-

Table 1. Calculated Values for the Thermodynamic Functions and Partition Coefficient of Iron Between *fp* and the A Site in Mg-*pv* at 70 GPa and 2500 K, and at 120 GPa and 3000 K. These are based on equation (11), combined with the numerical values given by in equations (10) and (12). At 1800 km depth, *K* increases by a factor of 8 at the onset of a spin transition, favouring the partitioning of iron in the LS phase (*fp*). At 2600 km depth, when both compounds are in the LS state, *K* is equal to 28, and remains 15 times higher than in the top part (fully HS) of the lower mantle.

P (GPa)	T (K)	Fe²⁺ in fp	Fe²⁺ in pv	ΔU (eV)	PΔV (eV)	TΔS (eV)	ΔG (eV)	K
70	2500	HS	HS	-0.108	0	0	-0.108	1.7
70	2500	LS	HS	-0.148	-1.00	-0.58	-0.558	13.3
120	3000	LS	HS	-0.824	-1.00	-0.70	-1.124	75
120	3000	LS	LS	-0.864	0	0	-0.864	28

tion coefficient is constant at the transition, and equal to 1.7 (Figure 10) as shown above. The effect on partitioning is only afterwards revealed. As pressure increases above the transition, we move into the field where iron occupies a lower volume in the LS state and partitions preferentially in the LS phase. This is driven once again by $P\Delta V$, except that P here is actually corrected for the transition pressure P_{tr} and the energetic term is $(P-P_{tr})\Delta V$.

This can be formally obtained from a Taylor development of the Gibbs free energy above the transition pressure. Given that $\Delta G = 0$ at the transition, this can be developed to the first order as follows:

$$\Delta G(P) = \left\{G_{LS}(P_{tr}) - G_{HS}(P_{tr})\right\} -$$

$$\left\{\left(\frac{\partial G_{LS}}{\partial P}\right) - \left(\frac{\partial G_{HS}}{\partial P}\right)\right\} \cdot \left(P - P_{tr}\right) + \vartheta^2 \qquad (13)$$

and since

$$\frac{\partial G}{\partial P} \equiv V,$$

this can be rewritten as $\Delta G(P) = \left\{G_{LS}(P_{tr}) - G_{HS}(P_{tr})\right\} - \left(V_{LS} - V_{HS}\right) \cdot \left(P - P_{tr}\right) + \vartheta^2$. The difference in the first bracket is identically

null by definition of thermodynamic equilibrium and therefore the relation becomes simply $\Delta G = \Delta V \cdot (P - P_{tr})$.

The partition coefficient can be written as:

$$\ln(K) = -\frac{\Delta G}{RT} = \ln(K_0) - \frac{\Delta V \cdot (P - P_{tr})}{RT}$$

or

$$K = K_0 \cdot \exp\left(-\frac{\Delta V \cdot (P - P_{tr})}{RT}\right) \qquad (14)$$

and is plotted in Figure 10. We see that the partition coefficient increases by 17-fold to reach 28.8 ($K_0 = 1.7$). After the second transition in perovskite, and within this thermodynamical framework, the partition coefficient should stabilize at that value.

3. Discussion. The partition coefficients calculated by both models are reported in Figure 10. Both models described above show that the spin transitions reported in both minerals should have the same effect on the partition coefficient of iron between these minerals, which is to enrich the oxide phase and deplete the silicate phase. They are in qualitative agreement in the intermediary regime and in perfect agreement at both transition pressures. In particular, they both

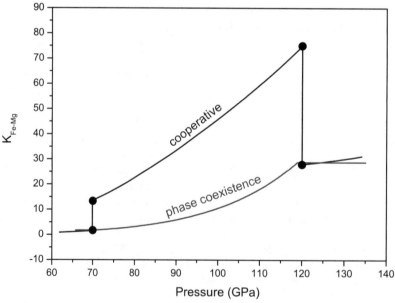

Figure 10. The predicted Fe–Mg partition coefficient between ferropericlase and magnesium silicate perovskite. Shown are the two transition models developed in the text. The lower curve is calculated considering a noncooperative phase transition (phase coexistence of HS and LS species). This is the standard thermodynamic theory. The four dots are calculated considering that the volume collapse in the LS state is not part of the equilibrium conditions (cooperative transition) and shows discontinuities in ΔG and therefore in K. The upper curve is also a guide to the eye. We see that both models predict an enrichment of iron in the oxide (ferropericlase) phase and depletion in the perovskite phase. Moreover, they both predict very close values at 120 GPa, with $K = 28$, that is, a 16-fold enrichment with respect to the standard lower mantle (below 70 GPa). In the intermediate region between 70 and 120 GPa, only a qualitative agreement is found. This effect and its magnitude still need to be confirmed experimentally or through more precise theoretical models.

show that $K_{Fe-Mg} \sim 30$ at 120 GPa, that is, at the onset of the D″ layer in the post-perovskite phase.

This is in very good agreement with recent experimental results [*Murakami et al.*, 2004b] obtained from an analytical transmission electron microscopy (ATEM) study of recovered samples, that shows a large increase of the partition coefficient between *fp* and the post-perovskite phase. These recent results, as well as our predictions, are at odds with the conclusions of previous work, whether based on ATEM studies [*Kesson et al.*, 2002] or on indirect assessment using equations of state [*Andrault*, 2001]. The latter was performed prior to the discovery of the spin transitions and assumes a constant compressibility and zero-pressure volume for the phases at high pressure. This assumption is not consistent with the observed spin transitions [*Badro et al.*, 2003, 2004] and more recently with the high-temperature equation of state measurement on *fp* [*Fei et al.*, 2004], which shows an important volume discontinuity at 60 GPa.

Obviously, more quantitative calculations based on quantum modelling of electronic interactions in both phases would be very useful for refining this qualitative prediction. Moreover, in a real mineralogical assemblage, the effect of partitioning will affect the spin-pairing transition pressures, and the whole interaction can be seen as a negative feedback loop: As *fp* is enriched in iron, its transition pressure increases and smooths and constrains the partition coefficient values between the phases. Again, state-of-the-art experimental investigation [*Murakami et al.*, 2004b] and modelling of these phenomena is required to obtain a self-consistent image of the processes occurring at these depths in the lower-mantle assemblage.

Nevertheless, we see qualitatively that such phenomena can have a major chemical impact. At depths around 1800 km, they indicate that lateral temperature variations can be responsible of tremendous large-scale chemical variations; indeed, because the spin-pairing transitions have different Clapeyron slopes in both compounds, it is possible for lateral (constant pressure) temperature heterogeneities on the order of a few hundred degrees to drive the *fp* phase in and out of the LS state. This in turn would induce the strong partitioning behaviour giving rise large-scale thermally induced lateral chemical heterogeneities, as suggested recently by tomographic modelling [*Trampert et al.*, 2004].

CONCLUSION

We have shown experimentally through high-pressure x-ray spectroscopy measurements on its major constituents (magnesium silicate perovskite and ferropericlase, accounting for 90–100% of its mass) that Earth's lower mantle is in a fundamentally different thermochemical

state at pressures exceeding 70 to 90 GPa [*Badro et al.*, 2003, 2004], corresponding to depths greater than 1700 to 2000 km. This is due to a change of the electronic properties of iron in these phases at those pressures. Indeed, iron in perovskite undergoes two successive spin transitions at about 70 and 120 GPa, to a partially and fully spin-paired (low-spin) state, respectively. On the other hand, iron in ferropericlase undergoes one transition between 60 and 70 GPa to a fully LS state. These changes can have major implications on the chemistry and dynamics of the lowermost mantle. They indicate that the lower mantle could be separated into three distinct regions in different thermochemical states:

- above 1700 km (below 70 GPa), the "normal" state, where iron is in the HS state in both lower-mantle compounds.
- between 1700 and 2600 km (between 70 and 120 GPa), the "transitional" state, where iron can be in the LS state in ferropericlase, and partially in the HS state in perovskite.
- below 2600 km and to the CMB (between 120 and 135 GPa), the "deep" state, where iron is in the LS state in both lower-mantle compounds.

Interestingly, these distinct regions have geophysical signatures that have been previously reported. The "transitional" region corresponds to depths where chemical heterogeneities have been observed by seismic tomography [*van der Hilst and Karason*, 1999]. Thermodynamic modelling of the partitioning of iron in this state indicates that the spin transitions at 70 GPa could promote large-scale chemical heterogeneities due to lateral temperature heterogeneities.

The depths and pressures of the "deep" region are in concordance with that of the D″ layer [*Lay*, 1983]. The transition pressure (120 GPa) is also in accord with a recently reported crystallographic transition in perovskite [*Murakami et al.*, 2004a], and so is our estimated Clapeyron slope of 8.5 MPa/ K [*Iitaka et al.*, 2004; *Oganov and Ono*, 2004; *Tsuchiya et al.*, 2004]. The latter value is also in agreement with the seismically derived value of the Clapeyron slope at the D″ layer of 6 MPa/K [*Sidorin et al.*, 1999].

All these concordances strengthen the suggestion that the spin-transition in perovskite at 120 GPa can very likely be linked to the pv-to-ppv transition [*Murakami et al.*, 2004a]. These studies do provide a mineral-physics basis for Earth's D″ layer. In the "deep" state, the lower-mantle mineralogical assemblage has an increased radiative conductivity, and iron in that assemblage has an increased affinity to the oxide phase (ferropericlase), as proposed here and shown very recently in a groundbreaking [*Murakami et al.*, 2004b] analysis of a KLB-1 composition synthetic peridotite recovered from these pressures and temperatures. That assemblage should be associated with a strong dynamical signature because the increase in thermal conductivity may hinder convection and favour layering in the lowermost mantle [*Anderson*, 2004; *Dubuffet and Yuen*, 2000], and the depletion of iron from the silicate phase should alter the electrical conductivity of the layer [*Dobson and Brodholt*, 2000; *Katsura et al.*, 1998; *Shankland et al.*, 1993]. These state-of-the-art measurements should stimulate discussions and further studies as to their geophysical consequences. Indeed, all of these recent observations pertaining to the properties of the minerals [*Badro, et al.*, 2003, 2004; *Dubrovinsky et al.*, 2000; *Iitaka et al.*, 2004; *Murakami et al.*, 2004a, 2004b; *Oganov and Ono*, 2004; *Tsuchiya et al.*, 2004] that constitute the D″ layer indicate that complex physical and chemical processes are undoubtedly taking place in this lowermost silicate envelope of the Earth.

Acknowledgments. We wish to thank Jean-Pascal Rueff, György Vankó, and Giulio Monaco, who participated in the experimental work. All experiments were carried out at beamline ID16 of the European Synchrotron Radiation Facility. Our thanks go to Tom Shankland, Ed Garnero, and Rob van der Hilst for stimulating discussions. We also thank Renata Wentzcovitch, Paul Asimow, Artem Oganov, and Philippe Sainctavit for fruitful discussions and comments concerning thermodynamic modelling and partitioning.

REFERENCES

Anderson, D. L. (2004), Simple scaling relations in geodynamics: the role of pressure in mantle convection and plume formation, *Chinese Science Bulletin, 49*, 2017-2021.

Andrault, D. (2001), Evaluation of (Mg,Fe) partitioning between silicate perovskite and magnesiowustite up to 120 GPa and 2300 K, *Journal of Geophysical Research-Solid Earth, 106*, 2079-2087.

Ashcroft, N. W., and N. D. Mermin (1976), *Solid State Physics*, Brooks Cole.

Badro, J., et al. (2003), Iron partitioning in Earth's mantle: Toward a deep lower mantle discontinuity, *Science, 300*, 789-791.

Badro, J., et al. (2004), Electronic transitions in perovskite: Possible nonconvecting layers in the lower mantle, *Science, 305*, 383-386.

Badro, J., et al. (1999), Magnetism in FeO at megabar pressures from X-ray emission spectroscopy, *Physical Review Letters, 83*, 4101-4104.

Birch, F. (1952), Elasticity and constitution of the Earth's interior, *J. Geophys. Res. (USA), 57*, 227-286.

Boehler, R. (2000), High-pressure experiments and the phase diagram of lower mantle and core materials, *Reviews of Geophysics, 38*, 221-245.

Brodholt, J. (2003), *Private Communication.*

Brown, J. M., and T. J. Shankland (1981), Thermodynamic parameters in the Earth as determined from seismic profiles, *Geophys. J. R. Astron. Soc. (UK), vol.66, no.3*, 579-596.

Burns, R. G. (1993), *Mineralogical Application of Crystal Field Theory*, Cambridge University Press.

Clark, S. P. (1957), *Trans. Am. Geophys. Union (USA), 38*, 931.

Dobson, D. P., and J. P. Brodholt (2000), The electrical conductivity of the lower mantle phase magnesiowustite at high temperatures and pressures, *Journal of Geophysical Research-Solid Earth, 105*, 531-538.

Dubrovinsky, L., et al. (2001), Stability of (Mg0.5Fe0.5)O and (Mg0.8Fe0.2)O magnesiowustites in the lower mantle, *European Journal of Mineralogy, 13*, 857-861.

Dubrovinsky, L. S., et al. (2000), Stability of ferropericlase in the lower mantle, *Science, 289*, 430-432.

Dubuffet, F., and D. A. Yuen (2000), A thick pipe-like heat-transfer mechanism in the mantle: nonlinear coupling between 3-D convection and variable thermal conductivity, *Geophysical Research Letters, 27*, 17-20.

Dubuffet, F., et al. (2000), Feedback effects of variable thermal conductivity on the cold downwellings in high Rayleigh number convection, *Geophysical Research Letters, 27*, 2981-2984.

Dziewonsky, A. M., and D. L. Anderson (1981), Preliminary reference Earth model, *Physics of the Earth and Planetary Interiors, 25*, 297-356.

Fei, Y. (1995), Thermal Expansion, in *Mineral Physics and Crystallography*, edited by T. J. Ahrens, pp. 29-44, American Geophysical Union.

Fei, Y., et al. (1994), Temperature-Dependent Electron Delocalization in (Mg,Fe)Sio3 Perovskite, *American Mineralogist, 79*, 826-837.

Fei, Y., et al. (2004), P-V-T equations of state of lower mantle minerals: Constraints on mantle composition models, *Eos Trans. AGU, 85*, Abstract MR14A-03.

Fiquet, G. (2001), Mineral phases of the Earth's mantle, *Zeitschrift Fur Kristallographie, 216*, 248-271.

Fiquet, G., et al. (2000), Thermoelastic properties and crystal structure of MgSiO3 perovskite at lower mantle pressure and temperature conditions, *Geophysical Research Letters, 27*, 21-24.

Fukao, Y., et al. (2001), Stagnant slabs in the upper and lower mantle transition region, *Reviews of Geophysics, 39*, 291-323.

Garnero, E. J., and T. Lay (2003), D '' shear velocity heterogeneity, anisotropy and discontinuity structure beneath the Caribbean and Central America, *Physics of the Earth and Planetary Interiors, 140*, 219-242.

Goto, T., et al. (1980), Absorption spectrum of shock-compressed Fe/sup 2+/-bearing MgO and the radiative conductivity of the lower mantle, *Physics of the Earth and Planetary Interiors, vol.22, no.3-4*, 277-288.

Grand, S. P. (2002), Mantle shear-wave tomography and the fate of subducted slabs, *Philosophical Transactions of the Royal Society of London Series a-Mathematical Physical and Engineering Sciences, 360*, 2475-2491.

Hermsmeier, B., et al. (1988), Direct evidence from gas-phase atomic spectra for an unscreened intra-atomic origin of outer-core multiplet splittings in solid manganese compounds, *Physical Review Letters*, vol.61, no.22, 2592-2595.

Iitaka, T., et al. (2004), The elasticity of the MgSiO3 post-perovskite phase in the Earth's lowermost mantle, *Nature, 430*, 442-445.

Jackson, I., and S. M. Rigden (1998), Composition and temperature of the Earth's mantle: Seismological models interpreted through experimental studies of Earth materials, in *The Earth's Mantle*, edited by I. Jackson, Cambridge University Press, Cambridge.

Jackson, J. M., et al. (2005), A synchrotron Mossbauer spectroscopy study of (Mg,Fe)SiO3 perovskite up to 120 GPa, *American Mineralogist, 90*, 199-205.

Jeffreys, H., and K. E. Bullen (1940), *Seismological Tables*, British Association for the Advancement of Science.

Jenkins, R. (1999), *X-Ray Fluorescence Spectrometry*, 2nd Edition ed., Wiley-Interscience.

Karato, S., and B. B. Karki (2001), Origin of lateral variation of seismic wave velocities and density in the deep mantle, *Journal of Geophysical Research-Solid Earth, 106*, 21771-21783.

Katsura, T., et al. (1998), Electrical conductivity of silicate perovskite at lower-mantle conditions, *Nature, 395*, 493-495.

Kellogg, L. H., et al. (1999), Compositional stratification in the deep mantle, *Science, 283*, 1881-1884.

Keppler, H., et al. (1994), Crystal-Field and Charge-Transfer Spectra of (Mg,Fe)Sio3 Perovskite, *American Mineralogist, 79*, 1215-1218.

Kesson, S. E., et al. (1998), Mineralogy and dynamics of a pyrolite lower mantle, *Nature, 393*, 252-255.

Kesson, S. E., et al. (2002), Partitioning of iron between magnesian silicate perovskite and magnesiowiistite at about 1 Mbar, *Physics of the Earth and Planetary Interiors, 131*, 295-310.

Knittle, E. (1995), Static Compression Measurements of Equations of State, in *Mineral Physics and Crystallography*, edited by T. J. Ahrens, pp. 98-142, American Geophysical Union.

Lay, T. (1983), Localized velocity anomalies in the lower mantle, *Geophys. J. R. Astron. Soc. (UK), vol.72, no.2*, 485-516.

Lennartsson, O. W., et al. (2004), Solar wind control of Earth's H+ and O+ outflow rates in the 15-eV to 33-keV energy range, *Journal of Geophysical Research-Space Physics, 109*.

Li, J., et al. (2004), Electronic spin state of iron in lower mantle perovskite, *Proceedings of the National Academy of Sciences of the United States of America, 101*, 14027-14030.

Lin, J., et al. (2005), Effects of the spin transition of iron in magnesiowüstite in earth's lower mantle, *Private Communication*.

Lin, J. F., et al. (2003), Stability of magnesiowustite in Earth's lower mantle, *Proceedings of the National Academy of Sciences of the United States of America, 100*, 4405-4408.

Liu, L. G. (1974), Silicate perovskite from phase transformations of pyrope garnet at high pressure and high temperature, *Geophysical Research Letters, 1*, 277.

Malavergne, V., et al. (1997), Partitioning of nickel, cobalt and manganese between silicate perovskite and periclase: A test of crystal field theory at high pressure, *Earth and Planetary Science Letters, 146*, 499-509.

Mao, H. K., et al. (1997), Multivariable dependence of Fe-Mg partitioning in the lower mantle, *Science, 278*, 2098-2100.

McCammon, C. (1997), Perovskite as a possible sink for ferric iron in the lower mantle, *Nature, 387*, 694-696.

McDonough, W. F., and S. S. Sun (1995), The Composition of the Earth, *Chemical Geology, 120*, 223-253.

Murakami, M., et al. (2004a), Post-perovskite phase transition in MgSiO3, *Science, 304*, 855-858.

Murakami, M., et al. (2004b), Mineralogy of the Earth's lowermost mantle, *Eos Trans. AGU, 85*, Abstract MR22A-01.

Oganov, A. R., and S. Ono (2004), Theoretical and experimental evidence for a post-perovskite phase of MgSiO3 in Earth's D'' layer, *Nature, 430*, 445-448.

Peng, G., et al. (1994), Spin selective X-ray absorption spectroscopy: demonstration using high resolution Fe K beta fluorescence, *Appl. Phys. Lett. (USA), vol.65, no.20*, 2527-2529.

Ringwood, A. E. (1975), *Composition and Petrology of the Earth's Mantle*, McGraw-Hill, New York.

Romanowicz, B. (2003), Global mantle tomography: Progress status in the past 10 years, *Annual Review of Earth and Planetary Sciences, 31*, 303-328.

Saxena, S. K., et al. (1996), Stability of perovskite (MgSiO3) in the Earth's mantle, *Science, 274*, 1357-1359.

Serghiou, G., et al. (1998), (Mg,Fe)SiO3-perovskite stability under lower mantle conditions, *Science, 280*, 2093-2095.

Shankland, T. J., et al. (1979), Optical absorption and radiative heat transport in olivine at high temperature, *J. Geophys. Res. (USA), vol.84, no.B4*, 1603-1610.

Shankland, T. J., et al. (1993), Electrical conductivity of the Earth's lower mantle, *Nature, vol.366, no.6454*, 453-455.

Sherman, D. M. (1991), The high-pressure electronic structure of magnesiowustite (Mg,Fe)O: applications to the physics and chemistry of the lower mantle, *J. Geophys. Res. (USA), vol.96, no.B9*, 14299-14312.

Shim, S. H., et al. (2001), Stability and structure of MgSiO3 perovskite to 2300-kilometer depth in Earth's mantle, *Science, 293*, 2437-2440.

Shim, S. H., and R. Jeanloz (2002), P-V-T equation of state of MgSiO3 perovskite and the chemical composition of the lower mantle, *Geochimica Et Cosmochimica Acta, 66*, A708-A708.

Sidorin, I., et al. (1999), Evidence for a ubiquitous seismic discontinuity at the base of the mantle, *Science, 286*, 1326-1331.

Tackley, P. J. (2002), Strong heterogeneity caused by deep mantle layering, *Geochemistry Geophysics Geosystems, 3*.

Trampert, J., et al. (2004), Probabilistic tomography maps chemical heterogeneities throughout the lower mantle, *Science, 306*, 853-856.

Tsuchiya, T., et al. (2004), Phase transition in MgSiO3 perovskite in the earth's lower mantle, *Earth and Planetary Science Letters, 224*, 241-248.

Tsutsumi, K., et al. (1976), X-ray Mn K beta emission spectra of manganese oxides and manganates, *Phys. Rev. B, Solid State (USA), vol.13, no.2*, 929-933.

van den Berg, A. P., et al. (2001), The effects of variable thermal conductivity on mantle heat-transfer, *Geophysical Research Letters*, *28*, 875-878.

van der Hilst, R. D., and H. Karason (1999), Compositional heterogeneity in the bottom 1000 kilometers of Earth's mantle: Toward a hybrid convection model, *Science, 283*, 1885-1888.

Vankó, G., et al. (2002), Molecular Spin Transitions Studied with X-ray Emission Spectroscopy, *ESRF Highlights 2002, 59.*

Wentzcovitch, R. M., et al. (1995), Ab-Initio Study of Mgsio3 and Casio3 Perovskites at Lower-Mantle Pressures, *Physics of the Earth and Planetary Interiors, 90*, 101-112.

Williams, Q., et al. (1990), Structural and electronic properties of Fe/sub 2/SiO/sub 4/-fayalite at ultrahigh pressures: amorphization and gap closure, *J. Geophys. Res. (USA), vol.95, no.B13,* 21549-21563.

James Badro, Guillaume Fiquet, and François Guyot, Laboratoire de Minéralogie Cristallographie de Paris (UMR CNRS 7590), Institut de Physique du Globe de Paris, Université Pierre et Marie Curie, 4 place Jussieu, 75252 Paris, France. (james.badro@lmcp.jussieu.fr)

Stability of MgSiO$_3$ Perovskite in the Lower Mantle

Sang-Heon Shim

Department of Earth, Atmospheric, and Planetary Sciences, Massachusetts Institute of Technology, Cambridge, Massachusetts

As it is expected to be the dominant phase in the lower mantle, any pressure-induced phase changes in MgSiO$_3$ could require significant modifications in current models of the dynamics and structures of Earth's deep mantle. Studies to date have yielded discrepant results regarding the high-pressure stability of MgSiO$_3$ perovskite. Understanding the source of discrepancy is essential, both to resolving the stability of perovskite and to developing more reliable techniques for understanding the Earth's deep interior. In this report, we give an overview of previous studies on the stability of MgSiO$_3$ perovskite and recent observations on the post-perovskite transition. We also summarize our measurements on MgSiO$_3$ perovskite to core–mantle boundary pressure–temperature (P–T) conditions using in-situ X-ray diffraction. Major peaks in our diffraction patterns are best explained by those of MgSiO$_3$ perovskite at 1200–2500-km depth conditions. No evidence of dissociation to MgO + SiO$_2$ and a transition to a cubic perovskite structure has been found to core–mantle boundary P–T conditions. We have also observed a new peak at 2.62–2.57 Å at 88–145 GPa, the existence of which may be relevant to a modification in perovskite crystal structure. However, the possibility that this peak may be from a chemical reaction among gasket, anvil materials, and sample cannot be ruled out. More significant changes are observed during heating above 2500 K at 135–145 GPa: appearance of new peaks, splitting of a peak, and intensity changes of some diffraction peaks. The recently proposed post-perovskite phase explains dominant new diffraction features. Based on the results available as of May 2005, the post-perovskite transition appears to be relevant to the D$''$ seismic discontinuity. Furthermore, depth of the post-perovskite transition may be very sensitive to variations in chemical composition as well as temperature.

1. INTRODUCTION

The paucity of the samples directly originating from the lower mantle and instability of most lower-mantle phases at ambient conditions make laboratory measurements at high pressures (P) and high temperatures (T), along with first-principles calculations, important means to understanding the

Earth's Deep Mantle: Structure, Composition, and Evolution
Geophysical Monograph Series 160

properties of the Earth's deep mantle. The pioneering work of *Liu* [1975] and *Liu and Ringwood* [1975] revealed that the major elements of the mantle, i.e., O, Si, Mg, Fe, and Ca, are hosted in perovskite-type structures at lower-mantle P–T conditions. In recent years, lines of evidence have been reported for their possible natural occurrences in meteorites [*Tomioka and Fujino*, 1997; *Langenhorst and Poirier*, 2000; *Tomioka and Kimura*, 2003] and xenoliths [*Collerson et al.*, 2000].

As a dominant mineral in the lower mantle, the phase relations and physical properties of magnesium silicate (MgSiO$_3$) perovskite have been of prime interest for understanding seismically observed structures and computationally predicted dynamics of the Earth's interior: Possible crystal structure changes in mantle minerals could play roles as seismic interfaces or flow regulators in the mantle, as is the case for the post-spinel boundary between the upper and lower mantle [*Ito and Takahashi*, 1989; *Ito et al.*, 1990; *Tackley et al.*, 1993; *van der Hilst et al.*, 1997; *Chundinovskikh and Boehler*, 2001; *Fukao et al.*, 2001; *Shim et al.*, 2001a; *Lebedev et al.*, 2002; *Fei et al.*, 2004b]. However, experimental and theoretical results of the past decade have yield discrepant findings regarding the stability of MgSiO₃ perovskite [*Knittle and Jeanloz*, 1987; *Meade et al.*, 1995; *Saxena et al.*, 1996, 1998; *Mao et al.*, 1997; *Serghiou et al.*, 1998; *Fiquet et al.*, 2000; *Andrault*, 2001; *Shim et al.*, 2001b, 2004]. Furthermore, a significant change in the crystal structure of (Mg,Fe)SiO₃ has been recently reported by several groups [*Murakami et al.*, 2004, 2005; *Shim et al.*, 2004; *Oganov and Ono*, 2004; *Iitaka et al.*, 2004; *Mao et al.*, 2004; *Tsuchiya et al.*, 2004a, b].

The phase relations for silicate perovskites are important issues to be resolved, owing to recent inferences about deep-mantle structures [e.g., *Lay et al.*, 1998; *Gurnis et al.*, 1998; *Sidorin et al.*, 1999; *Kellogg et al.*, 1999; *van der Hilst and Kárason*, 1999; *Albarède and van der Hilst*, 1999; *Trampert et al.*, 2004], and to the expected enhancement in resolution of seismic tomography for the lower mantle by the deployment of dense seismic arrays, e.g., USArray, J-array, and GEOSCOPE [*Levander et al.*, 1999; *Fischer and van der Hilst*, 1999]. In fact, developments in experimental techniques in the last decade [*Shen et al.*, 2001; *Andrault and Fiquet*, 2001; *Yagi et al.*, 2001; *Watanuki et al.*, 2001] enable experimentalists to study directly the phase relations at in-situ deep-mantle *P–T* conditions [e.g., *Fiquet et al.*, 1998, 2000; *Dewaele et al.*, 2000; *Shim et al.*, 2000, 2001b, 2004; *Murakami et al.*, 2003, 2004; *Oganov and Ono*, 2004; *Mao et al.*, 2004; *Shieh et al.*, 2005].

In this paper, we review recent experimental results for the stability and crystal structure of MgSiO₃ at mid- to lowermost mantle *P–T* conditions. Important issues in experimental technique at this extreme *P–T* condition are briefly introduced in order to discuss possible sources of discrepancy in the stability of MgSiO₃. A summary of our results [refer to *Shim et al.*, 2001b, 2004, for details] is presented in order to demonstrate experimental data used to examine the stability and crystal structure of MgSiO₃. We also consider potential implications of the observed changes in MgSiO₃ for the structure and dynamics of the lowermost mantle.

2. A BRIEF REVIEW

This section is not intended to provide a detailed technical review of all data bearing on the controversial issue of the stability of MgSiO₃ perovskite. Instead, we limit ourselves to a few important experimental issues for a general audience. Detailed technical reviews are available in the literature [e.g., *Dubrovinsky et al.*, 1999; *Serghiou et al.*, 1999; *Boehler*, 2000]. Knowledge about the post-perovskite transition is quickly growing at this moment. This summary for the post-perovskite transition is based on the results published in the literature as of May 2005.

2.1. Experimental Results

The data reported for pure MgSiO₃ so far in the literature are summarized in Figure 1. To facilitate the discussion, synthetic X-ray diffraction patterns of MgSiO₃ with different perovskite crystal structures are shown in Figure 2 (the method used to produce the synthetic diffraction patterns is presented in Appendix 1) .

The first data relevant to the stability of (Mg,Fe)SiO₃ perovskite at deep-mantle pressures were reported by *Knittle and Jeanloz* [1987]. They observed the synthesis of (Mg,Fe)SiO₃ perovskite from enstatite starting materials to 127 GPa by performing X-ray diffraction measurements for temperature-quenched samples.

The first in-situ study for (Mg,Fe)SiO₃ perovskite was carried out by *Meade et al.* [1995] (Figure 1a). They interpreted a decrease in peak width of the triplet (compare Pv3 in Figures 2a and b) during heating at 65 GPa as evidence of the phase transition from an orthorhombic (*Pbnm* space group) to a cubic structure in (Mg,Fe)SiO₃ perovskite. However, the method they used, energy-dispersive diffraction, does not provide sufficient resolution to resolve splitting of the triplet: Its disappearance could result from preferred orientation, in which diffraction peaks are observable only in certain directions for one-dimensional detectors, e.g., the solid-state detector. Later studies using two-dimensional detectors, which enable the recording of full diffraction rings, showed that the triplet splitting is still resolvable in this pressure range [e.g., *Fiquet et al.*, 2000]. *Meade et al.* [1995] also reported that (Mg,Fe)SiO₃ perovskite dissociates to (Mg,Fe)O + SiO₂ after heating for 1 h at 70 GPa.

On the basis of X-ray diffraction of heated products of two different crystalline starting materials (i.e., forsterite and enstatite) at high pressure, *Saxena et al.* [1996] reported that Mg end-member silicate perovskite (MgSiO₃) dissociates to MgO + SiO₂ at the similar conditions (Figure 1a). They claimed that longer heating (72 h) was required to observe the dissociation at 80 GPa and 1250 K, which is lower than the temperature normally reached in laser heating (>1500 K).

In contrast, *Mao et al.* [1997] confirmed the stability of MgSiO₃ perovskite to 85 GPa and 2000 K, using the energy-dispersive diffraction technique. They proposed that the

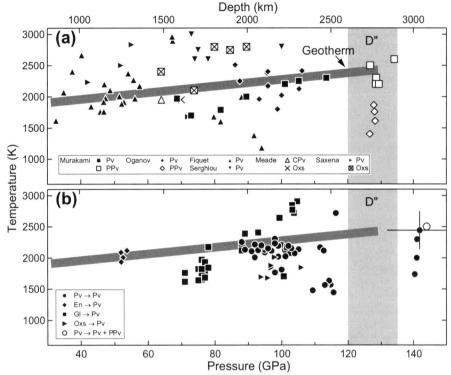

Figure 1. Pressure and temperature conditions for the data points reported for pure MgSiO$_3$: (a) previous measurements [*Meade et al.*, 1995; *Saxena et al.*, 1996, 1998; *Fiquet et al.*, 1998, 2000; *Serghiou et al.*, 1998; *Murakami et al.*, 2004] and (b) our measurements [*Shim et al.*, 2001b, 2004]. A representative error bar of our data is shown for pressure and temperature. A geotherm is obtained from *Brown and Shankland* [1981]. Different symbols in (a) represent data from different studies and different phases observed during or after heating. Different symbols in (b) represent different starting materials and phases observed during and after heating [Pv: perovskite, CPv: cubic perovskite, Oxs: MgO + SiO$_2$, PPv: post-perovskite, En: (Mg$_{0.9}$Fe$_{0.1}$)SiO$_3$ enstatite, Gl: MgSiO glass].

earlier observation of dissociation by *Meade et al.* [1995] may be due to incongruent melting or diffusion of MgSiO$_3$ perovskite in large thermal gradients.

Serghiou et al. [1998] published Raman spectra measured at 300 K and high pressure for heated products of MgSiO$_3$ glass, periclase (MgO) + quartz (SiO$_2$) mixtures, and MgSiO$_3$ enstatite. They confirmed the stability of MgSiO$_3$ perovskite to 100 GPa and 3000 K. They proposed that the possible existence of steep thermal and pressure gradients in the earlier experiments by *Meade et al.* [1995] and *Saxena et al.* [1996] might result in the dissociation of MgSiO$_3$ perovskite. In contrast, *Dubrovinsky et al.* [1999] argued that the glass and periclase + quartz mixture starting materials could lead to the synthesis of a metastable phase. They also claimed that pressure during heating could be significantly lower than the pressure measured after heating. Later, in-situ angle-dispersive X-ray diffraction patterns of MgSiO$_3$ perovskite were reported by *Fiquet et al.* [2000] to 90 GPa and 2000 K, and post-heating diffraction patterns of (Mg,Fe)SiO$_3$ (Fe ≤ 18 mol%) to 120 GPa and 2300 K were reported by *Andrault*

[2001]. Both studies confirmed the stability of (Mg,Fe)SiO$_3$ perovskite within their *P–T* range.

In the recent in-situ X-ray measurements by *Murakami et al.* [2004], the pressure and temperature range has been extended to 134 GPa and 2600 K for pure MgSiO$_3$. The stability of MgSiO3 perovskite was confirmed to 113 GPa and 2300 K. Above 126 GPa, they observed significant changes in diffraction patterns. To explain the observed changes, they proposed a phase transition in MgSiO$_3$ to a new phase with the CaIrO$_3$-type (*Cmcm*) structure at the base of the lower mantle. This result is confirmed in measurements carried out by two different groups [*Shim et al.*, 2004; *Oganov and Ono*, 2004].

2.2. Important Technical Issues

The key technical issues in this controversy include thermal gradients in the laser-heated diamond-cell samples, stress conditions of the samples, pressure determinations, transformation kinetics, and the possibility of incongruent

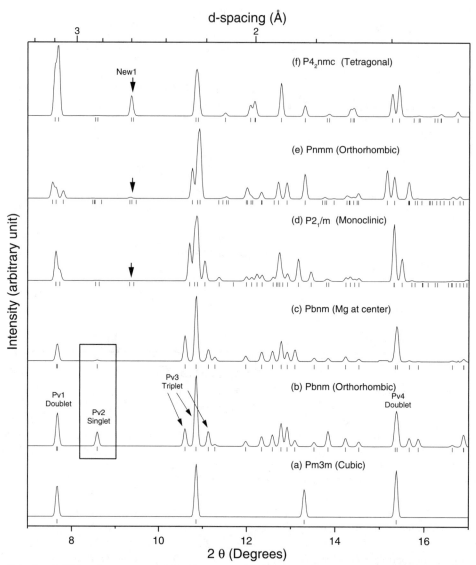

Figure 2. Calculated diffraction patterns of MgSiO₃ perovskite at 100 GPa and 300 K. (a) Cubic ($Pm\bar{3}m$) structure. (b) Orthorhombic (*Pbnm*) structure, which is a stable structure to at least 1800-km depth conditions, and (c) orthorhombic structure (*Pbnm*) without shift of Mg atom positions at the dodecahedral site centers. (d) Monoclinic ($P2_1/m$), (e) orthorhombic (*Pmmn*), and (f) tetragonal ($P4_2nmc$) structures, which are proposed to explain the new peak (New1) observed above 88 GPa (indicated by heavy arrows). The box in (b) and (c) highlights the intensity changes induced by the position shift of Mg atoms. Expected peak positions are shown as bars below the diffraction patterns for individual structures. "Pv1", "Pv2", "Pv3", and "Pv4" indicate a doublet (002+110), a singlet (111), a triplet (020+112+200), and a doublet (103+211), respectively—diagnostic diffraction features of perovskite. These notations are used throughout this paper. The calculation method is described in Appendix 1.

melting. Pressure variations in the diamond-anvil cell during laser heating is affected by several factors [*Andrault et al.*, 1998], including thermal pressure [*Heinz*, 1990] and thermal relaxation of the cell [*Kavner and Duffy*, 2001]. Whereas the thermal pressure increases total pressure, the thermal relaxation of the cell results in a pressure decrease.

Thus, the pressure change in the diamond cell is difficult to predict and this makes in-situ pressure measurement essential at high *P–T* . In fact, *Dubrovinsky et al.* [1999] argued that pressure during heating can be lower than what is measured after heating in *Serghiou et al.* [1998]'s experiments and thus their measurement conditions are not high enough to observe the dissociation. While in the earlier experiments pressure was determined after heating [*Saxena et al.*, 1996; *Serghiou et al.*, 1998], more recent in-situ X-ray diffraction measurements [*Fiquet et al.*, 2000; *Shim et al.*, 2001b, 2004; *Murakami et al.*, 2004] determine pressure during heating by using the equation of state for platinum [*Holmes et al.*, 1989], which was heated and compressed together with samples, at in-situ conditions.

One of the critical issues has been the stability of heating. Two different techniques have been used in previous studies: laser and resistance heating. Laser heating can generate temperatures directly relevant to the lower mantle (2000–4000 K). However, due to the limited size of laser beams, only a portion of the sample can be heated (typically 20–40 μm in diameter). This can introduce very steep thermal gradients radially in the sample. The other problem is the thermal gradients along the loading axes of the diamond-anvil cells due to the high thermal conductivity of diamond anvils. Several techniques have been developed to decrease the thermal gradients: Insulation materials, e.g., inert noble gases, can be loaded to separate the samples from diamond anvils (Figure 3). By directing laser beams on both sides of the sample, temperature gradients along the loading axis can be reduced [*Shen et al.*, 2001] (Figure 3). Temporal fluctuation of temperature in the samples can also be a source of the instability (Figure 4c).

Better temperature homogeneity is generally assumed for the resistance heating. In this technique, micro-heaters are placed near the sample chamber of the diamond-anvil cell. More precise temperature measurements are expected since thermocouples can be attached near the sample chamber. A major limitation of this technique is that temperatures are limited to 1200 K, which is much lower than those in the mantle. Also at this lower temperature, kinetics becomes important during phase transitions. It has been also suggested that high heat conduction through the diamond anvils can induce an extreme thermal gradient along the loading axis of the diamond-anvil cell in some type of resistance heating [see *Serghiou et al.*, 1999; *Dubrovinskaia and Dubrovinsky*, 2003, for a detailed discussion].

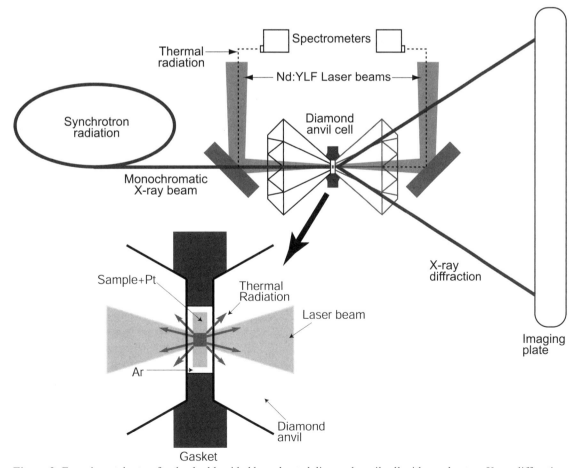

Figure 3. Experimental setup for the double-sided laser-heated diamond-anvil cell with synchrotron X-ray diffraction.

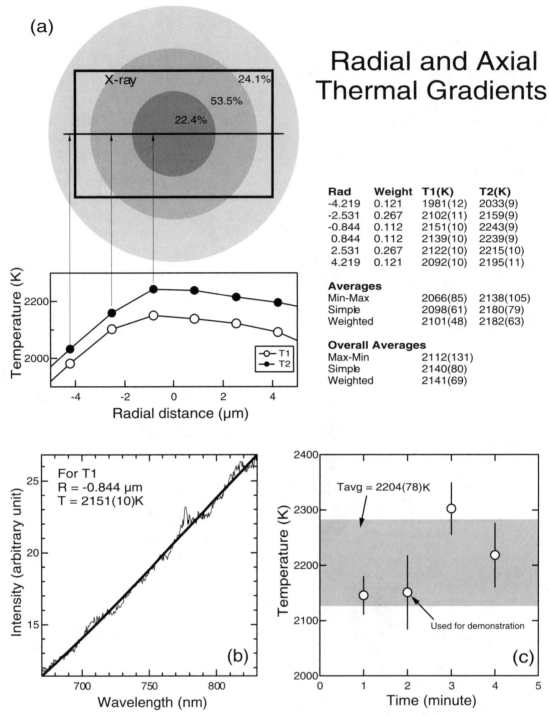

Rad	Weight	T1(K)	T2(K)
-4.219	0.121	1981(12)	2033(9)
-2.531	0.267	2102(11)	2159(9)
-0.844	0.112	2151(10)	2243(9)
0.844	0.112	2139(10)	2239(9)
2.531	0.267	2122(10)	2215(10)
4.219	0.121	2092(10)	2195(11)

Averages

Min-Max		2066(85)	2138(105)
Simple		2098(61)	2180(79)
Weighted		2101(48)	2182(63)

Overall Averages

Max-Min		2112(131)
Simple		2140(80)
Weighted		2141(69)

Figure 4. Temperature measurements at 100 ± 8 GPa and 2204 ± 78 K. (a) Schematic view of the heated spot (cylindrical symmetry is assumed), and radial thermal gradients for both sides of the sample (T1 and T2). A table is also shown, presenting temperatures measured along the solid line. The numbers in parentheses are 1σ uncertainties. The X-ray–probed area is indicated with a rectangle. Different areas represented by different temperature measurements are shown in different shades of gray. The numbers on those areas indicate the area fractions in the X-ray–probed region. (b) Fitting of a measured thermal radiation spectrum (thin line) to gray-body radiation (thick curve). (c) Temporal fluctuation of temperature during a diffraction pattern measurement. A temporal average and its uncertainty are shown by a gray area (refer to Appendix 3 for details).

There have been some reports that the thermal gradients can in fact result in element partitioning between the hot and cold spots of the samples [e.g., *Fei et al.*, 1996; *Andrault and Fiquet*, 2001]. Furthermore, *Kesson et al.* [1995] reported that $(Mg,Fe,Al)(Si,Al)O_3$ perovskite melts incongruently to SiO_2 and oxides in unevenly hot regions during laser heating. More technical development is warranted to ensure the stability and homogeneity of heating at lower-mantle *P–T* conditions. Use of the insulation media and larger homogeneous laser modes are among the methods that have been implemented.

Formation of a metastable phase has been observed in some high *P–T* measurements. For example, metastable $(Ca,Mg)SiO_3$ perovskite has been synthesized by heating glass starting material below 1300 K [*Kim et al.*, 1994]. Later, *Irifune et al.* [2000] found that $(Ca,Mg)SiO_3$ perovskite transforms to a stable phase assemblage ($CaSiO_3$ and $MgSiO_3$ perovskites) at sufficiently high temperature (>1500 K). A similar observation has also been made in $(Ca,Mg,Fe,Al)SiO_3$ [*Asahara et al.*, 2005]. Thus, it is very important to carry out measurements for different starting materials, such as glass, crystalline, and oxide component mixtures, and to heat samples to sufficiently high temperatures. *Serghiou et al.* [1998] confirmed the stability of $MgSiO_3$ perovskite by conducting experiments of different types of starting materials. A remaining issue is whether using quartz for the MgO + SiO2 mixture starting materials is appropriate [*Serghiou et al.*, 1998]. *Dubrovinsky et al.* [1999] argued that the use of a low-pressure polymorph of SiO_2, i.e., quartz, as a starting material could result in a synthesis of a metastable phase at high *P–T* where quartz is unstable.

Saxena et al. [1998] reported that longer heating is required to observe the dissociation. However, this was at lower temperature (1200 K), where kinetics may also play an important role and insufficient heating can cause synthesis of a metastable phase [*Irifune et al.*, 2000]. Furthermore, in their earlier experiment, *Saxena et al.* [1996] reported that they were able to observe the dissociation below 1500 K even with 10–15 min of laser heating above 83 GPa. This heating duration is similar or even less than those in other laser-heating studies [*Serghiou et al.*, 1998; *Fiquet et al.*, 2000; *Shim et al.*, 2001b, 2004; *Murakami et al.*, 2004]. Thus, it is unlikely that heating duration is the major factor to cause the discrepancy.

3. EXPERIMENTAL OBSERVATIONS

In this section, we summarize our observations (some of these results are published in *Shim et al.* [2001b, 2004]). We used three different starting materials: $(Mg_{0.91}Fe_{0.09})SiO_3$ enstatite, $MgSiO_3$ glass, and an equimolar MgO + SiO_2 mixture. The composition of the glass is very close to $MgSiO_3$ with a slight enrichment of Si (Table 1).

Table 1. Chemical Compositions of the Starting Materials, Expressed as the Molar Cation Fractions for a Formula Based on 3 Oxygens.

	Enstatite[a]	Glass[b]
Mg	0.91	0.98
Fe	0.09	< 0.01
Si	1.00	1.01
Na		< 0.01
Total	2.00	1.99

[a]Chemical analysis of enstatite obtained by energy-dispersive spectroscopy at Princeton Material Institute.
[b]Electron microprobe analyses of a glass starting material performed by J. Akins and T. Ahrens at Caltech.

In-situ X-ray diffraction was conducted at the GeoSoilEnviroCARS sector of the Advanced Photon Source with a double-sided laser-heating system (Figure 3 and *Shen et al.* [2001]). To minimize the temperature inhomogeneity in the X-ray–probed area, the X-ray beam was focused on less than $10 \times 10 \ \mu \ m^2$ on the sample and co-linearity was ensured among the incident lasers, X-ray beam, and diamond-anvil cell rotation axis. Nd:YLF laser beams with a TEM_{01} mode were focused on both sides of the sample in the diamond-anvil cells while diffraction patterns were recorded (Figure 4). Thermal radiation spectra, which are corrected for the system response, were fitted to Planck's equation in order to measure temperature [*Jephcoat and Besedin*, 1996; *Boehler*, 2000]. Including radial and axial thermal gradients, temporal fluctuations during the X-ray exposure, and fitting residual (e.g.), the temperature uncertainty is 100–300 K in these measurements (Figure 4, refer to Appendix 3 for details). The samples were heated for 30–45 min during a heating cycle.

Pressure is determined using the *P–V–T* relations in platinum [*Holmes et al.*, 1989] and argon [*Finger et al.*, 1981; *Ross et al.*, 1986], which are loaded together with the starting materials. The estimated uncertainty is ± 5 GPa at 300 K and ± 10 GPa at high temperature. The most significant error source is the overlap of diffraction peaks for platinum and argon above 70 GPa. Shear stresses and temperature uncertainty also contribute to this uncertainty. Detailed descriptions about the experimental technique can be found in *Shim et al.* [2001b, 2004].

3.1. Stability of MgSiO₃ Perovskite

From three different starting materials, we observe major diffraction peaks of perovskite during and after laser heating, including a triplet (Pv3) and two doublets (Pv1 and Pv4) at

50–120 GPa (Figure 5). Although the most intense diffraction peak of MgSiO₃ perovskite, Pv3 (Figure 2), is severely overlapped by a diffraction peak from the rhenium gasket, two weak shoulders of Pv3 are observed aligned off of the rhenium diffraction peak. A second doublet at a higher angle, Pv4, and a singlet, Pv2, are more readily observed throughout our experiments. No major diffraction peak from the gasket material, pressure standard, pressure medium, any proposed high-pressure silica phase [*Teter et al.*, 1998], or MgO overlaps these peaks (Figure 5). Furthermore, as shown in Figure 2a, Pv2 should have zero intensity for cubic perovskite and thus its presence can be diagnostic of lower-symmetry perovskite structures.

The stability of MgSiO₃ perovskite is thus confirmed at 50–118 GPa and 1500–2900 K (corresponding to 1200–2500-km depth) from the fact that the positions and intensities of major diffraction peaks remain consistent with those expected for MgSiO₃ perovskite without major changes (Figure 1b). The *P–T* conditions of our observations are close to the expected geotherm [e.g., *Brown and Shankland*, 1981] for the lower mantle and cover the entire *P–T* conditions expected for this region (Figure 1b).

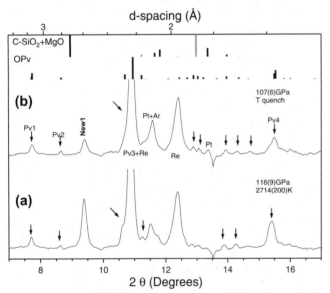

Figure 5. X-ray diffraction patterns measured (a) during and (b) after heating (the numbers in parentheses are 1σ uncertainties). The backgrounds are subtracted from the patterns. MgSiO₃ perovskite and the new peak (New1) are observed during and after heating. Calculated patterns for orthorhombic MgSiO₃ perovskite, *Pbnm* (black bars), periclase (gray bars), and the CaCl₂-type SiO₂ (black bars) are shown at the top. The major diffraction peaks of MgSiO₃ perovskite are identified with arrows [Pv: MgSiO₃ perovskite, OPv: orthorhombic (*Pbnm*) perovskite, C-SiO₂: CaCl₂-type SiO₂]. Other notations as in Figure 2. These patterns have also been reported in *Shim et al.* [2004].

A phase transition has been proposed for MgSiO₃ perovskite from orthorhombic to cubic structures at 65 GPa [*Meade et al.*, 1995]. However, we were able to resolve two or three of the distinct peaks of the perovskite triplet (Pv3) below 78 GPa (Figure 5a). Furthermore, Pv2, which is diagnostic of noncubic perovskite, is continuously observed. These suggest that MgSiO₃ perovskite has an orthorhombic unit cell and that no phase transition to a higher symmetry phase (tetragonal or cubic) exists to 78 GPa and 2100 K. Above this to core–mantle boundary *P–T* conditions, our angle-dispersive patterns contain Pv2 peaks during heating and the shoulder peaks of Pv3, which are diagnostic of noncubic perovskite structure (Figure 5). Thus, in our study there is no evidence for a phase transition to a cubic perovskite to the core–mantle boundary.

3.2. Possible Modifications in MgSiO₃ Perovskite Structure at 88 GPa

We observe a new peak at 2.62–2.57 Å (New1 in Figure 5) from 88 to 144 GPa in the patterns that also contain diffraction peaks of MgSiO₃ perovskite both during and after heating. This new peak appears in the runs with different starting materials (MgSiO₃ glass and MgO + SiO₂ oxide mixture) and in subsequent heating of materials already transformed to perovskite. This peak cannot be identified as arising from MgO, the CaCl₂-type SiO₂, or any other theoretically proposed post-stishovite phases, but this new peak is found to be indexed as a perovskite diffraction line. The pressure-induced shift of the New1 peak agrees well with the expected shift for peaks of MgSiO₃ perovskite. This observation associates New1 peak with MgSiO₃ perovskite or another material with very similar compressibility (refer to *Shim et al.* [2004] for details). From systematic absence of this peak in *Pbnm* (orthorhombic) perovskite and forward modeling, we have proposed that appearance of this peak may indicate a modification in crystal structure of MgSiO₃ perovskite to *P2₁/m* (monoclinic), *Pmmn* (orthorhombic), or *P4₂/nmc* (tetragonal) above 88 GPa (Figure 6; refer to Appendix 2 for details).

We calculated diffraction patterns of these three possible perovskite structures and compared them with our observed diffraction patterns (Figure 2). We found that the intensity ratio between New1 and Pv4 is 1.3 ± 0.3 and its sensitivity to pressure and temperature is minor at 88–144 GPa and 1500–2800 K from our angle-dispersive measurements. In Figure 2, except for the case of *P4₂/nmc*, the new peak intensity predicted by the space groups considered here is significantly lower than the observed peak intensity (Figure 2d,e,f).

However, direct comparison between observed and synthetic intensities is difficult in this case. Knowledge of atomic

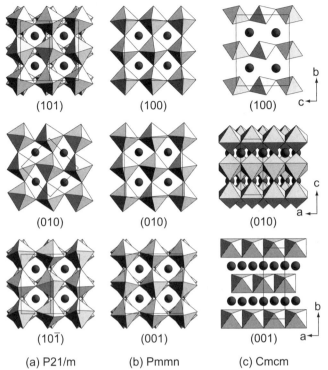

Figure 6. Crystal structure of (a) $P2_1/m$ and (b) $Pmmn$ perovskites, proposed for the observations above 88 GPa, and (c) the post-perovskite phase ($Cmcm$) of MgSiO$_3$ above 120 GPa, proposed by *Murakami et al.* [2004]. For perovskite structures, projections along three pseudo-cubic axes are shown for comparison. The crystal structures of $P2_1/m$ and $Pmmn$ have same tilting senses as $Pbnm$ and $P4_2/nmc$, respectively. The octahedra and the balls represent the SiO$_6$ structural units and Mg atoms, respectively. The unit cell is shown by solid lines. (a) and (b) have also been presented in *Shim et al.* [2001b]

positions is needed in order to calculate X-ray peak intensities. However, our forward modeling does not constrain the position shift of the cations from the dodecahedral site centers (e.g., Mg in MgSiO$_3$ perovskite) and the distortions of the octahedra (e.g., SiO$_6$ in MgSiO$_3$ perovskite) and these were not included in the calculation of Figure 2. To estimate the uncertainty in the calculated diffraction intensities in Figure 2, we calculate X-ray diffraction patterns of $Pbnm$ perovskite with and without the cation shifts from the center of dodecahedral sites. For some peaks, e.g., Pv2, the intensity is very sensitive to the cation shifts, whereas other peaks do not show significant changes in intensities. This indicates that the ignored factors could have a significant effect on the intensity of the new peak. Therefore, at this moment, the measured intensity does not provide sufficient information to distinguish the space group of the possible high-pressure phase of MgSiO$_3$ perovskite.

3.3. Post-Perovskite Transition in MgSiO$_3$

When the sample was heated at 135–145 GPa, we held the temperature below 2500 K for the first 20 min of heating. No significant changes were observed other than the diffraction peaks of MgSiO$_3$ perovskite and New1 at this condition. However, when the sample was heated above 2500 K, we observed several changes in the diffraction patterns: the appearance of other new features at 2.42 Å (New2), 2.93 Å (New3), and 2.75 Å (near Pv2) and intensity changes in the peaks near 13.7° and 14.1°, and more diffraction peaks in Pv4 (Figure 7). It is notable that major diffraction peaks of perovskite and the new peak observed from 88 GPa (New1) are retained during this change. These features remain after temperature quench.

We examined the possibility of diffraction from materials other than the sample, such as platinum (pressure standard), argon (pressure medium), and rhenium (gasket). None

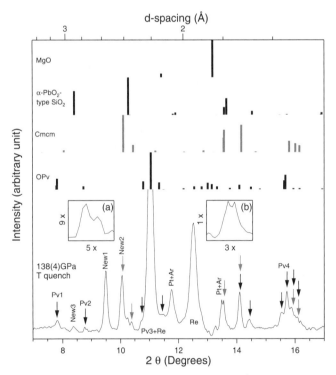

Figure 7. A diffraction pattern measured at 138 ± 4 GPa and 300 K after laser heating (bottom) together with calculated diffraction patterns of orthorhombic MgSiO$_3$ perovskite (OPv), the post-perovskite phase (CaIrO$_3$ type with a $Cmcm$ space group) proposed by *Murakami et al.* [2004], the α-PbO$_2$-type silica, and MgO. Inset (a) shows the observed peak-splitting of Pv2. Inset (b) shows the observed peak-splitting of a peak at 13.5°. The major diffraction peaks of MgSiO$_3$ perovskite and the post-perovskite phase are identified with black arrows and grey arrows, respectively. Notations are the same as in Figures 2 and 5.

of these materials can explain the newly observed peaks. Furthermore, none of the three possible perovskite structures can explain the new peak observed above 88 GPa predict peaks near New2 (Figure 2). Interestingly, the splitting of singlet Pv2 in *Pbnm* perovskite is predicted by all the candidate space groups we considered for New1 (Figure 2). However, the splitting has not been observed below 130 GPa. A possible explanation is that the splitting is induced by the transition at 88 GPa, and its magnitude becomes sufficiently large with compression that it can be observed at this higher pressure. We have also applied to New2 the geometrical modeling that we implemented for analysis of New1. However, no perovskite structure can explain the appearance of New2 merely with octahedral tilting.

A phase transition from perovskite to the CaIrO₃-type structure (*Cmcm*) at 120 GPa has been recently proposed by *Murakami et al.* [2004]. The peak position and intensity of New2 are very consistent with what is expected for the post-perovskite phase observed by *Murakami et al.* [2004] (Figure 7). Furthermore, the intensity increases in the peaks at 14–15° and the appearance of weak diffraction peaks at the higher angle side of Pv4 can be explained by the post-perovskite phase. Unlike MgSiO₃ perovskite, the post-perovskite phase has a relatively intense peak at 13.7° (Figure 7). However, the post-perovskite structure cannot explain New1 and New3. New1 appears above 88 GPa and remains even after the change at 145 ± 10 GPa. The fact that the major diffraction peaks observed in Figure 7 can be assigned to those of the modified MgSiO₃ perovskite and the post-perovskite phase may indicate that the observed pattern is due to an incomplete transition from the modified perovskite to the post-perovskite phase in our experiments.

We also investigate the possibility of a dissociation of MgSiO₃ during the heating run at 135–145 GPa. The peak positions of the new peaks do not match those of the CaC_2-type Si_2 diffraction peaks. We also calculate diffraction patterns of the α-PbO_2-type SiO_2 based on the unit-cell parameters reported by *Murakami et al.* [2003] (Figure 7). Although the most intense peak of the α-PbO_2-type SiO_2 is very close to New2, the most intense peak of the post-perovskite phase shows better agreement in position with New2. A new peak at 13.7° is quite close to one of the α-PbO_2-type SiO_2 peaks. However, the post-perovskite phase has a peak that better matches the observed peak position. New3 (8.36°) is very close to the second most intense peak of the α-PbO_2-type SiO_2 (8.42°). However, New3 appears to be a doublet, whereas a singlet is expected at this position for the α-PbO_2-type SiO_2. A structure (at 10.2°) between New2 and a weak peak of the post-perovskite phase at 10.4° may be assigned to the most intense peak of the α-PbO_2-type SiO_2 at 10.3° (Figure 7). However, if indeed these structures are the

diffraction from the α-PbO_2-type SiO_2, their intensities are very close to the background and significantly smaller than the intensities of the post-perovskite phase diffraction peaks. Furthermore, no significant evidence above noise level are observed for the presence of MgO. In summary, the observed diffraction patterns at 135–145 GPa are better explained by the mixture of the modified MgSiO₃ perovskite and the post-perovskite phase.

3.4. Observation of SiO₂

After heating MgSiO₃ perovskite at 107 GPa, in a pattern recorded for an area 10 μm away from the heated spot, we found a weak peak that can be assigned to the most intense diffraction peak of the CaCl₂-type silica phase (Figure 8). One possible interpretation of this is that the steep thermal gradients existing at the edge of the hot spot induce unstable conditions for MgSiO₃ perovskite, resulting in a dissociation as previously proposed [*Serghiou et al.*, 1998]. In fact, some of the previous studies [*Serghiou et al.*, 1999; *Boehler*, 2000] have proposed that steep thermal gradients in the laser-heated

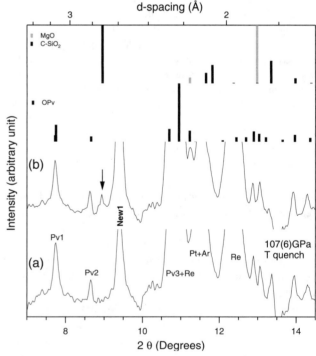

Figure 8. Integrated X-ray diffraction patterns measured at 107± 6 GPa after heating. (a) Heated spot and (b) a spot 10 μm away from the hot spot center. Calculated patterns for orthorhombic MgSiO₃ perovskite (*Pbnm*), periclase, and the CaCl₂-type SiO₂ are shown at the top. The observed SiO₂ diffraction is indicated by an arrow. The starting material of this sample is MgSiO₃ perovskite; other notations are as in Figure 5.

sample may be a source of discrepancy about the stability of $MgSiO_3$ perovskite in earlier studies. Unfortunately, diagnostic peaks of MgO overlap those of $MgSiO_3$ perovskite at this condition. We also note that excess SiO_2 has been widely observed in $MgSiO_3$ perovskite synthesis experiments at lower pressure [e.g., *O'Neill and Jeanloz*, 1994]. More study is needed before drawing further conclusions, as this was observed only once in our measurements and has yet to be reproduced.

4. DISCUSSION

4.1. Possible Sources for Discrepancy on the Stability

Comparison of the experimental techniques between our study and the other studies [*Meade et al.*, 1995; *Saxena et al.*, 1996; *Serghiou et al.*, 1998; *Fiquet et al.*, 1998, 2000; *Murakami et al.*, 2004] may provide some insights on the sources of the discrepancy on the stability. The main differences are: (1) Diffraction measurements are performed at in-situ high-P–T conditions relevant to those in the mantle. (2) The samples are loaded together with an inert pressure medium, argon, which insulates the sample from the very efficient thermal conductors, diamond anvils. Thus, the argon insulation layers surrounding the samples may reduce the axial thermal gradients in the diamond-anvil cell. (3) The TEM_{01} mode of Nd:YLF laser beams in the laser heating system [*Shen et al.*, 2001] used in our study provides a large homogeneous heating spot. Heating of both sides of the sample in this study may also reduce the thermal gradients along the loading axis of the diamond-anvil cells. These technical improvements provide a better sample environment for examining the stability of $MgSiO_3$ perovskite. In addition, the observation of SiO_2 diffraction at a spot 10 μm away from the center of the heated spot (20 μm in diameter) suggests that strong thermal gradients can lead to a dissociation of $MgSiO_3$ perovskite.

4.2. Uncertainty Associated With the New Peaks With $MgSiO_3$

It is worth discussing the uncertainties in the observation of the new peaks at 88 ± 5 GPa (New1). The New1 peak was not observed in recent measurements *by Murakami et al.* [2004]. Whereas *Murakami et al.* [2004] used only a gel starting material, we have confirmed the appearance of New1 when different crystalline starting materials were used. As shown in the case of $(Mg,Ca)SiO_3$ [*Kim et al.*, 1994; *Irifune et al.*, 2000], use of amorphous starting materials could result in different heating products. To insulate the sample + platinum foil, *Murakami et al.* [2004] separate their sample +

platinum mixture from diamond anvils by using pure sample foils. However, X-rays are scattered at both the sample + platinum foil and the pure sample foils, which are directly in contact with diamond anvils. Therefore, the measured diffraction is from both the sample with relatively less thermal gradients and the sample with extreme thermal gradients. Thus, the contamination with extreme thermal gradients in the sample cannot be ruled out in their experiments. However, we used argon to insulate the sample + platinum mixture foils throughout our experiments. This New1 peak also was not observed in *Mao et al.* [2004]. However, the difference in chemical composition between their starting materials, $(Mg,Fe)SiO_3$, and our starting materials, $MgSiO_3$, could result in the different observations.

The New1 peak was observed in *Oganov and Ono* [2004]. However, they assign the peak to platinum carbide, which is proposed to be a product of chemical reaction between diamond anvil and platinum. Interestingly, we do not observe the New1 peak in our $CaSiO_3$ perovskite measurements [*Shim et al.*, 2000], which were performed at similar P–T conditions with platinum in the diamond cell. If the New1 is from a reaction between platinum and diamond anvil, the New1 peak should have been observed in this experiment. Nevertheless, we note that there still exist uncertainties in evaluating the possibility of chemical reaction. More studies are warranted to resolve the problem.

4.3. Geophysical Implications of the Post-Perovskite Transition

Because the knowledge of the post-perovskite transition is rapidly growing and many more results are appearing in the literature, it may be too early to discuss implications of the post-perovskite transition at this moment. Yet it may be useful to summarize current knowledge for a general audience and discuss possible implications for the structure and dynamics near the core–mantle boundary. The discussion in this section is based on the results published as of May 2005.

The expected transition depth of the post-perovskite boundary is very close to the depth of the D'' discontinuity existing 200–300 km above the core–mantle boundary [*Lay and Helmberger*, 1983; *Wysession et al.*, 1998; *Sidorin et al.*, 1999]. According to measurements by *Murakami et al.* [2004] and *Oganov and Ono* [2004], the phase boundary may exist between 114 and 126 GPa at 2000–2500 K, which corresponds to 400 and 200 km above the core–mantle boundary in the preliminary reference Earth model (PREM) [*Dziewonski and Anderson*, 1981], respectively (Figure 1). The post-perovskite phase was observed at even higher pressure in our measurements, 140 GPa (Figure 1). With regard to a gap in our data set between 118 and 140 GPa, an appropriate interpretation is that the phase boundary exists in this range. Therefore, our data are not inconsistent with the results of *Murakami et al.* [2004] and *Oganov and Ono* [2004]. Furthermore, we found that the post-transition is very sluggish, which could result in overestimation of the transition pressure. This was confirmed in more recent measurement by *Mao et al.* [2004].

Note that uncertainties in the measured pressure and temperature are significant at these extreme experimental conditions. In synchrotron X-ray measurements, pressure is normally determined from the measured temperature and volume of the internal pressure standards (e.g., platinum and gold) using their equations of state. Volume can be precisely determined from the diffraction peak positions. Temperature uncertainty in typical laser heating [*Boehler*, 2000] is ±200–300 K at pressures greater than 100 GPa. Including these measurement uncertainties, ±5–10 GPa uncertainty in pressure (i.e., ±100–200 km uncertainty in depth) may be a reasonable estimation at 100–140 GPa and high temperature.

Uncertainties in the equations of state of the standard materials should also be considered. Including uncertainties in thermoelastic parameters, such as bulk modulus, pressure derivative of bulk modulus, and Grüneisen parameters, ±5 GPa uncertainty is a reasonable estimate for the pressure scales at 100–140 GPa and high temperature. However, as recently demonstrated by *Shim et al.* [2001a] and *Fei et al.* [2004a], different pressure scales may not be consistent with each other. The measurements performed by *Murakami et al.* [2004], *Shim et al.* [2004], and *Oganov and Ono* [2004] are based on the platinum scale [*Holmes et al.*, 1989; *Jamieson et al.*, 1982]. However, the accuracy of the platinum scale is not well known at core–mantle boundary *P–T* conditions. The gold pressure scale is one of the best studied among the existing internal standards [*Jamieson et al.*, 1982; *Heinz and Jeanloz*, 1984; *Anderson et al.*, 1989; *Shim et al.*, 2002; *Okube et al.*, 2002; *Tsuchiya*, 2003; *Matsui and Shima*, 2003]. As demonstrated by *Shim et al.* [2002], however, even the existing gold pressure scales are discrepant by about 10 GPa corresponding to 200 km in depth near core–mantle boundary *P–T* conditions (Figure 9).

First-principles calculations have provided constraints on the depth of the post-perovskite transition [*Oganov and Ono*, 2004; *Tsuchiya et al.*, 2004a]. Depending on used method, the value ranges from 108 to 125 GPa, which is comparable to the experimental results. Thus, a reasonable estimation of the transition depth based on the published data is 200–400 km above the core–mantle boundary in pure magnesium silicate (MgSiO$_3$) with an uncertainty of ±200 km.

Considering the chemical composition models of the lower mantle [e.g., *Ringwood*, 1975; *Taylor and McLennan*, 1985; *Anderson*, 1989], the post-perovskite phase should contains some amount of iron (maybe 5–15%). The effect of iron on

the transition pressure has been studied by *Mao et al.* [2004], who found that the post-perovskite transition occurs at 100 GPa in 10% iron ferromagnesian silicate, (Mg$_{0.9}$Fe$_{0.1}$)SiO$_3$. This is about 700 km above the core–mantle boundary. In other words, incorporation of 10% iron decreases the post-perovskite transition pressure by 15–25 GPa and elevates the height of the transition from the core–mantle boundary by 300–500 km. The sensitivity of the post-perovskite transition depth to iron is very high compared with that for the olivine-wadsleyite and the post-spinel transitions (both decrease by less than 1 GPa with 10% iron [*Akaogi et al.*, 1989; *Ito and Takahashi*, 1989]). There is some degree of uncertainty in comparing the results of *Mao et al.* [2004] with the data for pure MgSiO$_3$, as these two different sets of results are tied to different pressure scales (NaCl and platinum), for which the consistency is not well constrained at this pressure range.

Murakami et al. [2005] performed X-ray measurements for a pyrolitic starting material to 126 GPa and 2450 K and observed the post-perovskite transition between 103 and 113 GPa, i.e., 600 km and 400 km, respectively, above the core–mantle boundary. A direct comparison with the data that constrain the post-perovskite transition pressure in pure MgSiO$_3$ is difficult, because consistency between the gold (used in pyrolite measurements by *Murakami et al.* [2005]) and the platinum pressure scales (used in pure MgSiO$_3$ measurements by *Murakami et al.* [2004]; *Oganov and Ono* [2004]; *Shim et al.* [2004]) is not well known. *Murakami et al.* [2005] reported that the post-perovskite phase contains only 3% iron. From *Mao et al.* [2004], however, a 3% iron concentration is associated with a 5–8 GPa decrease in transition pressure. Thus, the decrease in the transition pressure measured in *Murakami et al.* [2005] can be mainly explained by the effect of iron. The effect of aluminum is not well understood yet. However, it is generally expected that aluminum may increase the post-perovskite transition pressure [*Caracas and Cohen*, 2005]. In summary, these two results demonstrate that the post-perovskite transition pressure may be very sensitive to minor elements, and the degree of the sensitivity may be much higher than those of the phase transitions that occur at the transition zone.

It is also important to note that the post-perovskite phase may have different element partitioning with magnesiowüstite compared to perovskite. As pointed out by *Murakami et al.* [2005], changes in iron partitioning could affect the viscosity [*Durham et al.*, 1979] and electrical conductivity [*Li and Jeanloz*, 1991; *Peyronneau and Poirier*, 1989] of the lowermost mantle. *Murakami et al.* [2005] reported that iron is still preferentially partitioned into magnesiowüstite after the post-perovskite transition; moreover, they observed that magnesiowüstite is more enriched in iron when it coexists with the post-perovskite than when it coexists with perovskite.

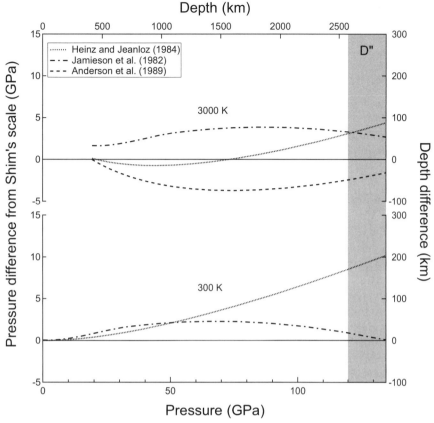

Figure 9. Pressure differences among different gold pressure scales [*Heinz and Jeanloz*, 1984; *Jamieson et al.*, 1982; *Anderson et al.*, 1989] at 300 K and 3000 K. A depth scale obtained from pressure–depth relations of PREM [*Dziewonski and Anderson*, 1981] is shown for comparison.

However, *Mao et al.* [2004] have proposed that iron is preferentially partitioned into the post-perovskite phase than into perovskite during the phase transition and predicted that the post-perovskite phase may have higher capacity for iron than any other phases at the lowermost mantle. This discrepancy may be related to the fact that the starting materials of *Mao et al.* [2004]'s contain only Mg, Fe, Si, and O, whereas *Murakami et al.* [2005] used starting materials having a pyrolitic composition with minor elements (e.g., aluminum and calcium), which may be more relevant to the chemical composition of the mantle. More measurements are necessary in order to understand element partitioning at the base of the mantle. Notably, both of the above results demonstrated that the iron partitioning among different phases in the lowermost mantle may be significantly different from that in the rest of the lower mantle.

Another important effect of iron observed in recent experiments [*Mao et al.*, 2004] is that iron increases the width of the post-perovskite transition significantly: the width of the transition for 10% iron is approximately 10 GPa, i.e., 200 km. This is a remarkably large width compared to the olivine-wadsleyite and post-spinel transitions (both less than 1 GPa for 10% iron [*Akaogi et al.*, 1989; *Ito and Takahashi*, 1989]). Assuming that 10% iron is a reasonable estimate in perovskite, this width seems rather large when compared to a seismically constrained transition width of the D″ discontinuity (less than 50–75 km [*Wysession et al.*, 1998]). However, there are many other factors to consider when comparing the laboratory measurements to the seismic observations. For example, as shown in the case of the olivine-wadsleyite transition [*Irifune and Isshiki*, 1998], transition width can be strongly affected by element partitioning with other phases, such as magnesiowüstite and CaSiO₃ perovskite. Furthermore, more precise in-situ measurements are necessary.

The Clapeyron slope of a phase transition is important for understanding the effect of the phase transition with regard to the mantle convection, as is well demonstrated for the phase boundaries at the transition zone [e.g., *Tackley et al.*, 1993]. If the post-perovskite transition exists above the core–mantle boundary, it will likely affect the mantle convection in the lowermost mantle. This is particularly interesting because the D″ layer has been believed to be the source region of the mantle plumes and the graveyard of the subducting slabs according to some models [e.g. *Morgan*, 1971; *Christensen and Hofmann*, 1994; *Garnero and Lay*, 2003].

Currently available measurements cannot constrain the sign and the magnitude of the Clapeyron slope of the post-perovskite boundary (Figure 1). Some first-principles calculations provide constraints on the slope: +9.56–9.85 MPa/K [*Oganov and Ono*, 2004] and +7.5 ± 0.3 MPa/K [*Tsuchiya et al.*, 2004a]. The sign and magnitude of the slope is comparable to a seismologic estimation for the D″ discontinuity, i.e., +6 MPa/K [*Sidorin et al.*, 1999]. It is notable that the magnitude of the slope is factor of 2–4 greater than those of the phase boundaries in the transition zone: +3.6 MPa/K for the olivine-wadsleyite transition [*Morishima et al.*, 1994], +2 MPa/K for the post-garnet transition [*Hirose et al.*, 2001], and −2.8 MPa/K for the post-spinel transition [*Ito and Takahashi*, 1989; *Irifune et al.*, 1998]. This makes the transition depth of the post-perovskite boundary very sensitive to temperature. For example, a 500 K decrease in temperature may elevate the transition depth by 100 km. At this moment, the effect of minor elements on the sign and magnitude of the slope has not been studied. However, from the high sensitivity of the transition depth to chemical composition and temperature, the post-perovskite boundary may develop significant topography if lateral variations in temperature and chemical composition exist near the core–mantle boundary. Interestingly, seismic studies have revealed that there exist lateral heterogeneities in the lowermost mantle [e.g., *Gaherty and Lay*, 1992; *Liu and Dziewonski*, 1998; *Kuo et al.*, 2000; *Garnero and Lay*, 2003; *Thomas et al.*, 2004a] and that the D″ discontinuity height varies laterally from 100 km to 450 km above the core–mantle boundary [*Wysession et al.*, 1998].

The effect of a phase transition with a positive Clapeyron slope near the core–mantle boundary has been studied in computer simulations by *Nakagawa and Tackley* [2004]. They showed that a positive slope boundary destabilizes the thermal boundary layer at the core–mantle boundary and will result in more vigorous plume activity. In future simulations, the finite transition width of the post-perovskite and the depth variation by lateral variations in chemical composition should be considered. Most of all, more knowledge about the post-perovskite transition is necessary to understand its effect on the mantle convection.

It is interesting to note that the estimated slope of the post-perovskite transition may be very close to the temperature gradient in the thermal boundary layer at the lowermost mantle. Although the lower-mantle geotherm is expected to be close to adiabatic and the gradient may be 0.3 K/km [*Brown and Shankland*, 1981; *Duffy and Hemley*, 1995], the mantle side of the core–mantle boundary should be strongly super-adiabatic, compared with the rest of the mantle, due to vertical heat

transfer across the core–mantle boundary. It is very difficult to estimate the temperature gradient in this layer because the necessary parameters are not well constrained, including the melting temperature of iron [*Williams et al.*, 1987; *Saxena et al.*, 1994; *Boehler*, 1993; *Williams et al.*, 1998; *Shen et al.*, 1998]. Using the PREM model [*Dziewonski and Anderson*, 1981], *Stacey and Loper* [1983] estimated that the thermal boundary layer extends to 73 km above the core–mantle boundary with a maximum thermal gradient of 11.2 K/km. However, according to seismic observations, the thickness of the thermal boundary layer could be much larger, i.e., 200–250 km [*Wysession et al.*, 1998]. Including existing uncertainties in important parameters, temperature increase across the D″ layer may lie in a range of 1000–2000 K [*Jeanloz and Morris*, 1986; *Boehler et al.*, 1995; *Williams et al.*, 1998]. Therefore, the temperature gradient of 5–10 K/km in the D″ layer would not be an unreasonable estimate. According to these estimations, the temperature gradient in the thermal boundary layer may be very close to the estimated Clapeyron slope of the post-perovskite boundary, i.e., 5–7 K/km. This similarity in slopes may result in very interesting changes in mineralogy in response to lateral variations in temperature and chemical composition at the lowermost mantle (Figure 10). We will demonstrate a few possible cases, but this discussion remains qualitative because uncertainties are not resolved for the post-perovskite boundary and the temperature gradient in the thermal boundary layer. To simplify the discussion, we shall treat the post-perovskite transition as a sharp boundary; however, as has been shown in some recent measurements [e.g., *Mao et al.*, 2004], minor elements, such as iron and aluminum, can increase the transition width. Also, although we represent the super-adiabatic thermal gradient in the thermal boundary layer with a line in Figure 10d, the gradient may be highly non-linear [e.g., *Stacey and Loper*, 1983].

Figure 10a shows a case where the post-perovskite transition exists above the thermal boundary layer. However, if the thermal boundary layer is extended to a higher elevation from the core–mantle boundary and/or if the post-perovskite boundary exists at a deeper depth (possibly due to the effect of minor elements), it is possible that perovskite is still stable to the core–mantle boundary and that the post-perovskite transition does not occur in the lowermost mantle (Figure 10b). If the post-perovskite transition exists in the thermal boundary layer with a higher slope than the temperature gradient in the thermal boundary layer (Figure 10c), the transition depth will be deeper than one expected for a mantle geotherm.

Very interesting changes in mineralogy at the lowermost mantle can be predicted when the temperature gradient in the thermal boundary layer is higher than the Clapeyron slope of the post-perovskite boundary and when the post-

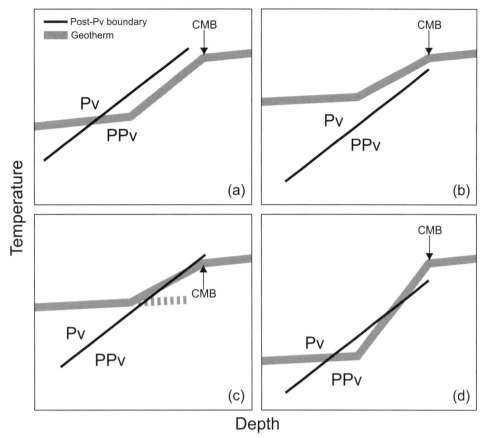

Figure 10. Schematic models of the post-perovskite (PPv) boundary and the mantle geotherm near the thermal boundary layer at the bottom of the mantle. (a) The post-perovskite transition exists above the thermal boundary layer. (b) The post-perovskite transition does not occur in the mantle. This is possible if minor elements move the boundary to a deeper depth and/or the temperature in the thermal boundary is sufficiently high. (c) The post-perovskite transition occurs in the thermal boundary layer. (d) The post-perovskite boundary crosses the geotherm more than once. This could result from more gentle slope of the post-perovskite boundary or a non-linear geotherm in the thermal boundary layer. This will result in alternating layers of perovskite (Pv) and the post-perovskite.

perovskite boundary exists near the thermal boundary layer. In Figure 10d, the post-perovskite boundary meets with a mantle geotherm at a shallower depth and with a temperature gradient in the thermal boundary layer at a deeper depth. At the first intercept, perovskite transforms to the post-perovskite. However, at the second intercept, the post-perovskite transforms back to perovskite. This will result in alternating layers of perovskite and the post-perovskite at the bottom of the mantle. A similar result is expected when the temperature gradient in the D″ layer is highly non-linear and the slope changes from lower to higher than the slope of the post-perovskite boundary. In fact, *Stacey and Loper* [1983] predicted that temperature in the thermal boundary layer exponentially increases with depth.

These models remain schematic due to the uncertainties discussed above. More precise determinations on the sensitiv-

ity of the post-perovskite transition to temperature and chemical composition will enable us to include or exclude some of these cases. In addition, although mantle geotherm and the post-perovskite boundary are assumed to be independent of each other in these schematic models, latent heat from the phase transition may perturb the geotherm near the transition boundary, as shown for the phase boundaries in the transition zone [e.g., *Agee*, 1998].

Nevertheless, these models may have some implications for seismic observations. An important point from these schematic models is that the expected sensitivity of the post-perovskite transition to temperature and chemical composition, a steep temperature gradient at the lowermost mantle, and lateral variations in temperature and chemical composition at the lowermost mantle as indicated in some seismic studies [e.g., *Gaherty and Lay*, 1992; *Liu and Dziewonski*,

1998; *Kuo et al.*, 2000; *Garnero and Lay*, 2003; *Thomas et al.*, 2004a] may result in significant lateral variations in the transition depth and mineralogy. For example, the D″ discontinuity becomes observable and not observable over distances of 10–100 km, and the discontinuity is sometimes coherently observed over a distance of 1000 km [*Wysession et al.*, 1998]. Furthermore, using seismic observations of beneath Eurasia [*Thomas et al.*, 2004b] and Caribbean [*Thomas et al.*, 2004a], *Hernlund et al.* [2005] has proposed that, depending on local temperature profile near the core–mantle boundary, the post-perovskite boundary can be crossed twice by the geotherm, which results in observation of a pair of seismic discontinuities near the core–mantle boundary.

According to the proposed crystal structure (CaIrO₃ type) for the post-perovskite phase [*Murakami et al.*, 2004; *Iitaka et al.*, 2004; *Oganov and Ono*, 2004; *Tsuchiya et al.*, 2004a], SiO₆ octahedra form layers in the post-perovskite structure, whereas they form a 3-dimensional network in perovskite (Figure 6c). This fundamental difference in the crystal structures may result in significantly different physical and chemical properties of the post-perovskite phase.

However, the post-perovskite phase is stable only at extreme *P–T* conditions and does not appear to be quenchable, which makes measurements of the physical properties of the post-perovskite phase difficult. First-principles calculations provide constraints on some of important physical properties of the post-perovskite phase. A density increase of 1.0–1.5% has been reported for the post-perovskite transition [*Murakami et al.*, 2004; *Oganov and Ono*, 2004; *Tsuchiya et al.*, 2004a]. Compared to other transitions related to seismic discontinuities in the transition zone, e.g., 4–6% for the post-spinel transition, this density increase is very moderate.

At this moment, elastic properties of the post-perovskite phase are constrained only by the first- principles calculations. The bulk modulus of the post-perovskite phase at 0 GPa is 221–232 GPa [*Oganov and Ono*, 2004; *Tsuchiya et al.*, 2004b], which is smaller than that of perovskite, i.e., 260 GPa. However, what makes the post-perovskite phase slightly more incompressible at the transition pressure is a higher pressure derivative of bulk modulus: 4.2–4.4 for the post-perovskite versus 3.7–4.0 for perovskite. Whereas *P*-wave velocity change is expected to be small [0.3%, *Oganov and Ono*, 2004] during the transition, *S*-wave velocity change is expected to be much higher (1.5% [*Tsuchiya et al.*, 2004b]). The magnitudes of the velocity changes in the post-perovskite transition are smaller than seismic estimation of the changes at the D″ discontinuity, which are approximately 3% for both *P* and *S* waves [*Wysession et al.*, 1998].

Shear-wave splitting has been observed in D″ below several regions, e.g., Caribbean [e.g, *Kendall and Silver*, 1996] and Alaska/North Pacific [e.g., *Lay and Young*, 1991], and

has been attributed to anisotropy in D″. *Kendall and Silver* [1998] reviewed the possibility of lattice-preferred orientation as the cause of the anisotropy. However, no satisfactory explanation was found from lattice-preferred orientation of most mantle minerals. The D″ anisotropy can be described as transverse isotropy with a symmetry axis along the vertical direction. Because of the 2-dimensional layered structure of the post-perovskite, some have proposed that the post-perovskite may be responsible for the observed D″ anisotropy with an expectation that a slip plane may be developed parallel to the SiO₆ octahedral layers, which are perpendicular to the *b*-axis, in the post-perovskite phase [Figure 6c, *Iitaka et al.*, 2004; *Oganov and Ono*, 2004]. First-principles calculations of *Iitaka et al.* [2004] have shown that the post-perovskite phase is elastically more anisotropic than is perovskite.

Oganov and Ono [2004] proposed that the shear-wave splitting with $V_{SH} > V_{SV}$ may be explained by the texturing of the post-perovskite phase. *Tsuchiya et al.* [2004b] argued that vertical alignment of the *c*-axis, which is parallel to the octahedral layers in the post-perovskite phase, could also explain the observed $V_{SH} > V_{SV}$. *Stackhouse et al.* [2005] studied the effect of temperature on the elastic anisotropy of perovskite and the post-perovskite and proposed that the elastic anisotropy of the post-perovskite phase is much less sensitive to temperature change than is perovskite, and $R_{S/P}$ $(= \partial \ln V_S / \partial \ln V_P)$ of the post-perovskite is much lower than that of perovskite and the seismically measured value of the lowermost mantle. Note that these calculations were performed for pure MgSiO₃; the effect of minor elements, e.g., Fe, Al, Ca, remains to be studied. Another important physical property change that may be related to the post-perovskite phase is a transition from high-spin to low-spin state of iron in (Mg,Fe)SiO₃ [*Badro et al.*, 2004] and aluminum bearing (Mg,Fe)SiO₃ [*Li et al.*, 2004] above 70 GPa. These investigators proposed that the transition will increase the intrinsic radiative thermal conductivity of this major mantle mineral and that this transition may increase contributions from radiative heat transfer.

Many other properties are yet to be determined for the post-perovskite phase, including the effect of minor elements, melting curve, element partitioning, rheology, elasticity, and chemical reaction with iron. This is particularly important for understanding the physical and chemical processes in the lowermost mantle, given that the post-perovskite phase seems to be the dominant phase in this region.

5. CONCLUSION

Although some earlier measurements [*Meade et al.*, 1995; *Saxena et al.*, 1996, 1998] have proposed a dissociation of

MgSiO$_3$ perovskite at mid-mantle P–T conditions, more recent measurements by many different groups [*Mao et al.*, 1997; *Serghiou et al.*, 1998; *Fiquet et al.*, 2000; *Andrault*, 2001; *Shim et al.*, 2001b, 2004] have shown that MgSiO$_3$ perovskite is stable to at least 118 GPa and 2900 K (2500-km depth). For example, in our in-situ X-ray diffraction measurements on MgSiO$_3$ at 50–144 GPa and 1400–2900 K, the dominant features observed in diffraction patterns during and after heating are identified as those of perovskite. This confirms the stability of MgSiO$_3$ perovskite at 1200–2500-km depth in the mantle.

However, a modification in the crystal structure of perovskite in deep mantle remains a viable explanation for our observation of a new peak that appears at 88 GPa. The source of this new peak remains inconclusive because of the possibility of alternative interpretations, such as chemical reaction or impurity.

A significant change in the crystal structure of MgSiO$_3$ has recently been observed by several groups, i.e., post-perovskite transition near core–mantle boundary P–T conditions [*Murakami et al.*, 2004; *Shim et al.*, 2004; *Oganov and Ono*, 2004]. According to recent studies, the post-perovskite transition depth is likely very sensitive to variations in chemical composition and temperature, which may be related to the observed large variation in depth of the D″ discontinuity. Furthermore, the proposed slope of the post-perovskite boundary seems very similar to the expected temperature gradient of the thermal boundary layer in the lowermost mantle. This may also result in a high sensitivity of mineralogy to variations in temperature and chemical composition. However, more studies are necessary to further investigate implications of the post-perovskite transition to the structure and dynamics of the lowermost mantle. The important measurements to be conducted include the transition depth, the Clapeyron slope, and the transition width in pure MgSiO$_3$ and "dirty" systems (the effect of minor elements). Interactions between the post-perovskite phase and other phases in the lowermost mantle are important: element partitioning between the post-perovskite phase and other major mantle minerals, and chemical reaction with iron and iron alloys. Elastic parameters, thermodynamic parameters, rheology, and transport properties of the post-perovskite are among the important parameters that should be measured.

APPENDIX 1: CALCULATION OF X-RAY DIFFRACTION PATTERNS

We calculate the expected positions and intensities for the diffraction peaks of MgSiO$_3$ perovskite, *Cmcm*-structured MgSiO$_3$, MgO, and SiO$_2$ phases for phase identification (e.g., Figures 2, 5, 7, and 8). Crystal-structure parameters of MgSiO$_3$ perovskite (orthorhombic, *Pbnm*) are obtained from recent X-ray measurements by *Fiquet et al.* [2000]. We calculate the diffraction pattern of the post-perovskite phase by using the CaIrO$_3$-type (*Cmcm*) structure proposed by *Murakami et al.* [2004]. Diffraction patterns of SiO$_2$ are calculated for all the phases proposed, based on recent experiments [*Dubrovinsky et al.*, 1997; *Andrault et al.*, 1998; *Murakami et al.*, 2003] and a first-principles calculation [*Teter et al.*, 1998], including the CaCl$_2$ and α-PbO$_2$ type structures. Pressure and temperature variations of unit-cell parameters are determined by using the published equations of state [*Fiquet et al.*, 2000; *Speziale et al.*, 2001; *Murakami et al.*, 2004]. We assume that the equations of state for all high-pressure forms of SiO$_2$ are the same as that for stishovite [*Hemley et al.*, 1994]. The predicted peak positions of the α-PbO$_2$-type SiO$_2$ under this assumption agree with those reported by *Murakami et al.* [2003].

To calculate X-ray diffraction patterns, it is necessary to know the space group, the unit-cell parameters, atomic positions, temperature factors for each atom, and other experimental parameters. To obtain unit-cell parameters at 100 GPa, we first calculate the expected volume of MgSiO$_3$ perovskite, using its equation of state [*Mao et al.*, 1991; *Jackson and Rigden*, 1996]. For non-cubic structures we need to know the axial ratios to obtain unit-cell parameters from the volume. For the *Pbnm* structure, we chose the axial ratios reported by *Fiquet et al.* [2000] at 79.7 GPa and 1681 K. For the proposed structures above 88 GPa, i.e., $P2_1/m$, $P4_2/nmc$, and *Pmmn*, the axial ratios are obtained from a geometric model (refer to Appendix 2).

Atomic positions are obtained from *Fiquet et al.* [2000] for *Pbnm* structure and from the geometrical modeling for $P2_1/m$, $P4_2/nmc$, and *Pmmn* structures. Temperature factors were assumed to be the same as those measured by the single-crystal X-ray diffraction study at lower pressure by *Ross and Hazen* [1990]. We assumed that Mg atoms are at the center of dodecahedral sites and that the SiO$_6$ octahedra are rigid and not distorted in $P2_1/m$, $P4_2/nmc$, and *Pmmn* structures, since no constraints are available for these. In fact, in *Pbnm* perovskite, the positions of Mg atoms are shifted from the center of the dodecahedral sites, and the SiO$_6$ octahedra are distorted [*Fiquet et al.*, 2000]. To demonstrate the effect of Mg atoms shifting from the dodecahedral site center, we calculate X-ray diffraction pattern for *Pbnm* structure with Mg atoms at the center (Figure 2c). For some of the peaks, e.g., Pv2, significant intensity changes are observed. Thus, there exists some degree of ambiguity in the calculated intensity of MgSiO$_3$ perovskite. However, peak positions are not affected by the Mg atom shifts. X-ray diffraction patterns were calculated with the GSAS program [*Larson and Dreele*, 1988].

APPENDIX 2: GEOMETRICAL MODELING FOR PEROVSKITE STRUCTURES

To investigate the possibility of associating the New1 peak observed above 88 GPa with perovskite, we performed geometrical calculations. We first assume that the distortion of perovskite is induced only by the tilting of SiO_6 octahedra and that the octahedra behave as rigid units. This distortion mechanism permits only a few modifications of the perovskite structures [*Glazer*, 1972; *Woodward*, 1997]. The constraints we implemented in the model are listed in *Shim et al.* [2001b]. We found that only three space groups can satisfy the constraints: $P2_1/m$ (monoclinic), $Pmmn$ (orthorhombic), and $P4_2/nmc$ (tetragonal).

The POPATO [*Woodward*, 1997] program enables us to calculate the atomic positions in the perovskite structures distorted only by the rigid octahedral tilting. The input parameters are unit-cell parameters and the distances between Si and O sites. We obtained the Si–O distance at 100 GPa from the equation of state, the axial compressibility of $MgSiO_3$ perovskite, and atomic positions reported at 79.7 GPa and 1681 K. Unit-cell parameters were obtained such that they can reproduce both the new peak (New1) and the observed perovskite diffraction peaks. The atomic positions obtained from this method are used for our X-ray diffraction calculations. The unconstrained parameters in this model are the atomic positions of Mg and the distortion of the SiO_6 octahedra.

APPENDIX 3: AVERAGING SCHEME FOR TEMPERATURE OF THE SAMPLE

In the double-sided laser heating system at GSECARS, two imaging spectrometers with charge-coupled devices measure thermal radiation spectra (Figure 4b) across the sample. These thermal radiation spectra are fitted to a gray-body equation after correcting for system response (Figure 4). As a result of this, a line profile of temperature can be obtained for both sides of the sample.

In each run, there exist several temperature determinations across the sample volume for which diffraction measurements are collected (rectangle in Figure 4a). Assuming cylindrical symmetry for the temperature distribution in the sample, each temperature measurement represents a different area, as shown in the schematic diagram of Figure 4a. Using this area as a weight for each data point, a weighted average can be calculated. In this particular case, the weighted averages for both sides (2101±48 K for T1 and 2182±63 K for T2) are not significantly different from simple averages (2098±61 K for T1 and 2180±79 K for T2) within the uncertainties. Axial thermal gradients can be estimated by taking the difference between the averages

for both sides. The thermal profiles are measured several times during X-ray diffraction measurements (Figure 4c). The uncertainty of measured temperature includes temporal fluctuations, axial and radial thermal gradients, and uncertainty in gray-body fitting.

It is important to note that the uncertainty calculated in this procedure should be viewed as an estimation of precision rather than absolute accuracy. The measured thermal gradients are contaminated by the interference of thermal radiation from different locations in the sample and by the transparency of the sample. Also, the wavelength-dependent emissivity of the sample is unknown at high P–T.

Acknowledgments. Discussions with T. S. Duffy, R. Jeanloz, S. Rekhi, J. Santillan, S. Lundin, J. Bass, and two anonymous reviewers improved this manuscript. I thank S. Speziale, S. Shieh, G. Shen, V. Prakapenka, and N. Sata for their assistant in the experiments. J. Akins and T. Ahrens kindly provided glass samples and their electron microprobe analysis results. This study is supported by NSF. The Miller Institute is acknowledged for support.

REFERENCES

Agee, C. B., Phase transformations and seismic structure in the upper mantle and transition zone, in *Ultrahigh-pressure mineralogy*, edited by R. J. Hemley, vol. 37 of *Reviews in Mineralogy*, pp. 165–203, Mineralogical Society of America, 1998.

Akaogi, M., E. Ito, and A. Navrotsky, Olivine-modified spinel-spinel transitions in the system Mg_2SiO_4–Fe_2SiO_4: Calorimetric measurements, thermochemical calculations and geophysical application, *Journal of Geophysical Research*, 94, 15,671–15,685, 1989.

Albarède, F., and R. D. van der Hilst, New mantle convection model may reconcile conflicting evidence, *Eos Transactions*, 80, 535–539, 1999.

Anderson, D. L., Composition of the earth, *Science*, 243, 367–370, 1989.

Anderson, O. L., D. G. Isaak, and S. Yamamoto, Anharmonicity and the equation of state for gold, *Journal of Applied Physics*, 65, 1534–1543, 1989.

Andrault, D., Evaluation of (Mg,Fe) partitioning between silicate perovskite and magnesiowustite up to 120 GPa and 2300 K, *Journal of Geophysical Research*, 106, 2079–2087, 2001.

Andrault, D., and G. Fiquet, Synchrotron radiation and laser heating in a diamond anvil cell, *Review of Scientific Instruments*, 72, 1283–1288, 2001.

Andrault, D., G. Fiquet, J. P. Itie, P. Richet, P. Gillet, D. Häusermann, and M. Hanfland, Thermal pressure in the laser-heated diamond-anvil cell: an X-ray diffraction study, *European Journal of Mineralogy*, 10, 931–940, 1998.

Asahara, Y., E. Ohtani, T. Kondo, T. Kubo, N. Miyajima, T. Nagase, K. Fujino, T. Yagi, and T. Kikegawa, Formation of metastable cubic-perovskite in high-pressure phase transformation of Ca(Mg,Fe,Al)Si₂O₆, *American Mineralogist*, 90, 457–462, 2005.

Badro, J., J.-P. Rueff, G. Vanko, G. Monaco, G. Fiquet, and F. Guyot, Electronic transitions in perovskite: possible nonconvecting layers in the lower mantle, *Science*, *305*, 383–386, 2004.

Boehler, R., Temperatures in the Earth's core from melting-point measurements of iron at high static pressures, *Nature*, *363*, 534–536, 1993.

Boehler, R., High-pressure experiments and the phase diagram of lower mantle and core materials, *Reviews of Geophysics*, *38*, 221–245, 2000.

Boehler, R., A. Chopelas, and A. Zerr, Temperature and chemistry of the core-mantle boundary, *Chemical Geology*, *120*, 199–205, 1995.

Brown, J. M., and T. J. Shankland, Thermodynamic parameters in the Earth as determined from seismic profiles, *Geophysical Journal of the Royal Astronomical Society*, *66*, 576–596, 1981.

Caracas, R., and R. E. Cohen, Prediction of a new phase transition in Al_2O_3 at high pressures, *Geophysical Research Letters*, *32*, L06,303, 2005.

Christensen, U. R., and A. W. Hofmann, Segregation of subducted oceanic crust in the convecting mantle, *Journal of Geophysical Research*, *99*, 19,867–19,884, 1994.

Chundinovskikh, L., and R. Boehler, High-pressure polymorphs of olivine and the 660-km seismic discontinuity, *Nature*, *411*, 574–577, 2001.

Collerson, K. D., S. Hapugoda, B. S. Kamber, and Q. Williams, Rocks from the mantle transition zone: Majorite-bearing xenoliths from Malaita, southwest pacific, *Science*, *288*, 1215–1223, 2000.

Dewaele, A., G. Fiquet, D. Andrault, and D. Häusermann, P–V–T equation of state of periclase from synchrotron radiation measurements, *Journal of Geophysical Research*, *105*, 2869–2877, 2000.

Dubrovinskaia, N., and L. Dubrovinsky, Whole-cell heater for the diamond anvil cell, *Review of Scientific Instruments*, *74*, 3433–3437, 2003.

Dubrovinsky, L. S., S. K. Saxena, P. Lazor, R. Ahuja, O. Eriksson, J. M. Wills, and B. Johansson, Experimental and theoretical identification of a new high-pressure phase of silica, *Nature*, *388*, 362–365, 1997.

Dubrovinsky, L. S., S. K. Saxena, and S. Rekhi, $(Mg,Fe)SiO_3$-perovskite stability and lower mantle conditions, *Science*, *285*, 983a, 1999.

Duffy, T. S., and R. J. Hemley, Some like it hot - the temperature structure of the earth, *Reviews of Geophysics*, *33 supplement*, 5–9, 1995.

Durham, W. B., C. Froideveaux, and O. Jaoul, Transient and steady-state creep of pure forsterite at low stress, *Physics of the Earth and Planetary Interiors*, *19*, 263–274, 1979.

Dziewonski, A. M., and D. L. Anderson, Preliminary reference Earth model, *Physics of the Earth and Planetary Interiors*, *25*, 297–356, 1981.

Fei, Y., Y. Wang, and L. W. Finger, Maximum solubility of FeO in $(Mg,Fe)SiO_3$-perovskite as a function of temperature at 26 GPa: Implication for FeO content in the lower mantle, *Journal of Geophysical Research*, *101*, 11,525–11,530, 1996.

Fei, Y., H. Li, K. Hirose, W. Minarik, J. V. Orman, W. V. Westrenen, T. Kmabayashi, and K. Funakoshi, A critical evaluation of pressure scales at high temperatures by in situ X-ray diffraction measurements, *Physics of the Earth and Planetary Interiors*, *143-144*, 515–526, 2004a.

Fei, Y., J. Van Orman, J. Li, W. van Westrenen, C. Sanloup, W. Minara, K. Hirose, T. Komabayashi, M. Walter, and K. Funakoshi, Experimentally determined postspinel transformation boundary in Mg_2SiO_4 using MgO as an internal pressure standard and its geophysical implications, *Journal of Geophysical Research*, *109*, B02,305, 2004b.

Finger, L. W., R. M. Hazen, G. Zou, H.-K. Mao, and P. M. Bell, Structure and compression of crystalline argon and neon at high pressure and room temperature, *Applied Physics Letters*, *39*, 92–894, 1981.

Fiquet, G., D. Andrault, A. Dewaele, T. Charpin, M. Kunz, and D. Häusermann, P–V–T equation of state of $MgSiO_3$ perovskite, *Physics of the Earth and Planetary Interiors*, *105*, 21–31, 1998.

Fiquet, G., A. Dewaele, D. Andrault, M. Kunz, and T. L. Bihan, Thermoelastic properties and crystal structure of $MgSiO_3$ perovskite at lower mantle pressure and temperature conditions, *Geophysical Research Letters*, *27*, 21–24, 2000.

Fischer, K. M., and R. D. van der Hilst, Perspectives: Geophysics—a seismic look under the continents, *Science*, *285*, 1365–1366, 1999.

Fukao, Y., S. Widiyantoro, and M. Obayashi, Stagnant slabs in the upper and lower mantle transition region, *Reviews of Geophysics*, *39*, 291–323, 2001.

Gaherty, J. B., and T. Lay, Investigation of laterally heterogeneous shear velocity structure in D″ beneath Eurasia, *Journal of Geophysical Research*, *97*, 417–435, 1992.

Garnero, E. J., and T. Lay, D″ shear velocity heterogeneity, anisotropy and discontinuity structure beneath the Caribbean and Central America, *Physics of the Earth and Planetary Interiors*, *140*, 219–242, 2003.

Glazer, A. M., The classification of tilted octahedra in perovskites, *Acta Crystallographica*, *B28*, 3384–3392, 1972.

Gurnis, M., M. E. Wysession, E. Knittle, and B. A. Buffett, *The core-mantle boundary region*, American Geophysical Union, Washington, DC, 1998.

Heinz, D. L., Thermal pressure in the laser-heated diamond anvil cell, *Geophysical Research Letters*, *17*, 1161–1164, 1990.

Heinz, D. L., and R. Jeanloz, The equation of state of the gold calibration standard, *Journal of Applied Physics*, *55*, 885–893, 1984.

Hemley, R. J., C. T. Prewitt, and K. J. Kingma, High pressure behavior of silica, in *Silica: physical behavior, geochemistry and material applications*, edited by P. J. Heaney, C. T. Prewitt, and G. V. Gibbs, pp. 41–82, Mineralogical Society of America, 1994.

Hernlund, J. W., C. Thomas, and P. J. Tackley, A doubling of the post-perovskite phase boundary and structure of the Earth's lowermost mantle, *Nature*, *434*, 882–886, 2005.

Hirose, K., Y. Fei, S. Ono, T. Yagi, and K. i. Funakoshi, In situ measurements of the phase transition boundary in $Mg_3Al_2Si_3O_{12}$: implications for the nature of the seismic discontinuities in the

Earth's mantle, *Earth and Planetary Science Letters*, *184*, 567–573, 2001.

Holmes, N. C., J. A. Moriarty, G. R. Gathers, and W. J. Nellis, The equation of state of platinum to 660 GPa (6.6 Mbar), *Journal of Applied Physics*, *66*, 2962–2967, 1989.

Iitaka, T., K. Hirose, K. Kawamura, and M. Murakami, The elasticity of the MgSiO₃ post-perovskite phase in the earth's lowermost mantle, *Nature*, *430*, 442–445, 2004.

Irifune, T., and M. Isshiki, Iron partitioning in a pyrolite mantle and the nature of the 410 km seismic discontinuity, *Nature*, *392*, 702–705, 1998.

Irifune, T., M. Miyashita, T. Inoue, J. Ando, K. Funakoshi, and W. Utsumi, High-pressure phase transformation in CaMgSi₂O₆ and implications for origin of ultra-deep diamond inclusions, *Geophysical Research Letters*, *27*, 3541–3544, 2000.

Irifune, T., et al., The postspinel phase boundary in Mg₂SiO₄ determined by in situ X-ray diffraction, *Science*, *279*, 1698–1700, 1998.

Ito, E., and E. Takahashi, Postspinel transformations in the system Mg₂SiO₄ - Fe₂SiO₄ and some geophysical implications, *Journal of Geophysical Research*, *94*, 10,637–10,646, 1989.

Ito, E., M. Akaogi, L. Topor, and A. Navrotsky, Negative pressure-temperature slopes for reactions forming MgSiO₃ perovskite from calorimetry, *Science*, *249*, 1275–1278, 1990.

Jackson, I., and S. M. Rigden, Analysis of P–V–T data: constraints on the thermoelastic properties of high-pressure minerals, *Physics of the Earth and Planetary Interiors*, *96*, 85–112, 1996.

Jamieson, J. C., J. N. Fritz, and M. H. Manghnani, Pressure measurement at high temperature in X-ray diffraction studies: gold as a primary standard, in *High-pressure research in geophysics*, edited by S. Akimoto and M. H. Manghnani, pp. 27–48, Center for Academic Publications Japan, Tokyo, 1982.

Jeanloz, R., and S. Morris, Temperature distribution in the crust and mantle, *Annual Review of Earth and Planetary Sciences*, *14*, 377–415, 1986.

Jephcoat, A. P., and S. P. Besedin, Temperature measurement and melting determination in the laser-heated diamond-anvil cell, *Philosophical Transactions of the Royal Society of London A*, *354*, 1333–1360, 1996.

Kavner, A., and T. S. Duffy, Pressure-volume-temperature paths in the laser-heated diamond anvil cell, *Journal of Applied Physics*, *89*, 1907–1914, 2001.

Kellogg, L. H., B. H. Hager, and R. D. van der Hilst, Compositional stratification in the deep mantle, *Science*, *283*, 1881–1884, 1999.

Kendall, J.-M., and P. G. Silver, Constraints from seismic anisotropy on the nature of the lowermost mantle, *Nature*, *381*, 409–412, 1996.

Kendall, J.-M., and P. G. Silver, Investigating causes of D" anisotropy, in *Gurnis et al.* [1998], pp. 97–118.

Kesson, S. E., J. D. Fitz Gerald, J. M. G. Shelley, and R. L. Withers, Phase relations, structure and crystal chemistry of some aluminous silicate perovskites, *Earth and Planetary Science Letters*, *134*, 187–201, 1995.

Kim, Y.-H., L. C. Ming, and M. H. Manghnani, High-pressure phase transitions in a natural crystalline diopside and a synthetic CaMgSi₂O₆ glass, *Physics of the Earth and Planetary Interiors*, *83*, 67–79, 1994.

Knittle, E., and R. Jeanloz, Synthesis and equation of state of (Mg,Fe)SiO₃ perovskite to over 100 gigaphascals, *Science*, *235*, 668–670, 1987.

Kuo, B. Y., E. J. Garnero, and T. Lay, Tomographic inversion of S–SKS times for shear velocity heterogeneity in D": Degree 12 and hybrid models, *Journal of Geophysical Research*, *105*, 28,139–28,157, 2000.

Langenhorst, F., and J.-P. Poirier, Anatomy of black veins in Zagami: clues to the formation of high-pressure phases, *Earth and Planetary Science Letters*, *184*, 37–55, 2000.

Larson, A. C., and R. B. V. Dreele, GSAS manual, *Tech. Rep. LAUR 86-748*, Los Alamos National Laboratory, 1988.

Lay, T., and D. V. Helmberger, A lower mantle *S*-wave triplication and the velocity structure of D", *Geophysical Journal of the Royal Astronomical Society*, *75*, 799–837, 1983.

Lay, T., and C. J. Young, Analysis of seismic SV waves in the core's penumbra, *Geophysical Research Letters*, *18*, 1373–1376, 1991.

Lay, T., Q. Williams, and E. J. Garnero, The core-mantle boundary layer and deep Earth dynamics, *Nature*, *392*, 461–468, 1998.

Lebedev, S., S. Chevrot, and R. van der Hilst, Seismic evidence for olivine phase changes at the 410- and 660-kilometer discontinuities, *Science*, *296*, 1300–1302, 2002.

Levander, A., E. Humphreys, G. Ekstrom, A. Meltzer, and P. Shearer, Proposed project would give unprecedented look under North America, *Eos Transactions*, *80*, 245, 1999.

Li, J., V. V. Struzhkin, H.-K. Mao, J. Shu, R. J. Hemley, Y. Fei, B. Mysen, P. Dera, V. Prakapenka, and G. Shen, Electronic spin state of iron in lower mantle perovskite, *Proceedings of the National Academy of Sciences of the United States of America*, *101*, 14,027–14,030, 2004.

Li, X., and R. Jeanloz, Effect of iron content on the electric conductivity of perovskite and magnesiowüstite assemblages at lower mantle conditions, *Journal of Geophysical Research*, *96*, 6113–6120, 1991.

Liu, L., and A. E. Ringwood, Synthesis of a perovskite-type polymorph of CaSiO₃, *Earth and Planetary Science Letters*, *14*, 209–211, 1975.

Liu, L.-G., Post-oxide phases of forsterite and enstatite, *Geophysical Research Letters*, *2*, 417–419, 1975.

Liu, X.-F., and A. M. Dziewonski, Global analysis of shear wave velocity anomalies in the lower-most mantle, in *Gurnis et al.* [1998], pp. 21–36.

Mao, H.-K., R. J. Hemley, Y. Fei, J. F. Shu, L. C. Chen, A. P. Jephcoat, and Y. Wu, Effect of pressure, temperature, and composition on lattice parameters and density of (Fe,Mg)SiO₃-perovskites to 30 GPa, *Journal of Geophysical Research*, *96*, 8069–8079, 1991.

Mao, H.-K., G. Shen, and R. J. Hemley, Multivariable dependence of Fe-Mg partitioning in the lower mantle, *Science*, *278*, 2098–2100, 1997.

Mao, W. L., G. Shen, V. B. Prakapenka, Y. Meng, A. J. Campbell, D. L. Heinz, J. Shu, R. J. Hemley, and H.-K. Mao, Ferromagnesian postperovskite silicates in the D" layer of the earth, *Proceedings of the National Academy of Sciences of the United States of America*, *101*, 15,867–15,869, 2004.

Matsui, M., and N. Shima, Electronic thermal pressure and equation of state of gold at high temperature and high pressure, *Journal of Applied Physics*, *93*, 9679–9682, 2003.

Meade, C., H.-K. Mao, and J. Hu, High-temperature phase transition and dissociation of (Mg,Fe)SiO$_3$ perovskite at lower mantle pressures, *Science*, *268*, 1743–1745, 1995.

Morgan, W. J., Convection plumes in the lower mantle, *Nature*, *230*, 42–43, 1971.

Morishima, H., T. Kato, M. Suto, E. Ohtani, S. Urakawa, W. Utsumi, O. Shimomura, and T. Kikegawa, The phase boundary between α-Mg$_2$SiO$_4$ and β-Mg$_2$SiO$_4$ determined by in-situ X-ray observation, *Science*, *265*, 1202–1203, 1994.

Murakami, M., K. Hirose, S. Ono, and Y. Ohishi, Stability of CaCl$_2$-type and α-PbO$_2$-type SiO$_2$ at high pressure and temperature determined by in-situ X-ray measurements, *Geophysical Research Letters*, *30*, 1207, 2003.

Murakami, M., K. Hirose, K. Kawamura, N. Sata, and Y. Ohishi, Post-perovskite phase transition in MgSiO$_3$, *Science*, *304*, 855–858, 2004.

Murakami, M., K. Hirose, N. Sata, and Y. Ohishi, Post-perovskite phase transition and mineral chemistry in the pyrolitic lowermost mantle, *Geophysical Research Letters*, *32*, L03304, 2005.

Nakagawa, T., and P. J. Tackley, Effects of a perovskite-post perovskite phase change near core-mantle boundary in compressible mantle convection, *Geophysical Research Letters*, *31*, L16,611, 2004.

Oganov, A. R., and S. Ono, Theoretical and experimental evidence for a post-perovskite phase of MgSiO$_3$ in Earth's D″ layer, *Nature*, *430*, 445–448, 2004.

Okube, M., A. Yoshiasa, O. Ohtaka, H. Fukui, Y. Katayama, and W. Utsumi, Anharmonicity of gold under high-pressure and high-temperature, *Solid State Communications*, *121*, 235–239, 2002.

O'Neill, B., and R. Jeanloz, MgSiO$_3$–FeSiO$_3$–Al$_2$O$_3$ in the earth's lower mantle—perovskite and garnet at 1200-km depth, *Journal of Geophysical Research*, *99*, 19,901–19,915, 1994.

Peyronneau, J., and J. P. Poirier, Electrical conductivity of the earth's lower mantle, *Nature*, *342*, 537–539, 1989.

Ringwood, A. E., *Composition and petrology of the Earth's mantle*, McGraw-Hill, New York, 1975.

Ross, M., H.-K. Mao, P. M. Bell, and J. A. Xu, The equation of state of dense argon: a comparison of shock and static studies, *Journal of Chemical Physics*, *85*, 1028–1033, 1986.

Ross, N. L., and R. M. Hazen, High-pressure crystal chemistry of MgSiO$_3$ perovskite, *Physics and Chemistry of Minerals*, *17*, 228–237, 1990.

Saxena, S. K., G. Shen, and P. Lazor, Temperatures in Earth's core based on melting and phase transformation experiments on iron, *Science*, *264*, 405–407, 1994.

Saxena, S. K., L. S. Dubrovinsky, P. Lazor, Y. Cerenius, P. Häggkvist, M. Hanfland, and J. Hu, Stability of perovskite (MgSiO$_3$) in the Earth's mantle, *Science*, *274*, 1357, 1996.

Saxena, S. K., L. S. Dubrovinsky, P. Lazor, and J. Z. Hu, In situ X-ray study of perovskite (MgSiO$_3$): phase transition and dissociation at mantle conditions, *European Journal of Mineralogy*, *10*, 1275–1281, 1998.

Serghiou, G., A. Zerr, and R. Boehler, (Mg,Fe)SiO$_3$-perovskite stability under lower mantle conditions, *Science*, *280*, 2093–2095, 1998.

Serghiou, G., A. Zerr, and R. Boehler, (Mg,Fe)SiO$_3$-perovskite stability and lower mantle conditions: response, *Science*, *285*, 983a, 1999.

Shen, G., H.-K. Mao, R. J. Hemley, T. S. Duffy, and M. L. Rivers, Melting and crystal structure of iron at high pressures and temperatures, *Geophysical Research Letters*, *25*, 373–377, 1998.

Shen, G., M. L. Rivers, Y. Wang, and S. R. Sutton, Laser heated diamond cell system at the Advanced Photon Source for in situ x-ray measurements at high pressure and temperature, *Review of Scientific Instruments*, *72*, 1273–1282, 2001.

Shieh, S. R., T. S. Duffy, and G. Shen, X-ray diffraction study of phase stability in SiO$_2$ at deep mantle conditions, *Earth and Planetary Science Letters*, *235*, 273–282, 2005.

Shim, S.-H., T. S. Duffy, and G. Shen, The stability and P–V–T equation of state for CaSiO$_3$ perovskite in the earth's lower mantle, *Journal of Geophysical Research*, *105*, 25,955–25,968, 2000.

Shim, S.-H., T. S. Duffy, and G. Shen, The post-spinel transformation in Mg$_2$SiO$_4$ and its relation to the 660-km seismic discontinuity, *Nature*, *411*, 571–574, 2001a.

Shim, S.-H., T. S. Duffy, and G. Shen, Stability and structure of MgSiO$_3$ perovskite to 2300-km depth conditions, *Science*, *293*, 2437–2440, 2001b.

Shim, S.-H., T. S. Duffy, and T. Kenichi, Equation of state of gold and its application to the phase boundaries near the 660-km depth in the mantle, *Earth and Planetary Science Letters*, *203*, 729–739, 2002.

Shim, S.-H., T. S. Duffy, R. Jeanloz, and G. Shen, Stability and crystal structure of MgSiO$_3$ perovskite to the core-mantle boundary, *Geophysical Research Letters*, *31*, L10,603, 2004.

Sidorin, I., M. Gurnis, and D. V. Helmberger, Evidence for a ubiquitous seismic discontinuity at the base of the mantle, *Science*, *286*, 1326–1331, 1999.

Speziale, S., C.-S. Zha, T. S. Duffy, R. J. Hemley, and H.-K. Mao, Quasi-hydrostatic compression of magnesium oxide to 52 GPa: implications for the pressure–volume–temperature equations of state, *Journal of Geophysical Research*, *106*, 515–528, 2001.

Stacey, F. D., and D. E. Loper, The thermal boundary-layer interpretation of D″ and its role as a plume source, *Physics of the Earth and Planetary Interiors*, *33*, 45–55, 1983.

Stackhouse, S., J. P. Brodholt, J. Wookey, J.-M. Kendall, and G. D. Price, The effect of temperature on the seismic anisotropy of the perovskite and post-perovskite polymorphs of MgSiO$_3$, *Earth and Planetary Science Letters*, *230*, 1–10, 2005.

Tackley, P. J., D. J. Stevenson, G. A. Glatzmaier, and G. Schubert, Effects of an endothermic phase transition at 670-km depth in a spherical model of convection in the Earth's mantle, *Nature*, *361*, 699–704, 1993.

Taylor, S. R., and S. M. McLennan, *The continental crust—Its composition and evolution—An examination of the geochemical record preserved in sedimentary rocks*, Blackwell Scientific, Oxford, 1985.

Teter, D. M., R. J. Hemley, G. Kresse, and J. Hafner, High pressure polymorphism in silica, *Physical Review Letters*, *80*, 2145–2148, 1998.

Thomas, C., E. J. Garnero, and T. Lay, High-resolution imaging of lowermost mantle structure under the Cocos plate, *Journal of Geophysical Research*, *109*, B08,307, 2004a.

Thomas, C., J. M. Kendall, and J. Lowman, Lower-mantle seismic discontinuities and the thermal morphology of subducted slabs, *Earth and Planetary Science Letters*, *225*, 105–113, 2004b.

Tomioka, N., and K. Fujino, Natural (Mg,Fe)SiO₃-ilmenite and -perovskite in the Tenham meteorite, *Science*, *277*, 1084–1086, 1997.

Tomioka, N., and M. Kimura, The breakdown of diopside to Ca-rich majorite and glass in a shocked H chondrite, *Earth and Planetary Science Letters*, *208*, 271–278, 2003.

Trampert, J., F. Deschamps, J. Resovsky, and D. Yuen, Probabilistic tomography maps chemical heterogeneities throughout the lower mantle, *Science*, *306*, 853–856, 2004.

Tsuchiya, T., First-principles prediction of the *P–V–T* equation of state of gold and the 660-km discontinuity in Earth's mantle, *Journal of Geophysical Research*, *108*, 2462, 2003.

Tsuchiya, T., J. Tsuchiya, K. Umemoto, and R. M. Wentzcovitch, Phase transition in MgSiO₃ perovskite in the earth's lower mantle, *Earth and Planetary Science Letters*, *224*, 241–248, 2004a.

Tsuchiya, T., J. Tsuchiya, K. Umemoto, and R. M. Wentzcovitch, Elasticity of post-perovskite MgSiO₃, *Geophysical Research Letters*, *31*, L14,603, 2004b.

van der Hilst, R. D., and H. Kárason, Compositional heterogeneity in the bottom 1000 kilometers of Earth's mantle: toward a hybrid convection model, *Science*, *283*, 1885–1888, 1999.

van der Hilst, R. D., S. Widiyantoro, and E. R. Engdahl, Evidence for deep mantle circulation from global tomography, *Nature*, *386*, 578–584, 1997.

Watanuki, T., O. Shimomura, T. Yagi, T. Kondo, and M. Isshiki, Contruction of laser-heated diamond anvil cell system for in situ X-ray diffraction study at SPring-8, *Review of Scientific Instruments*, *72*, 1289–1292, 2001.

Williams, Q., R. Jeanloz, J. Bass, B. Svendsen, and T. J. Ahrens, The melting curve of iron to 250 gigapascals: a constraint on the temperature at Earth's center, *Science*, *236*, 181–182, 1987.

Williams, Q., J. Revenaugh, and E. Garnero, A correlation between ultra-low basal velocities in the mantle and hot spot, *Science*, *280*, 546–549, 1998.

Woodward, P. M., Octahedral tilting in perovskites. I. geometrical consideration, *Acta Crystallographica*, *B53*, 32–43, 1997.

Wysession, M. E., T. Lay, J. Revenaugh, Q. Williams, E. J. Garnero, R. Jeanloz, and L. H. Kellogg, The D″ discontinuity and its implications, in Gurnis et al., *The core-mantle boundary region*, American Geophysical Union, Washington, DC, 1998, pp. 273–297.

Yagi, T., T. Kondo, T. Watanuki, O. Shimomura, and T. Kikegawa, Laser heated diamond anvil apparatus at the Photon Factory and SPring-8, *Review of Scientific Instruments*, *72*, 1293–1297, 2001.

S.-H. Shim, Department of Earth, Atmospheric, and Planetary Sciences, Massachusetts Institute of Technology, 77 Massachusetts Avenue, 54-514, Cambridge, Massachusetts 02139, USA. (sangshim@mit.edu)

Synthetic Tomographic Images of Slabs
From Mineral Physics

Y. Ricard, E. Mattern, and J. Matas

Laboratoire des Sciences de la Terre, UMR CNRS 5570, Lyon, France

The mantle structures observed by seismic tomography can only be linked with convection models by assuming some relationships between temperature, density, and velocity. These relationships are complex and nonlinear even if the whole mantle has a uniform composition. For example, the density variations are related not only to the depth dependent thermal expansivity and incompressibility, but also to the distribution of the mineralogical phases which are themselves evolving with temperature and pressure. In this paper, we present a stoichiometric iterative method to compute the equilibrium mineralogy of mantle assemblages by Gibbs energy minimization. The numerical code can handle arbitrary elemental composition in the system MgO, FeO, CaO, Al_2O_3, and SiO_2 and reaches the thermodynamic equilibrium by choosing the abundances of 29 minerals belonging to 13 possible phases. The code can deal with complex chemical activities for minerals belonging to solid state solutions. We illustrate our approach by computing the phase diagrams of various compositions with geodynamical interest (pyrolite, harzburgite, and oceanic basalt). Our simulations are in reasonable agreement with high-pressure and high-temperature experiments. We predict that subducted oceanic crust remains significantly denser than normal mantle, even near the core–mantle boundary. We then provide synthetic tomographic models of slabs. We show that properties computed at thermodynamic equilibrium are significantly different from those computed at fixed mineralogy. We quantify the three potential contributions of the seismic anomalies (intrinsic thermal effect, changes in mineralogy induced by temperature variations, changes in the bulk composition) and show that they are of comparable magnitudes. Although the accuracy of our results is limited by the uncertainties on the thermodynamic parameters and equations of state of each individual mineral, future geodynamical models will need to include these mineralogical aspects to interpret the tomographic results as well as to explain the geochemical observations.

1. TOMOGRAPHIC IMAGES OF SLABS

For the last 20 years the observation by seismic tomography of the large scale structure of the mantle has provided a fantastic tool to test, sometimes to contradict but often to confirm, the findings of geodynamical studies. It has also pointed to the necessity to go beyond simple qualitative agreements and have a closer look to mineralogical data that links temperature or density (the usual parameters of convective models) to velocity anomalies.

The general agreement on convection of fluids heated largely by internal radioactivity is that the regions of litho-

Earth's Deep Mantle: Structure, Composition, and Evolution
Geophysical Monograph Series 160
Copyright 2005 by the American Geophysical Union
10.1029/160GM17

spheric downwellings (the subduction zones) should be underlaid by cold descending plumes (the slabs). These structures should dominate the mantle heterogeneity structure whereas hot ascending plumes (presumably the hot-spots) should have a less important role [*Davies and Richards*, 1992; *Bercovici et al.*, 2000]. Indeed, the slab flow must extract the radioactive heat production plus the mantle and core secular cooling while the hot spot flow only carries the core cooling [*Davies*, 1988].

This view results in a testable consequence that the present-day mantle structure should, at first order, be the result of past subductions [*Richards and Engebretson*, 1992]. Starting from a compilation of plate reconstruction models spanning the last 200 Myrs [*Lithgow-Bertelloni et al.*, 1993], *Ricard et al.* [1993] compute the position that subducted slabs could have in the mantle. Their very simple model assume that each piece of slab sinks vertically and that the slab excess density is conserved through the whole mantle.

The quality of this geodynamical model based on paleo-plate reconstructions, lies in its robustness. Indeed there is basically no adjustable parameters; the density of the lithosphere at subduction is well known (i.e., *Turcotte and Schubert* [1982]) and the vertical velocity reduction at 660 km depth is easy to estimate. As large scale geoid modeling suggests a viscosity increase by a factor of 30–100 at 660 km depth, average mantle velocity should be decreased in the lower mantle by a factor of 3–5 [*Richards*, 1991].

The resulting density model provides a remarkable fit to the Earth's gravity field [*Ricard et al.*, 1993]. It also compares favorably to tomographic models as can be illustrated with the synthetic S model ("Smean model" of *Becker and Boschi* [2002]), which is a weighted average of previously published models [*Grand et al.*, 1997; *Ritsema and van Heijst*, 2000; *Masters et al.*, 2000]. The results are similar when other tomographic models are used but we hope that the synthetic Smean model emphasizes the common structures found by various tomographic approaches.

Plate 1 depicts the significance levels of the correlations between the slab and the tomographic models (this statistical quantity is an objective measure of the meaningfulness of the correlations). The horizontal axis is the spherical harmonic degree l of the heterogeneity (the inverse wavelength is ~20000/l in km). The vertical axis is depth. Only significance levels larger than 50% are plotted (i.e., when there is more than a 50% chance that the correlation is meaningful, the red colors indicate that the correlations are meaningful at 90% level). At degrees 2 and 3 there is a very good correlation between the subduction and the tomographic models. At higher degrees, the correlations are restricted to the upper half of the mantle but the upper-lower mantle transition is not associated with a significant decrease of these correlations. To precisely compute the correlations at high degree in the deep lower mantle would require a knowledge of the positions and velocities of Mesozoic subductions that we do not have.

The reader may or may not consider this plot to be a convincing indication that sinking slabs are the major ingredient of mantle anomalies. Taking into account the uncertainties of seismic topography and even more the extreme simplicity of the geodynamical model (neglecting heat diffusion, slab lateral advection, the necessary existence of hot plumes…) this first order correlation seems to confirm the fact that tomographic models are in agreement with geodynamical models based on mostly internally heated fluid dynamics, i.e. with a mantle dynamics dominated by cold plumes.

This overall correlation between seismic models and geodynamical models has also been discussed on a more regional scales (e.g., *van der Voo et al.* [1999]). These correlations, however, do not address the problem of the respective amplitudes of density and velocity anomalies. They do not even indicate that the observed velocity anomalies are of thermal rather than petrological origins.

In Plate 2, we depict a tomographic P wave model across Japan [*Karason and van der Hilst*, 2000]. In the top panel of Plate 2, this model has been plotted using the same color scale as in *Fukao et al.* [1992]. The lower panel of Plate 2 represents the exact same data set but with the color scale of *Bijwaard and Spakman* [2000]. Basically along the same great circle, these authors have published, with their own models, very similar results which indicate the good general agreement between models. Although Plate 2 may suggest slab penetrations, with significant thickening/deformation in the lower mantle in agreement with geodynamical models, the very significant decrease of the anomalies in the lower mantle must be explained. This can only be done after a closer look to temperature/velocity/density relationships and therefore to mineralogy. Otherwise the interpretation of tomographic images in term of non-penetration or straight penetration is affected by the non-objective choice of the color scale as illustrated in Plate 2.

In this paper we discuss the amplitudes of the density and velocity anomalies that may be expected for a sinking slab. To do this exercise, we first discuss the possible mineralogical composition at depth of a realistic slab, then the relationships between these mineralogical compositions and the associated thermal and chemical anomalies. We only address the direct problem, i.e. we discuss how the mineral physics can be used to predict realistic seismic anomalies. We analyze and evaluate the importance of the three fundamental contributions to lateral variations; thermal, mineralogical and petrological. We do not attempt to formulate the more complex inverse problem and do not try interpret any specific tomographic model in terms of temperature and compositional variations.

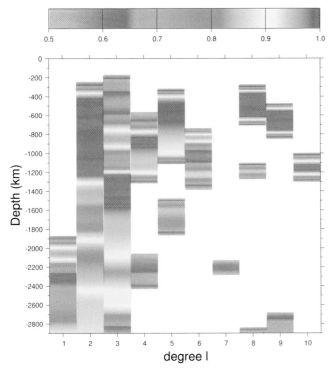

Plate 1. Comparison between a geodynamical model of Cenozoic and Mesozoic slab subduction and a tomographic model. The color panel represents, as a function of depth and harmonic degree, the significance level of the correlation between a slab model [*Ricard et al.*, 1993] and a synthetic tomographic Smean model [*Becker and Boschi*, 2002]. Only significance levels larger than 0.5 are depicted.

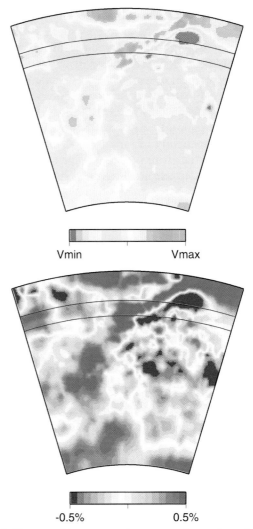

Plate 2. Tomographic cross-sections across Japan through the P-wave tomographic model of *Karason and van der Hilst* [2000]. The top map uses the same color scale as *Fukao et al.* [1992] and may suggest a stagnant slab lying on the upper–lower mantle interface. Using a color scale that saturates [*Bijwaard and Spakman*, 2000], the bottom map favors the interpretation of easy slab penetration.

2. MINERALOGICAL COMPOSITION OF THE MANTLE

The total elemental composition of the Earth can be esti-mated form geochemical considerations (e.g., *Hart and Zindler* [1989]). The direct natural samples (ophiolites, oro-genic peridotites, xenoliths) only represent the uppermost mantle (first 200 km). There are no samples of the lower mantle, the only information on its composition and structure comes from analyzes of propagation of seismic waves and of other geophysical observables (free oscillations, geoid modelling…). The mineralogical composition of the deeper mantle is thus obtained by indirect methods. Knowing the elastic properties of different minerals, it is indeed possible to test the compatibility of various bulk compositions and temperature profiles with various seismic and geophysical observations. An inverse problem can then be formulated to derive the most probable composition by fitting the seismic radial model (e.g. *Mattern et al.* [2005]) or the tomographic lateral variations (e.g. *Trampert et al.* [2004]). The elastic properties of each mantle minerals and the possible min-eralogical assemblages that need to be considered, can be approached by different ways:
  –experimental in-situ observations of high-pressure and high-temperature assemblages,
  –purely theoretical ab initio computations,
  –equilibrium thermodynamic computations based on single mineral properties either obtained from observations (e.g., calorimetry, Raman and Brillouin spectroscopy, high-P and high-T X-ray diffraction) or from theoretical models (e.g., vibrational approach, ab initio calculations).
Each method has some advantages and some limitations. Purely experimental works are providing the most important informations. However for practical reasons, they are limited to relatively simple mineralogical assemblages. Some of the major high-pressure mantle phases are not quenchable (e.g., $CaSiO_3$-perovskite) and the simultaneous HP-HT in situ analysis are still technically very difficult. The control of the sample oxidation state as well as the P-T conditions, is also another difficulty.

The ab initio calculation (i.e. directly solving a quantum mechanics problem) is very powerful since a precise control of pressure, temperature and oxidation conditions associated with the numerical experiment, is possible. The approach is however limited by various approximations needed to implement the quantum theory and by the computational cost necessary for the simulation of realistic mantle assemblages. It is usually used for single minerals and for a limited number of binary solid solutions (e.g., *Bukowinski and Akber-Knutson*, this volume).

The classical thermodynamic modeling stands in between the two previous methods. It combines fundamental thermody-namics (first and second principles) with accurate experimental observations of single mineral physical properties. It considers that the convecting Earth's mantle is locally in thermodynamic equilibrium. The proportions of minerals can therefore be cal-culated by a free-energy minimization technique and the global mantle properties are then obtained using the single mineral properties weighted by mineral proportions.

The assumption of thermodynamic equilibrium is question-able at least in the coldest part of subducting slabs [*Kirby et al.*, 1996]. A complete kinetic modeling of the mantle mineralogy is presently out of reach. Only a few kinetic reactions (e.g., between forsterite and Mg-wadsleyite) are reasonably well documented for a numerical modeling [*Kerschhofer et al.*, 2000; *Mosenfelder et al.*, 2000]. We are aware of the possible limitations of the thermodynamic approach that we use.

Over the past few decades several methods and techniques of equilibrium phase calculations have been proposed. In the early seventies, *Barron* [1968] has designed a computer pro-gram (GAPMIS) to determine phase equilibria in binary and ternary silicate systems. This code was designed to deal with complex non-ideal solid solutions in order to model various immiscibility gaps. It was however restricted to ambient pres-sure. The phase equilibrium calculations have been largely developed for low pressure igneous petrology. These studies were started by *Bottinga and Richet* [1978], *Ghiorso and Carmichael* [1980], *Berman and Brown* [1985]. A sophisti-cated computer code MELTS calculating phase equilibrium in silicate liquid systems was fine-tuned by Ghiorso in the nineties (see *Ghiorso* [1997], for a review).

The thermodynamic modeling of solid silicate systems under high pressures (i.e., at depths exceeding 300 km) has been developed by *Saxena and Eriksson* [1983], *Wood and Holloway* [1984] and *Bina and Wood* [1986]. The first stud-ies were limited to rather simple systems aiming essentially to study details of the 410-km discontinuity. The computa-tions in the MgO-FeO-SiO_2 multicomponent system were further extended by works by *Wood* [1987], *Kuskov and Panferov* [1991], *Fabrichnaya and Kuskov* [1991], *Saxena et al.* [1993], *Saxena* [1996].

Ita and Stixrude [1992] have considered realistic mantle compositions including CaO and Al_2O_3 components. However, the phase determination was discussed separately for the oliv-ine and the residual pyroxene-garnet parts. Their phase dia-grams were in reasonable agreement with experimental studies. However, having two separated subsystems prevent from modeling the details of 410-km and 660-km discontinuity. The interaction between olivine polymorphs, garnet and pyroxene around the depth of 410 km was carefully studied by *Stixrude* [1997] by analytical computational methods. The only com-plete free-energy minimization code for mantle like mineralo-gies was presented by *Sobolev and Babeyko* [1994].

The phase diagrams by *Ita and Stixrude* [1992] have been used in various later studies (e.g. *Vacher et al.* [1998], *Goes et al.* [2004]) as the baseline for the interpretation of observed lateral seismic anomalies. Such approach does not necessarily ensure the condition of thermodynamic equilibrium.

3. GIBBS MINIMIZATION

3.1. Basic Concepts

The Earth mantle mineralogical system can be represented by a collection of K simple oxides. We take into consideration MgO, FeO, CaO, Al_2O_3, SiO_2 labeled by the indices k, from 1 to $K = 5$. These five oxides allows us to describe 99 wt% of a peridotitic composition and about 95 wt% of an oceanic crust composition. Our mineralogical system does not include alcalins (Na, K,...), various transition metals, (Ti, Cr, Ni,...), ferric iron, hydrous phases, carbonates, metallic alloys, for which a complete set of thermodynamic parameters is unavailable. These elements can however be introduced for more specific studies [*Matas et al.*, 2000a, b; *Fiquet et al.*, 2002]. The oxides are distributed into J possible phases consisting in solid state solutions of I simple minerals also called end-members. In this paper we consider $I = 29$ minerals belonging to $J = 13$ phases (see Table 1).

Complex phases can include up to 5 simple end-members. The magnesium (index $k = 1$), with molar abundance $b_{Mg} = b_1$, is present in 11 different minerals, for example in forsterite, Mg_2SiO_4, an end-member with index i, belonging to the phase $\phi(i)$, olivine, $(Mg,Fe)_2SiO_4$. An end-member i, with molar abundance n_i, contains s_{ik} moles of the oxide k (e.g., Mg_2SiO_4 contains $s_{i1} = 2$ moles of MgO). We call S the matrix with coefficients s_{ik} of the linear operator that relates the end-member abundance (a space of dimension I) to the elemental abundances (a space of dimension K). Total elemental conservation simply writes

$$\sum_{i=1}^{I} s_{ik} n_i = b_k , \qquad (1)$$

where, of course, the molar abundances are positive

$$n_i \geq 0 \quad \forall i \in 1..I . \qquad (2)$$

The choice of minerals and phases shown in Table 1 is the result of a compromise between introducing the largest possible number of mantle minerals and avoiding minerals with too poorly known thermo-chemical parameters. This choice does not allow us to reproduce all minor complexities of the mantle mineralogy. For example, our choice of mineral species cannot account for the incorporation in small proportions

Table 1. Mineralogical Phases and Their End-Members. Each line gives the name, the crystallographic formula of an independent phase, and the names and chemical formulae of the end-members.

Phase, formula	Name	End-member
Olivine (α), $(Mg,Fe)_2SiO_4$	Forsterite	Mg_2SiO_4
	Fayalite	Fe_2SiO_4
Wadsleyite (β), $(Mg,Fe)_2SiO_4$	Mg-Wadsleyite	Mg_2SiO_4
	Fe-Wadsleyite	Fe_2SiO_4
Ringwoodite (γ), $(Mg,Fe)_2SiO_4$	Mg-Ringwoodite	Mg_2SiO_4
	Fe-Ringwoodite	Fe_2SiO_4
Magnesiowustite, $(Mg,Fe)O$	Periclase	MgO
	Wustite	FeO
Perovskite, (Mg,Fe,Al) $(Al,Si)O_3$	Mg-Perovskite	$MgSiO_3$
	Fe-Perovskite	$FeSiO_3$
	Al-Perovskite	Al_2O_3
Akimotoite, $(Mg,Fe)SiO_3$	Mg-Akimotoite	$MgSiO_3$
	Fe-Akimotoite	$FeSiO_3$
Orthopyroxene, $(Mg,Fe)SiO_3$	Orthoenstatite	$MgSiO_3$
	Orthoferrosilite	$FeSiO_3$
Clinopyroxene, (Ca,Mg,Fe) (Mg,Fe,Al) $(Al,Si)_2O_6$	Diopside	$CaMgSi_2O_6$
	Hedenbergite	$CaFeSi_2O_6$
	Ca-Tschermak	$CaAl_2SiO_6$
	Clinoenstatite	$Mg_2Si_2O_6$
	Clinoferrosilite	$Fe_2Si_2O_6$
Garnet, $(Ca,Mg,Fe)_3$ $(Mg,Fe,Al,Si)_2$ Si_3O_{12}	Pyrope	$Mg_3Al_2Si_3O_{12}$
	Almandin	$Fe_3Al_2Si_3O_{12}$
	Grossulaire	$Ca_3Al_2Si_3O_{12}$
	Mg-Majorite	$Mg_4Si_4O_{12}$
	Fe-Majorite	$Fe_4Si_4O_{12}$
Al-φ	Corundum	Al_2O_3
Coesite	Coesite	SiO_2
Stishovite	Stishovite	SiO_2
Ca-Perovskite	Ca-Perovskite	$CaSiO_3$

of Al in stishovite or $CaSiO_3$ perovskite, contrary to what is experimentally observed (e.g., *Ono et al.* [2001]; *Takafuji et al.* [2002]). In the absence of a complete set of parameters (i.e. the equation of state and a complete set of calorimetric parameters) the perovskite to post-perovskite phase change and high-spin/low-spin transition that could occur in the deep lower mantle are also not included [*Murakami et al.*, 2004; *Badro et al.*, 2003, 2004].

The Gibbs free energy G is a function of pressure, P, temperature, T and mineralogical composition:

$$G(P,T,n_i) = \sum_{i=1}^{I} n_i(P,T) \, \mu_i(P,T,n_j, \forall j \in \phi(i)) , \quad (3)$$

where μ_i are the chemical potentials of end-members obtained as:

$$\mu_i(P, T, n_j, \forall j \in \phi(i)) =$$

$$\mu_i^0(P, T) + RT \ln a_i(P, T, n_j, \forall j \in \phi(i)), \quad (4)$$

where μ_i^0 is the standard chemical potential of end-member i. The chemical activity a_i is a function of the abundances n_j of all end-members j belonging to the same phase, $\phi(i)$. The equilibrium composition at given pressure and temperature is obtained by minimizing the Gibbs energy (3), subject to constraints (1) and (2).

The numerical difficulty of solving this minimization problem comes from the fact that it is highly non-linear. For example for end-members in small quantities the activities are generally close to zero. This implies that their chemical potential that includes a logarithm 10 goes discontinuously from μ0 (when the phase is absent) to $-\infty$ (when the phase is in small proportions).

3.2. Standard Chemical Potentials and Activities

The standard chemical potentials are easily computed from the thermodynamic and elastic properties: enthalpy of formation, ΔH_f^0, entropy, S_0, molar volume, V_0, room pressure heat capacity, thermal expansion and isothermal incompressibility, $C_p(T)$, $\alpha(T)$ and $K_{T,0}(T)$, pressure dependence of the incompressibility, $K'_{T,0}$, and an equation of state (we use the classical 3rd order finite strain equation of state [Birch, 1952]). In order to compute realistic phase diagrams, the set of parameters must be thermodynamically self-consistent (i.e. obey the Maxwell identities with equalities of the second-order cross-derivatives) and also be in agreement with all accurate observations.

The data selection is a difficult and tedious task (see discussion on lower mantle minerals in Mattern et al. [2005]). For many minerals we had to come back to the original density measurements and fit them in order to obtain the complete set of parameters: K_T, K', $\partial K_T / \partial T$, $\alpha(T)$, etc. Indeed, using together parameters bluntly taken from various experimental studies quite systematically fails (i.e. using a molar volume from a given study and a compressibility from a second that did not use the same molar volume leads to thermodynamic inconsistencies). The calorimetry parameters were then refined by comparing the predicted univariant phase diagrams with accurate experimental observations. When adding any new element and solid solution, the predictions were checked as carefully as possible with available experimental data on phase diagrams. The table with all thermodynamic and elastic parameters used in this study is available on demand to the authors. Our choice of parameters can certainly, and should, be continu-

ously improved by new experiments. As demonstrated below, it however allows us to reproduce the major characteristics of various bulk compositions discussed in the present paper.

For minerals containing aluminum, a special care is taken for their structural and thermodynamic properties at high pressure. In the lower mantle, Al can be stored in a ortho-rhombic perovskite solid solution as well as in an exsolved Al-rich phase (for high Al concentrations like in mid oceanic ridge basalt, MORB). Several potential high-pressure phases have been proposed for the exsolved Al, with calcium-ferrite or hollandite-type structures, or with a new hexagonal phase (NAL) (see e.g., Fiquet [2001] for a detailed review). Some experiments on garnetite assemblage [Irifune et al., 1996; Kubo and Akaogi, 2000; Hirose et al., 2001] suggest that Al_2O_3 with corundum structure could be the potential host for the exsolved Al in the lower mantle. Due to the lack of precise thermodynamic data and equation of state parameters on the high pressure Al-rich phases, we choose the corundum phase as the candidate with the best known properties to evaluate the major effects associated with a possible Al exsolution.

In addition, for each mineralogical phase a mixing model is also required. In this paper, for binary solid state solutions, empirical activities with Margules coefficients are used [Fei et al., 1991]. For more complex assemblages, due to the lack of experimental data, we adopt an ideal mixing-on-site model [Spear, 1995]. This assumes that cations are randomly distributed on the crystallographic sites (as an example, in perovskite the dodecaedric site is randomly occupied by a Mg, Fe or Al cations, the octaedric site by Al or Si cations). This mixing model does not assume electro-neutrality at the level of the crystallographic unit cell (i.e., a Al^{3+} cation can replace a Mg^{2+} cation). All these activities (empirical or theoretical) take the forms of (often cumbersome) functions of n_j, $\forall j \in \phi(i)$. The choice of ideal mixing-on-site activities cannot account for all the mineralogical observations. For example, the demixion of clinopyroxenes into low-Ca and high-Ca components cannot be reproduced by our model, that therefore somewhat underestimates the stability of complex pyroxenes (especially for Al and Ca-rich compositions).

3.3. The Stoichiometric Algorithm

Our numerical code is adapted from a stoichiometric algorithm proposed by Smith and Missen [1991]. The input is a list of possible mineralogical phases and their corresponding end-members (see Table 1). We assume that the standard potential of all the end-members and their activities are known functions of the variables P, T and n_i.

The iterative procedure is initiated from a given elementary composition and an arbitrary starting composition, n_i^0 that verifies (1). Here, as the Earth mantle is represented by $I = 29$

different components (minerals) and $K = 5$ distinct chemical elements (five different oxides), there are $M = 29 - 5 = 24$ independent reactions that describe all possible evolutions of the system. Let us call ν_{im} the coefficients of these M vectors than span the kernel of S (equation 1). Any composition of the form

$$n_i = n_i^0 - \sum_{m=1}^{I-K} \xi_m \nu_{im}, \qquad (5)$$

satisfies the element conservation requirement. These coefficients ν_{im} are the stoichiometric coefficients of the end-members i in one of the M possible reactions (the stoichiometric coefficients of the equation $A \rightleftharpoons 2B$ are 1 and -2). From the definition of the chemical potentials and from the equation (5), we see that the coefficients ξ_m control the advancement of the M reactions as

$$\frac{\partial G}{\partial \xi_m} = \sum_{i=1}^{I} \frac{\partial G}{\partial n_i} \frac{\partial n_i}{\partial \xi_m} = - \sum_{i=1}^{I} \mu_i \nu_{im}. \qquad (6)$$

The M right-hand members are the affinities of the $I - K$ possible reactions. They are zero at equilibrium. To reach the minimum of G, we therefore advance the reactions according to these affinities. As each end-member i can be present in more than one chemical reaction, its total increment δn_i is therefore proportional to

$$\delta n_i = - \sum_{m=1}^{I-K} \nu_{im} \sum_{j=1}^{I} \mu_j \nu_{jm}. \qquad (7)$$

This advancement procedure (i.e., choosing the M independent reactions, computing their affinities and optimizing the reaction increments in the directions of δn_i) is done iteratively until all affinities are zero.

3.4. Numerical Difficulties

Although the previously described method may seem easy to implement, it is on the contrary a real numerical challenge. Without entering too much into technical details we want to summarize a few important tricks that have to be introduced in order to deal with the complex mantle mineralogies.

3.4.1. Component ordering. The choice of the independent reactions is not unique as any linear combination of such reactions is also acceptable. In the set of chemical equations, the left-hand member can be considered as a source and the right-hand members as products. Because of the positivity

constraint, equation (2), the advancement of each reaction is therefore limited by the molar abundance of the left-hand members (the sources). For an optimal iteration step the program chooses the set of independent chemical reactions that preferentially puts the most abundant end-members on the left side of these reactions.

3.4.2. Treatment of minor end-members of a phase. It sometimes happens that a phase of non-negligible abundance contains an end-member in a very small quantity. If this minor end-member stays on the source side of a chemical reaction the reaction increment decreases significantly. Below a certain threshold, the global properties are no more affected by the presence of a minor end-member. However it is generally not possible to simply assume that its abundance is zero because in that case, its potential (and sometimes those of the other end-members of the same phase) would jump to $-\infty$. The erased end-member would thus be recreated immediately in the next iterative step. In order to avoid such problems, at each iterations, we identify the reactions containing minor phases as sources and their affinities are arbitrarily and momentarily set to zero.

3.4.3. Phase introduction and phase elimination. At the beginning or at the end of a phase transformation, special care has to be taken in order to correctly eliminate and/or introduce the appropriate mineralogical phases. We eliminate a whole mineralogical phase if all its end-members are in very small quantities and if their reaction increments are all negative.

The problem of a phase introduction is more complicated. When a phase is not present the activities of its end-members are not yet defined. In other terms, for an absent phase, we only know the μ_i^0 not the μ_i that enters equation 7. At each iteration and for each absent phase, ϕ, the program must therefore find the virtual chemical potentials, μ_i^v, of all end-members i, that would possibly imply the stability of this phase. Then if the set of equations (see equation (4)),

$$\mu_i^v = \mu_i^0(P,T) + RT \ln a_i(P,T,n_j, \forall j \in \phi(i)), \qquad (8)$$

has a solution with $n_j \geq 0$ for all the end-members of the absent phase, then all these end-members must be simultaneously introduced in infinitesimal quantities proportional to n_j.

3.4.4. Elemental conservation. When a new phase is introduced or another eliminated, the global elemental composition has to remain constant. A slight correction of the abundances of major end-members is necessary.

4. VALIDATION OF THE METHOD

The predictions of our numerical code were first systematically tested with simple univariant, then bivariant phase transitions *Matas* [1999]. This was indeed the method to check and select our thermodynamic data set. Due to the non-linear behavior of the activity coefficients, it is not possible to formally prove that in the general cases with complex non-ideal solid solutions, the code reaches the absolute minimum rather than a secondary local minimum of the free-energy surface. We have however performed several tests varying the starting end-member composition at constant elementary composition (e.g. starting from MgO+SiO$_2$, oxide+coesite, or from MgSiO$_3$, perovskite, or from 1/2 Mg$_2$SiO$_4$ +1/2SiO$_2$, forsterite+stishovite,…) always gave the same final composition. The fact that the shape of the Gibbs minimum seems smooth is due to our simplified activity-composition relationships. A more refined modeling should in fact have local minima that are responsible for the immiscibility gaps in Ca- and Al-rich plagioclase or in clinopyroxene phases.

In order to validate our method for the mantle-like mineralogies, we first discuss the mineralogical and seismologic predictions for various bulk compositions along a given geotherm. We then consider compositions characteristic of a subducting slab.

Underlying the mid oceanic ridge basaltic crust there is a continuous range of depletion from the harzburgite layer to the lherzolite and the pyrolite [*Ringwood*, 1991]. In this paper, we only compute the phase relationships and bulk physical properties in pyrolite, harzburgite and MORB. Their molar fractions in the different oxides are summarized in Table 2.

Compared to pyrolite, harzburgite is characterized by a lower content in incompatible elements Al and Ca (less than 2%) and a higher Mg/Si ratio (increased by 20%) due to higher incompatibility of silicon. In MORB composition, SiO$_2$ represents over 50% of the whole molar budget whereas MgO only 15%. The amount of aluminum and calcium in basalt increases fivefold compared to pyrolite.

4.1. Mineral Proportions Along a Geotherm

The computed phase diagrams (in wt%) for three mantle assemblages, pyrolite, harzburgite and MORB, are shown in Plate 3 as a function of depth (vertical axis) (the depth-pressure relationship is taken from the PREM model). They are only a simplified summary of the results of our numerical code as the proportions of end-members of each phase are depth dependent. These diagrams are computed along a linear temperature profile with a gradient of 0.3 K km^{-1} crossing the upper-lower mantle interface at 1600°C. Our code can also compute the mineralogy along a self-consistent adiabat (i.e.

Table 2. Global Chemical Composition. Parental mantle (pyrolite) and subducting slab (pyrolite + harzburgite + MORB) in molar fractions of five major simple oxides.

	Pyrolite[a]	Harzburgite[a]	MORB[b]
SiO$_2$	38.9	36.4	52.2
Al$_2$O$_3$	2.2	0.7	10.2
CaO	3.1	0.9	14.8
MgO	50.0	56.6	15.8
FeO	5.8	5.4	7.0
Mg/Si	1.29	1.55	0.30
Fe/Si	0.15	0.15	0.13
Ca/Si	0.08	0.02	0.28
Al/Si	0.11	0.04	0.39

[a] [*Ringwood*, 1982].
[b] [*Ringwood and Irifune*, 1988].

a geotherm that accounts for the heat released or absorbed during phase changes and adiabatic compression). However, there is no profound physical reason for an adiabatic geotherm to be much more realistic (e.g., *Bunge et al.* [2001]). Moreover here we only show how mineralogical data can be used: no attempt has been made to precisely match the discontinuities of seismic reflectors or the exact compositions found in mineralogical experiments.

Pyrolite. In Plate 3 (top) the phase diagram can be divided into two major classes of minerals: 1) olivine and its high-pressure polymorphs and 2) pyroxene-garnet system. The transition from olivine to wadsleyite is predicted to occur at 430 km (14.5 GPa) over a narrow range (less than 10 km or 0.5 GPa). Wadsleyite dissociates to ringwoodite around 520 km (18 GPa) over a larger range (greater than 30 km). The breakdown of ringwoodite into perovskite and magnesiowustite at 640 km (23 GPa) is very sharp (~5 km) and can reasonably match the upper-lower mantle discontinuity. Pyroxenes and garnets constitute the second most abundant class of upper mantle minerals and contain almost all the Al and Ca budgets. It is by far the most difficult sequence to reproduce, since it requires highly accurate parameters as well as the precise knowledge of activity-composition relationships in complex solid solutions to correctly predict the immiscibility gap and the partitioning of elements between pyroxenes and garnet. As depth increases, clinopyroxenes progressively dissolve into the co-existing garnet phase. At 300 km depth, the computed garnet is 68 wt% pyrope, 23 wt% majorite and 9 wt% grossular. Complete conversion of clinopyroxenes is achieved at 460 km (16 GPa). At 600 km depth, the garnet phase is 56 wt% majorite-rich, with 29 wt% pyrope and 15 wt% grossular. The garnets are stable over a large pressure range in the pyrolite composition. Above 550 km (19 GPa), a CaSiO$_3$ perovskite phase exsolves from garnet. The garnet completely transforms to perovskite structure

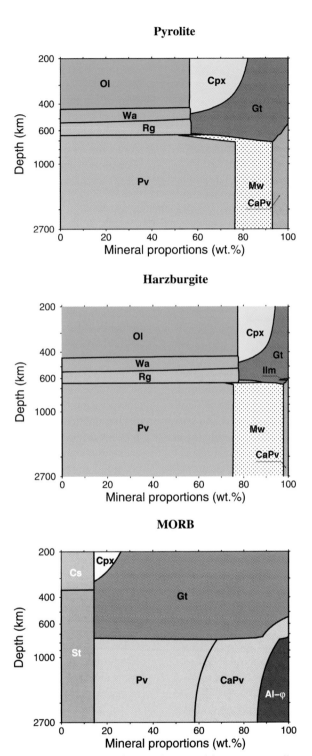

Plate 3. Computed phase diagram for three typical elemental compositions corresponding to pyrolite, harzburgite, and MORB (Ol: olivine, Wa: wadsleyite, Rg: ringwoodite, Pv: perovskite, Cpx: clinopyroxene, Gt: garnet, Mw: magnesio-wustite, CaPv: calcium perovskite, Ilm: ilmenite, Cs: coesite, St: stishovite, Al-φ: corundum). Notice that for a better visibility over the whole mantle, we use a non-linear scale for depth. The melt residue (harzburgite) is olivine-rich. Silica is an independent phase in MORB (mostly stishovite). In the lower mantle, the aluminum is present in two phases in the oceanic crust (Pv and Al-φ).

(calcium and magnesium silicate perovskites) above 720 km depth (26 GPa) whereas the Al_2O_3 content is entirely accommodated in Mg-rich perovskite. As a result, the 660 km discontinuity is made complex by the simultaneous dissociations of ringwoodite and garnet to perovskite phases. At greater depths, the pyrolite assemblage, composed of 76 wt% perovskite, 17 wt% magnesiowustite and 7 wt% calcium perovskite, is stable throughout the lower mantle, in agreement with a large number of experimental results on pyrolite composition (e.g. see review by *Fiquet* [2001]). This stability in the phase proportions is however associated with a gradual change in iron partitioning between magnesiowustite and perovskite that favors iron incorporation in perovskite as depth increases, in the presence of aluminum.

Harzburgite. In Plate 3 (middle) similarly to pyrolite, the harzburgite phase diagram may be divided into the olivine and pyroxene-garnet systems. The olivine sequence is rather similar to that of pyrolite. The olivine to wadsleyite and wadsleyite to ringwoodite transformations occur slightly deeper in harzburgite than in pyrolite due to a lower iron content. As already observed with pyrolite, the garnet/clinopyroxene ratio gradually increases with depth. The complete dissolution is achieved at similar depth, 460 km (16 GPa). Increasing depth stabilizes ilmenite over a very narrow region near the base of the transition zone (600 km). Ilmenite rapidly transforms over less than 20 km depth, into perovskite structure. The garnet to perovskite transformation is completed at around 650 km depth (23.5 GPa) whereas in pyrolite the garnet persists to 720 km (26 GPa). This is due to the stabilizing effect of aluminum on garnet. $CaSiO_3$-perovskite is exsolved from garnet at about 600 km (21 GPa). In the lower mantle, harzburgite is composed of 75 wt% perovskite, 23 wt% magnesiowustite and 2 wt% calcium perovskite. No other phase transition is present in our numerical simulation. These results are consistent with experimental observations of *Irifune and Ringwood* [1987].

MORB. In Plate 3 (bottom) the phase diagram is drastically different from that of harzburgite and pyrolite because the MORB is very silica rich with a Mg/Si molar ratio lower than 1 (see Table 2). No olivine polymorph is present. An exsolved SiO_2 phase coexists with clinopyroxene and garnet, either as coesite at depths shallower than 360 km (12 GPa) or as stishovite at larger depths. Notice that the computation are performed far above the stability field of plagioclase. Since we do not consider any other silica-rich phase such as high-pressure hollandite structures [*Fiquet*, 2001], stishovite is then found in the same proportion throughout the mantle. The dissolution of clinopyroxene into garnet is achieved at lower pressure than in pyrolite (around 310 km

or 10 GPa instead of 460 km or 16 GPa). The stability field of pyroxene is too shallow compared to experiments. This seems to be related to the difficulty of having realistic activity models for the end-members of pyroxenes. The proportion of garnet starts to decrease at 540 km (18.5 GPa) due to exsolving $CaSiO_3$-perovskite. Around 740 km (27 GPa), the Al-rich phase appears when garnet becomes unstable. Al is incorporated into both perovskite and corundum. The garnet-perovskite coexisting field is relatively narrow (less than 15 km) compared to that of the garnet-perovskite transformation in the pyrolite and harzburgite compositions. The garnet-perovskite transformation is completed around 760 km depth (28 GPa) which is consistent with experimental study by *Hirose et al.* [1999] and with the fact that high Al content stabilizes garnet in the lower mantle. Through the lower mantle after garnet vanishes, aluminum is exchanged between the two Al-bearing high-pressure phases (corundum/perovskite) favoring its incorporation into the corundum phase. In the lower mantle, the MORB composition crystallizes into an assemblage of 44 wt% Al-bearing perovskite, 28 wt% calcium perovskite, 14 wt% stishovite, and 14 wt% Al-rich phase. This is in agreement with recent experiments performed with similar MORB composition [*Perrillat et al.*, 2004; *Ricolleau et al.*, 2004].

Our thermodynamic approach allows us to reasonably reproduce the experimental phase diagrams. However, several observations are not well explained and underline the potential shortcoming of our approach. Our data base does not predict the presence of low-Ca pyroxenes in pyrolite and harzburgite as observed [*Irifune and Isshiki*, 1998; *Irifune and Ringwood*, 1987]. The computed stability field of clinopyroxene in MORB is too narrow compared to experiments that indicate the coexistence of garnet and clinopyroxene to 460 km (16 GPa) [*Aoki and Takahashi*, 2004]. A third Al-rich phase (NAL phase) seems to be locally present after garnet disappear in MORB [*Perrillat et al.*, 2004]. A more precise mineralogical modeling would be requested to resolve these discrepancies. The fundamental advantage of our method is that it however allows us to make predictions of phase diagrams when, composition, pressure and temperature are continuously varied, which is beyond practical possibility for experimental techniques.

4.2. Radial Density and Velocity Profiles

From the computed equilibrium mineralogies, we can deduce density ρ, isothermal incompressibility, adiabatic incompressibility K_S, and bulk sound velocity $v_\phi = \sqrt{K_S/\rho}$ of pyrolite (solid line), harzburgite (dashed-dotted) and MORB (dashed) compositions as a function of depth (see Figure 1). The bulk sound velocity shown in this figure corresponds to a Reuss average for an assemblage [*Watt et al.*, 1976]. The same arbitrary geotherm is used for the three mineralogies. For reference, PREM values are also depicted (dotted). Notice that we have not tried in this paper to fit PREM with any of the mineralogical models by varying the temperature profile or the bulk composition (see *Mattern et al.* [2005]).

Density. Figure 1 (top) shows that in the upper mantle, the ancient basaltic crust is denser than both pyrolite and harzburgite by around 0.1–0.2 g cm^{-3}. This is due to the large content of dense garnet and to the presence of the even denser stishovite. Between 645 and 760 km depth basaltic crust becomes on the contrary, lighter by about 0.2 g cm^{-3} in agreement with observations [*Irifune and Ringwood*, 1993; *Hirose et al.*, 1999]. These large buoyancy variations are due to the different sequences of garnet-perovskite as discussed above. Indeed, the garnet is ~10% lighter than perovskite. Once the transformation of garnet to perovskite is completed (at about 760 km depth) MORB is no longer buoyant and its density becomes again higher than that of pyrolite. The density excess is about 0.1 g cm^{-3} at the top of the lower mantle and decreases to about 0.03 g cm^{-3} near the CMB. This excess density is large compared to thermal density anomalies: as we predict a thermal expansivity of 0.7 10^{-5} K^{-1} in the lowermost mantle, a temperature increase of ~800 K is necessary for a subducted MORB to become neutrally buoyant. Harzburgite is by 0.02–0.05 g cm^{-3} slightly less dense than pyrolite throughout the mantle. The low FeO, Al$_2$O$_3$ and CaO contents lead to smaller amounts of garnets above 530 km depth of the very dense CaSiO$_3$-perovskite. However, between 640 and 700 km depth, harzburgite is slightly denser (by 0.04 g cm^{-3}) than pyrolite because of the sharp transformation of ringwoodite+garnet to perovskite+magnesiowustite assemblage at 640 km. This is in good agreement with the experimental results obtained by *Ringwood* [1991], *Irifune and Ringwood* [1987], and *Irifune and Ringwood* [1993]. On average, the crust and the corresponding depleted lithosphere have together the same average density as pyrolite, at the same temperature.

Bulk sound velocity. Figure 1 (bottom) shows that in the upper mantle, the MORB composition is faster than both pyrolite and harzburgite by about 0.2 km s^{-1}. This is due to the large content of garnet, particularly pyrope and grossular, as well as the presence of fast stishovite. Between 645 and 750 km depth basaltic crust becomes on the contrary slower by about 0.1 km s^{-1} due to the persistence of garnets. Below 800 km, the basaltic crust becomes faster than an average lower mantle, with a velocity profile which slightly diverges from that of pyrolite due to the progressive exchange of Al from perovskite to corundum phases. This last prediction

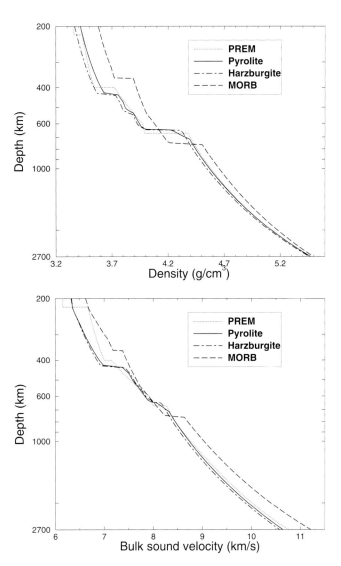

Figure 1. Computed density ρ and bulk sound velocity v_ϕ as a function of depth for pyrolite, harzburgite, and MORB compositions. Seismic profiles given by PREM [*Dziewonski and Anderson*, 1981] are also plotted. The mineralogical predictions have been computed assuming that the temperature increases linearly with depth. The choice of a non-linear scale induces an apparent curvature of the profiles. The pyrolite composition gives a close fit to PREM. The eclogitized MORB is very significantly denser and faster than the average mantle except in a limited range below the upper–lower mantle interface.

should however be taken with caution because it is based on our simplified model for the Al-rich phase in the lower mantle. More precise predictions of the velocity differences and of their behavior with depth would need accurate measurements of the elastic properties of the aluminum rich phase which are not available at present. The depleted mantle (harzburgite) has almost the same velocity as the parental mantle (pyrolite): the difference is between -0.06 km s^{-1} and +0.06 km s^{-1} in the upper mantle and is constant close to -0.06 km s^{-1} in the lower mantle.

5. APPLICATION TO SLAB PENETRATION MODELING

Our mineralogical model can be also used to compute density and bulk sound velocity of a hypothetical subducting slab. We consider a 100 km thick mature oceanic lithospheric plate far from an oceanic rift, across which there is a linear temperature difference of 1000 K. This initial thermal structure is then modified by heat diffusion once the plate enters vertically into the mantle with a constant vertical velocity of 5 cm yr^{-1}. We consider a uniform thermal diffusivity $\kappa = 10^{-6}$ m^2 s^{-1} and no heat source. We neglect vertical diffusion and only consider lateral thermal effect. In the mantle, far from the slab, we use a linear geotherm (a gradient of 0.3 K km^{-1} with a surface temperature of 1400 °C). We do not consider the possible thickening of the slab entering the lower mantle as suggested by geodynamical modeling. Our slab model is certainly very simple and we could have imposed a more realistic subduction angle but this would not have significantly affected our conclusions.

At each point, according to the local temperature and pressure, we compute the equilibrium mineralogy as discussed before. We consider two distinct cases: 1) the whole slab has pyrolite composition and 2) the slab is made of 50 km of pyrolite, 40 km of harzburgite and 10 km of MORB (the mixing of 4/5 harzburgite with 1/5 MORB gives approximately the pyrolite composition).

The thermo-chemical approach gives the density, the incompressibility and the bulk sound velocity, but not the shear modulus μ needed to compute the seismic velocities v_s and v_p. Although Brillouin and ultrasonic measurements of rigidity start to be available (e.g., *Sinogeikin and Bass* [2002]; *Kung et al.* [2002]; *Jackson et al.* [2004]), a complete data set for all the end-members introduced in our modeling is still missing. We therefore assume a simple semi-empirical model for the the rigidity. The lateral variations of rigidities have two components. The first one has a thermal origin at constant composition,

$$\Delta\mu/\mu = -(a + bz)10^{-5}\Delta T, \qquad (9)$$

where z, in km, is depth. In relative agreement with the few experimental studies we take $a = 21.3$ and $b = -0.0055$ km^{-1}. The second is related to changes in mineralogy,

$$\Delta\mu/\mu = c\Delta K_S/\mu, \qquad (10)$$

with $c = 0.631$ according to *Stacey* [1992], and where ΔK_S is the difference between the incompressibilities of the slab and of the normal mantle due to their differences of mineralogies. The shear velocities are certainly more affected than the compressional velocities by this poor modeling of rigidity. This is why we have chosen in the next paragraphs to only discuss our results in terms of sound velocity v_ϕ (independent of the rigidity model) and compressional velocity v_p for comparison with Plate 2.

5.1. Pyrolite Slab

In Plate 4 we show the computed temperature, the density excess $\Delta\rho/\rho$ and the equivalent tomographic images $\Delta v_\phi/v_\phi$, $\Delta v_p/v_p$ within a subducting slab and the surrounding mantle as a function of depth and horizontal distance. The temperature map (top left) shows the progressive re-heating of the slab with depth. Notice that with a vertical slab velocity is 5 cm yr^{-1}, at 2200 km depth, the slab is still not thermally re-assimilated (the coldest part of the slab is still colder by 350 K).

The relative density difference (in %) between subducting slab and surrounding mantle far from this cold region is shown in the top right panel of Plate 4. One can recognize two major contributions to the observed density heterogeneities. On the one hand, cold material is denser due to the intrinsic thermal contribution. As an example, the central part of the slab is about 2% denser at 200 km depth and 0.4% denser at 2200 km. This is both due to the diffusion of heat toward the center of the slab and to the decrease of thermal expansivity with depth (that decreases from 3.2 10^{-5} K^{-1} in the olivine stability field to 0.7 10^{-5} K^{-1} near the core mantle boundary). On the other hand, there are effects on density due to the phase transformations the occur in the slab at a shallower depth for exothermic phase changes with positive Clapeyron slope, and deeper for endothermic phase change with negative Clapeyron slope. Since the 410 km and 520 km discontinuities have positive Clapeyron slopes, they occur at lower depths in the cool slab inducing an additional increase of density of 5% (0.18 g cm^{-3}) at 430 km and of 2.8% (0.11 g cm^{-3}) at 500 km. This effect is not localized at one depth but forms a large zone of excess density from 390 to 520 km. The density contrast is opposite in the depth range from 630 to 690 km with a maximum difference of -5% (-0.22 g cm^{-3}) due to the endothermic decomposition of ringwoodite. The latter effect resists to convective downwellings but should not be large enough to impede slab penetration [*Christensen*, 1996].

The two bottom panels of Plate 4 depict the predicted seismic velocity perturbations in %. The lost of amplitude by a factor 3 when the slab enters the lower mantle is the main indication of this modeling in agreement with *Goes et al.* [2004]. In the upper mantle, we observe strong lateral variations (at least a few %) due to the combined thermal and phase change effects. The subducting slab becomes significantly less visible in the lower mantle. At 2000 km depth, lateral variations within the slab are at most 0.5% for v_p and 1% for v_s. The velocity anomalies are affected by the phase changes in the upper mantle but the details may not be realistic due to the poor quality of our modeling of rigidities.

5.2. Composite Slab

A more realistic slab petrology is considered, in Plate 5. We only represent the density and P-wave velocity perturbations in %. The temperature remains the same as in Plate 4 (top left panel). The thermal density excess Plate 5, (left) is partially balanced by the compositional effect in the harzburgitic lithosphere. The situation is very different in the 10-km thick MORB layer which is much denser than the pyrolitic mantle (4% above 640 km depth and ~2.2% below 780 km). In between, due to the persistence of garnet, the MORB layer becomes buoyant as discussed previously (−1.5%). Horizontally averaged, the excess mass and velocity of a composite slab is very similar to that of a slab of pure pyrolitic composition, in other words, the basalt together with the corresponding depleted lithosphere have properties roughly equivalent to those of the parental mantle. However the behavior in the convective mantle of a composite slab cannot be understood from a simple thermo convective model. During subduction, the dense crustal layer may delaminate from the rest of the lithosphere and segregate at depth. This dynamics is observed in numerical simulations and has been invoked to explain various geochemical observations [*Christensen and Hofmann*, 1994; *Coltice and Ricard*, 1999].

5.3. The Origin of Density and Velocity Anomalies

The lateral variations of any property in the mantle Δf (*f* standing for density, thermal expansivity, incompressibility, seismic velocity...) have potentially three contributions that can be summarized as

$$\Delta f = \left(\frac{\partial f}{\partial T}\right)_\phi \Delta T + \left(\frac{\partial f}{\partial \phi}\right)_T \left(\frac{\partial \phi}{\partial T}\right)_\chi \Delta T + \\ \left(\frac{\partial f}{\partial \phi}\right)_T \left(\frac{\partial \phi}{\partial \chi}\right)_T \Delta \chi. \quad (11)$$

This first term on the right-hand side, is the intrinsic thermal effect that is computed at constant mineralogical proportions (symbolized by the subscript ϕ). The second term is a thermo-chemical effect and corresponds to the changes in mineral proportions necessary to maintain the Gibbs equilibrium at constant elemental (bulk) composition. This term is responsible for a rise of the 410 km depth interface and deepens the 660 km depth interface in the presence of cold downwellings [*Irifune and Ringwood*, 1987]. However except for these two quasi univariant phase changes, previous studies have not considered this effect for larger phase transitions (e.g. olivine/ringwoodite), continuous phase changes (e.g. pyroxene/garnet and garnet/perovskite), or changes in partition coefficients with depth (e.g. perovskite/magnesiowustite). The last term of equation (11) is the intrinsic chemical effect (variations in the elemental budget, symbolized by χ, at constant temperature). Any change in the bulk elementary composition affects the mineralogical composition at constant temperature and can, therefore, contribute to lateral variations of *f*.

If the mantle average phase content is known the first contribution can be estimated. The last two contributions can however be precisely evaluated only by using a Gibbs free-energy minimization techniques. Computing the changes in mineralogy with temperature requires to take into account precisely various phase equilibria and to introduce, if necessary, new mineralogical phases not stable at the reference temperature. In the same way, assuming for example an enrichment in iron in the lower mantle would not affect the properties in the same way assuming that the iron goes preferentially into perovskite or into magnesiowustite. The effect of iron implicitly requires the knowledge of the *P*, *T* and Al-dependent partition coefficients (the chemical activity of iron in perovskite in indeed function of its aluminum content).

We depict in Plate 6 the three contributions of equation (11) for density (left column) and v_ϕ velocity (right column). The reference state is the pyrolitic mantle. The top row is the intrinsic thermal effect due to the presence of a cold slab while the mineralogy is held constant. The second row is the contribution due to phase changes and changes in mineralogical composition imposed by temperature variations within and around the slab. The sum of these two contributions (thermal and thermo-chemical) is depicted in the right column Plate 4.

Although a simple model that would move the main phase transition according to their Clapeyron slopes would catch some important mineralogical contributions, it is clear from Plate 6 that *all* phase changes contribute to the total properties: not only the olivine/wadsleyite and ringwoodite/perovskite + magnesiowustite but also the wadsleyite/ringwoodite phase change and all those involving pyroxene, garnet and Ca-perovskite. The dissolution of pyroxene into garnet seems to significantly affect the bulk sound velocity not the density. On the contrary, the phase transition between wadsleyite and ringwoodite around 520 km depth has a significant influence on the densities without affecting the velocities very much.

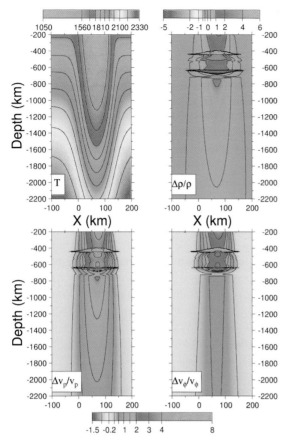

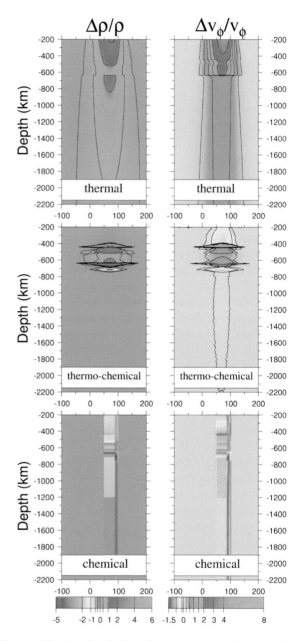

Plate 4. Temperature (K), lateral density, and velocity variations (%) associated with a slab sinking vertically in the mantle. The surface of the slab is initially at $x = 0$. The temperature variation have been simply computed assuming diffusive reheating for a 100-km-thick slab sinking at 5 cm/year. Far from the slab, we assume that the background temperature varies linearly with depth following a typical adiabat. We assume that the slab is composed of pure pyrolite. The density and velocities are deduced from our Gibbs minimization procedure.

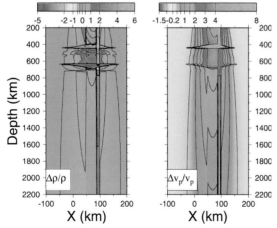

Plate 5. Lateral density and P-velocity variations associated with a slab sinking vertically in the mantle. Compared with Plate 4, the slab now includes a 10-km-thick oceanic crust (MORB composition) overlying a 40-km-thick depleted lithosphere (harzburgitic composition).

Plate 6. The density (left) and v_ϕ velocity (right) anomalies in a subducting slab have three origins: an intrinsic thermal origin (changes of properties with temperature at constant mineralogical content, top row), a thermochemical origin (phase changes and changes in phase composition induced by temperature anomalies, middle row), and an intrinsic chemical effect (changes in elementary composition, bottom row). The three contributions have basically similar amplitudes. In the two top rows, the slab has a pure pyrolitic composition and the same temperature as in Plate 4. The bottom row is the difference in properties at the same temperature between a composite slab and a pyrolite slab.

Notice also that even in the lower mantle there is a minute contribution to the velocity anomaly due to the progressive enrichment in iron of the perovskite phase (the level line in the middle right panel of Plate 6 is -0.05%). The bottom row represents the intrinsic chemical heterogeneities that are associated with a composite slab without any lateral temperature variations. Notice that the same color scale is used for the three contributions which emphasizes the fact that thermal, thermo-chemical and chemical anomalies are of the same order.

5.4. Effect of Elemental Variations on Density and Velocity Anomalies

The properties of subducted MORB are so different from those of pyrolite that we did not predict them by simply extrapolating from the pyrolite properties using the differential form (11). In other terms, the $\Delta\chi$ between MORB and pyrolite are not infinitesimal. We directly computed the MORB and pyrolite properties without explicitly using partial derivatives. However, it is also possible that the chemical composition of the lower mantle evolves continuously around its average composition with progressive, lateral or depth dependent chemical variations. Typically this is the case when the observed seismic tomography is to be interpreted in terms of smooth laterals variations of temperature and bulk chemical composition. To formulate properly the inverse problem one has to know precisely all the partial derivatives at given depth [*Trampert et al.*, 2004].

In Table 3, we show the values of $(\partial\ln f/\partial T)_\chi$ and $(\partial\ln f/\partial\chi)_T$ at two depths, 1030 and 2060 km. Here f stands for ρ, v_p or v_ϕ, χ is the mole abundance in oxides MgO, FeO, SiO_2, Al_2O_3 and CaO (except for the factor 2 in Al_2O_3, it is also the mole abundance in cations). These derivatives are computed under thermodynamic equilibrium. The reference state has a pyrolite composition and a reference temperature of 2000 K. One can see that the values of the individual partial derivatives with the composition are rather constant with depth. The values confirms that iron and temperature are the two most sensitive parameters in terms of density and velocity variations (e.g. *Mattern et al.* [2005]; *Trampert et al.* [2004]). In the case of the derivatives with respect to temperature and iron, our values are comparable but larger in amplitude to those of *Trampert et al.* [2004] and *Forte and Mitrovica* [2001], particularly for v_ϕ. The Table shows that the partial derivatives with respect to aluminum even though being lower than that with respect to iron, are not negligible specially for the density and have opposite signs. It follows that the effect of aluminum should be considered when interpreting the tomographic images. Finally, one can also see that the least sensitive parameter for the density is calcium.

Table 3. Partial Derivatives of $\ln\rho$, $\ln v_p$ or $\ln v_\phi$. Derivatives with respect to temperature (in K) and mole abundances of oxides. Two depths have been considered, 1030 and 2060 km; the reference case is pyrolite at 2000 K.

1030 km	$\ln\rho$	$\ln v_\phi$	$\ln v_p$
d/dT	-1.847×10^{-5}	-1.729×10^{-5}	-1.940×10^{-5}
d/d(MgO)	-0.087	-0.071	-0.080
d/d(FeO)	0.449	-0.342	-0.351
d/d(Al_2O_3)	-0.209	0.085	0.084
d/d(CaO)	-0.038	-0.097	-0.106
d/d(SiO_2)	0.060	0.145	0.159

2060 km	$\ln\rho$	$\ln v_\phi$	$\ln v_p$
d/dT	-1.137×10^{-5}	-1.684×10^{-5}	-1.969×10^{-5}
d/d(MgO)	-0.061	-0.060	-0.072
d/d(FeO)	0.475	-0.330	-0.342
d/d(Al_2O_3)	-0.208	0.087	0.084
d/d(CaO)	-0.015	-0.036	-0.042
d/d(SiO_2)	0.020	0.125	0.142

6. CONCLUSIONS

The simple visual inspection of tomographic sections across subduction zones is not a definitive argument for slab penetration unless a precise modeling of temperature-velocity relationship is done. The velocity amplitude reduction by a factor 3 across the upper-lower mantle interface observed locally in various tomographic model (see Plate 2) is however in good agreement with what can be estimated from mineral physics (see Plate 4, bottom left).

The computed phase diagrams do not explain all mineralogical details that have been experimentally observed. In spite of these problems it seems to us that the general agreement of the mineralogical predictions with various observations and the good fit with seismology is a remarkable success. It proves the high accuracy of the available thermochemical data and gives the exciting indication that these data will be soon integrated into complex geodynamical models. Understanding the dynamics of mantle reservoirs with different composition/petrology is indeed necessary to explain the complexities of geochemical observations [*van Keken and Ballentine*, 1998; *Ricard and Coltice*, 2004; *Tackley*, 2000]. Plate 6 however, illustrates the fact that this integration must be done on the basis of thermodynamical principle, not by assuming an a priori depth dependent mineralogy. The temperature, the trajectory, the possible thickening of slabs crossing the transition zone should not be guessed from a simple diffusive model as we did, but self consistently deduced from momentum and energy considerations.

The simulation of a complex slab mineralogy that we presented here is very crude. The depletion of the lithosphere is a continuous process without abrupt compositional transition between the parental fertile mantle and the most depleted harzburgitic layer. Although the uncertainties are probably large, our selection of the best available data does not show any crossover with depth between the densities of pyrolite, harzburgite and MORB. As seen in Plate 5 the crust remains denser while the cold depleted layer is basically neutrally buoyant. Although the variations in Fe and temperature are the major sources of seismic velocity variations, the other variations in bulk composition cannot be neglected.

Acknowledgments. We would like to thank Rob van der Hilst and Chang Li that made their tomographic model and the associated numerical tools available to us to draw Plate 2. This work also benefited from numerous discussions with François Guyot, Philippe Gillet, Guillaume Fiquet, Nicolas Coltice and Jay Bass. We also thank Saskia Goes, Jeannot Trampert and another anonymous colleague for their very constructive and detailed reviews.

REFERENCES

Aoki, I., and E. Takahashi (2004), Density of MORB in the upper mantle, *Phys. Earth Planet. Inter.*, *143–144*, 129–143.

Badro, J., G. Fiquet, and F. Guyot (2003), Iron partitioning in earth's mantle: Toward a deep lower mantle discontinuity, *Science*, *300*, 789–791.

Badro, J., J.-P. Rueff, H. Reniewicz, G. Fiquet, and F. Guyot (2004), Electronic transitions in perovskite: Possible nonconvecting layers in the lower mantle, *Science*, *305*, 383.

Barron, L. M. (1968), Thermodynamic, multicomponent silicate equilibrium phase calculations, *Am. Miner.*, *57*, 809–823.

Becker, T. W., and L. Boschi (2002), A comparison of tomographic and geodynamic mantle models, *Geochem. Geophys. Geosyst.*, *3*, 2001GC000,168.

Bercovici, D., Y. Ricard, and M. Richards (2000), The relation between mantle dynamics and plate tectonics: A primer, in: *History and Dynamics of Global Plate Motions, Geophys. Monogr. Ser.*, vol. 121, 5–46 pp., M.A. Richards, R. Gordon, and R. van der Hilst, AGU, Washington D.C.

Berman, R. G., and T. H. Brown (1985), A thermodynamic model for multicomponent melts, with application to the system CaO – Al_2O_3 – SiO_2, *Geochim. Cosmochim. Acta*, *49*, 613–614.

Bijwaard, H., and W. Spakman (2000), Non-linear global p-wave tomography by iterated linearized inversion, *Geophys. J. Int.*, *141*, 71–82.

Bina, C. R., and B. J. Wood (1986), The 400-km seismic discontinuity and the proportion of olivine in the earth's upper mantle, *Nature*, *324*, 449–451.

Birch, F. (1952), Elasticity and constitution of the earth's interior, *J. Geophys. Res.*, *57*, 227–286.

Bottinga, Y., and P. Richet (1978), Thermodynamics of liquid silicates, a preliminary report, *Earth Planet. Sci. Lett.*, *40*, 382–400.

Bunge, H. P., Y. Ricard, and J. Matas (2001), Non-adiabaticity in mantle convection, *Geophys. Res. Lett.*, *28*, 879–882.

Christensen, U. R. (1996), The influence of trench migration on slab penetration into the lower mantle, *Earth Planet. Sci. Lett.*, *140*, 27–39.

Christensen, U. R., and A. Hofmann (1994), Segregation of subducted oceanic crust in the convecting mantle, *J. Geophys. Res.*, *99*, 19,867–19,884.

Coltice, N., and Y. Ricard (1999), Geochemical observations and one-layer mantle convection, *Earth Planet. Sci. Lett.*, *174*, 125–137.

Davies, G. F. (1988), Ocean bathymetry and mantle convection, 1, large scale flow and hotspots, *J. Geophys. Res.*, *93*, 10,467–10,480.

Davies, G. F., and M. Richards (1992), Mantle convection, *J. Geol.*, *100*, 151–206.

Dziewonski, A. M., and D. Anderson (1981), Preliminary reference earth model, *Phys. Earth Planet. Inter.*, *25*, 297–356.

Fabrichnaya, O. B., and O. L. Kuskov (1991), Constitution of the mantle. 1. Phase-relations in the $FeO-MgO-SiO_2$ system at 10–30 GPa, *Phys. Earth Planet. Inter.*, *69*, 56–71.

Fei, Y., H.-K. Mao, and B. Mysen (1991), Experimental determination of element partitioning and calculation of phase relations in the $MgO-FeO-SiO_2$ system at high pressure and high temperature, *J. Geophys. Res.*, *96*(B2), 2157–2169.

Fiquet, G. (2001), Mineral phases of the earth's mantle, *Z. Kristallogr*, *216*, 248–271.

Fiquet, G., F. Guyot, M. Kunz, J. Matas, D. Andrault, and M. Hanfland (2002), Structural rafinements of magnesite at very high pressure, *Amer. Mineral.*, *87*, 1261–1265.

Forte, A. M., and J. Mitrovica (2001), Deep-mantle high-viscosity flow and thermochemical structure inferred from seismic and geodynamic data, *Nature*, *410*, 1049–1056.

Fukao, Y., M. Obayashi, H. Inoue, and M. Nenbai (1992), Subducting slabs stagnant in the mantle transition zone, *J. Geophys. Res.*, *97*, 4809–4822.

Ghiorso, M. S. (1997), Thermodynamic models of igneous preocesses, *Ann. Rev. Earth Planet. Sci.*, *25*, 221–241.

Ghiorso, M. S., and I. S. E. Carmichael (1980), Regular solution model for met-aluminous silicate liquids—applications to geothermometry, immiscibility, and the source regions of basic magmas, *Contrib. Mineral. Petrol.*, *71*, 323–342.

Goes, S., F. Cammarano, and U. Hansen (2004), Synthetic seismic signature of thermal mantle plumes, *Earth Planet. Sci. Lett.*, *218*, 403–419.

Grand, S. P., R. van der Hilst, and S. Widiyantoro (1997), Global seismic tomography a snapshot of convection in the earth, *GSA Today*, *7*(4), 1–7.

Hart, S. R., and A. Zindler (1989), Constraints on the nature and development of chemical heterogeneities in the mantle, in: *Mantle Convection*, W.R. Peltier, Gordon and Breach Science Publishers, New York.

Hirose, K., Y. Fei, Y. Ma, and H.-K. Mao (1999), The fate of subducted basaltic crust in the earth's lower mantle, *Nature*, *397*, 53–56.

Hirose, K., Y. Fei, S. Ono, T. Yagi, and K. I. Funakoshi (2001), In situ measurements of phase transformation boundary in $Mg_3Al_2Si_3O_{12}$, *Earth Planet. Sci. Lett.*, *184*, 567–573.

Irifune, T., and M. Isshiki (1998), Iron partitioning in a pyrolite mantle and the nature of the 410-km seismic discontinuity, *Nature*, *392*, 702–705.

Irifune, T., and A. Ringwood (1993), Phase transformations in subducted oceanic crust and buoyancy relationships at depths of 600–800 km in the mantle, *Earth Planet. Sci. Lett.*, *117*, 101–110.

Irifune, T., and A. E. Ringwood (1987), Phase transformations in a harzburgite composition to 26 GPa: Implications for dynamical behaviour of the subducting slab, *Earth Planet. Sci. Lett.*, *86*, 365–376.

Irifune, T., T. Koizumi, and J. Ando (1996), An experimental study of the garnet-perovskite transformation in the system $MgSiO_3$–$Mg_3Al_2Si_3O_{12}$, *Phys. Earth Planet. Inter.*, *96*, 147–157.

Ita, J., and L. Stixrude (1992), Petrology, elasticity, and composition of the mantle transition zone, *J. Geophys. Res.*, *97*(B5), 9849–6866.

Jackson, J. M., J. Zhang, and J. D. Bass (2004), Sound velocities and elasticity of aluminous $MgSiO_3$ perovskite: Implications for aluminum heterogeneity in earth's lower mantle, *Geophys. Res. J.*, *31*, doi:10:1029/2004GL019918.

Karason, H., and R. van der Hilst (2000), Constraints on mantle convection from seismic tomography, in: *The History and Dynamics of Global Plate Motion*, vol. 121, 277–288 pp., Richards, M. R., R. Gordon, and R. D. van der Hilst, Geophysical Monograph AGU, Washington, D.C.

Kerschhofer, L., D. C. Rubie, T. G. Sharp, J. D. C. McConnell, and C. Dupas-Bruzek (2000), Kinetics of intracrystalline olivine-ringwoodite transformation, *Phys. Earth Planet. Inter.*, *121*, 59–76.

Kirby, S. H., S. Stein, E. A. Okal, and D. Rubie (1996), Metastable mantle phase transformations and deep earthquakes in subducting oceanic lithosphere, *Rev. Geophys.*, *34*, 261–306.

Kubo, A., and M. Akaogi (2000), Post-garnet transitions in the system $Mg_4Si_4O_{12}$–$Mg_3Al_2Si_3O_{12}$ up to 28 GPa: Phase relations of garnet, ilmenite and perovskite, *Phys. Earth Planet. Inter.*, *121*, 85–102.

Kung, J., B. Li, D. J. Weidner, J. Zhang, and R. C. Liebermann (2002), Elasticity of $(Mg_{0.83}Fe_{0.17})O$ ferropericlase at high pressure: Ultrasonic measurements in conjunction with X-radiation techniques, *Earth Planet. Sci. Lett.*, *203*, 557–566.

Kuskov, O. L., and A. B. Panferov (1991), Phase-diagrams of the $FeO - MgO - SiO_2$ system and the structure of the mantle discontinuities, *Phys. Chem. Miner.*, *17*, 642–653.

Lithgow-Bertelloni, C., M. A. Richards, Y. Ricard, R. J. O'Connell, and D. C. Engebretson (1993), Toroidal-poloidal partitioning of plate motions since 120 ma, *Geophys. Res. Lett.*, *20*, 375–378.

Masters, G., G. Laske, H. Bolton, and A. Dziewonski (2000), The relative behavior of shear velocity, bulk sound speed, and compressional velocity in the mantle: Implications for chemical and thermal structure, in: *Earth's Deep Interior*, vol. 117, pp. 63–87, S. Karato, A. M. Forte, R. C. Liebermann, G. Masters and L. Stixrude, Geophysical Monograph AGU, Washington, D.C.

Matas, J. (1999), Modélisation thermochimique des propriétés de solides à hautes températures et hautes pressions. applications géophysiques., Ph.D. thesis, Ecole Normale Supérieure de Lyon, France.

Matas, J., P. Gillet, Y. Ricard, and I. Martinez (2000a), Thermodynamic properties of carbonates at high pressures from vibrational modelling, *Eur. J. Mineral.*, *12*, 703–720.

Matas, J., Y. Ricard, L. Lemelle, and F. Guyot (2000b), An improved thermodynamic model of metal-olivine-pyroxene stability domains, *Contrib. Mineral. Petrol.*, *140*, 73–83.

Mattern, E., J. Matas, Y. Ricard, and J. Bass (2005), Lower mantle composition and temperature from mineral physics and thermodynamic modelling, *Geophys. J. Inter.*, *160*, 973–990, doi:10:1111/j.1365-246X, 2004.02549.x.

Mosenfelder, J. L., J. A. D. Connolly, D. C. Rubie, and M. Liu (2000), Strength of $(Mg,Fe)_2SiO_4$ wadsleyite determined by relaxation of transformation stress, *Phys. Earth Planet. Inter.*, *120*, 63–78.

Murakami, T., K. Hirose, K. Kawamura, N. Sata, and Y. Ohishi (2004), Post-perovskite phase transition in $MgSiO_3$, *Science*, *304*, 855–858.

Ono, S., E. Ito, and T. Katsura (2001), Mineralogy of subducted basaltic crust (MORB) from 25 to 37 GPa and chemical heterogeneity of the lower mantle, *Earth Planet. Sci. Lett.*, *190*, 57–63.

Perrillat, J. P., A. Ricolleau, I. Daniel, G. Fiquet, M. Mezouar, and H. Cardon (2004), Phase transformations of MORB in the lower mantle, *Lithos*, *73*(1-2 Suppl. 1), S87–S87.

Ricard, Y., and N. Coltice (2004), Geophysical and geochemical models of mantle convection: Successes and future challenges, in: *The State of the Planet: Frontiers and Challenges in Geophysics*, vol. 150, 59–68 pp., Geophys. Monogr. Ser., AGU, Washington, D.C.

Ricard, Y., M. Richards, C. Lithgow-Bertelloni, and Y. L. Stunff (1993), A geodynamic model of mantle density heterogeneity, *J. Geophys. Res.*, *98*, 21,895–21,909.

Richards, M. (1991), Hotspots and the case for a high viscosity lower mantle, in: *Glacial Isostasy, Sea-Level and Mantle Rheology*, R. Sabadini et al., eds., 571–587 pp., Kluwer Academic Publishers, Dordrecht.

Richards, M. A., and D. C. Engebretson (1992), Large-scale mantle convection and the history of subduction, *Nature*, *355*, 437–440.

Ricolleau, A., G. Fiquet, J. Perillat, I. Daniel, N. Menguy, H. Cardon, A. Addad, C. Vanni, and N. Guignot (2004), The fate of subducted basaltic crust in the earth's lower mantle: an experimental petrological study, in *Eos Trans. AGU, Fall Meet. Suppl*, vol. 85(47), abstract U33B-02.

Ringwood, A. (1982), Phase transformations and differentiation in subducted lithosphere: Implications for mantle dynamics, basalt petrogenesis, and crustal evolution, *J. Geol.*, *90*(6), 611–642.

Ringwood, A. E. (1991), Phase transformations and their bearing on the constitution and dynamics of the mantle, *Geochim. Cosmochim. Acta*, *55*, 2083–2110.

Ringwood, A. E., and T. Irifune (1988), Nature of the 650-km seismic discontinuity: Implications for mantle dynamics and differentiation, *Nature, 331,* 131–136.

Ritsema, J., and H. J. van Heijst (2000), Seismic imaging of structural heterogeneity in earth's mantle: Evidence for large-scale mantle flow, *Sci. Progr., 83,* 243–259.

Saxena, S., N. Chatterjee, Y. Fei, and G. Shen (1993), *Thermodynamic Data on Oxides and Silicates,* Springer-Verlag.

Saxena, S. K. (1996), Earth mineralogical model: Gibbs free energy minimization computation in the system $MgO-FeO-SiO_2$, *Geochim. Cosmochim. Acta, 60,* 2379–2395.

Saxena, S. K., and G. Eriksson (1983), Theoritical computation of mineral assemblages in pyrolite and lherzolite, *J. Petrol., 24,* 538–555.

Sinogeikin, S. V., and J. D. Bass (2002), Elasticity of pyrope and majorite-pyrope solid solutions to high temperatures, *Earth Planet. Sci. Lett., 203.*

Smith, W. R., and R. Missen (1991), *Chemical Reaction Equilibrium Analysis: Theory and Algorithms,* Krieger Publishing Company, Malabar, Florida.

Sobolev, S. V., and A. Y. Babeyko (1994), Modeling of mineralogical composition, density and elastic-wave veleocities in anhydous magmatic rocks, *Surv. Geophys., 15,* 515–544.

Spear, F. S. (1995), *Metamorphic Phase Equilibria and Pressure-Temperature-Time Paths,* 1–799 pp., Mineralogical Society of America, Monograph.

Stacey, F. D. (1992), *Physics of the Earth,* 3rd edition, Brookfield Press.

Stixrude, L. (1997), Structure and sharpness of phase transitions and mantle discontinuities, *J. Geophys. Res., 102,* 14,835–14,852.

Tackley, P. (2000), Mantle convection and plate tectonics: Toward an integrated physical and chemical theory, *Science, 288,* 2002–2007.

Takafuji, N., T. Yagi, N. Miyajima, and T. Sumita (2002), Study on Al_2O_3 content and phase stability of aluminous-$CaSiO_3$ perovskite at high pressure and temperature, *Phys. Chem. Min., 29,* 532–537.

Trampert, J., F. Deschamp, J. Resovsky, and D. Yuen (2004), Probabilistic tomography maps chemical heterogeneities throughout the lower mantle, *Science, 306,* 853–856.

Turcotte, D. L., and G. Schubert (1982), *Geodynamics: Applications of Continuum Physics to Geological Problems,* 1–450 pp., Wiley, New York.

Vacher, P., A. Mocquet, and C. Sotin (1998), Computation of seismic profiles from mineral physics: The importance of the non-olivine components for explaining the 660 km depth discontinuity, *Phys. Earth Planet. Inter., 106,* 275–298.

van der Voo, R., W. Spakman, and H. Bijwaard (1999), Mesozoic subducted slabs under Siberia, *Nature, 397,* 246–249.

van Keken, P. E., and C. J. Ballentine (1998), Whole-mantle versus layered convection and the role of a high-viscosity lower mantle in terrestrial volatile evolution, *Earth Planet. Sci. Lett., 156,* 19–32.

Watt, J. P., G. F. Davies, and R. J. O'Connell (1976), Elastic properties of composite-materials, *Rev. Geophys., 14,* 541–563.

Wood, B. J. (1987), Thermodynamics of multicomponent systems containing several solid-solutions, *Rev. Mineral., 17,* 71–95.

Wood, B. J., and J. R. Holloway (1984), Thermodynamic model for subsolidus equilibrium in the system CaO; MgO; Al_2O_3, *Geochim. Cosmochim. Acta, 48,* 159–176.

J. Matas, E. Mattern, and Y. Ricard, Laboratoire des Sciences de la Terre, UMR CNRS 5570, Ecole Normale Supérieure de Lyon, 46 allée d'Italie, F069364 Lyon, Cedex 07, France. (jmatas@ens-lyon.fr; emattern@ens-lyon.fr; yanick.ricard@ens-lyon.fr)

Compositional Dependence of the Elastic Wave Velocities of Mantle Minerals: Implications for Seismic Properties of Mantle Rocks

Sergio Speziale

Department of Earth and Planetary Science, University of California, Berkeley, California

Fuming Jiang and Thomas S. Duffy

Department of Geosciences, Princeton University, Princeton, New Jersey

Using single-crystal elasticity data we constrain the effect of chemical substitutions on the elastic properties of mantle minerals and estimate the consequences for the seismic properties of mantle rocks. At ambient conditions the calculated relative variation of compressional and shear velocities $\partial \ln v_P / \partial X_{Fe}$ and $\partial \ln v_S / \partial X_{Fe}$ due to Fe–Mg substitution, range between -0.05 and -0.46 and between -0.08 and -0.74 respectively in the main mantle minerals. The corresponding heterogeneity ratios $R = \partial \ln v_S / \partial \ln v_P$ for Fe–Mg substitution range between 0.9 and 1.7 suggesting that the effect of this substitution is very different in different solid solutions systems. More limited experimental results for Ca–Mg substitution and Al enrichment in pyroxenes and garnets were also evaluated. Only Ca–Mg substitution in garnets is found to produce large (>2.0) values of R. Heterogeneity parameters at upper mantle and transition zone conditions can be substantially different from ambient P-T values in some cases. Using a first-order approximation of the effect of Fe–Mg substitution on the elastic properties of the most relevant upper mantle rocks, we find that the sensitivities of seismic velocities to Fe enrichment can vary as much as 2–3 times between the different rock types. We estimate that in the upper mantle the value of $\partial \ln v_S / \partial \ln v_P$ for pyrolite, piclogite and harzburgite decreases from 1.5 to 1.0 at the base of the transition zone, while it only decreases from 1.5 to 1.3 in mid ocean ridge basalt eclogite, which is enriched in garnet. We also estimate that the seismic effect of lateral changes in lithology from average mantle to subducted slab rocks decreases in intensity at upper mantle and transition zone depths, in agreement with seismic tomographic models. Information about the effects of Ca and Al enrichment are still too incomplete to make predictions of their effects on whole rocks, but they could be relevant based on our limited information.

1. INTRODUCTION

Earth's Deep Mantle: Structure, Composition, and Evolution
Geophysical Monograph Series 160
Copyright 2005 by the American Geophysical Union
10.1029/160GM18

The improvement of seismological models of the Earth mantle in recent years [e.g. *Kennett et al.*, 1998; *Masters et al.*, 2000; *Gung et al.*, 2003] has focused attention to the interpretation of a new level of details. The understanding of deviations from the radially averaged reference model requires more complete knowledge of the mechanical behavior of minerals at the conditions of the Earth interior. The majority of the heterogeneity present in global seismic tomography models can be interpreted as the effect of the thermal structure of the mantle [e.g. *Grand et al.*, 1997; *Antolik et al.*, 2003]. This interpretation has been confirmed by combined inversions of seismic and geodynamic data related to mantle flow [*Forte and Mitrovica*, 2001]. The presence of large compositional heterogeneities has been advocated based on the low correlation or anticorrelation of anomalies of bulk and shear seismic velocities at about 100 km depth and in the deepest part of the lower mantle [*Forte and Woodward*, 1997; *van der Hilst and Kárason*, 1999].

Enrichment and depletion in the average Fe content of lithospheric rocks is invoked as a mechanism of compositional stabilization of the thermal heterogeneity between the uppermost subcontinental and suboceanic mantle [*Jordan*, 1978]. This hypothesis is both supported by the available geochemical sampling of the mantle, represented by xenoliths [*Jordan*, 1979; *Lee*, 2003], and fits to geodynamical constraints [*Forte and Perry*, 2000]. The various interpretations ultimately depend on the correlation between seismic velocity anomalies and density or temperature derived from mineral physics [e.g. *Karato*, 1993; *Karato and Karki*, 2001] thereby stimulating a need for both improvement of the resolving power of tomographic models and a parallel development of a larger and more consistent base of data from mineral physics.

A major effort has been devoted to the development of in-situ high-pressure and high-temperature measurements of the thermoelastic properties of mantle minerals [*Zha et al.*, 2000; *Sinogeikin et al.*, 2004]. However, in spite of the technical advancement in high-pressure mineral physics, the basic question of the effects of compositional variations, high pressure, and high temperature on the elastic properties of the relevant minerals of the mantle has not yet been addressed comprehensively. Indeed the main components of mantle rocks are all solid solutions, often complex mixtures of three or more endmembers, and the effect of mixing is not fully taken into account in the extant mineralogical models [*Duffy and Anderson*, 1989; *Vacher et al.*, 1998; *Cammarano et al.*, 2003].

In this chapter we will evaluate the effect of composition on the elastic properties of upper mantle minerals and model rocks using an elasticity dataset derived solely from single-crystal experimental techniques. We will first calculate the sensitivity of sound wave velocities in olivine and aluminum-bearing garnets to Fe–Mg and Ca–Mg substitutions

by combining new experimental results by Brillouin spectroscopy [*Jiang et al.*, 2004a, 2004b; *Speziale et al.*, 2004] with extant data. We will then extend our calculation to the effect of Fe–Mg substitution, and where possible that of Al or hydrogen incorporation, in different families of minerals of the upper mantle. We will finally estimate the effect of Fe–Mg substitution in the rocks of the upper mantle. Our results give a picture of the strengths and the limitations of the currently available single-crystal experimental data on the elastic properties of mantle minerals. The inconsistency between our estimations and the results of other studies highlights the importance of a reasoned and consistent choice of the database to be used in constructing a mineralogical model of the Earth's mantle.

2. METHODS

2.1. Elasticity Dataset

We use an elasticity dataset based on a large number of experimental results from single-crystal techniques (e.g. Brillouin scattering, impulsive stimulated scattering, resonant ultrasound spectroscopy, GHz ultrasonic interferometry) in order to place constraints on the effects of Fe–Mg substitution in different families of mantle minerals. Single-crystal elasticity measurements allow us to determine the whole elastic tensor which completely defines the elastic anisotropy, gives information about interatomic interactions and integrates the information from crystallography. In selecting the database of elastic moduli, we systematically prefer single-crystal results with respect to measurements on polycrystalline samples (e.g. by ultrasonic techniques), because of potential grain size inhomogeneity, texture effects and porosity in the polycrystalline samples that can affect the validity of these results. Direct results from elasticity measurements are also emphasized over indirect determination of bulk moduli and their pressure derivatives from high-pressure x-ray diffraction because the latter are the results of fitting experimentally measured volumes to a pressure-volume equation of state, and can be affected by significant correlation [e.g. *Angel*, 2000].

2.2. Olivine and Ternary Garnets

We compute the sensitivity of the aggregate acoustic velocities to Fe–Mg substitution in olivine, $(Mg,Fe)_2SiO_4$ as well as the sensitivity to Fe–Ca–Mg substitutions in ternary garnets in the subsystem pyrope $(Mg_3Al_2Si_3O_{12})$, almandine $(Fe_3Al_2Si_3O_{12})$, grossular $(Ca_3Al_2Si_3O_{12})$, at upper mantle conditions. By combining our new single-crystal high-pressure Brillouin spectroscopy results for fayalite (Fe_2SiO_4) [*Speziale et al.*, 2004] and for three garnet compositions [*Jiang et al.*, 2004a, 2004b] with extant data for Mg-rich olivine [*Zha et al.*, 1996, 1998; *Abramson et al.*, 1997] and pyrope [*Sinogeikin and Bass*, 2000], we have a large coverage of the effect of chemical variation on the full elastic tensor up to upper mantle pressures for both systems.

We first derive the endmember properties of olivine by averaging available data (Table 1). In the case of ternary aluminum garnets in the subsystem $(Mg,Fe,Ca)_3Al_2Si_3O_{12}$ we proceeded by least square fitting of the large dataset of natural and synthetic garnets available in the literature. We extracted a complete set of properties for pyrope ($Mg_3Al_2Si_3O_{12}$), almandine ($Fe_3Al_2Si3O_{12}$), and grossular ($Ca_3Al_2Si3O_{12}$) reported in Table 2.

We generate mixture densities for olivine by averaging endmember densities using molar fractions as weighting factors. In the case of the ternary garnet (here treated as a mixture with complete disordered distribution of Fe, Mg, and Ca in the dodecahedral site) the densities are calculated using a regular solution model for volume mixing [*Geiger*, 1999]. We calculate thermal expansion coefficient, bulk and shear moduli, and their pressure and temperature derivatives by

Table 1. Thermoelastic Parameters Used in the Computations of Seismic Velocity of Olivine Mixtures.

Parameter	Forsterite		Fayalite	
A) ρ_0 (Mg/m³)	3.222	(7)	4.388	(9)
B) K_S (GPa)	128.7	(5)	137.6	(3)
C) $(\partial K_S/\partial P)_T$	4.2	(2)	4.85	(5)
D) G (GPa)	81.6	(3)	51.2	(2)
E) $(\partial G/\partial P)_T$	1.6	(2)	1.8	(1)
F) $\partial^2 G/\partial P^2$ (GPa⁻¹)	-0.021	(2)*	-0.11	(8)
G) γ_0	1.29	(2)	1.21	(3)
H) C_p (J/gK)	0.84		0.67	
I) $\partial K_S/\partial T$ (GPa/K)	-0.016	(1)	-0.022	(5)
J) $\partial G/\partial T$ (GPa/K)	-0.014	(1)	-0.013	(1)
K) α_1 (10^{-5} K⁻¹)	2.85		2.39	
L) α_2 (10^{-8} K⁻²)	1.01		1.15	
M) α_3 (K)	-0.384		-0.052	

Numbers in parentheses are 1 standard deviation uncertainty in the last digits.

$\alpha_1, \alpha_2, \alpha_3$, are coefficients to calculate the thermal expansion coefficient: $\alpha(T) = \alpha_1 + \alpha_2 T + \alpha_3/T^2$, where T is expressed in K.

* Required by truncation of the third-order Eulerian strain expansion.

References—Forsterite: A) *Anderson and Isaak* [1995]; B) *Isaak et al.* [1989], *Zha et al.* [1996]; C) *Zha et al.* [1996]; D) *Isaak et al.* [1989], *Zha et al.* [1996]; E) *Zha et al.* [1996]; F) calculated to be consistent with a third-order Eulerian strain fit; G) *Anderson and Isaak* [1995]; H) *Anderson and Isaak* [1995]; I) *Sumino et al.* [1977], *Isaak et al.* [1989]; J) *Sumino et al.* [1977], *Isaak et al.* [1989]; K–M) *Fei* [1995].
Fayalite : A–F) *Speziale et al.* [2004]; G–H) *Anderson and Isaak* [1995]; I) *Isaak et al.* [1993]; J) *Isaak et al.* [1993]; K–M) *Fei* [1995].

Table 2. Thermoelastic Parameters Used in the Computations of Seismic Velocity of Garnet Mixtures

Parameter	Pyrope		Almandine		Grossular	
A) ρ_0 (Mg/m³)	3.565	(1)	4.312	(2)	3.600	(1)
B) K_S (GPa)*	171	(3)	175	(2)	168	(1)
C) $(\partial K_S/\partial P)_T$*	4.1	(3)	4.9	(4)	3.9	(2)
D) G (GPa)*	94	(2)	96	(2)	109	(4)
E) $(\partial G/\partial P)_T$*	1.3	(2)	1.4	(1)	1.1	(1)
F) γ_0	1.16	(9)	0.90	(9)	1.04	(9)
G) C_p (J/gK)	0.814		0.714		0.736	
H) $\partial K_S/\partial T$ (GPa/K)*	-0.021	(2)	-0.022	(2)	-0.015	(1)
I) $\partial G/\partial T$ (GPa/K)*	-0.008	(1)	-0.012	(1)	-0.013	(1)
J) α_1 (10^{-5} K⁻¹)	2.30		1.78		1.95	
K) α_2 (10^{-8} K⁻²)	0.596		1.24		0.809	
L) α_3 (K)	-0.454		-0.507		-0.497	

$\alpha_1, \alpha_2, \alpha_3$: as in Table 1.
*The endmember values were determined by least-square fitting of the available single-crystal data for natural mixtures.
References—Pyrope: A) *Armbruster et al.* [1992]; B) *O'Neill et al.* [1991], *Sinogeikin and Bass* [2000]; C) *Sinogeikin and Bass* [2000]; D) *Isaak and Graham* [1976], *O'Neill et al.* [1991], *Sinogeikin and Bass* [2000]; E) *Sinogeikin and Bass* [2000]; F–G) *Watanabe* [1982]; H–I) *Suzuki and Anderson* [1983], *Isaak et al.* [1992]; J–L) *Skinner* [1956].
Almandine: A) *Armbruster et al.* [1992]; B) *Isaak and Graham* [1976], *Babuška et al.* [1978], *Jiang et al.* [2004b]; C) *Jiang et al.* [2004b]; D) *Isaak and Graham* [1976], *Babuška et al.* [1978], *Jiang et al.* [2004b]; E) *Jiang et al.* [2004b]; F–G) *Watanabe* [1982]; H–I) *Isaak and Graham* [1976], *Sumino and Nishizawa* [1978]; J–L) *Skinner* [1956].
Grossular: A) *Novack and Gibbs* [1971]; B) *Bass* [1986], *O'Neill et al.* [1989], *Isaak et al.* [1992], *Chai et al.* [1997]; *Jiang et al.* [2004b]; C) *Jiang et al.* [2004b]; D) *Bass* [1986], *O'Neill et al.* [1989], *Isaak et al.* [1992]; *Chai et al.* [1997]; *Jiang et al.* [2004b]; E) *Jiang et al.* [2004b]; F-G) *Anderson and Isaak* [1995]; H-I) *Suzuki and Anderson* [1983], *Isaak et al.* [1992]; J-L) *Skinner* [1956].

linear combination of the endmember properties using their molar fractions as weighting coefficients. We apply a linear mixing model because the available experimental data, both for natural and synthetic materials, do not show well resolved non-linearities.

We then compute acoustic velocities and densities of mixtures with arbitrary compositions along adiabatic pressure-temperatures paths with foot temperature of 1673 K, compatible with model geotherms for a radially averaged mantle. The zero-pressure density, bulk and shear modulus and pressure derivatives of the mixtures at 1673 K are calculated following the procedure outlined by *Duffy and Anderson* [1989]. Aggregate bulk and shear moduli along the adiabatic profiles are calculated with third order Eulerian finite strain equations [*Davies and Dziewonski*, 1975; *Duffy and Anderson*, 1989] except for the case of olivine where we use fourth order Eulerian strain equations (Table 1) because of the non-linear pressure dependence of the shear modulus G observed in fayalite [*Speziale et al.*, 2004]. We investigate selected paths in the compositional space of ternary garnets all intersecting at the "standard" composition for mantle garnets, which is 72 mol% pyrope, 14 mol% almandine, 14 mol% grossular [e.g. *McDonough and Rudnick*, 1998; *Lee*, 2003]. The results of the calculations are reported in Tables 3 and 4.

2.3. Other Mineral Families

In accordance with the philosophy of this study, we also use a dataset based on single-crystal elasticity data for other mantle mineral families. We limit our attention to ambient conditions because the single-crystal elasticity data at high pressure or high temperature available for these systems are largely incomplete. We calculate the sensitivities of aggregate acoustic velocity to Fe–Mg exchange in

orthopyroxens, clinopyroxenes of the diopside-hedenbergite series, periclase-wüstite, and in the high-pressure β- and γ- polymorphs of olivine. We determine the sensitivity of acoustic velocity to Al(Fe,Mg) substitution in orthopyroxenes and in majoritic garnets. We also estimate the effect of incorporation of water in different systems relevant for the Earth's upper mantle. The calculated sensitivities are reported in Tables 5 and 6.

2.4. Average Sensitivities of Mantle Rocks

We compute the effect Fe to Mg substitution on the seismic heterogeneity parameters in four relevant mantle rock types: pyrolite [*Ringwood*, 1975], piclogite [*Bass and Anderson*, 1984], harzburgite, and mid-ocean ridge basalt eclogite (MORB eclogite), along an isentropic pressure temperature path with a starting temperature of 1673 K. At each pressure, the bounds on the average rock heterogeneity parameters are calculated as an average, weighted by fractional volume, of the heterogeneity parameters of the single mineral

Table 3. Compositional Effect on the Compressional Velocity Heterogeneity Parameters $\partial \ln v_p / \partial X_i$, $\partial \ln v_S / \partial X_i$, $\partial \ln v_B / \partial X_i$, and the Ratio $R = \partial \ln v_S / \partial \ln v_p$ in Olivine and Ternary Aluminum Garnets.

Variable	Olivine Fe/(Fe+Mg)	Garnet Fe/(Fe+Mg) Ca/Fe = 1	Garnet Fe/(Fe+Mg) Ca/Mg = 0.19	Garnet Ca/(Ca+Mg) Fe/Mg = 0.19	Olivine + MgO Fe/(Fe+Mg) [K and K, 2001]	Whole rock Fe/(Fe+Mg) [Jordan, 1979]
$\partial \ln v_p / \partial X_i$						
1 bar, 300 K	-0.24 (1)	-0.05 (1)	-0.09 (1)	0.03 (1)	-0.4 (1)	-0.27
1 bar, 1673 K	-0.27 (1)	-0.08 (1)	-0.11 (1)	0.03 (1)		
4 GPa (high T)	-0.25 (1)	-0.08 (1)	-0.10 (1)	0.02 (1)		
8 GPa (high T)	-0.24 (1)	-0.08 (1)	-0.10 (1)	0.02 (1)		
14 GPa (high T)	-0.24 (1)	-0.08 (1)	-0.09 (1)	0.01 (1)		
$\partial \ln v_S / \partial X_i$						
1 bar, 300 K	-0.37 (1)	-0.08 (1)	-0.08 (1)	0.08 (1)	-0.5 (1)	-0.34
1 bar, 1673 K	-0.41 (1)	-0.08 (1)	-0.12 (1)	0.05 (1)		
4 GPa (high T)	-0.39 (1)	-0.08 (1)	-0.12 (1)	0.04 (1)		
8 GPa (high T)	-0.38 (1)	-0.08 (1)	-0.12 (1)	0.03 (1)		
14 GPa (high T)	-0.40 (1)	-0.09 (1)	-0.11 (1)	0.03 (1)		
$\partial \ln v_B / \partial X_i$						
1 bar, 300 K	-0.14 (1)	-0.08 (1)	-0.08 (1)	0.00 (1)	-0.2 (1)	-0.21
1 bar, 1673 K	-0.16 (1)	-0.08 (1)	-0.10 (1)	0.02 (1)		
4 GPa (high T)	-0.15 (1)	-0.08 (1)	-0.09 (1)	0.01 (1)		
8 GPa (high T)	-0.15 (1)	-0.08 (1)	-0.08 (1)	0.01 (1)		
14 GPa (high T)	-0.15 (1)	-0.07 (1)	-0.08 (1)	0.01 (1)		
$\partial \ln v_S / \partial \ln v_p$						
1 bar, 300 K	1.54 (8)	1.6 (4)	0.9 (1)	2.7 (6)	1.3 (4)	1.3
1 bar, 1673 K	1.52 (7)	1.0 (2)	1.1 (1)	1.7 (5)		
4 GPa (high T)	1.56 (7)	1.0 (2)	1.2 (2)	2.0 (7)		
8 GPa (high T)	1.58 (8)	1.0 (2)	1.2 (2)	1.5 (7)		
14 GPa (high T)	1.67 (8)	1.1 (2)	1.2 (2)	3 (2)		

High temperature values are calculated along an adiabat with 1673 K foot temperature. K and K: *Karato and Karki* [2001].

Table 4. Logarithmic Derivative of the Compressional Velocity with Respect to Density, $\partial \ln v_P / \partial \ln \rho$, $\partial \ln v_S / \partial \ln \rho$, and $\partial \ln v_B / \partial \ln \rho$ for Compositional Variations in Olivine and Ternary Aluminum Garnets.

Variable	Olivine Fe/(Fe+Mg)	Garnet Fe/(Fe+Mg) Ca/Fe = 1	Garnet Fe/(Fe+Mg) Ca/Mg = 0.19	Garnet Ca/(Ca+Mg) Fe/Mg = 0.19	Olivine + MgO Fe/(Fe+Mg) [K and K, 2001]	Whole rock Fe/(Fe+Mg) [Jordan, 1979]
$\partial \ln v_P / \partial \ln \rho$						
1 bar, 300 K	-0.67 (5)	-0.26 (3)	-0.52 (5)	-1.6 (2)	-0.8 (4)	-0.84
1 bar, 1673 K	-0.76 (6)	-0.40 (4)	-0.63 (6)	-1.7 (2)		
4 GPa (high T)	-0.69 (7)	-0.39 (4)	-0.58 (6)	-0.9 (1)		
8 GPa (high T)	-0.67 (6)	-0.38 (4)	-0.60 (6)	-0.7 (1)		
14 GPa (high T)	-0.66 (6)	-0.38 (4)	-0.50 (6)	-0.3 (1)		
$\partial \ln v_S / \partial \ln \rho$						
1 bar, 300 K	-1.1 (1)	-0.40 (4)	-0.48 (5)	-3.7 (4)	-1.0 (4)	-1.06
1 bar, 1673 K	-1.2 (1)	-0.38 (4)	-0.69 (7)	-3.1 (3)		
4 GPa (high T)	-1.1 (1)	-0.40 (4)	-0.72 (7)	-2.0 (2)		
8 GPa (high T)	-1.1 (1)	-0.41 (4)	-0.70 (7)	-1.7 (2)		
14 GPa (high T)	-1.0 (1)	-0.43 (4)	-0.62 (7)	-2.0 (2)		
$\partial \ln v_B / \partial \ln \rho$						
1 bar, 300 K	-0.37 (4)	-0.39 (5)	-0.44 (4)	0.00 (1)	-0.4 (3)	-0.66
1 bar, 1673 K	-0.44 (4)	-0.37 (4)	-0.57 (6)	-2.3 (2)		
4 GPa (high T)	-0.41 (4)	-0.38 (4)	-0.51 (5)	-0.8 (1)		
8 GPa (high T)	-0.40 (5)	-0.37 (4)	-0.47 (5)	-0.5 (1)		
14 GPa (high T)	-0.39 (4)	-0.34 (4)	-0.48 (5)	-0.3 (1)		

High temperature values are calculated along an adiabat with 1673 K foot temperature. K and K: *Karato and Karki* [2001].

phases. The modal compositions for the different rock types are inferred from experimental and computational studies [*Irifune and Ringwood*, 1987b; *Ita and Stixrude*, 1992] and are reported in Table 7.

We approximate the pressure and temperature effects on seismic heterogeneity induced by Mg-Fe substitution in the system ringwoodite–γ-Fe$_2$SiO$_4$ by combining isobaric temperature dependence and isothermal pressure dependence of the heterogeneity parameters from the available experimental data. Due to the absence of single-crystal high-pressure data on the effect of Fe–Mg substitution in pyroxenes and β-(Mg,Fe)$_2$SiO$_4$, we adopt the data available for these minerals groups at ambient conditions. The sensitivity of majoritic garnet to Fe–Mg substitution is assumed to be equivalent to that of Al-garnet.

We have first tested our averaging scheme in the case of arbitrary mixtures of olivine and ternary garnets in different proportions by comparing our calculations with heterogeneity parameters calculated as the relative variation of the average of Hashin–Shtrikmann bounds [*Hashin and Shtrikman*, 1961, 1963] for the velocities of the same mixtures in response to variation of the Fe content of the two component minerals. The parameters calculated at different pressures and different minerals proportions are in agreement between the two methods, within the estimated uncertainties.

3. RESULTS AND DISCUSSION

3.1. Olivine and Ternary Aluminum Garnets

The substitution of Fe for Mg in olivine results in a strong shear modulus weakening [*Graham et al.*, 1988; *Isaak et al.*, 1993], with 37% decrease of the shear modulus G, across the solid solution, and stiffening of the bulk modulus K corresponding to 7% increase from forsterite (α-Mg$_2$SiO$_4$) to fayalite (α-Fe$_2$SiO$_4$) [*Speziale et al.*, 2004]. The pressure derivative of the bulk modulus also increases with the increase of Fe content (Table 1), and the difference between the bulk moduli of forsterite and fayalite is increased to 9% at 10 GPa [*Speziale et al.*, 2004]. The pressure derivative of the shear modulus of olivine is less sensitive to the Fe content than that of the bulk modulus (Table 1). The difference of compressional sound velocity between forsterite and fayalite is 20.2% at ambient pressure and remains 17% at the pressures consistent with the bottom of the upper mantle. The variation of shear velocity is even larger: 32.1% at ambient pressure and 26% at 13 GPa.

The compositional effect on both the shear and compressional velocities at constant pressure can be represented by velocity heterogeneity parameters (also referred as compositional heterogeneity parameters in the following discussions) expressed as $(\partial \ln v_{P,S} / \partial X_i)_P$, [*Jordan*, 1979] where V$_P$ and V$_S$

306 MANTLE MINERALS' ELASTICITY

Table 5. Compositional Effect on the Compressional, Shear, and Bulk Velocity Heterogeneity Parameters in Different Solid Solution Series Calculated at Ambient Conditions.

Solid solution series	X	$\partial \ln v_P/\partial X$	$\partial \ln v_S/\partial X$	$\partial \ln v_B/\partial X$	$\partial \ln v_S/\partial \ln v_P$	$\partial \ln v_P/\partial \ln \rho$	$\partial \ln v_S/\partial \ln \rho$	$\partial \ln v_B/\partial \ln \rho$
Forsterite–fayalite	Fe/(Mg+Fe)	-0.24 (1)	-0.37 (1)	-0.14 (1)	1.54 (8)	-0.67 (5)	-1.1 (1)	-0.37 (4)
Al-garnet (Ca/Mg = constant)	Fe/(Mg+Fe)	-0.09 (1)	-0.08 (1)	-0.09 (2)	0.9 (1)	-0.52 (5)	-0.48 (5)	-0.44 (4)
Al-garnet (Ca/Fe = constant)	Fe/(Mg+Fe)	-0.05 (1)	-0.08 (1)	-0.08 (1)	1.6 (3)	-0.26 (3)	-0.40 (4)	-0.39 (5)
Orthoenstatite–orthoferrosilite	Fe/(Mg+Fe)	-0.21 (2)	-0.30 (4)	-0.14 (2)	1.4 (3)	-0.92 (9)	-1.3 (3)	-0.59 (6)
Diopside–hedenbergite	Fe/(Mg+Fe)	-0.09 (1)	-0.15 (2)	-0.04 (1)	1.7 (3)	-0.79 (8)	-1.4 (1)	-0.37 (4)
Wadsleyite–β-Fe$_2$SiO$_4$	Fe/(Mg+Fe)	-0.37 (5)	-0.52 (6)	-0.23 (6)	1.4 (2)	-1.2 (2)	-1.6 (2)	-0.8 (4)
Ringwoodite–γ-Fe$_2$SiO$_4$	Fe/(Mg+Fe)	-0.18 (3)	-0.28 (4)	-0.09 (5)	1.6 (3)	-0.48 (4)	-0.74 (5)	-0.26 (5)
Periclase–wüstite	Fe/(Mg+Fe)	-0.46 (5)	-0.74 (7)	-0.27 (2)	1.6 (2)	-1.0 (1)	-1.5 (2)	-0.35 (4)
Mg,Fe-Opx–Mg-Tschermak	Al^{VI}/(Al^{VI}+Mg+Fe)*	0.24 (4)	0.13 (4)	0.33 (6)	0.6 (5)	7 (2)	4 (2)	10 (4)
Majorite–pyrope	Al/(Al+Mg^{VI}+Si^{VI})*	0.03 (1)	0.04 (1)	0.02 (1)	1.3 (3)	1.2 (1)	1.5 (1)	0.8 (1)
Al-garnet (Fe/Mg = constant)	Ca/(Mg+Ca)	0.03 (1)	0.08 (1)	0.00 (1)	2.7 (6)	-1.6 (2)	-3.7 (4)	0.00 (3)

*Al^{VI}, Mg^{VI}, Si^{VI} are in octahedral coordination.

References—Orthoenstatite–orthoferrosilite: *Kumazawa* [1969]; *Frisillo and Barsch* [1972]; *Weidner et al.* [1978]; *Bass and Weidner* [1984]; *Duffy and Vaughan* [1988]; *Webb and Jackson* [1993]; *Jackson et al.* [1999]. Diopside–hedenbergite: *Kandelin and Weidner* [1988]; *Collins and Brown* [1998]; *Isaak and Ohno* [2003]. Wadsleyite–β-Fe₂SiO₄: *Sawamoto et al.* [1984]; *Sinogeikin et al.* [1998]; *Zha et al.* [1998]. Ringwoodite–γ-Fe₂SiO₄: *Weidner et al.* [1984]; *Rigden and Jackson* [1991]; *Rigden et al.* [1992]; *Sinogeikin et al.* [1997]; *Jackson et al.* [2000]; *Sinogeikin et al.* [2003a]. Majorite–pyrope: *Bass and Kanzaki* [1990]; *Yeganeh-Haeri et al.* [1990]; *O'Neill et al.* [1991]; *Sinogeikin and Bass* [2000]; *Sinogeikin et al.* [2002]. Periclase–wüstite: *Sinogeikin and Bass* [2000]; *Jacobsen et al.* [2002].

are the compressional and shear velocity and X_i represents the molar fraction of Fe, X_{Fe} = Fe/(Fe + Mg). Another important parameter is the compositional heterogeneity parameter ratio $R = (\partial \ln v_S/\partial \ln v_P)_P$, defined as the ratio of $(\partial \ln v_S/\partial X_i)_P$ and $(\partial \ln v_P/\partial X_i)_P$. In order to simplify the notation, all the compositional heterogeneity parameters and ratios will be hereafter considered as defined at constant pressure unless differently specified.

Heterogeneity parameters for olivine along an adiabatic path starting at 1673 K are presented in Figure 1. Consistent with our model of linear mixing of the endmember densities and aggregate moduli, we observe that the calculated effect of Fe substitution for Mg on shear velocity of olivine is slightly sensitive to the composition of the olivine itself (inset in Figure 1). However, within uncertainty, the second order coefficient is not constrained, and we limit our analysis to the linear term. Our estimation for $\partial \ln v_P/\partial X_{Fe}$ and $\partial \ln v_S/\partial X_{Fe}$ at ambient conditions are -0.24 ± 0.01 and -0.37 ± 0.01. The compositional heterogeneity parameters ratio $R = \partial \ln v_S/\partial \ln v_P$, is 1.54 ± 0.08. Our results are only in marginal agreement with the results presented by *Karato and Karki* [2001] and in disagreement with the estimates by *Jordan* [1979] for an average peridotite. We can also resolve the effect of pressure and temperature on the compositional heterogeneity parameters. At pressure above 8 GPa, $\partial \ln v_S/\partial X_{Fe}$ reverts its trend and it starts to increase (in absolute value), while $\partial \ln v_P/\partial X_{Fe}$ becomes pressure insensitive (Figure 1). This change of behavior at depth is the consequence of the very different pressure dependencies of both bulk and shear moduli of the two endmembers. In fact it also appears at pressures above 10 GPa at ambient

temperature. Metastability and incipient shear softening of the Fe-rich endmember [*Speziale et al.*, 2004] could be the explanation of the reversal. This observation emphasizes the need for direct measurements at high pressure and temperature to avoid possible complications associated with room temperature metastability. The effect on elastic velocity of density variations associated with chemical exchange is expressed by the logarithmic derivatives $\partial \ln v_{P,S}/\partial \ln \rho$. The logarithmic derivative $\partial \ln v_P/\partial \ln \rho$ and for Fe–Mg substitution in olivine is -0.67, lower but still in agreement within uncertainty with the estimation of *Karato and Karki* [2001]. The value of $\partial \ln v_S/\partial \ln \rho$ is -1.1, in excellent agreement with that reported by *Karato and Karki* [2001]. The logarithmic derivatives of compressional and shear velocity decrease 10% and 15% respectively with pressure along an adiabatic profile with foot temperature of 1673 K (Table 4).

The effect of Fe–Ca–Mg substitutions on the elastic velocities of garnets in the pyrope–almandine–grossular subsystem is more complex than the simple Fe–Mg substitution in olivine. Substitution of Fe for Mg causes as much as 8.5% and 8.2% decreases of compressional and shear velocity at ambient pressure, and 7.2% and 7.7% respectively, at 14 GPa. Substitution of Ca for Fe produces 10.2% and 14.9% increases of compressional and shear velocity at ambient pressure, and 7.3% and 12% increases at 14 GPa [*Jiang et al.*, 2004b].

In the case of ternary garnets we define two heterogeneity parameters X_{Fe} = Fe/(Fe + Mg) and X_{Ca} = Ca/(Ca + Mg) for Fe–Mg and for Ca–Fe substitutions, respectively. In addition to the heterogeneity parameter we have to specify a second ratio to completely determine the compositional path along which the heterogeneity parameter is defined. As an

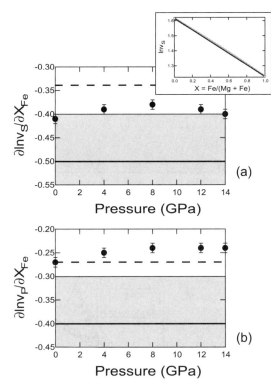

Figure 1. Compositional heterogeneity parameters of (a) shear and (b) compressional velocity for Fe–Mg substitution in olivine calculated along an adiabatic pressure / temperature path with foot temperature of 1673 K. Inset: the sensitivity of shear velocity, here plotted as $\ln v_S$, to Mg–Fe substitution calculated at ambient conditions (gray area: 1σ estimated uncertainty) suggests the existence of second order effects (see text). Dashed line: sensitivity of elastic wave velocity of lherzolite by Jordan [1979]. Continuous line: sensitivity for (Mg,Fe)O and olivine by *Karato and Karki* [2001] with uncertainty (shaded area).

example, a mixture of 70% Mg component, 10% Fe component and 20% of Ca component has the same X_{Fe} as a mixture of 77% Mg component, 11% Fe component and 12% Ca component, but it has different mixture properties because it is richer in the Ca component. In fact, the apparent effect of Fe enrichment is different in the case in which we maintain Ca/Fe = 1, compatible with the ratio for average peridotitic garnets (see section 2.2), with respect to the case in which Ca/Mg is fixed to the value 0.19 typical of average peridotitic garnets (Figure 2). The depth dependence of the heterogeneity parameters for the shear velocity ($\partial \ln v_S / \partial X_{Fe}$) along the adiabatic path have opposite sign in the two cases and the heterogeneity parameters tend to converge at pressures of the transition zone (Figure 2). The same effect can be also observed for $\partial \ln v_P / \partial X_{Fe}$.

Previous estimates of the effect of Fe to Mg substitution in an average peridotitic mineral assemblage [*Jordan*, 1979]

are in substantial disagreement with the results for garnets both for shear and compressional heterogeneity parameters (Figure 2). The disagreement cannot be compensated by olivine (Figure 2a). In garnets, we observe non-linearity in the calculated sensitivity of shear velocity to the Fe–Mg substitution, as in olivine. However, the overall uncertainty associated with the calculation does not allow us to resolve this second order effect.

The logarithmic derivatives of shear and compressional elastic wave velocities with respect to density, $\partial \ln v_i / \partial \ln \rho$, for Fe to Mg substitution in garnet are systematically smaller than those for olivine (Table 4) and both are smaller than the estimates presented by *Karato and Karki* [2001]. In addition,

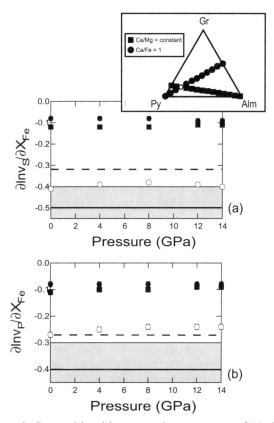

Figure 2. Compositional heterogeneity parameters of (a) shear and (b) compressional elastic wave velocity for Fe–Mg substitution in ternary aluminum-garnet in the case of Ca/Mg = constant (solid squares) and Ca/Fe = 1 (solid circles). The heterogeneity parameters are calculated along an adiabatic pressure / temperature path with foot temperature of 1673 K. Heterogeneity parameters of olivine (open circles) are reported for comparison. Dashed line: sensitivity of elastic wave velocity of lherzolite by Jordan [1979]. Continuous line: sensitivity for (Mg,Fe)O and olivine by *Karato and Karki* [2001] with uncertainty (shaded area). Inset: Compositional variation due to Fe–Mg substitution in ternary garnets in the cases examined in this study.

Table 6. Sensitivity of the Velocity Heterogeneity Parameters to Hydration in Different Solid Solution Series Calculated at Ambient Conditions.

Solid solution series	X	$\partial \ln v_P/\partial X$	$\partial \ln v_S/\partial X$	$\partial \ln v_B/\partial X$	$\partial \ln v_S/\partial \ln v_P$	$\partial \ln v_P/\partial \ln\rho$	$\partial \ln v_S/\partial \ln\rho$	$\partial \ln v_B/\partial \ln\rho$
Olivine humite group	$H^*/(H^*+Si)$	-0.2 (1)	-0.2 (1)	-0.2 (1)	1.0 (7)	1.7 (3)	1.1 (3)	2.2 (2)
Ringwoodite hydrous ringwoodite	$2H^*/(2H^*+Mg)$	-0.5 (1)	-0.5 (3)	-0.6 (2)	1.0 (6)	1.3 (3)	1.3 (6)	1.3 (4)
Mg,Fe-Opx–Mg-Tschermak	$H^*/(H^*+Si)$	-0.4	-0.4	-0.4	1.0	1	1	1
Periclase–wüstite brucite	$2H^*/(O-2H^*)$	-0.4	-0.5	-0.4	1.3	1	1	1

$(H^* = H/4.)$

References—Olivine–humite group: *Zha et al.* [1996]; *Beckman Fritzel and Bass* [1997]; *Sinogeikin and Bass* [1999]. Ringwoodite–γ-Fe$_2$SiO$_4$: *Weidner et al.* [1984]; *Rigden et al.* [1992]; *Inoue et al.* [1998]; *Jackson et al.* [2000]; *Wang et al.* [2003]. Grossular–hydrogrossular: *Bass* [1989]; *O'Neill et al.* [1989]; *Isaak et al.* [1992]; *O'Neill et al.* [1993]; *Chai et al.* [1997]; *Jiang et al.* [2004b]. Periclase–brucite: *Xia et al.* [1998]; *Sinogeikin and Bass* [2000].

under upper mantle pressures and temperatures, our calculations show that the values of $\partial \ln v_P/\partial \ln\rho$ and $\partial \ln v_S/\partial \ln\rho$ for garnets can vary as much as 50% relative to ambient conditions. In conclusion we cannot generalize the effect of Fe–Mg substitution between these two families of minerals.

We also analyze the effect of Ca enrichment or depletion in garnets at constant Fe/Mg ratio, compatible with that of peridotitic garnets (Figure 3). The effect of pressure is to decrease the value of the heterogeneity parameters both for compressional and shear velocity. At ambient conditions, the compositional heterogeneity parameters ratio R $= \partial \ln v_S/\partial \ln v_P$, due to Ca–Mg substitution, is equal to 2.7 ± 0.6, almost 2 times larger than the ratio related to Fe–Mg substitution in olivine and in garnet. This value is also larger than R in other solid solutions systems, as we will see in the following sections. A large value of R > 2.7 for Ca–Mg substitution has been inferred from comparison of theoretically calculated seismic velocities in perovskites by *Karato and Karki* [2001] who suggest that Ca variation can contribute to large R values in the deep mantle. Although also not a solid solution, the comparison of seismic velocities of orthoenstatite (MgSiO$_3$) and diopside (CaMgSi$_2$O$_6$), based on the available data at ambient conditions [*Weidner et al.*, 1978; *Jackson et al.*, 1999; *Isaak and Ohno*, 2003], yields R = 7 ± 4. The data for Ca–Mg substitution in these three systems all yield very large values of R at ambient conditions. However, our calculated value of R for Ca–Mg substitution in ternary Al-garnet averages at 1.7 ± 0.5 at upper mantle conditions and increases to 3 ± 2 in the transition zone (Table 3). A qualitative explanation of this result is that along our adiabatic path, the effect of temperature which reduces the value of R, is stronger at upper mantle depths while the effect of pressure (which increases R) becomes dominant at transition zone depths due to the shallow temperature gradient. In view of the complex combined effect of

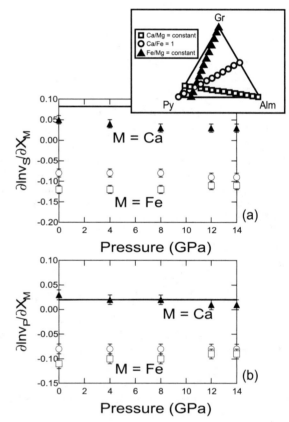

Figure 3. Compositional heterogeneity parameters of shear (a) and compressional velocity (b) for Ca–Mg substitution (solid triangles) in ternary aluminum–garnet calculated along an adiabatic pressure/temperature path with foot temperature of 1673 K. The heterogeneity parameters for Fe–Mg substitution in garnet are plotted for comparison as open symbols (cf. Figure 2). Line: sensitivity of elastic wave velocity for compositional variation between MgSiO$_3$-perovskite and CaSiO$_3$-perovskite by *Karato and Karki* [2001]. Inset: Compositional variation in ternary garnets for Ca–Fe–Mg substitutions examined in this study.

Table 7. Composition of Different Mantle Rock Types Used in Calculations.

Rock type	Bulk composition	Modal composition		Mineral composition		
		Mineral	(vol%)	Mg	Fe	Ca(mol.frac.)
Pyrolite						
SiO_2	44.31	olivine	59	0.88	0.12	
Al_2O_3	5.40	clinopyroxene	13	0.45	0.05	0.50
FeO	8.56	orthopyroxene	9	0.91	0.10	
MgO	37.82	garnet	19	0.82	0.14	0.04
CaO	3.48					
Piclogite						
SiO_2	44.69	olivine	41	0.88	0.12	
Al_2O_3	8.35	clinopyroxene	26	0.44	0.06	0.50
FeO	8.28	orthopyroxene	9	0.87	0.13	
MgO	30.24	garnet	24	0.62	0.19	0.19
CaO	8.43					
MORB eclogite						
SiO_2	47.20	olivine				
Al_2O_3	14.51	clinopyroxene	72	0.35	0.15	0.50
FeO	11.81	orthopyroxene				
MgO	12.10	garnet	28	0.45	0.35	0.25
CaO	13.19					
Harzburgite						
SiO_2	43.64	olivine	81	0.91	0.09	
Al_2O_3	0.65	clinopyroxene				
FeO	7.83	orthopyroxene	19	0.93	0.07	
MgO	46.36	garnet				
CaO	0.50*					

*Neglected in calculations. olivine $(Mg,Fe)_2SiO_4$; clinopyroxene $(Mg,Fe,Ca)_2Si_2O_6$; orthopyroxene $(Mg,Fe,Al)(Si,Al)SiO_3$; garnet $(Mg,Fe,Ca)_3Al_2Si_3O_{12}$.
References—Rock chemical and modal compositions: *Akaogi and Akimoto* [1979]; *Anderson and Bass* [1986]; *Irifune et al.* [1986]; *Irifune and Ringwood* [1987a,b, 1993]; *Ita and Stixrude* [1992, 1993]; *Irifune and Isshiki* [1998]; *Vacher et al.* [1998].

pressure and temperature, the available experimental results at ambient conditions must be used with caution in approximating the effects of Ca–Mg substitution at transition zone and lower mantle conditions.

3.2. Other Systems

The differences that we observed between the effect of Fe to Mg substitution in olivine and garnets highlight inherent risks involved in gross generalizations of compositional effects on the elastic properties of a limited subset of minerals to those of the rocks of the mantle as a whole [e.g. *Karato and Karki*, 2001]. Information about the chemical composition of the mantle should be accompanied by a precise petrological and mineralogical model. For this reason we calculate the heterogeneity parameters for Fe–Mg substitution, and where possible for other substitutions, for other minerals of the mantle based on single-crystal elasticity measurement consistent with the strategy adopted for

garnet and olivine. Unfortunately, the available elastic data for the members of the different families of mantle minerals are fragmentary. For this reason we limit our calculations to ambient conditions. All the computed logarithmic derivatives of compressional and shear velocity presented for the different systems discussed in the following sections are reported in Table 5.

3.2.1. Orthopyroxene. Al-free pyroxene, $(Mg,Fe)SiO_3$, is an abundant component in mantle rocks, with a compositional range varying from 98 to 82 mol% $MgSiO_3$ [*McDonough and Rudnick*, 1998]. Different studies have been performed to determine the whole elastic tensor of different compositions in the orthoenstatite $(MgSiO_3)$–orthoferrosilite $(FeSiO_3)$ solid solution series, at ambient conditions [e.g. *Webb and Jackson*, 1993; *Duffy and Vaughan*, 1988; *Jackson et al.*, 1999]. The available data are mainly limited to Mg-rich compositions with less than 10 mol% orthoferrosilite content augmented by a study of orthoferrosilite [*Bass and Weidner*,

1984]. The available data yield $\partial \ln v_P/\partial X_{Fe} = -0.21 \pm 0.02$, and $\partial \ln v_S/\partial X_{Fe} = -0.30 \pm 0.04$, which are comparable to olivine values ($\partial \ln v_P/\partial X_{Fe} = -0.24 \pm 0.01$, and $\partial \ln v_S/\partial X_{Fe} = -0.37 \pm 0.01$). However, the logarithmic derivatives of velocity with respect to density $\partial \ln v_P/\partial \ln \rho = -0.9 \pm 0.1$, and $\partial \ln v_S/\partial \ln \rho = -1.3 \pm 0.3$ are substantially different from those of olivine.

Chai et al. [1997] measured the elastic tensor of an orthorhombic pyroxene containing 2.6 weight% Al. By comparison with the calculated velocity for an Al-free orthopyroxene with the same Fe/Mg ratio, we calculate an average sensitivity of the heterogeneity parameters to Al enrichment in orthorhombic pyroxene: $\partial \ln v_P/\partial X_{Ts} = 0.24 \pm 0.04$, and $\partial \ln v_S/\partial X_{Ts} = 0.13 \pm 0.04$, where Ts = $MgAl_2SiO_6$ is the Mg-tschermak component. The presence of Al increases the acoustic velocity of orthorhombic pyroxene, in agreement with the observations for other silicate systems. The effect of Al enrichment on density is smaller than that on velocity, and the logarithmic derivatives $\partial \ln v_i/\partial \ln \rho$ are larger than unity albeit with large uncertainties (Table 5).

3.2.2. Clinopyroxene.

A second pyroxene is also present in fertile lherzolites. It is a Ca-rich monoclinic member of the diopside ($MgCaSi_2O_6$)–hedenbergite ($FeCaSi_2O_6$) solid solution series. The composition of Ca-rich clinopyroxene in mantle rocks ranges between 88 and 95 mol% diopside component [*McDonough and Rudnick*, 1998].

Because of the complexity of the monoclinic elastic tensor, only a few studies of the single-crystal elasticity of Ca-rich clinopyroxenes are available in the literature [*Kandelin and Weidner*, 1988; *Collins and Brown*, 1998; *Isaak and Ohno*, 2003]. However, the extant data cover a wide range of compositons, from pure Fe-endmember to almost pure (98 mol%) Mg-endmember.

The seismic heterogeneity parameters for Ca-rich clinopyroxenes are $\partial \ln v_P/\partial X_{Fe} = -0.09 \pm 0.01$, and $\partial \ln v_S/\partial X_{Fe} = -0.15 \pm 0.02$, generally smaller than those of olivine and orthopyroxene and closer to the range of values determined for ternary aluminum garnets. On the other hand, the logarithmic derivatives of velocity with respect to density are in close agreement with those of olivine, once again showing that the relationship between compositional effect on the elastic moduli and on density is structure-specific in the different solid-solution series (Table 5).

3.2.3. β-(Mg,Fe)₂SiO₄ and γ-(Mg,Fe)₂SiO₄.

Different studies of single-crystal elasticity of the high-pressure polymorphs of olivine have been published. Data for β-$(Mg,Fe)_2SiO_4$ (β-phase) are limited to Fe contents in the range 0–9 mol% [e.g. *Zha et al.*, 1998; *Sinogeikin et al.*, 1998]. The effect of Fe–Mg substitution is strong; in fact the heterogeneity parameters $\partial \ln v_P/\partial X_{Fe}$ and $\partial \ln v_S/\partial X_{Fe}$

of β-$(Mg,Fe)_2SiO_4$ are -0.37 ± 0.05 and -0.52 ± 0.06, larger than those of olivine and pyroxenes. Data for γ-$(Mg,Fe)_2SiO_4$ (γ-phase) span a larger compositional range, from pure ringwoodite (γ-Mg_2SiO_4) to γ-$(Mg_{0.75},Fe_{0.25})_2SiO_4$ [*Jackson et al.*, 2000; *Sinogeikin et al.*, 2003]. The heterogeneity parameters $\partial \ln v_P/\partial X_{Fe}$ and $\partial \ln v_S/\partial X_{Fe}$ of γ-$(Mg,Fe)_2SiO_4$ are -0.18 ± 0.03 and -0.28 ± 0.04. They are closer to olivine than those of β-phase (Table 5). However, because of the limited coverage of the compositional variation in the system β-$(Mg,Fe)_2SiO_4$, this result should be considered with caution.

3.2.4. Majorite–pyrope mixture.

Due to the high abundance of garnet–majorite solid solution in the transition zone in mineralogical models of the mantle [*Ringwood*, 1975; *Duffy and Anderson*, 1989], the binary solution series between majorite ($Mg_4Si_4O_{12}$) and pyrope has been the subject of extensive study. Data on the single-crystal elastic tensor of the endmembers and of intermediate compositions are available in the literature [*Sinogeikin and Bass*, 2000, 2002]. The calculated sensitivities of the seismic velocities to the pyrope content are both positive, $\partial \ln v_P/\partial X_{Al} = 0.03 \pm 0.01$, $\partial \ln v_S/\partial X_{Al} = 0.04 \pm 0.01$, where $X_{Al} = Al/(Al+Mg^{VI}+Si^{VI})$, and Mg^{VI} and Si^{VI} refer to atoms in octahedral coordination. This effect is about 5 times smaller than the effect of Al enrichment in orthopyroxenes. The logarithmic derivatives of velocity with respect to density related to Al enrichment in majoritic garnets are positive, $\partial \ln v_P/\partial \ln \rho = 1.2 \pm 0.2$, $\partial \ln v_S/\partial \ln \rho = 1.5 \pm 0.2$ (Table 5).

Recent results on polycrystalline Al-bearing Mg silicate perovskite [*Jackson et al.*, 2004] show that Al- enrichment produces a large R value at ambient conditions, in contrast with the results for orthopyroxenes and pyrope-majorite.

3.2.5. Periclase–wüstite mixture.

Periclase (MgO)–wüstite (FeO) solid solution represents the second most abundant mineral in the lower mantle in the reference mineralogical model of the Earth [e.g. *Jackson*, 1988]. A study of the single-crystal elasticity of the complete solid solution series has been performed at ambient conditions by *Jacobsen et al.* [2002] using GHz-ultrasound interferometry. Periclase has been subject of numerous single-crystal studies at both high-pressure and high-temperature [e.g. *Jackson and Niesler*, 1982; *Chen et al.*, 1998; *Sinogeikin and Bass*, 2000]. The strong nonlinearity in the compositional dependence of the shear modulus causes a compositional dependence of both compressional and shear velocity heterogeneity parameters. However, because the role of nonstoichiometry and defect density in the Fe-rich members of the series remains unclear [e.g. *Jacobsen et al.*, 2002], we limit our estimation to the first order effect only. Our computed heterogeneity parameters are $\partial \ln v_P/\partial X_{Fe} = -0.46 \pm 0.05$, and $\partial \ln v_S/\partial X_{Fe}$

= -0.74 ± 0.07. Both compressional and shear velocity heterogeneity parameters in periclase–wüstite are the largest among the whole set of minerals examined in this study. Moreover, they are much larger than $\partial \ln v_P / \partial X_{Fe}$ = -0.16, and $\partial \ln v_S / \partial X_{Fe}$ = -0.22, calculated by first principles for (Mg,Fe)SiO$_3$ perovskite, the other principal component of the lower mantle [*Kiefer et al.*, 2002].

On balance, we find that the heterogeneity parameter for a single substitution such as Fe–Mg can vary dramatically for different mantle mineral families. The logarithmic compositional heterogeneity parameters for v_P and v_S can vary by a factor of ~10, depending on the mineral structure involved (Table 5).

3.3. Temperature Versus Pressure Effects: Experimental Results

Reliable mineralogical models of the Earth interior must integrate high-temperature and high-pressure experimental results from mineral physics in a coherent thermodynamic and petrological model in order to compute single phase and rock aggregate properties along pressure–temperature–composition paths relevant to the Earth interior. The available high-quality high-temperature elasticity studies performed by resonant ultrasound spectroscopy (see *Anderson and Isaak* [1995] for reference) and by Brillouin spectroscopy [e.g. *Sinogeikin et al.*, 2003] combined with high-pressure results on comparable mineral phases, allow us to separately investigate temperature and pressure effects on the compositional sensitivity of acoustic velocity in three solid solution series: forsterite–fayalite, pyrope–grossular, and ringwoodite–γ-Fe$_2$SiO$_4$. The average relative uncertainties on the pressure and temperature derivatives of the elastic moduli from selected high-quality single-crystal measurements (Tables 1, 2) are always less than 10% and very often smaller than the uncertainties associated with results of other techniques, such as high-pressure and high-temperature polycrystalline ultrasonics and X-ray diffraction [e.g. *Liebermann*, 2000; *Jiang et al.*, 2004 b].

The temperature and pressure effects of Fe–Mg substitution in olivine and ringwoodite–γ-Fe$_2$SiO$_4$ and Ca–Mg substitution in pyrope–grossular are plotted in Figure 4 as relative variation with respect to the heterogeneity parameter, $\Delta v / (v \Delta X_{Fe})$ computed at ambient conditions (Table 5). The experimental results are extrapolated, when needed, using an approach based on Grüneisen theory [e.g. *Duffy and Anderson*, 1989].

In olivine, the pressure and temperature effects have an opposite sign, suggesting a near cancellation of pressure-temperature effects under mantle conditions (Figure 4a, b). In γ-phase (Figure 4c, d) pressure increase causes a negligible decrease of both $\partial \ln v_P / \partial X_{Fe}$ and $\partial \ln v_S / \partial X_{Fe}$, while the

effect of temperature is to decrease $\partial \ln v_P / \partial X_{Fe}$ as much as 35% and increase $\partial \ln v_S / \partial X_{Fe}$ up to 40% at T = 1800 K. The comparison between the two minerals suggests that the Fe to Mg substitution is more effective in decreasing the shear stiffness of the denser γ-phase than that of olivine.

In grossular–pyrope solid solutions, temperature increases $\partial \ln v_P / \partial X_{Ca}$ up to 20% at 1800 K and the effect on $\partial \ln v_S / \partial X_{Ca}$ is to decrease its value by 15% at 1800 K with a variation of the heterogeneity ratio R = $\partial \ln v_S / \partial \ln v_P$ from 2.2 at 300 K to 1.9 at 1800 K. The effect of pressure is that of decreasing both the heterogeneity parameters. However, the effect on the compressional velocity is two times larger than that observed on the shear velocity causing an increase of the heterogeneity ratio R to 2.5 at 10 GPa and 300 K (Figure 4 e,f). This behavior is in good agreement with our results for Ca–Mg substitution in ternary garnet with Fe contents consistent with peridotitic compositions (see section 3.1).

The available data suggest that temperature and pressure dependence of the heterogeneity parameters for the same chemical substitution can be radically different even in closely related systems such as olivine and its high-pressure polymorph γ-phase, suggesting caution in generalizing results obtained at ambient pressure and temperature by direct extrapolation to the conditions and the materials of the deep Earth's interior. The combination of high-quality data collected at high temperature and high pressure is the prerequisite to any reliable modeling or even qualitative estimation of the compositional effect on seismic velocity under mantle conditions. However, in general, the effects of P and T at upper mantle conditions on heterogeneity parameters are smaller than the changes in heterogeneity parameters observed across the different structure types.

3.4. The Effect of Water

The effect of water content on the elastic property of minerals of the mantle is far from being completely explored experimentally. The very few single-crystal elasticity studies of hydrated silicates give us a general idea limited to the olivine system, to γ-phase and grossular [e.g. *Wang et al.*, 2003]. Superhydrous phase B and phase D, two of the high-pressure magnesium silicate hydrous phases which are considered the most probable carriers of water in the deep interior of the Earth, have been subject of single-crystal elasticity measurements by *Pacalo and Weidner* [1996] and *Liu et al.* [2004] respectively. Here we report estimates of the average sensitivity of seismic velocity to water content in the cases of γ-phase and grossular, in which water incorporation does not involve structural modifications (Table 6). We also determine average sensitivities to water content by comparison of Mg-rich olivine with humite-group minerals

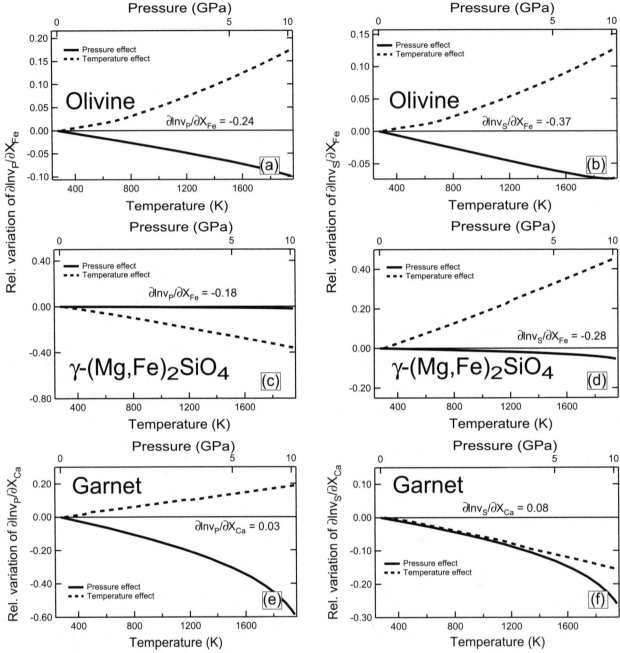

Figure 4. Temperature and pressure dependences of the elastic wave velocity compositional heterogeneity parameters from direct experiments for olivine (a, b), ringwoodite–γ-Fe2SiO4 (c, d), and grossular–pyrope garnets (e, f). Pressure and temperature effects are expressed as relative variations with respect to the heterogeneity parameters at ambient conditions. The pressure scale is not linear so as to match the temperature scale in agreement with the mantle geotherm by *Gasparik and Hutchison* [2000]. Data for olivine are from *Sumino* [1979], *Isaak et al.* [1989], *Isaak* [1992], *Zha et al.* [1996, 1998], *Abramson et al.* [1997], and *Speziale and Duffy* [2004]. Data for ringwoodite–γ-Fe2SiO4 are from *Jackson et al.* [2000] and *Sinogeikin et al.* [2003]. Data for pyrope–grossular are from *Suzuki and Anderson* [1983], *Isaak et al.* [1992], *Sinogeikin and Bass* [2000], and *Jiang et al.* [2004 b].

and of MgO with $Mg(OH)_2$, which instead represent systems where hydration involves structural modifications (Table 6). In this study all the different systems are examined at ambient conditions.

The softening caused by hydration affects in a similar way both compressional and shear velocity. As a consequence, the value of the ratio $R = \partial \ln v_S / \partial \ln v_P$ is equal to 1 for ring-woodite-hydrous ringwoodite, grossular-hydrogrossular, and olivine-humites (Table 6). This effect is different from that of Fe–Mg and Ca–Mg substitutions (Table 5). However, the amplitude of the effect is sensitive to the specific mechanism of water incorporation in the host crystallographic structure (H_4O_4 substitution for SiO_4, or OH substitution for O at the vertices of metal coordination polyhedra). Indeed the heterogeneity parameters in different systems are not immediately comparable in terms of total amount of water in the host mineral (Table 6). The complex interplay of crystallographic structure and density effect, already observed in the case of metal substitutions is also present in the case of OH enrichment, as shown in Table 6.

The results discussed here represent the effect of water content on the heterogeneity parameters in the elastic regime. However, it has been demonstrated that water plays a primary role in enhancing anelastic effects in silicates [e.g. *Karato and Jung*, 1998], which have to be fully taken into account when laboratory elasticity results (obtained using high frequency spectroscopic methods) are compared with seismological observations.

3.5. Applications to Mineralogical Models of the Mantle

3.5.1. Overview. In the prospective of understanding the Earth interior, mineral physics furnishes information about the elastic behavior of candidate mantle minerals at high pressures and temperatures in order to construct mineralogical and petrological models of the Earth that can satisfy both geophysical observation and geochemical constraints. In a first-order approximation, the rocks of the upper mantle contain mineral assemblages consistent with those observed in oceanic peridotites, mantle xenoliths and diamond inclusions [e.g. *Moore and Gurney*, 1985; *Menzies and Dupuy*, 1991; *Griffin et al.*, 1999], which give us a partial sampling mostly of regions shallower than 400 km depth.

Rocks of different origin and compositions contain substantially different minerals assemblages, which are subject to modifications at depth. For this reason, in absence of reliable mineral physics databases, general correlations between elemental enrichments and average sound velocities [e.g. *Lee*, 2003] could result in largely inaccurate model velocities, especially if extrapolated to mantle conditions. As an example of the complexity of the effects of variable rock

chemistry, in a pyrolitic average mantle, rich in $(Mg,Fe)_2SiO_4$ polymorphs, the volume fraction of garnet is 15% in the upper mantle and that of garnet-majorite solid solution is 40% in the transition zone [*Fei and Bertka*, 1999], a MORB eclogite, corresponding to subducted oceanic crustal material, contains 25% garnet at the top of the upper mantle but about 90% majorite-garnet solid solution in the transition zone [*Irifune and Ringwood*, 1993]. The relative contribution of garnets to the overall thermoelastic properties of the two rock types, will be dramatically different in varying temperature and pressure regimes. To model the behavior of the rock at mantle conditions the chemical composition has to be accompanied by a precise mineralogical and petrological model.

Single-crystal elastic properties need to be converted into effective or average elastic properties of mineral aggregates in order to compare mineral physics information with information from seismology. Simple bounding schemes for the elastic moduli of isotropic aggregates [*Hill*, 1963; *Hashin and Shtrikman*, 1961; *Watt et al.*, 1976] are often used in geophysical mineralogical models [*Duffy and Anderson*, 1989; *Lee*, 2003, *Cammarano et al.*, 2003]. In the same spirit, we use a very simple approach, described in section 2.4, to determine the average heterogeneity parameters for Fe–Mg substitution in pyrolite, piclogite, and harzburgite and in MORB eclogite (Table 7) along an adiabatic profile with starting temperature of 1673 K up to the bottom of the upper mantle, using information about the compositional sensitivity of seismic velocity in mantle minerals. As a test of our averaging scheme, we applied it to mixtures of garnet and olivine. The consistency of our results with a more rigorous approach based on Hashin–Shtrikman bounds (cf. Section 2.4) demonstrates the reliability of our computed heterogeneity parameters (Plate 1).

3.5.2. Average heterogeneity parameters. The heterogeneity parameters for Fe–Mg substitution calculated for MORB eclogite, which is strongly enriched in garnet-majorite component at high pressure, are $\partial \ln v_P / \partial X_{Fe} = -0.09 \pm 0.02$ and $\partial \ln v_S / \partial X_{Fe} = -0.14 \pm 0.02$ and they gradually decrease with depth to -0.08 ± 0.02 and -0.10 ± 0.02 at the conditions of the transition zone (Plate 2). Pyrolite, piclogite and harzburgite, all characterized by more than 40% olivine or its high-pressure polymorphs, show estimated heterogeneity parameters up to 3 times larger than MORB at ambient pressure (Plate 2). Both $\partial \ln v_P / \partial X_{Fe}$ and $\partial \ln v_S / \partial X_{Fe}$ for these rock types slowly decrease (in absolute value) with pressure within the stability field of olivine, and then more rapidly at the olivine–β-$(Mg,Fe)_2SiO_4$ transition and then increase in magnitude at the β- to γ-transition [*Irifune and Ringwood*, 1987a, b, 1993; *Ita and Stixrude*, 1992]. The ratio $R = \partial \ln v_S / \partial \ln v_P$ of pyrolite,

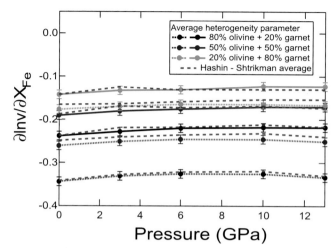

Plate 1. Heterogeneity parameters of binary mixtures of olivine (90 mol% forsterite) and garnet (72 mol% pyrope, 14 mol% almandine, 14 mol% grossular) in different volume ratios. The parameters calculated with our average scheme (see text) are compared with the results of a more formal approach based on Hashin–Shtrikman bounding of the elastic moduli of the mixtures. The results agree within uncertainties. Solid lines: $\partial \ln v_P / \partial X_{Fe}$. Dashed lines: $\partial \ln v_S / \partial X_{Fe}$.

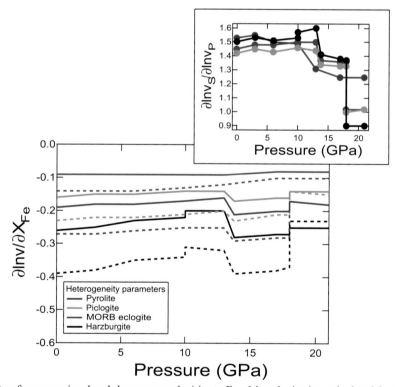

Plate 2. Sensitivity of compressional and shear wave velocities to Fe–Mg substitution calculated for pyrolite, piclogite, harzburgite, and MORB eclogite, along an isentropic pressure/temperature path with a foot temperature of 1673 K. The modal compositions and stability field boundaries for the different mineral phases in the four rock types were derived from *Irifune and Ringwood* [1987b, 1993], *Ita and Stixrude* [1992], and *Vacher et al.* [1998]. Solid lines: $\partial \ln v_P / \partial X_{Fe}$. Dashed lines: $\partial \ln v_S / \partial X_{Fe}$. Inset: variation of the heterogeneity parameters ratio $R = \partial \ln v_P / \partial \ln v_S$ for the four rock types along the same isentropic pressure/temperature path.

Table 8. Relative Velocity Changes Upon Transitions Between Different Mantle Rock Types.

Rock type	$\Delta v_P/v_P$	$\Delta v_S/v_S$	$\Delta v_B/v_B$	$[\Delta v_S/v_S]/[\Delta v_P/v_P]$
Pyrolite–piclogite	0.005 (2)	0.005 (1)	0.005 (2)	1.0 (3)
Pyrolite–harzburgite	-0.006 (2)	0.001 (3)	-0.012 (3)	-0.2 (3)
Pyrolite–MORB	-0.027 (2)	-0.038 (1)	-0.019 (3)	1.4 (1)
Piclogite–harzburgite	-0.011 (2)	-0.004 (1)	-0.016 (2)	0.4 (1)
Piclogite–MORB	-0.032 (2)	-0.043 (2)	-0.023 (3)	1.4 (1)
MORB–harzburgite	0.022 (2)	0.041 (1)	0.007 (2)	1.9 (2)

piclogite, and harzburgite is dominated by the properties of olivine and its polymorphs. It slowly increases with depth in the upper mantle, ranging between 1.45 ± 0.05 and 1.53 ± 0.05, and then it rapidly decreases in the transition zone to about 1.0 ± 0.1 at the β- to γ-$(Mg,Fe)_2SiO_4$ transition. MORB eclogite, which is enriched in garnet, shows a smaller decrease of the ratio $\partial \ln v_S/\partial \ln v_P$ from 1.5 ± 0.1 in the upper mantle to 1.3 ± 0.1 at the bottom of the transition zone. One reasonable interpretation of these results is that the seismic signature of bulk compositional variation in the olivine-rich rocks of the mantle fades with pressure in the transition zone, while it will remain visible in MORB eclogite.

3.5.3. Effect of elemental partitioning. We have carried out simple calculations of the effect of changes in Fe–Mg partitioning between the different minerals on the elastic velocities of these four rock-types at ambient conditions. The computations have been performed using Hashin–Shtrikman bounds to the rock elastic moduli to determine average rock velocities. The relative velocity variations $\Delta v_i/v_i$, where Δv_i is the velocity difference caused by enrichment of Mg up to 4 mol% in pyroxene at constant rock bulk composition in MORB eclogite, are $\Delta v_P/v_P = 0.002$ and $\Delta v_S/v_S = 0.003$. Similar enrichment in Mg in olivine results in $\Delta v_P/v_P = -0.002$ and $\Delta v_S/v_S = -0.002$ in harzburgite and $\Delta v_P/v_P = 0.002$ and $\Delta v_S/v_S = 0.003$ in pyrolite. We do not resolve a significant variation in piclogite, where the modal abundances of the different minerals are relatively close to one another (Table 7) and the compositional effects cancel out completely. The values of the calculated ratios $(\Delta v_S/v_S)/(\Delta v_P/v_P)$ for the different rock types are in the range 1.0–1.5, comparable to the heterogeneity ratio $R = \partial \ln v_S/\partial \ln v_P$ due to variations of the bulk Fe/Mg ratio in the same rocks. Based on these (indeed limited) results we cannot rule out that variations of the heterogeneity parameters with depth could represent an indicator of the sensitivity of seismic velocity to variations in the partition of Fe between the different mineral phases. Developing this interpretation we could use the heterogeneity parameters for bulk rock

compositional changes as qualitative estimates of the heterogeneity determined by transitions between the different rock types in the mantle, which also involve large changes in bulk rock chemical composition and different Fe/Mg ratios in the constituent minerals.

3.5.4. Effect of lateral variations of lithology. We have calculated the effect of transition from one to the other of the four examined rock compositions, as the relative velocity difference $\Delta v_i/v_i$, where v_i is elastic wave velocity at ambient conditions (Table 8). In the transition from pyrolite to MORB eclogite the relative variation $\Delta v_P/v_P$ is -0.027 ± 0.002 and $\Delta v_S/v_S$ is -0.038 ± 0.001 with $(\Delta v_S/v_S)/(\Delta v_P/v_P) = 1.4 \pm 0.1$. The relative changes in compressional and shear velocities between piclogite and MORB eclogite are $\Delta v_P/v_P = -0.032 \pm 0.002$ and $\Delta v_S/v_S = -0.043 \pm 0.002$, both larger than in the case of pyrolite, but the ratio $(\Delta v_S/v_S)/(\Delta v_P/v_P)$ is 1.4 ± 0.1, as for pyrolite. These results are in order of magnitude agreement with the range of calculated variations between different mantle rocks by *Goes et al.* [2000], which are based on a different mineral physics data selection.

The magnitude of the calculated heterogeneity ratios $(\Delta v_S/v_S)/(\Delta v_P/v_P)$ due to the change in Fe–Mg content at the transition between different types of upper mantle rocks are consistent with global seismic models [e.g. *Masters et al.*, 2000]. However, our results do not take explicitly into account the effect of Al substitution for Mg and Fe, whose effect cannot be modeled based on the limited available single-crystal elasticity data.

Using the heterogeneity parameters calculated for bulk rock variations of Fe/Mg ratio as an approximation of $(\Delta v_S/v_S)/(\Delta v_P/v_P)$ upon transitions between the different rocks (see discussion above), we estimate that this ratio decreases at depths of the transition zone, in agreement with the decreasing values of $\partial \ln v_S/\partial \ln v_P$ from 1.5 to 1.2 determined in the transition zone by a careful analysis of matching P and S data for global tomographic images [*Saltzer et al.*, 2001].

We finally compare our results with those of *Lee* [2003], by computing density, elastic wave velocity, and their logarithmic variation with respect to bulk Mg#, defined as the molar ratio Mg/(Mg + Fe), in "synthetic" garnet peridotite compositions that we calculated by small perturbations of the composition of natural peridotites, and their mineral phases selected from the databases used in that study. In order to obtain a stable fit of the compositional sensitivities of the different parameters, we computed 15,000 rock compositions, selected such that both the single mineral phases and the whole rock are within the ranges of Mg# of the dataset used by *Lee* [2003]. Our computed logarithmic dependence of shear velocity on Mg#, $\partial \ln v_S/\partial Mg\# = 3.45 \times 10^{-3}$, is 15% larger than the results of *Lee* [2003] (Table 9). Our results indicate that the compressional velocity has a logarithmic dependence $\partial \ln v_P/\partial Mg\# = 2.14 \times 10^{-3}$, similar to that of v_S, but with larger uncertainty. The larger scatter of longitudinal velocity was also observed by *Lee* [2003]. However, the average compositional dependence of v_P on Mg# is in disagreement with the results of *Lee* [2003], as it is demonstrated by the different compositional sensitivity of the ratio v_P/v_S, which is 45% smaller than that proposed by *Lee* [2003] (Table 9).

4. CONCLUSIONS

Global seismological and geodynamical models require an assessment not only of thermal effects but also of compositional effects on the elastic properties of the constituents of the Earth. Based on the available single-crystal elasticity data we have shown that it is possible to develop a dataset of compositional sensitivities of the main candidate minerals of the mantle. Our principal results are:

(1) Logarithmic velocity variations with respect to iron content are very sensitive to crystal structure, pressure, and temperature. At ambient conditions, $\partial \ln v_P/\partial X_{Fe}$ for garnets, olivines (including high-pressure polymorphs), pyroxenes, and periclase-wüstite ranges from -0.05 to -0.46, whereas $\partial \ln v_S/\partial X_{Fe}$ ranges from -0.08 to -0.74. The effects of pressure and temperature can change these values markedly under upper mantle conditions, but P-T effects vary greatly from structure to structure, and no general rules about the effects of pressure and temperature can be formulated.

(2) α- and γ- polymorphs of $(Mg,Fe)_2SiO_4$ exhibit broadly comparable values of $\partial \ln v/\partial X_{Fe}$ at ambient conditions, but the temperature dependences of these quantities are large and of opposite sign between the two polymorphs.

(3) Of the mineral families investigated for Fe–Mg substitution, the (Mg,Fe)O series is the most sensitive to chemical substitution showing $\partial \ln v_P/\partial X_{Fe} = -0.46 \pm 0.05$ and $\partial \ln v_S/\partial X_{Fe} = -0.74 \pm 0.07$, whereas garnets are the least

Table 9. Compositional Sensitivity of the Acoustic Wave Velocity in Garnet Peridotites.

Derivative	This study	*Lee* [2003]
$100 \times \partial \ln v_P/\partial Mg\#$	0.214 (5)	---
$100 \times \partial \ln v_S/\partial Mg\#$	0.345 (5)	0.30 (2)
$100 \times \partial(v_P/v_S)/\partial Mg\#$	-0.224 (5)	-0.41 (4)

---, parameter not determined.

sensitive, with $\partial \ln v_P/\partial X_{Fe} = -0.09 \pm 0.01$ and $\partial \ln v_S/\partial X_{Fe} = -0.08 \pm 0.01$.

(4) Al and Ca enrichments increase seismic velocities in orthopyroxenes, ternary garnets, and majorite garnets. The magnitude of the effect of Al incorporation in orthopyrene produces much larger changes in velocity ($\partial \ln v_P/\partial X_{Al} = 0.24 \pm 0.04$ and $\partial \ln v_S/\partial X_{Al} = 0.13 \pm 0.04$) than Al substitution in majorites ($\partial \ln v_P/\partial X_{Al}$ and $\partial \ln v_S/\partial X_{Al} \leq 0.04$) or Ca in garnets ($\partial \ln v_P/\partial X_{Fe}$ and $\partial \ln v_S/\partial X_{Fe} \leq 0.08$).

(5) Values of $\partial \ln v_S/\partial \ln v_P$ resulting from Fe–Mg substitution in different mineral families range from 0.9 to 1.7. A very large value of the heterogeneity ratio ($\partial \ln v_S/\partial \ln v_P = 2.7 \pm 0.6$) is found only for Ca–Mg substitution in ternary garnets at ambient conditions.

(6) The effect of hydration in grossular and ringwoodite and in the systems olivine–humites, and periclase–brucite is to decrease both compressional and shear velocity, with $\partial \ln v_S/\partial \ln v_P$ ranging between 1.0 and 1.3.

(7) In a simplified model of rock behavior, the very different response of garnets and olivines to Fe substitution gives rise to variations in the sensitivity of different rocks to iron concentration in accordance to their garnet content. For instance, velocities in mid-ocean ridge eclogites are about 2–3 times less sensitive to iron enrichment than those in harzburgites. Due to their enrichment in olivine component, values of $\partial \ln v_S/\partial \ln v_P$ of pyrolite, piclogite and harzburgite as a result of chemical substitution show a decrease from 1.5 ± 0.1 in the upper mantle to 1.0 ± 0.1 in the transition zone. MORB eclogite, which is extremely enriched in garnet component, shows a smaller decrease of $\partial \ln v_S/\partial \ln v_P$ from 1.5 ± 0.1 to 1.3 ± 0.1.

(8) The sensitivity of seismic velocities of pyrolite, harzburgite and MORB eclogite to change in Fe/Mg partitioning between the constituent minerals is one order of magnitude smaller than the effect of variation of the bulk rock composition. No clear effect is detected in piclogite. The heterogeneity parameters ratios $R = \partial \ln v_S/\partial \ln v_P$ are broadly comparable with those relative to bulk rock enrichment/depletion in Fe.

(9) Using the sensitivity to bulk compositional change in Fe content, and its pressure dependence, as a proxy for the

heterogeneity parameters upon transitions between the different mantle rock-types we estimate that the intensity of the seismic effect of the transition from average mantle to subducted slab rocks progressively fades at transition zone depths.

In order to have internal consistency in the selection of the whole set of parameters reported in this study, we have selected only single-crystal elasticity measurements, excluding many polycrystalline elasticity and X-ray diffraction results. The drawback of this choice is that it reduced our ability to give a complete description of the pressure/temperature dependence of the elastic properties of the examined minerals. On the other hand, the unavoidable inconsistency between the results from different techniques, the limitations of parameters deduced by approximate elasticity systematics, and the poor resolution of moduli constrained as fit parameters of standard equations of state (as in the case of bulk modulus and its pressure and temperature derivatives from X-ray diffraction experiments) suggest caution in interpreting the results based on the use of partially outdated or extensive but inconsistent elasticity databases. New perspectives in the interpretation of the properties of the Earth's mantle are represented by self-consistent thermodynamic models [e.g. Stixrude and Lithgow-Bertelloni, 2005] applied to a homogeneous (possibly single-crystal) elasticity database, and compatible with the mineralogical and petrological experimental constraints.

Acknowledgments. The authors thank two anonymous reviewers for their thoughtful comments and suggestions that helped to improve the original manuscript. S. S. thanks R. Jeanloz for the helpful discussions and the Miller Institute for Basic Research in Science that is supporting his work at the University of California at Berkeley.

REFERENCES

Abramson, E. H., J. M. Brown, L. J. Slutsky, and J. Zaug (1997), The elastic constants of San Carlos olivine to 17 GPa, *J. Geophys. Res., 102*, 12253–12263.

Akaogi, M., and S. Akimoto (1979), High-pressure phase equilibria in a garnet lherzolite, with special reference to Mg^{2+}-Fe^{2+} partitioning among constituent minerals, *Phys. Earth Planet. Inter., 19*, 31–51.

Anderson, D. L., and J. D. Bass (1986), Transition region of the Earth's upper mantle, *Nature, 320*, 321–328.

Anderson, O. L., and D. G. Isaak (1995), Elastic constants of mantle minerals at high temperature, in *Mineral physics and crystallography. A handbook of physical constants, AGU Reference shelf* 2, edited by T. J . Ahrens, pp.64–97, AGU, Washington D.C.

Angel, R. J. (2000), Equations of state, in *High-Temperature and High-Pressure Crystal Chemistry, Reviews in Mineralogy and Geochemistry*, vol. 41, edited by R. M. Hazen and R. T. Downs, pp. 35–59, Mineralogical Society of America, Washington D.C.

Antolik, M., Y. J. Gu, G. Ekström, and A. M. Dziewonski (2003), J362D28: a new joint model of compressional and shear velocity in the Earth's mantle, *Geophys. J. Inter., 153*, 443–466.

Armbruster, T., C. A. Geiger, and G. A. Lager (1992), Single-crystal X-ray structure study of synthetic pyrope almandine garnets at 100 and 293 K, *Am. Mineral., 77*, 512–521.

Babuška, V., J. Fiala, M. Kumazawa, I. Ohno, and Y. Sumino (1978), Elastic properties of garnet solid-solution series, *Phys. Earth Planet. Inter., 16*, 157–176.

Bass J. D. (1986), Elasticity of uvarovite and andradite garnets, *J. Geophys. Res., 91*, 7505–7516.

Bass, J. D. (1989), Elasticity of grossular and spessartine garnets by Brillouin spectroscopy, *J. Geophys. Res., 94*, 7621–7628.

Bass, J. D., and D. L. Anderson (1984), Composition of the upper mantle: geophysical tests of two petrological models, *Geophys. Res. Lett., 3*, 237–240.

Bass, J. D., and M. Kanzaki (1990), Elasticity of majorite-pyrope solid solution, *Geophys. Res. Lett., 97*, 4809–4822.

Bass, J. D., and D. J. Weidner (1984), Elasticity of single-crystal orthoferrosilite, *J. Geophys. Res., 89*, 4359–4371.

Beckman Fritzel, T. L., and J. D. Bass (1997), Sound velocity of clinohumite, and implications for water in Earth's upper mantle, *Geophys. Res. Lett., 24*, 1023–1026.

Cammarano, F., S. Goes, P. Vacher, and D. Giardini (2003), Inferring upper-mantle temperatures from seismic velocities, *Phys. Earth Planet. Inter., 138*, 197–222.

Chai, M, J. M. Brown, and L. J. Slutsky (1997), The elastic constants of an aluminum orthopyroxene to 12.5 GPa, *J. Geophys. Res., 102*, 14,779–14,785.

Chen, G., R. C. Liebermann, and D. J. Weidner (1998), Elasticity of single-crystal MgO to 8 gigapascal and 1600 K, *Science, 280*, 1913–1916.

Collins, M.D., and J. M. Brown (1998), Elasticity of an upper mantle clinopyroxene, *Phys. Chem. Minerals, 26*, 7–13.

Davies, G. F, and A. M. Dziewonski (1975), Homogeneity and constitution of the Earth's lower mantle and outer core, *Phys. Earth Planet. Inter., 10*, 336–343.

Duffy T. S., and D. L. Anderson (1989), Seismic velocities in mantle minerals and the mineralogy of the upper mantle, *J. Geophys. Res., 94*, 1895–1912.

Duffy, T. S., and M. T. Vaughan (1988), Elasticity of enstatite and its relationship to crystal structure, *J. Geophys. Res., 93*, 383–391.

Fei, Y. (1995), Thermal expansion, in *Mineral physics and crystallography. A handbook of physical constants, AGU Reference shelf* 2, edited by T. J. Ahrens, pp.29–44, AGU, Washington D.C. 64–97.

Fei, Y., and C. M. Bertka (1999), Phase transitions in the Earth's mantle and mantle mineralogy, in *Mantle Petrology: Field Observations and High Pressure Experimentation, Special Publication* No. 6, edited by Y. Fei, C. M. Bertka, and B. O. Mysen, pp.189–207, The Geochemical Society, Houston.

Forte, A. M., and J. X. Mitrovica (2001), Deep-mantle high-viscosity flow and thermochemical structure inferred from seismic and geodynamic data, *Nature, 410*, 1049–1056.

Forte, A. M., and R. Woodward (1997), Seismic-geodynamic constraints on three-dimensional structure, vertical flow, and heat transfer in the mantle, *J. Geophys. Res., 102*, 17,981–17994.

Forte, A. M., and H. K. C. Perry (2000), Geodynamic evidence for a chemical depleted continent tectosphere, *Science, 290*, 1940–1944.

Frisillo, A. L., and G. R. Barsch (1972), Measurement of single-crystal elastic constants of bronzite as a function of pressure and temperature, *J. Geophys. Res., 77*, 6360–6384.

Gasparik, T., and M. T. Hutchison (2001), Experimetnal evidence for the origin of two kinds of inclusions in diamonds from the deep mantle, *Earth Planet. Sci. Letters, 181*, 103–114.

Geiger, C. A. (1999), Thermodynamics of $(Fe^{2+},Mn^{2+},Mg^{2+},Ca)_3Si_3O_{12}$ garnet: An analysis and a review, *Mineral. Petrol., 66*, 271–299.

Goes, S., R. Govers, and P. Vacher (2000), Shallow mantletemperatures under Europefrom P and S wavetomography, *J. Geophys. Res., 105*, 11153–11169.

Graham, E. K., J. A. Schwab, S. M. Sopkin, and H. Takei (1988), The pressure and temperature dependence of the elastic properties of single-crystalfayalite Fe2SiO4, *Phys. Chem. Minerals*, 16, 186–198.

Grand, S. P, R.D. van der Hilst, and S. Widiyantoro (1997), Global seismic tomography: a snapshot of convection in the Earth, *GSA Today, 7*, 1–7.

Griffin, W. L., S. Y. O'Reilly, and C. G. Ryan (1999), The composition of sub-continental lithospheric mantle, in *Mantle Petrology: Field Observations and High Pressure Experimentation, Special Publlication* No. 6, edited by Y. Fei, C. M. Bertka, and B. O. Mysen, pp. 13–45, The Geochemical Society, Houston.

Gung, Y., M. Panning, and B. Romanowicz (2003), Global anisotropy and the thickness of continents, *Nature, 422*, 707–711.

Hashin, Z., and S. Shtrikman (1961), Note on a variational approach to the theory of composite elastic materials, *J. Franklin Inst., 271*, 336–341.

Hashin, Z., and S. Shtrikman (1963), A variational approach to the elastic behavior of multiphase materials, *J. Mech. Phys. Solids, 11*, 127–140.

Hill, R. (1963), Elastic properties of reinforced solids: Some theoretical principles, *J. Mech. Phys. Solids, 11*, 357–372.

Inoue, T., D. J. Weidner, P. A. Northrup, and J. B. Parise (1998), Elastic properties of hydrous ringwoodite (γ-phase) in Mg_2SiO_4, *Earth Planet. Sci. Lett., 160*, 107–113.

Irifune, T., and M. Isshiki (1998), Iron partitioning in a pyrolite mantle and the nature of the 410-km seismic discontinuity, *Nature, 392*, 702–705.

Irifune, T., and A. E. Ringwood (1987a), Phase transformations in primitive MORB and pyrolite compositions to 25 GPa and some geophysical implications, in *High Pressure Research in Geophysics*, edited by M. H. Manghnani and Y. Syono, pp. 231–242, Terrapub, AGU, Tokyo, Washington D.C.

Irifune, T., and A. E. Ringwood (1987b), Phase transformations in a harzburgite composition to 26 GPa: Implications for dynamical behavior of the subducting slab, *Earth Planet. Sci. Lett., 86*, 365–376.

Irifune, T., and A. E. Ringwood (1993), Phase transformation in subducted oceanic crust and buoyancy relationships at depths of 600–800 km in the mantle, *Earth Planet. Sci. Lett., 117*, 101–110.

Irifune, T., T. Sekine, A. E. Ringwood, and W. O. Hibberson (1986), The eclogite-garnetite transformation at high pressure and some geophysical implications, *Earth Planet. Sci. Lett., 77*, 245–256.

Isaak, D. G., O. L. Anderson, T. Goto, and I. Suzuki (1989), Elasticity of single-crystal forsterite measured to 1700 K, *J. Geophys. Res., 94*, 5895–5906.

Isaak, D. G., O. L. Anderson, and H. Oda (1992), High-temperature thermal expansion and elasticity of calcium-rich garnets, *Phys. Chem. Miner., 19*, 106–120.

Isaak, D. G., and E. K. Graham (1976), The elastic properties of an almandine-spessartine garnet and elasticity in the garnet solid solution series, *J. Geophys. Res., 81*, 2483–2489.

Isaak, D. G., E. K. Graham, J.D. Bass, and H. Wang (1993), The elastic properties of single-crystal fayalite as determined by dynamical measurements techniques, *Pure Appl. Geophys. 141*, 393–414.

Isaak, D. G., and I. Ohno (2003), Elastic constants of chrome-diopside: application of resonantultrasound spectroscopic to monoclinic single-crystals, *Phys. Chem. Minerals, 30, 430–439*.

Ita, J., and L. Stixrude (1992), Petrology, elasticity and composition of the mantle transition zone, *J. Geophys. Res., 97*, 6849–6866.

Ita, J., and L. Stixrude (1993), Density and elasticity of model upper mantle compositions and their implications for whole mantle structure, in *Evolution of the Earth and Planets, Geophysical Monograph* 74, edited by E. Takahashi, R. Jeanloz and D. Rubie, pp. 111–130, International Union of Geodesy and Geophysics and American Geophysical Union, Washington, D. C.

Jackson, I., and H. Niesler (1982), The elasticity of periclase to 3 GPa and some geophysical implications, in *High Pressure Research in Geophysics*, edited by S. I. Akimoto and M. H. Manghnani, pp.93–113, Cent. for Acad. Publ., Tokyo.

Jackson, J. M., S. V. Sinogeikin, and J.D. Bass (1999), Elasticity of $MgSiO_3$ orthoenstatite, *Am. Mineral., 84*, 677–680.

Jackson, J. M., S. V., S. V. Sinogeikin, and J.D. Bass (2000), Sound velocities and elastic properties of γ-Mg2SiO4 to 873 K by Brillouin spectroscopy, *Am. Mineral., 85*, 296–303.

Jackson, J. M., J. Zhang, and J.D. Bass (2004), Sound velocities and elasticity of aluminous MgSiO3 perovskite: Implications for aluminum heterogeneity in Earth's lower mantle, Geophys. Res. Lett., 31, L10614, doi:10.1029/2004GL019918.

Jacobsen, S.D., H. J. Reichmann, H. A. Spetzler, S. J. Mackwell, J. R. Smyth, R. J. Angel, and C. A. McCammon (2002), Structure and elasticity of single-crystal (Mg,Fe)O and a new method of generating shear waves for gigahertz ultrasonic interferometry, *J. Geophys. Res., 107*, Art. 2037.

Jiang, F., S. Speziale, R. S. Shieh, and T. S. Duffy (2004a), Single crystal elasticity of andradite garnet to 11 GPa, *J. Phys. Condens. Matter, 16*, S1041–S1052.

Jiang, F., S. Speziale, and T. S. Duffy (2004b), Single-crystal elasticity of grossular- and almandine-rich garnets to 12 GPa by Brillouin spectroscopy, *J. Geophys. Res., 109*, B10210, doi:10.1029/2004JB003081.

Jordan, T. H. (1978), Composition and development of the continental tectosphere, *Nature, 274*, 544–548.

Jordan, T. H. (1979), Mineralogies, densities and seismic velocities of garnet lherzolites and their geophysical implications, in *The Mantle Sample: Inclusions in Kimberlites and Other Volcanics, Proceedings of the Second Interantional Kimberlite Conference,* vol. 2, edited by F. R. Boyd, and H. O. Meyer, , pp. 1–14, AGU, Washington D.C.

Kandelin, J., and D. J. Weidner (1988), Elastic properties of hedenbergite, *J. Geophys. Res., 93*, 1063–1072.

Karato, S-I. (1993), Importance of anelasticity in the interpretation of seismic tomography, *Geophys. Res. Lett., 20*, 1623–1626.

Karato, S-I., and H. Jung (1988), Water, partial melting and the origin of the seismic low velocity and high attenuation zone in the upper mantle, *Earth Planet. Sci. Lett., 157*, 193–207.

Karato, S-I., and B. B. Karki (2001), Origin of lateral variation of seismic wave velocities and density in the deep mantle, *J. Geophys. Res., 106*, 21,771–21,783.

Kennett, B. L. N., S. Widiyantoro, and R.D. van der Hilst (1988), Joint seismic tomography for bulk sound and shear wave speed in the Earth's mantle, *J. Geophys. Res., 103*, 12,469–12,493.

Kiefer, B., L. Stixrude, and R. M. Wentzcovitch (2002), Elasticity of $(Mg,Fe)SiO_3$-perovskite at high pressure, *Geophys. Res. Lett., 29*, 10.1029/20002GL014683.

Kumazawa, M. (1969), The elastic constants of single-crystal orthopyroxene, *J. Geophys. Res., 74*, 5973–5980.

Lee, C-T. A. (2003), Compositional variation of density and seismic velocities in natural peridotites at STP conditions: Implications for seismic imaging of compositional heterogeneities in the upper mantle, *J. Geophys. Res., 108*, 2441, doi:10.1029/2003JB002413.

Liebermann, R. C. (2000), Elasticity of mantle minerals (experimental studies), in *Earth's Deep Interior: Mineral Physics and Tomography From the Atomic to the Global Scale, Geophys. Monogr. Ser.*, vol. 117, edited by S-I. Karato , A. Forte, R. Liebermann, G. Masters, and L. Stixrude, pp. 181–199, AGU, Washington D.C.

Liu, L. G., K. Okamoto, Y. J. Yang, C. C. Chen, and C. C. Lin (2004), Elasticity of single-crystal phase D (adense hydrous magnesiumsilicate)by Brillouin spectroscopy, Solid State Comm., 132, 517–520.

Masters, G., G. Laske, H. Bolton, and A. M. Dziewonski (2000), The relative behavior of shear velocity, bulk sound speed, and compressional velocity in the mantle: Implications for chemical and thermal structure, in *Earth's Deep Interior: Mineral Physics and Tomography From the Atomic to the Global Scale, Geophys. Monogr. Ser.*, vol. 117, edited by S-I. Karato , A. Forte, R. Liebermann, G. Masters, and L. Stixrude, pp. 63–87, AGU, Washington D.C.

McDonough, W. F., and R. L. Rudnick (1998), Mineralogy and composition of the upper mantle, in *Ultrahigh-Pressure Mineralogy: Physics and Chemistry of the Earth's Deep Interior, Reviews in Mineralogy, vol. 37*, edited by R. J. Hemley, pp. 139–164, Min. Soc. Am., Washington, D.C.

Menzies, M. A., and C. Dupuy (1991), Orogenic massifs: protolith, process and provenance, *J. Petrol. Spec. Lherzolites Issue,* edited by M. A. Menzies, C. Dupuy, and A. Nicolas, pp. 1–16, Oxford University Press, New York.

Moore, R. O., and J. J. Gurney (1985), Pyroxene solid solution in garnets included in diamond, *Nature, 335*, 784–789.

Novack, G. A., and G. V. Gibbs (1971), The crystal chemistry of the silicate garnets, *Am. Mineral., 56*, 791–825.

Pacalo, R. E. G., and D. J. Weidner (1996), Elasticity of superhydrous B, *Phys. Chem. Minerals, 23*, 520–525.

O'Neill, B., J.D. Bass, J. R. Smyth, and M. T. Vaughan (1989), Elasticity of a grossular-pyrope-almandine garnet, *J. Geophys. Res., 94*, 17,819–17,824.

O'Neill, B., J.D. Bass, G. R. Rossman, C. A. Geiger, and K. Langer (1991), Elastic properties of pyrope, *Phys. Chem. Miner., 17*, 617–621.

O'Neill, B., J.D. Bass, and G. M. Rossman (1993), Elastic properties of hydrogrossular garnet and and implications for water in the upper mantle, *J. Geophys. Res., 98*, 20,031–10,037.

Rigden, S. M., and I. Jackson (1991), Elasticity of germanate and silicate spinels at high pressure, *J. Geophys. Res., 96*, 9999–10,006.

Rigden S. M., G.D. Gwanmesia, I. Jackson, R. C. Liebermann (1992), Progress in high-pressure ultrasonic interferometry, the pressure dependence of elasticity of Mg_2SiO_4 polymorphs and constraints on the composition of the transition zone of the Earth mantle, in *High-Pressure Research: Application to Earth and Planetary Science, Geophys. Monogr. Ser.*, vol. 67, edited by Y. Syono and M. H. Manghnani, pp.167–182, AGU, Washington D.C.

Ringwood, A. E. (1975), *Composition and Petrology of the Earth Mantle*, McGraw-Hill, New York.

Saltzer, R. L., R.D. van Der Hilst, and H. Kárason (2001), Comparing P and S wave heterogeneity in the mantle, *Geophys. Res. Lett., 28*, 1335–1338.

Sawamoto, H., D. J. Weidner, S. Sasaki, and M. Kumazawa (1984), Single-crystal elastic properties of the modified spinel(β) phase of magnesium orthosilicate, *Science, 224*, 749–751.

Sinogeikin, S. V., J.D. Bass, A. Kavner, R. Jeanloz (1997), Elasticity of natural majorite and ringwoodite from the Catherwood meteorite, *Geophys. Res. Lett., 24*, 3265–3268.

Sinogeikin, S. V., and J.D. Bass (1999), Single-crystal elastic properties of chondrodite,: implications for water in the upper mantle, *Phys. Chem. Minerals, 26*, 297–303.

Sinogeikin, S. V., J.D. Bass (2000), Single-crystal elasticity of pyrope and MgO to 20 GPa by Brillouin scattering in the diamond anvil cell, *Phys. Earth Planet. Inter. 120*, 43–62.

Sinogeikin, S. V., and J.D. Bass (2002), Elasticity of majorite and a majotite-pyrope solid solution to high pressure: implications for the transition zone, Geophys. Res. Lett., 29, 10.1029/2001GL013937.

Sinogeikin, S. V., J.D. Bass, and T. Katsura (2003), Single-crystal elasticity of ringwoodite to high pressures and high temperatures: implications for 520 km seismic discontinuity, *Phys. Earth. Planet. Inter., 136*, 41–66.

Sinogeikin, S. V., D. L. Lakshtanov, J.D. Nicholas, and J.D. Bass (2004), Sound velocitymeasurements on laser-heated MgO and Al2O3, *Phys. Earth Planet. Inter., 143–144*, 575–586.

Sinogeikin, S. V., T. Katsura, and J.D. Bass (1998), Sound velocities and elastic properties of Fe-bearing wadsleyite and ringwoodite, *J. Geophys. Res., 103*, 20,819–20,825.

Skinner, B. J. (1956), Physical properties of end-members of the garnet group, *Am. Mineral., 41*, 428–436.

Speziale, S. and T. S. Duffy (2002), Single-crystal elastic constants of fluorite (CaF_2) to 9.3 GPa, *Phys. Chem. Miner., 29*, 465–472.

Speziale, S., T. S. Duffy, and R. J. Angel (2004), Single-crystal elasticity of fayalite to 12 GPa, *J. Geophys. Res., 109*, B12, B12202, doi:10.1029/2004JB003162.

Stixrude, L., and C Lithgow-Bertelloni (2005), Mineralogy and elasticity of the oceanic upper mantle: Origin of the low velocity zone, J. Geophys. Res., in press.

Su, W-J., and A. M. Dziewonski (1997), Simultaneous inversion for 3D variations in shear and bulk velocity in the mantle, *Phys. Earth Planet. Inter., 100*, 135–156.

Sumino, Y., and O. Nishizawa (1978), Temperature variation of elastic constants of of pyrope-almandine garnets, *J. Phys. Earth, 26*, 239–252.

Sumino, Y., O. Nishizawa, T. Goto, I. Ohno, and M. Ozima (1977), Temperature variation of elastic constants of single-crystal forsterite between -190 and 400° C, *J. Phys. Earth, 25*, 377–392.

Suzuki, I., and O. L. Anderson (1983), Elasticity and thermal expansion ofa natural garnet up to 1,000 K, *J. Phys. Earth, 31*, 125–138.

van der Hilst, R.D., and H. Kárason (1999), Compositional heterogeneity in the bottom 100 kilometers of the Earth's mantle: toward a hybrid convection model, *Science, 283*, 1885–1888.

Vacher, P., A. Moquet, and C. Sotin (1998), Computation of seismic profiles from mineral physics: the importance of the non-olivine components for explaining the 660 depth discontinuity, *Phys. Earth. Planet. Inter., 196*, 275–298.

Wang, J., S. V. Sinogeikin, T. Inoue, and J.D. Bass (2003), Elastic properties of hydrous rongwoodite, *Am. Mineral., 88*, 1608–1611.

Watanabe, H. (1982), Thermochemical properties of synthetic high-pressure compounds relevant to the earth's mantle, in *High pressure Research in Geophysics*, edited by S. Akimoto and M. H. Manghnani, pp. 441–464, Center for Academic Publications, Tokyo, Japan.

Watt, J. P., G. F. Davies, and R. J. O'Connell (1976), The elastic properties of composite materials, *Rev. Geophys. Space Phys., 14*, 541–563.

Webb, S. and I. Jackson (1993), The pressure dependence of the elastic moduli of single –crystal orthopyroxene ($Mg_{0.8}Fe_{0.2})SiO_3$, *Eur. J. Mineral., 5*, 1111–1119.

Weidner, D. J., H. Wang, and J. Ito (1978), Elasticity of orthoenstatite, *Phys. Earth Planet. Inter., 17*, P7–P13.

Weidner D. J., H. Sawamoto, S. Sasaki, and M. Kumazawa (1984), Single-crystal elastic properties of the spinel phase of Mg2SiO4, *J. Geophys. Res., 89*, 7852–7860.

Xia, X., D. J. Weidner, and Y. Zhao (1998), Equation of state of brucite: Single-crystal Brillouin spectroscopy study and polycrystalline pressure-volume-temperature measurement, *Am. Mineral., 83*, 68–74.

Yeganeh-Haeri, A., D. J. Weidner, and E. Ito (1990), Elastic properties of the pyroxene-majorite solid solution series, *Geophys. Res. Lett., 17*, 2453–2456.

Yoneda, A., and M. Morioka (1992), Pressure derivatives of elastic constants of single crystal forsterite, in *High-pressure research: Application to Earth and planetary sciences*, edited by Y. Syono and M. H. Manghnani, pp. 207–214. Terra Scientific Company (TERRAPUB), Tokyo / American Geophysical Union, Washington, D.C.

Zha, C-S., T. S. Duffy, R. T. Downs, H-K. Mao, and R. J. Hemley (1996), Sound velocity and elasticity of single-crystal forsterite to 16 GPa, *J. Geophys. Res., 101*, 17535–17545.

Zha, C-S., T. S. Duffy, H-K. Mao, R. T. Downs, R. J. Hemley, and D. J. Weidner (1997), Single-crystal elasticity of β-Mg_2SiO_4 to the pressure of the 410 km seismic discontinuity in the Earth's mantle, *Earth and Planet. Sci. Lett., 147*, E9–E15.

Zha, C-S., T. S. Duffy, R. T. Downs, H-K. Mao, and R. J. Hemley (1998), Brillouin scattering and X-ray diffraction of San Carlos olivine: direct pressure determination to 32 GPa, *Earth Planet. Sci. Lett., 159*, 25–33.

Zha, C-S., H-K. Mao, and R. J. Hemley (2000), Elasticity of MgO and a primary pressure scale to 55 GPa, *Proc. Natl. Acad. Sci. USA, 97*, 13,494–13,499.

T. S. Duffy and F. Jiang, Department of Geosciences, Guyot Hall, Princeton University, Princeton, New Jersey 08544-1003, USA. (duffy@princeton.edu)

Sergio Speziale, Department of Earth and Planetary Science, 307 McCone Hall, University of California, Berkeley, California 94720-4767 USA.

Recent Progress in Experimental Mineral Physics: Phase Relations of Hydrous Systems and the Role of Water in Slab Dynamics

Eiji Ohtani

Institute of Mineralogy, Petrology, and Economic Geology, Tohoku University, Japan

Water is stored in various hydrous minerals in subducting slabs. The series of hydrous minerals, which are stable in the peridotite layer of the slabs, can bring water into the transition zone in the cold slabs with temperature profiles below the choke point. Even along the geotherms passing above the choke point, slabs can bring water into the deeper mantle by several mechanisms, such as transport by hydrous phases in the basalt and sediment layers of slabs, K-bearing hydrous phases at the base of the mantle wedge, and the isolated fluid pockets. The water storage capacity is very low in the upper and lower mantles, whereas the transition zone has a large water storage capacity due to high water solubilities in wadsleyite and ringwoodite. Recent seismological observations suggest existence of hydrous slabs in the transition zone depth. The seismic reflectors observed in the upper part of the lower mantle beneath the Mariana subduction zone may be accounted for by the existence of the fluid-bearing basaltic layer with perovskite lithology, where the fluid may be generated by dehydration of the hydrous phase D (phase G) in the underlying peridotite layer of the slab.

1. INTRODUCTION

Water is transported into the mantle by subducting slabs. The important carriers of water in the slabs are the high-pressure hydrous minerals. The hydrous minerals such as chlorite and serpentine have been considered to be stable only at relatively shallow upper mantle depths, and most of these hydrous phases are unstable at depths of around 150–200 km, which is often called a choke point [*Kawamoto et al.*, 1996]. Subducting slabs passing though a choke point in the lower portions of the upper mantle may cause dehydration of the slabs. It has been considered that volatiles stored in hydrous minerals in the slabs are dehydrated to trigger the island arc volcanism. Recent studies on the stability of high-pressure hydrous phases revealed that several high-pressure

hydrous phases are stable at the very high pressure and high temperature conditions of the transition zone [e.g., *Shieh et al.*, 1998; *Schmidt et al.*, 1998]. Thus, it is likely that water is transported to the transition zone and the lower mantle by slab subduction.

It has been clarified that the major constituent minerals in the transition zone, wadsleyite and ringwoodite, can accommodate a large amount of water of more than 2 wt.% [*Inoue et al.*, 1995; *Kohlstedt et al.*, 1996]. Therefore, the transition zone is an important water reservoir in the earth, although the real water content in the transition zone is a matter of great debate. Geophysical studies in seismology and electrical sounding are now undertaken to clarify the actual amount of water in the mantle. In this manuscript, I use the word "water" in a rather crude definition: i.e., it includes not only the molecular form of liquid water H_2O, but also the H and O atoms in various forms such as solid H_2O, vapor, supercritical fluid, hydroxyl and hydrogen ions in minerals and magmas.

Earth's Deep Mantle: Structure, Composition, and Evolution
Geophysical Monograph Series 160
Copyright 2005 by the American Geophysical Union
10.1029/160GM19

Mass balance calculations [e.g., *Peacock*, 1990] suggest that the amount of water transported into the mantle is greater than that degassed and supplied into the surface of the Earth by the magmatism in the island arcs, ocean islands, and mid-oceanic ridges. Therefore, water in the ocean at the surface of the earth may decrease with time. The mass balance calculations suggest that approximately half of the current ocean water can be transported into the mantle by slab subduction every billion years [*Peacock*, 1990; *Williams and Hemley*, 2001].

The phase relations of water bearing silicate systems relevant to slabs provide the first order petrological constrains on whether "water" is transported into the mantle or not. In this paper, I will summarize the recent studies on phase relations of the slab compositions under the hydrous conditions, and estimate the mechanisms for transport and storage capacity of water in the deep mantle. I also discuss possibilities of a hydrous transition zone and dehydration in the lower mantle based on the recent petrological phase relations, mineral physics data, and seismological observations. The topography of the 410 and 660 km discontinuities, and physical properties of some seismic reflectors in the lower mantle may be consistent with the effect of water in the transition zone and the lower mantle.

2. EFFECT OF WATER FOR THE PHASE RELATIONS IN THE MANTLE AND STABILITY OF HIGH-PRESSURE HYDROUS PHASES IN THE SLABS

2.1. Peridotite-Water System

Table 1 shows the hydrous minerals expected to exist in the peridotite bulk compositions. The phase relations of hydrous peridotite to 30 GPa in the CaO-MgO-Al$_2$O$_3$-SiO$_2$ (CMAS) system with 2 wt.% of water are shown in Figure 1 [*Litasov and Ohtani*, 2002, 2003; *Ohtani et al.*, 2004]. The details of the experimental conditions and the chemical compositions of minerals are given in *Litasov and Ohtani* [2002, 2003]. The bulk composition of the hydrous peridotite is: SiO$_2$, 45.15; Al$_2$O$_3$, 5.20; MgO, 43.46; CaO, 4.19; H$_2$O, 2.0 wt.%, which is obtained by simplification of the primitive mantle composition [*Jagoutz et al.*, 1979] to a CMASH five component system by adding 2 wt.% water. The phase relations at pressures above 10 GPa are based on the experimental results by *Litasov and Ohtani* [2002, 2003], whereas those below 10 GPa are based on the results by *Schmidt and Poli* [1998] and *Poli and Schmidt* [2002]. The peridotite layer of the slabs is depleted in the basaltic component, therefore it may correspond to herzolite or harzburgite depending on the degree of depletion by magma extraction [e.g., *Schmidt and Poli*, 1998].

The peridotite layer of slabs contains water due to serpentinization or chloritization, although there is little direct information on its water content. Serpentine is observed along the fracture zones of the ocean floor. *Schmidt and Poli* [1998] estimated that the upper 5 km of the peridotite layer suffers about 10% serpentinization, whereas *Kesson and Ringwood* [1989] suggested more extensive serpentinization. Thus, under the upper mantle conditions, chlorite and serpentine are the major water reservoirs in the cold slabs. Recently, *Fumagalli et al.* [2001] reported that 10Å phase is stable at temperatures above the stability of serpentine at around 6 GPa in the peridotite composition, and it can be a candidate for the water carrier in the peridotite layer of the slabs (Figure 1). Serpentine or 10Å phase is replaced by hydrous phase A at the depths greater than 200 km (~7 GPa). The maximum capacity of water storage in peridotite by these hydrous minerals is estimated to be in the range of 4.1~6.5 wt.% by *Schmidt and Poli* [1998] up to a pressure of 8 GPa.

The mineral assemblage of a serpentine peridotite containing more than 4.6 wt.% water transforms to the assemblage of phase A + Ca-poor clinopyroxene + Ca-rich clino-

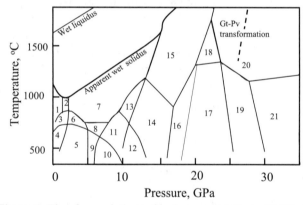

Figure 1. The phase relations of hydrous peridotite to 30 GPa in the CaO-MgO-Al$_2$O$_3$-SiO$_2$ (CMAS) system with 2 wt.% of water. The mineral assemblages of the numbered regions of the phase diagram are as follows: 1, ol+opx+sp+amp; 2. ol+opx+amp+gt; 3, ol+opx+chl+amp; 4, ol+tc+chl+amp; 5, ol+cpx+chl+serp; 6, ol+opx+cpx+chl; 7, ol+cen+gt+cpx+fl; 8, ol+cpx+gt+10Å; 9, ol+cpx+gt+serp; 10, ol+cpx+gt+serp+A; 11, ol+cen+cpx+gt+A; 12, wd+cen+cpx+gt+A; 13, ol+cen+cpx+gt+E; 14, wd+E+gt+cpx; 15, wd+gt+fl(+cpx); 16, wd+rw+gt+cpx+sb; 17, rw+sb+gt+cpv(+cpx); 18, rw+gt+cpv+fl; 19, mpv+cpv+pc(+gt)+sb; 20, mpv+pc+cpv(+gt)+fl; 21, mpv+cpv+pc+D. ol, olivine; opx, orthopyroxene; sp, spinel; amp, amphibole; chl, chlorite; tc, talc; serp, serpentine; cen, clinoenstatite; cpx, clinopyroxene; gt, garnet; fl, fluid; 10Å, 10 angstrom phase; A, hydrous phase A; wd, wadsleyite; E, hydrous phase E; rw, ringwoodite; sb, superhydrous phase B; cpv, CaSiO$_3$ perovskite; mpv, MgSiO$_3$ perovskite; pc, magnesiowustite; D, hydrous phase D (= hydrous phase G).

Table 1. The Hydrous Minerals Expected to Exist in the Peridotite, Basalt, and Sediment Bulk Compositions.

name	formula	density g/cm³	Mg/Si	H₂O wt.%	reference
chlorite	$Mg_5Al_2Si_3O_{10}(OH)_8$	2.6–3.4	1.67	13	1
serpentine	$Mg_3Si_2O_5(OH)_4$	2.55	1.5	14	2
10 Å phase	$Mg_3Si_4O_{14}H_6$	2.65	0.75	13	3
phase A	$Mg_7Si_2O_8(OH)_6$	2.96	3.5	12	4
phase B	$Mg_{12}Si_4O_{19}(OH)_2$	3.38	3	2.4	4
superhydrous phase B = phase C	$Mg_{10}Si_3O_{14}(OH)_4$	3.327	3.3	5.8	4, 5
phase D(= phase F = phase G)	$Mg_{1.14}Si_{1.73}H_{2.81}O_6$	3.5	0.66	14.5–18	6
phase E	$Mg2.3Si1.25H2.4O6$	2.88	1.84	11.4	7
wadsleyite	Mg_2SiO_4	3.47	2	≤3	8, 9
ringwoodite	Mg_2SiO_4	3.47–3.65	2	1.0–2.2	9
phlogopite	$K_2(Mg,Fe)_6Si_6Al_2O_{20}H_2$	2.78	~1	2.3	10
K-richterite	$K_2Ca(Mg,Fe)_5Si_8O_{22}(OH)_2$	3.01	~0.6	2.1	10
zoisite	$Ca_2Al_3Si_3O_{12}OH$	3.15	-	1.9	11
phase X	$K_4Mg_8Si_8O_{25}(OH)_2$	2.95–3.28	1	1.9	12
talc	$Mg_3Si_4O_{10}(OH)$	3.15	0.75	2.3	13
phengite	$K(Al,Mg)AlSi_3O_{10}(OH)_2$	2.83	0.33	4.6	14
lawsonite	$CaAl_2Si_2O_{10}H_4$	3.09	-	11.5	15
topaz-OH	$Al_2SiO_4(OH)_2$	3.37	-	10	16
diaspore	$AlOOH$	2.38	-	15	16
phase pi	$Al_3Si_2O_7(OH)_3$	3.23	-	9	16
phase Egg	$AlSiO_3OH$	3.84	-	7.5	17
phase δ	$AlOOH$	3.533	-	15	18

1, *Pawley* [1996]; 2, *Evans et al.* [1976]; 3, *Fumagalli et al.* [2001]; 4, *Ringwood and Major* [1967]; 5, *Gasparik et al.*[1990]; 6, *Kudoh et al.* [1997]; 7, *Kanzaki* [1991]; 8, *Inoue et al.*[1995]; 9, *Kohlstedt et al.*[1996]; 10, *Konzett and Ulmer* [1999]; 11, *Poli* [1993]; 12, *Konzett and Fei* [1998]; 13, *Bose and Gangley* [1995]; 14, *Domanik and Holloway* [1996]; 15, *Schmidt* [1980]; 16, *Wunder et al.* [1993]; 17, *Eggelton et al.* [1979]; 18, *Suzuki et al.* [2000].

pyroxene + garnet + fluid which is characterized by absence of olivine even in a peridotite bulk composition [*Schmidt and Poli*, 1998]. Since the water content is lower than that of *Schmidt and Poli* [1998], olivine can coexist with the hydrous phases in our hydrous CMAS peridotite phase relation given in Figure 1. Hydrous peridotite with 2 wt.% water transforms to the assemblage containing olivine, Ca-poor and Ca-rich clinopyroxenes, garnet, and phase A.

In the hydrous peridotite bulk compositions, the assemblage containing phase A changes to that containing hydrous phase E. The fluid appears at higher temperatures above about 1200°C in the upper part of the transition zone by decomposition of phase E (region 15 in Figure 1), whereas it appears above 1300°C at around 16 GPa by decomposition of superhydrous phase B (18 in Figure 1). Appearance of super-hydrous phase B in the transition zone depends on the water content. The upper bound of the water content for appearance of superhydrous phase B may be around 1.2 wt.%, since

ringwoodite, composing about 60 wt.% of the transition zone, can accommodate water up to 2 wt.% assuming that the water storage capacities of the other phases such as majorite and Ca-perovskite are negligible. The water content of appearance of superhydrous phase B in the transition zone could be lower, since water is partitioned between superhydrous phase B and ringwoodite [e.g., *Ohtani et al.*, 2000]. This phase appears also as a decomposition product of ringwoodite at pressures close to the 660 km discontinuity (18 in Figure 1). The mineral assemblage of superhydrous phase B, Mg-perovskite, majorite, and Ca-perovskite were observed in peridotite containing 2 wt.% water under the lower mantle conditions below 1400°C. Periclase appears in addition to this assemblage in the peridotite compositions with a lower water content.

Superhydrous phase B decomposes into periclase + Mg-perovskite + phase D (= phase G by *Kudoh et al.* [1997]) in the uppermost part of the lower mantle at around 30 GPa and temperatures below 1200°C, and it further decomposes into periclase + Mg-perovskite + fluid at higher temperature at 30 GPa [Ohtani et al., 2003]. Therefore, superhydrous phase B exists up to a depth of 900 km in the upper part of the lower mantle.

Water can be transported by the slabs into the deep upper mantle and transition zone, provided that the slab geotherm passes through the temperature below the choke point [Kawamoto et al., 1996] of 600°C at 6 GPa; when the temperature of the slab is lower than this temperature, the stability field of serpentine overlaps with that of phase A, which is one of the most important dense hydrous magnesium silicate (DHMS) phases stable in the deep upper mantle [Ulmer and Trommsdorff, 1995; Luth, 1995]. The high temperature stability limit of serpentine is placed by formation of hydrous phase A at 600°C and 6 GPa, i.e., serpentine + phase A = enstatite + fluid. However, the recent studies of the phase relations of water-bearing silicate systems revealed that the choke point is not so effective as a barrier for water transport because several mechanisms does exist for water transport into the deep upper mantle and transition zone even along the geotherms passing through the temperatures above the choke point as is discussed in the succeeding sections.

Figure 1 shows the phase relation of peridotite mantle in the CMASH system, which is a simplified model of the natural peridotite. Although it is a FeO-free model system, it can well model the phase relations of the real hydrous peridotite mantle. The effect of FeO on the phase boundary of peridotite is rather small, i.e., difference in the stability field of hydrous phases in FeO-free and FeO-bearing systems of the model mantle may be negligible as is shown in the location of the choke point and the stability of phase A [Schmidt and Poli, 1998; Luth, 1995; Ulmer and Trommsdorf, 1995].

2.2. Basalt-Water System

The phase relations of the basalt-water system have been studied by several authors [e.g., Schmidt and Poli, 1998; Ono, 1998]. They showed that lawsonite is stable up to the deep upper mantle of 8 GPa, whereas no hydrous phases are observed at higher pressures except for the K-bearing basalt composition. K-bearing hydrous phase, phengite, can exist at higher pressures up to about 10 GPa. The phase relation of the basalt-water system [Schmidt and Poli, 1998] is summarized in Figure 2. The oceanic crust composed of the basalt component does not have a water storage capacity at pressures above 10 GPa after decomposition of lawsonite, although some amount of water may be trapped in the grain boundary due to a large interfacial energy between garnet and fluid [e.g., Ono et al., 2002] or in phengite for the K-bearing basaltic composition. The hydrous phases appearing in the

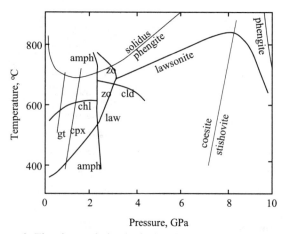

Figure 2. The phase relation in the basalt-water system. The phase relation is based on the results by Schmidt and Poli [1998]. zo, zoicite; cld, chloritoid; law, lawsonite. The other abbreviations are the same as those used in Figure 1. Lawsonite is the major water reservoir in hydrous basalt. In K-bearing compositions, water is stored in phengite and stabilized to 10 GPa and high temperatures, after decomposition of lawsonite.

basalt-water systems are given in Table 1. The phase relations of the basalt-water system at pressures above 10 GPa are clearly needed to clarify the water storage in the oceanic crust in the transition zone and the lower mantle.

2.3. Sediment-Water System

The stability conditions of the hydrous phases in siliceous sediments are given in Figure 3 as a phase diagram [e.g., Ono, 1998; Schmidt and Poli, 1998]. The hydrous phases in sediments are also listed in Table 1. Phengite is stable up to about 8 GPa, which decomposes to the assemblage containing a hydrous phase, topaz-OH. It is replaced by phase Egg [Eggelton et al., 1979] in the transition zone conditions. The stability field of phase Egg has been studied at the pressures above 10 GPa [Ono, 1999; Schmidt and Poli, 2001]. According to these authors, phase Egg is stable up to 1400°C in the transition zone conditions when the sediment components containing water are transported to the depths.

Sano et al. [2004] clarified that this phase is stable up to the top of the lower mantle, and it decomposes into a new hydrous phase δ-AlOOH and stishovite at 30 GPa and the temperature below 1200°C. It decomposes into corundum + stishovite + fluid at the same pressure and above 1200°C. These phase relations indicate that δ-AlOOH is a candidate for water reservoirs in the lower mantle. The elastic properties of δ-AlOOH have been studied by Vanpeteghem et al. [2002]. They showed that this phase has a high bulk

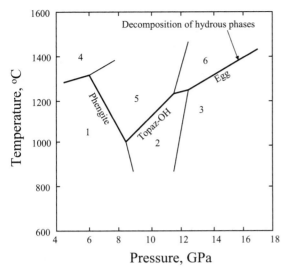

Figure 3. The stability of hydrous phases in sediments [modified from *Ono*, 1998]. 1, gt+cpx+co+phen; 2, gt+cpx+st+ho+ox+to; 3, gt+cpx+st+ho+ox+eg; 4, gt+cpx+coe+melt; 5, gt+cpx+ho+ox+ky+co/st; 6, gt+cpx+st+ho+ox+fl. The assemblage contains fluid in all PT conditions, i.e., the system is saturated with water in the experiments by *Ono* [1998]. coe, coesite; phen, phengite; ox, FeTi oxide; st, stishivite; ho, hollandite; to, topaz-OH; ky, kayanite; eg, phase Egg; the other abbreviations are the same as those in Figures 1 and 2.

modulus close to that of corundum, K=253 GPa, assuming K'=4, although it contains a large amount of hydrogen in the lattice. An important nature of this phase is a high solid solubility of MgO and SiO_2 components in the structure. *Suzuki et al.* [2000] reported the phase containing a significant amount of Mg and Si substituting Al in the lattice with a composition of $Al_{0.86}Mg_{0.07}Si_{0.07}OOH$. δ-AlOOH containing Si and Mg may exist in the basalt-water system although its stability in the basalt bulk composition is not yet studied. δ-AlOOH phase together with phase D is a potential water carrier into the lower mantle. The hydrous phases existing in the hydrous sediment compositions at high pressures and temperatures are also given in Table 1. The phase relations of the sediment-water systems are clarified up to ca. 18 GPa. We need further studies of the phase relations of these systems at higher pressures in order to discuss the role of the sediment components in the transition zone and the lower mantle.

3. WATER SOLUBILITY OF NOMINALLY ANHYDROUS MINERALS AND WATER STORAGE CAPACITY IN THE MANTLE

In the upper mantle, olivine can accommodate only limited amounts of water along the normal mantle geotherm,

although the amount of water stored in olivine increases with pressure [*Kohlstedt et al.*, 1996]. DHMS phases are stable at lower temperatures corresponding to the descending slabs. The hydrous phases stable in peridotite at the depths from the transition zone to the lower mantle are phase E, superhydrous phase B, and phase D. The other nominal anhydrous minerals such as garnet and pyroxenes constituting the oceanic crust also contain some water. Clinoenstatite contains about 580 ppm wt. H_2O at 15 GPa and 1300°C, whereas majorite garnet contains about 680 ppm H_2O at 17.5 GPa and 1500°C [*Bolfan-Casanova*, 2000]. Therefore these minerals could also be important water carriers in basaltic component of subducting slabs for transporting water into the deep upper mantle.

Wadsleyite and ringwoodite, the major constituent minerals in the transition zone, can accommodate water up to about 3 wt.% and 2 wt.%, respectively, at the temperature conditions of the cold slabs [e.g., *Inoue et al.*, 1995; *Kohlstedt et al.*, 1996]. Thus, the transition zone is a potential water reservoir in the Earth. The maximum water solubility in wadsleyite at 15 GPa decreases with increasing temperatures from 2.0 wt.% at 1000°C to 0.5 wt.% at 1600°C, and to 0.3 wt.% with increasing pressure to 18 GPa at 1600°C [*Litasov and Ohtani*, 2002]. Thus the shallower transition zone has a larger water storage capacity than the deeper region. Water solubility in ringwoodite also decreases with increasing temperature, but it has a weak pressure dependence. Since wadsleyite and ringwoodite can contain H_2O up to about 0.5 wt.% along the normal mantle geotherm in the transition zone, the total amount of water which can be accommodated in the transition zone may be as large as 2×10^{24} g, which is about twice the present ocean mass.

Water storage capacity in the lower mantle is a matter of great debates. In the cold subducting slabs (<1200°C), superhydrous phase B and hydrous phase D can store and transport water into the lower mantle. Ilmenite $MgSiO_3$, which is stable in the depleted harzburgite layer of subducting slabs, can also accommodate water up to 0.2 wt.% at about 1600–1750°C near the ilmenite-perovskite phase boundary [*Bolfan-Casanova*, 2002]. The main constituent of the lower mantle is Mg-perovskite, magnesiowustite and Ca-perovskite. *Bolfan-Casanova et al.* [2000] showed absence of water (<1 ppm wt. H_2O) in $MgSiO_3$-perovskite. New data revealed that the water solubility in aluminous perovskite observed in the peridotite bulk composition can accommodate a significant amount of water, more than 0.1 wt.% H_2O [e.g., *Murakami et al.*, 2002; *Litasov et al.*, 2003]. On the other hand, Fe-bearing aluminous perovskite observed in the basalt component in the slabs can contain a limited amount of water (100–400 ppm wt. H_2O) [*Litasov et al.*, 2003]. *Murakami et al.* [2002] reported that Ca-perovskite and magnesiowustite accommodate about 0.3–0.4 wt.% and 0.2 wt.% water, respectively. However, *Bolfan-Casanova et al.* [2002] reported that periclase can accommodate very little water up to 2 ppm wt.%, whereas magnesiowustite contains only 20 ppm wt.% H_2O at 25 GPa.

The average lower mantle peridotite contains 80 wt.% Mg-perovskite, 15 wt.% magnesiowustite, and 5 wt.% Ca-perovskite. Summarizing the water solubility of the lower mantle minerals, we can estimate that the average lower mantle peridotite can contain 0.15–0.2 wt.% water, i.e., the water storage potential of the lower mantle can be estimated to be $3.4–4.5 \times 10^{24}$ g. This is about 2–3 times of the present ocean mass, which is comparable to the amount of water stored in the transition zone. We can generally conclude that there is a layered structure in the water storage capacity in the mantle; i.e., the upper and lower mantles have relatively low water storage capacities, whereas the transition zone has a high capacity, although there are ambiguities in the water storage capacity in perovskite and magnesiowustite. A relatively low water solubility of the lower mantle minerals indicates that the subducting hydrous slabs might dehydrate at the top of the lower mantle.

The water content in the transition zone might be large because the dehydrated water from the subducted slabs together with the primordial water in the lower mantle have been stored in the transition zone during the global circulation of water in the mantle. Although there is a huge water storage capacity in the transition zone and the lower mantle, i.e., the total water storage capacity in whole mantle will be about 4–5 times the ocean mass, actual amounts of water stored in these regions are a matter of debate. *Dixon et al.* [2002] suggested that no evidence for water concentration in the deep mantle, i.e., water content in basalts originating from the enriched mantle containing a recycled component in the plume, is lower than the MORB, suggesting effective extraction of water during subduction. Geophysical data such as elevation of the 410 km discontinuity, seismic wave velocity and electrical conductivity profiles [e.g., *Koyama et al.*, 2004] indicate that some amounts of water exist heterogeneously in the mantle transition.

4. MECHANISM FOR WATER TRANSPORT INTO THE TRANSITION ZONE AND LOWER MANTLE

4.1. Choke Point

Water may be transported into the transition zone by the cold subducting slabs. It has been argued that subducting slabs will pass though a choke point in the lower parts of the upper mantle where the absence of stable hydrous phases causes dehydration of the slabs, thus preventing water from

being subducted into the transition zone [e.g., *Kawamoto et al.*, 1995]. *Kawamoto et al.* [1996] placed the base of the choke point at 6 GPa and 600°C which is an invariant point coexisting with serpentine, phase A, enstatite, and fluid. *Fumagalli et al.* [2001] recently reported that the 10Å phase is stable at temperatures above the choke point at around 6 GPa even in the peridotite composition, and can be a candidate for the water carrier in the peridotite layer of the slabs.

The upper bound of the geotherm of the wet peridotite layer of the slab can be placed by the stability field of 10Å phase, which is stable up to about 700°C at 7 GPa [*Poli and Schmidt*, 2002]. The slabs with this geotherm passing below the temperature of the invariant point of phase A, 10Å phase, enstatite, and water at around 7 GPa and 700 °C can transport water into the deeper upper mantle. The temperature of the choke point, i.e., 700°C at 7 GPa is higher than that estimated by earlier authors [*Kawamoto et al.*, 1996; *Schmidt and Poli*, 1998, 6 GPa and 600°C]. Thus, the important water carriers in the peridotite layer of subducting slabs in the upper mantle are chlorite, serpentine, 10Å phase, and phase A. At higher pressures, the stability field of phase A expands with a positive slope of its dehydration boundary as shown in Figure 4. Cold subducting slabs, i.e., old subducting slabs such as the slab beneath the Tonga subduction zone, can transport water into the deep upper mantle without dehydration from the slabs. Although higher temperature slabs may release volatiles at the

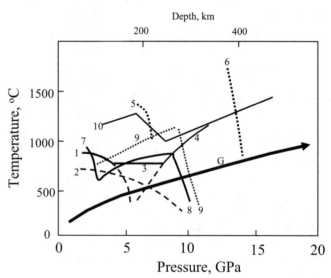

Figure 4. The stability limit of hydrous phases in the peridotite-water, basalt-water, and siliceous sediment-water systems. 1, chlorite; 2, serpentine; 3, 10Å phase; 4, phase A and phase E; 5, phlogopite in K-bearing peridotite; 6, K-richterite in K-bearing peridotite and appearance of phase X; 7, zoisite in basalt; 8, lawsonite in basalt; 9, phengite in K-bearing basalt; 10, phengite, topaz-OH, phase Egg in sediment; G, the cold slab geotherm [e.g., *Peacock*, 1990].

choke point in the upper mantle by dehydration of serpentine, 10Å phase, or phase A, the released water can be trapped in the slabs by several mechanisms, and further transported into the transition zone and lower mantle. The choke point may not work effectively as a barrier for the water transport into the deep upper mantle.

4.2. Mechanisms for Transport of Water Into the Deep Upper Mantle and Transition Zone

Several mechanisms have been proposed to transport water into the deep mantle even along the geotherms passing above the choke point. High pressure hydrous phases exist in basalt and sediment systems [*Ono*, 1998]. The wetting angle between garnet and fluid is greater than 60°, indicating that water can be trapped in the grain boundary of the oceanic crust and transported into the deeper mantle [*Ono et al.*, 2002]. Effect of potassium is also important to stabilize hydrous phases in basalts and peridotites at temperatures above the choke points [e.g., *Konzett and Fei*, 1998].

High-pressure hydrous phases in basalts and sediments. The stability fields of hydrous phases in basalt and pelagic sediment components are shown in Figure 4. *Ono* [1998] argued that lawsonite and phengite are stable in the hydrous basalt at the temperatures higher than the dehydration temperatures of serpentine and chlorite, and thus water generated by dehydration of the peridotite layer of the slab can be trapped and stored in the overlying lawsonite or phengite in the basalt layer at least up to the pressure of 8 GPa [*Ono*, 1998].

The phase relation of the sediment-water system implies that there are various hydrous phases, including phengite, topaz-OH, phase Egg, and phase δ in the conditions of the cold subducting slabs. These hydrous phases in sediments are stable even under the conditions of dehydration of the hydrous phases in the peridotite and basalt layers of the slabs. Water released by decomposition of lawsonite and/or phengite at around 8 GPa can be trapped in the hydrous minerals in sediments, such as topaz-OH and phase Egg, and further transported to the transition zone. The phase relations among the peridotite-water, basalt-water, and siliceous sediment-water systems (Figure 4) indicate that water dehydrated from the peridotite layer or the basalt layer of the slabs can be absorbed by the sediment layer, i.e., the siliceous sediment can work as an absorber of water in the slabs in the deep upper mantle and the transition zone, although the amount of sediments is limited compared to the basalt and peridotite components in the slab.

Water generated by high pressure decomposition of lawsonite and/or phengite may also be stored as phase A in the mantle wedge, since the stability field of phase A expands to the pressures higher than the stability field of lawsonite and phengite. Therefore, water can be tapped by the mantle wedge and transported further into the deep upper mantle by the drag flow due to the slab subduction. The stability conditions of hydrous phases for the thermal structure of the typical subducting slab (subducting rate, 7.2 cm/year and dipping angle of 60°, *Davies and Stevenson* [1992]) are schematically shown in Figure 5.

Wetting angle between garnet and fluid. The wetting angle is also an important factor for transport of water into the deep mantle. *Ono et al.* [2002] showed that the dihedral angle between garnet crystals and fluid is greater than 60°, which suggests that the fluid pockets are isolated in the oceanic slabs composed mainly of majorite garnet crystals. Thus, the dehydrated fluid can be trapped in the grain boundaries in the descending oceanic crust even at high temperatures above the stability of lawsonite and phengite. Water can be transported even if the slab geotherm passes above the choke

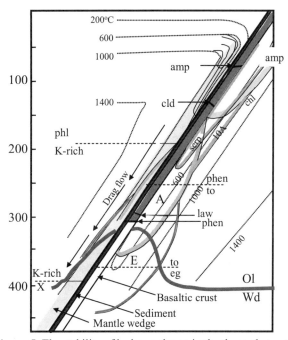

Figure 5. The stability of hydrous phases in the thermal structure of the subducting slab (horizontal axis is the distance; dipping angle of 60°; subduction rate of 7.2 cm/year, *Davies and Stevenson* [1992]). The shaded areas in the slab and mantle wedge indicate existence of hydrous phases. Phengite, topaz-OH, and phase Egg are the hydrous phases expected in sediment. amp, amphibole; chl, chlorite; cld, chloritoid; serp, serpentine; phl, phlogopite; K-rich, K-richterite; phen, phengite; to, topaz-OH; law, lawsonite; A, phase A; E, phase E; eg, phase Egg [*Eggelton et al.*, 1979]; X, hydrous X-phase [*Luth*, 1997; *Inoue et al.*, 1998]; ol, olivine; wd, wadsleyite.

point. The choke point is not an effective barrier for water transport into the transition zone and lower mantle.

Effect of potassium-bearing hydrous phases. K-bearing assemblages may be more relevant to the mantle wedge as a result of lower pressure dehydration of the slab and removal of K into the over lying mantle. Existence of potassium can affect water storage and transport by subducting slabs and dragged mantle wedge, since potassium-bearing hydrous silicates such as phlogopite and K-amphibole (K-richiterite) are stable at higher pressure and temperature even in the peridotite composition. In peridotite bulk compositions, phlogopite is stable to at least at 6 GPa and 1000°C coexisting with orthopyroxene + clinopyroxene + olivine [e.g., *Konzett and Ulmer*, 1999], and to at least 12 GPa and 1350°C coexisting with clinopyroxene [*Luth*, 1997].

K-amphibole with a composition close to $KNaCaMg_5Si_8O_{22}(OH)_2$ can be formed as a high-P breakdown product of phlogopite in the peridotite compositions at P > 6 GPa [*Konzett and Ulmer*, 1999]. K-amphibole and garnet react to form pyroxenes and another hydrous phase, phase X, $K_4Mg_8Si_8O_{25}(OH)_2$, and water at high pressures between 12 and 14 GPa corresponding to 340–400 km depth [*Luth*, 1997; *Inoue et al.*, 1998]. At higher pressures above 20 GPa, phase X dehydrates to form K-hollandite [*Konzett and Fei*, 1998]. The stability conditions of these hydrous phases in the mantle wedge are summarized in Figure 4 and schematically shown in Figure 5.

In the hot subducting slabs, phengite dehydration, K-bearing basalt melting, or phlogopite breakdown at <3 GPa can provide K and H_2O into the peridotite mantle wedge, and can stabilize K-bearing hydrous phases, which are dragged down into the transition zone (Figure 5). The succession of the K-bearing hydrous minerals in the mantle wedge is phlogopite (+Ca-amphibole) →K-amphibole (K-richterite) →phase X, with a final dehydration product of K-hollandite at pressures greater than 20 GPa. Thus it is inevitable that water can be transported into the deep upper mantle and the transition zone, if the slab contains water at shallow levels.

5. GEOPHYSICAL IMPLICATIONS

5.1. Subduction of the Hydrated Oceanic Crust Into the Lower Mantle

A large density crossover between the basalt layer and the underlying peridotite layer of the slab exists at relatively high temperatures due to a positive slope of the garnet-perovskite phase boundary and a negative slope of the post-spinel phase boundary [e.g., *Oguri et al.*, 2000; *Katsura et al.*, 2003, *Litasov et al.* 2004]. Thus, subducting oceanic

crust component in the hot Archean time may not penetrate into the lower mantle, and it might be stored in the mantle transition zone as was suggested by *Ringwood* [1994]. The kinetics of the phase transformations may enhance this tendency, i.e., *Kubo et al.* [2002] showed that the reaction kinetics of the garnet-perovskite transformation under the dry condition is very sluggish and the metastable garnet could survive on the order of 10 My even at 1600 K, although the metastable ringwoodite quickly breaks down at 1000 K. The sluggish reaction rate of the garnet-perovskite transformation and the fast reaction rate of the post-spinel transformation provide a large pressure interval of the density crossover between the basaltic oceanic crust layer and the underlying peridotite mantle, creating a buoyancy force in the oceanic crust to resist its penetration into the lower mantle [*Kubo et al.*, 2002]. Hot and heterogeneous mantle plumes containing the oceanic crust component, on the other hand, can easily be transported from the lower mantle into the transition zone due to buoyancy in the basaltic component.

In the wet subducting slabs, on the other hand, the garnet-perovskite phase transformation is kinetically enhanced by water [e.g., *Litasov et al.*, 2004]. A strong positive slope of the garnet-perovskite transformation [e.g., *Oguri et al.* 2000, *Litasov et al.*, 2004] and rapid reaction kinetics under the hydrous conditions result in absence of the density crossover in the cold subducting slabs. Thus, slabs can easily penetrate into the lower mantle without buoyancy of the density crossover under the conditions of the hydrous subduction. The oceanic crust descending into the lower mantle could cause the heterogeneity in the lower mantle.

The stability of high pressure hydrous phases and the temperature profiles in the slabs indicate that there are several dehydration regions in the slabs as was discussed by *Ohtani et al.* [2004]; the major regions of dehydration are the top of the lower mantle where dehydration of hydrous ringwoodite and/or superhydrous phase B is expected, and the base of the upper mantle where dehydration melting is expected in the ascending plumes due to difference in water content in olivine and wadsleyite.

Revenaugh and Sipkin [1994] suggested an evidence for existence of melt at the base of the upper mantle or the top of the transition zone beneath eastern China. P-wave tomography data beneath Japan [*Zhao*, 2001, 2004; *Fukao et al.*, 2001] suggest the existence of low velocity anomalies at the base of the upper mantle and the top of the lower mantle. A low P-wave anomaly in the transition zone and the base of the upper mantle is also reported beneath the Tonga trench, the coldest one among subducted slabs [*Zhao et al.*, 1997]. The low velocity anomaly may originate from the presence of water as a fluid or hydroxyle in wadsleyite or ringwoodite in the transition zone [e.g., *Inoue et al.*, 1998; *Smyth et al.*, 2004].

5.2. Seismic Evidence for Hydrous Subducting Slabs

A large elevation of the 410 km discontinuity to 60–70 km in Izu-Bonin slab was observed by *Collier et al.* [2001]. They argued that the discontinuity can be explained by the equilibrium boundary of the olivine-wadsleyite transformation in the cold slab, and a large elevation of the 410 km discontinuity corresponds to the temperature difference of about 1000°C compared to the surrounding normal mantle. This indicates the temperature of the 410 km discontinuity is around 500°C. Recent studies on the phase transformation kinetics [e.g., *Rubie and Ross*, 1994] indicate that a very low temperature transformation, such as that proposed by *Collier et al.* [2001], would be kinetically inhibited and therefore unlikely to occur in the dry slabs.

A small amount of water, about 0.12–0.5 wt.%, enhances dramatically the olivine-wadsleyite phase transformation kinetics, which corresponds to the temperature elevation of about 150°C [*Kubo et al.*, 1998; *Ohtani et al.*, 2004]. The elevation of the 410 km discontinuity of the Izu-Bonin slab reported by *Collier et al.* [2001] may be explained as the equilibrium phase boundary under the hydrous conditions. *Koper et al.* [1998] also observed no evidence for depression of the 410 km discontinuity of the coldest Tonga slab originating from metastable olivine wedge in the slab.

The depression of the 660 km discontinuity in the subduction zone may be consistent with the hydrated transition zone [*Litasov et al.*, 2005]. Recent in situ X-ray diffraction studies of the phase boundary of the decomposition of ringwoodite [*Katsura et al.*, 2003; *Fei et al.*, 2004] under the dry condition indicate that the slope of the boundary, dP/dT, is around −0.4 to −1.3 MPa/K, which is significantly smaller than that determined previously (−2.5 MPa/K [*Irifune et al.*, 1998]). A decrease of the temperature by 1500 K is needed to account for the depression of the discontinuity about 40 km observed in the some subduction zones [e.g., *Collier et al.*, 2001], suggesting anomalously cold subducting slabs. *Litasov et al.* [2005] showed that the slope of the phase boundary becomes greater by existence of water due to high water contents in ringwoodite. Thus, a large depression of the 660 km discontinuity beneath some subduction zones is consistent with wet subducting slabs.

Seismic reflectors have been reported in the upper part of the lower mantle by several authors [e.g., *Kaneshima and Helffrich*, 1998; *Niu et al.*, 2003]. Recent study by *Niu et al.* [2003] indicates that the physical properties of the reflector observed at the depth of 1115 km beneath the Mariana subduction zone show a decrease in shear wave velocity by 2–6%, an increase in density by 2–9% within the reflector, whereas there is almost no difference in P-wave velocity (<1%) between the reflector and the surrounding mantle. The

estimated thickness of the reflector is around 12 km. Origin of the seismic reflectors is one of the most interesting issues in understanding the nature of heterogeneity in the lower mantle. It is likely that the seismic reflectors may correspond to the subducted oceanic crust as was suggested by some authors [e.g., *Kaneshima and Helffrich*, 1999]. In order to consider this possibility, we need to clarify whether the oceanic crust in the lower mantle can explain the observed seismological properties of the reflectors. There may be three possible states for the oceanic crust in the lower mantle for explaining the seismic properties of the reflector. First, it may have an equilibrium lower mantle lithology composed of Mg-perovskite, Ca-perovskite, stishovite, and calcium ferrite type aluminous phase (CF). CF may be replaced by Na-aluminous phase (NAL). Second, it may have the lower mantle lithology containing metastable majorite garnet due to the sluggish transformation kinetics of the garnet-perovskite transformation [*Kubo et al.*, 2002]. There is also a possibility that the physical properties of the seismic reflectors may not be accounted for by the above two lithologies. In such a case, we may need additional factors to explain the properties of the seismic reflectors.

In order to find a plausible explanation for the properties of the seismic reflectors, we need to estimate physical properties of the oceanic crust at high pressure and temperature. Table 2 summarizes the physical properties of high-pressure minerals in the oceanic crust and mantle. Table 3 gives compressional (V_p) and shear (V_S) velocities, and density at 30 GPa and 1600°C for the stable and metastable oceanic crust calculated from the data given in Table 2. V_p and V_S of the oceanic crust and pyrolite mantle are calculated using the averaged bulk modulus, K_S, and rigidity, μ, by the following equations; $V_p = (K_S(P,T)+(4/3)\mu(P,T))^{1/2}/\rho(P,T)$, $V_S = (\mu(P,T)/\rho(P,T))^{1/2}$, where $\rho(P,T)$ is the density at pressure P and temperature T. The $K_S(P,T)$ and $\mu(P,T)$ of the oceanic crust and mantle are estimated by averaging those of the constituent minerals by the Voigt-Reuss-Hill average scheme. $K_T(P,T)$, isothermal bulk modulus, was converted to the adiabatic bulk modulus $K_S(P,T)$ by the following relation, $K_S(P,T) = K_T(P,T)(1 + \alpha\gamma T)$, where α and γ are thermal expansion coefficient and Grüneisen parameter at P and T, respectively. The detailed procedure of calculation was based on *Akaogi et al.* [2002]. There is a large uncertainty in estimation of V_p and V_S of the constituent minerals. The zero pressure density, bulk moduli, and its pressure derivative may have uncertainties typically <1%, 2–3%, and 5–10%, respectively [e.g., *Weidner and Wang*, 1998; *Wang et al.*, 2004]. The shear modulus G may have an uncertainty of the same order as the bulk modulus, although its pressure and temperature derivative are not yet reported. The resulting uncertainties in density and the velocity (V_p and V_S) are 1–2% and ±5%, respectively [e.g., *Wang et* al., 2004] for the minerals of which elastic properties are available. The uncertainty in the shear modulus of CF phase is larger, since there are no available data for shear modulus. We calculated the shear modulus based on density ρ, bulk modulus K_S, and V_P estimated from Birch's law [Anderson, 1989]. If the shear modulus G of CF phase is similar to that of Ca-perovskite or majorite, it could vary from 90 to 164 GPa, resulting in the uncertainty of V_P and V_S around 10–20 %. Figure 6 shows the differences between density, V_P, and V_S of MORB (basaltic crust in the slab) and the surrounding mantle peridotite calculated in Table 3. The density and velocity (V_P and V_S) differences between the reflector and surrounding mantle observed seismologically [Niu et al., 2003] are also superimposed in Figure 6. The properties of MORB with a metastable garnet-bearing lithology are also shown.

A decrease in V_S and an increase in density are the characteristic properties of the reflector which can not be accounted for by the perovskite lithology nor metastable

majorite lithology of MORB, i.e., by a complete or incomplete phase transformation from majorite to perovskite lithologies, although there is a large uncertainty in seismic velocity, especially in V_S for MORB. We need additional factors to reduce V_S, although the density increase can be explained by the lower mantle lithology; i.e., the complete phase transformation from majorite lithology to perovskite lithology shows positive jumps in density, V_P, and V_S relative to the surrounding mantle, whereas the metastable majorite lithology shows the physical properties (ρ, V_P, and V_S) smaller than those of the surrounding mantle. We may be able to explain a drastic decrease of V_S as the effect of fluid or melt films in the subducted oceanic lithosphere in the lower mantle [e.g., Eshelby, 1957; Williams and Garnero, 1996]. Recent experiments on stability of hydrous phase D [e.g., Shieh et al., 1998] indicated that this phase dehydrates at pressures around 40–50 GPa. Therefore, the properties of the seismic reflector might be explained

Table 2. Physical Properties of the Constituent Minerals in the Lower Mantle for the Basaltic Crust and Peridotite Mantle Systems.

		Mg-perovskite (in peridotite, O=12)	Magnesiowustite	Ca-perovskite (O=3)	Mg-perovskite (in MORB, O=12)	Garnet	Stishovite[1]	CF phase[2]
$V(0,298)$	cm³/mol	97.76 (I)	11.25(H)	27.45(W)	101.68(I,F,K)	117.13(Wa,M,He)	14.09	37.13
$d(\ln V)/d(X\mathrm{Fe})$		0.036 (F)	0.089(H)					
$d(\ln V)/d(X\mathrm{Al})$		0.01(K)						
$K_T(0,298)$	GPa	261(Fu)	157(Fe)	232(W)	261(Fu)	160(P)	291	243(O)
dK_T/dP		4(Fu)	4(Fe)	4.8(W)	4(Fu)	4(Ak)	4.3	4(O)
dK_T/dT	GPa/K	-0.028(Fu)	-0.027(Fe)	-0.033(W)	-0.028(Fu)	-0.024(And)	-0.041	-0.033
	($\alpha T = a_0 + a_1 T + a_2 T^{-2}$, K^{-1})							
$a_0 \times 10^5$	K⁻¹	1.982 (Fu)	3.681(Su)	3.01(W)	1.982 (Fu)	2.874(SA)	1.40E-05	3.01E-05
$a_1 \times 10^9$	K⁻²	8.18(Fu)	9.283(Su)	0(W)	8.18(Fu)	2.886(SA)	1.09E-08	0.00E+00
$a_2 \times 10$	K	-4.74(Fu)	-7.445(Su)	0(W)	-4.74(Fu)	-0.5443(SA)	0.00E+00	0.00E+00
$G(0,298)$	GPa	175(Y)	132(An)	164(K)	175(Y)	90(P)	220	142.9*
dG/dP		1.8(S)	2.5(D)	1.8(Ak)	1.8(S)	2(Ri)	1.8	1.8
dG/dT	GPa/K	-0.029(S)	-0.024(An)	-0.014(Ak)	-0.029(S)	-0.014(Ak)	-0.018	-0.014
gamma(0)		1.4(A)	1.5(F)	1.7(WW)	1.4(A)	1.2(Wa)	1.35	1.7
Debye T	K	1030(A)	500(F)	1100(Ak)	1030(A)	800(Ri)	1152	1100

[1] Anderson [1989]; [2] assumed to be the same as Ca-perovskite.
(I), Ito and Yamada [1982]; (F), Fei et al. [1996]; (K), Kubo and Akaogi [2000]; (Fu), Funamori et al. [1996]; (Y), Yeganeh-Haeri [1994]; (S), Sinelnikov et al. [1998]; (A), Akaogi and Ito [1993]; (O), Ono et al. [2002]; (H), Hazen and Jeanloz [1984]; (Fe), Fei et al. [1992]; (Su), Suzuki et al. [1975]; (An), Anderson et al. [1992]; (D), Duffy and Ahrens [1995]; (W), Wang et al. [1996]; (K), Karki and Crain [1998]; (Ak), assumed (Akaogi et al., 2002); (WW), Wang and Weidner [1994]; (Wa), Wang et al. [1998]; Matsubara et al. [1990]; (He), Heinemann et al. [1997]; (P), Pacalo and Weidner [1997]; (And), Anderson et al. [1991]; (SA), Suzuki and Anderson [1983]; (Ri), Rigden et al. [1991].
* calculated by K_s, ρ, and V_P estimated from the Birch's law [e.g., Anderson, 1989].

Table 3. Density, V_P, and V_S of the Basaltic Crust and Peridotite and Their Constituent Minerals at 30 GPa and 1600°C.

30 GPa 1600°C	Density* g/cm³	V_P* km/sec	V_S* km/sec	Density difference from LM, %	V_P difference from LM %	V_S difference from LM %
Lower mantle (PREM)	4.47	11.24	6.28			
Mg-perovskite	4.48	11.43	6.29			
Ferropericlase	4.39	10.29	6.12			
Ca-perovskite	4.56	11.43	6.51			
MORB	4.58	11.54	6.52	2.41	2.64	3.87
MORB-perovskite	4.7	11.27	6.24	5.25	0.28	-0.60
Garnet	4.2	10.02	5.52	-5.97	-10.89	-12.12
Ca-perovskite	4.57	11.5	6.55	2.18	2.27	4.30
Stishovite	4.52	12.43	7.37	1.21	10.59	17.34
CF-phase	4.46	11.29	6.26	-0.30	0.47	-0.26

*, Uncertainties of density, V_P, and V_S are discussed in the text.

by existence of fluid in the oceanic crust generated by dehydration of this phase in the hydrous peridotite layer of the slab in the lower mantle. Although existence of fluid film in the oceanic crust component of the slab is a plausible

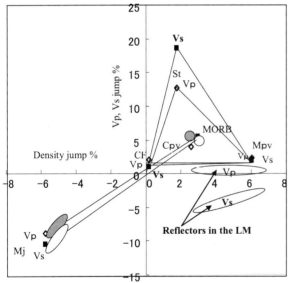

Figure 6. Differences (%) in density, V_P, and V_S between MORB (oceanic crust) and peridotite mantle at 30 GPa and 1600°C. MORB with the lower mantle assemblage (composed of Mg-perovskite, Ca-perovskite, stishovite, and CF phase) and metastable majorite assemblage (composed of majorite and stishovite) are shown in this figure. The differences between the reflectors and surrounding mantle observed seismologically [*Niu et al.*, 2003] are also shown in this figure. Open diamonds represent the V_P jump, whereas solid squares are the V_S jump from the properties of the peridotite mantle. Mj, majorite garnet; Mpv, Mg-perovskite; Cpv, Ca-perovskite; CF, CF phase; St, stishovite.

explanation for the physical properties of the reflector, existence of minor metallic iron formed by the garnet-perovskite transformation in the oceanic crust [*Miyajima et al.*, 1999; *Frost et al.*, 2004] could also cause the reduction of shear wave velocity and to increase the density of the reflectors. Following explanations may also be possible to account for the unusual properties of the seismic reflector, i.e., existence of elastic anomaly associated with the phase transformation of stishivite to post-stishovite $CaCl_2$ phase [e.g., *Stixrude*, 1998; *Carpenter et al.*, 2000], or the ortho-rhombic to cubic transformation in Al_2O_3-bearing $CaSiO_3$ perovskite [*Kurashina et al.*, 2004], although the elastic anomalies associated with these phase transformations are not yet confirmed experimentally.

Acknowledgments. I appreciate T. Kondo, K. Litasov, T. Kubo, A. Suzuki, and A Sano for collaborating the research project on the effect of water in mantle dynamics. This work was partially supported by the 21st Century Center-of-Excellence program of Tohoku University, Advanced Science and Technology Center for the Dynamic Earth and the grant-in-aid of Scientific Research for Priority Area from Monbusho (no. 16075202) to Eiji Ohtani.

REFERENCES

Akaogi, M., Ito, E., 1999. Calorimetric study of majorite–perovskite transition in the system $Mg_4Si_4O_{12}$–$Mg_3Al_2Si_3O_{12}$: transition boundaries with positive pressure–temperature slopes. *Phys. Earth Planet. Inter.* 114, 129–140. 416–422.

Akaogi, M., A. Tanaka, E. Ito, Garnet-ilmenite-perovskite transitions in the system $Mg_4Si_4O_{12}$-$Mg_3A_{12}Si_3O_{12}$ at high pressures and high temperatures: phase equilibria, calorimetry and implications for mantle structure, *Phys. Earth Planet. Inter.*, 132, 303–324, 2002.

Anderson, D. L., Theory of the Earth. Blackwell Scientific Publ., Boston, 366p, 1989.

Anderson, O. L., D. Isaak, H Oda, 1991. Thermoelastic parameters of six minerals at high temperatures. *J. Geophys. Res.* 96, 18037–18046, 1991.

Anderson, O. L., D. Isaak, H Oda, High-temperature elastic constant data on minerals relevant to geophysics. *Rev. Geophys.* 30, 57–90, 1992.

Bolfan-Casanova N, H. Keppler, D. C. Rubie, Water partitioning between nominally anhydrous minerals in the $MgO-SiO_2-H_2O$ system up to 24 GPa: implications for the distribution of water in the Earth's mantle, *Earth Planet. Sci. Let.* 182, 209–221, 2000.

Bolfan-Casanova N, S. Mackwell, H. Kepple., C. McCammon, D. C. Rubie, Pressure dependence of H solubility in magnesiowustite up to 25 GPa: implications for the storage of water in the Earth's lower mantle, *Geophys. Res. Lett.*, 10.1029, 2002.

Bose, K. and J. Gangley, Experimental and theoretical studies of the stabilies of talc, antigolite, and phase A at high pressures with application to subduction processes, *Earth Planet. Sci. Lett.*, 136, 109–121, 1994.

Carpenter, M. A., R. J. Hemley, H-K. Mao, High pressure elasticity of stishovite and the P42/mnm-Pnnm phase transition, *Geophys. Res.*, 105, 10807–10816, 2000.

Collier, J. D., G. R. Helffrich, and B. J. Wood, Seismic discontinuities and subduction zones, *Phys. Earth Planet. Inter.*, 127, 35–49, 2001.

Davies, J. H. and D. J. Stevenson, Physical model for the source region of subduction zone volatiles, *J. Geophys. Res.*, 97, 2037–2070, 1992.

Dixon, J. E., L. Leist, C. Langmuir, J. G. Schilling, Recycled dehydrated lithosphere observed in plume-influenced mid-ocean-ridge basalt, *Nature*, 420, 385–389, 2002.

Domanik, K. and J. R. Holloway, The stability and composition of phengitic muscovite and associated phases from 5.5 to 11 GPa: implications for deeply subducted sediments, *Geochim. Cosmochim. Acta,* 60, 4233–4150, 1996.

Duffy, T. S., Ahrens, T. J., Compressional sound velocity, equation of state, and constitutive response of shock-compressed magnesium oxide. *J. Geophys. Res.* 100, 529–542, 1995.

Eggelton, RA, N. J. Boland, A. E. Ringwood, High pressure synthesis of a new aluminum silicate: $Al_5Si_5O_{17}(OH)$, *Geochem. J.*, 12, 191–194, 1979.

Eshelby, D., The determination of the elastic field of an ellipsoidal inclusion, and related problem, *Proc. Roy. Soc. London, Ser. A*, 241, 376–396, 1957.

Evans, B. W., W. Johannes, H. Oterdoom, V. Trommsdorff, Stability of chrysotile and antigorite in their serpentine multisystem, Schweiz, *Mineral. Petrogr. Mitt.*, 50, 481–492, 1976.

Fei, Y., Y. Wang, L. W. Finger, Maximum solubility of FeO in (Mg,Fe)SiO_3 perovskite as a function of temperature at 26 GPa: Implication for FeO content in the lower mantle. *J. Geophys. Res.*, 101, 11525–11530, 1996.

Fei, Y., Mao, H. K., Shu, J., Hu, J., P–V–T equation of state of magnesiowustite ($Mg_{0.6}Fe_{0.4}$)O. *Phys. Chem. Minerals* 18, 416–422, 1992.

Fei, Y., Van Orman, J., Li, J., van Westrenen, W., Sanloup, C., Minarik, W., Hirose, K., Komabayashi, T., Walter, M., and Funakoshi, K., 2004, Experimentally determined postspinel transformation boundary in Mg_2SiO_4 using MgO as an internal pressure standard and its geophysical implications: *J. Geophys. Res.*, v. 109, doi: 10.1029/2003JB002562.

Frost D. J., C. Liebske, F. Langenhorst, C. McCammon, R. G. Tronnes, D. C. Rubie, Experimental evidence of iron-rich metal in the Earth's lower mantle, *Nature*, 428, 409–411, 2004.

Fukao, Y, S. Widiyantoro, M. Obayashi, Stagnant slabs in the upper and lower mantle transition region, *Rev. Geophys.*, 39, 291–323, 2001.

Fumagalli P, L. Stixrude, S. Poli, and L. Snyder, The 10Å phase: a high pressure expandable sheet silicate during subduction of hydrated lithosphere, *Earth Planet. Sci. Lett.*, 186, 125–141, 2001.

Funamori, N., T. Yagi, W. Utsumi, T. Kondo, T. Uchida, M. Funamori, Thermoelastic properties of $MgSiO_3$ perovskite determined by in situ X-ray observations up to 30 GPa and 2000 K, *J. Geophys. Res.* 101, 8257–8269,1996.

Gasparik, T., Phase relations in the transition zone, *J. Geophys. Res.*, 95, 15751–15769, 1990.

Hazen, R. M., R. T. Downs, L. W. Finger, and Ko, J. Crystal chemistry of ferromagnesian silicate spinels: evidence for Mg–Si disorder. *Am. Mineral.* 78, 1320–1323, 1993.

Heinemann, S., T. G. Sharp, F. Seifert, D. C. Rubie, The cubic–tetragonal phase transition in the system majorite ($Mg_4Si_4O_{12}$)–pyrope ($Mg_3Al_2Si_3O_{12}$), and garnet symmetry in the earth's transition zone. *Phys. Chem. Minerals* 24, 206–221, 1997.

Inoue T, H. Yurimoto, and Y. Kudoh, Hydrous modified spinel, $Mg_{1.75}SiH_{0.5}O_4$: A new water reservoir in the mantle transition region, *Geophys. Res. Lett.*, 160, 117–120, 1995.

Inoue, T., D. J. Weidner, P. A. Northrup, and J. B. Parise, Elastic properties of hydrous ringwoodite (γ-phase) in Mg_2SiO_4, *Earth Planet. Sci. Lett.*, 160, 107–113, 1998.

Irifune, T., N. Nishiyama, K. Kuroda, T. Inoue, M. Isshiki, W. Utsumi, K. Funakoshi, S. Urakawa, T. Uchida, T. Katsura, O. Ohtaka, The postspinel phase boundary in Mg_2SiO_4 determined by *in situ* X-ray diffraction, *Science* 279 (1998) 1698–1700.

Ito, E., Yamada, H., Stability relations of silicate spinels, ilmenites, and perovskites. In: Akimoto, S., Manghnani, M. H. (Eds.), *High-Pressure Research in Geophysics*. Center for Academic Publication, Japan, pp. 405–419, 1982.

Jagoutz E, H. Palme,H. Baddenhausen, K. Blum, M. Cendales, G. Dreibus, B. Spettel, V. Lorenz, H. Wanke, The abundances of major, minor and trace elements in the Earth's mantle as derived from primitive ultramafic nodules. *Proc 10th Lunar Planet Sci Conf,* Tucson, Arizona, pp 2031–2050, 1979.

Kaneshima, S. and G. Helffrich, Detection of lower mantle scatterers northeast of the Mariana subduction zone using short-period array data, *J. Geophys. Res.*, 103, 4825–4838, 1998.

Kaneshima, S. and G. Helffrich, Dipping lower-velocity layer in the mid-lower mantle: evidence for geochemical heterogeneity, *Science, 283*, 1888–1891, 1999.

Kaneshima, S. Small-scale heterogeneity at the top of the lower mantle around the Mariana slab, *Earth Planet. Sci. Lett.,* 209, 85–101, 2003.

Kaneshima, S. and G. Helffrich, Subparallel dipping heterogeneities in the mid-lower mantle, *J. Geophys. Res.,* 108, doi: 10.1029/2001jb001596, 2003.

Kanzaki, M., Stability of hydrous magnesium silicates in the mantle transition zone, *Phys. Earth Planet. Inter.,* 66, 307–316, 1991.

Karki, B. B., J. Crain, First-principles determination of elastic properties of CaSiO$_3$ perovskite at lower mantle pressures. *Geophys. Res. Lett.* 25, 2741–2744. 5–24, 1998.

Katsura, T., H. Yamada, T. Shinmei, A. Kubo, S. Ono, M. Kanzaki, A. Yoneda, M. J. Walter, S. Urakawa, E. Ito, K. Funakoshi, and W. Utsumi, Post-spinel transition in Mg$_2$SiO$_4$ determined by in situ X-ray diffractometry, *Phys. Earth Planet. Inter.,* 136, 11–24, 2003.

Kawamoto, T., K. Leinenweber, R. L. Hervig, J. R. Holloway, Stability of hydrous minerals in H$_2$O-saturated KLB-1 peridotite up to 15 GPa, in *Volatiles in the Earth and Solar System,* edited by K. A. Farley, pp.229–239, Amer. Inst. of Phys., New York, 1995.

Kawamoto, T., R. L. Hervig, and R. Holloway, Experimental evidence for a hydrous transition zone in the Earth's early mantle, *Earth Planet. Sci. Lett.,* 142, 587–592, 1996.

Kesson, S. and A. E. Ringwood. Slab-mantle interactions. I. Sheared and refertillised garnet peridotite xenoliths-samples of wadachi-benioff zone? *Chem. Geol.,* 78, 83–96, 1989.

Kohlstedt, D. L., H. Keppler, and D. C. Rubie, Solubility of water in the α–β–γ phases of (Mg,Fe)$_2$SiO$_4$, *Contrib. Mineral. Petrol.,* 123, 345–357, 1996.

Konzett, J. and Y. Fei, Hydrous potassic phases at high pressures: the stability of potassium amphibole and phase X and a new ordered hydrous pyribole, *EOS Trans, Amer. Geophys. Union,* 79 (17), S161, 1998.

Konzett, J. and P. Ulmer, The stability of hydrous potassic phases in lherzolitic mantle—an experimental study to 9.5 GPa in simplified and natural bulk compositions, *Jour. Petrol.,* 40, 629–652, 1999.

Koper, K. D., D. A. Wiens, L. M. Dorman, J. A. Hildebrand, and S. C. Webb, Modeling the Tonga slab: can travel time data resolve a metastable olivine wedge? *J. Geophys. Res.* 103: 30079–30100, 1998.

Koyama, T., H. Shimizu, H. Utada, E. Ohtani, and R. Hae, Water content in the mantle transition zone beneath north Pacific derived from the electrical conductivity anomaly, *Eos,* 85 (47), Fall meeting suppl., abstract, T31F-05, 2004.

Kubo, A. and M. Akaogi, Post-garnet transitions in the system Mg$_4$Si$_4$O$_{12}$–Mg$_3$Al$_2$Si$_3$O$_{12}$ up to 28 GPa: phase relations of garnet, ilmenite and perovskite. *Phys. Earth Planet. Inter.* 121, 85–102, 2000.

Kubo, T, E. Ohtani, T. Kato, T. Shinmei, and K. Fujino, Effect of water on the α–β transformation kinetics in San Carlos Olivine, *Science,* 281, 85–87, 1998.

Kubo, T., E. Ohtani, T. Kondo, T. Kato, M. Toma, T. Hosoya, A. Sano, T. Kikegawa, and T. Nagase, Metastable garnet in oceanic crust at the top of the lower mantle, Nature, 420, 803, 2002.

Kudoh Y., T. Nagase, H. Mizobata, E. Ohtani, S. Sasaki, and M. Tanaka, Structure and crystal chemistry of phase G, a new hydrous magnesium silicate synthesized at 22 GPa and 1050°C, *Geophys. Res. Lett.,* 24, 1051–1054, 1997.

Kurashina, T., K. Hirose, S. Ono, N. Sata, and Y. Ohishi, Phase transition in Al-bearing CaSiO$_3$ perovskite: implications for seismic discontinuities in the lower mantle, *Phys. Earth and Planet. Inter.,* 145, 67–74, 2004.

Litasov, K. and E. Ohtani, Phase relations and melt compositions in CMAS pyrolite-H$_2$O system up to 25 GPa, *Phys. Earth Planet. Inter.,* 134, 105–127, 2002.

Litasov K. and E. Ohtani, CMAS pyrolite-H$_2$O system up to 25 GPa, *Phys. Chem. Minerals,* 30, 147–156, 2003.

Litasov, K., E. Ohtani, A. Suzuki, T. Kawazoe, K. Funakoshi, Absence of density crossover between basalt and peridotite in the cold slabs passing through 660 km discontinuity, *Geophys. Res. Lett.,* 31, L24607, doi:10.1029/2004GL21306, 2004.

Litasov, K. and E. Ohtani, Phase Relations in Hydrous MORB at 18–25 GPa: Implications for Heterogeneity of the Lower Mantle, *Phys. Earth Planet. Inter.,* 150, 239–263, 2005.

Litasov K., E. Ohtani, F. Langenhorst, H. Yurimoto, T. Kubo, and T. Kondo, Water solubility in Mg-perovskites and water storage capacity in the lower mantle, *Earth Planet. Sci. Lett.,* 211, 189–203, 2003.

Litasov, K., E. Ohtani, A. Sano, A. Suzuki, and K. Funakoshi, 2005, Wet Subduction versus Cold Subduction: *Geophys. Res. Lett.,* 32, L13312, doi:10.1029/2005GL022921, 2005.

Luth, R. W., Is phase A relevant to the Earth's mantle, *Geochim. Cosmochim. Acta,* 59, 679–682, 1995.

Luth, R. W., Experimental study of the system phlogopite-diopside from 3.5 to 17 GPa, *Am. Mineral.,* 82, 1198–1209, 1997.

Matsubara, R., H. Toraya, S. Tanaka, H. Sawamoto, Precision lattice parameter determination of (Mg, Fe)SiO$_3$ tetragonal garnet, *Science* 247, 697–699, 1990.

Miyajima N, K. Fujino, N. Funamori, T. Kodo, and T. Yagi, Garnet-perovskite transformation under conditions of the Earth's lower mantle: an analytical transmission electron microscopy study. *Phys. Earth Planet. Inter.,* 116, 117–131, 1999.

Murakami M, K. Hirose, H. Yurimoto, S. Nakashima, and N. Takafuji, Water in Earth's lower mantle, *Science* 295, 1885–1887, 2002.

Niu F, H. Kawakatsu, and Y. Fukao, A slightly dipping and strong seismic reflector at the mid-mantle depth beneath the Mariana subduction zone, *J. Geophys. Res.,* 108,, 2419, doi:10.1029/ 2002JB002384, 2003.

Oguri, K., N. Funamori, T. Uchida, N. Miyajima, T. Yagi, and T. Fujino, Post-garnet transition in a natural pyrope: a multi-anvil study based on in situ X-ray diffraction and transmission electron microscopy, *Phys. Earth Planet. Inter.,* 122, 175–186, 2000.

Ohtani, E., H. Mizobata, H. Yurimoto, Stability of dense hydrous magnesium silicate phases in the systems Mg$_2$SiO$_4$-H$_2$O and MgSiO$_3$-H$_2$O at pressures up to 27 GPa, *Phys. Chem. Minerals,* 27, 533–544, 2000.

Ohtani, E., M. Toma, T. Kubo, and T. Kondo, T. Kikegawa, *Geophys. Res. Lett.,* 30, 1029, doi:10.1029/2002GL015549, 2003.

Ohtani, E., K. Litasov, T. Hosaya, T. Kubo, and T. Kondo, Water transport into the deep mantle and formation of a hydrous transition zone, *Phys. Earth Planet. Inter.*, 143–144, 201–213, 2004.

Ono, S., Stability limits of hydrous minerals in sediment and mid-ocean ridge basalt compositions: Implications for water transport in subduction zones, *J. Geophys. Res.*, 103,18253–18264, 1998.

Ono, S., High temperature stability of phase egg, AlSiO$_3$(OH). *Contrib. Mineral. Petrol.*, 137, 83–89, 1999.

Ono, S., K. Hirose, M. Isshiki, K. Mibe, Y. Saito, Equation of state of hexagonal aluminous phase in basaltic composition to 63 GPa at 300 K, *Phys. Chem. Mineral.*, 29, 527–531, 2002.

Pacalo, R. E. G. and D. J. Weidner, Elasticity of majorite, MgSiO$_3$ tetragonal garnet. *Phys. Earth Planet. Inter.* 99, 145–154, 1997.

Pawley, A. R., High pressure stability of chlorite: a source of H$_2$O for subduction magmatism, *Terra*, 8, Abstr. Suppl.,1, 50, 1996.

Peacock, S. M., Fluid processes in subduction zone, *Science,* 248, 329–337, 1990.

Poli, S., The amphiborite-eclogite transformation: an experimental result on basalt, *Am. J. Sci.*, 293, 1061–1107, 1993.

Poli, S and M. W. Schmidt, Petrology of subducted slabs, *Ann. Rev. Earth Planet. Sci.*, 30, 207–235, 2002.

Revenaugh J and S. A. Sipkin, Seismic evidence for silicate melt atop the 410-km mantle discontinuity, *Nature*, 369, 474–476, 1994.

Rigden, S. M., G. D. Gwanmesia, J. D. Fitz Gerald, I. Jackson, R. C. Liebermann, Spinel elasticity and seismic structure of the transition zone of the mantle. *Nature,* 354, 143–145, 1991.

Ringwood AE., Role of the transition zone and 660 km discontinuity in mantle dynamics, *Phys. Earth Planet. Inter.,* 86, 5–24, 1994.

Ringwood A. E. and A. Major, High pressure reconnaissance investigation in the system Mg$_2$SiO$_4$-MgO-H$_2$O, *Earth Planet. Sci. Lett.*, 2, 130–133, 1967.

Rubie, D. C. and C. R. Ross II, Kinetics of olivine-spinel transformation in subducting lithosphere: experimental constraints and implications on deep slab processes, *Phys. Earth Planet. Inter.* 86, 223–241, 1994.

Sano A., E. Ohtani, T. Kubo, and K. Funakoshi, In situ X-ray observation of decomposition of hydrous aluminum silicate AlSiO$_3$OH and aluminum oxide hydroxide δ-AlOOH at high pressure and temperature, *J. Phys. Chem. Solids*, 65, 1547–1554, 2004.

Schmidt, M. W., Lawsonite: Upper pressure stability and formation of higher density hydrous phases, *Am. Mineral.*, 80, 1286–1292, 1995.

Schmidt, M. W. and S. Poli, Experimental base water budgets for dehydrating slabs and consequences for arc magma generation. *Earth Planet. Sci. Lett.*, 163, 361–379, 1998.

Schmidt, M. W., L. W. Finger, R. J. Angel, and R. E. Dinnebier, Synthesis, crystal structure, and phase relations of AlSiO$_3$(OH), a high-pressure hydrous phase, *Am. Miner.*, 83, 881–888, 1998.

Shieh, S. R., H. K. Mao, and J. C. Ming, Decomposition of phase D in the lower mantle and the fate of dense hydrous silicates in subducting slabs, *Earth Planet. Sci. Lett.*, 159, 13–23, 1998.

Sinelnikov, Y. D., G. Chen, D. R. Neuville, M. T. Vaughan, R. C. Liebermann, Ultrasonic shear wave velocities of MgSiO$_3$

perovskite at 8 GPa and 800 K and lower mantle composition, *Science* 281, 677–679, 1998.

Smyth, R., C. M. Holl, D. J. Frost, and S. D. Jacobsen, High pressure crystal chemistry of hydrous ringwoodite and water in the earth's interior, *Phys. Earth Planet. Inter.*, 143–144, 271–278, 2004.

Stixrude, L., Elastic constants and anisotropy of MgSiO$_3$ perovskite, periclase, and SiO$_2$ at high pressure, in Core-mantle boundary region. Edited by M. Gurnis, M. E. Wysession, E. Knittle, and B. A. Buffet, *Geodynamics Series*, 28, AGU, pp83–96, 1998.

Suzuki, I., Thermal expansion of periclase and olivine, and their anharmonic properties. *J. Phys. Earth*, 23, 145–159, 1975.

Suzuki, I. and O. L. Anderson, Elasticity and thermal expansion of a natural garnet up to 1000 K, *J. Phys. Earth,* 31, 125–138, 1983.

Suzuki, A., E. Ohtani, and T. Kamada, A new hydrous phase δ-AlOOH synthesized at 20.9 GPa and 1000°C, *Phys. Chem. Miner.* 27, 689–693, 2000.

Ulmer, P and V. Trommsdorff, Serpentine stability to mantle depths and subduction related magmatism, *Science*, 268, 858–861, 1995.

Vanpeteghem, C. B., E. Ohtani, and T. Kondo, Equation of state of the hydrous phase δ-AlOOH at room temperature up to 22.5 GPa, *Geophys. Res. Lett.*, 29, 7,10.1029, 2002.

Wang, Y. and D. J. Weidner, Thermoelasticity of CaSiO$_3$ perovskite and implications for the lower mantle. *Geophys. Res. Lett.* 21, 895–898, 1994.

Wang, Y., D. J. Weidner., F. Guyot, Thermal equation of state of CaSiO$_3$ perovskite. *J. Geophys. Res.* 101, 661–672, 1996.

Weidner, D. J., H. Sawamoto, S. Sasaki, M. Kumazawa, Single-crystal elastic properties of the spinel phase of Mg$_2$SiO$_4$. *J. Geophys. Res.* 89, 7852–7860, 1984.

Williams, Q, and R. J. Hemley, Hydrogen in the deep earth, *Ann. Rev. Earth Planet. Sci.*, 29, 365–418, 2001.

Williams, Q. and E. J. Garnero, Seismic evidence for partial melting at the base of the lower mantle, *Science,* 273, 1528–1530, 1996.

Wunder, B., D. C. Rubie, C. R. Ross II, O. Medenbach, F. Seifert, W. Schreyer, Synthesis, stability, and properties of Al$_2$SiO$_4$(OH)$_2$: a fully hydrated analogue of topaz, *Am. Mineral.*, 78, 285–297, 1997.

Yeganeh-Haeri, A., Synthesis and re-investigation of elastic properties of single-crystal magnesium silicate perovskite. *Phys. Earth Planet. Inter.* 87, 111–121, 1994.

Zhao, D., Seismological structure of subduction zones and its implications for arc magmatism and dynamics, *Phys. Earth Planet. Inter.*, 127, 197–214, 2001.

Zhao, D., Global tomographic images of mantle plumes and subducting slabs: Insight into deep Earth dynamics, *Phys. Earth Planet. Inter.*, 146, 3–34, 2004.

Zhao, D., Y. Xu, D. Wiens, L. Dorman, J. Hildebrand, S. Webb, Depth extent of the Lau back-arc spreading center and its relation to subduction processes, *Science*, 278, 254–257, 1997.

Eiji Ohtani, Institute of Mineralogy, Petrology, and Economic Geology, Faculty of Science, Tohoku University, Sendai 980-8578, Japan. (ohtani@mail.tains.tohoku.ac.jp)